T0269132

Advances in Intelligent Systems and Computing

Volume 403

Series editor

About this Series

The series "Advances in Intelligent Systems and Computing" contains publications on theory, applications, and design methods of Intelligent Systems and Intelligent Computing. Virtually all disciplines such as engineering, natural sciences, computer and information science, ICT, economics, business, e-commerce, environment, healthcare, life science are covered. The list of topics spans all the areas of modern intelligent systems and computing.

The publications within "Advances in Intelligent Systems and Computing" are primarily textbooks and proceedings of important conferences, symposia and congresses. They cover significant recent developments in the field, both of a foundational and applicable character. An important characteristic feature of the series is the short publication time and world-wide distribution. This permits a rapid and broad dissemination of research results.

More information about this series at http://www.springer.com/series/11156

Robert Burduk · Konrad Jackowski
Marek Kurzyński · Michał Woźniak
Andrzej Żołnierek
Editors

Proceedings of the 9th International Conference on Computer Recognition Systems CORES 2015

Springer

Editors
Robert Burduk
Department of Systems and Computer Networks
Wrocław University of Technology
Wrocław
Poland

Konrad Jackowski
Department of Systems and Computer Networks
Wrocław University of Technology
Wrocław
Poland

Marek Kurzyński
Department of Systems and Computer Networks
Wrocław University of Technology
Wrocław
Poland

Michał Woźniak
Department of Systems and Computer Networks
Wrocław University of Technology
Wrocław
Poland

Andrzej Żołnierek
Department of Systems and Computer Networks
Wrocław University of Technology
Wrocław
Poland

ISSN 2194-5357 ISSN 2194-5365 (electronic)
Advances in Intelligent Systems and Computing
ISBN 978-3-319-26225-3 ISBN 978-3-319-26227-7 (eBook)
DOI 10.1007/978-3-319-26227-7

Library of Congress Control Number: 2015959588

Printed on acid-free paper

This Springer imprint is published by SpringerNature
The registered company is Springer International Publishing AG Switzerland

Preface

The goal of the CORES series of conferences is the development of theories, algorithms, and applications of pattern recognition methods. These conferences have always served as a very useful forum where researchers, practitioners, and students working in different areas of pattern recognition can come together and help each other keeping up with this active field of research.

This book is collection of 79 carefully selected works, which have been carefully reviewed by the experts from the domain and accepted for presentation during the 9th International Conference on Computer Recognition Systems CORES 2015.

We hope that the book can become the valuable source of information on contemporary research trends and the most popular areas of application.

The chapters are grouped into seven parts on the basis of the main topics they dealt with:

1. *Features, Learning, and Classifiers* consists of the works concerning new classification and machine learning methods;
2. *Biometrics* presents innovative theories, methodologies, and applications in the biometry;
3. *Data Stream Classification and Big Data Analytics* section concentrates on both data stream classification and massive data analytics issues;
4. *Image Processing and Computer Vision* is devoted to the problems of image processing and analysis;
5. *Medical Applications* presents chosen applications of intelligent methods into medical decision support software;
6. *Applications* describes several applications of the computer pattern recognition systems in the real decision problems;
7. *RGB-D Perception: Recent Developments and Applications* presents pattern recognition and image processing algorithms aimed specifically at applications in robotics.

Editors would like to express their deep thanks to authors for their valuable submissions and all reviewers for their hard work. Especially we would like to thank Dr. Tomasz Kornuta, Prof. Włodzimierz Kasprzak, Warsaw University of Technology, and Prof. Piotr Skrzypczyński, Poznań University of Technology, who organized a special session entitled "RGB-D Perception: Recent Developments and Applications." We would like to also thank Prof. Jerzy Stefanowski from Poznań University of Technology who helped Prof. Michał Woźniak and Bartosz Krawczyk, Wroclaw University of Technology to organize a special session on data stream classification and big data analytics.

We believe that this book could be a reference tool for scientists who deal with the problems of designing computer pattern recognition systems.

CORES 2015 enjoyed outstanding keynote speeches by distinguished guest speakers:

- Prof. Nitesh Chawla—University of Notre Dame, USA,
- Prof. Krzysztof J. Cios—Virginia Commonwealth University, USA,
- Prof. João Gama—University of Porto, Portugal,
- Prof. Francisco Herrera—University of Granada, Spain.

Last but not least, we would like to give special thanks to local organizing team (Robert Burduk, Kondrad Jackowski, Dariusz Jankowski, Bartosz Krawczyk, Maciej Krysmann, Jose Antonio Saez, Alex Savio, Paweł Trajdos, Marcin Zmyślony, and Andrzej Żołnierek) who did a great job.

This edition of the CORES was organized under the framework the ENGINE project, and thus the authors of the selected, best papers did not pay conference fee. ENGINE has received funding from the European Union's the Seventh Framework Programme for research, technological development, and demonstration under grant agreement no 316097. We would like to give our special thanks to the management of the ENGINE project—Prof. Przemysław Kazienko and Dr. Piotr Bródka—for this valuable sponsorship.

Also we would like to fully acknowledge the support from the Wrocław University of Technology, especially Prof. Andrzej Kasprzak—Chairs of Department of Systems and Computer Networks and vice Rector of the Wroclaw University of Technology, Prof. Jan Zarzycki—Dean of Faculty of Electronics, and Prof. Zdzisław Szalbierz—Dean of Faculty of Computer Science and Management, which has also supported this event.

We believe that this book could be a great reference tool for scientists who deal with the problems of designing computer pattern recognition systems.

Wrocław
July 2015

Robert Burduk
Konrad Jackowski
Marek Kurzyński
Michał Woźniak
Andrzej Żołnierek

Contents

Part VI Application

Part I
Features, Learning and Classifiers

New Ordering-Based Pruning Metrics for Ensembles of Classifiers in Imbalanced Datasets

Mikel Galar, Alberto Fernández, Edurne Barrenechea, Humberto Bustince and Francisco Herrera

Abstract The task of classification with imbalanced datasets have attracted quite interest from researchers in the last years. The reason behind this fact is that many applications and real problems present this feature, causing standard learning algorithms not reaching the expected performance. Accordingly, many approaches have been designed to address this problem from different perspectives, i.e., data preprocessing, algorithmic modification, and cost-sensitive learning. The extension of the former techniques to ensembles of classifiers has shown to be very effective in terms of quality of the output models. However, the optimal value for the number of classifiers in the pool cannot be known a priori, which can alter the behaviour of the system. For this reason, ordering-based pruning techniques have been proposed to address this issue in standard classifier learning problems. The hitch is that those metrics are not designed specifically for imbalanced classification, thus hindering the performance in this context. In this work, we propose two novel adaptations for ordering-based pruning metrics in imbalanced classification, specifically the margin distance minimization and the boosting-based approach. Throughout a complete

M. Galar (✉) · E. Barrenechea · H. Bustince
Departamento de Automática y Computación, ISC (Institute of Smart Cities), Universidad Pública de Navarra, Pamplona, Spain
e-mail: mikel.galar@unavarra.es

E. Barrenechea
e-mail: edurne.barrenechea@unavarra.es

H. Bustince
e-mail: bustince@unavarra.es

A. Fernández
Department of Computer Science, University of Jaén, Jaén, Spain
e-mail: alberto.fernandez@ujaen.es

F. Herrera
Department of Computer Science and Artificial Intelligence, University of Granada, Granada, Spain
e-mail: herrera@decsai.ugr.es

R. Burduk et al. (eds.), *Proceedings of the 9th International Conference on Computer Recognition Systems CORES 2015*, Advances in Intelligent Systems and Computing 403, DOI 10.1007/978-3-319-26227-7_1

experimental study, our analysis shows the goodness of both schemes in contrast with the unpruned ensembles and the standard pruning metrics in Bagging-based ensembles.

Keywords Imbalanced datasets · Ensembles · Ordering-based pruning · Bagging

1 Introduction

The unequal distribution among examples of different classes in classification tasks is known as the problem of imbalanced datasets [9, 22]. The use of standard algorithms in this framework lead to undesirable solutions as the model is usually biased towards the most represented concepts of the problem [13]. Therefore, several approaches have been developed for addressing this issue, which can be divided into three large groups including preprocessing for resampling the training set [3], algorithmic adaptation of standard methods [2], and cost-sensitive learning [25]. Additionally, all these schemes can be integrated into an ensemble learning algorithm, increasing the capabilities and performance of the baseline approach [7, 8, 13]. An ensembles is a set of classifiers where its components are supposed to complement each other, so that the learning space is completely covered and the generalization capability is enhanced with respect to the single baseline learning classifier [18, 21]. When classifying a new instance, all individual members are queried and their decision is obtained in agreement. The total number of classifiers that compose an ensemble is not a synonym of its quality and performance [27], since several issues that can degrade its behavior must be taken into account: (1) the time elapsed in the learning and prediction stages; (2) the memory requirements; and (3) contradictions and/or redundancy among components of the ensemble. In accordance with the above, several proposals have been developed to carry out a pruning of classifiers within the ensemble [26]. Specifically, ordering-based pruning is based on a greedy approach that adds classifiers iteratively to the final set with respect to the maximization of a given heuristic metric, until a preestablished number of classifiers are selected [10, 16]. In this contribution, we aim at developing an adaptation of two popular metrics towards the scenario of classification with imbalanced datasets, i.e., Margin Distance Minimization (MDM) and Boosting-Based pruning (BB) [6, 15]. Specifically, we consider that the effect of each classifier in both classes must be analyzed after the construction of the classifier and not only before (for example, rebalancing the dataset). The goodness of this novel proposal is analyzed by means of a thorough experimental study, including a number of 66 different imbalanced problems. We have selected SMOTE-Bagging [23] and Under-Bagging [1] as ensemble learning schemes which, despite of being simple approaches, have shown to achieve a higher performance than many other more complex algorithms [7]. As in other related studies, we have selected the well-known C4.5 algorithm as baseline classifier [20]. Finally, our results are supported by means of non-parametric statistical tests [5]. In order to do so, this work is organized as follows. Section 2 briefly introduces the problem of imbalanced datasets. Then, Sect. 3 presents ordering-based pruning methodology, in which we describe standard metrics for performing this process and our adaptations to imbalanced classification. Next, the details about the experimental framework are provided in Sect. 4. The analysis and discussion of the experimental results are carried out in Sect. 5. Finally, Sect. 6 summarizes and concludes the work.

2 Basic Concepts on Classification with Imbalanced Datasets

Classification with imbalanced datasets appears when the distribution of instances between the classes of a given problem is quite different [13, 19]. Therefore, this classification task needs a special treatment in order to carry out an accurate discrimination between both concepts, independently of their representation. The presence of classes with few data can generate sub-optimal classification models, since there is a bias towards the majority class when the learning process is guided by the standard accuracy metric. Furthermore, recent studies have shown that other data intrinsic characteristics have a significant influence for the correct identification of the minority class examples [13]. Some examples are overlapping, small-disjuncts, noise, and dataset shift. Solutions developed to address this problem can be categorized into three large groups [13]: (1) *data level solutions* [3], (2) *algorithmic level solutions* [2], and (3) *cost-sensitive solutions* [25]. Additionally, when the former approaches are integrated within an ensemble of classifiers, their effectiveness is enhanced [7, 13]. Finally, in order to evaluate the performance in such a particular classification scenario, the metrics used must be designed to take into account the class distribution. One commonly considered alternative is the Area Under the ROC curve (AUC) [11]. In those cases where the used classifier outputs a single solution, this measure can be simply computed by the following formula:

$$AUC = \frac{1 + TP_{rate} - FP_{rate}}{2} \tag{1}$$

where $TP_{rate} = \frac{TP}{TP+FN}$ and $FP_{rate} = \frac{FP}{TN+FP}$.

3 A Proposal for Ordering-Based Pruning Scheme for Ensembles in Imbalanced Domains

Ensemble-based classifiers [18] are composed by a set of so-called *weak learners*, i.e., low changes in data produce big changes in the induced models. Diversity is quite significant in the performance of this type of approach, implying that individual classifiers must be focused on different parts of the problem space [12]. There are mainly two types of ensemble techniques: Bagging [4] and Boosting [6]. In this work, we will focus on the first scheme, due to the simplicity for the integration of data preprocessing techniques [7]. In this methodology, an ensemble of classifiers is trained with different sets of random instances from the original training data. When classifying a new sample, all individual classifiers are fired and a majority or weighted vote is used to infer the class. The first parameter to take into account when building these types of models is the number of classifiers considered in the ensemble. In this sense, pruning methods were designed to obtain the "optimal" number of classifiers by carrying out a selection from a given pool of components of the ensemble. The hypothesis is that accuracy generally increases monotonically as more elements are added to the ensemble [10, 15, 16]. Most of pruning techniques make use of an heuristic function to seek for the reduced set of classifiers. In the case of *ordering-based pruning*, a metric that measures the goodness of adding each classifier to the ensemble is defined and the classifier with the highest value is added to the final sub-ensemble. The same process is performed until the size of the sub-ensemble reaches the specified parameter value. In this work, we study two popular pruning metrics MDM and BB [6, 15]. We describe both schemes and our adaptation to imbalanced classification below:

- *MDM* is based on certain distances among the output vectors of the ensembles. These output vectors have the length equal to the training set size, and their value at the ith position is either 1 or -1 depending on whether the ith example is classified or misclassified by the classifier. The signature vector of a sub-ensemble is computed as the sum of the vectors of the selected classifiers. To summarize, the aim is to add those classifiers with the objective of obtaining a signature vector of the sub-ensemble where all the components are positive, i.e., all examples are correctly predicted. For a wider description please refer to [16]. This method selects the classifier to be added depending on the closest Euclidean distance between an objective point (where every components are positive) and the signature vector of the sub-ensemble after adding the corresponding classifier. As a consequence, every example has the same weight in the computation of the distance, which can bias the selection to those classifiers favoring the majority class. Therefore, we compute the distance for the majority class examples and minority class examples independently. Then, distances are normalized by the number of examples used to compute them and added afterwards. That is, the same weight is given to both classes in the distance. This new metric is noted as *MDM-Imb*.
- *BB* selects the classifier that minimizes the cost with respect to the boosting scheme. This means that boosting algorithm is applied to compute the weights (costs) for each example in each iteration, but instead of training a classifier with these weights, the one that obtains the lowest cost from those in the pool is added to the sub-ensemble and weights are updated accordingly. Hence, it makes no difference whether classifiers were already learned using a boosting scheme or not. Different from the original boosting method, when no classifier has a weighted training error below 50 %, weights are reinitialized (equal weights for all the examples) and the method continues (whereas in boosting it is stopped). Once classifiers are

selected the scores assigned to each classifier by boosting are forgotten and not taken into account in the aggregation phase. It is well-known that boosting by itself is not capable of managing class imbalance problem [7]. For this reason, we have also adapted this approach in a similar manner as in the case of *MDM*. In boosting, every example has initially the same weight and these are updated according to whether they are correctly classified or not. Even though minority class instances should get larger weights if they are misclassified, these weights can be negligible compared with those of the majority class examples. Hence, before finding the classifier that minimizes the total cost, we normalize the weights of the examples of each class by half of their sum, so that both classes has the same importance when selecting the classifier (even though each example of each class would have a different weight). This is only done before selecting the classifier, and then weights are updated according to the original (non-normalized ones). This working procedure tries to be similar to that successfully applied in several boosting models such as EUS-Boost [8]. This second weighting approach is noted as *BB-Imb*.

4 Experimental Framework

Table 1 shows the benchmark problems selected for our study, in which the name, number of examples, number of attributes, and IR (ratio between the majority and minority class instances) are shown. Datasets are ordered with respect to their degree of imbalance. Multi-class problems were modified to obtain two-class imbalanced problems, defining the joint of one or more classes as positive and the joint of one or more classes as negative, as defined in the name of the dataset. A wider description for these problems can be found at http://www.keel.es/datasets.php. The estimates of AUC measure are obtained by means of a Distribution Optimally Balanced Stratified Cross-Validation (DOB-SCV) [17], as suggested in the specialized literature for working in imbalanced classification [14]. Cross-validation procedure is carried out using five folds, aiming to include enough positive class instances in the different folds. In accordance with the stochastic nature of the learning methods, these five folds are generated with five different seeds, and each one of the fivefold cross-validation is run five times. Therefore, experimental results are computed with the average of 125 runs. As ensemble techniques, we will make use of SMOTE-Bagging [23] and Under-Bagging [1]. In order to apply the pruning procedure, we will learn a number of 100 *classifiers* for each ensemble, choosing a subset of only 21 *classifiers* as suggested in the specialized literature [16]. The baseline ensemble models for comparison will use 40 classifiers as recommended in [7]. For *SMOTE-Bagging*, SMOTE configuration will be the standard with a 50 % class distribution, 5 neighbors for generating the synthetic samples, and Heterogeneous Value Difference Metric for computing the distance among the examples. Finally, both learning approaches include the C4.5 decision tree [20] as baseline classifier, using a confidence level at

Table 1 Summary of imbalanced datasets used

Name	#Ex.	#Atts.	IR	Name	#Ex.	#Atts.	IR
Glass1	214	9	1.82	Glass04vs5	92	9	9.22
Ecoli0vs1	220	7	1.86	Ecoli0346vs5	205	7	9.25
Wisconsin	683	9	1.86	Ecoli0347vs56	257	7	9.28
Pima	768	8	1.87	Yeast05679vs4	528	8	9.35
Iris0	150	4	2.00	Ecoli067vs5	220	6	10.00
Glass0	214	9	2.06	Vowel0	988	13	10.10
Yeast1	1484	8	2.46	Glass016vs2	192	9	10.29
Vehicle2	846	18	2.52	Glass2	214	9	10.39
Vehicle1	846	18	2.52	Ecoli0147vs2356	336	7	10.59
Vehicle3	846	18	2.52	Led7digit02456789vs1	443	7	10.97
Haberman	306	3	2.78	Ecoli01vs5	240	6	11.00
Glass0123vs456	214	9	3.19	Glass06vs5	108	9	11.00
Vehicle0	846	18	3.25	Glass0146vs2	205	9	11.06
Ecoli1	336	7	3.36	Ecoli0147vs56	332	6	12.28
Newthyroid2	215	5	4.92	Cleveland0vs4	1771	13	12.62
Newthyroid1	215	5	5.14	Ecoli0146vs5	280	6	13.00
Ecoli2	336	7	5.46	Ecoli4	336	7	13.84
Segment0	2308	19	6.01	Shuttle0vs4	1829	9	13.87
Glass6	214	9	6.38	Yeast1vs7	459	8	13.87
Yeast3	1484	8	8.11	Glass4	214	9	15.47
Ecoli3	336	7	8.19	Pageblocks13vs4	472	10	15.85
Pageblocks0	5472	10	8.77	Abalone918	731	8	16.68
Ecoli034vs5	200	7	9.00	Glass016vs5	184	9	19.44
Yeast2vs4	514	8	9.08	Shuttle2vs4	129	9	20.50
Ecoli067vs35	222	7	9.09	Yeast1458vs7	693	8	22.10
Ecoli0234vs5	202	7	9.10	Glass5	214	9	22.81
Glass015vs2	506	8	9.12	Yeast2vs8	482	8	23.10
Yeast0359vs78	172	9	9.12	Yeast4	1484	8	28.41
Yeast0256vs3789	1004	8	9.14	Yeast1289vs7	947	8	30.56
Yeast02579vs368	1004	8	9.14	Yeast5	1484	8	32.73
Ecoli046vs5	203	6	9.15	Yeast6	1484	8	41.40
Ecoli01vs235	244	7	9.17	Ecoli0137vs26	281	7	39.15
Ecoli0267vs35	244	7	9.18	Abalone19	4174	8	129.44

0.25, with 2 as the minimum number of item-sets per leaf, and the application of pruning will be used to obtain the final tree. Reader may refer to [7] in order to get a thorough description of the former ensemble methods. Finally, we will make use of Wilcoxon signed-rank test [24] to find out whether significant differences exist between a pair of algorithms.

5 Experimental Study

Our analysis is focused on determining whether the new proposed metrics, specifically designed for dealing with class imbalance, are well-suited for this problem with respect to the original metrics, i.e., *BB* and *MDM*. Additionally, we will analyze the improvement in the performance results with respect to the original ensemble model. The average values for the experimental results are shown in Table 2, whereas full results are shown in Table 3. Regarding the comparison between the pruning schemes, in the case of *BB* and *BB-Imb* we find that for SMOTE-Bagging the metric adapted for imbalanced classification achieves a higher average performance. Regarding Under-Bagging, the relative differences are below 1 % in favour of the standard approach. On the other hand, the analysis for *MDM* and *MDM-Imb* metrics shows the need for the imbalanced approach, as it stands out looking at the experimental results. Finally, the robustness of the imbalanced metrics must be stressed in accordance with the low standard deviation shown with respect to the standard case. In order to determine statistically the best suited metric, we carry out a Wilcoxon pairwise test in Table 4. We have included a symbol for stressing whether significant differences are found at 95 % confidence degree (*) or at 90 % (+). Results of these tests agree with our previous remarks. The differences in the case of MDM are clear in favour of the imbalanced version. In the case of BB, the behaviour vary depending on the ensemble technique, where significant differences are obtained for SMOTE-Bagging whereas none are found for Under-Bagging. Finally, when we contrast these results versus the standard ensemble approach, we also observe a two-fold behaviour: in the case of SMOTE-Bagging the pruning approach enables the definition of a simpler ensemble with a low decrease of the performance, especially when the imbalanced metric is selected. On the other hand, for Under-Bagging we observe a notorious improvement of the results in all cases when the ordering-based pruning is applied,

Table 2 Average test results for the standard ensemble approach (Base) and the ordering-based pruning with the original (BB and MDM) and imbalanced pruning metrics (BB-Imb and MDM-Imb)

Ensemble	Base	BB	BB-Imb	MDM	MDM-Imb
SMOTE-Bagging	0.8645 ± 0.0587	0.8602 ± 0.0632	**0.8635 ± 0.0610**	0.8596 ± 0.0629	**0.8625 ± 0.0622**
Under-Bagging	0.8647 ± 0.0516	**0.8755 ± 0.0564**	0.8734 ± 0.0544	0.8653 ± 0.0563	**0.8699 ± 0.0558**

Table 3 Test results for the standard ensemble (Base) and ordering-based pruning schemes (BB, BB-Imb, MDM, and MDM-Imb) using AUC metric

	SMOTE-Bagging					Under-Bagging				
Dataset	Base	BB	BB-Imb	MDM	MDM-Imb	Std.	BB	BB-Imb	MDM	MDM-Imb
Glass1	0.7675	0.8021	0.7925	0.7866	0.7900	0.7686	0.7979	0.7927	0.7918	0.7928
Ecoli0vs1	0.9812	0.9750	0.9763	0.9802	0.9788	0.9805	0.9806	0.9764	0.9826	0.9809
Wisconsin	0.9707	0.9692	0.9700	0.9662	0.9666	0.9691	0.9698	0.9704	0.9678	0.9672
Pima	0.7568	0.7451	0.7546	0.7500	0.7558	0.7598	0.7561	0.7548	0.7532	0.7539
Iris0	0.9880	0.9888	0.9880	0.9880	0.9880	0.9900	0.9900	0.9900	0.9900	0.9900
Glass0	0.8347	0.8517	0.8464	0.8413	0.8430	0.8264	0.8469	0.8438	0.8399	0.8352
Yeast1	0.7312	0.7192	0.7321	0.7301	0.7315	0.7304	0.7333	0.7310	0.7331	0.7307
Vehicle2	0.9723	0.9752	0.9734	0.9686	0.9691	0.9704	0.9750	0.9744	0.9680	0.9686
Vehicle1	0.7848	0.7691	0.7918	0.7898	0.7934	0.8016	0.8020	0.7983	0.7959	0.7985
Vehicle3	0.7784	0.7593	0.7827	0.7795	0.7808	0.8060	0.7979	0.7976	0.7966	0.7974
Haberman	0.6627	0.6517	0.6476	0.6500	0.6498	0.6627	0.6616	0.6486	0.6488	0.6620
Glass0123vs456	0.9405	0.9318	0.9357	0.9308	0.9378	0.9335	0.9432	0.9379	0.9264	0.9337
Vehicle0	0.9635	0.9630	0.9636	0.9609	0.9614	0.9492	0.9558	0.9595	0.9539	0.9544
Ecoli1	0.9053	0.8988	0.9067	0.9044	0.9107	0.8988	0.8981	0.9101	0.9043	0.9123
Newthyroid2	0.9642	0.9540	0.9586	0.9567	0.9577	0.9605	0.9572	0.9696	0.9614	0.9692
Newthyroid1	0.9558	0.9460	0.9486	0.9456	0.9467	0.9490	0.9479	0.9550	0.9594	0.9613
Ecoli2	0.9145	0.9153	0.9128	0.9131	0.9099	0.9054	0.9057	0.8996	0.9017	0.8996
Segment0	0.9917	0.9917	0.9924	0.9922	0.9926	0.9866	0.9881	0.9887	0.9872	0.9878
Glass6	0.9291	0.9164	0.9213	0.9157	0.9203	0.9096	0.9277	0.9248	0.9228	0.9190
Yeast3	0.9330	0.9308	0.9325	0.9315	0.9329	0.9311	0.9326	0.9305	0.9311	0.9295
Ecoli3	0.8462	0.8508	0.8560	0.8506	0.8514	0.8830	0.8702	0.8670	0.8793	0.8707

(continued)

Table 3 (continued)

Dataset	SMOTE-Bagging					Under-Bagging				
	Base	BB	BB-Imb	MDM	MDM-Imb	Std.	BB	BB-Imb	MDM	MDM-Imb
Pageblocks0	0.9580	0.9552	0.9585	0.9572	0.9581	0.9610	0.9631	0.9626	0.9612	0.9615
Ecoli034vs5	0.9129	0.9032	0.9018	0.9029	0.8948	0.8922	0.9148	0.9203	0.8701	0.9037
Yeast2vs4	0.9277	0.9192	0.9155	0.9123	0.9223	0.9445	0.9408	0.9482	0.9383	0.9536
Ecoli067vs35	0.8576	0.8651	0.8626	0.8653	0.8630	0.8582	0.8624	0.8578	0.8670	0.8523
Ecoli0234vs5	0.9007	0.9008	0.9036	0.8935	0.8939	0.8641	0.9053	0.9027	0.8404	0.8784
Glass015vs2	0.7041	0.7004	0.7015	0.7052	0.7025	0.7412	0.7117	0.7604	0.7553	0.7628
Yeast0359vs78	0.7173	0.7023	0.7174	0.7016	0.7134	0.7373	0.7414	0.7386	0.7394	0.7387
Yeast02579vs368	0.8028	0.7982	0.7995	0.7927	0.7993	0.8159	0.8090	0.8068	0.8136	0.8075
Yeast0256vs3789	0.9183	0.9173	0.9176	0.9150	0.9185	0.9149	0.9136	0.9099	0.9140	0.9098
Ecoli046vs5	0.9132	0.9086	0.9114	0.9046	0.9083	0.8869	0.9188	0.9238	0.8666	0.9123
Ecoli01vs235	0.8988	0.8665	0.8815	0.8789	0.8883	0.8815	0.9031	0.9047	0.8893	0.8942
Ecoli0267vs35	0.8617	0.8544	0.8611	0.8664	0.8642	0.8573	0.8623	0.8556	0.8662	0.8483
Glass04vs5	0.9910	0.9836	0.9879	0.9876	0.9869	0.9940	0.9900	0.9940	0.9940	0.9940
Ecoli0346vs5	0.8921	0.8888	0.8929	0.8762	0.8884	0.8799	0.8961	0.9051	0.8618	0.8956
Ecoli0347vs56	0.8595	0.8701	0.8707	0.8590	0.8643	0.8762	0.8875	0.8897	0.9009	0.8800
Yeast05679vs4	0.8177	0.8152	0.8133	0.8088	0.8124	0.8209	0.8287	0.8189	0.8018	0.8182
Ecoli067vs5	0.8897	0.8894	0.8888	0.8909	0.8886	0.8820	0.8883	0.8888	0.9028	0.8779
Vowel0	0.9878	0.9874	0.9880	0.9838	0.9853	0.9588	0.9671	0.9684	0.9689	0.9685
Glass016vs2	0.7009	0.7083	0.7176	0.7168	0.7214	0.7025	0.7185	0.7291	0.7265	0.7323
Glass2	0.7425	0.7390	0.7436	0.7458	0.7458	0.7569	0.7394	0.7691	0.7452	0.7702
Ecoli0147vs2356	0.8685	0.8637	0.8719	0.8673	0.8793	0.8328	0.8625	0.8536	0.8665	0.8468
Led7digit02456789vs1	0.8466	0.8547	0.8407	0.8500	0.8383	0.8268	0.8397	0.8322	0.8449	0.8399
Ecoli01vs5	0.8881	0.8786	0.8782	0.8688	0.8755	0.8726	0.9142	0.9174	0.8795	0.8937

(continued)

Table 3 (continued)

Dataset	SMOTE-Bagging					Under-Bagging				
	Base	BB	BB-Imb	MDM	MDM-Imb	Std.	BB	BB-Imb	MDM	MDM-Imb
Glass06vs5	0.9926	0.9954	0.9954	0.9916	0.9912	0.9151	0.9910	0.9940	0.9940	0.9940
Glass0146vs2	0.6961	0.7161	0.7295	0.7189	0.7254	0.7214	0.7335	0.7336	0.7323	0.7434
Ecoli0147vs56	0.8703	0.8848	0.8804	0.8682	0.8750	0.8738	0.9035	0.8870	0.8819	0.8756
Cleveland0vs4	0.7894	0.7933	0.8004	0.7815	0.7835	0.8492	0.8714	0.8305	0.7917	0.8069
Ecoli0146vs5	0.8875	0.9037	0.9022	0.8828	0.8994	0.8933	0.9197	0.9273	0.8639	0.8988
Ecoli4	0.9245	0.9220	0.9247	0.9094	0.9135	0.8952	0.9357	0.9349	0.9017	0.8969
Shuttle0vs4	0.9999	0.9999	0.9999	0.9999	0.9999	1.0000	1.0000	1.0000	1.0000	1.0000
Yeast1vs7	0.7458	0.7354	0.7349	0.7368	0.7303	0.7661	0.7869	0.7852	0.7463	0.7824
Glass4	0.9069	0.8795	0.8788	0.8716	0.8675	0.9065	0.9182	0.8903	0.8943	0.8882
Pageblocks13vs4	0.9952	0.9932	0.9964	0.9963	0.9963	0.9804	0.9937	0.9946	0.9928	0.9928
Abalone9vs18	0.7120	0.7140	0.7076	0.7090	0.7085	0.7560	0.7490	0.7388	0.7222	0.7354
Glass016vs5	0.9865	0.9493	0.9747	0.9675	0.9674	0.9429	0.9698	0.9675	0.9670	0.9663
Shuttle2vs4	1.0000	1.0000	1.0000	1.0000	1.0000	1.0000	1.0000	1.0000	1.0000	1.0000
Yeast1458vs7	0.6330	0.6175	0.6144	0.6059	0.6153	0.6374	0.6530	0.6263	0.6009	0.6315
Glass5	0.9769	0.9533	0.9619	0.9586	0.9626	0.9488	0.9596	0.9639	0.9631	0.9621
Yeast2vs8	0.8064	0.7916	0.7946	0.8014	0.8068	0.7526	0.7846	0.7608	0.7579	0.7629
Yeast4	0.8211	0.8117	0.8114	0.8046	0.8124	0.8420	0.8534	0.8537	0.8416	0.8543
Yeast1289vs7	0.7046	0.6818	0.6905	0.6831	0.7004	0.7370	0.7194	0.7392	0.6918	0.7433
Yeast5	0.9622	0.9536	0.9581	0.9525	0.9585	0.9593	0.9689	0.9673	0.9623	0.9625
Yeast6	0.8375	0.8354	0.8446	0.8369	0.8431	0.8673	0.8736	0.8570	0.8706	0.8514
Ecoli0137vs26	0.8347	0.8273	0.8336	0.8363	0.8400	0.7807	0.8774	0.7874	0.8060	0.7789
Abalone19	0.5432	0.5380	0.5447	0.5375	0.5462	0.7121	0.7034	0.7251	0.7213	0.7307
Average	0.8645	0.8602	0.8635	0.8596	0.8625	0.8647	0.8755	0.8734	0.8653	0.8699

Table 4 Wilcoxon test for pruning metrics: standard [R^+] and imbalanced [R^-]

Ensemble	Comparison	R^+	R^-	p-value
SMOTE-Bagging	BB versus BB-Imb	540.0	1671.0	0.00028*
	MDM versus MDMimb	436.0	1775.0	0.00002
Under-Bagging	BB versus BB-Imb	1277.0	934.0	0.27939
	MDM versus MDMimb	831.5	1379.5	0.07246+

Table 5 Wilcoxon test to compare the standard ensemble approach (Std.) [R^+] and the one with imbalanced ordering-based pruning [R^-]

Ensemble	Comparison	R^+	R^-	p-value
SMOTE-bagging	Std. versus BB-Imb	1261.5	883.5	0.215579
	Std. versus MDMimb	1386.5	758.5	0.039856*
Under-bagging	Std. versus BB-Imb	502.0	1709.0	0.000114*
	Std. versus MDMimb	637.0	1574.0	0.002735*

showing a better behaviour for MDM-Imb and especially in BB-Imb (see Tables 2 and 3). These findings are complemented by means of a Wilcoxon test (shown in Table 5), for which we observe significant differences in favour of the ordering-based pruning for the Under-Bagging approach.

6 Concluding Remarks

Ordering-based pruning in ensembles of classifiers consists of carrying out a selection of those elements of the ensemble set that are expected to work with better synergy. The former process is guided by a given metric of performance which is focused on different capabilities of the ensemble. However, they have not been previously considered within been developed within the scenario of imbalanced datasets. In this work, we have proposed two adaptations of metrics for ordering-based pruning in imbalanced classification, namely BB-Imb and MDM-Imb. The experimental analysis has shown the success of these novel metrics with respect to their original definition, especially in the case of the SMOTE-Bagging approach. Additionally, we have pointed out that a significant improvement in the behaviour of the Under-Bagging ensemble is achieved by means of the application of the ordering-based pruning, outperforming the results with respect to the original model. As future work, we plan to include a larger number of pruning metrics and ensemble learning methodologies, aiming at giving additional support and strength to the findings obtained in this contribution.

Acknowledgments This work was supported by the Spanish Ministry of Science and Technology under projects TIN-2011-28488, TIN-2012-33856, TIN2013-40765-P; the Andalusian Research Plans P11-TIC-7765 and P10-TIC-6858; and both the University of Jaén and Caja Rural Provincial de Jaén under project UJA2014/06/15.

References

1. Barandela, R., Valdovinos, R., Sánchez, J.: New applications of ensembles of classifiers. Pattern Anal. Appl. **6**(3), 245–256 (2003)
2. Barandela, R., Sánchez, J.S., García, V., Rangel, E.: Strategies for learning in class imbalance problems. Pattern Recognit. **36**(3), 849–851 (2003)
3. Batista, G.E.A.P.A., Prati, R.C., Monard, M.C.: A study of the behaviour of several methods for balancing machine learning training data. SIGKDD Explor. **6**(1), 20–29 (2004)
4. Breiman, L.: Bagging predictors. Mach. Learn. **24**(2), 123–140 (1996)
5. Demšar, J.: Statistical comparisons of classifiers over multiple data sets. J. Mach. Learn. Res. **7**, 1–30 (2006)
6. Freund, Y., Schapire, R.: A decision-theoretic generalization of on-line learning and an application to boosting. J. Comput. Syst. Sci. **55**(1), 119–139 (1997)
7. Galar, M., Fernández, A., Barrenechea, E., Bustince, H., Herrera, F.: A review on ensembles for class imbalance problem: bagging, boosting and hybrid based approaches. IEEE Trans. Syst., Man Cybern. Part C: Appl. Rev. **42**(4), 463–484 (2012)
8. Galar, M., Fernández, A., Barrenechea, E., Herrera, F.: Eusboost: enhancing ensembles for highly imbalanced data-sets by evolutionary undersampling. Pattern Recognit. **46**(12), 3460–3471 (2013)
9. He, H., Garcia, E.A.: Learning from imbalanced data. IEEE Trans. Knowl. Data Eng. **21**(9), 1263–1284 (2009)
10. Hernández-Lobato, D., Martínez-Muñoz, G., Suárez, A.: Statistical instance-based pruning in ensembles of independent classifiers. IEEE Trans. Pattern Anal. Mach. Intell. **31**(2), 364–369 (2009)
11. Huang, J., Ling, C.X.: Using AUC and accuracy in evaluating learning algorithms. IEEE Trans. Knowl. Data Eng. **17**(3), 299–310 (2005)
12. Kuncheva, L.I.: Diversity in multiple classifier systems. Inf. Fusion **6**(1), 3–4 (2005)
13. López, V., Fernández, A., García, S., Palade, V., Herrera, F.: An insight into classification with imbalanced data: empirical results and current trends on using data intrinsic characteristics. Inf. Sci. **250**(20), 113–141 (2013)
14. López, V., Fernández, A., Herrera, F.: On the importance of the validation technique for classification with imbalanced datasets: addressing covariate shift when data is skewed. Inf. Sci. **257**, 1–13 (2014)
15. Martínez-Muñoz, G., Suárez, A.: Using boosting to prune bagging ensembles. Pattern Recognit. Lett. **28**(1), 156–165 (2007)
16. Martínez-Muñoz, G., Hernández-Lobato, D., Suárez, A.: An analysis of ensemble pruning techniques based on ordered aggregation. IEEE Trans. Pattern Anal. Mach. Intell. **31**(2), 245–259 (2009)
17. Moreno-Torres, J.G., Sáez, J.A., Herrera, F.: Study on the impact of partition-induced dataset shift on k-fold cross-validation. IEEE Trans. Neural Netw. Learn. Syst. **23**(8), 1304–1313 (2012)

18. Polikar, R.: Ensemble based systems in decision making. IEEE Circuits Syst. Mag. **6**(3), 21–45 (2006)
19. Prati, R.C., Batista, G.E.A.P.A., Silva, D.F.: Class imbalance revisited: a new experimental setup to assess the performance of treatment methods. Knowledge and Information Systems, 1–25 (2014, in press)
20. Quinlan, J.: C4.5: Programs for Machine Learning. Morgan Kauffman, San Mateo (1993)
21. Rokach, L.: Ensemble-based classifiers. Artif. Intell. Rev. **33**(1), 1–39 (2010)
22. Sun, Y., Wong, A.K.C., Kamel, M.S.: Classification of imbalanced data: a review. Int. J. Pattern Recognit. Artif. Intell. **23**(4), 687–719 (2009)
23. Wang, S., Yao, X.: Diversity analysis on imbalanced data sets by using ensemble models. In: Proceedings of the 2009 IEEE Symposium on Computational Intelligence and Data Mining (CIDM'09), 324–331 (2009)
24. Wilcoxon, F.: Individual comparisons by ranking methods. Biom. Bull. **1**(6), 80–83 (1945)
25. Zadrozny, B., Langford, J., Abe, N.: Cost-sensitive learning by cost-proportionate example weighting. In: Proceedings of the 3rd IEEE International Conference on Data Mining (ICDM'03), 435–442 (2003)
26. Zhang, Y., Burer, S., Street, W.N.: Ensemble pruning via semi-definite programming. J. Mach. Learn. Res. **7**, 1315–1338 (2006)
27. Zhou, Z.H., Wu, J., Tang, W.: Ensembling neural networks: many could be better than all. Artif. Intell. **137**(1–2), 239–263 (2002)

A Variant of the K-Means Clustering Algorithm for Continuous-Nominal Data

Aleksander Denisiuk and Michał Grabowski

Abstract The core idea of the proposed algorithm is to embed the considered dataset into a metric space. Two spaces for embedding of nominal part with the Hamming metric are considered: Euclidean space (the classical approach) and the standard unit sphere $\mathbb{S}$ (our new approach). We proved that the distortion of embedding into the unit sphere is at least 75 % better than that of the classical approach. In our model, combinations of continuous and nominal data are interpreted as points of a cylinder $\mathbb{R}^p \times \mathbb{S}$, where p is the dimension of continuous data. We use a version of the gradient algorithm to compute centroids of finite sets on a cylinder. Experimental results show certain advances of the new algorithm. Specifically, it produces better clusters in tests with predefined groups.

1 Introduction

From the very beginning we define a dissimilarity function or a metric on combinations of continuous and nominal (categorical) data. There is a huge collection of dissimilarity functions on vectors of nominal data, used in data exploration. For example, the Hamming distance, the Jaccard distance, the distance defined after the Bayesian numerical codding of nominal values, and other concepts [6, 8]. In this paper we follow the approach with the Hamming distance. Let (x, n) be a record of continuous (x) and nominal (n) data, where $x \in \mathbb{R}^p$. We define metric on the space of such records as $\operatorname{dist}\big((x_1, n_1), (x_2, n_2)\big) = K\big(d(x_1, x_2), H(n_1, n_2)\big)$, where

A. Denisiuk (✉)
University of Warmia and Mazury in Olsztyn, Olsztyn, Poland
e-mail: denisiuk@matman.uwm.edu.pl

M. Grabowski
Warsaw School of Computer Science, Warsaw, Poland
e-mail: mgrabowski@poczta.wwsi.edu.pl

R. Burduk et al. (eds.), *Proceedings of the 9th International Conference on Computer Recognition Systems CORES 2015*, Advances in Intelligent Systems and Computing 403, DOI 10.1007/978-3-319-26227-7_2

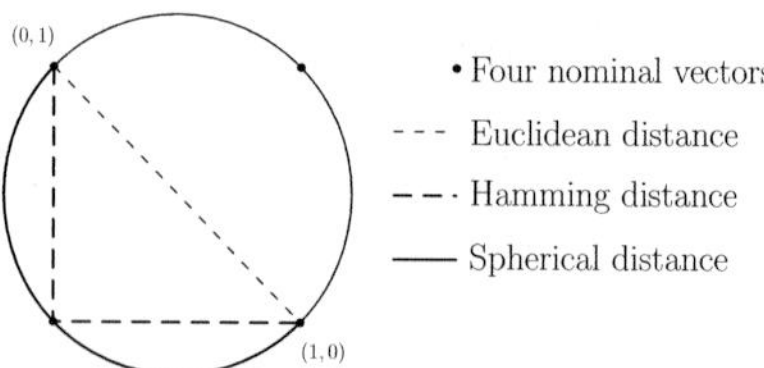

Fig. 1 The classical embedding of the Hamming metric of the Cartesian product of two-elements nominal domain $\{0, 1\} \times \{0, 1\}$ into the Euclidean plane. One can see that normalized spherical distance coincides with the Hamming distance

$d(x_1, x_2)$ is the standard Euclidean distance, $H(n_1, n_2)$ is the Hamming distance, and $K : \mathbb{R}^2 \to \mathbb{R}^+$ is an appropriate function. It can be observed [5] that such an approach produces classification results that are not worse than those of Bayesian numerical codding of nominal values. Thus, the Hamming metric may be considered as a sufficiently strong option. We accept a metric of this form and suggest a suitable function K in Sect. 2. To perform the k-means algorithm, one should be able to measure a distance between two records of data and to compute a centroid of a finite set of data records. The embedding of the considered dataset into a metric space equipped with a method of computing centroids is the core idea of this paper. We search for a relevant space by embedding the Hamming metric space of nominal data into a Riemannian manifold with possibly small distortion. The classical approach, representing nominal values as equidistant vertexes of a simplex, can be considered as embedding of the Hamming metric space into the Euclidean space. In general, isometric embedding of the Hamming metric into the Euclidean space is not possible [7, 9]. In Sect. 3 we analyze the distortion of this embedding. Two- and three-dimensional examples suggest that embedding of nominal values into a sphere has a distortion that is less than distortion of classical embedding into Euclidean space (Fig. 1). In Sect. 4 we prove that this is a general phenomenon: every Hamming metric space $(A_1 \times \cdots \times A_s, d_H)$ can be embedded into a suitable multidimensional sphere $\mathbb{S}$ with distortion that is better than the distortion of embedding it into the Euclidean space. We give a quantitative measure of this distortion improvement (Theorems 3 and 4), which is at least 75 % better than the distortion of embedding into the Euclidean space. We represent combinations of continuous and nominal data (x, n) as points on a cylinder $\mathbb{R}^p \times \mathbb{S}$, where p is the dimension of continuous data. To compute centroids of a finite subset of the cylinder, a version of the gradient algorithm with respect to the cylinder metrics is proposed. The experimental results (Sect. 5) show that the within index of clusters, produced by both algorithms are comparable. On the other hand, our algorithm has an advantage with respect to the degree of covering of predefined groups. The authors thank the participants of the CORES2015 conference for interesting discussions, related to the article.

2 Technical Preliminaries

In this section the basic notations, definitions, and algorithms are introduced. Specifically, we formulate a general scheme of the k-means clustering algorithm in an abstract metric space. This scheme is instanced in Sects. 3 and 4 by embedding the Hamming metric into the Euclidean space and the unit sphere respectively.

Definition 1 Let $A_1, \ldots, A_s$ be finite sets of nominal values, $H = A_1 \times \cdots \times A_s$. The Hamming metric on H is defined as follows: for each two vectors $n, n' \in H$,

$$d_H(n, n') = s^{-1} \Big| \{ i = 1, \ldots, s | n_i \neq n'_i \} \Big|, \tag{1}$$

where $n = (n_1, \ldots, n_s)$, $n' = (n'_1, \ldots, n'_s)$.

Definition 2 (*cf.* [9]) Let (U, d) and (U', d') be two metric spaces. A mapping $f : U \to U'$ is said to have a distortion of at most c, if there exists a constant $\nu \in (0, +\infty)$ such that for all $x, y \in U$

$$\nu \cdot d(x, y) \le d'(f(x), f(y)) \le c \cdot \nu \cdot d(x, y). \tag{2}$$

In other words, the mapping f can magnify the relative distances between points not greater than c times. In this paper we consider datasets combining both continuous and nominal data. To represent continuous data we use points from the standard p-dimensional Euclidean metric space $(\mathbb{R}^p, d_E)$. The nominal part of record is represented as a point at the Hamming metric space $H = A_1 \times \cdots \times A_s$. The set $X = \mathbb{R}^p \times H$ is considered as a set of all records. Every record $r \in X$ is a pair of continuous and nominal data, $r = (x, n)$. We admit the following metric on X:

$$\text{dist}^2\big((x, n), (x', n')\big) = d_E^2(x, x') + d_H^2(n, n')$$

We end this section with a general scheme of the k-means clustering algorithm. Let (M, d_M) be a metric space.

Definition 3 A point $m \in M$ is said to be a *centroid* of a finite set $\{ x_1, \ldots, x_N \} \subset M$, if $\sum_{j=1}^{N} d_M^2(m, x_j) = \min_{x \in M} \sum_{j=1}^{N} d_M^2(x, x_j)$.

We assume that we have some algorithm for computation of centroids of finite subsets of M. Let $\iota : X \to M$ be an embedding map. The (M, ι) algorithm is defined in algorithm 1. In what follows we use for $y, y' \in \mathbb{R}^p$ the following: notations $\langle y, y' \rangle = \sum_{j=1}^{p} y_j y'_j$, $\|y\|^2 = \sum_{j=1}^{p} y_j^2$.

Algorithm 1 A general k-means algorithm

Require: k is the number of clusters, $T \subset X$ is a finite set of records
Ensure: The resulting clusters C_1, ..., C_k in the data set T
  $MT = i(T) \subset M$
  Choose randomly initial centroids $m_1, \ldots, m_k \in M$
  {Compute initial clusters $MC_1, \ldots MC_k \subset M$}
  for all $x \in MT$ **do**
    x is classified to the cluster MC_j, if $d_M(x, m_j) = \min(d_M(x, m_1), \ldots, d_M(x, m_k))$
  end for
  while The current set of clusters $MC_1, \ldots, MC_k$ is not stabilized **do**
    for all Current clusters $MC_1, \ldots, MC_k$ **do**
      Compute new centroids $m_1, \ldots, m_k \in M$
      {Compute new clusters $MC_1, \ldots MC_k \subset M$}
      for all $x \in MT$ **do**
        x is classified to MC_j, if $d_M(x, m_j) = \min(d_M(x, m_1), \ldots, d_M(x, m_k))$
      end for
    end for
  end while
Return: $C_1 = \imath^{-1}(MC_1), \ldots, C_k = \imath^{-1}(MC_k)$

3 Euclidean Embedding of the Hamming Metric

The Hamming metric space cannot be isometrically mapped into the Euclidean space (v. [7]). However, it can be embedded with a particular distortion. Distortion analysis of embeddings of finite metric spaces into Euclidean spaces is actually hard [7, 9]. Bourgaine in [1] proved that every m-point metric space can be embedded into the $O(\log^2 m)$-dimensional Euclidean space with distortion $O(\log m)$. Consider the following well-known, folklore-type idea: map each of the m values of the nominal domain into a vertex of a simplex of equidistant points in $\mathbb{R}^{m-1}$. We expand this mapping to the space of nominal records:

$$\phi_{sim} : A_1 \times \cdots \times A_s \to \mathbb{R}^{a_1-1} \times \cdots \times \mathbb{R}^{a_s-1},$$

where $|A_j| = a_j$, $j = 1, \ldots, s$.

Theorem 1 *The distortion of ϕ_{sim} is not greater than $\sqrt{\sum(a_j - 1)}$.*

The simple proof is omitted. The above estimation of distortion is optimal. We can deduce Theorem 1 as well as optimality of this estimation from the following theorem by Enflo [2].

Theorem 2 *The standard embedding of s-dimensional unit cube with graph metric has optimal distortion, which is equal to $\sqrt{s}$.*

Let $q = \sum_{j=1}^{s}(a_j - 1)$. We map a data record $r = (x, n) \in X$ into the Euclidean space $\mathbb{R}^p \times \mathbb{R}^q$ as follows: $\phi_E(x, n) = (x, \phi_{sim}(n))$. Considering this embedding as $\imath$ and $\mathbb{R}^p \times \mathbb{R}^q$ with the standard Euclidean metric as M (with obvious computation of centroids as means), we instance algorithm 1. Thus we arrive at Euclidean k-means clustering of combinations of continuous and nominal data by embedding the Hamming metric into the Euclidean space. The main purpose of this paper is to compare this approach with embedding of the Hamming metric into a sphere, with better distortion.

4 Spherical Embedding of the Hamming Metric

In this section we consider another metric space M, the cylinder $\mathbb{R}^p \times \mathbb{S}$, where $\mathbb{S}$ is the unit sphere. We define embedding of the Hamming metric space into the sphere and analyze the distortion of this embedding (Theorems 3 and 4). Theorems 2 and 3 imply that the distortion of the spherical embedding is much better than the distortion of Euclidean embedding. We end this section with an algorithm of calculation of centroids. We show that binary Hamming s-dimensional set can be embedded into $s-1$ dimensional sphere with distortion less than that of Enflo's theorem.

Theorem 3 *For any $s > 1$ an s-dimensional cube $A = \{-1, 1\}^s$ with Hamming distance can be embedded into the standard unit sphere $\mathbb{S}^{s-1}$ with distortion less than $\lambda^{-1}\sqrt{s}$, where $\lambda^{-1} < 3/4$ is the constant, defined in (5).*

Proof Place elements of A into corresponding vertexes of the standard cube $[-1, 1]^s$ in $\mathbb{R}^s$. Consider the circumscribed sphere and project it to the unit sphere $\mathbb{S}^{s-1}$, centered at the origin. The spherical distance between embedded vertexes is the angle measure between corresponding cube vertexes. Using the Euclidean scalar product, we obtain for $\theta = \widehat{e, f}$

$$\cos\theta = \frac{\sum e_i f_i}{s} = \frac{s - 2|\{i|e_i \neq f_i\}|}{s} = 1 - 2d_H(e, f),$$

since all the coordinates of e and f are ± 1. Therefore,

$$d_s(e, f) = \arccos(1 - 2d_H(e, f)), \tag{3}$$

where d_s is the spherical distance and d_H is the Hamming distance. To estimate the distortion of this embedding, note that possible values for the Hamming distances are: $\frac{1}{s}, \frac{2}{s}, \ldots, \frac{s-1}{s}, 1$. Therefore, possible values for the spherical distances are: $\arccos(1 - 2 \cdot \frac{1}{s})$, $\arccos(1 - 2 \cdot \frac{2}{s})$, …, $\arccos(1 - 2 \cdot \frac{s-1}{s})$, π. So, the distortion σ satisfies inequality $\sigma < M/m$, where

$$M = \max_{k=1,\ldots,s} \frac{\arccos(1 - 2 \cdot \frac{k}{s})}{k/s}, \quad m = \min_{k=1,\ldots,s} \frac{\arccos(1 - 2 \cdot \frac{k}{s})}{k/s}. \tag{4}$$

To estimate M and m in (4), consider a function $f(t) = \arccos(1 - t)/t, t \in (0, 2)$. By analyzing its first and second derivatives, one can prove that f is convex on $(0, 2)$

and decreases for $t \in (0, 1)$. Since $f(1) = f(2) = \frac{\pi}{2}$, $f(t)$ has the only minimum, belonging to $(1, 2)$. Hence $M = s \arccos(1 - 2/s)$, $m > 2\lambda$, where

$$\lambda = \min_{t \in [1,2]} \arccos(1 - t)/t. \tag{5}$$

To estimate M, we make use of the inequality

$$\theta < 2 \tan \frac{\theta}{2} = 2\sqrt{\frac{1 - \cos\theta}{1 + \cos\theta}},$$

where $\theta = \arccos(1 - 2/s)$, and therefore $\cos\theta = 1 - 2/s$. So, $M < 2\frac{s}{\sqrt{s+1}} < 2\sqrt{s}$. Therefore $\sigma < \lambda^{-1}\sqrt{s}$, where λ is defined in (5). Since $f(t)$, defining λ, is strictly convex, one can numerically estimate its value: $\lambda \approx 1.380$, therefore $\lambda^{-1} < 3/4$.

Now let us consider a general case for any finite set.

Theorem 4 *Let $|A_j| = a_j$ for $j = 1, \ldots, s$. Then $A = A_1 \times \cdots \times A_s$ with the standard Hamming metrics d_h (1) can be embedded into the standard unit sphere $\mathbb{S}^{q-1}$ ($q = \sum_{j=1}^{s} a_j - s$) with distortion less than*

$$\frac{1}{\lambda}\sqrt{2\frac{a_{\min}}{a_{\min} - 1}\frac{a_{\max} - 1}{a_{\max}}}\sqrt{s}, \tag{6}$$

where $a_{\min} = \min_{j=1,\ldots,s} a_j$, $a_{\max} = \max_{j=1,\ldots,s} a_j$, λ is the constant (5), $\lambda^{-1} < 3/4$.

Proof The set A_j can be embedded into $\mathbb{R}^{a_j - 1}$ by placing its elements into vertexes of a simplex, centered at the origin. Let $(e_1, \ldots, e_{a_j - 1})$ be the coordinates of certain vertex of the simplex. We will assume that $\sum e_i^2 = 1$,. The desired embedding of $A_1 \times \cdots \times A_s$ to $\mathbb{R}^q$ is defined as follows:

$$\phi_s : n = (n_1, \ldots, n_s) \mapsto (e_1^{(1)}, \ldots, e_{k_1}^{(1)}, \ldots, e_1^{(s)}, \ldots, e_{k_s}^{(s)}), \tag{7}$$

where $k_j = a_j - 1$ and $e_1^{(j)}, \ldots, e_{k_j}^{(j)}$ are the coordinates of n_j on the corresponding simplex ($j = 1, \ldots, s$). One can see that the image of n belongs to the sphere of radius $\sqrt{s}$, i.e. $\|y\|^2 = s$. The distance between two points y and y' on the sphere is the angular measure between corresponding vertexes. For the Euclidean scalar product we obtain: $\langle y, y' \rangle = \lambda_1 + \cdots + \lambda_s$, where

$$\lambda_j = \begin{cases} 1, & \text{if } y_j = y'_j, \\ \frac{1}{1 - a_j}, & \text{if } y_j \neq y'_j. \end{cases}$$

Here $\frac{1}{a_j - 1}$ is the cosine of the angle between corresponding simplex vertexes. If y and y' differ in k places, $d_H(y, y') = k/s$, then the possible values for $d_s(y, y')$ are

$$d_s(y, y') = \arccos\left(1 - \frac{1}{s}\left(k + \sum\nolimits_{i=1}^{k} \frac{1}{a_{j_i} - 1}\right)\right). \tag{8}$$

To estimate the distortion, we should estimate the minimum and maximum of $d_s(y, y')/k$ for all $k = 1, \ldots, s$ and all k-tuples $j_1, \ldots, j_k$. Since $\arccos(1 - t)$ increases, one can substitute a_j in (8) by a minimal value for estimation of $M = \max d_s(y, y')/k$, and by a maximal value for estimation of $m = \min d_s(y, y')/k$. Then, repeating the estimates from Theorem 3, we get

$$M \le \sqrt{2s\frac{a_{\min}}{a_{\min} - 1}} \quad m \ge \lambda\frac{a_{\max}}{a_{\max} - 1},$$

These estimates immediately give (6).

The embedding of a record $r = (x, n)$ into the cylinder $\mathbb{R}^p \times \mathbb{S}^{q-1}$ is defined as follows: $(x, n) \mapsto (x, y) = (x, \phi_s(n)) = (x, y)$, where ϕ_s is defined in (7), $\|y\| = 1$. The distance between two records on the cylinder is defined by the following formula: $\mathrm{dist}^2\big((x, y), (x', y')\big) = \sum_{j=1}^{p}\big(x_j - x_j'\big)^2 + \kappa \arccos^2\big(\langle y, y'\rangle\big)$, where κ is a scaling coefficient. Let a finite set X of records, embedded into the cylinder be given as $(x^{(j)}, y^{(j)}) = (x_1^{(j)}, \ldots, x_p^{(j)}, y_1^{(j)}, \ldots, y_q^{(j)})$, where $\|y^{(j)}\| = 1$, and $j = 1, \ldots, N$. To compute a centroid $(\bar{x}, \bar{y})$ we should minimize the following expression:

$$\sum_{j=1}^{N}\sum_{k=1}^{p}\big(x_k - x_k^{(j)}\big)^2 + \kappa\sum_{j=1}^{N}\arccos^2\big(\langle y, y^{(j)}\rangle\big).$$

So, the minimization splits into two independent problems. Minimization with respect to x gives the standard mean. For the second term we suggest to use a gradient method. This is summarized in Algorithm 2.

Algorithm 2 Calculation of centroid of the finite set on a sphere

Require: $(x^{(j)}, y^{(j)}) = (x_1^{(j)}, \ldots, x_p^{(j)}, y_1^{(j)}, \ldots, y_q^{(j)})$, where $\|y^{(j)}\| = 1$, $j = 1, \ldots, N$
Ensure: $(\bar{x}, \bar{y})$ is a centroid with the accuracy ε
  $\bar{x} = N^{-1}\sum_{j=1}^{N} x^{(j)}$
  Zero approximation: $\bar{y} = \left(\sum_{j=1}^{N} y^{(j)}\right) / \left\|\sum_{j=1}^{N} y^{(j)}\right\|$.
  if $\sum_{j=1}^{N} y^{(j)} = 0$ **then**
    choose one of $y^{(j)}$ (of maximal multiplicity).
  end if
  repeat {The next approximation}
    $y' = \frac{y - \omega\Delta y}{\|y - \omega\Delta y\|}$, where ω is a relaxation parameter, $\Delta y = \nabla\left(\sum \arccos^2\big(\langle y, y^{(j)}\rangle\big)\right)$
  until $\|y' - y\| > \varepsilon$

5 Experimental Results

Clustering validation is a very subtle issue. There is a clear measure of success in the classification problem: cross-validation as estimation of expected loss over the joint probability distribution on data × decision. In the context of clustering, we have no such direct measure of success. As Hastie, Tibshirani, Friedman put it (v. [6]): *This uncomfortable situation has led to heavy proliferation of proposed methods, since effectiveness is a matter of opinion and cannot be verified directly.* The approach with predefined clusters seems to be credible (to some extent). Some predefined groups of data are considered as intended clusters. The mean covering degree of predefined groups by computed clusters may be considered as the clustering validity index. The degree of covering A by B is defined as $|A \cap B|/|B|$. We analyze two training sets with decision categories as intended clusters [4]: the *Heart Disease* (6 continuous, 7 nominal attributes, 370 records, 2 decision categories) and the *Australian Credit Approval* (6 continuous, 8 nominal attributes, 690 records, 2 decision categories). The following artificial datasets are analyzed as well:

ADS1: The set of records $X = \mathbb{R} \times \{0, 1\}$—one continuous attribute and one nominal attribute with two-elements domain. We define ADS1 as the union of two intended clusters $A \times \{0\} \cup B \times \{1\}$, where the elements of A and B have been randomly chosen according to the normal distributions $N(170, 10)$ and $N(155, 10)$ respectively. Each of A and B contains 300 real numbers.

ADS2: The set of records $X = \mathbb{R}^4 \times A_1 \times A_2 \times A_3 \times A_4 \times A_5$, where nominal domains are of arbitrarily chosen cardinalities: 3, 2, 4, 5 and 2 respectively. Let $\mathbb{B}(x, r) = \{ y \in X | \text{dist}(x, y) \leq r \}$ be a ball with a center x and radius r. Choose randomly two elements $x_1, x_2 \in X$, $r = \text{dist}(x_1, x_2)/2$. Let $A, B \subset X$ be two sets of 200 uniform randomly chosen elements from balls $\mathbb{B}(x_1, r)$ and $\mathbb{B}(x_2, r)$ respectively. The disjoint sets A, B are considered as two intended, not well separated, clusters. We define ADS2 as $A \cup B$. The dataset ADS2 was randomly generated 10 times in experiments.

The following two clustering validity indexes appear to make sense in the context of k-means clustering: the number of iterations stabilizing clustering and the within index W, which, for $C_1, \ldots, C_k \subset X$ is defined as $W = 1/2 \cdot \sum_j \sum_{r,r' \in C_j} \text{dist}^2(r, r')$. We compare the following variants of k-means clustering:

MC: k-medoids clustering: the centroids are records from the analyzed dataset.

EC: Euclidean k-means: with classical embedding of the Hamming metric into the Euclidean space.

CC: Cylindrical k-means: the Hamming metric is embedded into the sphere.

In order to maintain a balance between continuous and Hamming parts of the metric, all the continuous attributes were renormalized to the normal Gaussian distribution $N(0, 1)$ or to the segment $[0, 1]$. The results of numerical experiment are presented in Table 1. We do not present results for the within index W and for number of iterations, stabilizing clustering. These results do not indicate any advantage of one

Table 1 The average covering degree for tested datasets

Dataset	MC	EC	CC
ADS1	0.93	0.98	1
ADS2–I	0.78	0.84	0.91
ADS2–II	0.88	1	1
ADS2–III	0.92	0.95	0.96
ADS2–IV	1	0.99	1
ADS2–V	0.9	0.9	1
ADS2–VI	0.79	0.84	0.93
ADS2–VII	0.85	0.88	0.97
ADS2–VIII	0.85	0.94	1
ADS2–IX	0.92	0.91	0.95
ADS2–X	0.95	0.93	0.95
Heart Disease	0.72	0.80	0.86
Australian Credit Approval	0.69	0.79	0.83

Each algorithm was executed 30 times with two randomly chosen initial centroids on each dataset

clustering (EC, CC) over another, with respect to these indexes. The only observed phenomenon is that MC clustering on datasets ADS2 is visibly worse with respect to the index W. We can see that the cylindrical clustering has better covering degree in all datasets. Note that it is very stable on ADS1 with respect to choice of initial centroids; in this case the normalized spherical metric coincides with the Hamming metric.

6 Conclusions and Final Remarks

The explored datasets showed the advantage of cylindrical clustering over two other analyzed methods with respect to the degree of covering of predefined groups. However, we do not see a hard relation between the distortion improvement and the resulting values of the within index W. None of the considered clustering algorithms (Euclidean, cylindrical) has a definite advantage over the other with respect to the within index W. The k-medoids algorithm produces visibly worse results compared to the Euclidean and the cylindrical k-means algorithms. At least on the analyzed datasets and with respect to the here accepted clustering validation indexes. In our opinion, the cylindrical k-means clustering is worth considering as one of the alternatives, when several algorithms are tested in order to find relevant, interpretable clusters of continuous-nominal data. Our approach is applicable to those datasets where information hidden in nominal data influences relevant clusters by the Hamming metric. Many other ways of such nominal information influence are possible, for instance, via leading approaches by probabilistic models like attribute information role in the classical version of COBWEB algorithm [3]. If real clusters are not (roughly) spherical with respect to the metric of the data space, then the resulting clusters, produced by our approach, may not be credible. The same objection is valid for the Euclidean embedding. Perhaps it is worth to analyze the embedding of Hamming metric into the Euclidean space using the idea of Bourgaine's theorem [1], since the dimension of this space is significantly less than the dimension of Euclidean spaces and of spheres used by our approach. This is a possible future work as well as a deeper experimental study. We have analyzed only two geometrical structures: Euclidean space and sphere. There could be other possibilities, with distortion that is significantly better than that of spherical embedding. In general, the following hard question is opened up: does there exist a Riemannian manifold providing really good k-means clustering of continuous-nominal data with respect to the within index W? Note that Riemannian metric works well in some data exploration contexts, for instance, a version of adaptive k-nearest neighbors algorithm based on Riemannian metric [10].

References

1. Bourgaine, J.: On lipschitz embeddings of finite metric spaces in Hilbert space. Isr. J. Math. **52**, 46–52 (1985)
2. Enflo, P.: On the nonexistence of uniform homeomorphisms between Lp spaces. Ark. Mat. **8**, 5–103 (1969)
3. Fisher, D.H.: Knowledge acquisition via incremental conceptual clustering. Mach. Learn. **2**, 139–172 (1987)
4. Frank, A., Asuncion, A.: UCI Machine Learning Repository http://archive.ics.uci.edu/ml (2010). University of California, Irvine, School of Information and Computer Science
5. Grabowski, M., Korpusik, M.: Metrics and similarities in modeling dependencies between continuous-nominal data. Zeszyty Naukowe WWSI. 10/7 (2013)
6. Hastie, T., Tibshirani, R., Friedman, J.: The Elements of Statistical Learning. Springer Series in Statistics. Springer, Berlin (2001)
7. Indyk, P., Matoušek, J.: Low-distortion embeddings of finite metric spaces. Handbook of Discrete and Computational Geometry. CRC Press LLC, Boca Raton (2004)
8. Krzanowski, W.J.: Principles of Multivariate Analysis: A User's Perspective. Clarendon Press, Oxford (1998)
9. Linial, N.: Finite metric spaces: combinatorics, geometry and algorithms. In: Symposium on Computational Geometry. ACM, Barcelona (2002)
10. Peng, J., Heisterkamp, D., Dai, H.: Adaptative quasiconformal kernel nearest neighbor classification. IEEE Trans. Pattern Anal. Mach. Intell. **26**(5), 656–661 (2004)

Combining One-vs-One Decomposition and Ensemble Learning for Multi-class Imbalanced Data

Bartosz Krawczyk

Abstract Learning from imbalanced data poses significant challenges for machine learning algorithms, as they need to deal with uneven distribution of examples in the training set. As standard classifiers will be biased toward the majority class there exist a need for specific methods than can overcome this single-class dominance. Most of works concentrated on binary problems, where majority and minority class can be distinguished. But a more challenging problem arises when imbalance is present within multi-class datasets, as relations between classes tend to complicate. One class can be a minority class for some, while a majority for others. In this paper, we propose an efficient method for handling such scenarios that combines the problem decomposition with ensemble learning. According to divide-and-conquer rule, we decompose our multi-class data into a number of binary subproblems using one-versus-one approach. To each simplified task we delegate a ensemble of classifiers dedicated to binary imbalanced problems. Then using a dedicated classifier fusion approach, we reconstruct the original multi-class problem. Experimental analysis backed-up with statistical testing clearly proves that such an approach is superior to state-of-the art ad hoc and decomposition methods used in the literature.

Keywords Ensemble classifiers · Imbalanced data · Multi-class classification · Pairwise learning · Binary decomposition

1 Introduction

Usage of machine learning algorithms for pattern classification is a still rapidly developing field, despite several decades of research conducted. This can be contributed to continuous emergence of novel and challenging topics that require specific types of methods to be tackled. One of the vital fields is related with the problem of imbal-

---

B. Krawczyk (✉)
Department of Systems and Computer Networks, Wroclaw University of Technology, Wroclaw, Poland
e-mail: bartosz.krawczyk@pwr.edu.pl

R. Burduk et al. (eds.), *Proceedings of the 9th International Conference on Computer Recognition Systems CORES 2015*, Advances in Intelligent Systems and Computing 403, DOI 10.1007/978-3-319-26227-7_3

anced data [15]. Here, we deal with a situation where the number of examples for each class in the training set differs significantly. Additionally, there are a number of distinct properties embedded in the nature of the imbalanced data that accompanies the uneven distributions. Canonical machine learning algorithms assume that the distribution of objects among classes is roughly equal—therefore, they tend to become biased toward the class with the highest number of examples. Additionally, they cannot deal with the mentioned difficulties embedded in the nature of data. This has lead to development of a family of models dedicated to handling imbalanced data. Such methods work on data level (e.g., preprocessing schemes) or classifier level (e.g., changing the focus of the training procedure). They have been proven to offer an efficient aid for mining imbalanced datasets, however often require careful parameter tuning. Recently, a combination of such techniques and multiple classifier systems was proposed [10]. Ensemble learning is considered as one of the most efficient tracks in machine learning, due to its ability to extract useful information from a group of classifiers [18]. Ensembles can allow us to outperform each base classifier by combining their areas of competence [13], improve the generalization abilities [4, 5], and remove redundant learners from the pool [3]. Therefore, a combination of classifier committees and methods for data preprocessing is currently considered as one of the most efficient solution for mining imbalanced sets. These solutions were mainly developed for binary imbalanced problems. There is still a lack of works dealing with the presence of imbalance in multi-class datasets [17]. Here we deal with a more complicated situation, as the relations between classes are no longer obvious. One class may be a majority one when compared to some other classes, but a minority or well balanced for rest of them. Therefore developing efficient solutions to deal with this scenario is a needed research direction. In this paper, we present a solution for handling multi-class imbalanced datasets with the combination of binary decomposition and ensemble learning. We propose to apply a divide-and-conquer principle in a form of one- -one decomposition, in which we create an exhaustive two-class combination of all available classes. This way we simplify our task and revert it to canonical binary imbalanced problem. We propose to solve each subproblem with the usage of specific ensemble classifiers dedicated to imbalanced classification, as they return excellent performance for standard binary problems. To reconstruct the original multi-class problem from binary outputs of local ensemble classifiers, we apply the pairwise coupling technique.

2 Learning from Imbalanced Data

The performance and quality of machine learning algorithms is conventionally evaluated using predictive accuracy. However, this is not appropriate when the data under consideration is strongly imbalanced, since the decision boundary may be strongly biased toward the majority class, leading to poor recognition of the minority class. Disproportion in the number of class examples makes the learning task more complex [15], but it is not the sole source of difficulties for machine learning algorithms.

It is usually accompanied by difficulties embedded in the structure of data such as small sample size (very limited availability of minority examples), small disjuncts (minority class can consist of several subconcepts), or class overlapping. Over the last decade there was developed a number of dedicated techniques for handling such difficulties [1, 6]. They can be divided into the major categories. First one consists of data-level approaches that in the preprocessing stage aim at rebalancing the original distribution. Classifier-level approaches try to adapt existing algorithms to the problem of imbalanced datasets and alleviate bias toward the majority class. Third group relies on cost-sensitive classification and assign higher misclassification cost for minority class, while classification is performed so as to reduce the overall learning cost. Ensemble systems have also been successfully applied to this domain, and mainly combine a committee learning algorithm (such as Bagging or Boosting) with one of the above-mentioned methods [10]. For example, one may combine Bagging with under- or oversampling to balance the distributions of objects in each of bags [2, 16]. One may also propose a hybrid training procedure for such a combined classifier that will apply cost-sensitive learning locally (for each base classifier) and globally (in the ensemble pruning step) [14]. These methods however are specific to binary imbalanced datasets. Multi-class problems arFi.we far more complicated, as relationships between classes are no longer obvious. Therefore, on may easily lose performance on one class while trying to gain it on other. There are still only few approaches for handling this task. Static-SMOTE [7] applied resampling procedure in M steps, where M is the number of classes. In each iteration, the resampling procedure selects the minimum size class, and duplicates the number of instances of the class in the original dataset. An ensemble learning algorithm for multi-majority and multi-minority cases was proposed in [17]. Authors combine AdaBoost with negative correlation learning, where starting weights of examples are calculated in inverse proportion to the number of objects in this class. Finally, a combination of binary decomposition and pre-processing methods was proposed as an efficient solution when the number of classes is high [8].

3 Proposed Approach

According to divide-and conquer rule, we should aim at solving each complex problem by dividing it into a series of subproblems, each easier to solve than the original task. This strategy can be easily applied in machine learning, where dealing with complex, multi-class datasets is a common practice. The most popular approach is to decompose the original dataset into a number of binary problems [9]. With this approach we obtain locally specialized classifiers that are trained over simplified task. This leads to lower complexity of the classifier training step and obtaining a pool of locally accurate models. However, we still need to reconstruct the original multi-class decision from a set of binary outputs. Two most common decomposition schemes are one-versus-one (OVO) and one-versus-all (OVA). The former creates all possible pairwise combinations of classes, while the latter selects one class as the positive class and uses all the remaining ones as the negative class. One should note drawbacks of these methods: OVO creates a larger number of base classifiers (especially cumbersome for dataset with numerous classes), while OVA introduced an artificial imbalance into data (as it combines all but one classes into an aggregated collection). Recent studies show that OVO outperforms in most cases OVA approach [11], and thus we can observe a shift toward this approach. We will also use it in this work. Additionally, most of the available multi-class imbalanced datasets do not have an extremely numerous umber of classes, so the number of base classifiers is so far not a problem in this domain. OVO divides an M-class dataset into $M(M-1)/2$ binary pairwise problems. Each classifier is trained on the basis of the reduced training dataset, which consists only of two corresponding classes. Example of such decomposition is given in Fig. 1. In the testing phase the new object is presented to each of the classifiers from the pool. The possible output of the classifier is given by $y_{ij} \in [0, 1]$. This stands for a classifier discriminating between classes i and j in favor of the former. The outputs are stored in the confidence matrix for each possible pairwise combination. The OVO combination methods work directly on this matrix. In this work, we will use the **Pairwise Coupling (PC)** method [12]. It is based on the estimation of the joint probability for M classes from the pairwise probabilities of all possible binary combinations. Therefore, for given Prob($Class_i|Class_i$ or $Class_j$), the fuser approximates the posteriori probabilities $\hat{p}(x) = (\hat{p_1}(x), \ldots, \hat{p_M}(x))$, based on the individual classifier outputs.

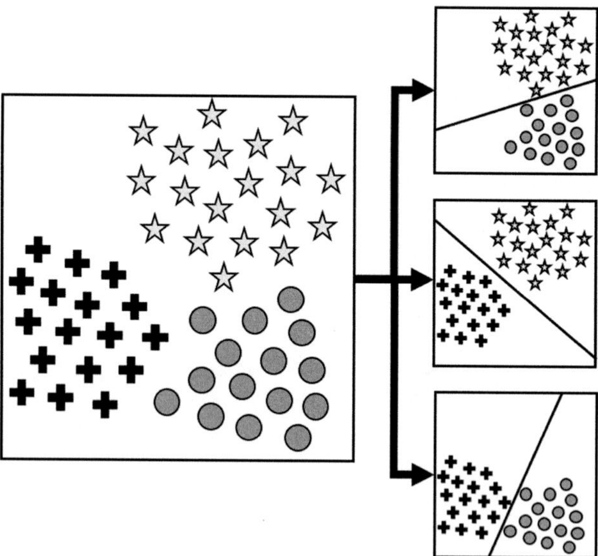

Fig. 1 An example of one-versus-one (OVO) decomposition of a three-class problem into three two-class problems

The class with the highest posteriori probability is selected as the final output of the system:

$$Class_x = arg \max_{m=1,2,\ldots,M} \hat{p_m}. \tag{1}$$

In order to calculate the posteriori probabilities, the Kullback–Leibler distance between r_{ij} and μ_{ij} is minimized:

$$l(p) = \sum_{1 \leq i \neq j \leq M} n_{ij} r_{ij} log \frac{r_{ij}}{\mu_{ij}} = \sum_{i<j} n_{ij} \left(r_{ij} log \frac{r_{ij}}{\mu_{ij}} + (1 - r_{ij}) log \frac{1 - r_{ij}}{1 - \mu_{ij}} \right), \tag{2}$$

where $\mu_{ij} = p_i/(p_i + p_j)$ and n_{ij} stands for the number of objects in the ith and jth classes. The positive impact of decomposition techniques on multi-class imbalanced data mining was recently reported by Fernández et al. [8]. However authors used there only simple preprocessing techniques. We propose to improve this by using ensemble learning methods dedicated specifically to binary imbalanced problems for each decomposed subproblem. This way we take advantage of efficient combined classifiers and show how easily extend their area of usability onto multi-class imbalanced problems. We propose to investigate the performance of two ensemble classifiers based on Bagging algorithm, but utilizing a different preprocessing method:

- **UnderBagging** [2] applies undersampling of the majority class conducted independently in each bag of the ensemble. This way we obtain balanced training sets for each base classifier and take advantage of the bagging scheme for diversifying classifier's input. Additionally, one may use a resampling with replacement of the minority class in order to further increase the diversity of the ensemble.
- **SMOTEBagging** [16] uses SMOTE algorithm for artificial introduction of samples into the minority class in a more directed way than random oversampling. It aims at creating a balanced distribution in each bag, as the number of instances contributed to each class is equal to the number of instances in majority class N_{maj}. However, in each iteration we set a different resampling rate $a\%$ raging from 10 % to 100 %. This defines the number of minority class instanced randomly resampled with replacement ($a\% \cdot N_{maj}$) in each bag. Remaining minority instances are generated by SMOTE.

As we can see, one of the ensembles works with reducing the size of the majority class, while the other concentrates on increasing the number of examples from the minority class. Examples of such approaches for decomposed multi-class imbalanced problem are presented in Fig. 2. One should note that there is a possibility that some of pairs of classes after a decomposition will have roughly equal number of examples. For such cases both UnderBagging and SMOTEBagging converge to normal Bagging method, which is desirable for such cases. Thus imbalanced pairs of classes will be balanced, while those with similar number of objects will be processed like standard cases.

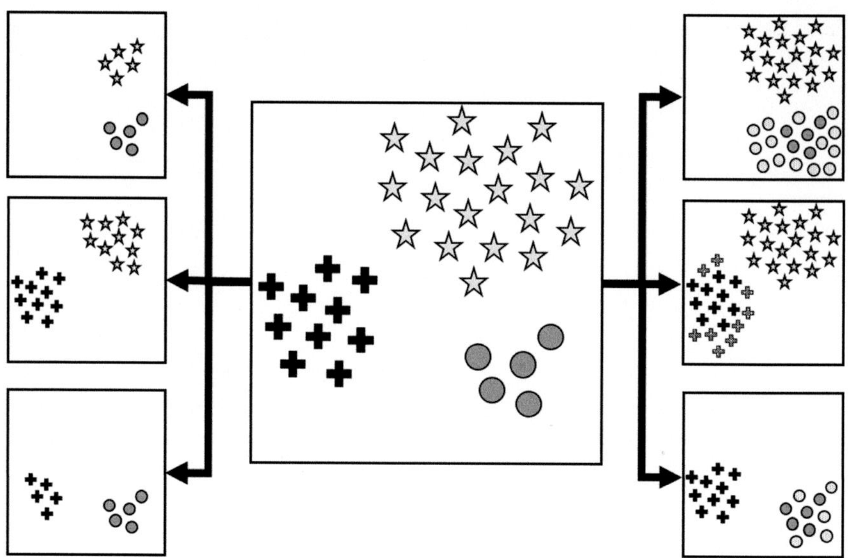

Fig. 2 An example of preprocessing methods applied over a decomposed multi-class imbalanced dataset. *Left* Undersampling applied to each subproblem. *Right* Oversampling applied to each subproblem

4 Experimental Study

The aim of the experiments was to evaluate the usefulness of the proposed combination of OVO decomposition and ensemble learning for mining multi-class imbalanced data, and to compare it with state-of-the-art approaches.

4.1 Datasets and Set-Up

We have selected 10 multi-class imbalanced datasets taken from KEEL repository.[1] Their details are given in Table 1. As a base classifier, we have applied a C4.5 decision tree. As reference methods we use two binary preprocessing techniques (SMOTE and Random Undersampling), and two ad hoc techniques (ST-SMOTE and AdaBoost.NC). C4.5 uses confidence level of 0.25, the minimum number of item-sets per leaf equal to 2, and post-pruning applied. SMOTE uses the 5-nearest neighbors of the minority class to generate the synthetic samples, and balance both classes to the 50 % of distribution. UnderBagging and SMOTEBagging use 40 base classifiers as suggested by previous experimental studies [10]. AdaBoost.NC uses penalty parameter $\alpha = 2$ and uses 51 classifiers in the committee as suggested by

[1] http://sci2s.ugr.es/keel/datasets.php.

Table 1 Details of datasets used in the experimental investigations, with the respect to no. of objects, no. of features, no. of classes, and imbalance ratio (IR)

No.	Name	Objects	Features	Classes	IR
1.	Autos	159	25	6	16.00
2.	Cleveland	467	13	5	12.62
3.	Contraceptive	1473	9	3	1.89
4.	Ecoli	336	7	8	71.50
5.	New-thyroid	215	5	3	5.00
6.	Page-blocks	5472	10	5	175.46
7.	Satimage	6435	36	7	2.45
8.	Thyroid	7200	21	3	40.16
9.	Wine-Quality-White	4898	11	11	439.60
10.	Yeast	1484	8	10	92.60

authors [17]. We use 5x2 CV F-test for training / testing and pairwise statistical analysis, while Friedman ranking test and Shaffer post hoc tests are applied for statistical comparison over multiple datasets.

4.2 *Results and Discussion*

The results are given in Table 2. The output of Shaffer post hoc test is reported in Table 3. From the obtained results one may see that the proposed combination of OVO decomposition and ensemble learning outperforms all other methods in 6 out of 10 cases. What is highly interesting it never leads to a drop of accuracy in comparison with combinations of OVO and preprocessing techniques. This can be contributed to highly efficient ensemble learning techniques that have been proven to deliver excellent performance for binary imbalanced data. Therefore, when considering decomposition-based approaches it is worthwhile to examine the performance of multiple classifier systems, which is further confirmed by Shaffer post hoc test. When comparing with ad hoc procedures for multi-class imbalance one may see that they always outperform ST-SMOTE. What is interesting is that in most cases a combination of OVO and preprocessing techniques also delivers better performance. This leads to a conclusion that it is more efficient to decompose the problem and use SMOTE locally than to use the modification for multi-class problem. AdaBoost.NC provides a more challenging reference and is able to outperform the proposed methods for four datasets. However for the remaining cases the decomposition-based solution returns superior results. This can be explained by the fact that multi-class ensemble learning method may often not be able to sufficiently capture the relations between classes and falsely label some situations as multi-majority or multi-minority.

Table 2 Results according to accuracy [%] for each examined methods over 10 datasets

Data	ST-SMOTE[1]	AdaBoost.NC[2]	OVO-RUS[3]	OVO-SMOTE[4]	OVO-UB[5]	OVO-SMB[6]
1.	82.13	83.26	81.56	80.36	84.46 *ALL*	83.18 3,4
2.	24.91	25.33	29.27	34.08	33.21 1,2,3,4	**36.04** *ALL*
3.	47.22	50.36	49.02	50.04	**53.18** 1,2,3,4	52.31 1,2,3,4
4.	65.38	74.28	73.04	71.13	**75.48** *ALL*	72.45 1,4
5.	90.65	**95.02**	90.20	93.04	93.18 1,3	93.87 1,3,4
6.	85.61	90.72	91.80	90.34	**93.05** *ALL*	90.98 1
7.	83.19	**88.78**	84.57	84.09	86.19 1,3,4	85.92 1,3,4
8.	98.24	99.02	98.88	99.29	99.19 1	**99.42** 1,3
9.	38.84	**49.73**	41.96	42.63	46.14 1,3,4,6	44.73 1,3,4
10.	51.72	56.03	51.58	52.73	**56.08** 1,3,4,6	54.22 1,3,4
Rank	5.65	2.40	4.40	3.90	1.60	3.05

OVO-UB stands for UnderBagging with OVO, OVO-SMB for SMOTEBagging with OVO. Small numbers under proposed methods stands for the indexes of reference classifiers that were statistically inferior in the pairwise 5 × 2 CV F-test. Last row represents the ranks after Friedman test

Table 3 Shaffer test for comparison between the proposed OVO ensemble learning and reference methods over multiple datasets

hypothesis	p-value
OVO-UB vs ST-SMOTE	+ (0.0147)
OVO-UB vs AdaBoost.NC	+ (0.0378)
OVO-UB vs OVO-RUS	+ (0.0191)
OVO-UB vs OVO-SMOTE	+ (0.0188)
OVO-SMB vs ST-SMOTE	+ (0.0192)
OVO-SMB vs AdaBoost.NC	− (0.0688)
OVO-SMB vs OVO-RUS	+ (0.0411)
OVO-SMB vs OVO-SMOTE	+ (0.0258)
OVO-UB vs OVO-SMB	+ (0.0208)

Symbol '+' stands for a situation in which the proposed method is superior, '−' for vice versa, and '=' represents a lack of statistically significant differences

When using the binary decompositions the relations between two classes become much more obvious and thus it is easier to properly analyze them. When comparing two used ensemble methods one may see a significant dominance of UnderBagging. This confirms the observations from the binary cases that undersampling-based ensembles tend to return excellent accuracy for imbalanced datasets, regardless of their simplicity.

5 Conclusions and Future Works

Mining multi-class imbalanced data is a challenging task and poses significantly higher difficulty than binary problems. So far only few methods were proposed for this domain, thus there is a need for further development in dedicated classification methodologies. This paper proposed a new approach for handling multi-class imbalanced datasets with a combination of pairwise OVO decomposition and ensemble learning. We proposed to divide the original set into a number of binary subproblems in oder to simplify the classification task. Then for each of such simplified subconcepts we proposed to delegate a ensemble classifier originally designed for handling binary problems. Outputs of individual ensembles were combined with pairwise coupling method. This way we were able to utilize highly efficient binary classification techniques in multi-class scenario. Experimental study, backed-up with a thorough statistical analysis had proven the quality of this proposal. We were able to achieve a boost in accuracy for all of analyzed problems, and for six of them we outperformed all of reference methods used in literature for multi-class imbalanced analysis. In future, we plan to examine the performance of other aggregation methods and ensemble classifiers for this task, and to propose dynamic classifier selection scheme for multi-class imbalanced datasets.

Acknowledgments This work was supported by the Polish National Science Center under the grant no. DEC-2013/09/B/ST6/02264.

References

1. Antonelli, M., Ducange, P., Marcelloni, F.: An experimental study on evolutionary fuzzy classifiers designed for managing imbalanced datasets. Neurocomputing **146**, 125–136 (2014)
2. Barandela, R., Valdovinos, R., Sánchez, J.: New applications of ensembles of classifiers. Pattern Anal. Appl. **6**(3), 245–256 (2003)
3. Burduk, R.: Classifier fusion with interval-valued weights. Pattern Recognit. Lett. **34**(14), 1623–1629 (2013)
4. Cyganek, B.: Recognition of road signs with mixture of neural networks and arbitration modules. In: Advances in Neural Networks—ISNN 2006, Third International Symposium on Neural Networks, Chengdu, China, 28 May–1 June, 2006, Proceedings, Part III. pp. 52–57 (2006)

5. Cyganek, B., Woźniak, M.: Pixel-based object detection and tracking with ensemble of support vector machines and extended structural tensor. In: Proceedings of the 4th International Conference on Computational Collective Intelligence Technologies and Applications, ICCCI, Part I, pp. 104–113. Ho Chi Minh City, Vietnam, 28–30 Nov 2012
6. Czarnecki, W.M., Tabor, J.: Multithreshold entropy linear classifier: theory and applications. Expert Syst. Appl. **42**(13), 5591–5606 (2015)
7. Fernández-Navarro, F., Hervás-Martínez, C., Gutiérrez, P.A.: A dynamic over-sampling procedure based on sensitivity for multi-class problems. Pattern Recognit. **44**(8), 1821–1833 (2011)
8. Fernández, A., López, V., Galar, M., del Jesús, M.J., Herrera, F.: Analysing the classification of imbalanced data-sets with multiple classes: binarization techniques and ad-hoc approaches. Knowl.-Based Syst. **42**, 97–110 (2013)
9. Galar, M., Fernandez, A., Barrenechea, E., Bustince, H., Herrera, F.: An overview of ensemble methods for binary classifiers in multi-class problems: experimental study on one-versus-one and one-versus-all schemes. Pattern Recognit. **44**(8), 1761–1776 (2011)
10. Galar, M., Fernandez, A., Barrenechea, E., Bustince, H., Herrera, F.: A review on ensembles for the class imbalance problem: bagging-, boosting-, and hybrid-based approaches. IEEE Trans. Syst., Man Cybern. Part C: Appl. Rev. **42**(4), 463–484 (2012)
11. Galar, M., Fernández, A., Barrenechea, E., Herrera, F.: DRCW-OVO: distance-based relative competence weighting combination for one-versus-one strategy in multi-class problems. Pattern Recognit. **48**(1), 28–42 (2015)
12. Hastie, T., Tibshirani, R.: Classification by pairwise coupling. Annal. Stat. **26**(2), 451–471 (1998)
13. Jackowski, K., Woźniak, M.: Algorithm of designing compound recognition system on the basis of combining classifiers with simultaneous splitting feature space into competence areas. Pattern Anal. Appl. **12**(4), 415–425 (2009)
14. Krawczyk, B., Woźniak, M., Schaefer, G.: Cost-sensitive decision tree ensembles for effective imbalanced classification. Appl. Soft Comput. **14**, 554–562 (2014)
15. Sun, Y., Wong, A.K.C., Kamel, M.S.: Classification of imbalanced data: a review. Int. J. Pattern Recognit. Artif. Intell. **23**(4), 687–719 (2009)
16. Wang, S., Yao, X.: Diversity analysis on imbalanced data sets by using ensemble models. In: Proceedings of 2009 IEEE Symposium on Computational Intelligence and Data Mining, CIDM 2009. pp. 324–331 (2009)
17. Wang, S., Yao, X.: Multiclass imbalance problems: analysis and potential solutions. IEEE Trans. on Syst., Man, Cybern., Part B: Cybern. **42**(4), 1119–1130 (2012)
18. Woźniak, M., Graña, M., Corchado, E.: A survey of multiple classifier systems as hybrid systems. Inf. Fusion **16**, 3–17 (2014)

Combining One-Versus-One and One-Versus-All Strategies to Improve Multiclass SVM Classifier

Wiesław Chmielnicki and Katarzyna Stąpor

Abstract Support Vector Machine (SVM) is a binary classifier, but most of the problems we find in the real-life applications are multiclass. There are many methods of decomposition such a task into the set of smaller classification problems involving two classes only. Two of the widely known are one-versus-one and one-versus-rest strategies. There are several papers dealing with these methods, improving and comparing them. In this paper, we try to combine theses strategies to exploit their strong aspects to achieve better performance. As the performance we understand both recognition ratio and the speed of the proposed algorithm. We used SVM classifier on several different databases to test our solution. The results show that we obtain better recognition ratio on all tested databases. Moreover, the proposed method turns out to be much more efficient than the original one-versus-one strategy.

1 Introduction

Support vector machine (SVM) proposed by Vapnik [28] is widely known and robust binary classifier. However, most of the tasks in the real-world applications are the classification problems that involve more than two classes. How to effectively extend a binary to a multiclass classifier is still an ongoing research problem. There are many methods proposed to deal with this issue. The most popular is to construct the multiclass classifier by combining several binary classifiers. To use this approach, we have to find the method of the decomposition multiclass task into the set of the smaller classification problems involving two classes only. There are many papers addressing benefits obtained from the decomposition of such task for example [1, 10, 25, 30].

W. Chmielnicki (✉)
Faculty of Physics, Astronomy and Applied Computer Science,
Jagiellonian University, Łojasiewicza 11, 30-348 Krakow, Poland
e-mail: wieslaw.chmielnicki@uj.edu.pl

K. Stąpor
Institute of Computer Science, Silesian University of Technology,
Akademicka 16, 44-100 Gliwice, Poland
e-mail: katarzyna.stapor@polsl.pl

R. Burduk et al. (eds.), *Proceedings of the 9th International Conference on Computer Recognition Systems CORES 2015*, Advances in Intelligent Systems and Computing 403, DOI 10.1007/978-3-319-26227-7_4

Additionally, some interesting review considering this topic can be found in [21]. We can also look at the problem of the decomposition from the efficiency point of view [7] or we can investigate how the problem properties can be employed to the construction of the decomposition scheme [22]. It is also worth mentioning an approach to solve the multiclass problem by combining several one-class classifiers [20]. The one-versus-one strategy is well-known and successfully used in many applications. In general, its principle is to separate each pair of the classes ignoring the remaining ones. Then all objects are tested against these classifiers and then a voting scheme is used. A closer look at this strategy shows the problem which impacts the final result of the combined classifier. Each binary classifier vote for each object even if it does not belong to one of the two classes on which it is trained on. This problem is addressed in our strategy. In our solution, additional classifiers are used to select the objects which will be considered by the binary classifiers. The similar solution has been proposed in our other paper [6] but the method has some limitations which we will address later in the next section. The proposed solution has been tested on several data sets. We used two real-life databases: proteins database [8, 17] and gesture database [14] and two other databases from the UCI Machine Learning Repository [27]. Obtained results show that our strategy outperforms original OVO algorithm, pairwise coupling (PWC), and the solution proposed by Moreira and Mayoraz [23]. The difference is more significant when the number of the classes in the problem is growing.

2 Related Work

There are many methods of the decomposition of the multiclass problems into a set of the binary classification tasks described in the literature. We can list some of the most popular such as OVR (One-Versus-Rest), OVO (One-Versus-One) strategies, DAG (Directed Acyclic Graph), ADAG (Adaptive Directed Acyclic Graph) methods [18, 24], BDT (Binary Decision Tree) approach [11], DB2 method [29], PWC (PairWise Coupling) [15], or ECOC (Error-Correcting Output Codes) [9]. One of the best known and widely used method of the decomposition is one-versus-one strategy where the input vector x is presented to the binary classifiers trained against each pair of the classes. We can assume that each classifier discriminates between class ω_i and class ω_j and computes the estimate $\hat{p}_{ij}$ of the probability:

$$p_{ij} = P(x \in \omega_i | x, x \in \omega_i \cup \omega_j), \tag{1}$$

then the classification rule is defined as below:

$$\arg \max_{1 \leq i \leq K} \sum_{j \neq i} f(\hat{p}_{ij}), \tag{2}$$

where K is the number of the classes and $f(\hat{p}_{ij})$ is defined as:

$$f(\hat{p}_{ij}) = \begin{cases} 1 & \hat{p}_{ij} > 0.5 \\ 0 & otherwise \end{cases} \quad (3)$$

This approach has been proposed by Friedman and we call it max-voting scheme [12]. Another approach has been suggested in [15, 23]. We can take into consideration that the outputs $\hat{p}_{ij}$ of the binary classifiers are representing the class probabilities. So these values can be used as the estimates $\hat{p}_i$ of a posteriori probabilities:

$$p_i = P(x \in \omega_i | x), \quad (4)$$

Assuming that we have a square matrix $K \times K$ of $\hat{p}_{ij}$ for $i, j = 1 \ldots K$ and $\hat{p}_{ji} = 1 - \hat{p}_{ij}$ then we can calculate the value of $\hat{p}_i$ as:

$$\hat{p}_i = \frac{2}{K(K-1)} \sum_{j \neq i} \sigma(\hat{p}_{ij}), \quad (5)$$

and then we can use the classification rule:

$$\arg \max_{1 \le i \le K} \hat{p}_i, \quad (6)$$

where σ takes the form a threshold function at 0.5 for the max-voting scheme and the identity function for the solution proposed by [15]. Some other σ functions are considered in [23]. If we look closer at this decomposition scheme, we will see that in all approaches we are using values of $\sigma(\hat{p}_{ij})$ for a given vector x which belongs neither to the class ω_i nor to ω_j. Looking at (5), we see that the estimation of p_i takes into account all classifiers even if they are not trained on the samples of the class to which x belongs to. This problem has been addressed in several papers, for example in [13] where a dynamic classifier selection strategy was proposed based on the neighborhood of each instance. Another procedure to overcome this problem has been proposed in [23], which consists of training additional correcting classifiers separating the classes i and j from all the other classes. These classifiers are producing the outputs $\hat{q}_{ij}$. The $\hat{q}_{ij}$ provides us with an estimate of the probability that sample x belongs to the class i or to the class j. Therefore, we can modify the Eq. (5), which now becomes:

$$\hat{p}_i = \frac{2}{K(K-1)} \sum_{j \neq i} \sigma(\hat{p}_{ij}) \hat{q}_{ij}, \quad (7)$$

The use of correcting classifiers causes the incompetent classifiers have no significance and it improves the quality of the estimation $\hat{p}_i$. Another approach using different classifier (RDA) and the special weights to reduce the influence of these classifiers has been proposed in [4, 5]. However, all these solutions have one common drawback. All samples have to be tested using all $K(K-1)/2$ binary classifiers. When the number of the classes grows it becomes very inefficient.

3 The Comparison of OVO and OVR Strategies

The OVO strategy is employing the binary classifiers between each pair of the classes. The OVR approach uses the binary classifiers distinguishing between each class and the all other classes. Both strategies have their advantages and disadvantages which may impact the final result of the combined classifier. They will be shortly discussed in this section. When we use OVO strategy, we have to train a set of $K(K-1)/2$ binary classifiers between each pair of the classes. Then all objects representing all classes are tested against these classifiers. It brings us to the problem of the incompetent classifiers which has been mentioned in the previous section. Look at the Fig. 1 how the samples from other classes are treated by the 8 versus 9 classifier. Especially, samples from classes 4 and 1. Some of them are classified as 8s some as 9s. Another problem can be seen when the number of the classes increases. The number of the binary classifiers rises quadratically and all the samples have to be tested against each classifier during the testing phase. So 1000 classes means about half a million of the binary classifiers. There are several methods to deal with the issue. Some solutions for this problem have been proposed in [5]. The OVR strategy uses the samples of all the classes to train each binary classifier. The samples from one distinguished class are treated as the class ω_1 (we will call it the "one" class) and all the other samples are considered as to belong to the class ω_r (we will call it the "rest" class). When comparing with the OVO strategy, the number of the binary classifiers, which we have to train is quite small. We need K binary classifiers only and we see that this number will increase linearly with the number of the classes. However, when the number of the classes increases another problem can be seen. As it was stated earlier in this section, almost all the samples representing the classes

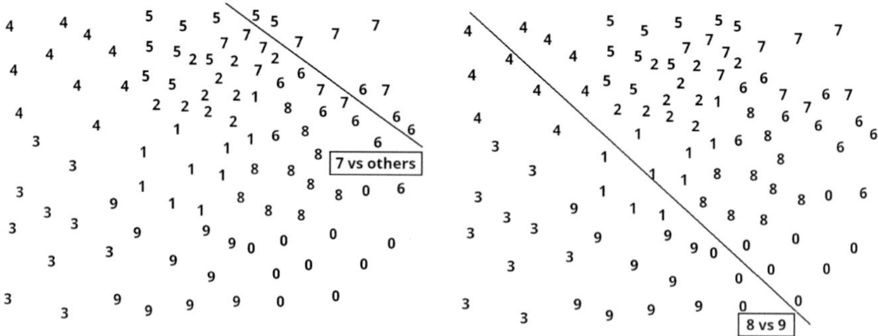

Fig. 1 One-versus-rest approach (*left*) and one-versus-one approach (*right*)

except one distinguished class are treated as one big class. It causes the problem of overrepresenting the "rest" class. Therefore, the results obtained by these binary classifiers could be very biased. It is clear, if we consider the problem with 1000 classes. Assuming that each class is represented by the same number of the samples, then we have 999 times more samples of the "rest" class than the samples of the "one" class. We can observe this problem in the Fig. 1. As much as 6 from 10 samples of the 7 class have been misclassified because the class "rest" is highly overrepresented. We can also notice that at the training phase we have much more samples in the training data set especially when the number of the classes is big. On the contrary, when we use OVO strategy we train our binary classifiers using data sets which consist of the samples from two classes only. When we use OVR strategy, we have to use all the samples. It impacts the training time which might be a problem in some applications, for example, when we have to often retrain our classifier online using new samples.

4 Proposed Method

As we stated in the previous section, one of the major weakness of the OVR strategy is the problem of the imbalanced training data sets. That is the number of the samples representing the "one" class is much smaller than the number of the samples from the "rest" class. Moreover, the problem is more and more visible when the number of the classes grows. It causes that the result of the OVR classifier can be very biased [16]. It is quite obvious, if 99 % of the samples come from one class, then the algorithm which marks all the samples as belonging to this one class achieves 99 % accuracy. We address this problem in another paper [6] proposing a way to balance the number of the samples representing "one" and the "rest" class. The results are promising but the solution have one deficiency. There is a problem how to select the samples for the "rest" data set. We proposed some random procedure which works quite well but it have to be repeated several times to get satisfactory results. It impacts the speed of the combined classifier. We had looked for a better procedure which might improve the results obtained by the correcting classifiers. You can notice that the procedure of the finding the optimal hyperplane and the support vectors on which it is based should be very robust even if the data are unbalanced. However, this is no longer true when we implement the soft margin mechanism [3]. The problem is that the C value which represents the punishment for wrong classification of the sample has the value for the samples representing the "rest" and the "one" class. Let us consider the following solution: for the samples representing the "one" class, we choose C_1 value and for the all other samples we choose different C_r value. Let us consider the behavior of the classifier when we set the penalty for the wrong classification of the sample belonging to the "rest" class to zero $C_r = 0$. Such classifiers would classify more samples as belonging to the "one" class. However, in this case, it will become very fragile to the outliers, see the Fig. 2. So there have to be some penalty for the

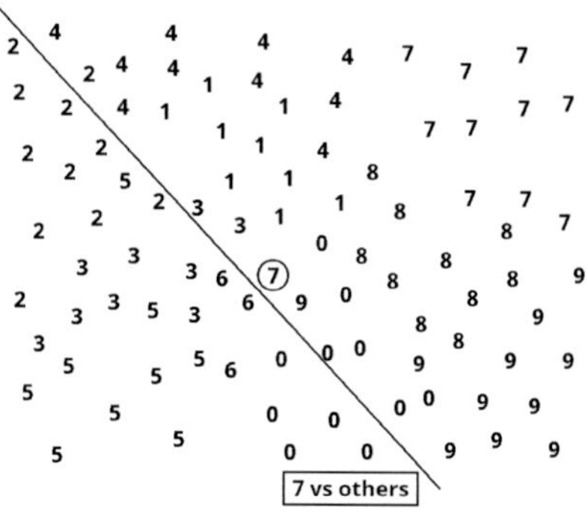

Fig. 2 OVR approach with $C_r = 0$

wrong classification of the sample from the "rest" class. The values of C_r and C_1 equal to

$$C_1 = N/K, \quad C_r = 1 \tag{8}$$

seems to be the good candidates, where C_1 is the penalty for wrong classification of the sample representing the "one" class and C_r is the penalty for wrong classification of the sample representing the "rest" class. N is number of the samples in the data set and K is the number of the classes. It is visible that the proposed classifier will be more eager to wrongly classify the sample representing the "rest" class than the sample representing the "one" class. It will of course impact the result of the correcting classifier but this time the number of the wrong classified samples of the "one" class is much more smaller—we avoid the effect of the poor performance on the minority ("one") class [19]. Considering the original one-versus-one strategy with max-voting scheme [12], this approach has two main advantages. First, we eliminate the majority of the samples which do not belong to the classes on which the classifier has not been trained. Second, the OVO classifiers are not run against all these samples and as we can recall it there is as much as $K(K-1)/2$ OVO classifiers. Let us consider the number of the algorithm runs in the different approaches: for OVO and PWC algorithm, we have $NK(K-1)/2$ runs (each binary classifier tests each sample). It is even worse for the PWC-CC algorithm because we have $2NK(K-1)/2 = NK(K-1)$ runs (for each pair we have to use one correcting classifier). In our solution, the number of the runs is much smaller, i.e., $NK + PK(K-1)/2$ runs where $P < N$ and P is the average number of the samples selected by OVR correcting classifiers. It is hard to estimate the value of the P. It is obvious that it should not be less that N/K, i.e., the average number of the samples representing the class. Assuming that we have the same number of the samples representing each class and assuming that each OVR classifier has 100 % accuracy, then the number of the runs will be: $NK + N/KK(K-1)/2 = NK + N(K-1)/2 = N(3K-1)/2$. The real number of the runs depends on concrete case, on which database we use our algorithm and how accurate are our OVR correcting classifiers. In our experiments, the values of P's were in the range $N/K < P < 2N/K$. For example, for ACRS database it means that we have 25 times less OVO classifier runs.

5 Results of the Experiments

Several experiments have been conducted to test the proposed method. We used an SVM classifier because it manages quite efficiently even if we have a lot of the samples in the training data set. We used the modified implementation of the SVM classifier [2]. The modification enables the possibility to use different values of C for the “one” and the “rest” classes. We need several databases with different characteristics to test our solution. Some of the databases we found in the UCI Machine Learning Repository [27]. We also used the protein database [8], gesture database created at the Institute of Theoretical and Applied Informatics of the Polish Academy of Sciences [14], and the leafs database [26]. In the Table 1, we show the number of classes and the size of the feature vector for all databases used in our experiments. On each database, the four algorithms have been tested; i.e., one-versus-one OVO with max-voting scheme, pairwise coupling PWC, the improved version of PWC pairwise coupling with corrected classifiers (PWC-CC) proposed by Moreira and Mayoraz [23], and our new solution modified OVO (MOVO) proposed in this paper. The results are presented in the Table 2. We can notice that the obtained results are better than achieved by the OVO, PWC, and the PWC-CC classifiers. If we compare the speed of the algorithms, we can see the full advantage of the proposed method. The MOVO algorithm is much more efficient than all other methods. It is obvious when we consider the average number of the binary classifier runs we have to use to obtain the results. In all the experiments, the procedure of k-crossvalidation has been used to avoid biased results. We use $k = 7$ in our experiments. we choose this value because one class in the protein database was represented by seven samples only. The average values of the recognition ratios obtained using k-crossvalidation are shown in the Table 2. We can observe that our solution overcomes all other algorithms on all databases no matter which classifier is used.

Table 1 The databases used in the experiments

Name	Classes	Samples	Features
Gestures IITiS	22	1320	256
ISOLET	26	7797	617
Proteins	27	698	126
Leafs	36	340	13
ACRS	50	1500	10000

Table 2 The results using SVM classifier

DB name	OVO (%)	PWC (%)	PWC-CC (%)	MOVO (%)
Gestures	81.3	81.6	82.0	82.3
ISOLET	96.5	96.6	96.5	96.7
Proteins	57.4	57.9	57.9	58.2
Leafs	79.1	79.4	79.0	80.8
ACRS	73.2	73.4	72.9	74.5

6 Conclusions

The OVO strategy has two main drawbacks: the number of the binary classifiers we have to use to obtain the result and the usage of the incompetent classifiers. Both these drawbacks are addressed in our solution. We avoided the use of incompetent classifiers by using OVR correcting classifiers and additionally we limit the number of the runs of the binary OVO classifiers, what is even more important. We show that the computational complexity is reduced from $O(NK^2)$ to $O(NK)$. It means that our solution is even more efficient when the number of the classes grows. The experimental results also confirm that our algorithm obtains better recognition ratios than other presented in the literature. In our experiments, our algorithm overcomes OVO, PWC, and PW-CC methods on all databases. It was also much more efficient if we consider the execution time. The results and the analysis of the proposed methods suggest that it should perform even better when the number of the classes is growing. Considering the fact that the average number of the classes in the databases from UCI Machine Learning Repository [27] increases from 5.6 in the years 1988–1992 to 35.3 in the years 2008–2012, it is very important result. Some future experiments in this area might be interesting. It is clear that the efficiency of our algorithm is dependent from the recognition ratio of the OVR classifiers in two aspects: i.e., if we misclassify the sample belonging to ω_1 class, then OVO classifiers would not have a chance to correct this error; so, we deteriorate the recognition ratio of the final classifier. When we misclassify the sample from the ω_r class, then we deteriorate the speed of the final classifier but the OVO binary classifiers can be able to correct the error. We can experiment with the values of C_1 and C_r parameters to find the optimal settings to maximize the recognition ratio or to maximize the speed of the proposed algorithm.

References

1. Allwein, E., Schapire, R., Singer, Y.: Reducing multiclass to binary: a unifying approach for margin classifiers. J. Mach. Learn. Res. **1**, 113–141 (2001)
2. Chang, C.C., Lin, C.J.: LIBSVM: a library for support vector machines. Software available at http://www.csie.ntu.edu.tw/cjlin/libsvm (2001)
3. Chen, D.R., Wu, Q., Ying, Y., Zhou, D.X.: Support vector machine soft margin classifiers: error analysis. J. Mach. Learn. Res. **5**, 1143–1175 (2004)
4. Chmielnicki, W., Stapor, K.: Protein fold recognition with combined SVM-RDA classifier. Lect. Notes Artif. Intell. **6076**, 162–169 (2010)
5. Chmielnicki, W., Stapor, K.: A hybrid discriminative/generative approach to protein fold recognition. Neurocomputing **75**(1), 194–198 (2012)
6. Chmielnicki, W., Stapor, K.: A modification of the pairwise coupling algorithm to solve multiclass problem. In: Proceedings of XLIII Conference on Mathematics Applications, in Polish (2014)
7. Chmielnicki, W., Roterman-Konieczna, I., Stapor, K.: An improved protein fold recognition with support vector machines. Expert Syst. **20**(2), 200–211 (2012)
8. Ding, C.H., Dubchak, I.: Multi-class protein fold recognition using support vector machines and neural networks. Bioinformatics **17**, 349–358 (2001)

9. Dietterich, T.G., Bakiri, G.: Solving multiclass problems via error-correcting output codes. J. Artif. Intell. Res. **2**, 263–286 (1995)
10. Dubchak, I., Muchnik, I., Holbrook, S.R., Kim, S.H.: Prediction of protein folding class using global description of amino acid sequence. Proc. Natl. Acad. Sci. USA **92**, 8700–8704 (1995)
11. Fei, B., Liu, J.: Binary tree of SVM: a new fast multiclass training and classification algorithm. IEEE Trans. Neural Netw. **17**(3), 696–704 (2006)
12. Friedman, J.H.: Another approach to polychotomous classification. Stanford Department of Statistics (1996)
13. Galar, M., Fernandez, A., Tartas, E.B., Sola, B., Herrera, F.: Dynamic classifier selection for one-vs-one strategy: avoiding non-competent classifiers. Pattern Recognit. **46**(12), 3412–3424 (2013)
14. Glomb, P., Romaszewski, M., Opozda, S., Sochan, A.: Choosing and modeling hand gesture database for natural user interface. In: Proceedings of the 9th International Conference on Gesture and Sign Language in Human-Computer Interaction and Embodied Communication, pp. 24–35 (2011)
15. Hastie, T., Tibshirani, R.: Classification by pairwise coupling. Ann. Stat. **26**(2), 451–471 (1998)
16. He, H., Garcia, E.A.: Learning from imbalanced data. IEEE Trans. Knowl. Data Eng. **21**(9), 1263–1284 (2009)
17. Hobohm, U., Scharf, M., Schneider, R., Sander, C.: Selection of a representative set of structures from the Brookhaven protein bank. Protein Sci. **1**, 409–417 (1992)
18. Kijsirikul, B., Ussivakul, N.: Multiclass support vector machines using adaptive directed acyclic graph. In: Proceedings of the International Joint Conference on Neural Networks, pp. 980–985 (2002)
19. Kotsiantis, S., Kanellopoulos, D., Pintelas, P.: Handling imbalanced datasets: a review. GESTS Int. Trans. Comput. Sci. Eng. **30**(1), 25–36 (2006)
20. Krawczyk, B., Wozniak, M., Cyganek, B.: Clustering-based ensembles for one-class classification. Inf. Sci. **264**, 182–195 (2014)
21. Lorena, A.C., Carvalho, A.C., Gama, J.M.: A review on the combination of binary classifiers in multiclass problems. Artif. Intell. Rev. **30**(1–4), 19–37 (2008)
22. Lorena, A.C., Carvalho, A.C.: Building binary-tree-based multiclass classifiers using separability measures. Neurocomputing **73**(16–18), 2837–2845 (2010)
23. Moreira, M., Mayoraz, E.: Improved pairwise coupling classification with correcting classifiers. In: Proceedings of Tenth European Conference on Machine Learning ECML, Chemmitz, Germany, pp. 160–171 (1998)
24. Platt, J.C., Cristianini, N., Shawe-Taylor, J.: Large margin DAGs for multiclass classification. In: Proceedings of Neural Information Processing Systems, pp. 547–553 (2000)
25. Sáez, J.A., Galar, M., Luengo, J., Herrera, F.: A first study on decomposition strategies with data with class noise using decision trees. In: Proceedings of the 7th International Conference on Hybrid Artificial Intelligent Systems—Volume Part II, pp. 25–35 (2012)
26. Silva, P.F.B., Marcal, A.R.S., Almeida da Silva, R.M.: Evaluation of Features for Leaf Discrimination. Springer Lecture Notes in Computer Science, vol. 7950, pp. 197–204. Springer, Heidelberg (2013)
27. UCI Machine Learning Repository. http://archive.ics.uci.edu/ml/ (2014)
28. Vapnik, V.: The Nature of Statistical Learning Theory. Springer, New York (1995)
29. Vural, V., Dy, J.G.: A hierarchical method for multi-class support vector machines. In: Proceedings of the XXI ICML, pp. 831–838 (2004)
30. Windeatt, T., Ghaderi, R.: Coding and decoding for multiclass learning problems. Inf. Fusion **4**(1), 11–21 (2003)

A Wrapper Evolutionary Approach for Supervised Multivariate Discretization: A Case Study on Decision Trees

Sergio Ramírez-Gallego, Salvador García, José Manuel Benítez and Francisco Herrera

Abstract The main objective of discretization is to transform numerical attributes into discrete ones. The intention is to provide the possibility to use some learning algorithms which require discrete data as input and to help the experts to understand the data more easily. Due to the fact that in classification problems there are high interactions among multiple attributes, we propose the use of evolutionary algorithms to select a subset of cut points for multivariate discretization based on a wrapper fitness function. The algorithm proposed has been compared with the best state-of-the-art discretizers with two decision trees-based classifiers: C4.5 and PUBLIC. The results reported indicate that our proposal outperforms the rest of the discretizers in terms of accuracy and requiring a lower number of intervals.

Keywords Discretization · Numerical attributes · Evolutionary algorithms · Data preprocessing · Classification

1 Introduction

Data preprocessing [10] is a crucial research topic in Data Mining (DM) since most real-world databases are highly influenced by negative elements such as the presence of noise, missing values, inconsistent, and superfluous data. Discretization, as one of the basic reduction techniques, has received increasing research attention in

---

S. Ramírez-Gallego (✉) · S. García · J.M. Benítez · F. Herrera
Department of Computer Science and Artificial Intelligence,
University of Granada, 18071 Granada, Spain
e-mail: sramirez@decsai.ugr.es

S. García
e-mail: salvagl@decsai.ugr.es

J.M. Benítez
e-mail: J.M.Benitez@decsai.ugr.es

F. Herrera
e-mail: herrera@decsai.ugr.es

R. Burduk et al. (eds.), *Proceedings of the 9th International Conference on Computer Recognition Systems CORES 2015*, Advances in Intelligent Systems and Computing 403, DOI 10.1007/978-3-319-26227-7_5

recent years [9] and has become one of the most broadly used techniques in DM. The objective of a discretization process is to transform numerical attributes into discrete ones by producing a finite number of intervals, and by associating a discrete, numerical value to each interval [14]. Although real-word DM tasks often involve numerical attributes, many algorithms can only handle categorical attributes, such as Naive Bayes [19]. Among the most important advantages of using discretized data is that it is simpler and more reduced than numerical data. For example, some kind of decision trees yield more compact, shorter, and more accurate results than the derived ones using numerical values [14]. Discretization of data besides has the effect of improving the speed and accuracy of DM algorithms. Cut points selection problem for discretization is formed by all the singleton values present in each input attribute (candidate points). As this space can become very complex, specially when the data grow (both instances and features); we resorted to the use of a subset of attribute values or cut points, considering only those points that fall in the class borders (boundary points). Since this problem can be considered as an optimization problem with a binary search space, evolutionary algorithms (EAs) can be applied. EAs have been used for data preparation with promising results [8]. In discretization, few evolutionary approaches can be found in the literature. For instance, in [11], a multivariate proposal is presented based on finding hidden association patterns using genetic algorithms and clustering. In this contribution, we present an evolutionary-based discretization algorithm with binary representation called evolutionary multivariate discretizer (EMD), which selects the most adequate combination of boundary cut points to create discrete intervals. We compare our approach with the discretizers emphasized as the best performing according to [9]. We consider two of the most influential decision trees that benefit from discretization: C4.5 and PUBLIC. The empirical study consists of 45 data sets, 6 discretizers for comparison and an analysis based on nonparametric statistical testing. The rest of the contribution is organized as follows: Sect. 2 defines some basic concepts. In Sect. 3, our proposal is explained. In Sect. 4, we provide the experimental framework, the results obtained and an analysis based on nonparametric statistical testing. Finally, Sect. 5 concludes the contribution.

2 The Problem of Discretization

Considering a classification problem with C target classes, a set with N instances and M attributes, we can define the discretization as follows. A discretization algorithm would partition a continuous attribute A into k_A discrete and disjoint intervals:

$$D_A = \{[d_0, d_1], (d_1, d_2], \ldots, (d_{k_A-1}, d_{k_A}]\} \tag{1}$$

where d_0 and d_{k_A}, respectively, are the minimum and maximal value, the values in D_A are arranged in ascending order. Such set of discrete intervals D_A is called a discretization scheme on attribute A and $P_A = \{d_1, d_2, \ldots, d_{k_A-1}\}$ is the set of

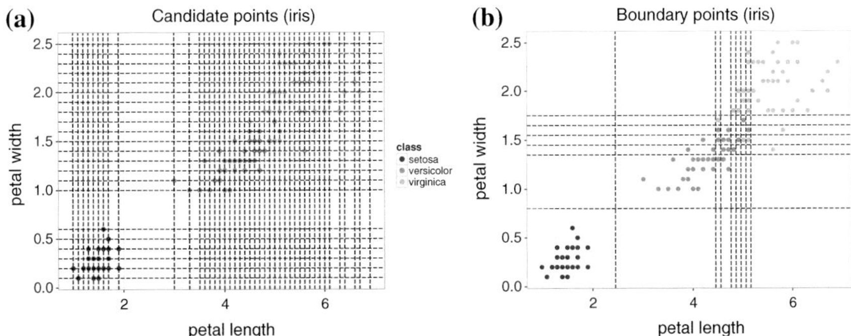

Fig. 1 Cut points selection problem

cut points of attribute A. Finally, P denotes the complete set of cut points for all the continuous attributes in M. Therefore, the search space is formed by all the *candidate cut points* for each attribute, which is basically all the different existing values in the training set, considering each attribute separately. In order to reduce the complexity of the initial search space, we resorted to the use of a reduced subset formed by the *boundary points* of each attribute. Partitioning of a numerical attribute A begins by sorting its values into ascending order. Let $Dom(A)$ denote the domain of the attribute A and $val_A(s)$ denote the value of the attribute A in a sample $s \in S$. If a pair of samples $u, v \in S$ exists, having different classes, such that $val_A(u) < val_A(v)$, and another sample $w \in S$ does not exist such that $val_A(u) < val_A(w) < val_A(v)$. We define, a boundary point $bp \in Dom(A)$ as the midpoint value between $val_A(u)$ and $val_A(v)$. Thus, the set of boundary points for attribute A is denoted as BP_A, and BP denotes the complete set. Boundary points are known to form the optimal intervals for most of the evaluation measures used, as inconsistency or information gain [5]. Hence, using only the boundary points in the search space, we can obtain significant savings in complexity and time consumption. Additionally, the above definition of boundary points allow us to achieve the maximum separability between classes, thus obtaining better discretization results. A representation of the candidate and boundary points for the *iris* data set is showed in Fig. 1.

3 Evolutionary Multivariate Discretizer

In this section, we describe EMD, an evolutionary-based algorithm with binary representation for discretization. EMD uses a wrapper fitness function based on a trade-off between the classification error provided by the straightforward application of two classifiers, and the number of intervals produced. As backbone of our EA, we have used the CHC model [6]. Despite it being well known that multivariate approaches improve the discretization process on supervised learning, they have been less exploited in the literature. The proposed algorithm follows a multivariate approach, taking advantage of the existing interactions and dependencies among the attributes to improve the underlying discretization process. Furthermore, to tackle larger problems and, in general, to speed up the discretization process on all data sets; we include a chromosome reduction mechanism.

3.1 Representation

Classically, an individual is represented in a binary array of 0's and 1's. Let us define the search space of the problem as all the possible discretization schemes offered by their set of boundary points *BP*. In this way, the common binary encoding is suitable to represent these schemes so that each boundary point corresponds to a gene. A chromosome thus consists of $|BP|$ genes with two possible values: 0, if the associated cut point is not included in *P*; and 1, if it is. The chromosome structure associated with the boundary points approach is presented in Fig. 2a and an example of a possible solution along with its spatial representation are depicted in Fig. 2b.

3.2 Fitness Function

To select the most appropriate discretization scheme from the population, it is necessary to define a suitable fitness function to evaluate the solutions. Let *Q* be a subset of cut points selected from *BP* and be coded by a chromosome. We define a fitness

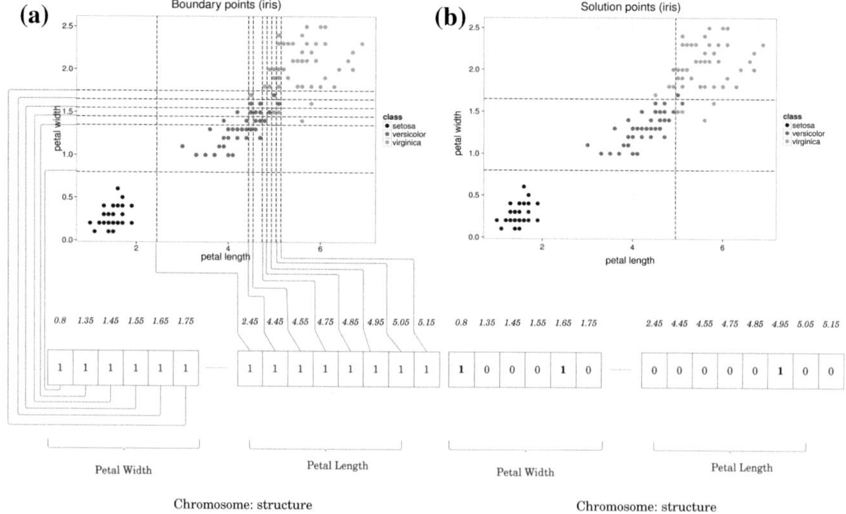

Fig. 2 Chromosome representation and solution

function as the aggregation of two objectives, namely the classification error of the discretized data and the minimization of the number of cut points.

$$Fitness(Q) = \alpha \cdot \frac{|Q|}{|BP|} + (1 - \alpha) \cdot \Delta \tag{2}$$

where $|Q|$ is the number of boundary cut points currently selected in the chromosome, $|BP|$ is the total number of cut points, Δ is the classification error of the discretized data, and α is the weight factor specified as input parameter. With this fitness design, we can obtain more effective discretization schemes (through the classifier evaluation), while maintaining solutions as simple as possible. The *classification error* is a supervised measure used to compute the number of misclassified instances. In our case, the classification error is computed as the aggregated error obtained using two classifiers [18]: an unpruned version of C4.5 [15] and Naive Bayes [3]. In this manner, the algorithm has several criteria to evaluate the solutions. This part is defined as follows:

$$\Delta = \frac{\delta_{C45} + \delta_{NB}}{2} \tag{3}$$

where δ_{C45} is the the total number of instances misclassified by C4.5 divided by the total number of instances, and δ_{NB} represents the same computation for Naive Bayes. This measure is basically the arithmetic mean of the classification error committed by these two classifiers with the domain [0, 1]. Regarding the time complexity of our approach, this is mainly conditioned by the number of cut points (L) and the number of instances (I) of the data set, as well as, the maximum number of evaluations (E) until the stopping criterion is reached. According to these variables, we can state that the time complexity of the whole algorithm strongly depends on the evaluation phase, which in turn depends on the time complexity of the two classifiers used in the wrapper function. Naive Bayes classifier provides an efficient time complexity $O(LI)$ similar to the consistency measures presented in [4]. For C4.5, it is known that the tree pruning introduces a time complexity of $O(LIlog(I)^2)$ [15]. To mitigate it, we use an unpruned C4.5 version that improves the time complexity ($O(LIlog(I))$) and becomes it in a evaluation measure more efficient and sensitive to the error. Therefore, EMD obtains a time complexity $O(ELIlog(I))$. The objective of the GA is to minimize the fitness function defined; therefore, to obtain consistent and simple discretization schemes with the minimum possible number of cut points, but always keeping a fair classification accuracy through the error counterpart.

3.3 Reduction

Whenever an EA faces problems with a significant size, its application can become unsatisfying due to the growth of the chromosomes. A great number of attributes and examples cause an increment in complexity because the number of boundary points represented increases. As in any optimization problem, the cut points selection problem offers multiple solutions (local optima) which can be considered as valid, although not the best ones (global optima) [16]. Agreeing that the size of the problem becomes troublesome, it is desirable to reach some local optima to avoid a long delay or even inability in obtaining the global optima. In our case, we propose a chromosome reduction process to speed up the convergence by reducing the number of boundary cut points to be considered. For that, in each generation we only preserve the most selected points in previous evaluations (for each reduction stage). The reduction process is applied at the beginning of each generation. Given a counter of the times each point is selected C_p, the current best chromosome bc, the current size of the chromosomes C_s (initially equal to $|BP|$), the number of reductions accomplished n_r, the maximum size for chromosomes M_s, the number of evaluations accomplished up to now n_e, the maximum number of evaluations M_e, the reduction rate R_{rate} and the reduction percentage R_{perc}. EMD applies a distinct reduction process each $M_e * R_{rate}$ evaluations, if and only if the maximum size allowed M_s is not exceeded. The process described as follows:

1. Calculate the new reduced size for the chromosomes: $|bc| * (1 - R_{perc})$.
2. Initialize the set of new points T with all the points marked as 1 in the best previous chromosome bc. The remaining points are ordered (by ranking) and stored in RP.
3. According to the counter C_p, add the most selected points from the rest of individuals to the new structure until completing the new chromosome size C_s.
4. Once completed T, we reorder the points by value also considering the original order of the attributes, form the new chromosome structure (as in Fig. 2a).
5. Apply the new structure to each individual, removing all non-selected points.
6. Reinitialize the counter C_p according to the new structure.
7. Restart the new population $newPop$ using the best previous solution bc as reference.
8. Evaluate the new population.

4 Experimental Framework and Results

Next, we describe the methodology followed in the experimental study which compares the proposed technique with the best discretization algorithms in the experimental review [9] using a wide range of classifiers. Here our method is denoted by EMD.

4.1 Experimental Framework

The discretizers involved in the comparison for the first two parts are: CAIM [13], ChiMerge [12], FUSINTER [20], MDLP [7], Mod-Chi2 [17], and PKID [19].

Implementations of the discretizers as well as the classifiers used in these experiments can be found under the KEEL data mining tool [1]. Performance of the algorithms is analyzed by using 45 data sets taken from the UCI machine learning database repository [2].[1] The data sets considered are partitioned using the *ten fold cross-validation (10-fcv)* procedure. The recommended parameters of the discretizers and the classifiers are set according to their authors' specification and they are the same used in [9]. For our method, GA parameters (population size and number of evaluations) has been established to standard values that have been demonstrated to perform well in most of cases where GAs have been applied on data preprocessing tasks [10]. The alpha weight factor is set to 0.7 in favor to the removal of points for two reasons: first, the more points removed during the cycle, the faster the algorithm converges. On the other hand, the selection of candidate points is a multimodal problem in which many subsets of points achieve similar performances, making more difficult the reduction subobjective.

4.2 Analysis and Empirical Results

To compare the performance of these methods, test classification accuracy obtained by C4.5 and PUBLIC using all discretizers is presented in Table 1. The best case in each data set is highlighted in bold. Likewise, Fig. 3 shows the average cut points required by each discretizer. Observing the results, the following analysis can be stated:

- According to the classification accuracy, EMD is the best alternative when using C4.5 as classifier on average (0.7852). Our proposal outperforms 14/45, being the most promising algorithm out of all those used.
- Better results are obtained using PUBLIC, where EMD represents the best alternative with the most promising mean (0.7713), outperforming on 19/45 data sets.
- The lowest number of cut points on average (26/45) is held by our method. EMD yields simpler discretization schemes in almost half of the cases.

Statistical analysis will be carried out by means of the Wilcoxon nonparametric statistical tests. Table 2 collects the results offered by the Wilcoxon test considering a level of significance equal to $\alpha = 0.05$. This table is divided into three parts, each one associated with columns: in the first and second parts, the measure of accuracy classification in the test is used for C4.5 and PUBLIC, respectively. In the third part, we accomplish the Wilcoxon test by using as a performance measure the number of cut points produced by the discretizers. The table indicates, for each method in the rows, the number of discretizers outperformed by using the Wilcoxon test under the column represented by the '+' symbol. The column with the '±' symbol indicates the number of wins and ties obtained by the method in the row. The maximum value

[1] They are specified in Table 1.

Table 1 Average accuracy obtained for C4.5 and PUBLIC

	CAIM		ChiMerge		FUSINTER		MDLP		Modified Chi2		PKID		EMD	
	C4.5	*PUBLIC*	*C4.5*	*PUBLIC*	*C4.5*	*PUBLIC*	*C4.5*	*PUBLIC*	*C4.5*	*PUBLIC*	*C4.5*	*PUBLIC*	*C4.5*	*PUBLIC*
Abalone	0.2432	0.2549	0.2245	0.254	0.253	0.2568	0.2537	0.2585	0.1749	0.2554	0.2307	0.2252	0.2343	**0.2595**
Appendicitis	0.8336	0.8427	0.8336	0.8327	0.8236	0.8145	0.8336	0.8609	0.7855	0.8336	0.8018	0.7927	0.87	**0.87**
Australian	0.8725	0.8377	0.8565	0.8377	0.8681	**0.8551**	0.8638	0.8478	0.858	**0.8551**	0.8493	0.8522	0.8536	0.8478
Autos	0.7263	0.5956	0.7474	0.6275	0.7937	0.6045	0.7691	0.6353	0.7897	0.6347	0.767	0.5462	0.7644	0.5759
Balance	0.7472	0.7073	0.7777	0.7619	0.7282	0.7456	0.6992	0.6959	0.664	0.7489	0.6482	0.7265	0.8047	**0.7697**
Banana	0.6387	0.6387	0.6253	0.6253	0.8791	0.8674	0.7485	0.7492	0.6392	0.5517	0.7043	0.5517	0.873	**0.8704**
Bands	0.6458	**0.6901**	0.6346	0.679	0.5975	0.6864	0.5378	0.5529	0.6642	0.6568	0.6197	0.6753	0.6494	0.6734
Banknote	0.8892	0.8899	0.8863	0.8833	0.9584	0.9613	0.9446	0.9424	0.9526	0.9476	0.8484	0.6472	0.9657	**0.9694**
Bupa	0.6065	0.598	0.6361	**0.6395**	0.6198	0.5789	0.5715	0.5715	0.5704	0.5789	0.5789	0.5789	0.6814	0.6382
Cleveland	0.5484	0.5444	0.5482	0.541	0.564	0.528	0.5377	**0.5609**	0.5471	0.5409	0.5345	0.5444	0.548	0.5577
Climate	0.9278	0.9222	0.9	0.9148	0.937	0.9148	0.9333	**0.9333**	0.9167	0.9037	0.9148	0.9148	0.9333	0.9222
Contraceptive	0.5105	0.5384	0.5506	**0.5425**	0.5221	0.4942	0.5045	0.4915	0.5045	0.465	0.4875	0.4691	0.5261	0.5092
Crx	0.8739	0.8348	0.8667	0.8435	0.8797	0.8449	0.8681	0.8377	0.8768	0.8551	0.8522	0.8551	0.8638	**0.8551**
Dermatology	0.9318	**0.9619**	0.9424	0.9536	0.9508	0.9592	0.9589	0.9536	0.9589	0.9592	0.9454	0.9563	0.9482	0.9509
Ecoli	0.7469	**0.8063**	0.7708	0.7534	0.7503	0.7949	0.7771	0.8038	0.7381	0.7059	0.6603	0.6966	0.7472	0.7623
Flare	0.6782	**0.6754**	0.6782	**0.6754**	0.6735	**0.6754**	0.6754	**0.6754**	0.6754	**0.6754**	0.6754	**0.6754**	0.6726	0.6754
Glass	0.6761	0.6782	0.6814	0.6335	0.6679	0.6471	0.7579	0.6744	0.628	0.6092	0.5788	0.4701	0.737	**0.7184**
Haberman	0.7512	**0.7512**	0.7347	0.7186	0.7251	0.7482	0.7253	0.7285	0.7353	0.7353	0.7353	0.7353	0.7415	0.7447
Hayes	0.802	**0.83**	0.802	**0.83**	0.7338	0.7113	0.5202	0.5202	0.7201	0.7415	0.7107	0.6604	0.7426	0.7476
Heart	0.7852	0.7556	0.8037	0.7556	0.8074	0.7704	0.7963	0.7593	0.7889	0.7667	0.7926	0.7593	0.8222	**0.8074**
Hepatitis	0.8271	0.7554	0.7883	0.7808	0.8517	0.7683	0.8325	0.7742	0.8008	0.7808	0.8258	0.7804	0.8071	**0.8008**
Iris	0.9333	0.9333	0.9333	0.9333	0.94	0.94	0.9333	0.9333	0.9333	0.9467	0.9267	0.8333	0.9533	**0.9533**

(continued)

Table 1 (continued)

	CAIM		ChiMerge		FUSINTER		MDLP		Modified Chi2		PKID		EMD	
	C4.5	*PUBLIC*	*C4.5*	*PUBLIC*	*C4.5*	*PUBLIC*	*C4.5*	*PUBLIC*	*C4.5*	*PUBLIC*	*C4.5*	*PUBLIC*	*C4.5*	*PUBLIC*
Mammographic	0.8284	**0.8315**	0.8325	0.8169	0.8263	0.8315	0.8315	0.8294	0.82	0.8242	0.8117	0.8315	0.8315	0.8294
Movement	0.475	0.35	0.3972	0.3722	0.6194	**0.575**	0.6056	0.55	0.6306	0.5528	0.3222	0.1917	0.6361	0.5417
Newthyroid	0.9351	0.9255	0.9444	0.9398	0.9165	0.9255	0.9444	0.9299	0.9398	0.9442	0.9398	0.9028	0.9494	**0.9494**
Pageblocks	0.9618	0.9627	0.9666	0.9642	0.9618	0.9611	0.9684	0.9638	0.9474	0.9476	0.9496	0.9472	0.9693	0.9653
Penbased	0.8877	0.9171	0.8882	0.9131	0.8947	0.9218	0.8866	0.9124	0.8904	0.9142	0.67	0.7865	0.9491	**0.9519**
Phoneme	0.7913	0.7877	0.782	0.7794	0.8205	0.7889	0.8124	0.795	0.7533	0.7065	0.7681	0.7065	0.8447	**0.8151**
Pima	0.7345	0.7346	0.7292	0.7266	0.7318	0.7003	0.7344	0.7305	0.7203	0.7019	0.7307	0.655	0.7435	**0.7499**
Saheart	0.7055	**0.7078**	0.708	0.6949	0.645	0.6797	0.6817	0.671	0.6971	0.6816	0.658	0.6386	0.6839	0.6818
Satimage	0.8541	**0.8493**	0.8637	0.8421	0.8449	0.8448	0.8454	0.8252	0.8387	0.8413	0.802	0.7277	0.8483	0.8449
Segment	0.9468	0.9329	0.9524	0.9459	0.9502	0.9398	0.9385	0.939	0.8831	0.8502	0.8476	0.7519	0.9606	**0.9468**
Seismic	0.9342	**0.9342**	0.9342	**0.9342**	0.9342	**0.9342**	0.9342	**0.9342**	0.9342	**0.9342**	0.9342	**0.9342**	0.9334	0.9334
Sonar	0.74	0.7243	0.7455	**0.7643**	0.6967	0.7162	0.7638	0.7398	0.7398	0.74	0.6962	0.5338	0.7731	0.7014
Spambase	0.9356	0.9232	0.9384	**0.9278**	0.9206	0.9102	0.9273	0.9147	0.8764	0.8601	0.8869	0.903	0.9252	0.9147
Specfheart	0.7785	0.7793	0.7829	0.7942	0.7496	0.7942	0.7268	0.7792	0.7752	0.7979	0.7942	0.7942	0.8204	**0.8017**
Tae	0.4583	**0.5838**	0.5433	0.4379	0.5113	0.3379	0.3442	0.3442	0.5296	0.3313	0.4708	0.3379	0.5317	0.4504
Thoracic	0.8468	**0.8511**	0.8468	**0.8511**	0.8468	**0.8511**	0.8468	**0.8511**	0.8511	**0.8511**	0.8511	**0.8511**	0.8468	**0.8511**
Titanic	0.7774	0.7783	0.7733	0.7751	0.776	0.7751	0.7715	0.7678	0.776	0.7751	0.7892	0.7906	0.7906	**0.7906**
Transfusion	0.7727	0.77	0.758	0.7487	0.7619	0.7554	0.7621	0.7514	0.7621	0.7621	0.7621	0.7621	0.7728	0.7554
Vehicle	0.6739	0.682	0.6667	0.6668	0.6832	0.6726	0.6832	0.6679	0.6831	0.6572	0.6454	0.5899	0.6831	0.6832
Vowel	0.696	0.6121	0.6677	0.5818	0.7253	**0.6717**	0.7323	0.6182	0.7121	0.6576	0.4848	0.3667	0.7071	0.6515
Wine	0.9101	0.8925	0.9042	0.9271	0.9379	0.9438	0.8984	**0.9608**	0.9268	0.949	0.7974	0.8082	0.9212	0.9157
Wisconsin	0.9385	0.9385	0.9514	0.9471	0.9456	**0.9528**	0.9442	0.9471	0.9471	0.9427	0.9384	0.9399	0.9499	0.9456
Yeast	0.5304	0.5358	0.5021	0.5249	0.5674	0.5567	0.5722	**0.5661**	0.4434	0.4266	0.3862	0.4455	0.5243	0.558
Mean	*0.7624*	*0.7566*	*0.7622*	*0.7532*	*0.7732*	*0.7602*	*0.7600*	*0.7500*	*0.7556*	*0.7422*	*0.7250*	*0.6921*	***0.7852***	***0.7713***

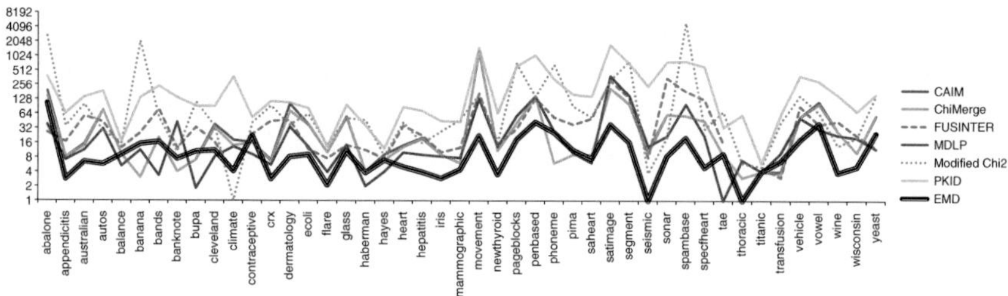

Fig. 3 Number of cut points required for each discretizer

Table 2 Wilcoxon test results for accuracy (C45 and PUBLIC) and number of cut points

Algorithms	C4.5		PUBLIC		# Cut Points	
	+	±	+	±	+	±
CAIM	1	6	1	6	3	4
ChiMerge	1	6	1	6	4	6
FUSINTER	2	6	1	6	2	5
MDLP	1	6	1	6	5	6
Modified Chi2	1	5	1	6	1	1
PKID	0	0	0	0	0	0
EMD	7	7	7	7	7	7

for each column is highlighted by a shaded cell. According to the statistical analysis, we can assert that:

- Statistically, no method can be considered to be better than our proposal in any of the measures used.
- The nonparametric tests confirms that EMD is better than all the other discretizers for all measures.
- Considering the trade-off accuracy/simplicity, we can establish EMD as the best option. It needs to make a lower number of cut points than the discretizers which are similar in accuracy, specially for the largest data sets.
- EMD obtains the best results considering both classifiers.

5 Conclusions

In this contribution, we have presented a new evolutionary-based discretization algorithm called EMD which selects the most adequate combination of boundary cut points to create discrete intervals. For this purpose, EMD uses a wrapper fitness function based on the classification error provided by two important classifiers, and the number of cut points produced. The proposed algorithm follows a multivariate approach, being able to take advantage of the existing interactions and dependencies among the set of input attributes and the class output to improve the discretization process. It also includes a chromosome reduction mechanism to tackle larger problems and, in general, to speed up its performance on all kinds of data sets. A large experimental study performed allows us to show that EMD is a suitable method for discretization in small and large problems.

Acknowledgments This work was partially supported by the Spanish Ministry of Science and Technology under project TIN2011-28488 and the Andalusian Research Plans P11-TIC-7765, P10-TIC-6858.

References

1. Alcalá-Fdez, J., Sánchez, L., García, S., del Jesus, M.J., Ventura, S., Garrell, J.M., Otero, J., Romero, C., Bacardit, J., Rivas, V.M., Fernández, J.C., Herrera, F.: KEEL: a software tool to assess evolutionary algorithms for data mining problems. Soft Comput. **13**(3), 307–318 (2009)
2. Bache, K., Lichman, M.: UCI machine learning repository (2013). http://archive.ics.uci.edu/ml
3. Cios, K.J., Pedrycz, W., Swiniarski, R.W., Kurgan, L.A.: Data Mining: A Knowledge Discovery Approach. Springer, New York (2007)
4. Dash, M., Liu, H.: Consistency-based search in feature selection. Artif. Intell. **151**(1–2), 155–176 (2003)
5. Elomaa, T., Rousu, J.: General and efficient multisplitting of numerical attributes. Mach. Learn. **36**, 201–244 (1999)
6. Eshelman, L.J.: The CHC adaptive search algorithm: how to have safe search when engaging in nontraditional genetic recombination. In: FOGA, pp. 265–283 (1990)
7. Fayyad, U.M., Irani, K.B.: Multi-interval discretization of continuous-valued attributes for classification learning. In: Proceedings of the 13th International Joint Conference on Artificial Intelligence (IJCAI), pp. 1022–1029 (1993)
8. Freitas, A.A.: Data Mining and Knowledge Discovery with Evolutionary Algorithms. Springer, New York (2002)
9. García, S., Luengo, J., Sáez, J.A., López, V., Herrera, F.: A survey of discretization techniques: taxonomy and empirical analysis in supervised learning. IEEE Trans. Knowl. Data Eng. **25**(4), 734–750 (2013)
10. García, S., Luengo, J., Herrera, F.: Data Preprocessing in Data Mining. Springer, New York (2015)
11. He, Z., Tian, S., Huang, H.: EMVD-BDC: an evolutionary multivariate discretization approach for association rules. J. Comput. Inf. Syst. **2**(4), 1343–1350 (2006)
12. Kerber, R.: ChiMerge: discretization of numeric attributes. In: National Conference on Artificial Intelligence American Association for Artificial Intelligence (AAAI92), pp. 123–128 (1992)
13. Kurgan, L.A., Cios, K.J.: CAIM discretization algorithm. IEEE Trans. Knowl. Data Eng. **16**(2), 145–153 (2004)
14. Liu, H., Hussain, F., Tan, C.L., Dash, M.: Discretization: an enabling technique. Data Min. Knowl. Discov. **6**(4), 393–423 (2002)
15. Quinlan, J.R.: C4.5: Programs for Machine Learning. Morgan Kaufmann Publishers Inc., San Mateo (1993)
16. Sheng, W., Liu, X., Fairhurst, M.C.: A niching memetic algorithm for simultaneous clustering and feature selection. IEEE Trans. Knowl. Data Eng. **20**(7), 868–879 (2008)
17. Tay, F.E.H., Shen, L.: A modified Chi2 algorithm for discretization. IEEE Trans. Knowl. Data Eng. **14**, 666–670 (2002)

18. Wu, X., Kumar, V. (eds.): The Top Ten Algorithms in Data Mining. Chapman & Hall/CRC Data Mining and Knowledge Discovery, Boca Raton (2009)
19. Yang, Y., Webb, G.I.: Discretization for Naive-Bayes learning: managing discretization bias and variance. Mach. Learn. **74**(1), 39–74 (2009)
20. Zighed, D.A., Rabaséda, S., Rakotomalala, R.: FUSINTER: a method for discretization of continuous attributes. Int. J. Uncertain. Fuzziness Knowl.-Based Syst. **6**, 307–326 (1998)

Extreme Learning Machine as a Function Approximator: Initialization of Input Weights and Biases

Grzegorz Dudek

Abstract Extreme learning machine is a new scheme for learning the feedforward neural network, where the input weights and biases determining the nonlinear feature mapping are initiated randomly and are not learned. In this work, we analyze approximation ability of the extreme learning machine depending on the activation function type and ranges from which input weights and biases are randomly generated. The studies are performed on the example of approximation of one variable function with varying complexity. The ranges of input weights and biases are determined for ensuring the sufficient flexibility of the set of activation functions to approximate the target function in the input interval.

Keywords Extreme learning machine · Function approximation · Activation functions · Feedforward neural networks

1 Introduction

Feeforward neural networks (FNNs) have been successfully applied to solve many complex and diverse tasks. They are widely used in regression and classification problems due to their adaptive nature and excellent approximation properties (FFN is an universal approximator, i.e., it is capable of approximating any nonlinear function). As a learning machine FNN can learn from observed data and generalize well in unseen examples. All inner parameters of the networks (weights and biases) are adjustable. Due to the layered structure of FNN the learning process is complicated, inefficient, and requires the activation functions of neurons to be differentiable. The training usually employ some form of gradient descent method, which is generally

G. Dudek (✉)
Department of Electrical Engineering, Czestochowa University of Technology, Al. Armii Krajowej 17, 42-200 Czestochowa, Poland
e-mail: dudek@el.pcz.czest.pl

R. Burduk et al. (eds.), *Proceedings of the 9th International Conference on Computer Recognition Systems CORES 2015*, Advances in Intelligent Systems and Computing 403, DOI 10.1007/978-3-319-26227-7_6

time-consuming and converges to local minima. Moreover some parameters, such as number of hidden neurons or learning algorithm parameters, have to be tuned manually.

The Extreme Learning Machine (ELM) is an alternative learning algorithm proposed for training single-hidden-layer FNNs [1]. The learning process does not require iterative tuning of weights. The input weights (linking the inputs with hidden layer) and biases of hidden neurons need not to be adjusted. They are randomly initiated according to any continuous sampling distribution without the knowledge of the training data. The only parameters need to be learned are the output weights (between the hidden and output layers). Thus ELM can be simply considered as a linear system in which the output weights can be analytically determined through simple generalized inverse operation of the hidden layer output matrices. As theoretical studies have shown, even with randomly generated hidden nodes, ELM with wide type of activation functions can work as an universal approximator. Numerous experiments and applications have demonstrated that ELM and its variants are efficient, accurate, and easy to implement. The learning speed of ELM can be thousands of times faster than traditional gradient descent-based learning.

In this work, we analyze approximation ability of ELM depending on the activation function type and ranges from which input weights and biases are randomly generated. To visualize results the studies are performed on the example of approximation of one variable function with varying complexity. The ranges of input weights and biases are determined for ensuring the sufficient flexibility of ELM in the input interval.

2 Basic Extreme Learning Machine

ELM originally proposed by Huang et al. [1] learns in three steps. Given a training set $\Phi = \{(\mathbf{x}_k, t_k) \mid (\mathbf{x}_k \in \mathbb{R}^n, t_k \in \mathbb{R}, k = 1, 2, \ldots, N\}$, hidden node activation function type $h(\mathbf{x})$, and the number of hidden nodes L,

1. Randomly initiate according to any continuous sampling distribution hidden node parameters, i.e., input weights and biases: $\mathbf{a}_i = [a_{i,1}, a_{i,2}, \ldots, a_{i,n}]^T$ and b_i, $i = 1, 2, \ldots, L$. Usually uniform distribution is used for this: $a_{i,j} \sim \mathrm{U}(a_{\min\, i,j}, a_{\max\, i,j})$, $b_i \sim \mathrm{U}(b_{\min\, i,j}, b_{\max\, i,j})$. (The ranges from which weights and biases are generated, $a_{\min}$, $a_{\max}$, $b_{\min}$ and $b_{\max}$, are the main subject of this work.)
2. Calculate the hidden layer output matrix $\mathbf{H}$:

$$\mathbf{H} = \begin{bmatrix} \mathbf{h}(\mathbf{x}_1) \\ \vdots \\ \mathbf{h}(\mathbf{x}_N) \end{bmatrix} = \begin{bmatrix} h_1(\mathbf{x}_1) & \cdots & h_L(\mathbf{x}_1) \\ \vdots & \vdots & \vdots \\ h_1(\mathbf{x}_N) & \cdots & h_L(\mathbf{x}_N) \end{bmatrix}, \tag{1}$$

where $h_i(\mathbf{x})$ is an activation function of the ith neuron, which is nonlinear piecewise continuous function, e.g., the sigmoid:

$$h_i(\mathbf{x}) = \frac{1}{1 + \exp(-(\mathbf{a}_i \cdot \mathbf{x} + b_i))}, \tag{2}$$

$\mathbf{a}_i \cdot \mathbf{x}$ denotes the inner product of $\mathbf{a}_i$ and $\mathbf{x}$.
The ith column of $\mathbf{H}$ is the ith hidden neuron output vector with respect to inputs $\mathbf{x}_1, \mathbf{x}_2, \ldots, \mathbf{x}_N$. Hidden neurons maps the data from n-dimensional input space to the L-dimensional feature space H, and thus, $\mathbf{h}(\mathbf{x})$ is a nonlinear feature mapping. The most popular activations functions are: sigmoid, Gaussian, multiquadric, hard-limit, triangular, and sine functions. Different activation functions can be used in different hidden neurons.
The output matrix $\mathbf{H}$ remains unchanged because parameters of the activation functions, $\mathbf{a}_i$ and b_i, are fixed.

3. Calculate the output weights β_i:

$$\boldsymbol{\beta} = \mathbf{H}^{+}\mathbf{T}, \tag{3}$$

where $\boldsymbol{\beta} = [\beta_1, \beta_2, \ldots, \beta_L]^T$ is the vector of the output weights, $\mathbf{T} = [t_1, t_2, \ldots, t_N]^T$ is the training data output matrix, and $\mathbf{H}^{+}$ is the Moore–Penrose generalized inverse of matrix $\mathbf{H}$.
The above equation for $\boldsymbol{\beta}$ results from the minimizing the approximation error:

$$\min \|\mathbf{H}\boldsymbol{\beta} - \mathbf{T}\|. \tag{4}$$

The output function of ELM is of the form (one output case):

$$f_L(\mathbf{x}) = \sum_{i=1}^{L} f_i(\mathbf{x}) = \sum_{i=1}^{L} \beta_i h_i(\mathbf{x}) = \mathbf{h}(\mathbf{x})\boldsymbol{\beta}, \tag{5}$$

where $f_i(\mathbf{x}) = \beta_i h_i(\mathbf{x})$ is the weighted output of the ith hidden node.

The output function $f_L(\mathbf{x})$ is a linear combination of the activation functions $h_i(\mathbf{x})$. Characteristically for ELM the hidden nodes parameters, $\mathbf{a}_i$ and b_i, are randomly generated instead of being explicitly trained. This process is independent of the training data and provide random feature mapping.

To improve generalization performance of ELM its regularized version was proposed [2]. Recent developments in theoretical studies and applications of ELM are reported in [3].

3 Approximation Capability of ELM: Simulation Study

In this section we analyze the approximation capability of ELM depending on the activation function types and the way of initialization of the input weights and biases. For brevity, we use the following acronyms:

- TF: target function $g(x)$,
- FC: fitted curve $f_L(x)$,
- AF: activation function $h_i(x)$,
- II: input interval, i.e., the interval to which inputs are normalized.

The simulation tests were performed in MATLAB R2010b environment. We used Matlab implementation of ELM: function `elm` created by the authors of ELM algorithm (downloaded from http://www.ntu.edu.sg/home/egbhuang/elm_random_hidden_nodes.html). The input weights and biases in this implementation are generated randomly from the uniform distribution: weights from the range of $[-1, 1]$ and biases form the range of $[0, 1]$. There are five types of AFs in `elm` to choose from. Each AF gets linear combination of ELM inputs: $\mathbf{a} \cdot \mathbf{x} + b$ as an argument. The coefficients of this combination are input weights and bias of the ith neuron. The types of AFs implemented in `elm` function are shown in Table 1 in the single input version.

In [4] we analyze the impact of ranges from which the input weights and biases are randomly generated on the fitted curve complexity when sigmoid AFs are used. In this work, we consider ELM with other AFs. To illustrate results the single variable TF is used of the form:

$$g(x) = \sin(20 \cdot e^x) \cdot x^2, \tag{6}$$

where $x \in [0, 1]$.

The complexity of function (6) increases along the interval $[0, 1]$. TF is flat at the left border of the interval, while at the right border its variability is the highest. To express TF variability we use the percentage slope function [4]:

$$s_{g\%}(x) = 100 \cdot \left| \frac{dg(x)}{dx} \right| \cdot \left(\max_{x \in \text{II}} \left| \frac{dg(x)}{dx} \right| \right)^{-1}. \tag{7}$$

The training set includes 5000 points (x_k, y_k), where x_k are uniformly randomly distributed on $[0, 1]$ and y_i are distorted by adding the uniform noise distributed in $[-0.2, 0.2]$. The testing set is created similarly but without noise. The outputs are normalized into the range $[-1, 1]$. These settings are the same as in [1], where ELM performance was evaluated on the SinC function benchmark.

In Fig. 1 the results of approximation using ELM with 20 hidden neurons with Gaussian AFs are shown. As can be seen from this figure the FC fluctuates in the flat part of TF and is underfitted in the complex part. More neurons in the hidden layer (up to 1000) does not improve the result. If we look at the fragments of AFs, $h_i(x)$, in the input window (II), we notice their low variability, which does not correspond

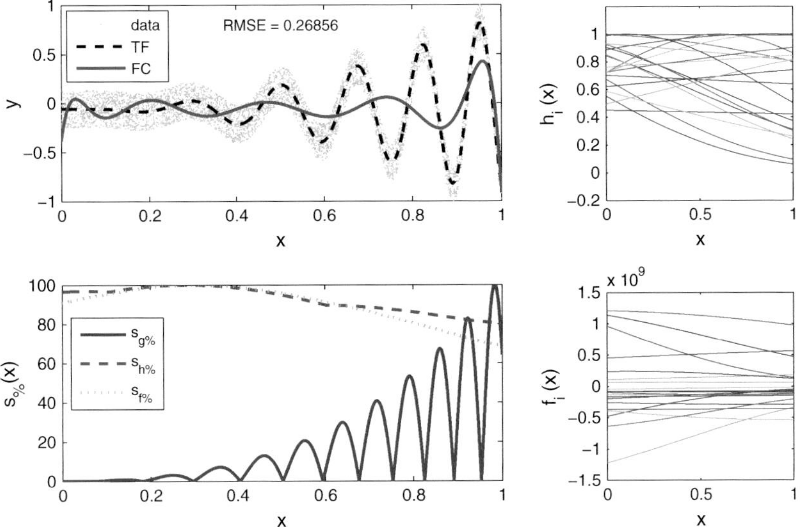

Fig. 1 Results of approximation using ELM with 20 Gaussian neurons generated from the default intervals

to the variability of TF. The AF fragments in II compose the set of basis functions from which the FC is constructed by their linear combination. The components of this combination, $f_i(x)$, i.e., AFs weighted by the output weights β_i, are also shown in Fig. 1. To measure the variability of the set of AFs we use the percentage slope functions of AFs, $s_{h\%}(x)$, and the weighted AFs, $s_{f\%}(x)$, defined as follows [4]:

$$s_{h\%}(x) = \frac{100 \cdot s_h(x)}{\max_{x \in \mathrm{II}} s_h(x)}, \text{ where } s_h(x) = \frac{1}{L}\sum_{i=1}^{L} \left| \frac{dh_i(x)}{dx} \right|, \tag{8}$$

$$s_{f\%}(x) = \frac{100 \cdot s_f(x)}{\max_{x \in \mathrm{II}} s_f(x)}, \text{ where } s_f(x) = \frac{1}{L}\sum_{i=1}^{L} \left| \frac{df_i(x)}{dx} \right| = \frac{1}{L}\sum_{i=1}^{L} \left| \beta_i \frac{dh_i(x)}{dx} \right| \tag{9}$$

The plots of these functions, expressing the AF set variability along II, in Fig. 1 are shown, together with function (7) expressing TF variability. Note that the highest variability of the AF set is at the left border of II, whereas the highest variability of TF is at the right border.

To improve approximation capability of ELM let us increase the AF set variability in the II of [0, 1]. This will be reached in two ways:

- by increasing the input weights determining the slopes of the Gaussian AFs and
- by adjusting the biases to the II so that the maxima of AFs are inside the II.

The first requirement is not very important because the AF slopes are regulated by output weights (see (9)). But to avoid large output weights necessary for providing steep weighted AFs to model the steep TF fragments, let us assume that $a_i \in [0, 10]$. (For one input case a_i can be limited to the positive values.)

According to the second requirement the maximum of AF should be for $x \in [0, 1]$. When the maximum is in the left border of our II we get:

$$h_i(0) = \exp(-(a_i \cdot 0 + b_i)^2) = 1 \rightarrow b_i = 0, \tag{10}$$

and when it is in the right border we get:

$$h_i(1) = \exp(-(a_i \cdot 1 + b_i)^2) = 1 \rightarrow b_i = -a_i. \tag{11}$$

Thus the bias of the ith neuron should be randomly generated within the range:

$$b_i \in [-a_i, 0], \text{ where } a_i \geq 0. \tag{12}$$

For the lowest value of the input weight $a_i = 0$, $b_i = 0$ and AF is a constant function: $h_i(x) = 1$. The higher the value of a_i, the steeper the AF is.

The results of approximation for weights and biases randomly generated in the intervals proposed above in Fig. 2 are presented. Here 100 Gaussian neurons are used in the hidden layer (RMSE for 20 neurons was 0.088). Note higher variability of AFs in the II than outside this interval. The slope function $s_{f\%}(x)$ corresponds better to the variability of TF than in the previous example. Similar results were achieved

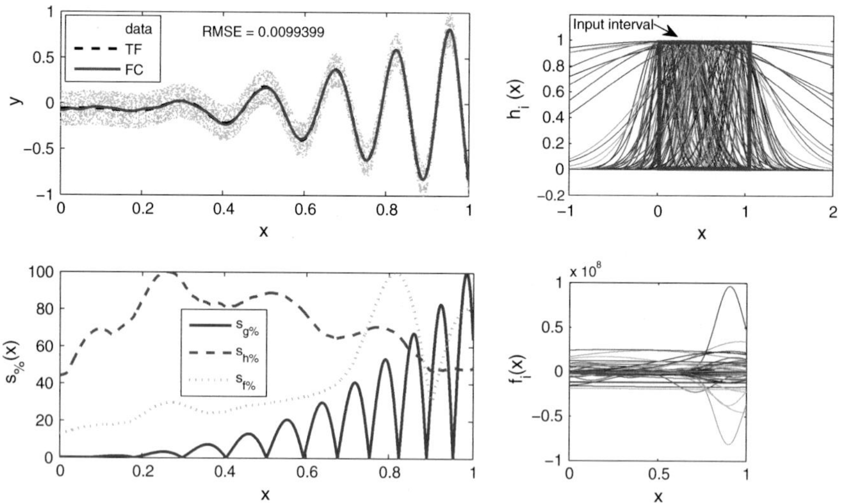

Fig. 2 Results of approximation using ELM with 100 Gaussian neurons generated from the proposed intervals

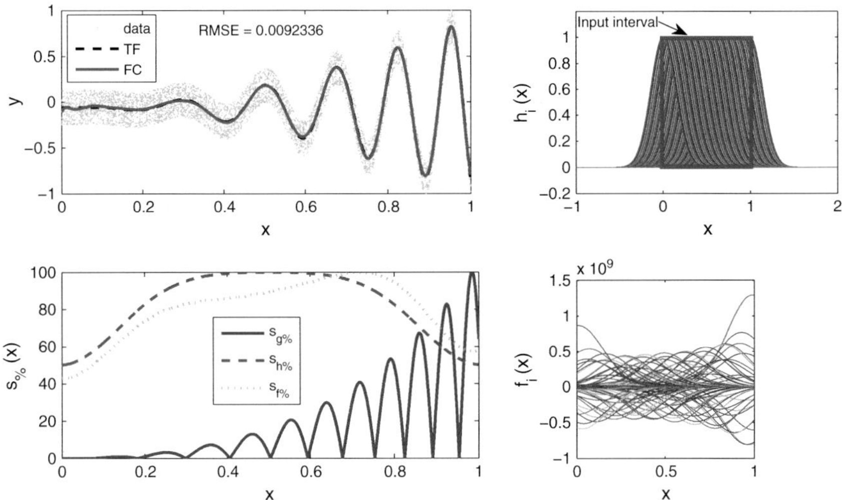

Fig. 3 Results of approximation using ELM with 100 Gaussian neurons evenly distributed in the II

when input weights were all set to constant values $a_i = 5$, and biases were uniformly distributed in the interval $[-a_i, 0]$. This is illustrated in Fig. 3. In the next experiment we replace Gaussian AFs by triangular AFs. When using default intervals for random generation of input weights and biases ($[-1, 1]$ and $[0, 1]$, respectively) the results are not satisfactory. The problem of underfitting appears. To improve approximation capability in this case we use the same approach as for Gaussian AFs. First, we increase the input weight values defining their range as $[0, 10]$. Then we assume that the maxima of AFs are in II. This leads to the same range for biases (12) as in the case of Gaussian AFs. Results of approximation in Fig. 4 are shown. Note that FC is piecewise linear and unsmooth. Combining triangular basis functions results in the "jagged" FC. When instead of random weights and biases constant input weights were used ($a_i = 5$) and biases were uniformly distributed in $[-a_i, 0]$, the results were similar (RMSE = 0.016). In the case of hard-limit AFs the FC is a linear combination of unit step functions. When the step position (jump from 0 to 1 or vice versa) is outside the II we get constant fragment of AF in the II. Such fragments are useless for modeling TF fragments of nonzero slopes. When using default settings for ranges of input weights and biases, many AFs have their jumps outside the II of $[0, 1]$. The jump positions can be calculated from: $x = -b/a$. For b randomly generated from the uniform distribution on the interval $[0, 1]$ and a generated similarly on $[-1, 1]$ about 75 % of AFs have jumps outside our II. To bring the jumps into II the biases should be randomly generated from the intervals: $[-a_i, 0]$, if $a_i \geq 0$ or $[0, -a_i]$, if $a_i < 0$. The value of a_i is not important in this case. (In other types of AFs presented in Table 1, a_i regulates the slope of AF, but not in hard-limit function.) Only its sign deciding about the hard-limit function direction, i.e., from 0 to 1 or from 1 to 0, is important.

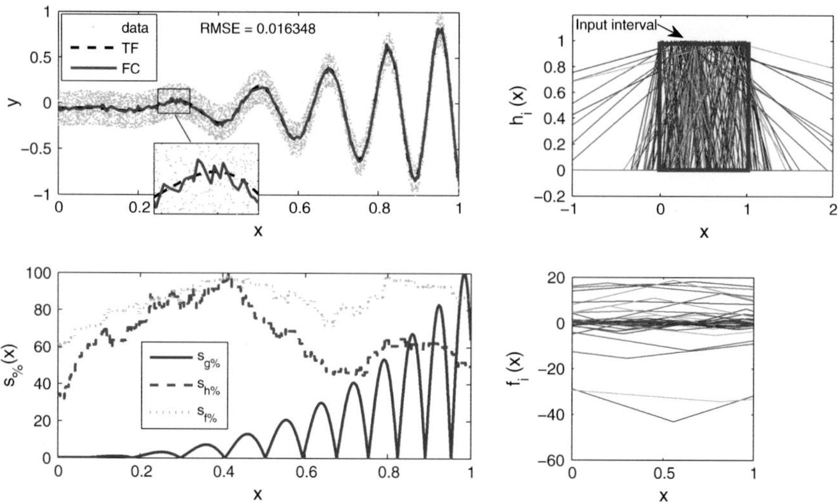

Fig. 4 Results of approximation using ELM with 100 triangular neurons generated from the proposed intervals

So we can assume $a_i \in \{-1, 1\}$, and draw its value with the same probability. In Fig. 5 the results of approximation using ELM composed of 100 hidden neurons with hard-limit AFs are presented. Note that FS is a step function. When we use constant input weights $a_i = +1$ for even-numbered neurons and $a_i = -1$ for odd-numbered ones, and biases evenly distributed in $[-1, 0]$, for $a_i = +1$ or in $[0, 1]$, for $a_i = -1$, the results were similar (RMSE $= 0.046$).

Now we test ELM with sine functions as AFs. For default ranges for input weights and biases the results of approximations look similar to results for Gaussian AFs presented in Fig. 1. To increase variability of AFs in the II we change the bounds of the interval from which input weights are generated to $[0, 30]$. The biases are generated from the range $[0, 2\pi]$. This range for b_i ensures uniform distribution of a single sine AF in II. In such a case, at each point x from II, each value of the AF from its codomain is achievable. The results of approximation for ELM with 100 sine AFs with parameters randomly generated from the above intervals in Fig. 6 are shown. Note that due to the periodicity of AFs their variability is the same inside and outside the II. When instead of random weights and biases we use constant input weights and biases evenly distributed in $[0, 2\pi]$, the resulting FC does not fit to TF. It has a form of a sine function with changing amplitude, and with decreasing period with a_i.

The ranges from which input weights and biases should be randomly generated for a single variable function approximation in Table 1 are summarized. Sigmoid AF was analyzed in [4]. The border value A of the interval for input weights depends on the AF type and variability of the TF. When TF is flat, lower border values can be used. This prevents overfitting. For TF with high variability, higher values of

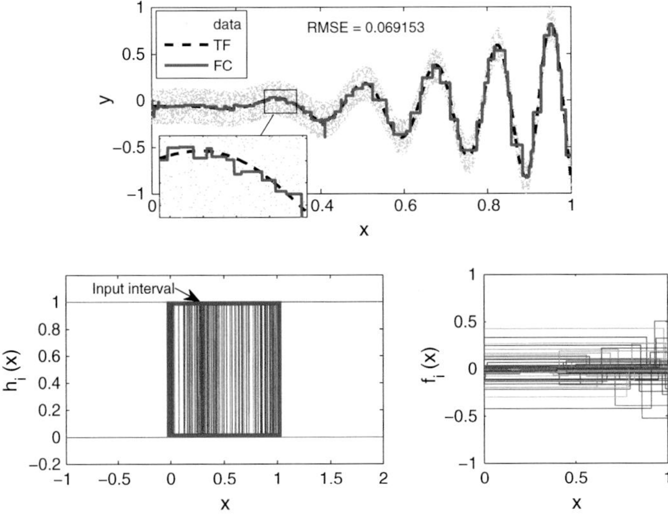

Fig. 5 Results of approximation using ELM with 100 hard-limit neurons generated from the proposed intervals

Table 1 Activation functions and ranges for their parameters

Activation function	Weights $a_i \in$	Biases $b_i \in$
Sigmoid $[1+\exp(-(a_i x+b_i))]^{-1}$	$[-A, A]$	$\left[-\ln\left(\frac{q}{1-q}\right)-a_i, -\ln\left(\frac{1-q}{q}\right)\right]$, if $a_i \geq 0$ $\left[-\ln\left(\frac{q}{1-q}\right), -\ln\left(\frac{1-q}{q}\right)-a_i\right]$, if $a_i < 0$
Gaussian $\exp(-(a_i x+b_i)^2)$	$[0, A]$	$[-a_i, 0]$
Triangular $\begin{cases} 1-\lvert a_i x+b_i\rvert, & \text{if } \lvert a_i x+b_i\rvert \leq 1 \\ 0, & \text{otherwise} \end{cases}$	$[0, A]$	$[-a_i, 0]$
Hard-limit $\begin{cases} 1, & \text{if } a_i x+b_i \leq 0 \\ 0, & \text{otherwise} \end{cases}$	$\{-1, 1\}$	$[-1, 0]$, if $a_i = 1$ $[0, 1]$, if $a_i = -1$
Sine function $\sin(a_i x+b_i)$	$[0, A]$	$[0, 2\pi]$

where: $A > 0$, $q \in [0.5, 1]$ (see [4])

A are needed to prevent underfitting. Generally, parameter A controls bias-variance tradeoff of the ELM. In all cases except sine AF the intervals for biases are dependent on the input weights. So for each neuron first input weight is randomly chosen, and next bias is randomly generated from the appropriate interval.

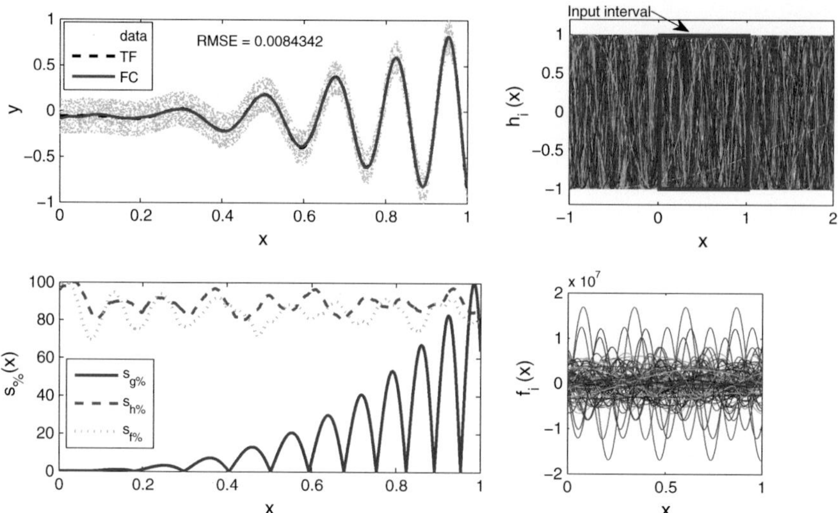

Fig. 6 Results of approximation using ELM with 100 sine neurons generated from the proposed intervals

4 Conclusions

The fitted curve in ELM is a linear combination of the basis functions, i.e., the activation functions of the hidden neurons. The basis function is a simple nonlinear piecewise continuous function which parameters are randomly generated in ELM. The set of basis functions should have the sufficient flexibility to ensure the best fitting to the target function in the input interval. In the classical learning scheme, such as gradient descent-based learning, the input weights and biases are adjusted during learning. This results in the modifications of the basis functions: they change their slopes and slide along the x-axis. So the flexibility of the set of basis functions is adapted to the complexity of the target function. In ELM such a mechanism does not work. Therefore, the ELM designer should ensure the flexibility of the basis function set in the input interval.

The aim of this work is to find the ranges from which basis function parameters should be randomly generated. These ranges are dependent on the basis function type, moreover, the ranges for biases are dependent on the input weights. In the future work, the results achieved here for approximation of the single argument function we will try to generalize for multiple argument functions. It is worth examining also the ability of ELM generalization depending on the input weight ranges determining the slopes of the activation functions.

References

1. Huang, G.-B., Zhu, Q.-Y., Siew, C.-K.: Extreme learning machine: theory and applications. Neurocomputing **70**, 489–501 (2006)
2. Huang, G.-B., Zhou, H., Ding, X., Zhang, R.: Extreme learning machine for regression and multiclass classification. IEEE Trans. Syst. Man Cybern.—Part B: Cybern. **42**(2), 513–529 (2012)
3. Huang, G., Huang, G.-B., Song, S., You, K.: Trends in extreme learning machines: a review. Neural Netw. **61**(1), 32–48 (2015)
4. Dudek, G.: Extreme learning machine for function approximation—interval problem of input weights and biases. In: Proceedings of the 2nd IEEE International Conference on Cybernetics (CYBCONF), pp. 62–67 (2015). http://dx.doi.org/10.1109/CYBConf.2015.7175907

Electron Neutrino Classification in Liquid Argon Time Projection Chamber Detector

Piotr Płoński, Dorota Stefan, Robert Sulej and Krzysztof Zaremba

Abstract Neutrinos are one of the least known elementary particles. The detection of neutrinos is an extremely difficult task since they are affected only by weak subatomic force or gravity. Therefore, large detectors are constructed to reveal neutrino's properties. Among them the Liquid Argon Time Projection Chamber (LAr-TPC) detectors provide excellent imaging and particle identification ability for studying neutrinos. The computerized methods for automatic reconstruction and identification of particles are needed to fully exploit the potential of the LAr-TPC technique. Herein, the novel method for electron neutrino classification is presented. The method constructs a feature descriptor from images of observed event. It characterizes the signal distribution propagated from vertex of interest, where the particle interacts with the detector medium. The classifier is learned with a constructed feature descriptor to decide whether the images represent the electron neutrino or cascade produced by photons. The proposed approach assumes that the position of primary interaction vertex is known. The method's performance in dependency to the noise in a primary vertex position and deposited energy of particles is studied.

Keywords Electron neutrino · Classification · Image descriptor · Liquid argon · Time projection chambers

P. Płoński (✉) · K. Zaremba
Institute of Radioelectronics, Warsaw University of Technology,
Nowowiejska 15/19, 00-665 Warsaw, Poland
e-mail: pplonski@ire.pw.edu.pl

K. Zaremba
e-mail: zaremba@ire.pw.edu.pl

D. Stefan
Instituto Nazionale di Fisica Nucleare, Sezione di Milano e Politecnico,
Via Celoria 16, I-20133 Milano, Italy
e-mail: dorota.stefan@ifj.edu.pl

R. Sulej
National Center for Nuclear Research, A. Soltana 7, 05-400 Otwock/Swierk, Poland
e-mail: Robert.Sulej@cern.ch

R. Burduk et al. (eds.), *Proceedings of the 9th International Conference on Computer Recognition Systems CORES 2015*, Advances in Intelligent Systems and Computing 403, DOI 10.1007/978-3-319-26227-7_7

1 Introduction

Neutrinos are one of the fundamental particles as well as one of the least understood. They exist in three flavors: electron, muon, and tau. There is a hypothesis about the existence of the fourth type of flavor, namely sterile [9]. The detection of neutrinos is an extremely difficult task since they are affected only by weak subatomic force or gravity. Therefore, large detectors are constructed to reveal neutrino's properties. Among them the Liquid Argon Time Projection Chamber (LAr-TPC) detector, proposed by C. Rubbia in 1977 [8], provides excellent imaging ability of charged particles, making it ideal for studying neutrino oscillation parameters, sterile neutrinos existence [9], charge-parity violation, violation of baryonic number conservation, and dark matter searches. The LAr-TPC technique is used in several projects around the world [1, 2, 10–12]. Among them, the ICARUS T600 [10] was the largest working detector located at Gran Sasso in underground Italian National Laboratory operating on CNGS beam (CERN[1] Neutrinos to Gran Sasso). In this study, the T600 parameters will be used since other existing or planned LAr-TPC detectors have the same or similar construction and settings. A neutrino which passes through the LAr-TPC detector can interact with nuclei of argon. Density of argon in liquid form makes the rate of interactions practical for experimental study. Charged particles, created in interaction, produce both scintillation light and ionization electrons along its path in the LAr-TPC detector. The scintillation light, which is poor compared to ionization charge, is detected by photomultipliers which trigger the read-out process. Free electrons from ionizing particles drift in a highly purified liquid argon in an uniform electric field toward the anode. Electrons diffusion approximate value $4.8\,\mathrm{cm}^2/\mathrm{s}$ is much slower than electron drift velocity $1.59\,\mathrm{mm}/\mu\mathrm{s}$, therefore they can drift to macroscopic distances preserving high resolution of track details. The anode consists of three wire planes, so-called Induction1, Induction2, Collection. A signal is induced in a nondestructive way on the first two wire planes, which are practically transparent to the drifting electrons. The signal on the third wire plane (Collection) is formed by collecting the ionization charge. The wires in consecutive planes are oriented in three different degrees with respect to the horizontal with 3 mm spacing between wires in the plane. This allows to localize the signal source in the XZ plane, whereas Y coordinate is calculated from wire signal timing and electron drift velocity.[2] Signal on wires is amplified and digitized with 2.5 MHz sampling frequency which results in 0.64 mm spatial resolution along the drift coordinate. The digitized waveforms from consecutive wires placed next to each other form 2D projection images of an event, with resolution $0.64\,\mathrm{mm} \times 3\,\mathrm{mm}$. One of the common aims of proposed and future neutrino experiments is to study the appearance of electron neutrinos in the muon neutrino beam. Fundamental requirement for a such study is the method for classification of the interacting neutrino flavor and, in the case of detectors placed on surface, the method for the cosmogenic background rejection. Selection of ν_e interaction among other ν interactions and background events should

[1] CERN—European Organization for Nuclear Research.

[2] Coordinate system labeling is given for reference.

involve analysis of the primary interaction vertex (PIV) features, including detection of a single electron and the presence of hadronic activity. These features may allow to eliminate the majority of events that can mimic signal, namely:

- ν interactions with π_0 produced in the vertex, which decays into gammas immediately, and one or more gammas converts to e+/e− in close vicinity of the vertex;
- gammas produced by cosmogenic sources, converting in the detector volume within the data taking time window with production of e+/e− pair.

In this paper, a novel method for automatic classification of ν_e from cosmogenic sources is presented. The considered range of energy deposited by an event in the detector is 0.2–1.0 GeV. We expect that appearance of interesting ν_e events within this range and therefore we are preparing the method for rejection of background events resulting in similar energy deposit. In the proposed method for each event a feature descriptor is constructed. It describes the signal distribution in Induction2 and Collection views. The method assumes that the localization of the PIV, where the particle starts interaction with detector, is known. The classifier is learned with a created feature descriptor. Herein, the different settings used in the feature vector construction are examined. The settings with the best performance are selected. The impact of noise in PIV localization on the method's performance is analyzed. Additionally, the classifier performance is assessed on various energy ranges of classified events.

2 Methods

2.1 Dataset Creation

The dataset was generated with the FLUKA software [4] and T600 detector parameters. There were generated 7090 events with energy from 0.2 to 1.0 GeV equally distributed. Among them, there were 3283 events from electron neutrino (positive class label) and 3807 events from cosmogenic sources (negative class label). For each event the position of primary interaction vertex is assumed to be known. All the events have PIV located at least 5 cm from anode or cathode. All images were deconvoluted with impulse response of the wire signal read-out chain. The segmentation procedure [7] was applied to remove detector's noise from the images. There are considered two views for each event, namely Induction2 and Collection. From each view the image chunk with size 101 wires x 505 samples and center in the PIV was considered. The used chunk's size is sufficiently large for analysis conducted in this paper. The images were downsampled to 101 x 101 pixels size to provide similar resolution on both the axes, where 1 wire corresponds to 1 pixel in the x-axis and 5 samples corresponds to 1 pixel in the y-axis.

2.2 Event's Feature Descriptor

The events from different classes have charge amplitudes propagated in different ways starting from the PIV. Herein, the event's feature descriptor is proposed to describe this property. The event observed in the detector is described as two images from Induction2 and Collection views. Each image is converted into the polar coordinate system with radius R and number of bins B spaced to each other with $360/B$ degrees and center in the PIV. The total charge in each bin is summed and creates a charge histogram for the considered view. Additionally, for each view the statistics variables are computed, which describe minimum, maximum, standard deviation, mean, and total sum of charge in the histogram. The histogram values and statistics variables from both the views form a feature vector. This results in $2(B + 5)$ features describing each event. In Fig. 1 are presented images of example events from positive and negative classes, with corresponding images after conversion into polar coordinates and charge histograms. It can be observed that the histograms from the event with negative class (Fig. 1c, f) have one peak, whereas histograms for positive class (Fig. 1i, l) have more than one peak. This is the main difference between positive and negative classes. What is more, the tracks in the images from negative class events have broader peaks in the histograms, contrary to tracks from positive class events which appear as lines in the image and narrow peaks in the histograms. The classifier algorithm is learned to distinguish these properties coded in the feature vector.

2.3 Classification Framework

In the proposed approach, the created feature descriptor of the event is an input for the classifier, which response is a probability whether image represents interaction of electron neutrino. The Random Forest [3] algorithm was used as classifier. To asses the classifier's performance the Receiver Operating Characteristic (ROC) [5] curve was used, where

$$\text{True Positive Rate} = \frac{TP}{TP + FN}, \tag{1}$$

$$\text{False Positive Rate} = \frac{FP}{FP + TN}. \tag{2}$$

The TP stands for true positives—correctly classified positive samples, TN are true negative—properly classified negative samples, FP are false positives—negative samples incorrectly classified, and FN are false negatives, which are positive samples improperly classified as negatives. Additionally, the Area Under ROC Curve (AUC) and accuracy (the number of all correctly classified samples) were used. They were computed on fivefold cross-validation (CV) repeated 10 times for stability of the obtained results.

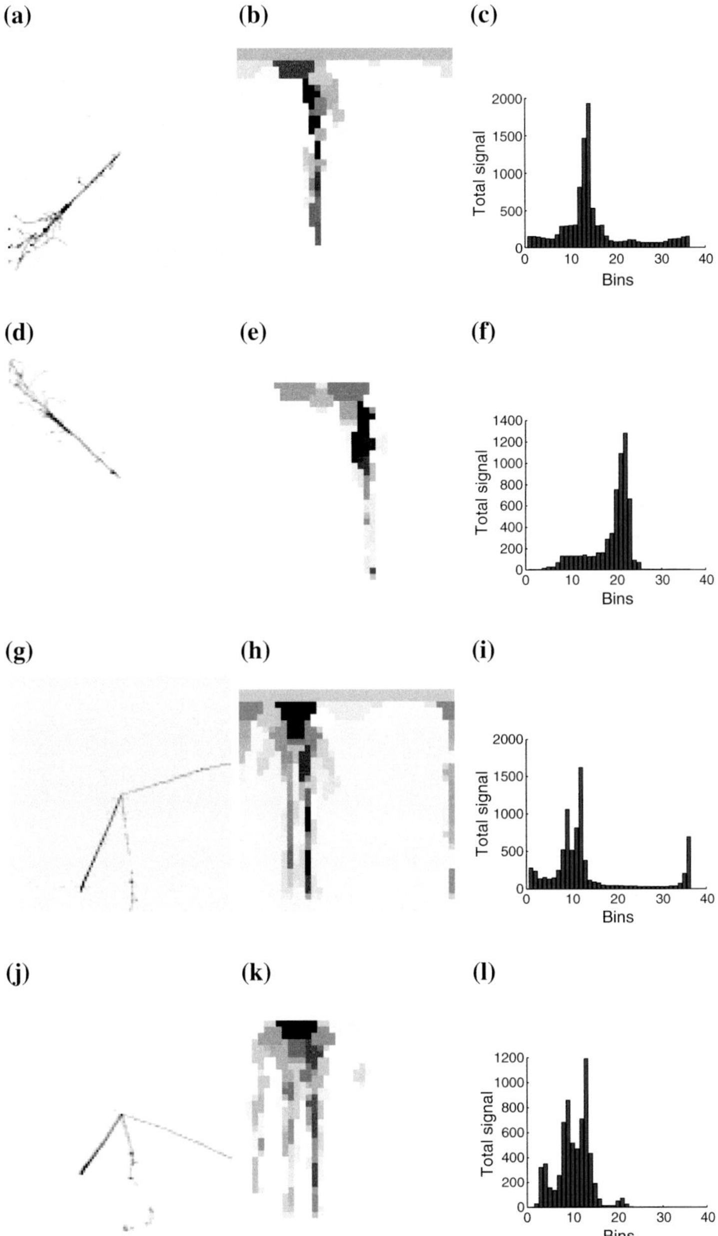

Fig. 1 The example of negative (bkg.) and positive (sig.) events are presented in different perspectives, namely as a raw image observed in the detector, the image in polar coordinates, and charge distribution with 36 bins and radius equal 10 pixels. Each event is presented in Induction2 (Ind2) and Collection (Coll) views. Each row describes one event in a selected view. **a** Raw, Ind2, bkg. **b** Polar, Ind2, bkg. **c** Distribution, Ind2, bkg. **d** Raw, Coll, bkg. **e** Polar, Coll, bkg. **f** Distribution, Coll, bkg. **g** Raw, Ind2, sig. **h** Polar, Ind2, sig. **i** Distribution, Ind2, sig. **j** Raw, Coll, sig. **k** Polar, Coll, sig. **l** Distribution, Coll, sig.

3 Results

The feature vector depends on two parameters: the number of bins and the length of the radius. In order to construct the most discriminative feature vector, the various parameter combinations were checked. There were considered a number of bins: {18, 36, 72, 180}, radius length: {2, 5, 10, 20, 50} pixels and presence of signal statistics. To asses the discriminative power of feature vector, performance of the Random Forest (RF) classifier with 1000 trees was measured. The results are presented in Fig. 2. There can be observed that performance of the classifier is higher when statistics variables are included in the feature vector. The best performance was obtained for the feature vector with 36 bins and radius length 10 pixels and signal statistics included, the AUC is 0.9893 ± 0.0001 and accuracy is equal 0.9535 ± 0.0007. From this point, in further experiments the feature vector is constructed with 36 bins, radius length equal 10 pixels and included signal statistics variables.

The performance of the proposed method in the dependency to the number of trees used in the RF classifier is presented in Fig. 3. The AUC of the method increases with

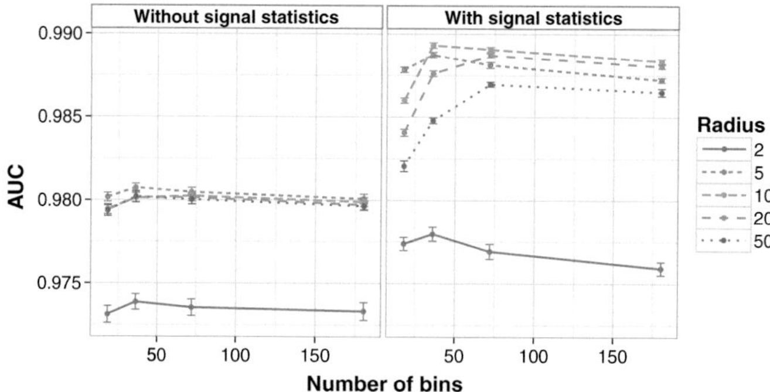

Fig. 2 The performance of the method computed on fivefold CV repeated 10 times for different settings used for event's feature descriptor construction. There were used different number of bins, various length of the radius and presence of additional variables with signal statistics

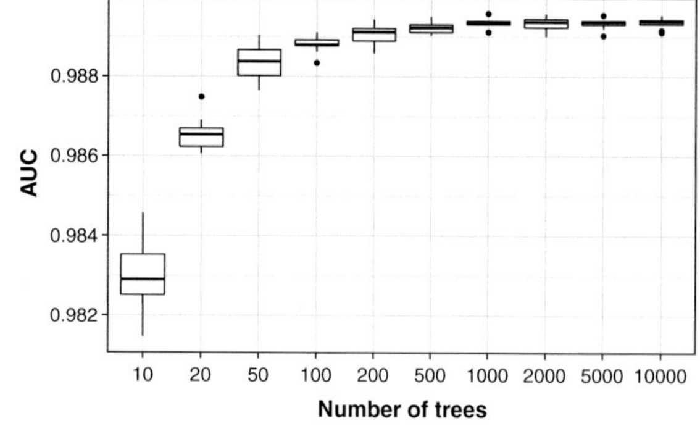

Fig. 3 The performance of the method computed on 5 fold CV repeated 10 times for different tree numbers for 36 bins and radius length equal 10 pixels and signal statistics included in feature vector

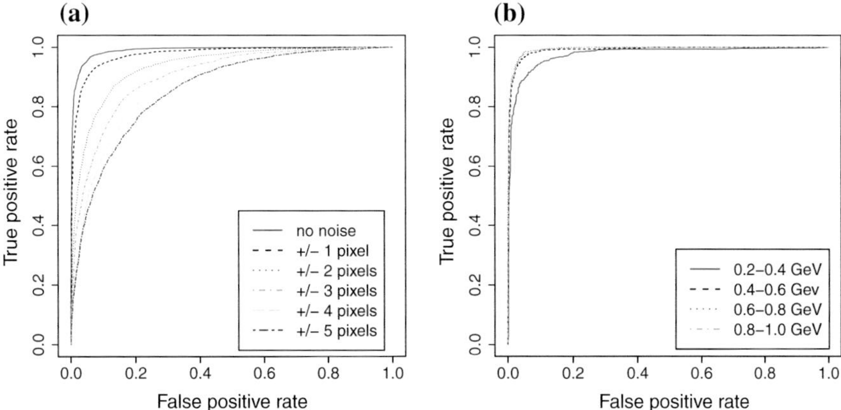

Fig. 4 The performance of the proposed method for different **a** noise levels in PIV's location and **b** different energies of observed event

increasing tree number in the RF. However, the performance for 1000 or more trees in the forest is almost stable. In further analysis the 1000 trees in the RF were used. The proposed method assumes that PIV position is known. It should be designated before classification by another algorithm, for instance with algorithm presented in [6] and additional logic rules about the PIV position. Therefore, the performance of the method for different noise levels in PIV's position is examined. The noise levels were generated by drawing a random number of pixels and adding to the true PIV's location. The ROC curves for different noise levels in PIV are presented in Fig. 4a and the AUC values for fivefold CV repeated 10 times are in Table 1. The performance of the method decreases with more noise in PIV's location. However, the method's AUC is 0.9360 for ±2 pixels noise in PIV's location, which is considered as expected upper bound noise level of the algorithm which designates the PIV's location.

The image content depends on the observed event's energy. Therefore, the method performance was tested for different events' energy ranges (0.2–0.4, 0.4–0.6, 0.6–0.8, 0.8–1.0) GeV. The ROC curves are presented in Fig. 4b and AUC values for fivefold CV with 10 times repetition are in Table 2. It can be observed that for very low energies (0.2–0.4) GeV the performance of the method decreases slightly. However, for events with energy greater than 0.4 GeV the accuracy of the method is almost stable.

Table 1 The performance of the proposed method for different noise levels in PIV's location computed with fivefold CV repeated 10 times

Noise level	AUC
Zero noise	0.9893 ± 0.0001
1 pixel	0.9771 ± 0.0004
2 pixels	0.9360 ± 0.0006
3 pixels	0.9031 ± 0.0009
4 pixels	0.8800 ± 0.0009
5 pixels	0.8601 ± 0.0014

Table 2 The performance of the proposed method for different energies of observed event computed with fivefold CV repeated 10 times

Energy range	AUC
0.2–0.4	0.9770 ± 0.0005
0.4–0.6	0.9906 ± 0.0002
0.6–0.8	0.9944 ± 0.0003
0.8–1.0	0.9934 ± 0.0002

4 Conclusions

The fundamental requirement in the case of surface neutrino detectors is an ability for rejection of cosmogenic background events. Herein, the novel method for classification of neutrino with electron flavor based on raw images from LAr-TPC detector is presented. The method constructs the feature vector for an observed event. It describes the distribution of the charge starting from primary interest vertex, in which position is assumed to be known. The classifier makes a decision whether the detected event is an electron neutrino based on a feature descriptor. The best combination of parameters used for a feature vector construction was selected. The method has AUC 0.9893 and accuracy 0.9535. The experiments for different noise levels in PIV locations show that even with large noise (2 pixels, which corresponds to 2 wires on x-axis and 10 samples on y-axis randomly added in any direction) the AUC of the method is 0.9360. The performance of the method is slightly lower for low-energy events (<0.4 GeV) and for events with energy higher than 0.4 GeV is almost constant with AUC greater than 0.99. The future work will focus on combining the proposed method with algorithm for PIV position estimation.

Acknowledgments PP and KZ acknowledge the support of the National Science Center (Harmonia 2012/04/M/ST2/00775). Authors are grateful to the ICARUS Collaboration and Polish Neutrino Group for useful suggestions and constructive discussions during a preliminary part of this work.

References

1. Anderson, C., et al.: The ArgoNeuT detector in the NuMI low-energy beam line at Fermilab. J. Instrum. **7**, P10019 (2012)
2. Autiero, D., et al.: Large underground, liquid based detectors for astro-particle physics in Europe: scientific case and prospects. J. Cosmol. Astropart. Phys. **11**, 011 (2007)
3. Breiman, L.: Random forests. Mach. Learn. **45**, 5–32 (2001)
4. Ferrari, A., Sala, P.R., Fasso, A., Ranft, J.: FLUKA: a multi-particle transport code, CERN-2005-10, INFN TC 05 11, SLAC-R- 773 (2005)
5. Hastie, T., Friedman, J., Tibshirani, R.: The Elements of Statistical Learning. Springer, New York (2009)
6. Morgan, B.: Interest point detection for reconstruction in high granularity tracking detectors. J. Instrum. **5**, P07006 (2010)
7. Płoński, P., Stefan, D., Sulej, R., Zaremba, K.: Image Segmentation in Liquid Argon Time Projection Chamber Detector. Lecture Notes in Artificial Intelligence, vol. 9119, pp. 606–615. Springer, Heidelberg (2015)

8. Rubbia, C.: The liquid-argon time projection chamber: a new concept for neutrino detectors. CERN report (1977)
9. Sulej, R.: Sterile neutrino search with the ICARUS T600 in the CNGS beam. XV Workshop on Neutrino Telescopes, PoS Neutel (2013)
10. The ICARUS Collaboration, Design, construction and tests of the ICARUS T600 detector, Nuclear Instruments and Methods in Physics Research, vol. A527 (2004)
11. The LBNE Collaboration, The long-baseline neutrino experiment–exploring fundamental symmetries of the universe, FERMILAB-PUB-14-022 (2014). arXiv:1307.7335
12. The MicroBooNE Collaboration, Proposal for a new experiment using the booster and NuMI neutrino beamlines: MicroBooNE, FERMILAB-PROPOSAL-0974 (2007)

Stroke Tissue Pattern Recognition Based on CT Texture Analysis

Grzegorz Ostrek, Artur Nowakowski, Magdalena Jasionowska, Artur Przelaskowski and Kazimierz Szopiński

Abstract The main objective of this paper is a texture-based solution to the problem of acute stroke tissue recognition on computed tomography images. Our proposed method of early stroke indication was based on two fundamental steps: (i) segmentation of potential areas with distorted brain tissue (selection of regions of interest), and (ii) acute stroke tissue recognition by extracting and then classifying a set of well-differentiating features. The proposed solution used various numerical image descriptors determined in several image transformation domains: 2D Fourier domain, polar 2D Fourier domain, and multiscale domains (i.e., wavelet, complex wavelet, and contourlet domain). The obtained results indicate the possibility of relatively effective detection of early stroke symptoms in CT images. Selected normal or pathological blocks were classified by LogitBoost with the accuracy close to 75 % with the use of our adjusted cross-validation procedure.

Keywords Hypodensity recognition · Stroke recognition · Multiscale domain

1 Introduction

Stroke is one of the most frequent and most devastating events among human diseases, the first cause of permanent disability (for people over 40 years old) worldwide, and the third most frequent cause of death. Detection of hyperacute (<6 h after onset)

G. Ostrek (✉) · A. Nowakowski · A. Przelaskowski
Faculty of Mathematics and Information Science, Warsaw University of Technology,
ul. Koszykowa 75, 00-662 Warsaw, Poland
e-mail: ostrekg@mini.pw.edu.pl

M. Jasionowska
Institute of Radioelectronics, Warsaw University of Technology,
ul. Nowowiejska 15/19, 00-665 Warsaw, Poland

K. Szopiński
Department of Dental and Maxillofacial Radiology, Faculty of Medicine and Dentistry,
Medical University of Warsaw, ul. Nowogrodzka 59, 02-006 Warsaw, Poland

R. Burduk et al. (eds.), *Proceedings of the 9th International Conference on Computer Recognition Systems CORES 2015*, Advances in Intelligent Systems and Computing 403, DOI 10.1007/978-3-319-26227-7_8

stroke on noncontrast computed tomography (NCCT) images is a challenging task for both neuroradiologists and pattern recognition techniques. CT availability makes it still the most wide-spread imaging diagnostics tool in a time-demanding acute stroke management. NCCT radiological assessment is reported to reach low sensitivity (22–87%) and interobserver agreement ($\kappa = 0.14-0.78$) [1] for the detection of early infarction signs. However, the pathology appearances are specific and some information on viability of tissue can be obtained from NCCT [2] in 50–60% cases prior to 12 h post stroke. The presented approach is a pattern recognition scheme used in neuroimaging [3] and related to a previously developed method [4, 5]. In this work some essential steps of the method have been improved, additional texture features were added, and new classifiers were used. Specifically, an analysis of extensive boosting-based classifiers was performed and directional texture features, dedicated to local tissue density differentiation, were introduced. Testing procedure was adjusted for a *one-patient-leave-out* cross-validation method, which tends to generate more reliable results than previously presented experiments with selected training and two testing sets [5]. Most of the developed stroke detection algorithms utilize an inter-hemispheres relationship or involve texture features or volumetric statistics only. Our approach extends the applied descriptors set, extracted from selected multiscale wavelet decomposition domains.

1.1 Stroke

Typically, stroke symptoms occur suddenly and require fast and emphatic activities. The *time is brain* concept [6] means that stroke treatment should be considered as a medical emergency state by the patient, bystanders, and medical services. Early diagnosis opens the possibility of early treatment. Thus, the avoiding delay should be the major goal in the pre-hospital proceedings of acute stroke care, i.e., recognition of signs and symptoms of stroke by the patient, relatives or bystanders, further emergency services, and emergency physicians, until 'stroke code' pathway at the hospital stroke unit [7]. Among imaging techniques, NCCT remains the method of choice during a routine emergency procedure because it allows to exclude intracerebral hemorrhage and other cerebral contusions in a short amount of time. Intracerebral bleeding is a main contraindication to introduce a recommended intra-venous tissue plasminogen activator (tPA) therapy which remains the only efficient cure. The therapy must be introduced at least 3–4.5 h after the stroke onset. Otherwise, its application can cause serious consequences like intracerebral bleeding or hemorrhage. Magnetic resonance imaging (MRI) and diffusion weighted imaging (DWI) are considered superior to NCCT in ischemia imaging. However, these methods are less sensitive to bleeding and may be not so widely available during out-of-emergency hours. Moreover, MRI and DWI have more contraindications and usually take more time to perform. Some stroke units prefer magnetic resonance imaging for stroke diagnostics. Image reading support schemes, such like Alberta Stroke Program Early CT score (ASPECTS), were introduced to improve diagnosis and quantification of ischemic lesion. The size of hypodensity area of middle cerebral artery is an important tPA therapy contraindication.

1.2 Stroke Automatic Detection

For experienced neurologists, stroke is considered to be easy to recognize but seems to be difficult for general physicians or the patient's family. Telemedicine is recommended by European Stroke Organisation in the rural settings. Thus, it allows to employ computer-assisted methods. The authors of this paper are aware of over a dozen image-based stroke detection algorithms (e.g., [8–16]) in which the classification task is based on density measures, spatial or value distribution, histogram, texture, or multiscale analysis. In most of the published papers there is calculated an inter-hemisphere relation to detect abnormality. Many use statistical tests to differentiate normal from abnormal tissues. Other approaches include thresholding or using classifiers, i.e., support vector machine (SVM), neural networks, decision trees, or linear discrimination analysis. These methods use various validation procedures: selected train and test sets, *n*-fold, or one-leave-out cross-validation. Many image-based stroke detection reports had a database of several patients; however, a cross-validation scheme was applied only in a few publications [14–16]. Based on our experience, the patient-specific approach, *one-patient-leave-out*, cross-validation which considers all samples of a single patient, should be preferred.

2 Materials and Methods

Hyperacute NCCT studies, performed <6 h since stroke symptoms have occurred with follow-up imaging, were analyzed by an experienced radiologist who created infarct lesions outline based on observed progression in a follow-up study. Series with strong image artifacts (e.g., caused by bones or patient movement), a significant asymmetry, and brain stem strokes were excluded. The latter type of stroke is too small for detection by our algorithm, and frequently contaminated by bone artifacts. Finally, eight normal and thirty-two stroke CT studies were selected for further analysis. All slices presenting lateral ventricles were manually selected from the studies.

2.1 Stroke Slicer Method Design

The fundamental steps of the Stroke Slicer [4, 5] method are as follows: (a) automatic segmentation of the CT study, (b) multiscale analysis, (c) semantic maps expression and visualization, and (d) automatic patch-based stroke recognition. The observed stroke lesion size deviates from a focal (observed as $\sim 10 \times 10$ pixels in one slice) hypoattenuation to a vast region in a hemisphere (on a few slices). Based on our experiments and the real anatomical structures size, the 32×32 pixels blocks were used in our procedures of feature calculation. In order to obtain stroke-susceptible blocks, we probed CT images with an automatic block selection procedure, which (a) selects the first block which belongs to the segmented stroke-susceptible region of interest (ROI), (b) selects the next block with selected step in x and y resolution which belongs to ROI, (c) tests whether the selected block overlaps with other blocks with an area less than a specified factor. Negative blocks selected in the training set creation procedure could not overlap with any other negative block. Moreover, negative blocks could not overlap with the outlined hypodensity area. Blocks overlapping over 80 % of an outlined lesion were considered positive.

2.2 *Image Texture Characteristic*

Automatic differentiating between normal and hyperacute stroke CT images is still an ongoing problem. The set of well-differentiating features has been searched among previously developed sets of features. *New added directional and calculated in complex wavelets domain texture features, and image acquisition parameters.* Various features, estimated in both the image and transformation domains, applied for recognition of suspicious regions (blocks of 32×32 pixels) were as follows:

2.2.1 Features in the Image Domain

(a) statistical standard deviation, skewness, entropy, energy; (b) weighted sum of standard deviation and mean of entropy-based filtration; (c) based on co-occurrence matrix: correlation, energy, homogeneity (with and without normalization).

2.2.2 Statistic Features in the 2D Fourier Domain

(*2D FFT calculated in 8-connected neighborhood of around 0 frequency*) (a) standard deviation of squared coefficients (FNSDSC); (b) absolute value of averaged coefficients (FNAVAC); (c) averaged coefficients (FNAE); (d) energy distribution, average energy (MHH), energy standard deviation (SDHH) in selected subbands.

2.2.3 Directional Features in the Polar 2D Fourier Domain

(a) estimated for the angular spectrum which is calculated with the use of soft thresholding [17], i.e., (i) contrast, correlation, energy, entropy, and homogeneity based on co-occurrence matrix (only for $\alpha = 0$); (ii) statistic features (variance, standard deviation, kurtosis, skewness, energy, entropy); (iii) slope and point of intersection with y-axis computed with the use of approximation of angular spectrum; (b) joint entropy, contrast, homogeneity, and the product of contrast and homogeneity for co-occurrence matrix of the magnitude of the polar 2D Fourier transform.

2.2.4 Directional Features in the Wavelet Domain

(2-level DWT-2D decomposition with near-symmetric *5,7 tap* filters) and the *complex wavelet domain* (2-level DWT-2D decompostion using near-symmetric *5,7 tap* filters and *Q-shift 14,14 tap* filters) [18], i.e., (a) statistical features—maximum values of different statistic features calculated from transform coefficients among all directions (three direction ranges $i = 1, 2, 3$ in the wavelet domain, six direction ranges $j = 1, \ldots, 6$ in the complex wavelet domain) and two scales: variance, standard deviation, kurtosis, and energy of selected subbands; (b) co-occurrence matrix features based on the variance matrix of wavelet coefficients among all directions (three in wavelet domain, six in complex wavelet domain) and two scales: correlation, contrast, energy,homogeneity; (c) co-occurrence matrix features based on the image reconstructions with gained third and fourth levels (containing lower frequency): correlation, contrast, energy, homogeneity; (d) three vectors of tissue directionality distribution—calculated as a column-wise sum of binary map obtained by thresholding of selected statistical features, respectively (variance, standard deviation, and energy); (e) three vectors of tissue directionality distribution—calculated as a logical column-wise sum of binary map obtained by thresholding of selected statistical features, respectively (variance, standard deviation, and energy).

2.2.5 Features in the Contourlet Domain

(*5/3* for the pyramidal decomposition step, *pkva6* for the directional decomposition step) (a) mean of coefficients of lowpass subband; (b) correlation for co-occurrence matrix of coefficients of the lowpass subband and the detail subband; (c) mean of contrast for co-occurrence matrix of coefficients, calculated for four directions of the coefficients of the coarse subband; (d) standard deviation of coefficients of the detail subband.

2.2.6 Features in the Wavelet Domain

(2-level of decomposition using *symlet2* from nearly symmetrical wavelets) (a) energy of approximation related to the energy of details, (b) joint entropy of maximum magnitude distribution for successive scales of details, (c) energy for co-occurrence matrix of quantized magnitude details.

2.3 Stroke Tissue Pattern Recognition

Contemporary supervised classification methods used in the image analysis includes decision trees, support vector machines (SVM), neural networks, random forest, and boosting methods. In many applications the boosting methods are the most accurate tool [19] for data classification. Additional advantage of the boosting methods is the fact that these methods do not need a preselection of an initial feature set. Existence of non-relevant features does not affect a final accuracy. The first practical boosting algorithm, called AdaBoost, was introduced by Freund and Shapire in 1996 [20]. Since then many different variations of this solution as well as new boosting algorithms have been introduced. Among all of them, a particular attention should be paid to GentleBoost, RobustBoost, and especially to LogitBoost. These classifiers were reported to be more robust than the original AdaBoost in different cases, especially whereas the image noise was present [21–24]. In this research, performance of these four boosting methods has been investigated in comparison to two other efficient classifiers: SVM and random forest. Boosting is an example of an ensemble classification, where the final strong classifier is built from many weak ones. In general, it is an iterative algorithm where every weak classifier is evaluated on weighted training samples in each iteration and the best one is chosen and next added to the strong classifier. In the same iteration, the algorithm increases the weights of samples which have been wrongly classified. As a weak classifier a simple one-feature thresholding has been chosen in order to reduce computational costs and to gain direct information about the relevancy of the used features.

3 Results and Discussion

3.1 Classification Assessment

In order to assess the accuracy in a real operational situation, all blocks of the same case were included in training or testing sets exclusively (for each patient the same number of positive and negative blocks was randomly selected). The *one-patient-leave-out* cross-validation approach was used, where all blocks of one patient were chosen for testing while remaining cases constituted the training set in each iteration. ROC curves for selected classifiers with the use of all features and for seven radiologists[1] with different levels of expertise (3–15 years experience) are presented in Fig. 1. The performance of all the considered classification methods reach a similar level of accuracy.[2] However, the LogitBoost method is slightly better than the rest

---

[1]Calculated with DBM MRMC 2.5 software tool.

[2]Definition of measures: $Accuracy = \frac{TP+TN}{N_{pos}+N_{neg}}$, $Sensitivity = \frac{TP}{N_{pos}}$, $Specificity = \frac{TN}{N_{neg}}$, $Precision = \frac{TP}{TP+FP}$.

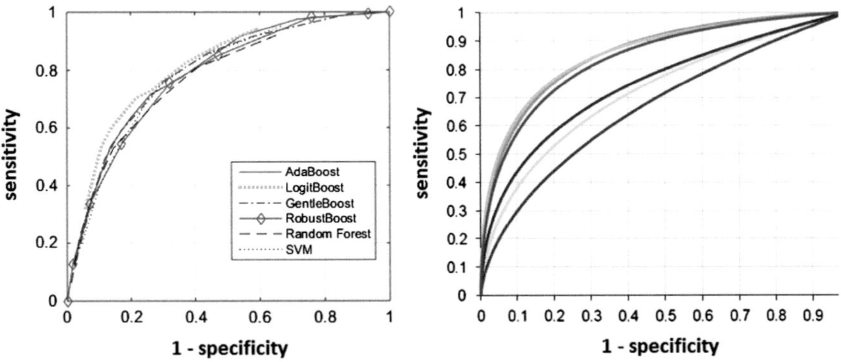

Fig. 1 ROC curves for selected classifiers (*left*) and 7 radiologists (*right*)

of the methods. The best overall accuracy obtained for the LogitBoost was 74.6 % with a sensitivity of 64.2 %, specificity 82.6 %, and precision 69.6 %.

3.2 Feature Relevancy

We prepared three analyses to answer the question of features relevancy. First, we investigated which features classified the training sets the most accurately. The result of this simulation is presented in Table 1. Central moment and skewness calculated on histogram-enhanced CT images seemed to be the most relevant features. The second analysis was aimed at finding the features which improve classification obtained

Table 1 Feature relevancy

Feature	Occurence (%)
Central moment	97
Skewness	3

Table 2 Features improving classification

Feature	Occurrence (%)
Central moment	9.6
Exposure	6.9
FNAE SEGM	5.8
MHH	4.9
SDHH	4.5
FNAE MUPP	4.4
Absolute deviation	4.3
FNAE MUDE	3.3

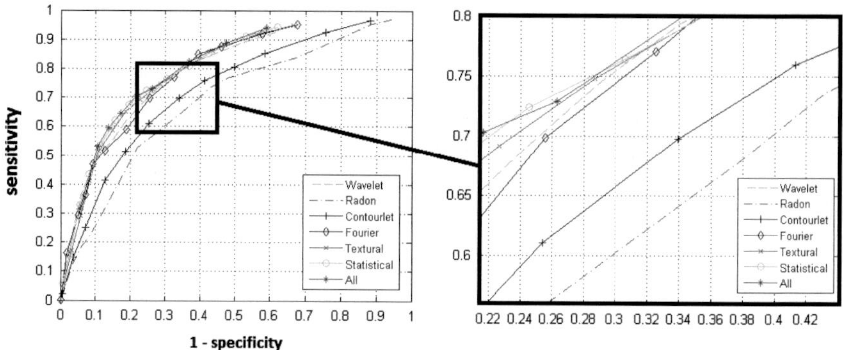

Fig. 2 ROC curves for various groups of features

with the use of only the most relevant features. In other words, we were searching for the most relevant features which allow to classify the previously wrongly classified samples from the training set with the highest precision. Out of a total of 40 weak classifiers the ones with highest occurrence averaged by iterations are listed in Table 2. The number of weak classifiers was selected basing on observed strong classifier performance. The most relevant features from Table 1 occur also in the top of Table 2, that makes central moment even more important in the classification problem. Other significant features were extracted with the use of multiscale signal decomposition, i.e., Fourier (FNAE) and wavelet (MHH an SDHH) selected coefficients statistics, and image statistics. It is worth noting that the acquisition parameter, namely exposure also contributed beside image features. The last analysis was questioning the usefulness of a particular features groups' relevancy. These groups namely were: wavelet, Radon, contourlet, Fourier, textural, and statistical features. ROC curves for the groups, obtained with the use of the LogitBoost classifier were considered as the best in our experiments, are presented in Fig. 2. The differences in results for groups are significant, the highest rates are achieved for four groups: wavelet, Fourier, textural, and statistical features.

4 Conclusion

The achieved results confirmed the usefulness of automatic stroke tissue recognition based on the recognized procedure. The procedure of normal and ischemic tissue blocks differentiation effectively recognized the hypothetical penumbra patches with the accuracy close to 75 %. The cases used in automatic classification were evaluated in our previous work by radiologists with the mean accuracy of 61.0 % (range 47–84%), sensitivity of 56 % (range 40–81%) with 92 % specificity (range 66–100%), and 97 % precision (range 91–100%). Comparing the results of patch-based analysis and radiologists, diagnostics is not straightforward. However, one can be convinced of the usefulness of our method. A wide set of features was analyzed and then boosting-based feature selection was performed. Both histogram and wavelet domains energy distributions were considered useful for features calculation. Moreover, the classical image moments were found useful, whereas exposure seems to contribute to image features extraction. Directional contourlets and Radon features showed their limited usage; however, rotation normalization was not performed. Our experiment expands the set of known useful and well-differentiating features and classifiers in CT based hyperacute stroke detection.

Acknowledgments This publication was funded by the National Science Centre (Poland) based on the decision DEC-2011/03/B/ST7/03649.

References

1. Wardlaw, J.M., Mielke, O.: Early signs of brain infarction at CT: observer reliability and outcome after thrombolytic treatment-systematic review. Radiology **235**(2), 444–453 (2005)
2. Muir, K.W., et al.: Can the ischemic penumbra be identified on noncontrast CT of acute stroke? Stroke **38**(9), 2485–2490 (2007)
3. Orrù, G., et al.: Using support vector machine to identify imaging biomarkers of neurological and psychiatric disease: A critical review. Neurosci. Biobehav. R. **36**(4), 1140–1152 (2012)
4. Przelaskowski, A. et al.: Stroke slicer for CT-based automatic detection of acute ischemia. In: Kurzynski, Marek, Wozniak, Michal (eds.) Comput. Recognit. Syst. 3. Advances in Intelligent Systems and Computing, vol. 57, pp. 447–454. Springer, Heidelberg (2009)
5. Ostrek, G., Przelaskowski, A.: Automatic early stroke recognition algorithm in CT images. In: Piętka, E., Kawa, J. (eds.) Inf. Technol. Biomed. Lecture Notes in Computer Science, vol. 7339, pp. 101–109. Springer, Heidelberg (2012)
6. Ragoschke-Schumm, W.S., et al.: Translation of the 'time is brain' concept into clinical practice: focus on prehospital stroke management. Int. J. Stroke **9**(3), 333–340 (2014)
7. The European Stroke Organisation (ESO) Executive committee: Guidelines for management of ischaemic stroke and transient ischaemic attack 2008. Cerebrovasc. Dis. **25**(5), 457–507 (2008)
8. Hudyma, E., Terlikowski, G.: Computer-aided detecting of early strokes and its evaluation on the base of CT images. In: Proceedings of the International Multiconference on Computer Science and Information Technology (IMCSIT 2008), pp. 251–254 (2008)
9. Chawla, M., et al.: A method for automatic detection and classification of stroke from brain CT images. In: Proceedings of the Annual International Conference of the IEEE Engineering in Medicine and Biology Society (EMBC 2009), pp. 3581–3584 (2009)
10. Yongbum, L., Noriyuki, T., Du-Yih T.: Computer-aided diagnosis for acute stroke in CT images, Dr. L. Saba (ed.) Computed Tomography—Clinical Applications (2012). ISBN: 978-953-307-378-1
11. Noriyuki, T., et al.: Computer-aided detection scheme for identification of hypoattenuation of acute stroke in unenhanced CT. J. Radiol. Phys. Tech. **5**(1), 98–104 (2012)
12. Hema Rajini, N., Bhavani, R.: Computer aided detection of ischemic stroke using segmentation and texture features. Measurement **46**(6), 1865–1874 (2013)
13. Nowinski, W.L., et al.: Automatic detection, localization, and volume estimation of ischemic infarcts in noncontrast computed tomographic scans: method and preliminary results. Invest. Radiol. **48**(9), 661–670 (2013)
14. Tang, F.-H., et al.: An image feature approach for computer-aided detection of ischemic stroke. Comp. Biol. Med. **41**(7), 529–536 (2011)

15. Takahashi, N., et al.: An automated detection method for the MCA dot sign of acute stroke in unenhanced CT. Radiol. Phys. Technol. **7**(1), 79–88 (2014)
16. Nowinski, W.L., et al.: Population-based stroke atlas for outcome prediction: method and preliminary results for ischemic stroke from CT. PLoS ONE **9**(8), e102048 (2014)
17. Jasionowska, M., et al.: A two-step method for detection of architectural distortions in mammograms. Inf. Technol. Biomed., Adv. Soft Comput. **69**, 73–84 (2010)
18. Jasionowska, Magdalena, Przelaskowski, Artur: Subtle directional mammographic findings in multiscale domain. In: Piętka, Ewa, Kawa, Jacek (eds.) Information Technologies in Biomedicine. Lecture Notes in Computer Science, vol. 7339, pp. 77–84. Springer, Heidelberg (2012)
19. Caruana, R., Niculescu-Mizil, A.: An empirical comparison of supervised learning algorithms. In: Proceedings of the 23rd International Conference on Machine Learning, ACM (2006)
20. Freund, Y., Schapire, R.E., Experiments with a new boosting algorithm, ICML 96, (1996)
21. Friedman, J., Hastie, T., Tibshirani, R.: Additive logistic regression: a statistical view of boosting. Ann. Stat. **38**, 337–374 (2000)
22. McDonald, R.A., Hand, D.J., Eckley, I.A.: An empirical comparison of three boosting algorithms on real data sets with artificial class noise. MCS, LNCS **2709**, 35–44 (2003)
23. Torralba, A., Murphy, K., Freeman, W.: Sharing features: efficient boosting procedures for multiclass object detection. CVPR04 **2**, 762–769 (2004)
24. Valmianski, I., et al.: Automatic identification of fluorescently labeled brain cells for rapid functional imaging. J. Neurophysiol. **104**(3), 1803–1811 (2010)

Conversion of Belief Networks into Belief Rules: A New Approach

Teresa Mroczek and Zdzisław S. Hippe

Abstract This paper shows a new method of explaining Bayesian networks by creating descriptions of their properties in a manner closer to the human perceptual abilities, i.e., decision rules in the IF…THEN form (called by us *belief rules*). The conversion method is based on the cause and effect analysis of the Bayesian network quantitative component (the probability distribution). Proposed analysis of the quantitative component leads to a deeper insight into the structure of knowledge hidden in the analyzed data set.

Keywords Belief networks · Belief rules · *Belief* SEEKER

1 Introduction

Among various fields of artificial intelligence, the automation of extraction of knowledge hidden in data, especially when data is incomplete or even inconsistent, gained recently special attention. One of the possible approach to such data is probabilistic methods; among them the Bayesian theorem provides the inference using probabilistic models (hypotheses). Such approach makes it possible to estimate the reliability of the proposed hypothesis basing on the probability of observation, supported by the conviction of matching analyzed data to the hypothesis [1]. This is the foundation for the construction of so-called Bayesian networks, being a complex representation of knowledge about casual relationships between properties of objects [2]. The complexity of the network is determined by the combination of the probability distribution concept and the learning model of the considered problem, forming a directed,

T. Mroczek (✉) · Z.S. Hippe
Department of Expert Systems and Artificial Intelligence, University of Information Technology and Management, 2 Sucharskiego Str., 35-225 Rzeszow, Poland
e-mail: tmroczek@wsiz.rzeszow.pl

Z.S. Hippe
e-mail: zhippe@wsiz.rzeszow.pl

R. Burduk et al. (eds.), *Proceedings of the 9th International Conference on Computer Recognition Systems CORES 2015*, Advances in Intelligent Systems and Computing 403, DOI 10.1007/978-3-319-26227-7_9

acyclic graph in which nodes represent objects and arcs—probability relationships. Both the structure of the Bayesian network and the associated data analysis methodology, allow inference about the probability of a new observation belonging to the particular category, basing on previously estimated set of probabilities. Although in the past few years, the Bayesian method of probabilistic reasoning has gained many supporters, it is rarely used by researchers from outside of the narrow circle of people involved in data mining and artificial intelligence. This may be due to the fact that this methodology is based on difficult concepts and requires a good knowledge of the probability theory foundations. Hence, the study conducted by us have been focused on developing a new methodology for explaining the graphical learning model of a Bayesian network with the help of the more commonly used, closer to human perceptual abilities, declarative forms of knowledge representation—decision rules in the **IF…THEN** form (hereinafter referred to as *belief rules*). Analysis of the current state of knowledge regarding discussed issue, made across many areas of science shows that advantageous features of belief networks, especially the possibility of inference under uncertainty, resulted in attempts at conversion of various knowledge representation formalism into such networks. Most studies dealt with the conversion of *decision rules* → *belief network* [3–5], while the conversion in the opposite direction, i.e., the transition from belief network to decision rules, is little known. It may be said that recently a method of automatic conversion has been developed, which is based largely on the interpretation of the qualitative component of the belief network model—the directed, acyclic graph [6–8]. A different conversion method, based on the cause and effect analysis of the quantitative component (probability distribution) proposed by us, leads to a deeper insight into the structure of knowledge hidden in the analyzed data set. For the sake of keeping the size of the article in recommended bounds, a detailed discussion of the method of Bayesian network explanation has been omitted.

2 The Belief Networks → Belief Rule Conversion Algorithm

Assume that the Bayesian network is a *pair (G, P)* where G is an acyclic directed graph and P is a set of conditional probability distributions over random variables represented by the graph nodes [9]. Two types of variables can be present in the network—independent variables $(X_1, X_2, \ldots, X_n)$, where X_i is any descriptive variable belonging to the set X and dependent variable Y, also referred as decision attribute. A set of older generation variables of X_i is depicted as $parents(X_i)$, while set of its descendants—$children(X_i)$. Every variable characterizes complete set of mutually exclusive states called also a domain $DX_i = x_1, \ldots, x_h^1$ and $DY = y_1, \ldots, y_k$. Additionally, there is conditional probability table $T \in P$ given for any dependent variable Y. If it is possible to select subset Q from set G of graph nodes, such that

$Q \in X \wedge Q \in parents(Y)$ and $|Q| = q$, then it is possible to define rules in the following form:

$$\textbf{IF } X_q \ is \ x_h^q \textbf{ THEN } Y \ is \ y_k : P, \tag{1}$$

where X_q is a node belonging to the subset Q, having values x_h^q defined in the set DX_q, i.e., domain of node X_q, whereas y_k is a value of decision attribute such that $y_k \in DY$, finally P is conditional probability value read from table T. From all classes of dependent variable Y, the class y_k is chosen, having the greatest value of probability P in given row of table T. The conditional part of the new rule is built from values of descriptive variables from the given row. In a case where a single class of dependent variable Y cannot be chosen (more than one class having the same, maximal value) the rule is rejected. Due to that it is possible to eliminate inconsistent cases, characterized by the same conditional frequency for different decision classes. Predefined set of rules is then passed through selection process in order to choose rules for which $P >= PA$, where PA is so-called *acceptation level*—a criterion taking values of conditional probability—allowing to segregate rules and choose those of them that achieve the best classification results. An outline of the mentioned conversion algorithm is shown on Algorithm 1. In the first stage of the developed algorithm the conditional probability table T of the decision variable, consisting of m rows (where m is the number of possible combinations of values of attributes directly influencing the decision) and $q + k$ columns (where q is the number of the mentioned attributes and k is the number of decision classes) is being analyzed. Each row of the table T is analyzed in order to find the maximum value of probability that fulfills the specified *acceptation level*. The number of such maximum

Algorithm 1 The algorithm for converting belief networks to a set of belief rules (detailed description given in the text)

```
R := ∅ for i = 1...m do
  max := −1 max_count := 0 for j = 1...k do
    if T(i, q + j) ≥ PA and (T(i, q + j) ≥ max) then
      if T(i, q + j) = max then
        | max_count := max_count + 1
      else
        | max := T(i, q + j)
      end
      c := j;
    end
  end
  if max_count = 1 then
    r := CreateRule(c) for j = 1...q do
      | AddStatement(r, j)
    end
    R := R ∪ r
  end
end
returnR;
```

values is counted and if it is equal to one then the rule r is created, otherwise the analyzed row is ignored. Two functions take part in the process of rule creation—*CreateRule* and *AddStatement*. The first of them creates the decision part and the second—the condition part of the rule. Both functions employ the table T as a source of values used during rule construction. For the probability value derived in the previous step of the algorithm the values of descriptive attributes and the value of decision attribute is retrieved from the table T. The created rule r is then added to the set of rules R. The algorithm finishes when the entire table is analyzed and the created set of rules R is returned as a result. There is a possibility to extend the set of rules by recursively repeating the described conversion method for subsequent generations of the decision attribute. This means that the developed methodology first addresses cause and effect relationships between the decision node and its parents and then between the parents of the decision node and their ancestors. Thus the set of rules may be expanded in a controlled manner by specifying the number of generations taken into account. In order to perform the complexity analysis [10] of the developed *belief network* $\rightarrow$ *belief rules* conversion algorithm it is necessary to consider its dependency on conditional probability tables (CPT), the number of descriptive attributes (m) and the number of generations (p). For the first generation the computational complexity is linear O(n), where n is the number of cells in the conditional probability table. While the complexity with respect to the number of descriptive attributes is polynomial $O(m^P)$, where m is the number of descriptive attributes and p is a number of attribute generations taken into consideration (both m and p are small natural numbers). Thus it can be concluded that the computational complexity with respect to the number of generations is exponential.

3 Example of Usage of the Bayesian Network → Belief Rules Conversion Method

In order to illustrate the developed conversion method a known database (Table 1) from machine learning domain was used. The database describes three well-known fruits (concepts: $<$ *Banana* $>$, $<$ *Apple* $>$, $<$ *Grape* $>$) using five symbolic and numeric attributes (*Color, Size, Shape, Taste, Weight*). For the selected database the Bayesian network presented in Fig. 1 has been generated.[1] Graphical structure of a belief network allows to define the importance hierarchy of descriptive attributes used in analyzed learning model and a set of the most important attribute—namely attributes directly influencing the decision variable. These attributes are shown in Fig. 1 in bold arrows. The conversion process involves precisely these attributes in the first place. It should be noted that the developed conversion mechanism is based on the analysis of the quantitative network component—the *Conditional Probability*

[1]The Bayesian network has been generated using ***Belief*SEEKER** system [11]. The system generates learning models in a form of belief networks, using a heuristic algorithm utilizing Bayesian fitness function—matching the network structure to the probability distribution—as a metric.

Table 1 Example of decision table

Color	Size	Shape	Taste	Weight	Fruit
Green	Small	Round	Sweet	0.1	Banana
Green	Medium	Round	Sweet	0.5	Apple
Green	Big	Round	Sweet	0.4	Apple
Green	Medium	Round	Sour	0.3	Apple
Green	Small	Round	Sweet	0.1	Grape
Yellow	Medium	Elongated	Sweet	0.3	Banana
Green	Medium	Elongated	Sweet	0.2	Banana
Green	Big	Elongated	Sweet	0.3	Banana
Maroon	Small	Round	Sweet	0.2	Grape
Green	Small	Round	Sour	0.1	Grape
Red	Medium	Round	Sweet	0.3	Apple
Red	Medium	Round	Sour	0.4	Apple
Maroon	Small	Round	Sweet	0.2	Grape

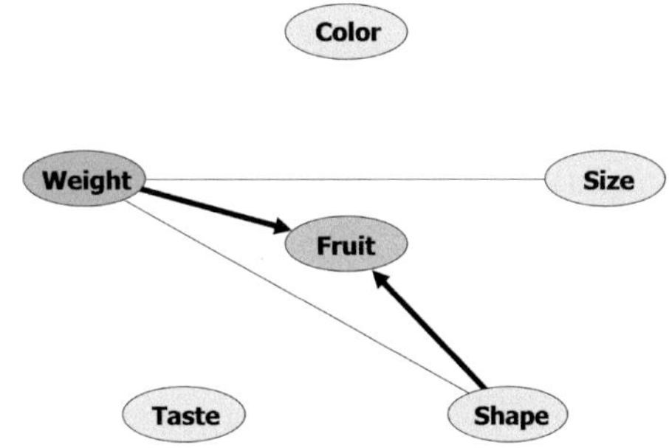

Fig. 1 The learning model in a form of Bayesian network generated by ***Belief*SEEKER** system

Tables (*CPT*) for the dependent variable *Fruit*. The conversion process is controlled by two parameters: the already mentioned parameter *number of generations* and *acceptation level* (PA). The number of generations is determined by the qualitative component—i.e., the Bayesian network structure. In the presented example there are two generations of attributes with respect to the decision attribute. The first generation consists of the most important attributes (*Weight* and *Shape*) while the second generation is the *Size* attribute. For the presentation purposes the value **PA = 0.3** has been chosen.[2]

The algorithm starts with the analysis of *CPT* of the dependent variable (Fig. 2). Just in the first row of that table from all classes of the dependent variable, the class having the highest probability value is chosen and since there is only one such value

[2]The conversion process allows for controlled modification of the acceptation level (PA) parameter. The research conducted in this direction shows that such development of the set of rules does not always help to improve the classification results. Is it therefore necessary to carry out studies evaluating the classification effectiveness of each developed set.

Fruit					
Weight	**Shape**	**Apple**	**Banana**	**Grape**	
0.100..0.233	**elongated**	0.048	0.905	0.048	**1**
0.233..0.367	**elongated**	0.026	0.949	0.026	**2**
0.367..0.500	**elongated**	0.333	0.333	0.333	**3**
0.100..0.233	**round**	0.011	0.204	0.785	**4**
0.233..0.367	**round**	0.949	0.026	0.026	**5**
0.367..0.500	**round**	0.965	0.018	0.018	**6**

Fig. 2 The conditional probability table of the dependent variable fruit. Rows have been additionally numbered on the right (description given in the text)

Fig. 3 The belief rule obtained by the analysis of the row numbered 1 from the CPT of the dependent variable *Fruit*

RULE 1
IF Weight >= 0.100
AND Weight < 0.233
AND Shape IS elongated
THEN Fruit IS Banana

a new rule is created. The process of rule creation is as follows: first the decision part is created by reading out the category of the dependent variable—in this case $< Banana >$—then the condition part is constructed basing on the values of the descriptive attributes readout from the currently analyzed row of the table—thus the attribute *Weight*, whose values are in the range $< 0.100 \dots 0.233 >$, and the attribute Shape taking the value $< elongated >$. The result of this step is the rule shown in Fig. 3. In the same way the subsequent rows of the *CPT* are analyzed, so the rows numbered from 2 to 6. And thus in rows 2 the maximum value of the probability have been obtained for the concept $< Banana >$, in row 4 for the concept $< Grape >$, while in rows 5 and 6 for the concept $< Apple >$. The rules generated in this stage are shown in Fig. 5. Particularly noteworthy is the 3rd row of the CPT. In this row the probability values of all concepts of the dependent variable *Fruit* are the same and are 0.333. In such a case, when more than one class of the dependent variable has a maximum value of the probability for given row of a CPT, the row is not taken into consideration, because it introduces ambiguity. In the presented example, as a result of the 6th row conversion, three belief rules would be created having the same condition part but different indications of the decision variable—therefore—inconsistent rules. Consequently, as a result of the *bayesian network* $\rightarrow$ *belief rules* conversion method for the first generation of attributes, a set of five rules has been obtained (Figs. 3 and 4). In the Bayesian network being described (Fig. 1) there are two generations of attributes relative to the decision attribute. The network transformation (Fig. 5) allowed a better insight into the relationships between attributes. In the second generation, the attribute *Size* indirectly participates in the identification of the *Fruit* category. Hence, at this stage of the algorithm, the analysis of the *CPT* (Fig. 6) of the attribute *Weight*—parent of the attribute *Size*—will be performed. It should be noted that the formation of the probability distribution of the attribute Weight is also influenced by the attribute *Shape*—already included in the process of rule creation for the first generation. This is a special case, which during the enlargement of the set

Fig. 4 Belief rules obtained as a result of the analysis of rows 2–6 of the CPT of the dependent variable *Fruit*

RULE 2
IF Weight >= 0.233
AND Weight < 0.367
AND Shape IS elongated
THEN Fruit IS Banana

RULE 3
IF Weight >= 0.100
AND Weight < 0.233
AND Shape IS round
THEN Fruit IS Grape

RULE 4
IF Weight >= 0.233
AND Weight < 0.367
AND Shape IS round
THEN Fruit IS Apple

RULE 5
IF Weight >= 0.367
AND Weight <= 0.500
AND Shape IS round
THEN Fruit IS Apple

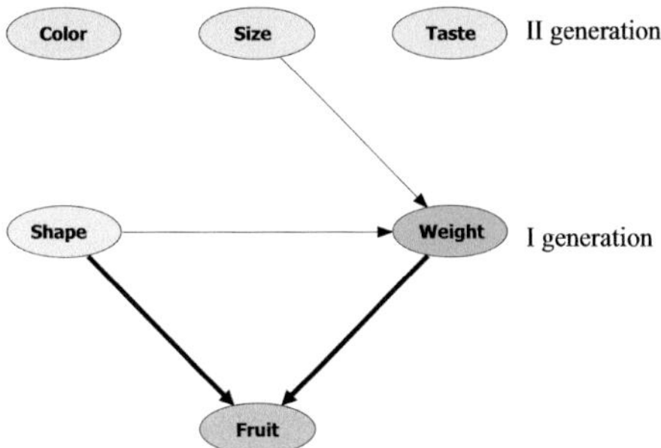

Fig. 5 The Bayesian network—the hierarchy of attributes

of rules by the second generation requires the control of both attributes *Weight* and *Shape*. Thus, a new rule is created if the value of probability readout from the *CPT* (Fig. 6) satisfies the given acceptation level for the values of the attributes *Weight* and *Shape* obtained in the first generation. In particular, the conversion algorithm in the second generation analyzes columns of the *CPT* searching for maximum values of the conditional probability. And so in the column number 1 in Fig. 6 the maximum probability value which exceeds the given acceptation level (PA $= 0.3$) is 0.978. It was achieved for the following values of attributes: $Weight =< 0.100 \ldots 0.233 >$, $Shape =< round >$ and $Size =< small >$. In the next step the rules obtained in the first generation are verified by checking if for the obtained values of the attributes Weight and Shape a rule was created. The rule number 1 fulfills these criteria. Then the attribute *Weight* in Rule 3 will be replaced by a descendant—the attribute *Size*, and a new rule as the presented in Fig. 7 will be added to the set of rules of the first generation.[3] Analysis of the remaining columns of CPT, so the columns number 2 and 3, expanded the set of rules by the additional two rules shown in Fig. 8. Consequently, the presented belief network → belief rules conversion algorithm allowed the explanation of the Bayesian network in the form of 8 rules.

[3]The lack of the rule number is due to the fact, that during the development of the rule set, the rules are automatically numbered after applying the operation of grouping by the concept of the dependent variable and sorted within the concepts. In the result set the rule will be numbered 2. There is already a rule having the number in the set of rules presented in Fig. 5. To avoid ambiguity the authors deliberately omitted the numbering of the second generation rules.

Fig. 6 The conditional probability table of the attribute *Weight*

Weight		1	2	3
Size	**Shape**	**0.100..0.233**	**0.233..0.367**	**0.367..0.500**
big	elongated	0.048	0.905	0.048
medium	elongated	0.487	0.487	0.026
small	elongated	0.333	0.333	0.333
big	round	0.048	0.048	0.905
medium	round	0.013	0.493	0.493
small	round	0.978	0.011	0.011

RULE #
IF Size IS small
AND shape IS round
THEN Fruit IS Grape

Fig. 7 The belief rule created for the second generation of the dependent variable *Fruit*

RULE #
IF Size IS big
AND shape IS elongated
THEN Fruit IS Banana

RULE #
If Size IS big
AND Shape IS round
THEN Fruit IS Apple

Fig. 8 The belief rules obtained from the analysis of columns 2 and 3 of the CPT of the attribute *Weight*

Table 2 Classification results of belief rules

Data set	Error rate (%)	
	Equal width	Equal quantity
Australian	7.39	7.39
Bupa	13.04	45.22
Echocardiogram	4.17	41.67
Glass	7.04	7.04
Hepatitis	13.51	13.51
Iris	2.00	2.00
Pima	2.34	22.66
Wine Recognition	3.39	3.39
Yeast	1.21	0.81

The research was performed under the same system settings. For each data set two working sets were randomly created for learning (2/3 of cases) and for testing (1/3 of cases). For numerical attributes two methods of discretization were used: equal width of each range and equal quantity of instances in each range

The research on the efficiency of the developed conversion algorithm was carried out applying various data sets available in the repository of databases for the machine learning (a few results are presented in Table 2), as well as in the neonatal infection [12] and the brain stroke [13, 14] databases. Furthermore, the recent applications of the methodology relate to classification of laryngopathies [15] and to investigation of the problem of hybridization and optimization of the knowledge base for the Copernicus system [16].

4 Summary

The research described in this paper consternates on a new method of explaining Bayesian networks by creating descriptions of their properties in a manner closer to the human perceptual abilities, i.e., decision rules in the **IF...THEN** form (called by us *belief rules*). Proposed analysis of the quantitative component (probability distribution) leads to a deeper insight into the structure of knowledge hidden in the analyzed data set. We assume that application of the developed method offers a reliable possibility to explain learning models usually difficult for perception, like traditional Bayesian networks. It was found that understanding of various Bayesian models is significantly facilitated by conversion of them into **belief IF...THEN rules**, thus by conversion to knowledge representation formalism easier and better understandable by human beings. It should also be emphasized that numerous and extended experiments, discussed in detail in papers mentioned above, have shown that the main measure of network quality, i.e., the level of classification error, has been always preserved.

References

1. Pearl, J.: Probabilistic Reasoning in Intelligent Systems: Networks of Plausible Inference. Morgan Kaufmann Publishers, San Mateo (1988)
2. Klopotek, M.: Intelligent Web Search Engine. EXIT, Warsaw (2001)
3. Duda, R.O., Hart, P.E., Nilsson, N.J.: Subjective bayesian methods for rule-based inference systems, Technical Note 124, SRI Project 4763 (1976)
4. Korver, M., Lucas, P.: Converting a rule-based expert system into a belief network. Med. Inform. **18**(3), 219–241 (1993)
5. Shwe, M., Middleton, B., Heckerman, D., Henrion, M., Horvitz, E., Lehmann, H.: Probabilistic diagnosis using a reformulation of the INTERNIST-1/QMR knowledge base I. The probabilistic model and inference algorithms. Meth. Inf. Med. **30**(4), 241–255 (1991)
6. Sniezynski, B.: Converting a naive bayes model into a set of rules. Intelligent Information Processing and Web Mining. Advances in Soft Computing, pp. 221–229. Springer, Berlin (2006)
7. Sniezynski, B.: Converting a naive bayes models with multi-valued domains into sets of rules. Database and EXpert Systems Applications, pp. 634–643. Springer, Berlin (2006)
8. Sniezynski, B.: Algorithm for converting bayesian network into set of rules: initial results. Technical Report TR-1/2007, AGH University of Science and Technology (2007)
9. Jensen, F.: Bayesian Networks and Decision Graphs. Springer, New York (2001)
10. Bachmann, P.: Die Analytische Zahlentheorie. Teubner, Leipzig (1984)
11. Grzymała-Busse, J.W., Hippe, Z.S., Mroczek, T.: Deriving belief networks and belief rules from data: a progress report. Transactions on Rough Sets VII. Lecture Notes in Computer Science, vol. 4400, pp. 53–69. Springer, Heidelberg (2007)

12. Grzymala-Busse, J.W., Hippe, Z.S., Kordek, A., Mroczek, T., Podraza, W.: Neonatal infection diagnosis using constructive induction in data mining. Rough Sets, Fuzzy Sets, Data Mining and Granular Computing. Lecture Notes in Computer Science (Lecture Notes in Artificial Intelligence), vol. 4482, pp. 289–296. Springer, Heidelberg (2007)
13. Grzymala-Busse, J.W., Hippe, Z.S., Mroczek, T., Paja, W., Bucinski, A., Strepikowska, A., Tutaj, A.: Brain stroke database—evaluation of glasgow outcome scale and the rankine scal. INFOBAZY (2008) Systems, Applications, Services. Academic Computer Centre in Gdansk—TASK 127–131(2008)
14. Grzymala-Busse, J.W., Hippe, Z.S., Mroczek, T., Roj, E., Skowronski, B.: Two Rough Set Approaches to Mining Hop Extraction Data. Information Science Reference, New York (2008)
15. Mroczek, T., Pancerz, K., Warchol, J.: Belief Networks in Classification of Laryngopathies Based on Speech Spectrum Analysis. Lecture Notes in Artificial Intelligence, pp. 222–231. Springer, Berlin (2012)
16. Gomuła, J., Paja, W., Pancerz, K., Mroczek, T., Wrzesień, M.: Experiments with hybridization and optimization of the rules knowledge base for classification of MMPI profiles. Advances in Data Mining. Lecture Notes in Computer Science, vol. 6870, pp. 121–133. Springer, Heidelberg (2011)

Semi-supervised Naive Hubness Bayesian k-Nearest Neighbor for Gene Expression Data

Krisztian Buza

Abstract Classification of gene expression data is the common denominator of various biomedical recognition tasks. However, obtaining class labels for large training samples may be difficult or even impossible in many cases. Therefore, semi-supervised classification techniques are required as semi-supervised classifiers take advantage of the unlabeled data. Furthermore, gene expression data is high dimensional which gives rise to the phenomena known under the umbrella of the *curse of dimensionality*, one of its recently explored aspects being the presence of hubs or *hubness* for short. Therefore, hubness-aware classifiers were developed recently, such as Naive Hubness Bayesian k-Nearest Neighbor (NHBNN). In this paper, we propose a semi-supervised extension of NHBNN and show in experiments on publicly available gene expression data that the proposed classifier outperforms all its examined competitors.

Keywords Semi-supervised classification · Gene expression data · High dimensionality

1 Introduction

Proteins play essential role in almost all biological processes at the cellular level. Genes are particular subsequences of the DNA that code for proteins. While each cell of the organism has the same DNA, activation levels of genes may vary in different tissues: informally speaking, the expression level of a gene means how frequently the corresponding DNA fragment is transcribed to RNA and translated to proteins. Various tissues are characterized by different gene expression patterns, furthermore, diseases such as cancer may be associated with characteristic gene expression pat-

K. Buza (✉)
BioIntelligence Lab, Institute of Genomic Medicine and Rare Disorders,
Semmelweis University, Budapest, Hungary
e-mail: buza@biointelligence.hu
URL: http://www.biointelligence.hu

R. Burduk et al. (eds.), *Proceedings of the 9th International Conference on Computer Recognition Systems CORES 2015*, Advances in Intelligent Systems and Computing 403, DOI 10.1007/978-3-319-26227-7_10

terns. Classification of gene expression data may contribute to diagnosis of various diseases such as colon cancer, lymphoma, lung cancer, and subtypes of breast cancer [7]. However, the classification task is challenging for several reasons. Usually, the expression levels of several thousands of genes are measured, therefore, the data is high dimensional which gives rise to the phenomena known under the umbrella of the *curse of dimensionality*. While well-studied aspects of the curse are the sparsity and distance concentration, see, e.g., [17], a recently explored aspect of the curse is the presence of hubs [13], i.e., instances that are similar to surprisingly many other instances. A hub is said to be bad if its class label differs from the class labels of those instances that have this hub as one of their k-nearest neighbors. In the context of k-nearest neighbor classification, bad hubs were shown to be responsible for a surprisingly large portion of the total classification error. Therefore, hubness-aware classifiers were developed, such as the Naive Hubness Bayesian k-Nearest Neighbor, or NHBNN for short [21]. Hubness-aware classifiers were shown to work well with various types of noise [18] which are particularly relevant from the point of view of the current study as gene expression data is often noisy due to measurement uncertainty. Furthermore, it may be expensive (or in case of rare diseases even impossible) to collect *large* amount of labeled data, therefore, we have to account for the fact that only relatively few labeled instances are available which may not reflect the structure of the classes well enough. Therefore, besides learning from labeled data, the classification algorithm should be able to use unlabeled data too in order to discover the structure of the classes. Therefore, in this paper we introduce a semi-supervised hubness-aware classifier. In particular, our approach is an extension of the aforementioned NHBNN. As we will show, straight forward incorporation of semi-supervised classification techniques with NHBNN leads to suboptimal results, therefore, we develop a hubness-aware inductive semi-supervised classification schema. To our best knowledge, this paper is the first that studies hubness-aware semi-supervised classification of gene expression data.

2 Background

Semi-supervised classification, often in a general data mining context, i.e., without special focus on the analysis of genetic data, has been studied intensively, see, e.g., [5, 10] and the references therein for related works on semi-supervised classification. Although the difficulties related to the analysis of high-dimensional data are often referred to as the *curse of dimensionality* [3], and some results even suggest that the notion of distances between instances of a dataset becomes meaningless in high-dimensional spaces [17], algorithms developed recently under the umbrella of hubness-aware data mining try to address the curse of dimensionality, see, e.g., [4, 11, 12, 14, 19, 20, 22, 23] for a survey. Hubs were observed in gene expression data [8, 14] and hubness was brought into relation with the performance of the SUCCESS semi-supervised time series classifier [9], however, none of the aforementioned works focused on hubness-aware classifiers in semi-supervised mode, i.e., when the classifier is allowed to learn both from labeled and unlabeled instances. In order to ensure that our study is self-contained, next, we review the Naive Hubness Bayesian k-Nearest Neighbor (NHBNN) classifier [21] and the self-training semi-supervised learning technique. The presentation of NHBNN and self-training is based on [19] and [10], respectively.

2.1 NHBNN: Naive Hubness Bayesian k-Nearest Neighbor

We aim at classifying instance x^*, i.e., we want to determine its unknown class label y^*. We use $\mathcal{N}_k(x^*)$ to denote the set of k-nearest neighbors of x^*. For each class C, Naive Hubness Bayesian k-Nearest Neighbor (NHBNN) estimates $P(y^* = C|\mathcal{N}_k(x^*))$, i.e., the probability that x^* belongs to class C given its nearest neighbors. Subsequently, NHBNN selects the class with highest probability. NHBNN follows a Bayesian approach to assess $P(y^* = C|\mathcal{N}_k(x^*))$. For each labeled training instance x, one can estimate the probability of the event that x appears as one of the k-nearest neighbors of any labeled training instance belonging to class C. This probability is denoted by $P(x \in \mathcal{N}_k|C)$. While calculating nearest neighbors, throughout this paper, an instance x is *never* treated as the nearest neighbor of itself, i.e., $x \notin \mathcal{N}_k(x)$. Assuming conditional independence between the nearest neighbors given the class, $P(y^* = C|\mathcal{N}_k(x^*))$ can be assessed as follows:

$$P(y^* = C|\mathcal{N}_k(x^*)) \quad \propto \quad P(C) \prod_{x_i \in \mathcal{N}_k(x^*)} P(x_i \in \mathcal{N}_k|C). \tag{1}$$

where $P(C)$ denotes the prior probability of the event that an instance belongs to class C. From the labeled training data, $P(C)$ can be estimated as $P(C) \approx \frac{|\mathcal{D}_C^{lab}|}{|\mathcal{D}^{lab}|}$, where $|\mathcal{D}_C^{lab}|$ denotes the number of labeled training instances belonging to class C and $|\mathcal{D}^{lab}|$ is the total number of labeled training instances. The maximum likelihood estimate of $P(x_i \in \mathcal{N}_k|C)$ is the fraction

$$P(x_i \in \mathcal{N}_k|C) \approx \frac{N_{k,C}(x_i)}{|\mathcal{D}_C^{lab}|}, \tag{2}$$

where $N_{k,C}(x_i)$ denotes how many times x_i occurs as one of the k-nearest neighbors of labeled training instances belonging to class C.

2.2 Example

Figure 1 shows a simple two-dimensional example, i.e., instances correspond to points of the plane. In this example, we use $k = 1$. In Fig. 1, a directed edge points from each labeled training instance to its first nearest neighbor among the labeled training instances. In other words: the nearest neighbor relationships shown in Fig. 1 are calculated solely on the *labeled training data*. Out of the 10 labeled training

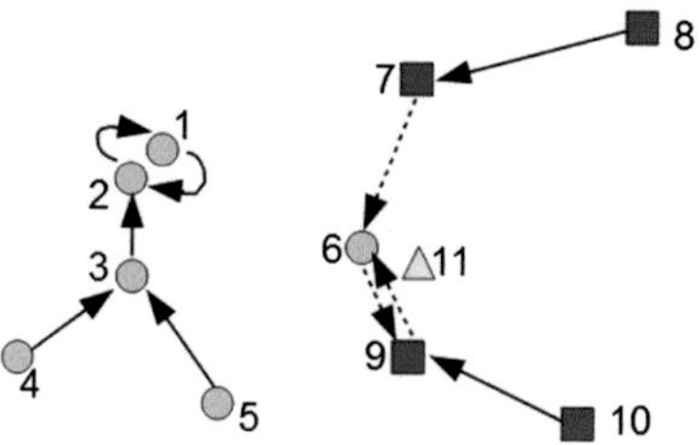

Fig. 1 Running example used to illustrate NHBNN. Labeled training instances belong to two classes, denoted by *circles* and *rectangles*. From each labeled training instance, a directed edge points to its first nearest neighbor among the labeled training instances. The *triangle* is an instance to be classified. For details, see the description of NHBNN

instances, 6 belong to the class of circles (C_1) and 4 belong to the class of rectangles (C_2). Thus: $|\mathcal{D}^{lab}_{C_1}| = 6$, $|\mathcal{D}^{lab}_{C_2}| = 4$, $P(C_1) = 0.6$, and $P(C_2) = 0.4$. Next, we calculate $N_{k,C}(x_i)$ for both classes and classify instance 11 using its first nearest neighbor, i.e., x_6. In particular, Eq. (2) leads to $P(x_6 \in \mathcal{N}_1|C_1) \approx \frac{N_{1,C_1}(x_6)}{|\mathcal{D}^{lab}_{C_1}|} = \frac{0}{6} = 0$ and $P(x_6 \in \mathcal{N}_1|C_2) \approx \frac{N_{1,C_2}(x_6)}{|\mathcal{D}^{lab}_{C_2}|} = \frac{2}{4} = 0.5$. According to Eq. (1) we calculate $P(y_{11} = C_1|\mathcal{N}_2(x_{11})) \propto 0.6 \times 0 = 0$ and $P(y_{11} = C_2|\mathcal{N}_2(x_{11})) \propto 0.4 \times 0.5 = 0.2$. As $P(y_{11} = C_2|\mathcal{N}_2(x_{11})) > P(y_{11} = C_1|\mathcal{N}_2(x_{11}))$, instance 11 will be classified as rectangle. The previous example also illustrates that estimating $P(x_i \in \mathcal{N}_k|C)$ according to (2) may simply lead to zero probabilities. In order to avoid it, we can use a simple Laplace estimate for $P(x_i \in \mathcal{N}_k|C)$ as follows:

$$P(x_i \in \mathcal{N}_k|C) \approx \frac{N_{k,C}(x_i) + m}{|\mathcal{D}^{lab}_C| + mq}, \tag{3}$$

where $m > 0$ and q denotes the number of classes. Informally, this estimate can be interpreted as follows: we consider m additional pseudo-instances from each class and we assume that x_i appears as one of the k-nearest neighbors of the pseudo-instances from class C. We use $m = 1$ in our experiments. Even though k-occurrences are highly correlated, as shown in [19, 21], NHBNN offers improvement over the basic kNN. This is in accordance with other results from the literature that state that Naive Bayes can deliver good results even in cases with high independence assumption violation [15].

2.3 *Self-training*

Self-training is one of the most commonly used semi-supervised algorithms. Self-training is a wrapper method around a supervised classifier, i.e., one may use self-training to enhance various classifiers. To apply self-training, for each instance x^* to be classified, besides its predicted class label, the classifier must be able to output a certainty score, i.e., an estimation of how likely the predicted class label is correct. Self-training is an iterative process during which the set of labeled instances is grown

```
SELF-TRAINING(L, U)
L_0 = L
U_0 = U
t = 0
repeat
    M = SUPERVISED-LEARNING(L_t)
    x_best = arg max_{x∈U_t} CERTAINTY(M, x)
    ŷ = CLASSIFY(M, x_best)
    L_{t+1} = L_t ∪ {(x_best, ŷ)}
    U_{t+1} = U_t \ {x_best}
    t = t + 1
until |U_t| == 0
return M
```

Fig. 2 Simple self-training algorithm

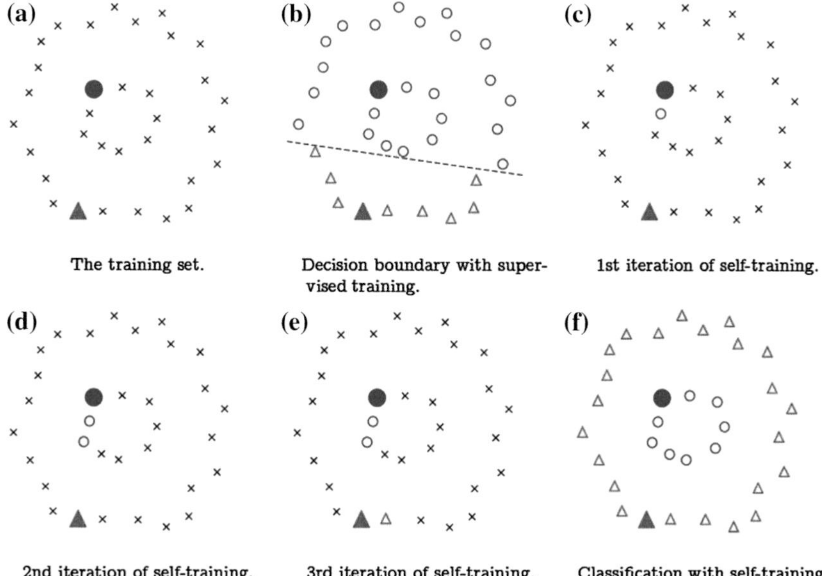

Fig. 3 Self-training with nearest neighbor. There are two classes, *circles* and *triangles*. *Bold* symbols correspond to instances of the initially labeled training set L_0, while unlabeled instances are marked with *crosses*, see (**a**). **c**–**e** Shows the first three iterations of self-training. The final output of self-training is shown in (**f**)

until all the instances become labeled. Let L_t denote the set of labeled instances in the tth iteration ($t \geq 0$) while U_t shall denote the set of unlabeled instances in the tth iteration. L_0 denotes the instances that are labeled initially, i.e., the labeled training data, while U_0 denotes the set of initially unlabeled instances. In each iteration of self-training, the base classifier is trained on the labeled set L_t. Then, the base classifier is used to classify the unlabeled instances. Finally, the instance with highest certainty score is selected. This instance, together with its predicted label $\hat{y}$, is added to the set of labeled instances, in order to construct L_{t+1} the set of labeled instance for the next iteration. The pseudocode of this algorithm is shown in Fig. 2. In context of nearest neighbor classification, the algorithm is illustrated in Fig. 3. If an unlabeled instance is classified incorrectly and this instance is added to the training data of the subsequent iterations, this may cause a chain of classification errors. Therefore, as noted in [6], it may be worth to stop self-training after a moderate number of iterations and use the resulting model to label all the remaining unlabeled instances.

3 Certainty Estimation for NHBNN

In order to allow NHBNN to be used in self-training mode, we only need to define an appropriate certainty score. A straightforward certainty score may be based on the probability estimates as follows:

$$certainty(x^*) = \frac{P(C') \prod_{x_i \in \mathcal{N}_k(x^*)} P(x_i \in \mathcal{N}_k | C')}{\sum_{C_j \in \mathcal{C}} \left(P(C_j) \prod_{x_i \in \mathcal{N}_k(x^*)} P(x_i \in \mathcal{N}_k | C_j) \right)}. \tag{4}$$

where C' denotes the class with maximal estimated probability and $\mathcal{C}$ denotes the set of all the classes. In the example shown in Fig. 1, the above certainty estimate gives $0.2/(0 + 0.2) = 1$ when classifying instance 11. However, this certainty estimate does not take into account that, usually, unlabeled instances appearing as nearest neighbors of many labeled instances can be classified more accurately as these instance are expected to be located "centrally" in the dataset, i.e., they appear in relatively dense regions of the data, see, e.g., [22]. Therefore, we propose to use the following hubness-aware certainty score:

$$hc(x^*) = \frac{N_k(x^*) P(C') \prod_{x_i \in \mathcal{N}_k(x^*)} P(x_i \in \mathcal{N}_k | C')}{\sum_{C_j \in \mathcal{C}} \left(P(C_j) \prod_{x_i \in \mathcal{N}_k(x^*)} P(x_i \in \mathcal{N}_k | C_j) \right)}, \tag{5}$$

where $N_k(x^*)$ denotes how many times instance x^* appears as nearest neighbors of other instances when considering the labeled training data $\mathcal{D}^{lab}$ together with the unlabeled instance x^*, i.e., $\mathcal{D}^{lab} \cup \{x^*\}$. Please note that in order to calculate $hc(x^*)$, we do not take other unlabeled instances into account. In the example shown in Fig. 1, the above certainty estimate gives $(2 \times 0.2)/(0 + 0.2) = 2$ when classifying instance 11, as instance 11 appears as nearest neighbor of instance 6 and instance 9 when considering all the eleven instances for the computation of the nearest neighbor relationships (we assume that the distance between instance 11 and instance 9 is lower that the distance between instance 9 and instance 6, therefore, instance 11 will be the nearest neighbor of instance 9 when considering all the instances).

4 Experimental Evaluation

4.1 Datasets

We used publicly available gene expression data of breast cancer tissues [16], colon cancer tissues [1], and lung cancer tissues [2]. In these datasets, the expression levels of 7650, 6500, and 12,600 genes have been measured for 95, 62, and 203 patients in the breast cancer, colon cancer, and lung cancer datasets, respectively. The breast and colon cancer datasets had two classes, while the lung cancer dataset had five classes. In all the cases, classes correspond to subtypes of the disease or healthy tissues, see [7] for details. Out of the five classes of the lung cancer dataset, we ignored one because extraordinarily few instances (in particular, only six instances) belonged to that class.

4.2 Experimental Protocol

In order to simulate scenarios in which the available training data is not fully representative, we considered five randomly selected instances per class as labeled training data. This results in balanced distribution of classes in the labeled training data whereas the entire datasets were class-imbalanced [7]. We repeated all the experiments 100 times with 100 different initial random selection of the labeled training instances. We measured the performance of the classifiers in terms of classification accuracy, i.e., the fraction of correctly classified "unlabeled instances." Note that the true class labels of the "unlabeled instances" were given in the datasets, however, these true class labels were used for evaluation purposes only, i.e., the labels of the "unlabeled instances" were unknown to the classifier. We report the average and standard deviation of the accuracies achieved in the aforementioned 100 runs. Additionally, we used t-test at significance level of 0.01 to judge if the differences between our approach and the baselines are statistically significant.

4.3 Compared Methods

We compared the following approaches:

- NHBNN-HS, i.e., NHBNN in self-training mode with the proposed hubness-aware certainty score according to Formula (5),
- NHBNN-Simple, i.e., NHBNN in self-training mode with the straightforward certainty score according to Formula (4),
- k-NN in self-training mode with the proposed hubness-aware certainty score according to Formula (5),
- NHBNN-SV, i.e., supervised NHBNN that uses only the labeled training instances but does not learn from the unlabeled data.

In accordance with [20], by default, we used $k = 5$, for all the aforementioned variants of NHBNN and k-NN. Note, however, that we performed experiments with other k values as well and we observed similar trends. As distance measure, we used the Cosine distance. In order to avoid the propagation of errors, in accordance with [6], in case of semi-supervised classifiers, we performed 20 iterations of self-training, i.e., 20 instances were labeled and added to the training set iteratively and then the model resulting after the 20th iteration was used to label all the remaining unlabeled instances. We performed experiments with other number of self-training iterations as well and we observed similar trends regarding the order of the semi-supervised approaches.

4.4 Results

Our results are summarized in Table 1. The results show that our approach, NHBNN-HS, consistently outperforms the baselines on all the three datasets. As we can see, both the choice of the algorithm and the certainty score matters: both NHBNN in self-training mode with the straight forward certainty score and k-NN with the hubness-aware certainty score achieve suboptimal accuracy compared with our approach NHBNN-HS. Furthermore, as we expected, semi-supervised classification outperforms supervised classification as it can be seen from the comparison against NHBNN-SV. Figure 4 shows that NHBNN-HS systematically outperforms its competitors for various k values, except for $k = 1$. Additionally, we tried support vector machines from the Weka software package with polynomial and RBF kernels with various settings of the complexity constant and the exponent of the polynomial kernel. According to our observations, self-training was not able to substantially improve the performance of SVMs overall: SVMs without self-training performed as well as (or sometimes even better than) SVMs with self-training. More importantly, NHBNN-HS was competitive to SVMs too: for example on the Breast Cancer and Colon Cancer datasets, best performing SVMs achieved classification accuracy of 0.781 and 0.705, respectively. We also note that the model built by NHBNN is more interpretable to human experts than the model built by an SVM.

Table 1 Accuracy ± standard deviation of our approach, NHBNN-HS and the baselines averaged over 100 runs

	Breast Cancer	Colon Cancer	Lung Cancer
NHBNN-HS	**0.840 ± 0.044**	**0.794 ± 0.073**	**0.784 ± 0.152**
NHBNN-Simple	0.835 ± 0.049 ○	0.790 ± 0.082 ○	0.679 ± 0.114 •
k-NN	0.649 ± 0.155 •	0.650 ± 0.162 •	0.674 ± 0.329 •
NHBNN-SV	0.756 ± 0.103 •	0.637 ± 0.139 •	0.617 ± 0.125 •

Bold font denotes the best approach for each dataset. The symbol •/○ denotes if the difference between NHBNN-HS and its competitor is statistically significant (•) or not (○) according to t-test at significance level of 0.01

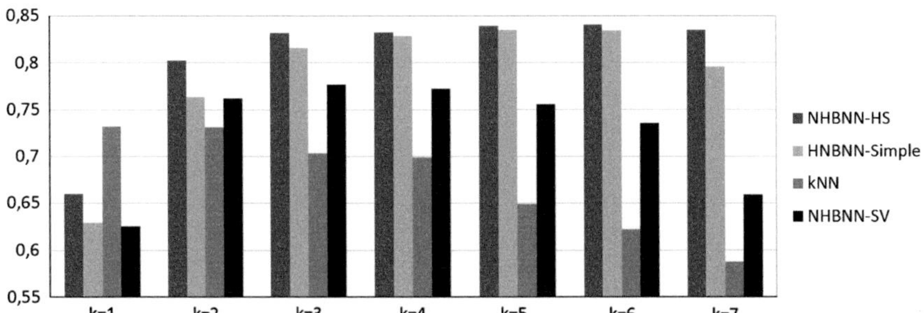

Fig. 4 Accuracy of our approach, NHBNN-HS, and its competitors for various k values on the Breast Cancer dataset

5 Conclusions and Outlook

In many applications, obtaining reliable class labels for large training samples may be difficult or even impossible. Therefore, semi-supervised classification techniques are required as they are able to take advantage of unlabeled data. Some of the most prominent recent methods developed for the classification of high-dimensional data follow the paradigm of hubness-aware data mining. However, hubness-aware classifiers have not been used for semi-supervised classification tasks previously. Therefore, in this paper, we introduced a semi-supervised hubness-aware classifier and we showed that it outperforms all the examined relevant baselines on the classification of gene expression data. While classification of gene expression data is highly relevant due to its biomedical applications, we expect that hubness-aware semi-supervised classifiers will also be utilized in various other classification tasks in the future.

Acknowledgments This research was performed within the framework of the grant of the Hungarian Scientific Research Fund—OTKA 111710 PD. This paper was supported by the János Bolyai Research Scholarship of the Hungarian Academy of Sciences.

References

1. Alon, U., Barkai, N., Notterman, D.A., Gish, K., Ybarra, S., Mack, D., Levine, A.J.: Broad patterns of gene expression revealed by clustering analysis of tumor and normal colon tissues probed by oligonucleotide arrays. Proc. Natl. Acad. Sci. **96**(12), 6745–6750 (1999)
2. Bhattacharjee, A., Richards, W.G., Staunton, J., Li, C., Monti, S., Vasa, P., Ladd, C., Beheshti, J., Bueno, R., Gillette, M., et al.: Classification of human lung carcinomas by mrna expression profiling reveals distinct adenocarcinoma subclasses. Proc. Natl. Acad. Sci. **98**(24), 13790–13795 (2001)
3. Bishop, C.M.: Pattern Recognition and Machine Learning (Information Science and Statistics). Springer, New Jersey (2006)
4. Buza, K., Nanopoulos, A., Schmidt-Thieme, L.: INSIGHT: Efficient and effective instance selection for time-series classification. In: Huang, J.Z., Cao, L., Srivastava, J. (eds.) Advances

in Knowledge Discovery and Data Mining. Lecture Notes in Computer Science, vol. 6635, pp. 149–160. Springer, Heidelberg (2011)
5. Chapelle, O., Schölkopf, B., Zien, A., et al.: Semi-Supervised Learning. MIT Press, Cambridge (2006)
6. Guillaumin, M., Verbeek, J., Schmid, C.: Multimodal semi-supervised learning for image classification. In: Proceedings of the IEEE Conference on Computer Vision and Pattern Recognition (CVPR 2010), pp. 902–909 (2010)
7. Lin, W.J., Chen, J.J.: Class-imbalanced classifiers for high-dimensional data. Br. Bioinform. **14**(1), 13–26 (2013)
8. Marussy, K.: The curse of intrinsic dimensionality in genome expression classification. In: Proceedings of the Students' Scientific Conference, Budapest University of Technology and Economics (2014)
9. Marussy, K., Buza, K.: Hubness-based indicators for semi-supervised time-series classification. In: Proceeding of the 8th Japanese-Hungarian Symposium on Discrete Mathematics and Its Applications. pp. 97–108 (2013)
10. Marussy, K., Buza, K.: SUCCESS: A new approach for semi-supervised classification of time-series. In: Rutkowski, L., Korytkowski, M., Scherer, R., Tadeusiewicz, R., Zadeh, L.A., Zurada, J.M. (eds.) Artificial Intelligence and Soft Computing. Lecture Notes in Computer Science, vol. 7894, pp. 437–447. Springer, Heidelberg (2013)
11. Radovanović, M., Nanopoulos, A., Ivanović, M.: Nearest neighbors in high-dimensional data: the emergence and influence of hubs. In: Proceedings of the 26rd International Conference on Machine Learning (ICML). pp. 865–872. ACM (2009)
12. Radovanović, M., Nanopoulos, A., Ivanović, M.: Hubs in space: popular nearest neighbors in high-dimensional data. J. Mach. Learn. Res. (JMLR) **11**, 2487–2531 (2010)
13. Radovanović, M., Nanopoulos, A., Ivanović, M.: Time-series classification in many intrinsic dimensions. In: Proceedings of the 10th SIAM International Conference on Data Mining (SDM). pp. 677–688 (2010)
14. Radovanović, M.: Representations and Metrics in High-Dimensional Data Mining. Izdavačka knjižarnica Zorana Stojanovića, Novi Sad, Serbia (2011)
15. Rish, I.: An empirical study of the naive Bayes classifier. In: Proceedings of the IJCAI Workshop on Empirical Methods in Artificial Intelligence (2001)
16. Sotiriou, C., Neo, S.Y., McShane, L.M., Korn, E.L., Long, P.M., Jazaeri, A., Martiat, P., Fox, S.B., Harris, A.L., Liu, E.T.: Breast cancer classification and prognosis based on gene expression profiles from a population-based study. Proc. Natl. Acad. Sci. **100**(18), 10393–10398 (2003)
17. Tan, P.N., Steinbach, M., Kumar, V.: Introduction to Data Mining. Addison Wesley, Boston (2005)
18. Tomašev, N., Buza, K.: Hubness-aware knn classification of high-dimensional data in presence of label noise. Neurocomputing **160**, 157–172 (2015)
19. Tomašev, N., Buza, K., Marussy, K., Kis, P.B.: Hubness-aware classification, instance selection and feature construction: survey and extensions to time-series. Feature Selection for Data and Pattern Recognition, pp. 231–262. Springer, Heidelberg (2015)
20. Tomašev, N., Mladenić, D.: Nearest neighbor voting in high dimensional data: learning from past occurrences. Comput. Sci. Inf. Syst. **9**, 691–712 (2012)
21. Tomašev, N., Radovanović, M., Mladenić, D., Ivanovicć, M.: A probabilistic approach to nearest neighbor classification: naive hubness Bayesian k-nearest neighbor. In: Proceeding of the CIKM Conference (2011)
22. Tomašev, N., Radovanović, M., Mladenić, D., Ivanović, M.: The role of hubness in clustering high-dimensional data. In: Huang, J.Z., Cao, L., Srivastava, J. (eds.) Advances in Knowledge Discovery and Data Mining. Lecture Notes in Computer Science, vol. 6634, pp. 183–195. Springer, Heidelberg (2011)
23. Tomašev, N., Radovanović, M., Mladenić, D., Ivanović, M.: Hubness-based fuzzy measures for high-dimensional k-nearest neighbor classification. Int. J. Mach. Learn. Cybern. **5**(3), 79–84 (2013)

The Multi-Ranked Classifiers Comparison

Norbert Jankowski

Abstract Is it true that everybody knows how to compare classifiers in terms of reliability? Probably not, since it is so common that just after reading a paper we feel that the classifiers' performance analysis is not exhaustive and we would like to see more information or more trustworthy information. The goal of this paper is to propose a method of multi-classifier comparison on several benchmark data sets. The proposed method is trustworthy, deeper, and more informative (multi-aspect). Thanks to this method, we can see much more than overall performance. Today, we need methods which not only answer the question whether a given method is the best, because it almost never is. Apart from the general strength assessment of a learning machine we need to know when (and whether) its performance is outstanding or whether its performance is unique.

1 Introduction

The proposed method of classifiers comparison is based on known statistical elements like accuracy, statistical tests, and rankings in general. For clarity, let us define *accuracy* as the fraction of correctly classified instances to the whole number of instances m:

$$acc(m, D) = 1 - err(m, D) = \frac{1}{m} \sum_{\langle \mathbf{x}, y\rangle \in D,\, y=m(\mathbf{x})} 1 \qquad (1)$$

In highly unbalanced cases (when the numbers of class instances differ significantly), it is recommended to use a balanced version of accuracy:

N. Jankowski (✉)
Department of Informatics, Nicolaus Copernicus University, Toruń, Poland
e-mail: norbert@is.umk.pl

R. Burduk et al. (eds.), *Proceedings of the 9th International Conference on Computer Recognition Systems CORES 2015*, Advances in Intelligent Systems and Computing 403, DOI 10.1007/978-3-319-26227-7_11

$$bacc(m, D) = \frac{1}{K}\sum_{k=1}^{K}\frac{\sum_{x \in D^k, m(\mathbf{x}_i)=k} 1}{|D^k|}, \tag{2}$$

where K is the number of classes, and $D^k = \{\langle \mathbf{x}_i, y_i\rangle : \langle \mathbf{x}_i, y_i\rangle \in D \wedge y_i = k\}$ is a set of pairs belonging to the kth class. As it can be seen that an error in classification of an instance of a smaller class is more strongly weighted, accordingly to class counts' proportions. The most common testing tool is the *cross-validation* test which divides randomly a given data set D into p equally counted subsets D_i. In consequence, we obtain p training–testing data set pairs $[D'_i, D_i]$, where $D'_i = D \backslash D_i$. Next, we have p phases of classifier learning and testing. The average accuracy over the testing part defines estimated accuracy ($eacc = 1/p \sum_{i=1}^{p} acc(m_i, D_i)$), where m_i is a classifier learnt on the D'_i data. However, such estimation should not be considered trustworthy, and it is recommended to repeat the cross-validation process q times (usually 10 times). For more about parametrization of cross-validation and their statistical relations see [2]. Now, the accuracy estimation is based on much more tests. Assume that

$$Eacc_m^D = [acc_1, \ldots, acc_{pq}] \tag{3}$$

is a vector of accuracies for all p test parts and for all q repetitions of cross-validation ($p * q$ single tests). It is highly recommended to use a *stratified* version of cross-validation. This test additionally keeps the proportions of classes in subsequent D_i sets to be close to the proportions of class counts in D. To keep the process of classifier comparison as trustworthy as possible it is also recommended to control the seed in the drawing process of training and testing parts. This means that each classifier should be trained and tested on the same training data and testing data. This is even more important when we use statistical tests like paired tests (e.g., paired t-test). In case of paired tests, it is obligatory to train and test all classifiers on the same data. Except the accuracy and the error, the reader should in some cases consider usage of other factors like recall, precision, specificity or confusion matrix, for more see [3, 6]. Statistical tests can serve as an important tool in classifier comparison. The reason for that is quite simple. Let us consider an example of two tests where average accuracies were equal to 0.87 and 0.879. In such case, we cannot directly claim that one of those classifiers is significantly better than the other one, however they differ in accuracy values. It is because the variances of classification of test data for both classifiers have a crucial role. Thanks to the statistical tests, the significance of results can be calculated. For detailed description on how to calculate statistical test, see [2, 8]. The most oftenly used statistical test in the context of machine learning is the t-test, which goal is to check whether the accuracy mean of the first classifier is significantly greater than the accuracy mean of the second classifier (in such case we use the *one tail* test version). The null hypothesis is that the first classifier is not better than the second one. Another possibility is to check whether the accuracy means of two populations are significantly different or not (this is two tail test version). The one tail is slightly more advised, as we usually have to choose the better one. And to calculate this test for two classifiers m and m' the accuracy differences $Eacc_m^D - Eacc_{m'}^D$ are used (the paired version of test). For classifiers' comparison the t-test can be used in one of two versions: paired or unpaired. The unpaired version has to be used only if the classifiers learned on different draws from data set. Practically, for a machine learning task it is not difficult to use the same distribution for learning and testing. If we used the same data distribution in cross-validation (or Monte Carlo as well), then the paired version is a more reliable test and we should avoid using the unpaired version if possible. The necessary condition to use one of the t-tests is that the test samples are approximately normally distributed. In a case where test samples are not normally distributed, there are two other interesting options. For the paired case, the Mann–Whitney test can be used and for unpaired, the Wilcoxon test. The last two test are ranked tests (accuracies are first transformed to ranks and are then further analyzed). Another resourceful test for computational intelligence which is rarely used is the McNemmar test. This test is designed to analyze whether the correctnesses of instance vectors for two classifiers are not statistically different. In that case, the test is not based on accuracies (or equivalently on errors) but on the correctness of each data instance. Even if two classifiers are characterized by the same means and similar variations they may differ in classification of appropriate instances. To compare more than two classifiers, the Anova test can be used, but its usability is somewhat limited. The base goal of this test is to calculate whether any two classifiers in a group are significantly unequal. The limitation of this test stems from the fact that in a case of comparing several classifiers, we are usually sure that same classifiers will differ, but we are interested in how they differ, not just whether they differ.

1.1 Common Traps in Learning Machine Comparison

The description below mostly concerns classification testing *traps*, but indeed most of the *traps* are of universal behavior. The ultimate solution or a trap? In so many cases a seemingly trustworthy comparison can be easily misleading. There are some types of commonly repeated errors in numerous articles. To avoid those problems in the future we can enumerate some of them.

1. The overall average accuracy as a measure of classifier performance. In some papers average accuracies are averaged over several data sets for given learning machines. Of course this information can be useful, however if we try to compare such averages obtained from two (or more) learning machines, such comparison is not trustworthy. It happens that in case of one or few data sets in a tested group the average accuracies differ strongly between classifiers and then, even if one classifier has a better overall accuracy, realistically, in case of most data sets, its performance may be significantly worst.
2. Another commonly observed scheme of classifier comparison is calculating how many of the given classifiers were not worse (the average accuracy was not smaller significantly) than others. The results of such calculation is the number of wins for all classifiers over several data sets. Such information is really interesting because

winners are certainly positive. However, it is somewhat risky in the following case: assume the first classifier wins a few times more than the second classifier and each win was significantly better, but just by a bit. A problem arises if for the second classifier the wins are much more than just *a bit* significant.

3. In some cases, benchmark data set was originally divided into two parts: the training and testing. If for a given classifier's configuration we train the classifier just once using the training part and then test use the testing part, then the test result is trustworthy. A problem arises when we repeatedly: learn a classifier, test the classifier, then basing on the test results we tune the configuration of the classifier. In such scenarios, researchers do not test different configurations but learn with validation using the testing data as validation data. Presentation of such results means a presentation of unreliable data.
4. One of the very typical errors in comparing classifiers is when the cross-validation testing is prepared after supervised[1] data transformation/preprocessing. Any supervised preprocessing must be embedded inside each cross-validation fold—before every classifier learning, first the data must be preprocessed for each cv-fold. Without following this scenario, the results can differ too strongly and are unreliable.
5. In some articles authors propose a new method but the conclusions are sometimes based just on a few data set benchmarks. However, the authors claim that the method works always and is universal (really?). Such scenario should also raise suspicions—just a few data sets should mean 'sometimes', not almost 'always'. A close problem to the above one is when authors claim that a method is *scalable* while the results are presented only on small data sets and the computational complexity is not investigated and is probably far from linear. Although it happens that a new method is proposed just in context of one problem (one data set) and this scenario can be correct.
6. Another unfair type of construction of comparisons is based on consciously misleading testing procedures. One of the most common examples of such problem is the usage of atypical parametrization of a test. For example, the usage of monte carlo randomization in place of commonly used cross-validation for given benchmarks. Generally, monte carlo randomization is correct, but if in context of the results for benchmark data sets all previous authors used cross-validation, then if we see monte carlo randomization, we can be sure of one thing: we cannot compare those results with the previous ones. They should be considered negligible. Another way of erring in the test procedure is to select different error measures, even though the author knows which measure was selected in previous articles about the considered problem (benchmark). Again, new results will not be trustworthily compared. Of course some of traps are entered unconsciously while others, consciously. I hope the above examples will help to avoid some mistakes in the future. The goal of the following part is to present how to plan a classifier comparison to be clear, trustworthy, informative and deep.

[1] Supervised process (learning of data transformation) means to use the class labels.

2 The Multi-ranked Classifier Comparison

The main goal of this section is to present the multi-ranked classifiers comparison which will analyze a series of classifiers over a sequence of benchmark data sets. Except the standard mean accuracies and the number of wins, we plan to present additional supporting information which significantly simplify the estimation of the role of a given classifier compared to others. Let us assume we have a sequence of benchmark data sets D_i $(i \in \{1, \ldots d\})$ and a series of classifiers m_j $(j \in \{1, \ldots, T\})$. First, as usually, we need the accuracy vectors from cross-validation tests for every classifier and for every benchmark data set. This gives us a matrix of $Eacc_{m_j}^{D_i}$ (remember that $Eacc_{m_j}^{D_i}$ is a vector, not a scalar). The matrix of mean test accuracies $\bar{a}_j^i$ for a machine m_j and benchmark D_i is the base of further presentation. Strictly, the $\bar{a}_j^i$ is the mean of $Eacc_{m_j}^{D_i}$ vector accuracies. Additionally, we define the σ_j^i to be the standard deviation of $Eacc_{m_j}^{D_i}$.

2.1 Machine Ranks and Significance Groups

For given data D_i, we can group machines in accordance with their mean accuracies using the paired t-test.[2] Such groups will be assigned to a rank. We can define the rank assignment for machines as follows:

- the machine with highest accuracy mean is ranked 1,
- all machines whose accuracy means are not significantly smaller (measured with t-test) are also ranked with 1,
- rank 2 is assigned to the machine of highest accuracy amongst those whose accuracy is significantly smaller than the machine's first ranked with 1,
- rank 2 is assigned to all machines which have not been ranked yet, whose accuracies are insignificantly smaller than the first machine's ranked with 2,
- the following ranks are assigned in the same way.

All machines with the same rank compose a *significance group*. Let's define r_j^i as the rank of machine m_j obtained for benchmark data D_i. Such ranks forms rank groups, each group is composed of machines which are characterized by the same (insignificantly different) performance. This feature is so important because for different benchmark data sets the spread between accuracies varies. Additionally, ranks are independent of the differences between mean accuracies of different benchmark data sets. This feature is important for comparing the results for two (or more) benchmarks, basing on ranks, instead of comparing mean accuracies for different

---

[2] To compute paired t-test for machine s and t and data D_i use the vector of differences: $Eacc_{m_s}^{D_i} - Eacc_{m_t}^{D_i}$.

benchmarks. Additionally, let us define $\bar{r}_j$ to be a mean rank for machine m_j across benchmarks and σ_j^r as the standard deviation of ranks for a given machine m_j. The mean ranking is the best estimate of overall performance. If it is close to 1, it means that the machine is usually the winning one. And standard deviation informs us of the changes across benchmark data sets. Winners versus ranking: Typically, authors of classifier comparisons use the division into two parts for a given benchmark: winners (machines with best accuracies-insignificantly different) and losers (machines which perform significantly worse than the best one). Such binary spread is sometimes not adequate—in some cases the mean accuracies naturally form more than two groups of performance and division into two groups in fact hides some information. The reader will able to observe in the example below that in case of some benchmarks, the ranks form several groups of performance which reflect several levels of performance degradation. And across several benchmarks we can simply observe how frequently the performance of a given classifier degraded and how deeply.

2.2 Winners and Unique Winners

Observation of machines which win for given data sets is important, but apart from the observation of winners we should also observe machines which are unique winners. A unique winner is a machine which is the best for a given data set and no other machine is insignificantly worse. Such machines are not redundant in contrary to nonunique winners, which can be substituted by another machine(-s). Define the w_i to be the number of wins for machine m_i (win means that machine is the best one for given data or insignificantly worse). Define the u_i to be a count of unique wins of machine m_i, which is the number of wins while no other machine has the same rank (unique win means that only one machine has rank 1 for given benchmark).

2.3 Multi-ranked Classifiers Comparison

The above part of this section has presented all necessary definitions to present the proposed classifier comparison. This comparison will consist of

- mean accuracy and standard deviation for each machine and each benchmark with its rank,
- overall mean accuracy per machine with its standard deviation,
- overall mean rank per machine with its standard deviation,
- wins count and unique wins count.

All this information is nested in the matrix below:

	m_1	m_2	$\cdots$	m_p
D_1	$\bar{a}_1^1 \pm \sigma_1^1(r_1^1)$	$\bar{a}_2^1 \pm \sigma_2^1(r_2^1)$	$\cdots$	$\bar{a}_p^1 \pm \sigma_p^1(r_p^1)$
D_2	$\bar{a}_1^2 \pm \sigma_1^2(r_1^2)$	$\bar{a}_2^2 \pm \sigma_2^2(r_2^2)$	$\cdots$	$\bar{a}_p^2 \pm \sigma_p^2(r_p^2)$
$\cdots$	$\cdots$	$\cdots$	$\cdots$	$\cdots$
D_q	$\bar{a}_1^q \pm \sigma_1^q(r_1^q)$	$\bar{a}_2^q \pm \sigma_2^q(r_2^q)$	$\cdots$	$\bar{a}_p^q \pm \sigma_p^q(r_p^q)$
Mean Accuracy	$\bar{a}_1^* \pm \sigma_1^*$	$\bar{a}_2^* \pm \sigma_2^*$	$\cdots$	$\bar{a}_p^* \pm \sigma_p^*$
Mean Rank	$\bar{r}_1 \pm \sigma_1^r$	$\bar{r}_2 \pm \sigma_2^r$	$\cdots$	$\bar{r}_p \pm \sigma_p^r$
Wins[unique wins]	$w_1[u_1]$	$w_2[u_2]$	$\cdots$	$w_p[u_p]$

(4)

2.4 An Example of Multi-ranked Comparison

Probably, the best way to see the attractiveness of the presented comparison method is to analyze a real world example. 40 benchmark data sets from the UCI machine learning repository [7] were selected to present the comparison below. Two neural networks, k Nearest neighbor [1], and two types of Support Vector Machines (linear and gaussian) [9] were selected to compare with the proposed method. The first neural network is a simple linear model (no hidden layer) learned by pseudo-inverse matrix (via singular values decomposition). The linear model $g(\mathbf{x}) = \mathbf{w}^T\mathbf{x}$ is learned by:

$$\mathbf{w} = (\mathbf{X}^T\mathbf{X})^{-1}\mathbf{X}^T\mathbf{y} = \mathbf{X}^\dagger\mathbf{y} \tag{5}$$

where $\mathbf{X}$ is a matrix of input data, $\mathbf{y}$ label (class) vector and $\mathbf{X}^\dagger$ is pseudo-inverse matrix. The above equation is a solution for the goal:

$$J_s(\mathbf{w}) = ||\mathbf{X}\mathbf{w} - \mathbf{y}||^2 = \sum_{i=1}^{m}(\mathbf{w}^T\mathbf{x}_i - y_i)^2 \tag{6}$$

obtained by zeroing the gradient. The next neural network is a nonlinear model generated by a set of gaussian kernels $(k_1, \ldots, k_l)$, and learned in a similar way as above networks after transforming the original space into the space obtained by kernels. It means that instead of $\mathbf{X}$ in Eq. 5 the matrix F is used:

$$F_{ij} = k_j(\mathbf{x}_{z_j}; \mathbf{x}_i), \tag{7}$$

where $\mathbf{x}_{z_j}$ are randomly selected between all data vectors. Such construction of neural networks is equivalent to Extreme learning machines [4, 5]. Note that the parameters of all learning machines were not optimized because the goal of this paper is not to achieve optimal performance of given machines, but to present the attractiveness of the comparison method.

Table 1 The multi-ranked classifiers comparison

	NN-linear	NN-Gauss	kNN	L-SVM	SVM
Autos	71.06±9.11(2)	**77.37±9.37(1)**	64.75±10.8(3)	53.31±10.4(5)	59.2±9.61(4)
Balance-scale	49.44±5.06(5)	**90.8±1.56(1)**	88.81±2.45(3)	84.48±2.84(4)	89.5±1.89(2)
Breast-cancer-diagnostic	43.01±2.91(5)	90.65±3.29(4)	96.8±2.26(2)	**97.4±1.93(1)**	95.9±2.47(3)
Breast-cancer-original	85.18±3.52(3)	95.74±2.34(2)	**96.82±2.01(1)**	**96.65±2.21(1)**	**96.97±1.98(1)**
Breast-cancer-prognostic	76.59±4.36(2)	75.87±5.79(2)	76.39±8.11(2)	**80.14±8.54(1)**	76.38±4.04(2)
Breast-tissue	**64.43±13.3(1)**	52.45±15.7(2)	**66.77±14.3(1)**	43.42±7.92(3)	42.45±8.79(3)
Car-evaluation	75.01±2.11(5)	89.02±1.63(2)	**93.79±1.31(1)**	82.9±2.31(4)	88.49±1.53(3)
Cardiotocography-1	73.31±3(3)	74.02±2.62(2)	**76.09±3.08(1)**	58.79±2.53(5)	71.65±2.3(4)
Cardiotocography-2	40.84±15.4(5)	88.93±1.69(3)	**91.15±1.5(1)**	87.6±1.52(4)	90.57±1.51(2)
Chess-rook-versus-pawn	66.45±9.28(5)	70.4±1.96(4)	94.86±1.17(3)	96.92±0.855(2)	**98.42±0.726(1)**
CMC	49±3.8(2)	**55.35±3.44(1)**	49.45±4.21(2)	19.24±2.6(4)	30.67±3.07(3)
Vongressional-voting	70.25±7.49(4)	94.53±4.37(2)	91.91±5.07(3)	95.04±3.97(2)	**96.3±3.22(1)**
Connectionist-bench-sonar	52.43±5.05(5)	57.59±4.91(4)	**82.65±6.79(1)**	75.86±7.94(3)	78.55±6.79(2)
Connectionist-bench-vowel	53.66±5(4)	87.48±4.25(2)	**94.72±2.65(1)**	26.66±3.77(5)	64.32±4.69(3)
Cylinder-bands	73.42±8.3(2)	36.98±2.22(5)	64.93±8.44(4)	**76.43±7.71(1)**	67.14±2.55(3)
Dermatology	**93.43±4.1(1)**	**94.07±3.99(1)**	92.82±3.87(2)	**93.4±3.99(1)**	86.69±5.23(3)
Ecoli	**86.13±5.05(1)**	**85.46±5.17(1)**	**85.75±5.29(1)**	76.06±6.62(3)	83.25±5.23(2)
Glass	54.46±8.39(4)	**67.83±9.13(1)**	65.68±7.65(2)	35.67±6.52(5)	57.22±8.07(3)
Habermans-survival	**73.78±2.82(1)**	71.65±6.4(2)	70.99±6.44(3)	72.46±2.77(2)	**73.77±3.93(1)**
Hepatitis	83.75±10.9(3)	**91.25±9.32(1)**	87.38±12.1(2)	81.63±10.3(3)	88.25±9.03(2)
Ionosphere	73.3±4.54(4)	66.55±3.4(5)	84.67±4.56(3)	88.15±4.75(2)	**94.87±3.65(1)**

(continued)

Table 1 (continued)

	NN-linear	NN-Gauss	kNN	L-SVM	SVM
Iris	51.73±8.79(4)	94.33±5.22(2)	**95±5.79(1)**	77.8±7.58(3)	**96.13±5.03(1)**
Libras-movement	65.81±6.7(3)	67.75±7.37(2)	**77.11±5.99(1)**	50.33±6.39(4)	47.97±5.97(5)
Liver-disorders	43.34±2.37(4)	67.55±8.25(2)	61.02±7.74(3)	69.22±5.68(2)	**71.13±6.41(1)**
Lymph	80.1±9.6(2)	**85.65±10(1)**	80.47±10(2)	80.17±9.75(2)	79.97±10.1(2)
Monks-problems-1	50±0.81(5)	96.42±2.7(3)	99.66±0.838(2)	74.64±3.5(4)	**100±0(1)**
Monks-problems-2	**65.73±0.877(1)**	64.01±4.55(2)	56.18±4.47(4)	**65.73±0.877(1)**	60.88±4.22(3)
Monks-problems-3	60.3±3.69(3)	98.09±1.93(2)	**98.77±1.2(1)**	**98.92±1.12(1)**	**98.92±1.12(1)**
Parkinsons	82.67±4.86(4)	**92.41±5.48(1)**	**91.82±6.37(1)**	87.79±6.29(3)	89.81±5.19(2)
Pima-Indians-diabetes	36.13±1.66(5)	74.95±4.78(3)	73.97±4.52(4)	**77.24±4.76(1)**	76.47±4.65(2)
Sonar	52.43±5.05(5)	57.59±4.91(4)	**82.65±6.79(1)**	75.86±7.94(3)	78.55±6.79(2)
Spambase	43.2±0.807(5)	68.12±1.92(4)	90.92±1.29(3)	**92.79±0.962(1)**	91.65±1.2(2)
Spect-heart	79.42±2.05(2)	**82.13±6.66(1)**	**81.84±7.21(1)**	**80.86±7.43(1)**	**82.58±6.89(1)**
Spectf-heart	79.27±2.17(2)	**80.61±5.93(1)**	72.73±7.47(3)	79±6.96(2)	78.22±5.25(2)
Statlog-Australian-credit	62.04±3.76(5)	71.64±4.4(4)	79.61±4.44(3)	**84.74±4.42(1)**	83±4.32(2)
Statlog-German-credit	70.59±1.04(4)	70.74±2.77(4)	72.4±3.7(3)	**76.55±3.91(1)**	75.21±2.82(2)
Statlog-heart	78.3±7.16(2)	79.15±7.73(2)	**82.15±7.9(1)**	**83.67±6.7(1)**	**83.44±7.53(1)**
Statlog-vehicle	59.65±3.99(5)	**75.87±4.47(1)**	72.95±4.06(2)	68.4±4.89(3)	66.05±3.84(4)
Teaching-assistant	**54.99±11.7(1)**	**56.82±10.3(1)**	42.4±11.5(3)	54.09±12(2)	40.44±12.4(3)
Thyroid-disease	14.76±26.3(5)	95.25±0.584(2)	94.94±0.499(3)	93.7±0.35(4)	**95.41±0.526(1)**
Mean accuracy	63.48±5.92	77.33±5.06	80.59±5.4	74.84±5.09	78.16±4.61
Mean rank	3.35±0.24	2.25±0.2	2.1±0.16	2.525±0.22	2.175±0.17
Wins [unique]	6[0]	13[8]	15[7]	13[7]	12[6]

Table 2 The multi-ranked classifiers comparison—version II

	NN-linear	NN-Gauss	kNN	L-SVM	SVM
Autos	71.06±9.11(2)	**83.92±9.1(1)**	64.75±10.8(3)	53.31±10.4(5)	59.2±9.61(4)
Balance-scale	49.44±5.06(5)	**90.21±1.93(1)**	88.81±2.45(3)	84.48±2.84(4)	89.5±1.89(2)
Breast-cancer-diagnostic	43.01±2.91(5)	92.13±3.01(4)	96.8±2.26(2)	**97.4±1.93(1)**	95.9±2.47(3)
Breast-cancer-original	85.18±3.52(3)	95.48±2.49(2)	**96.82±2.01(1)**	**96.65±2.21(1)**	**96.97±1.98(1)**
Breast-cancer-prognostic	76.59±4.36(2)	76.55±6.55(2)	76.39±8.11(2)	**80.14±8.54(1)**	76.38±4.04(2)
Breast-tissue	**64.43±13.3(1)**	48.94±16(2)	**66.77±14.3(1)**	43.42±7.92(3)	42.45±8.79(3)
Car-evaluation	75.01±2.11(5)	92.56±1.61(2)	**93.79±1.31(1)**	82.9±2.31(4)	88.49±1.53(3)
Cardiotocography-1	73.31±3(3)	**78.21±2.37(1)**	76.09±3.08(2)	58.79±2.53(5)	71.65±2.3(4)
Cardiotocography-2	40.84±15.4(4)	**90.93±1.57(1)**	**91.15±1.5(1)**	87.6±1.52(3)	90.57±1.51(2)
Chess-rook-versus-pawn	66.45±9.28(5)	77.16±1.94(4)	94.86±1.17(3)	96.92±0.855(2)	**98.42±0.726(1)**
CMC	49±3.8(2)	**54.26±3.59(1)**	49.45±4.21(2)	19.24±2.6(4)	30.67±3.07(3)
Congressional-voting	70.25±7.49(4)	**95.66±3.58(1)**	91.91±5.07(3)	95.04±3.97(2)	**96.3±3.22(1)**
Connectionist-bench-sonar	52.43±5.05(5)	60.96±5.04(4)	**82.65±6.79(1)**	75.86±7.94(3)	78.55±6.79(2)
Connectionist-bench-vowel	53.66±5(4)	**96.82±2.31(1)**	94.72±2.65(2)	26.66±3.77(5)	64.32±4.69(3)
Cylinder-bands	73.42±8.3(2)	38.49±2.39(5)	64.93±8.44(4)	**76.43±7.71(1)**	67.14±2.55(3)
Dermatology	93.43±4.1(2)	**95.13±3.43(1)**	92.82±3.87(2)	93.4±3.99(2)	86.69±5.23(3)
Ecoli	**86.13±5.05(1)**	82.55±5.32(2)	**85.75±5.29(1)**	76.06±6.62(3)	83.25±5.23(2)
Glass	54.46±8.39(3)	**64.51±7.74(1)**	**65.68±7.65(1)**	35.67±6.52(4)	57.22±8.07(2)
Habermans-survival	**73.78±2.82(1)**	70.57±6.32(3)	70.99±6.44(3)	72.46±2.77(2)	**73.77±3.93(1)**
Hepatitis	83.75±10.9(3)	**91.25±9.32(1)**	87.38±12.1(2)	81.63±10.3(3)	88.25±9.03(2)
Ionosphere	73.3±4.54(4)	67.27±4.3(5)	84.67±4.56(3)	88.15±4.75(2)	**94.87±3.65(1)**
Iris	51.73±8.79(4)	94.47±5.36(2)	**95±5.79(1)**	77.8±7.58(3)	**96.13±5.03(1)**
Libras-movement	65.81±6.7(3)	**80.92±6.12(1)**	77.11±5.99(2)	50.33±6.39(4)	47.97±5.97(5)
Liver-disorders	43.34±2.37(5)	64.4±7.5(3)	61.02±7.74(4)	69.22±5.68(2)	**71.13±6.41(1)**

(continued)

Table 2 (continued)

	NN-linear	NN-Gauss	kNN	L-SVM	SVM
Lymph	80.1±9.6(2)	**84.84±9.33(1)**	80.47±10(2)	80.17±9.75(2)	79.97±10.1(2)
Monks-problems-1	50±0.81(5)	99.87±0.526(2)	99.66±0.838(3)	74.64±3.5(4)	**100±0(1)**
Monks-problems-2	**65.73±0.877(1)**	64.02±5.9(2)	56.18±4.47(4)	**65.73±0.877(1)**	60.88±4.22(3)
Monks-problems-3	60.3±3.69(3)	98.61±1.31(2)	**98.77±1.2(1)**	**98.92±1.12(1)**	**98.92±1.12(1)**
Parkinsons	82.67±4.86(5)	**94.84±4.47(1)**	91.82±6.37(2)	87.79±6.29(4)	89.81±5.19(3)
Pima-Indians-diabetes	36.13±1.66(4)	74.17±4.65(3)	73.97±4.52(3)	**77.24±4.76(1)**	76.47±4.65(2)
Sonar	52.43±5.05(5)	60.96±5.04(4)	**82.65±6.79(1)**	75.86±7.94(3)	78.55±6.79(2)
Spambase	43.2±0.807(5)	71.94±1.94(4)	90.92±1.29(3)	**92.79±0.962(1)**	91.65±1.2(2)
Spect-heart	79.42±2.05(2)	**82.17±6.19(1)**	**81.84±7.21(1)**	**80.86±7.43(1)**	**82.58±6.89(1)**
Spectf-heart	**79.27±2.17(1)**	**79.89±6.18(1)**	72.73±7.47(3)	**79±6.96(1)**	78.22±5.25(2)
Statlog-Australian-credit	62.04±3.76(5)	75.29±4.47(4)	79.61±4.44(3)	**84.74±4.42(1)**	83±4.32(2)
Statlog-German-credit	70.59±1.04(4)	71.72±3.3(3)	72.4±3.7(3)	**76.55±3.91(1)**	75.21±2.82(2)
Statlog-heart	78.3±7.16(2)	78.37±7.44(2)	**82.15±7.9(1)**	**83.67±6.7(1)**	**83.44±7.53(1)**
Statlog-vehicle	59.65±3.99(5)	**79.22±3.69(1)**	72.95±4.06(2)	68.4±4.89(3)	66.05±3.84(4)
Teaching-assistant	54.99±11.7(2)	**63.65±10.3(1)**	42.4±11.5(3)	54.09±12(2)	40.44±12.4(3)
Thyroid-disease	14.76±26.3(5)	**95.77±0.595(1)**	94.94±0.499(3)	93.7±0.35(4)	95.41±0.526(2)
Mean accuracy	63.48±5.92	78.97±4.86	80.59±5.4	74.84±5.09	78.16±4.61
Mean rank	3.35±0.23	2.1±0.2	2.2±0.15	2.5±0.21	2.2±0.16
Wins [unique]	5[0]	18[13]	12[3]	13[7]	11[4]

All results in accordance with the above definitions was presented in Tables 1 and 2 which have the same form as matrix in the Eq. 4. The difference between tables lies in the number of kernels used to learn the NN-Gauss neural network—the numbers of kernels were equal to 80 and 160, respectively. Starting from the top of Table 1 we can analyze significance groups for selected benchmark data. In contrary to a presentation based only on wins and defeats here we can observe that in case of several benchmarks data the numbers of significance groups spread from 2 even up to 5 (5 is the maximum of course). The number of significance groups is very often relatively huge—it is close to the maximum—and is directly related to the diversity of model's performance. Divergent performance of quality is directly correlated with the numbers of significance groups. In case of a presentation based on wins and defeats, this feature is invisible. After the rows which present accuracies statistics and significance groups we come to the sum-up information. The first row informs about commonly used average accuracies over all benchmark data sets. Next row presents the information about the average rank for each classifier. The best ranking informs us about the best classifier over all benchmarks. Note that the best average accuracy over all benchmarks may not be as good an estimation of the best classifier as the machine with the best average rank. It is because the magnitudes of average accuracy for machines are independent, which can significantly bias the rank, and this can be seen in Table 2. The last row informs us about the number of wins and the number of unique wins. The best number of wins is quite closely related to the best rank. But more special information is captured by the unique wins. This informs us about the uniqueness of a given machine. Larger number of unique wins means a more unique and more significant machine. If the number of unique wins is really small, it means that such machine can be simply substituted by another machine. It shows that machine redundancy can be very easily analyzed. Compare the two tables to see how the redundancy can change. Additionally, all non-small unique win counters are connected with nonredundant winning machines.

3 Summary

Model selection is one of most important tasks in machine learning and computational intelligence. The comparison of classifiers should as informative as possible, and should not hide any important information. Typically, we observe that classifiers comparisons are oversimplified and in consequence, to select a model, we need another results which comment the behavior of learning and the obtained results. The proposed scheme of multi-ranked classifiers comparison bases on the same statistical tools but calculates more and different features for the prepared test. Thanks to the proposed scheme, we can easily analyze information like

- significance groups which describe difference in performance for a given benchmark without the bias of variance,
- the overall best classifier information is based mostly on the averaged ranks, which may be additionally compared with the win counts,
- machine uniqueness and machine redundancy,
- the best winner machine (the machine with the most wins),
- detailed information about performance for a given machine and a given benchmark.

Such classifier comparison significantly simplifies the process of results analysis and the model(-s) selection is simplified.

References

1. Cover, T.M., Hart, P.E.: Nearest neighbor pattern classification. Inst. Electr. Electron. Eng. Trans. Inf. Theory **13**(1), 21–27 (1967)
2. Dietterich, T.G.: Approximate statistical tests for comparing supervised classification learning algorithms. Neural Comput. **10**(7), 1895–1923 (1998)
3. Friedman, J., Hastie, T., Tibshirani, R.: The Elements of Statistical Learning: Data Mining, Inference, and Prediction. Springer, New York (2001)
4. Huang, G.-B., Zhu, Q.-Y., Siew, C.K.: Extreme learning machine: a new learning scheme of feedforward neural networks. In: International Joint Conference on Neural Networks, pp. 985–990. IEEE Press (2004)
5. Huang, G.-B., Zhu, Q.-Y., Siew, C.-K.: Extreme learning machine: theory and applications. Neurocomputing **70**, 489–501 (2006)
6. Larose, D.: Discovering Knowledge in Data. An Introduction to Data Mining. Wiley, New York (2005)
7. Merz, C.J., Murphy, P.M.: UCI repository of machine learning databases (1998). http://www.ics.uci.edu/~mlearn/MLRepository.html
8. Montgomery, D.C., Runger, G.C.: Applied Statistics and Probability for Engineers. Wiley, New York (2002)
9. Vapnik, V.: The Nature of Statistical Learning Theory. Springer, New York (1995)

Using a Genetic Algorithm for Selection of Starting Conditions for the EM Algorithm for Gaussian Mixture Models

Wojciech Kwedlo

Abstract This paper addresses the problem of initialization of the expectation-maximization (EM) algorithm for maximum likelihood estimation of Gaussian mixture models. In order to avoid local maxima of the likelihood function, a genetic algorithm (GA) which searches for best initial conditions of the EM algorithm is proposed. In the GA, a chromosome represents a set of initial conditions, in which initial mean vectors of mixture components are feature vectors chosen from the training set. The chromosome also encodes variances of initial spherical covariance matrices of mixture components. To evaluate each chromosome in the GA we run the EM algorithm until convergence and use the obtained log likelihood as the fitness. In computational experiments our approach was applied to clustering problem and tested on two datasets from the image processing domain. The results indicate that our method outperforms the standard multiple restart EM algorithm and is at least comparable to the state-of-the art random swap EM method.

1 Introduction

A finite Gaussian mixture model (GMM) [7, 15] assumes that a probability density function is a weighted sum of K multivariate Gaussian components

$$p(\mathbf{x}|\Theta) = \sum_{m=1}^{K} \alpha_m \mathcal{N}(\mathbf{x}|\boldsymbol{\mu}_m, \boldsymbol{\Sigma}_m). \tag{1}$$

In the above equation, $\alpha_1, \alpha_2, \ldots, \alpha_K$ are the component weights, also called the mixing proportions. $\mathcal{N}(\mathbf{x}|\boldsymbol{\mu}_m, \boldsymbol{\Sigma}_m)$ is the probability density function of a d-dimensional Gaussian distribution with a mean vector $\boldsymbol{\mu}_m$ and a covariance matrix $\boldsymbol{\Sigma}_m$. The mixing proportions must be positive and add up to one. $\Theta = \{\boldsymbol{\mu}_1, \boldsymbol{\Sigma}_1, \ldots, \boldsymbol{\mu}_K,$

W. Kwedlo (✉)
Faculty of Computer Science, Bialystok University of Technology, Wiejska 45A, 15-351 Bialystok, Poland
e-mail: w.kwedlo@pb.edu.pl

R. Burduk et al. (eds.), *Proceedings of the 9th International Conference on Computer Recognition Systems CORES 2015*, Advances in Intelligent Systems and Computing 403, DOI 10.1007/978-3-319-26227-7_12

$\mathbf{\Sigma}_K, \alpha_1, \ldots, \alpha_K\}$ is the complete set of parameters needed to define a mixture. The number of components K is either known as a priori or has to be determined during the mixture learning process. In the paper we assume that K is known. GMMs are widely used in such applications as data clustering [7], data classification [8], speaker recognition [21], image segmentation and classification [18]. Estimation of the parameters of a GMM can be performed using the maximum likelihood approach. Given a set of independent and identically distributed feature vectors $X = \{\mathbf{x}_1, \mathbf{x}_2, \ldots, \mathbf{x}_N\}$, called the learning set, the log likelihood corresponding to a K-component GMM is given by:

$$\log p(X|\Theta) = \log \prod_{i=1}^{N} p(\mathbf{x}_i|\Theta) = \sum_{i=1}^{N} \log \sum_{m=1}^{K} \alpha_m \mathcal{N}(\mathbf{x}_i|\boldsymbol{\mu}_m, \mathbf{\Sigma}_m). \tag{2}$$

Unfortunately, for $K > 1$ the solution which maximizes 2 cannot be obtained in a closed form [2]. For this reason a numerical optimization algorithm has to be employed. Because of its simplicity, the EM (expectation-maximization) algorithm [5, 14] is the most popular tool for maximizing 2. This procedure, however, is highly sensitive to initialization and easily gets trapped in a local maximum. Therefore, the quality of the final solution is strongly dependent on the initial guess of the mixture parameters. The problem can be to some degree alleviated by performing multiple runs of the algorithm, each run starting from different initial conditions, and returning the result with the highest $\log p(X|\Theta)$. This approach is called multiple restart EM (MREM). Researchers investigating the problem of local maxima of the log likelihood have recently started to apply population based stochastic global optimization algorithms such as evolutionary algorithms (EAs) [1, 19]. However, a random nature of search operators used by these algorithms makes it difficult to optimize in a solution space consisting of GMM parameters. For instance, it is very likely [3] that a result of applying a mutation operator of an EA to a covariance matrix will not be positive definite, and thus will be an invalid covariance matrix. To overcome these difficulties parametrizations of covariance matrix based on Givens rotation angles [3] or Cholesky decomposition [11] were proposed. This paper investigates another line of research, in which, instead of using an EA to search a solution space consisting of all possible GMM parameters we search a more restricted space of *initial* solutions of the EM algorithm. To obtain the final solution and to compute the fitness of an individual in EA we run the EM algorithm, using an initial GMM encoded in a chromosome as the starting condition, until convergence. We test our approach, which we call GAIEM (GAIEM, for Genetic Algorithm for Initialization of EM) on model-based clustering [7] problem. This problem is an important application of GMMs. The aim of clustering can be defined as dividing a set of objects into K disjoint groups (clusters), in such a way that the objects within the same cluster are close to each other, whereas the objects in different clusters are very distinct. In applications of GMMs to clustering we assume each feature vector was generated from one of K mixture components. Clustering of feature vectors can be performed by identifying, for each feature vector, the mixture component from which it was generated. The feature vectors generated from the same component are then assigned to the same cluster. If we are able to estimate the mixture parameters Θ, this assignment can be performed using maximum a posteriori (MAP) rule, by allocating a feature vector to a cluster (mixture component) with the highest posterior probability. From the Bayes theorem, this probability for the mixture component m can be expressed as:

$$h_m(\mathbf{x}_i) = \frac{\alpha_m \mathcal{N}(\mathbf{x}_i | \boldsymbol{\mu}_m, \boldsymbol{\Sigma}_m)}{p(\mathbf{x}_i | \Theta)}. \tag{3}$$

Maximization of 3 is equivalent to finding the mixture index m with the highest value $\alpha_m \mathcal{N}(\mathbf{x}_i | \boldsymbol{\mu}_m, \boldsymbol{\Sigma}_m))$. The rest of the paper is organized as follows. In the next section, the EM algorithm for GMMs is outlined. Section 3 contains description of the GAIEM method. The results of initial computational experiments with two real-life datasets are reported in Sect. 4. The last section concludes the paper.

2 EM Algorithm for Gaussian Mixture Models

The EM algorithm [5, 14] is an iterative method which, starting from an initial guess of mixture parameters $\Theta^{(0)}$, generates a sequence of estimates $\Theta^{(1)}, \Theta^{(2)}, \ldots, \Theta^{(j)}, \ldots$ with increasing log likelihood (i.e., $\log p(X|\Theta^{(j)}) > \log p(X|\Theta^{(j-1)})$). Each iteration j of the algorithm consists of two steps called the expectation step (E-step) and the maximization step (M-step). For GMMs, these steps are defined as follows [20]:

- E-Step: Given the set of mixture parameters $\Theta^{(j-1)}$ from the previous iteration, for each $m = 1, \ldots, K$ and $i = 1, \ldots, N$, the posterior probability that a feature vector $\mathbf{x}_i$ was generated from the mth component is calculated as:

$$h_m^{(j)}(\mathbf{x}_i) = \frac{\alpha_m^{(j-1)} p_m(\mathbf{x}_i | \theta_m^{(j-1)})}{\sum_{k=1}^{K} \alpha_k^{(j-1)} p_k(\mathbf{x}_i | \theta_k^{(j-1)})}, \tag{4}$$

  where $\theta_m^{(j-1)}$ and $\theta_k^{(j-1)}$ denote parameters of components m and k, in the iteration $j-1$, respectively.
- M-Step: Given the posterior probabilities $h_m^{(j)}(\mathbf{x}_i)$, the set of parameters $\Theta^{(j)}$ is calculated as:

$$\alpha_m^{(j)} = \frac{1}{N} \sum_{i=1}^{N} h_m^{(j)}(\mathbf{x}_i) \tag{5}$$

$$\boldsymbol{\mu}_m^{(j)} = \frac{\sum_{i=1}^{N} h_m^{(j)}(\mathbf{x}_i) * \mathbf{x}_i}{\sum_{i=1}^{N} h_m^{(j)}(\mathbf{x}_i)} \tag{6}$$

$$\boldsymbol{\Sigma}_m^{(j)} = \frac{\sum_{i=1}^N h_m^{(j)}(\mathbf{x}_i)(\mathbf{x}_i - \mu_m^{(j)})(\mathbf{x}_i - \boldsymbol{\mu}_m^{(j)})^T}{\sum_{i=1}^N h_m^{(j)}(\mathbf{x}_i)} \tag{7}$$

The E-steps and M-steps are applied alternately until a convergence criterion is met.

3 Genetic Algorithm for Initialization of EM

A genetic algorithm in GAIEM uses tournament selection with the tournament size $ts = 5$ and the population size $np = 64$. Its main loop is implemented using three standard steps [16]: the selection step, crossover step and mutation step.

3.1 Search Space

The genetic algorithm searches the space of initial solutions of the EM algorithm. In GAIEM this search space is restricted, i.e., not every GMM can be a valid initial solution. We impose three following restrictions on the space of initial solutions:

- An initial mean vector must be a feature vector from the learning set X.
- An initial covariance matrix must be spherical, i.e., the form $\sigma^2 I$, where I is the $d \times d$ identity matrix.
- Mixing proportions must be equal, i.e., $\alpha_m^{(0)} = 1/K$.

A chromosome (member of EA population) in GAIEM is able to represent any GMM fulfilling the above restrictions. More specifically, the chromosome takes the form of an unordered set $S = \{s_1, s_2, \ldots, s_K\}$. Each $s_j \in S$ is used to initialize a single component (the mean and covariance) of a GMM. s_j takes the form of a pair $(\mathbf{x}_j, \sigma_j^2)$ where $x_j \in X$ is a feature vector used to initialize the mean of the component and σ_j^2 is the variance of spherical covariance matrix $\sigma_j^2 I$ of the initial component. At the start of the EA, all chromosomes are initialized as follows.

$$\mathbf{x}_j = \mathbf{x}_r, \text{ where } x_r \in X \text{ and } r \text{ is randomly chosen from } \{1, 2, 3, \ldots, N\} \tag{8}$$

and

$$\sigma_j^2 = 1/d * \text{trace}(\boldsymbol{\Sigma}) * 0.1^{3r+1}, \tag{9}$$

where $\boldsymbol{\Sigma}$ is the covariance matrix of the learning set X and r is a random number from the uniform distribution on $[0, 1)$, which is drawn independently for each chromosome in the population.

3.2 Mutation and Crossover

The mutation operator in GAIEM is applied with a small probability p_m to each pair $s_j = (\mathbf{x}_j, \sigma_j^2)$ in a chromosome. It first randomly decides whether to mutate $\mathbf{x}_j$ or σ_j^2. If $\mathbf{x}_j$ is chosen for mutation, it is replaced by a new feature vector from the learning set X. The new vector is selected as follows. At the beginning tc (tc is a user-supplied parameter, in our experiments we used $tc = 20$) unique candidate feature vectors is randomly chosen from the learning set X. Next $\mathbf{x}_j$ is replaced by the candidate vector closest to the component represented by a pair $(\mathbf{x}_j, \sigma_j^2)$. As a measure of distance between a candidate feature vector and an initial Gaussian component we use the Mahalanobis distance. If σ_j^2 is chosen for mutation, it is replaced by σ'^2_j according to

$$\sigma'^2_j = \sigma_j^2 * \exp(\mathcal{N}(0, 2)), \tag{10}$$

where $\mathcal{N}(0, 2)$ is a random number from the univariate Gaussian distribution with mean 0 and standard deviation 2. The crossover operator in GAIEM uses two parents and generates one offspring. It first copies into the offspring the more fit parent. Next it selects randomly in the offspring a pair $(\mathbf{x}_j, \sigma_j^2)$. This pair is replaced by the closest pair from the less fit parent. As a measure of distance between two pairs we use the Bhattacharyya distance [10] between the corresponding multivariate Gaussian distributions.

3.3 Fitness Function

To compute the fitness in the chromosome we run the EM algorithm, as outlined in Sect. 2, until the convergence. The GMM encoded by the chromosome becomes an initial solution of the EM algorithm. The log likelihood 2 of the final solution found by the EM is regarded as the chromosome fitness. It should be noted that the restrictions on the form of initial GMMs described in Sect. 3.1 are not enforced during the EM run. After the termination of the algorithm, the final EM solution of the chromosome with the highest fitness is considered as the result of GAIEM.

4 Experimental Results

In this section, the results of the computational experiments performed on two real-life datasets are presented. The experiments used GMMs for model-based clustering, in which, after the learning of model parameters, the feature vectors were clustered using the MAP rule, as explained in Sect. 1. Apart from GAIEM, two other parameter estimation methods were used in the experiments. The former was multiple restart EM algorithm (MREM), as outlined in the Sect. 1. The latter was recently proposed [22] random swap EM (RSEM) method. RSEM uses a random swap operation, which perturbs the current solution by assigning a randomly chosen feature vector to a randomly chosen component mean. After this operation a run of the EM algorithm, starting from a modified solution, is performed. The solution found by the EM is accepted as a new current solution if its log likelihood is higher; otherwise it is discarded. The random swaps and EM runs are applied until a termination condition (e.g., based on the maximum number of swaps is met). Experiments have demonstrated [22], that RSEM is able to find a high quality solution in terms of log likelihood. The results found by the three methods were compared using two criteria. The first of them was the clustering accuracy, which compares the partitions of data discovered in the course of clustering with the original partitions (the ground truth). The clustering accuracy is equivalent of the classification accuracy in supervised learning. Let r_i and s_i denote the obtained cluster label and the label provided by the ground truth, respectively. The clustering accuracy a is defined as [9]:

$$a = \frac{\sum_{i=1}^{N} \delta(s_i, \mathrm{map}(r_i))}{N}, \quad \text{where} \quad \delta(x, y) = \begin{cases} 1 & \text{if } x = y \\ 0 & \text{otherwise} \end{cases}. \tag{11}$$

In the above equation, $\mathrm{map}(r_i)$ is the optimal permutation function, which maps each cluster label r_i to a corresponding ground truth label from the dataset. The optimal mapping can be found in the polynomial time by using the Kuhn–Munkres (Hungarian) algorithm [13]. The clustering accuracy assumes the value in the interval [0, 1]. A higher value of a indicates a higher similarity between cluster labels and the ground truth; the maximum value of 1 means that partitions of the data defined by the cluster labels and the ground truth are identical. The second criterion used in the experiments was the log likelihood 2. of the GMM solution. It was the objective function, which all three algorithms were trying to maximize. The higher log likelihood the better a given GMM fits the data. To assure a fair comparison the three clustering methods were compared on the basis of equal CPU runtime. Each of them was allocated one hour (3600 s) of CPU time, and the best solution found in this time was considered as the final result. Since all three methods are probabilistic we repeated this experiment 50 times, each time starting with a different seed of the random number generator. The results shown in Tables 1, 2 and Figs. 1, 2 are averages over the 50 independent runs. To assess statistical significances of differences in a and $\log p(X|\Theta)$ the Wilcoxon rank sum test [4] was performed.

Table 1 Comparison of three GMM parameter estimation methods for color texture recognition dataset

Method	$\log p(X\|\Theta)$	p_L	a	p_a
GAIEM	285816.3 ± 3.67	–	$0.8916 \pm 8.4e{-}4$	–
RSEM	285809 ± 12.2	0.12	0.8804 ± 0.014	2.9e−06
MREM	285514.5 ± 133	7.1e−18	0.8557 ± 0.03	9.3e−17

The results are based on 50 independent 3600 s runs

Table 2 Comparison of three GMM parameter estimation methods for image recognition dataset

Method	$\log p(X\|\Theta)$	p_L	a	p_a
GAIEM	18318.8 ± 643	–	0.7817 ± 0.02	–
RSEM	18948.1 ± 819	1.4e−4	0.7452 ± 0.038	1.4e−7
MREM	9249.2 ± 467	7.1e−18	0.5991 ± 0.037	7.0e−18

The results are based on 50 independent 3600 s runs

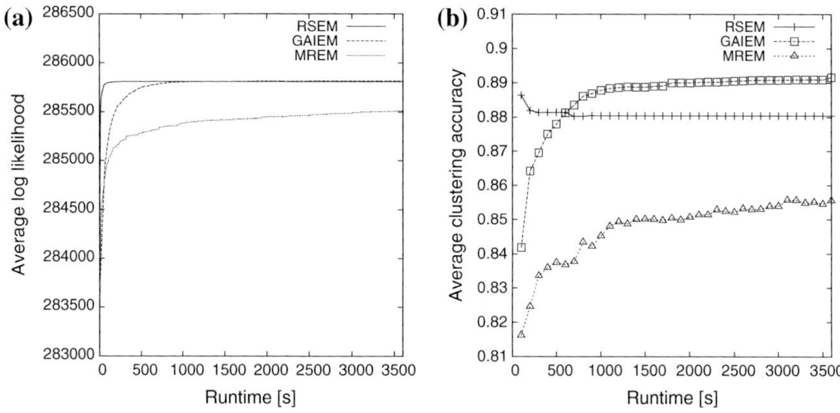

Fig. 1 Progress towards the final solution of three GMM parameter estimation algorithms for color texture recognition dataset: **a** log likelihood. **b** Clustering accuracy. The *curves* are averages over 50 independent runs

4.1 Hardware and Software

All the algorithms were implemented in C++ language on a system with 16 GB of RAM and eight-core AMD FX 8150 processor, running Ubuntu Linux 12.04. The programs were compiled using gcc 4.9.2 compiler using optimizing options (-Ofast -flto -March=bdver1 -mprefer-avx128) recommended for this processor. The algorithms were utilizing all eight cores of this system, using a parallel version of the EM algorithm described in [12].

4.2 Clustering of Color Texture Patches

Two real-life datasets were used in the clustering experiments. The first dataset was based on subset of 24 textured color images from MIT Vision Texture database.[1] The same subset of 24 color texture images was used in [18] for supervised (classification) learning. The size of each image was 256 × 256 pixels. We first rescaled RGB values

[1] http://vismod.media.mit.edu/vismod/imagery/VisionTexture/vistex.html.

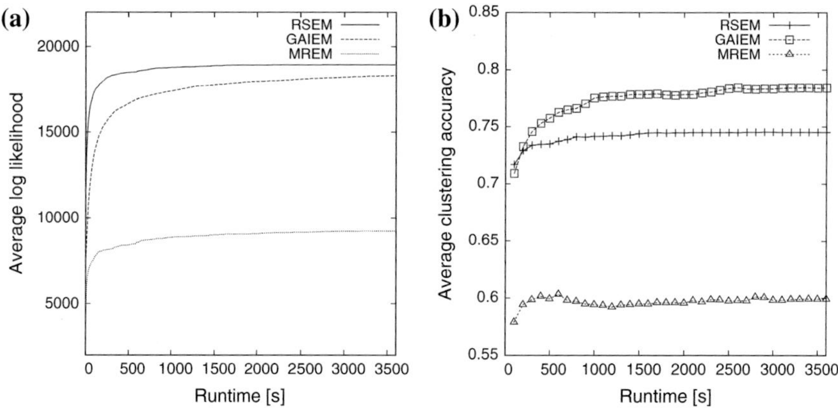

Fig. 2 Progress towards the final solution of three GMM parameter estimation algorithms for image recognition dataset: **a** log likelihood. **b** Clustering accuracy. The *curves* are averages over 50 independent runs

in each image into the [0, 1) interval. Next, we divided each image into 256 16×16 blocks. From each block, a feature vector was extracted, similarly as in [18], by taking the mean and covariance of its pixel RGB values. Since the covariance matrix is symmetric, we used its upper triangular part, including diagonal. Thus, a feature vector consisted of 9 elements, 6 coming from the covariance and 3 from the mean. The training set consisted of $24 * 256 = 6144$ feature vectors. For each feature vector we also stored its image number as ground truth. We expected the vectors extracted from the same image to form clusters in the feature space. Figure 1 illustrates the progress towards final solution of the three methods. The curves in Fig. 1a were obtained by measuring the clustering accuracy of best (in the sense of highest log likelihood) solution found so far by the given algorithm in time steps of 200 s. The curves in Fig. 1b were updated after each EM run. Table 1 shows comparison of the final solutions of the three methods based on 50 independent runs. The first column of the table indicates the name of the method. The second column shows the average (over 50 experiments) log likelihood obtained by the method and the standard deviation after $\pm$ sign. The third column shows the p-value for a statistical comparison of log likelihood (obtained using the Wilcoxon rank sum test) of the method with GAIEM. The fourth column presents average clustering accuracy and the standard deviation. The last column shows the p-value for a statistical comparison of clustering accuracy (obtained using the Wilcoxon rank sum test) of the method with GAIEM.

4.3 *Clustering of COIL20 Images*

The second dataset was based on COIL20 image library from Columbia University [17]. It consists of 1440 gray scale images of 20 objects (72 images per object). The images of each object were taken against a black background, at pose intervals of 5 degrees as the object was rotated on a turntable through 360°. The size of original images was 128×128 pixels. We used a preprocessed version of the database [9], in which size of each image was reduced to 32×32 pixels, giving a 1024-dimensional feature vector, and gray levels were scaled to the interval [0,1). For this dataset, it was not possible to fit a GMM directly, because the number of dimensions was much larger than the number of data points in each class, which results in linear dependencies between features and singular covariance matrices. To get rid of these dependencies we used the principal component analysis (PCA) to reduce the dimensionality. We kept 23 first principal components which together captured 85 % of total data variance. Figure 2 illustrates the progress towards final solution of the three methods. Table 2 shows comparison of the final solutions of the three methods based on 50 independent runs.

5 Conclusions and Future Work

The initial results allow us to conclude that GAIEM outperforms the standard multiple restart EM algorithm by a large margin and is at least as good as the recently proposed RSEM approach. Although it was not able to statistically significantly dominate RSEM with respect to the objective function 2, its clustering results were significantly better with respect to the clustering accuracy a. Several directions of future work exist. We are going to apply GAIEM to the problem of modeling class-conditional densities by GMMs in classification. This method is known as the mixture discriminant analysis [8]. An important limitation of the current version GAIEM is the requirement of knowing the number of components K in advance. We plan to extend GAIEM to be able to dynamically adjust the number of components in the initial mixture. For this purpose we are going to incorporate into GAIEM new search operators, which will add (and remove) components from the mixture. Since global optimum of log likelihood 2 increases monotonically with increase of K, a version of GAIEM with variable number of components will have to use a different fitness function, for instance based on the minimal-description length principle (MDL, similarly to [19]) or minimum message length principle [6].

Acknowledgments This work was supported by the Bialystok University of Technology grant S/WI/2/2013.

References

1. Andrews, J.L., McNicholas, P.D.: Using evolutionary algorithms for model-based clustering. Pattern Recognit. Lett. **34**(9), 987–992 (2013)
2. Bishop, C.M.: Pattern Recognition and Machine Learning. Springer, New York (2006)
3. Caglar, A., Aksoy, S., Arikan, O.: Maximum likelihood estimation of Gaussian mixture models using stochastic search. Pattern Recognit. **45**(7), 2804–2816 (2012)

4. Conover, W.J.: Practical Nonparametric Statistics. Wiley, New York (1999)
5. Dempster, A.P., Laird, N.M., Rubin, D.B.: Maximum likelihood from incomplete data via the EM algorithm. J. R. Stat. Soc. Ser. B **39**(1), 1–38 (1977)
6. Figueiredo, M., Jain, A.: Unsupervised learning of finite mixture models. IEEE Trans. Pattern Anal. Mach. Intell. **24**(3), 381–396 (2002)
7. Fraley, C., Raftery, A.E.: Model-based clustering, discriminant analysis, and density estimation. J. Am. Stat. Assoc. **97**(458), 611–631 (2002)
8. Hastie, T., Tibshirani, R.: Discriminant analysis by Gaussian mixtures. J. R. Stat. Soc. Ser. B **58**(1), 155–176 (1996)
9. He, X., Cai, D., Shao, Y., Bao, H., Han, J.: Laplacian regularized Gaussian mixture model for data clustering. IEEE Trans. Knowl. Data Eng. **23**(9), 1406–1418 (2011)
10. Kailath, T.: The divergence and Bhattacharyya distance measures in signal selection. IEEE Trans. Commun. Technol. **15**(1), 52–60 (1967)
11. Kwedlo, W.: Estimation of parameters of Gaussian mixture models by a hybrid method combining a self-adaptive differential evolution with the EM algorithm. Adv. Comput. Sci. Res. **14**, 109–123 (2014)
12. Kwedlo, W.: A parallel EM algorithm for Gaussian mixture models implemented on a NUMA system using OpenMP. In: Proceedings of the 22nd Euromicro International Conference on Parallel, Distributed, and Network-Based Processing PDP 2014, pp. 292–298. IEEE CPS (2014)
13. Lovasz, L., Plummer, M.D.: Matching Theory. American Mathematical Society, Providence (2009)
14. McLachlan, G., Krishnan, T.: The EM Algorithm and Extensions. Wiley, New York (2008)
15. McLachlan, G., Peel, D.: Finite Mixture Models. Wiley, New York (2000)
16. Michalewicz, Z.: Genetic Algorithms + Data Structures = Evolution Programs. Springer, Berlin (1996)
17. Nene, S.A., Nayar, S.K., Murase, H.: Columbia object image library (COIL-20). Technical report, CUCS-005-96, Columbia University (1996)
18. Permuter, H., Francos, J., Jermyn, I.: A study of Gaussian mixture models of color and texture features for image classification and segmentation. Pattern Recognit. **39**(4), 695–706 (2006)
19. Pernkopf, F., Bouchaffra, D.: Genetic-based EM algorithm for learning Gaussian mixture models. IEEE Trans. Pattern Anal. Mach. Intell. **27**(8), 1344–1348 (2005)
20. Redner, R.A., Walker, H.F.: Mixture densities, maximum likelihood and the EM algorithm. SIAM Rev. **26**(2), 195–239 (1984)
21. Reynolds, D., Quatieri, T., Dunn, R.: Speaker verification using adapted Gaussian mixture models. Digit. Signal Process. **10**(1), 19–41 (2000)
22. Zhao, Q., Hautamäki, V., Kärkkäinen, I., Fränti, P.: Random swap EM algorithm for Gaussian mixture models. Pattern Recognit. Lett. **33**(16), 2120–2126 (2012)

On the Combination of Pairwise and Granularity Learning for Improving Fuzzy Rule-Based Classification Systems: GL-FARCHD-OVO

Pedro Villar, Alberto Fernández and Francisco Herrera

Abstract Fuzzy rule-based systems constitute a wide spread tool for classification problems, but several proposals may decrease its performance when dealing with multi-class problems. Among existing approaches, the FARC-HD algorithm has excelled as it has shown to achieve accurate and compact classifiers, even in the context of multi-class problems. In this work, we aim to go one step further to improve the behavior of the former algorithm by means of a "divide-and-conquer" approach, via binarization in a one-versus-one scheme. Besides, we will contextualize each binary classifier by adapting the database for each subproblem by means of a granularity learning process to adapt the number of fuzzy labels per variable. Our experimental study, using several datasets from KEEL dataset repository, shows the goodness of the proposed methodology.

Keywords Multi-class classification · Fuzzy association rules · One-vs-One decomposition · Genetic algorithms · Granularity learning

1 Introduction

Fuzzy rule-based systems is a commonly used tool for classification problems. An advantage of fuzzy systems is the interpretability of the generated model, especially when fuzzy sets are represented by linguistic labels [12]. We will make use of

P. Villar (✉)
Department of Software Engineering, University of Granada, Granada, Spain
e-mail: pvillarc@ugr.es

A. Fernández
Department of Computer Science, University of Jaén, Jaén, Spain
e-mail: alberto.fernandez@ujaen.es

F. Herrera
Department of Computer Science and Artificial Intelligence, University of Granada, Granada, Spain
e-mail: herrera@decsai.ugr.es

R. Burduk et al. (eds.), *Proceedings of the 9th International Conference on Computer Recognition Systems CORES 2015*, Advances in Intelligent Systems and Computing 403, DOI 10.1007/978-3-319-26227-7_13

linguistic fuzzy association rule-based classification systems, being one of the most robust method proposed in this field the FARC-HD algorithm (Fuzzy Association Rule-based Classification model for High Dimensional problems) [1], that obtains models with high prediction ability and low complexity. This method is based on fuzzy association rules for classification, where the antecedent of the rule is a combination of fuzzy labels and the consequent is a class label. The possible values for the antecedents are sets of fuzzy labels of linguistic partitions defined over the attribute's universe of discourse. FARC-HD considers a fuzzy Database (DB) with a fixed number of labels for all fuzzy partitions. In the last step of the algorithm the DB definition is adjusted by a lateral tuning, but maintaining the predefined granularity level for variable. This is the usual way to proceed in the majority of fuzzy systems learning methods, i.e., it is necessary to select a priori a number of labels for all fuzzy partitions. However, the granularity level has a significant influence on the Fuzzy Systems performance as it has been analyzed in [7]. Some learning methods proposed in fuzzy modeling and fuzzy classification include the granularity level learning [8, 22]. Classification problems with more than two classes (multi-class problems) are known to present more difficulties than binary-class problems. A robust solution to cope with the former problem is to use a decomposition approach [13]. Its main strategy is to reduce the multi-class problem to several binary-class problems [3], where the One-vs-One (OVO) technique is widely used [16]. This method divides the original problem by confronting all pairs of classes against them. Then, an independent classifier is built for each pair of classes and it is necessary to combine the outputs of these classifiers to obtain the final predicted class label for a given instance [13, 14]. The main purpose of this contribution is to improve the performance of FARC-HD in multi-class problems, using an OVO scheme built with fuzzy association classifiers generated by learning an appropriate granularity level for each attribute in the FARC-HD method. We think that the optimal granularity level for learning each pair of classes will be different and so, looking for a good number of labels for each binary classifier can contribute to obtain better prediction ability in this decomposition scheme (OVO) for multi-class problems. To do so, we employ an approach to derive the Fuzzy classifier that involves the use of two different (and independent) learning processes, in which a DB definition process wraps the FARC-HD algorithm. We use a Genetic Algorithm (GA) for the granularity learning. A similar learning scheme was performed in [22] to design Fuzzy Rule-Based Classification Systems for binary-class problems with imbalanced datasets. In order to illustrate the good performance of the proposed scheme of an OVO strategy with the learning process mentioned for FARC-HD, we will contrast the obtained results with the FARC-HD algorithm with the application of an OVO decomposition using the Cohen's kappa measure, that equilibrates the performance in each individual class, independently of the number of examples of every class. We have selected a collection of multi-class datasets from KEEL dataset repository [2] for developing our experimental analysis. Furthermore, we will perform a statistical analysis using non-parametric tests [15] to find significant differences among the obtained results.

This paper is organized as follows. First, Sect. 2 introduces the preliminary concepts used in this paper. Next, in Sect. 3 we will describe our proposal, an OVO strategy for FARC-HD designed using a GA for granularity learning. Then, Sect. 4 describes the experimental study. Finally, in Sect. 5, some conclusions will be pointed out.

2 Preliminaries

This section provides a description of the working procedure of FARC-HD joint with some basic concepts about fuzzy association rules (Sect. 2.1). Next, a brief introduction of the OVO scheme is given (Sect. 2.2). Finally, we present the metric of performance used in this paper, i.e., the kappa metric (Sect. 2.3).

2.1 FARC-HD Learning Method

In this paper, we have make use of a robust fuzzy model known as FARC-HD [1]. This algorithm is based on association discovery, a commonly used technique in data mining for extract interesting knowledge from large datasets by means of finding relationships between the different items in a database. The integration between association discovery and classification leads to precise and interpretable models. FARC-HD is aimed at obtaining an accurate and compact fuzzy rule-based classifier with a low computational cost. The usual data set of classification examples used for learning a Classifier consists of m training patterns $x_p = (x_{p1}, \dots, x_{pn})$, $p = 1, 2, \dots, m$ from M classes where x_{pi} is the ith attribute value ($i = 1, 2, \dots, n$) of the pth training pattern. In this work we use the Fuzzy Association rules used in FARC-HD: $R_i : If\ A\ then\ Class = C_k$, where R_i is the label of the ith rule, A is a set of labels of the attribute's fuzzy partitions and C_k is a class label. We use triangular membership functions as antecedent fuzzy sets as it is used in FARC-HD. In short, this method is based on the following three stages:

Stage 1. *Fuzzy association rule extraction for classification*: A search tree is employed to list all possible frequent fuzzy item sets and to generate fuzzy association rules for classification, limiting the depth of the branches in order to find a small number of short (i.e., simple) fuzzy rules.

Stage 2. *Candidate rule pre-screening*: Afterwards the rule generation, the size of the rule set can be too large to be interpretable by the end user. Therefore, a preselection of the most interesting rules is carried out by means of a "subgroup discovery" mechanism based on an improved weighted relative accuracy measure (wWRAcc').

Stage 3. *Genetic rule selection and lateral tuning*: Finally, in order to obtain a compact and accurate set of rules within the context of each problem, an evolutionary process will be carried out in a combination for the selection of the rules with a tuning of membership function.

2.2 One-vs-One Decomposition

The most common approaches for decomposition a multi-class problem into a binary-class problem are OVO [16] and OVA (One-vs-All) [5]. The former learns a binary classifier for each possible pair of classes, whereas the latter constructs a binary classifier considering each single class and all the other classes joined. OVA approaches are easier to apply and they have shown to be an interesting scheme for one-class classifiers [9, 18]. However, in a general framework the goodness of the OVO scheme versus the former have been experimentally proven [13]. OVO divides a m-class problem into $m(m-1)/2$ independent binary subproblems by contrasting all classes among them, each of which is learnt by a single classifier. In the classification stage, the input instance is presented to all classifiers, so that each one of them outputs a confidence degree r_{ij} and $r_{ji} \in [0, 1]$ in favor of their couple of classes C_i and C_j (usually $r_{ji} = 1 - r_{ij}$). Then, these confidence degrees are set within a score-matrix:

$$R = \begin{pmatrix} - & r_{12} & \cdots & r_{1m} \\ r_{21} & - & \cdots & r_{2m} \\ \vdots & & & \vdots \\ r_{m1} & r_{m2} & \cdots & - \end{pmatrix} \tag{1}$$

It is necessary an additional phase to combine the confidence degrees of each single classifier. Different aggregation methods have been proposed in order to determine the final class [13]. The simplest aggregation is the voting strategy, where each classifier contributes with a vote for its predicted class. However, in our case, we aim to benefit from the characteristics of fuzzy classifiers to make use of the framework of fuzzy preference relations for classification [17] as it will be explained in Sect. 3.2.

2.3 Performance Metric: Cohen's Kappa Index

Cohen's kappa is an alternative measure to *classification rate*, since it compensates for random hits [4, 6]. In contrast to classification rate, kappa evaluates the portion of hits that can be attributed to the classifier itself (i.e., not to mere chance), relative to all the classifications that cannot be attributed to chance alone. An easy way of computing Cohen's kappa is by making use of the resulting confusion matrix (Table 1) in a classification task. With the expression (2), we can obtain Cohen's kappa as

$$kappa = \frac{n \sum_{i=1}^{m} h_{ii} - \sum_{i=1}^{m} T_{ri} T_{ci}}{n^2 - \sum_{i=1}^{m} T_{ri} T_{ci}}, \tag{2}$$

where h_{ii} is the cell count in the main diagonal (the number of true positives for each class), n is the number of examples, m is the number of class labels, and T_{ri}, T_{ci} are the rows' and columns' total counts, respectively ($T_{ri} = \sum_{j=1}^{m} h_{ij}$, $T_{ci} = \sum_{j=1}^{m} h_{ji}$). Cohen's kappa ranges from -1 (total disagreement) through 0 (random

Table 1 Confusion matrix for an n-class problem

Correct class	Predicted class				
	C_1	C_2	…	C_m	Total
C_1	h_{11}	h_{12}	…	h_{1m}	T_{r1}
C_2	h_{12}	h_{22}	…	h_{2m}	T_{r2}
$\vdots$			$\ddots$		$\vdots$
C_m	h_{1m}	h_{2m}	…	h_{mm}	T_{rm}
Total	T_{c1}	T_{c2}	…	T_{cm}	T

classification) to 1 (perfect agreement). Being a scalar, it is less expressive than the ROC curves applied to binary-class cases. However, for multi-class problems, kappa is a very useful, yet simple, meter for measuring a classifier's classification rate while compensating for random successes.

3 OVO Strategy Using FARC-HD with Granularity Learning: GL-FARCHD-OVO

In this section, we describe the proposed method for learning the Fuzzy association rule-based model of each binary classifier that form the set of classifiers of the OVO scheme and the aggregation method used for compute the final class prediction. We denote our proposal as GL-FARCHD-OVO (Granularity Learning for FARC-HD in an OVO scheme).

3.1 Genetic Algorithm for Granularity Learning in FARC-HD

Any optimization/search algorithm can be used for our learning approach. In our case, we have considered a GA, and more specifically, a integer-coded CHC algorithm [10] as a robust model in accordance with its tradeoff between exploration and exploitation. The individuals of the GA codify the granularity level of each feature. For evaluating every individual, first, the fuzzy partitions are built considering the number of labels codified in the chromosome. Uniform partitions with triangular membership functions are chosen as in FARC-HD. As we employ a GA for determining a good granularity level for variable, we need to run the FARC-HD algorithm in the evaluation of each chromosome, that also includes a GA in the last tuning stage. In order to decrease the computational cost of the proposed method, only the two first two stages of FARC-HD are executed in the chromosome evaluation. The last stage of FARC-HD is executed once, over the best granularity level configuration found by the GA of our proposal. Next, we describe the components of the GA for granularity learning.

3.1.1 Coding Scheme

An integer coding approach is considered, with a chromosome length equal to the number of features in the data set. Each value stands for the number of fuzzy partitions to be used in each input variable. In this contribution, the possible values considered are taken from the set $\{2, \ldots, 7\}$. If g_i is the value that represents the granularity of variable i, a graphical representation of the chromosome is: $C = (g_1, g_2, \ldots, g_N)$.

3.1.2 Initial Gene Pool

The initial population is composed of two parts. In the first group, all the chromosomes have the same granularity in all its variables. This group is composed of v chromosomes, with v being the cardinality of the significant term set, in our case $v = 6$, corresponding to the six possibilities for the number of labels, $\{2 \ldots 7\}$. For these six possible granularity levels, one individual is created. The second part is composed for the remaining chromosomes, and all of their components are randomly selected among the possible values.

3.1.3 Evaluation of the Chromosome

Composed of three steps:

1. Define the DB using the granularity level encoded in the chromosome. For all the features, a uniform fuzzy partition with triangular membership functions is built considering the specific number of labels of the variable (g_i).
2. Generate the fuzzy association rules by running the FARC-HD method using the DB obtained. We must remark that only the two first stages of FARC-HD are executed, providing a rule base composed of "interesting" rules.
3. Calculate the fitness value, that it is the kappa index of the rule base obtained in the previous step over the training data set.

3.1.4 Selection

This GA makes use of a mechanism of "Selection of Populations". M parents (population size) and their corresponding offspring are put together to select the best M individuals to take part in the next population.

3.1.5 Crossover

This operator combines two chromosomes of the population to generate their offspring. The standard crossover operator in one point is applied, which works as follows. A crossover point p is randomly generated (the possible values for p are $\{2, \ldots, N\}$) and the two parents are crossed at the pth variable.

3.1.6 Incest Prevention

It promotes diversity among solutions (which is important to properly search the whole search space). Two parents are crossed if their distance divided by two is above a predetermined threshold T, which is initially computed as $N/4$, being N the length of the chromosome. If no individuals are recombined, then the threshold value is reduced by one. If C_1 and C_2 are the two chromosome to recombine: $C_1 = (g_1, g_2, \ldots, g_N)$, $C_2 = (h_1, h_2, \ldots, h_N)$, the distance measure used in this paper $(Dist)$ is calculated by

$$Dist = \sum abs(g_i - h_i) \quad i : 1 \ldots N$$

3.1.7 Restarting Approach

The mutation operator is replaced by this mechanism in order to get away from local optima. When the threshold value T is zero, the best chromosome is maintained and used as a template from generate at random new chromosomes by randomly changing the 35 % of the genes.

3.2 Aggregation Method for the OVO Decomposition

As mentioned in the previous section, we make use of the fuzzy preference relations for aggregating the outputs of each binary classifier. In this scheme, the classification problem is translated into a decision making problem for determining the final predicted class among all predictions for the binary classifiers. Specifically, in this paper we consider the use of a maximal *Non-Dominance Criterion* (*ND*) [11] for the final decision process. This method predicts the class which is less dominated by all the remaining classes:

$$Class = \arg \max_{i=1,\ldots,m} \left\{ 1 - \sup_{j \in C} r'_{ji} \right\} \tag{3}$$

where r'_{ji} corresponds to the normalized and strict score-matrix.

Table 2 Summary description of datasets

Data-set	#Ex.	#Atts.	#Cl.	Data-set	#Ex.	#Atts.	#Cl.
Balance	625	4	3	Page-blocks	548	10	5
Contraceptive	1473	9	3	Autos	159	25	6
Hayes-Roth	132	4	3	Shuttle	5800	9	7
Iris	150	4	3	Glass	214	9	7
Newthyroid	215	5	3	Satimage	643	36	7
Tae	151	5	3	Segment	2310	19	7
Thyroid	720	21	3	Ecoli	336	7	8
Wine	178	13	3	Penbased	1100	16	10
Vehicle	846	18	4	Yeast	1484	8	10
Cleveland	297	13	5	Vowel	990	13	11

Table 3 Parameters of FARC-HD

Conjunction operator	Product T-norm	Parameter K of the prescreening	2
Fuzzy reasoning method	Additive combination	Maximum evaluations	15000
Minimum support	0.05	Population size	50
Maximum confidence	0.8	Parameter alpha	0.15
Depth of the trees	3	Bits per gen:	30

4 Experimental Study

We have used twenty multi-class datasets from KEEL dataset repository[1] [2]. In order to correct the dataset shift [20], situation in which the training data set and the test data set do not follow the same distribution, we do not use the commonly used cross-validation scheme. We will employ a recently published partitioning procedure called Distribution Optimally Balanced Cross Validation [19] with five different partitions for each dataset. Table 2 summarizes the characteristics of these datasets: number of examples, number of attributes and number of classes. There are different imbalance ratios, from totally balanced datasets to highly imbalanced ones, besides the different number of classes. Some of the largest datasets (page-blocks, penbased, satimage, shuttle and thyroid) were stratified sampled at 10 % in order to reduce the computational time required for training. In the case of missing values (autos and cleveland), we removed those instances from the dataset before doing the partitions. We will analyze the influence of granularity learning by means of a comparison between the performance of GL-FARCHD-OVO and the original FARC-HD method used in an

[1] http://www.keel.es/dataset.php.

Table 4 Experimental results in training and test with the kappa metric

Dataset	FARCHD-OVO		GL-FARCHD-OVO		Dataset	FARCHD-OVO		GL-FARCHD-OVO	
	tra	tst	tra	tst		tra	tst	tra	tst
Balance	0.846	0.682	0.808	**0.710**	Page-blocks	0.774	0.554	0.787	**0.563**
Contraceptive	0.468	0.268	0.425	**0.287**	Autos	0.986	0.708	0.984	**0.737**
Hayes	0.826	0.663	0.868	**0.672**	Shuttle	0.827	0.824	0.994	**0.990**
Iris	0.975	0.920	0.988	**0.930**	Glass	0.850	**0.571**	0.830	0.560
Newthyroid	0.993	0.861	0.998	**0.900**	Satimage	0.832	0.717	0.849	**0.751**
Tae	0.697	0.337	0.680	**0.344**	Segment	0.941	0.920	0.959	**0.936**
Thyroid	0.530	0.368	0.485	**0.401**	Ecoli	0.921	**0.771**	0.926	0.769
Wine	1.000	0.906	1.000	**0.932**	Penbased	0.990	0.899	0.989	**0.903**
Vehicle	0.811	**0.636**	0.820	0.600	Yeast	0.590	0.478	0.585	**0.484**
Cleveland	0.936	**0.325**	0.858	0.311	Vowel	0.979	0.918	0.988	**0.924**
Average	FARCHD-OVO					GL-FARCHD-OVO			
	tra		tst			tra		tst	
	0.839		0.666			0.841		**0.685**	

Table 5 Results obtained by the Wilcoxon test for algorithm GL-FARCHD-OVO

VS	R^+	R^-	p-value	Hypothesis
FARCHD-OVO	172.0	38.0	0.010688	Rejected for GL-FARCHD-OVO

OVO strategy. The original fitness function of the GA performed in the third stage of FARC-HD has been modified changing the accuracy rate for the kappa index. The configuration and parameters for FARC-HD are the ones suggested in its seminal paper [1] and they are presented in Table 3 being "Conjuction operator" the operator used to compute the compatibility degree of the example with the antecedent of the rule. FARC-HD needs also a predefined number of labels for all the fuzzy partitions, we have used 5 as granularity level, as suggested in [1]. In the execution of the two first stages of FARC-HD performed in the GA for learning the granularity level, we have used the same parameters except the depth of the trees, that it reduced to 2, in order to go down the computational cost of the GA proposed. We remark that the final step of GL-FARCHD-OVO is the execution of the stage 3 of FARC-HD over the best individual found by the GA. The specific parameters setting for the GA of GL-FARCHD-OVO are 60 individuals and $100 \cdot N$ number of evaluations, being N the number of variables. In order to carry out the comparison of the classifiers appropriately, nonparametric tests should be considered, according to the recommendations made in [15]. We will use the Wilcoxon paired signed-rank test [21] to perform comparisons between the two algorithms executed. Table 4 shows the results in performance (using the performance metric) for GL-FARCHD-OVO and FARCHD-OVO, being *tra* the kappa index over the training dataset and tst the kappa index over the test dataset. The highest performance value for each test dataset is stressed in boldface. As it can be observed, the values obtained by GL-FARCHD-OVO are higher than the obtained for FACHD-OVO, showing the influence of the granularity level in the behavior of the classifier regarding to the classical way to proceed (with a predefined number of labels, the same for all the attributes). In order to validate these results, we show the ranking on precision of the different models. Table 5 presents the results obtained in by applying Wilcoxon test. The p-value obtained shows significative differences between our proposed method (GL-FARCHD-OVO) and FARCHD-OVO.

5 Conclusions

This contribution has described a learning process for multi-class problems following the OVO decomposition strategy that aggregates the outputs of the binary classifiers obtained for each pair of classes. We have used FARC-HD as learning method to build the classifiers. A stationary GA based on the well-known CHC algorithm is used for granularity learning. Our proposal uses a divide-and-conquer strategy and aims at finding a good granularity level for each pair of classes that outperform the prediction ability of the classifier and it is compared with an OVO scheme using the original FARC-HD algorithm, that is, considering a fixed granularity level. The proposed method obtains better results in performance rate in the majority of datasets considered, showing significative differences according the non-parametric statistical test. In future works, we will try to adjust the learning process in order to improve the results and to decrease the computational time of the GA.

Acknowledgments This work was partially supported by the Spanish Ministry of Science and Technology under project TIN-2012-33856, the Andalusian regional projects P10-TIC-06858 and P11-TIC-7765 and both the University of Jaén and Caja Rural Provincial de Jaén under project UJA2014/06/15.

References

1. Alcalá-Fdez, J., Alcalá, R., Herrera, F.: A fuzzy association rule-based classification model for high-dimensional problems with genetic rule selection and lateral tuning. IEEE Trans. Fuzzy Syst. **19**(5), 857–872 (2011)
2. Alcalá-Fdez, J., Fernández, A., Luengo, J., Derrac, J., García, S., Sánchez, L., Herrera, F.: KEEL data-mining software tool: data set repository, integration of algorithms and experimental analysis framework. J. Multi-Valued Log. Soft Comput. **17**(2–3), 255–287 (2011)
3. Allwein, E.L., Schapire, R.E., Singer, Y.: Reducing multiclass to binary: a unifying approach for margin classifiers. J. Mach. Learn. Res. **1**, 113–141 (2000)
4. Ben-David, A.: A lot of randomness is hiding in accuracy. Eng. Appl. Artif. Intell. **20**, 875–885 (2007)
5. Clark, P., Boswell, R.: Rule induction with cn2: some recent improvements. In: Kodratoff, Y. (ed.) EWSL. Lecture Notes in Computer Science, vol. 482, pp. 151–163. Springer, Berlin (1991)
6. Cohen, J.A.: Coefficient of agreement for nominal scales. Educ. Psychol. Meas. **20**, 37–46 (1960)
7. Cordón, O., Herrera, F., Villar, P.: Analysis and guidelines to obtain a good uniform fuzzy partition granularity for fuzzy rule-based systems using simulated annealing. Int. J. Approx. Reason. **25**(3), 187–215 (2000)
8. Cordón, O., Herrera, F., Villar, P.: Generating the knowledge base of a fuzzy rule-based system by the genetic learning of the data base. IEEE Trans. Fuzzy Syst. **9**(4), 667–674 (2001)
9. Cyganek, B.: One-class support vector ensembles for image segmentation and classification. J. Math. Imaging Vis. **42**(2–3), 103–117 (2012)
10. Eshelman, L.J.: The CHC adaptive search algorithm: how to have safe search when engaging in nontraditional genetic recombination. Foundations of Genetic Algorithms, pp. 265–283. Morgan Kaufman, Burlington (1991)
11. Fernandez, A., Calderon, M., Barrenechea, E., Bustince, H., Herrera, F.: Solving multi-class problems with linguistic fuzzy rule based classification systems based on pairwise learning and preference relations. Fuzzy Sets Syst. **161**(23), 3064–3080 (2010)
12. Gacto, M.J., Alcalá, R., Herrera, F.: Interpretability of linguistic fuzzy rule-based systems: an overview of interpretability measures. Inf. Sci. **181**(20), 4340–4360 (2011)
13. Galar, M., Fernández, A., Barrenechea, E., Bustince, H., Herrera, F.: An overview of ensemble methods for binary classifiers in multi-class problems: experimental study on one-vs-one and one-vs-all schemes. Pattern Recognit. **44**(8), 1761–1776 (2011)
14. Galar, M., Fernandez, A., Barrenechea, E., Bustince, H., Herrera, F.: Dynamic classifier selection for one-vs-one strategy: avoiding non-competent classifiers. Pattern Recognit. **46**(12), 3412–3424 (2013)

15. García, S., Herrera, F.: An extension on "statistical comparisons of classifiers over multiple data sets" for all pairwise comparisons. J. Mach. Learn. Res. **9**, 2607–2624 (2008)
16. Hastie, T., Tibshirani, R.: Classification by pairwise coupling. Ann. Stat. **26**(2), 451–471 (1998)
17. Hüllermeier, E., Brinker, K.: Learning valued preference structures for solving classification problems. Fuzzy Sets Syst. **159**(18), 2337–2352 (2008)
18. Krawczyk, B., Wozniak, M., Cyganek, B.: Clustering-based ensembles for one-class classification. Inf. Sci. **264**, 182–195 (2014)
19. Moreno-Torres, J.G., Raeder, T., Aláiz-Rodríguez, R., Chawla, N.V., Herrera, F.: A unifying view on dataset shift in classification. Pattern Recognit. **45**(1), 521–530 (2012)
20. Moreno-Torres, J.G., Sáez, J.A., Herrera, F.: Study on the impact of partition-induced dataset shift on k-fold cross-validation. IEEE Trans. Neural Netw. Learn. Syst. **23**(8), 1304–1313 (2012)
21. Sheskin, D.: Handbook of Parametric and Nonparametric Statistical Procedures. Chapman & Hall/CRC, Boca Raton (2006)
22. Villar, P., Fernández, A., Carrasco, R.A., Herrera, F.: Feature selection and granularity learning in genetic fuzzy rule-based classification systems for highly imbalanced data-sets. Int. J. Uncertain. Fuzziness Knowl.-Based Syst. **20**(3), 369–397 (2012)

Measures for Combining Prediction Intervals Uncertainty and Reliability in Forecasting

Vânia Almeida and João Gama

Abstract In this paper we propose a new methodology for evaluating prediction intervals (PIs). Typically, PIs are evaluated with reference to confidence values. However, other metrics should be considered, since high values are associated to too wide intervals that convey little information and are of no use for decision-making. We propose to compare the error distribution (predictions out of the interval) and the maximum mean absolute error (MAE) allowed by the confidence limits. Along this paper PIs based on neural networks for short-term load forecast are compared using two different strategies: (1) dual perturb and combine (DPC) algorithm and (2) conformal prediction. We demonstrated that depending on the real scenario (e.g., time of day) different algorithms perform better. The main contribution is the identification of high uncertainty levels in forecast that can guide the decision-makers to avoid the selection of risky actions under uncertain conditions. Small errors mean that decisions can be made more confidently with less chance of confronting a future unexpected condition.

Keywords Load forecasting · Prediction intervals · Neural networks · Conformal prediction · Uncertainty assessment

1 Introduction

In time series forecasting, most research focuses around producing and evaluating point forecasts. Point forecasts are a topic of first-order importance, being easy to compute and understand. However, prediction intervals (PIs) are assuming increasing importance comparatively to conventional techniques. By definition, a PI is an

V. Almeida (✉) · J. Gama
LIAAD/INESC TEC, University of Porto, Porto, Portugal
e-mail: vania.g.almeida@inescporto.pt

J. Gama
Faculty of Economics, University of Porto, Porto, Portugal
e-mail: jgama@fep.up.pt

R. Burduk et al. (eds.), *Proceedings of the 9th International Conference on Computer Recognition Systems CORES 2015*, Advances in Intelligent Systems and Computing 403, DOI 10.1007/978-3-319-26227-7_14

estimate of an interval in which a future observation will fall, with a certain probability called confidence level. Similarly to point forecasts, error measures play an important role in calibrating or refining a PI model [1, 2]. Typically, PIs evaluation is focused on the calibration of confidence intervals that indicates the probability for correct predictions. But, confidence values cannot be considered individually. High probability values are associated to intervals that can include extreme prediction errors. These too wide intervals convey little information and is of no use for decision-making. Sharpness and resolution are also considered as added value, i.e., the average size and the variability of intervals, respectively. The literature offers some metrics for PIs evaluation. However, a reliable representation based on the error distribution has not yet been studied. This paper aims to describe a useful methodology for evaluating PIs. The prediction errors are computed as the distance to the upper and lower bounds. Additionally, the Mean Absolute Error (MAE) range is computed, considering that the prediction values are contained within the lower and upper prediction bounds. This value represents the range of "acceptable" errors, and it is correlated with the interval width. Since many of intervals are asymmetric, and the forecast is not the midpoint of the estimated interval [8], the evaluation of the cost associated to the underestimation or overestimation is also considered. For the purpose of experimental evaluation a case study in electrical load forecast is presented. This is a challenging topic where PIs assume major importance. In order to increase sustainability and optimize resource consumption, electric utilities are constantly trying to adjust power supply to the demand. However, more than providing accurate forecasts, reliable interval predictions are fundamental. The paper is organized as follows. Section 2 describes related work. The proposed PIs evaluation methodology is formulated in Sect. 3. Case study description and results are presented in Sect. 4. Finally, Sect. 5 concludes this paper and provides guidelines for future work.

2 Related Work

2.1 Models Used in This Study

Several strategies can be used to provide PIs. Two strategies are adopted: (1) dual perturb and combine (DPC) algorithm [4] which produces PIs based on the perturbed predictions, and (2) conformal prediction (CP), one of the most promising strategies used to determine precise levels of confidence. Other strategies can be used, such as detailed in [7, 11].

2.1.1 Dual Perturb and Combine Method

The DPC algorithm is an efficient method that allows the reduction of the variance exhibited by neural networks (NNs), but also the estimation of the confidence values associated to the predictions. It consists of perturbing each test example several times, adding white noise to the attribute values, and predicting each perturbed version of the test examples. The final prediction is obtained by aggregating all the predictions, implemented as follows:

1. For each input variable in the test set x, k perturbations are performed, $i = 1, \ldots, k$.
$$x_i = x + \delta_i \tag{1}$$
with δ_i white noise N$(0, \sigma_i^2)$, where σ_i and k are user-defined parameters.
2. k predictions $\hat{y}_i$ are obtained, and the final prediction $\hat{y}$ is:
$$\hat{y} = \frac{\sum \hat{y}_i}{k} \tag{2}$$
3. The lower and upper bounds are defined as: $[\min(\hat{y}_i), \max(\hat{y}_i)]$.

2.1.2 Conformal Prediction

CP uses the past experience to determine precise levels of confidence in new predictions, assuming that the data is identically and independently distributed (*i.i.d*). CPs have been developed based on several algorithms, such as Support Vector Machines [14], k-Nearest Neighbors [10], or Neural Networks Regression [9]. In this paper a Neural Networks Regression based on Inductive Conformal Prediction (NNR-ICP) is implemented as proposed by Papadopoulos and Haralambous [9]:

1. The training and the calibration sets are represented as:
Training: $\{(x_1, y_1), (x_2, y_2), \ldots, (x_m, y_m)\}$ where $m < l$
Calibration: $\{(x_{m+1}, y_{m+1}), (x_{m+2}, y_{m+2}), \ldots, (x_l, y_l)\}$ with $k = l - m$ elements
2. A nonconformity score is associated with every pair (x_{m+i}, y_{m+i}) in the calibration set. It evaluates how strange the pair is for the trained NNR rule, being defined as:
$$\alpha_i = \left|\hat{y}_{m+1} - y_{m+1}\right| \tag{3}$$
where $\hat{y}_{m+1}$ is the predicted value.
3. Assuming *i.i.d.* distribution, these α's are sort in descending order
$$\alpha_{m+1}, \ldots, \alpha_{m+k} \tag{4}$$
4. Finally, the lower and upper bounds are computed according to
$$(\hat{y}_{l+1} - \alpha_{m+s}, \hat{y}_{l+1} + \alpha_{m+s}) \tag{5}$$
where $s = \delta(k + 1)$.

Assuming that a confidence level, $1 - \delta$, is given a priori, where $\delta > 0$ is a small constant (e.g., 5 %). It means that for a $\alpha = 0.05$ and a confidence of 95 %, the interval width is given by $\alpha_{m+0.05(k+1)}$, where k is the calibration set length.

2.2 PIs Metrics

The literature offers a variety of methods for the evaluation of the performance of point prediction methods, e.g., mean square error (MSE), mean absolute error (MAE), or mean absolute percentage error (MAPE). However, there is no well-established error measure dedicated to PIs assessment. Typically, the evaluation is only made based on the PI coverage probability (PICP) that can be interpreted as the probability that target values will be covered by the interval bounds. It is defined as

$$\text{PICP} = \frac{1}{N}\sum_{i=1}^{N} c(i) \tag{6}$$

where N is the number of samples and $c(i) = 1$, if $\hat{y}(i) \in [L(i), U(i)]$, $L(i)$ is the lower bound, and $U(i)$ is the upper bound, otherwise $c(i) = 0$. Ideally, the PICP should be as close as possible to its nominal value $(1 - \alpha)$ %, the confidence level for which PIs have been constructed. However, without considering its length the PI evaluation sound more subjective than objective. Therefore, the computation of width-based indices is essential. Typically, the intervals are normalized to the number of intervals, PI normalized average width (PINAW).

$$\text{PINAW} = \frac{1}{N}\sum_{i=1}^{N} (U(i) - L(i)) \tag{7}$$

Typically, very narrow PIs with a low coverage probability are not very reliable. On the other hand, very wide PIs with a high coverage probability are not very useful to use practically. The combination of both PI aspects can be performed by the use of different criteria, like coverage-length-based criterion (CLC) [12].

$$\text{CLC} = \text{NPINAW}\left(1 + e^{-\eta(PICP-\mu)}\right) \tag{8}$$

where μ and η are two controlling parameters. The CLC tries to compromise between informativeness and correctness of a PI [7]. PIs should be as narrow as possible from the informativeness perspective. However, the narrowness tends to result in a low coverage probability.

3 PIs Evaluation

3.1 Min–Max Error

We propose to compute errors with reference to the lower bound (min error) and upper bound (max error). They result from the distance from the predicted value to the respective limit ($\min_{dist}$) or ($\max_{dist}$). If the predictions are within PI limits, $\min_{dist}$ is negative and $\max_{dist}$ is positive.

$$\min_{dist} = \hat{y}_{\min} - y \tag{9}$$

$$\max_{dist} = \hat{y}_{\max} - y \tag{10}$$

Error values are represented in histograms. The bins associated to the prediction outside of the respective bound are identified as

- Bins placed in $x > 0$ in $\min_{dist}$ histogram.
- Bins placed in $x < 0$ in $\max_{dist}$ histogram.

The key points are: (1) the area associated to the error bins (min–max error), and (2) the drift from zero (min–max drift). PIs can range from point predictions (very narrow intervals, $\hat{y}_{\max} = \hat{y}_{\min}$) to very wide intervals. The key properties are presented in Table 1, assuming an *i.i.d* distribution.

Table 1 Main properties of min–max distribution

Interval	Min–max distance	Min–max error	Min–max drift
Very narrow	$\min_{dist} \approx \max_{dist}$	$\min_{area} \approx \max_{area}$	≈ 0
mindrift max drift Overestimation	$\min_{dist} < \max_{dist}$	$\min_{area} < \max_{area}$	$\min_{drift} < \max_{drift}$
min drift max drift Subestimation	$\min_{dist} > \max_{dist}$	$\min_{area} > \max_{area}$	$\min_{drift} > \max_{drift}$
Very wide	$\min_{dist} \approx \max_{dist} \approx \infty$	≈ 0	$\approx \infty$

3.2 Mean Absolute Error of the Interval

The calculation of MAE for a point forecast is relatively simple. It involves summing the magnitudes (absolute values) of the errors to obtain the total error and then dividing the total error by N. In the case of a PI, a single measure is not possible, since the prediction belongs to a range of values. We propose to compute the maximum MAE allowed for an interval (considering a hit, $\hat{y} \in [\hat{y}_{\min}, \hat{y}_{\max}]$). To facilitate, the central value of the interval is taken as the forecast value:

$$\hat{y} = \frac{\hat{y}_{\max} - \hat{y}_{\min}}{2} \tag{11}$$

So, MAE is computed as

$$\text{MAE} = \frac{1}{n} \sum_{i=1}^{n} \left| y - \frac{\hat{y}_{\max} - \hat{y}_{\min}}{2} \right| \tag{12}$$

The MAE limits are verified when the real value is $y = \hat{y}_{\max}$ or $y = \hat{y}_{\min}$. In this case, MAE ranges between

$$\frac{1}{n} \sum_{i=1}^{n} \left| \hat{y}_{\min} - \frac{R}{2} \right| \leq \text{MAE} \leq \frac{1}{n} \sum_{i=1}^{n} \left| \hat{y}_{\max} - \frac{R}{2} \right| \tag{13}$$

considering $\left|\hat{y}_{\max} - \frac{R}{2}\right| > \left|\hat{y}_{\min} - \frac{R}{2}\right|$, where R is the interval width. MAE range is the absolute value of the difference between limits.

4 Case Study and Experimental Setup

4.1 Data

The dataset includes historical data from April 1 to November 31, 2014, collected in the Customer Load Active System Services (CLASS) Project run by the UK Distribution Network Operator Electricity North West Limited.[1] The data consist of 30 MV substations. Each one is treated individually, being that 70 % of data used for learning the global model, and the remaining 30 % for prediction. All of the experiments were repeated 5 times.

[1]https://www.enwclass.nortechonline.net/data#substation-group/31.

4.2 Horizon Forecasting

NNs are one of the most popular options in the electric load forecasting [6]. The predictive model for the next 24h is a multilayer perceptron (MLP) network, with 3 inputs and a linear output. The choice of the network topology and inputs was motivated by previous work [3, 5, 13]. It is constituted by *Inputs:*

- 24 values of the load curve $[L(d-1)1, L(d-1)2, \ldots, L(d-1)24]$ of day $d-1$ (day before the forecasting day d).
- Day of week, entered as two different variables, in the form of sines and cosines, by means of $\sin[(2\pi\ d)/7]$ and $\cos[(2\pi\ d)/7]$, for each one of the days: Sunday($d=0$), Monday($d=1$), Tuesday($d=2$), Wednesday($d=3$), Thursday($d=4$), Friday($d=5$), Saturday($d=6$).

Output: 24 values of the load curve $[Ld1, \ldots, Ld24]$.

4.3 Calibration

The NNR-ICP model is calibrated through Eq. 4. In Fig. 1 the expected and the observed reliability values are plotted, considering all the substations. As shown, the calibration fit is linear with predictions falling near to the line of equality for the predicted and expected values. It can be concluded that NNR-ICP is well calibrated in the case of the database considered in this study. In opposition to the NNR-ICP model, the DPC method is not calibrated a priori. The jit added to the input variables follows $x_i \approx N(0, \sigma_i^2), i = 1, \ldots, 10$. The calibration curve is shown in Fig. 2. Results are presented individually for each substation, considering five independent trials. Calibration curves for three of the substations are depicted, evidencing the inter-substation variability. In Fig. 3 the NNR-ICP and DPC are compared at two confidence levels. As expected, as the confidence level increases, the corresponding interval width is enlarged. Additionally, NNR-ICP produces symmetric intervals, while DPC intervals are asymmetric.

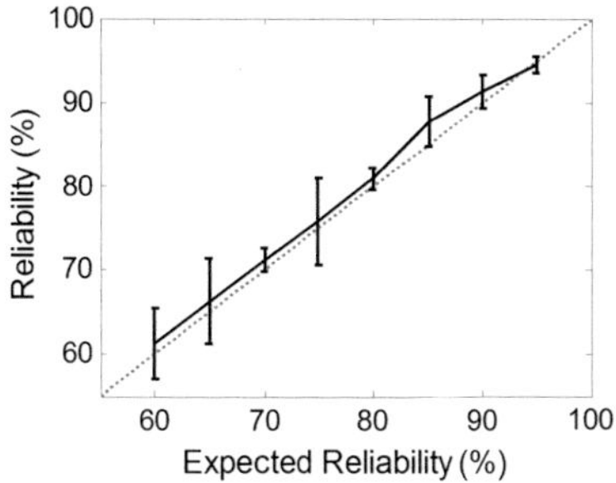

Fig. 1 NNR-ICP calibration considering all the substations

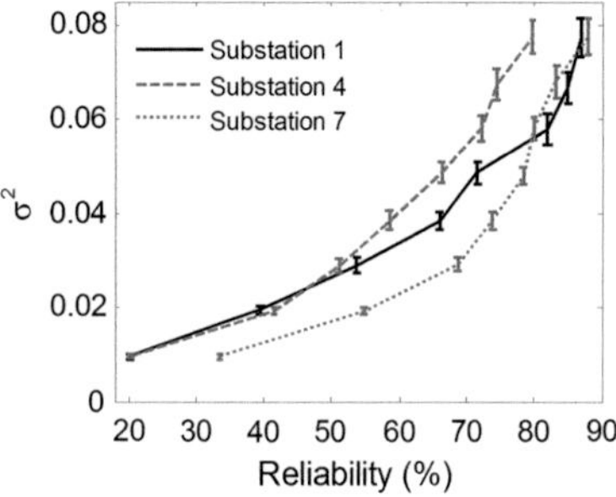

Fig. 2 DPC calibration for three different substations

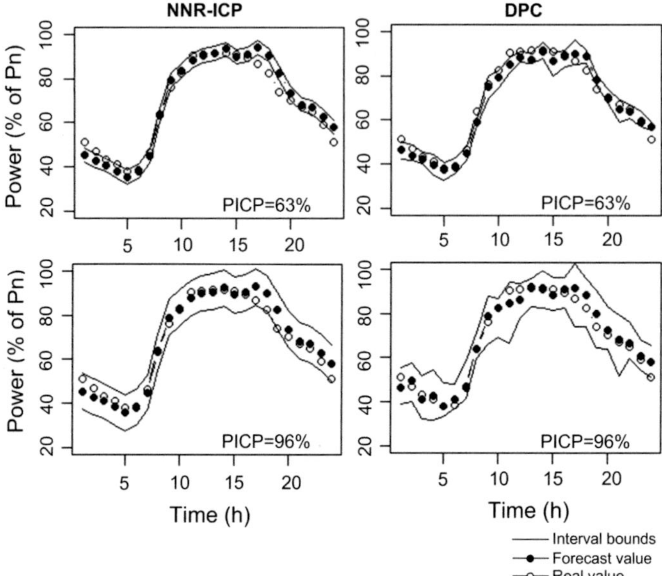

Fig. 3 PIs for the NNR-ICP and DPC methods at different confidence values

4.4 Min–Max Evaluation

An unreliable PI can lead to the underestimation or overestimation of the real value. In load forecast, the trend to underestimate or overestimate depends on the location, season, or even time of the day. The min–max error is computed as described in Sect. 3. Three different situations are depicted in Fig. 4 (upper panel), representing the forecast errors in different time schedules. In (a) the overlapped red/blue area is minimized due to the low number of predictions that fall out of the interval. The error increases to maximum values (b), and in (c) turns to decrease. At lower panel, the underestimation/overestimation ratio for both algorithms at confidence level 82 %

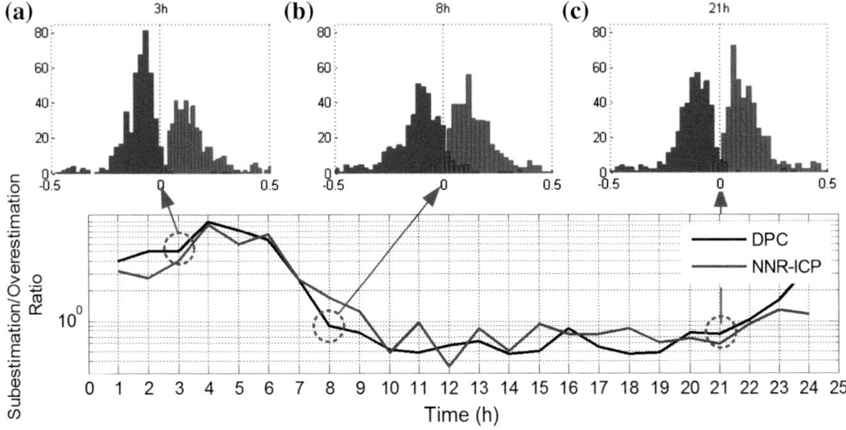

Fig. 4 Forecast errors (*upper panel*) and underestimation/overestimation errors along day (*lower panel*) for both algorithms

is presented. It is visible that during the periods [1–8 h] and [21–24 h] the forecasts tend to overestimate the real value, while during the period [9–21 h] the forecats are underestimated.

4.5 PIs Comparison

The MAE limits for the NNR-ICP and DPC algorithms at different confidence levels are presented in Fig. 5. The interval width increases at faster or slower speeds depending on the method used to built it. DPC method often leads to PIs whose lengths are significantly larger than PIs constructed using the NNR-ICP method. The paid cost

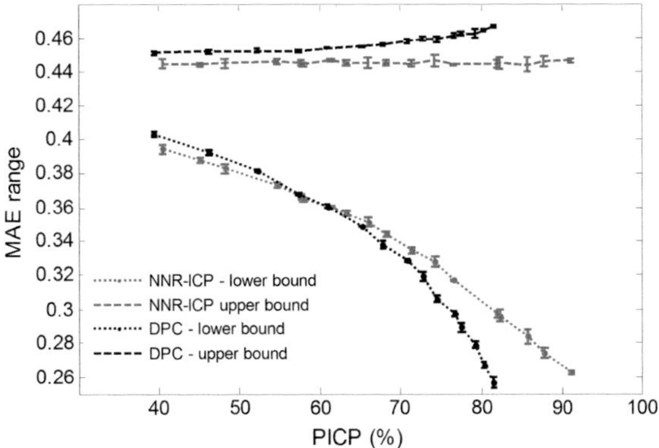

Fig. 5 MAE limits for the NNR-ICP and DPC algorithms

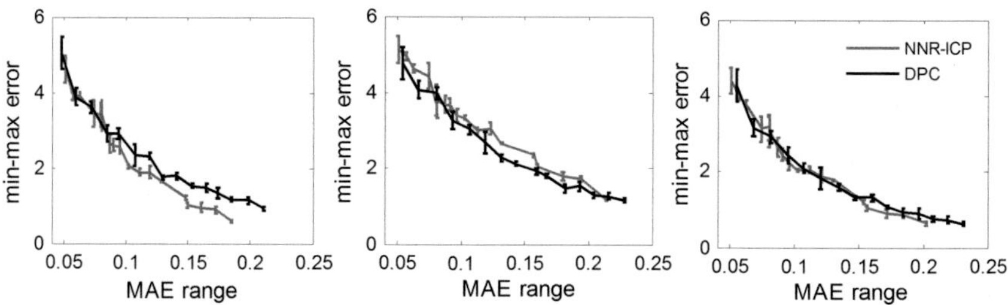

Fig. 6 Min–max error versus MAE at three different periods: (1) [1–4 h], (2) [11–14 h], and (3) [21–24 h]

is a wide MAE range and less informative intervals. This effect is visible for PICP values above 50 %. Finally, we compare the min–max error versus MAE for three different periods: (1) [1–4 h], (2) [11–14 h], and (3) [21–24 h]. A model is better as the min–max error is minimized. In (a) NNR-ICP presents a superior performance. In (b) the min–max error is minimized for the DPC method. This means, that PIs obtained with DPC are more robust during this period of the day. In (c) both methods are comparable. Min–max error measured during the period [11–14 h] is superior. This period is associated with higher forecast uncertainty, and so the selection of the DPC algorithm at this schedule is justified due to its wider and asymmetric interval bounds (Fig. 6).

5 Conclusions

A new methodology for evaluating PIs is addressed. From the methodological aspect, we have adopted two innovative approaches. One is the exploration of prediction errors distribution (the min–max error), while the other is the quantification of the MAE values associated to the interval bounds. The comparison consists in analyzing the contrast of the maximum MAE allowed, and the minimization of the min–max error for these values. It aims to guide model selection for PIs with the shortest length and the lowest error dispersion. Large errors are an indication of the presence of higher levels of uncertainty in forecasts. This information can guide the decision-makers to avoid the selection of risky actions. Small errors mean that decisions can be made more confidently with less chance of confronting an unexpected condition in the future. First, the calibration issue was addressed. Results indicate that NNR-ICP is well calibrated in the case of the database considered in this study. In the case of DPC (non-calibrated method) the inter-substation variability is evident. The min–max error depends on the time of day for both algorithms. Additionally, we can conclude that during periods of day associated with higher levels of uncertainty, DPC tends to have a better performance. This is justified due to their wider and asymmetric interval bounds, in comparison to the shorter and symmetric NNR-ICP limits. However, in general NNR-ICP tends to have a superior performance.

Acknowledgments This work was supported by NORTE-07-0124-FEDER-000056 financed by ON.2−O Novo Norte, under the National Strategic Reference Framework, through the Development Fund, and by national funds, through FCT. Additionally, it was funded by European Commission through MAESTRA (ICT-2013-612944).

References

1. Armstrong, J., Collopy, F.: Error measures for generalizing about forecasting methods: empirical comparisons. Int. J. Forecast. **8**(1), 69–80. http://www.sciencedirect.com/science/article/pii/016920709290008W (1992)
2. Christoffersen, P.F.: Evaluating interval forecasts. Int. Econ. Rev. **39**(4), 841–862 (1998)
3. Gama, J., Rodrigues, P.: Stream-based electricity load forecast. In: Knowledge Discovery in Databases: PKDD 2007. Lecture Notes in Computer Science, vol. 4702, pp. 446–453. Springer, Berlin. http://dx.doi.org/10.1007/978-3-540-74976-9_45 (2007)
4. Geurts, P., Wehenkel, L.: Closed-form dual perturb and combine for tree-based models. In: Raedt, L.D., Wrobel, S. (eds.) Proceedings of the Twenty-Second International Conference on Machine Learning (ICML 2005), Bonn, Germany, August 7–11, 2005. ACM International Conference Proceeding Series, vol. 119, pp. 233–240. ACM. http://doi.acm.org/10.1145/1102351.1102381 (2005)
5. Hernández, L., Baladrón, C., Aguiar, J.M., Carro, B., Sánchez-Esguevillas, A., Lloret, J.: Artificial neural networks for short-term load forecasting in microgrids environment. Energy **75**(0), 252–264. http://www.sciencedirect.com/science/article/pii/S0360544214008871 (2014)
6. Hippert, H., Pedreira, C., Souza, R.: Neural networks for short-term load forecasting: a review and evaluation. IEEE Trans. Power Syst. **16**(1), 44–55 (2001)
7. Khosravi, A., Nahavandi, S., Creighton, D.C., Atiya, A.F.: Comprehensive review of neural network-based prediction intervals and new advances. IEEE Trans. Neural Netw. **22**(9), 1341–1356 (2011)
8. OâConnor, M., Remus, W., Griggs, K.: The asymmetry of judgemental confidence intervals in time series forecasting. Int. J. Forecast. **17**(4), 623–633. http://www.sciencedirect.com/science/article/pii/S0169207001001030 (2001)
9. Papadopoulos, H., Haralambous, H.: Reliable prediction intervals with regression neural networks. Neural Netw. **24**(8), 842–851. http://www.sciencedirect.com/science/article/pii/S089360801100150X (2011). Artificial Neural Networks: Selected Papers from ICANN 2010
10. Proedrou, K., Nouretdinov, I., Vovk, V., Gammerman, A.: Transductive confidence machines for pattern recognition. ECML 2002, pp. 381–390. Springer, Berlin (2001)
11. Quan, H., Srinivasan, D., Khosravi, A.: Short-term load and wind power forecasting using neural network-based prediction intervals. IEEE Trans. Neural Netw. Learn. Syst. **25**(2), 303–315 (2014)
12. Quan, H., Srinivasan, D., Khosravi, A.: Uncertainty handling using neural network-based prediction intervals for electrical load forecasting. Energy **73**(0), 916–925. http://www.sciencedirect.com/science/article/pii/S0360544214008032 (2014)
13. Rodrigues, P.P., Gama, J.: A system for analysis and prediction of electricity-load streams. Intell. Data Anal. **13**(3), 477–496 (2009)
14. Saunders, C., Gammerman, A., Vovk, V.: Transduction with confidence and credibility. Proc. Int. Jt. Conf. Artif. Intell. **16**, 722–726 (1999)

Detection of Elongated Structures with Hierarchical Active Partitions and CEC-Based Image Representation

Arkadiusz Tomczyk, Przemysław Spurek, Michał Podgórski, Krzysztof Misztal and Jacek Tabor

Abstract In this paper, a method of elongated structure detection is presented. In general, this is not a trivial task since standard image segmentation techniques require usually quite complex procedures to incorporate the information about the expected shape of the segments. The presented approach may be an interesting alternative for them. In its first phase, it changes the representation of the image. Instead of a set of pixels, the image is described by a set of ellipses representing fragments of the regions of similar color. This representation is obtained using cross-entropy clustering (CEC) method. The second phase analyses geometrical and spatial relationships between ellipses to select those of them that form an elongated structure within an acceptable range of its width. Both phases are elements of hierarchical active partition framework which iteratively collects semantic information about image content.

Keywords CEC · Hierarchical active partition · Structural description

1 Introduction

Detection of different structures in the images is a crucial part of almost any system analyzing image content. A typical example of such systems are tools supporting medical diagnosis. Elongated objects are frequent structures that need to be detected in those systems. Some good examples are not only arteries and veins, because

A. Tomczyk (✉)
Institute of Information Technology, Lodz University of Technology,
Wólczańska 215, 90-924 Łódź, Poland
e-mail: arkadiusz.tomczyk@p.lodz.pl

P. Spurek · K. Misztal · J. Tabor
Faculty of Mathematics and Computer Science, Jagiellonian University,
Łojasiewicza 6, 30-348 Kraków, Poland
e-mail: przemyslaw.spurek@ii.uj.edu.pl

M. Podgórski
Department of Radiology and Diagnostic Imaging, Medical University of Lodz,
Kopcinskiego 22, 90-159 Łódź, Poland
e-mail: michal.podgorski@umed.lodz.pl

R. Burduk et al. (eds.), *Proceedings of the 9th International Conference on Computer Recognition Systems CORES 2015*, Advances in Intelligent Systems and Computing 403, DOI 10.1007/978-3-319-26227-7_15

of their natural shape, but also different types of tissues are visible in such a way if different sections of 3D structures are analyzed separately. In medical images, detection of such structures is usually not a trivial task. The first problem is the noise being a result of image acquisition process. Second, and maybe even more important, is the fact that different, adjacent structures have similar colors of the pixels representing them which consequently result in blurred borders between those structures. There are different methods that can be used to detect such structures. Typical segmentation techniques have, however, problems with this task [2, 4, 12]. Application of region-based techniques with standard pixel similarity criteria would lead to region leaks. Edge-based methods would require precise contour tracing which usually fails in areas with blurred borders between structures. To overcome those problems, an additional knowledge about expected shape must be used. In the aforementioned approaches, it is not impossible but very troublesome. A natural solution to this problem is the active contour techniques where additional knowledge may be incorporated either in energy function or in contour evolution constraints [1, 7]. This, however, also requires additional, sometimes complex procedures like, for example, training of a shape model [1]. In this work, an alternative method is proposed which, instead of focusing on complex methods, uses a hierarchical approach with a sequence of simpler techniques and specific representation of image content. The results of the described method are illustrated with a problem of articular cartilage detection that can be of use in a process of osteoarthritis (OA) diagnosis. OA is a chronic, degenerative disease that leads to loss of articular cartilage and joint deterioration [13]. It is a leading cause of disability worldwide [10]. Knee is most commonly affected, especially in elderly and obese individuals [8, 13].

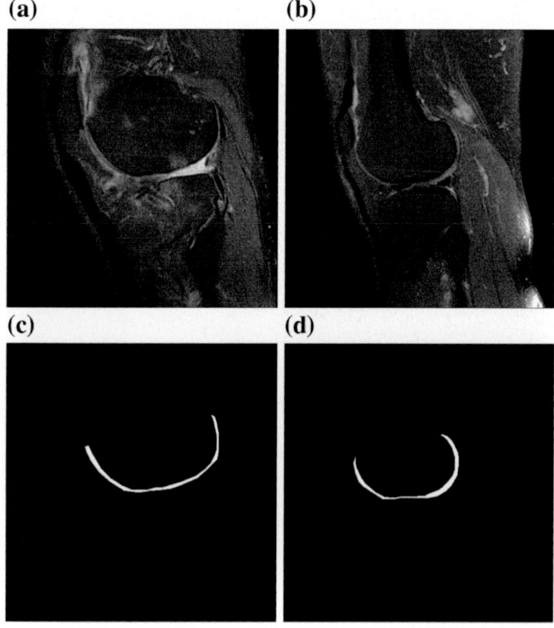

Fig. 1 Sample sections of knee MRI examinations: **a**, **b** images, **c**, **d** fragments of articular cartilage

Although, plain radiography is traditionally used to diagnose OA, the joint space narrowing is observed typically at the late-stage disease [17]. Due to the fact that evaluation of structural changes in articular cartilage is important for assessment of the progression and the effect of OA treatment, more sensitive methods are required [5]. MRI enables high-resolution visualization of the cartilage, is noninvasive, and does not use ionizing radiation; thus, it was applied for quantitative evaluation of knee join articular cartilage [5], see Fig. 1a, b. By application of automated or semi-automated segmentation methods, analysis can be performed comprehensively and easily [5]. The paper is organized as follows: in the second section, the hierarchical active partition approach is described as a framework of image analysis; in the third section, the CEC method and resulting image representation is presented; the fourth section focuses on detection of elongated structures, and finally the last two sections are devoted to presentation and discussion of the results.

2 Hierarchical Active Partitions

The concept of active partitions originates in active contour techniques [16]. Those methods look for an optimal contour describing object in the image. The search objective is defined by an energy function and as a search procedure any appropriate optimization technique can be used. As it was mentioned in the previous section, one of the biggest advantages of this approach is that it is prepared to incorporate any kind of additional, expert knowledge expressing expectations about sought structures. In [16] and earlier works of the same authors it was shown that contours can be considered as classifiers of pixels as they allow to discriminate pixels representing object and background. In other words, they partition the whole set of pixels into two subsets. As, in general, classification techniques can be used to recognize almost any kind of objects, the idea of active partitions was proposed where instead of pixels other objects representing image content are considered. Also in this case to find an optimal partition, any reasonable optimization process can be used. If this process is an iterative algorithm, then partitions change in every iteration and consequently they can be called active. In practice, different types of objects can be considered. They can be line segments, circular regions of the same color, etc. In this work, CEC-based image representation is proposed where image content is described by a set of ellipses. The process of object detection can be considered as a hierarchical problem where consecutive phases deliver additional information about image content. Such an interpretation has strong biological foundations in human vision system [3, 9] as well as in conscious process of image analysis. All the mentioned so-far segmentation techniques, including active partitions, can be elements of such a hierarchy. As an example of the problem where such an approach can be considered is the cartilage detection task which is discussed in this work. Instead of complex algorithm it focuses first on localization of regions with similar color and representation of those regions by a set of ellipses, then elongated structures are extracted which later can be used to detect specific tissues based on medical knowledge about knee structure.

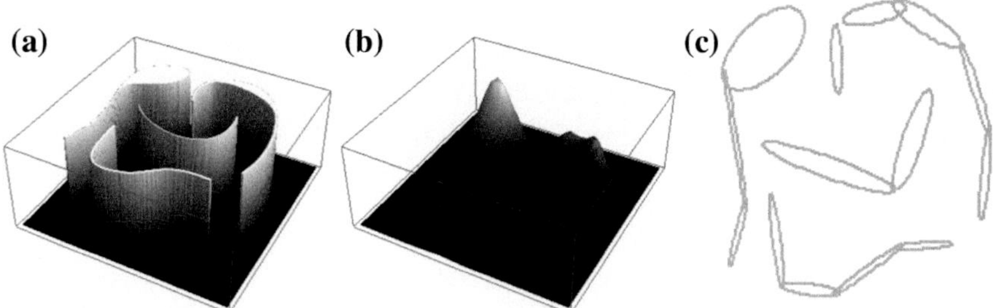

Fig. 2 CEC method applied to white regions of Fig. 3a: **a** uniform distribution, **b** several Gaussian functions approximating given distribution, **c** ellipses corresponding with those functions

3 CEC-Based Image Representation

In this section, we present a method which uses CEC[1] algorithm to construct a new type of image representation. It can be interpreted as an approximation of uniform distribution by several Gaussian densities. Since level-sets of Gaussian densities are ellipses, it can be understood as covering a data set by ellipses. The simple example is presented in Figs. 2 and 3. Image of a given set depicted in Fig. 3a can be interpreted as a uniform distribution on white parts of this set, see Fig. 2a. We can approximate the distribution by several Gaussian functions, see Fig. 2b. Finally, each of Gaussian component we can interpret as a ellipses which is presented in Fig. 2c. More precisely, we represent Gaussian densities by one level-set. Before we present how to choose the optimal one, let us recall that the normal random variable with the mean equal to zero and the covariance matrix Σ has a density:

$$g_{\Sigma}(x) := \frac{1}{2\pi\sqrt{\det(\Sigma)}} \exp(-\frac{1}{2}\|x\|^2_{\Sigma}) \tag{1}$$

where by $\|x\|^2_{\Sigma} := x^T \Sigma^{-1} x$ we denote the square of the Mahalanobis norm. The 2D ellipse generated by the positive definite matrix Σ (of size 2×2) with its center in zero is defined as follows:

$$\mathbb{B}_{\Sigma} := \left\{(x_1, x_2) \in \mathbb{R}^2 \colon \|(x_1, x_2)\|^2_{\Sigma} < 1\right\}. \tag{2}$$

The eigenvectors of Σ define the principal directions of the ellipse and the eigenvalues of Σ are the squares of the semi–axes: a^2, b^2. On the other hand, the covariance matrix of uniform density of an ellipse $\left\{(x_1, x_2) \in \mathbb{R}^2 \colon \frac{x_1^2}{a^2} + \frac{x_2^2}{b^2} < 1\right\}$ is given by $\begin{bmatrix} \frac{a^2}{4} & 0 \\ 0 & \frac{b^2}{4} \end{bmatrix}$. Therefore, we represent a Gaussian distribution by ellipse with radii $2\sqrt{\lambda_1}, 2\sqrt{\lambda_2}$ and

[1] Implementation of the CEC algorithm for the Project R—a free software environment for statistical computing and graphics—is available at [6].

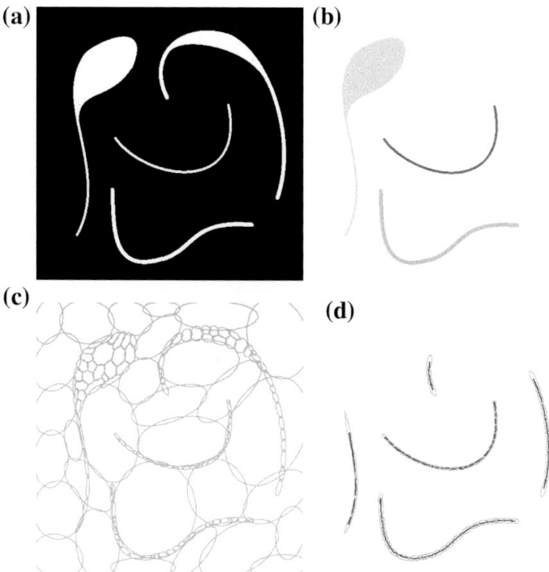

Fig. 3 Detection of elongated structures: **a** sample image, **b** regions of similar characteristic, **c** CEC-based representation, **d** detected structures

the principal directions v_1, v_2, where λ_1, λ_2 are eigenvalues and v_1, v_2 are eigenvectors of the covariance matrix. There are few possible methods for extracting Gaussian-like clusters in the data. In this paper, we use cross-entropy clustering (CEC) [15] algorithm, which is a modification of the classical EM approach. In general, CEC aims to find parameters:

$$p_1, \ldots, p_k \geq 0 : \sum_{i=1}^{k} p_i = 1, \tag{3}$$

and $f_1, \ldots, f_k$ Gaussian densities such that the convex combination:

$$f := \max(p_1 f_1, \ldots, p_k f_k) \tag{4}$$

optimally approximates the scattering of the data under consideration $X = \{x_1, \ldots, x_n\}$. The optimization is taken with respect to cost function:

$$\mathrm{CEC}(f, X) := -\frac{1}{|X|} \sum_{j=1}^{n} \ln\left(\max(p_1 f_1(x_j), \ldots, p_k f_k(x_j))\right), \tag{5}$$

where all p_i for $i = 1, \ldots, k$ satisfy the condition 3. It occurs, see [15], that the above formula implies that it is profitable to reduce some clusters (as each cluster has its cost). Consequently, after the stop of the procedure some parameters p_i for $i \in \{1, \ldots, k\}$ can equal zero, which implies that the clusters they represented have disappeared. Consequently, CEC determines optimal number of clusters, and therefore reduces the complication of the model. Moreover, it is easy to adapt method to the various type of Gaussian model. In particular, we can use diagonal (Gaussians with diagonal covariances) or spherical models (Gaussians with covariances proportional to identity matrix)—compare with [11, 14].

4 Detection of Elongated Structures

CEC algorithm can be used to approximate a given segment of the image with ellipses. However, since usually we are interested in regions of similar characteristic, in this work the first step of changing image representation is to find connected components of the image that have similar color, see Fig. 3b. For that purpose, region growing algorithm is used preceded by median filter to remove noise. Region growing is applied many times as long as there are pixels that are not assigned to any segment. Each execution starts from the brightest unassigned pixel. The similarity criterion accepts points with tolerance 20 of intensity difference between current and seed pixels. Too small regions are removed if necessary. In the next step, CEC algorithm is executed separately for each segment. The algorithm is executed 10 times to find possibly the best value of the cost function 5. It starts with at most $k = 30$ Gaussian functions and the precise number depends on the size of the considered segment (the smaller region the less number of functions is considered). The sample results of this phase are presented in Fig. 3c. The obtained set of ellipses describing whole image is a input for the second phase—elongated structures detection. It constructs first a graph where the ellipses are vertexes and edges indicates which ellipses are close to each other. Thus neighboring ellipses for a given vertex can be easily found. Next, starting from every ellipse its neighbors are examined to check if, in the direction determined by a line connecting centers of those ellipses, their width do not differ too much (in this work the tolerance of 2 pixels was considered). If this condition is satisfied, the procedure is continued recursively comparing the width of successive ellipses with the width of the starting ellipse. The width is always considered in a direction defined by current and previous ellipse. Additionally, starting from the third ellipse, it is also checked if the angle between the lines connecting the last two pairs of ellipses is not too small (it should be larger than 150°). The recursion is stopped if one of the described criteria is not satisfied. As there can be more than one ellipse satisfying those conditions at each level of the recursion as a result for a single ellipse, we can obtain a tree where branches represent a chain of ellipses reflecting elongated structures. When the number of ellipses is huge, it is computationally inefficient to consider all the paths in the graph as some paths will be analyzed many times. An acceptable solution requires remembering which vertex has been already visited. This implies, however, that to obtain the final result additional postprocessing must be performed. It connects some of the paths that starts with the same or neighboring ellipses. The sample results with paths connecting more than 4 ellipses are presented in Fig. 3d.

5 Results

The method proposed in this work was applied to a set of 103 selected images coming from MRI examinations. For all of them the fragment of articular cartilage was manually pointed out by the radiologist, see Fig. 1c, d. Figure 4a, b present the detected connected components of the image. The complexity of the considered images causes that the number of regions is quite large. The CEC-based representation of the images based on those segments is depicted in Fig. 4c, d. Finally, the structures detected with

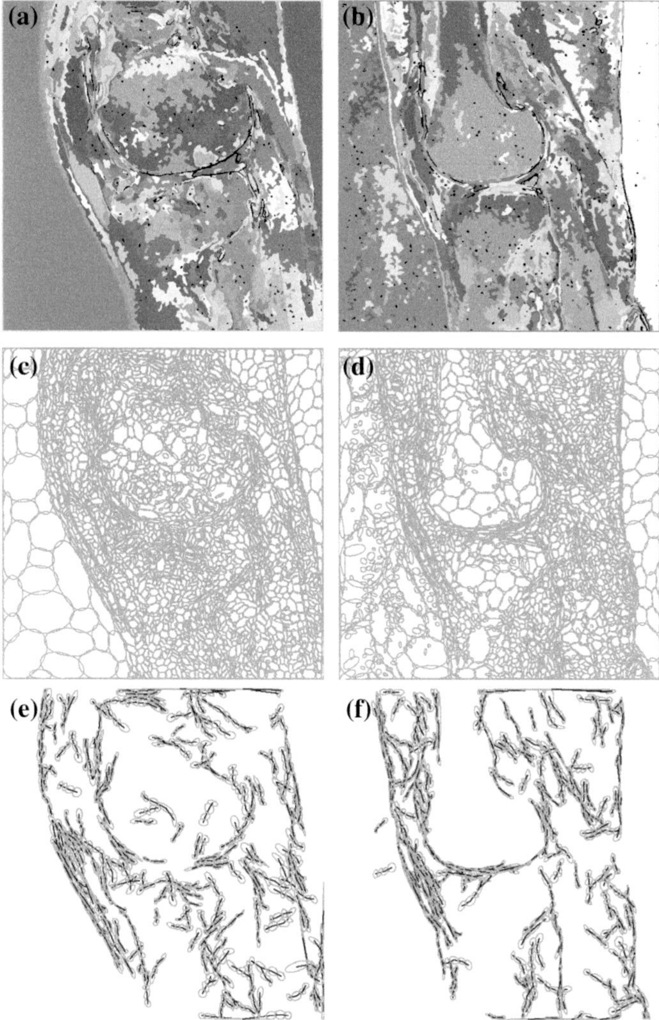

Fig. 4 Sample results for MRI images presented in Fig. 1, **a**, **b** regions of similar characteristics, **c**, **d** CEC-based representations, **e**, **f** detected structures

a discussed method are presented in Fig. 4e, f. Those results reveal that a large number of elongated structures is detected. It is, however, an expected outcome. There are two main reasons explaining this effect. Firstly, the analyzed images contain many structures with the sought characteristic as there are many layers of different tissues of human body visible in selected sections of MRI examinations. Secondly, in this work only geometrical and spatial aspects of structures were considered and no information about color of the region represented by ellipses was taken into account. It can lead to detection of structures lying in regions of different colors. Moreover, it can also be observed that not all the fragments of articular cartilage are always detected. The reason maybe its actual loss or blurred borders between different tissues of the knee. This problem can be overcome only by additional medical knowledge about its precise localization or shape. This would also significantly reduce the number of detections. Such information can be used in the next phase of analysis where from among the elongated structures their subset will be selected. This phase, although it is an element of hierarchical active partition approach, is out of the scope of this work. To measure objectively the effectiveness of the proposed method the regions manually indicated by physician were used. As they only represent a fragment of articular cartilage and there are other elongated structures in the images, we could only check what percentage of those regions was covered be ellipses. Precisely, in the conducted experiments, we checked how many pixels from those regions was covered by a smallest rectangles containing those ellipses. For a considered set of images it was on average 60.6 % which, taking into account that there are usually blurred borders in the center of the fragments in question and no additional, medical knowledge was considered, is a satisfactory result.

6 Summary

In this paper, a method for elongated structures in the images was presented. It utilizes a CEC-based image representation with a set of ellipses covering regions of similar color. The presented approach can be used in hierarchical articular cartilage detection with three phases: detection of ellipses, detection of elongated structures and detection of sought tissue. The last phase was out of the scope of this work but will be investigated in future. What should be emphasized is fact that each phase increases the semantic knowledge about image content reducing the number of details that had to be considered if bare pixel representation was used. Moreover, this approach resembles the human process of such detection as first the regions of different colors are identified, then connected to compose elongated structures, and finally those regions are combined using anatomical knowledge to find the whole articular cartilage even in those areas where information in the image is not sufficient (blurred borders). The conducted experiments revealed that the crucial element of the proposed method is proper extraction of regions with similar colors which will be suitable for images obtained by different MRI devices. The method itself has also a great potential which can be used for other tasks. First of all, CEC allows to obtain image elements of different characteristics: only circles, only flattened ellipses, etc. Second, the method of looking for an optimal subsets of the ellipses can be chosen freely depending on specific tasks. It can not only consider geometrical properties of those ellipses but it can also take under consideration color of the circumscribed region, the neighboring ellipses, and almost any external knowledge. The results presented in this work were not compared to other, mentioned techniques that can be applied to articular cartilage detection as the main goal of the authors was to present an alternative approach to image analysis. Such comparison is also, however, under further investigation.

Acknowledgments This project has been funded with support from the National Science Centre, Republic of Poland, decision number DEC-2012/05/D/ST6/03091.

– The work of Przemysław Spurek was supported by the National Centre of Science (Poland) [grant no. 2013/09/N/ST6/01178].
– The work of Krzysztof Misztal was supported by the National Centre of Science (Poland) [grant no. 2012/07/N/ST6/02192].
– The work of Jacek Tabor was supported by the National Centre of Science (Poland) [grant no. 2014/13/B/ST6/01792].

References

1. Cootes, T., Taylor, C., Cooper, D., Graham, J.: Active shape models - their training and application. CVGIP Image Underst. **61**(1), 38–59 (1994)
2. Davies, E.R.: Machine Vision: Theory, Algorithms, Practicalities. Morgan Kaufmann, San Francisco (2004)
3. Frisby, J., Stone, J.: Seeing: The Computational Approach to Biological Vision. The MIT Press, Cambridge (2010)
4. Gonzalez, R.C., Woods, R.E.: Digital Image Processing. Prentice Hall, Upper Saddle River (2002)
5. Iranpour-Boroujeni, T., Watanabe, A., Bashtar, R., Yoshioka, H., Duryea, J.: Quantification of cartilage loss in local regions of knee joints using semi-automated segmentation software: analysis of longitudinal data from the osteoarthritis initiative (OAI). Osteoarthr. Cartil. **19**(3), 309–314 (2011)
6. Kamieniecki, K., Spurek, P.: CEC: cross-entropy clustering. http://CRAN.R-project.org/package=CEC, (2014), R package version 0.9.2
7. Kass, M., Witkin, A., Terzopoulos, D.: Snakes: active contour models. Int. J. Comput. Vis. **1**, 321–331 (1988)
8. Laberge, M., Baum, T., Virayavanich, W., Nardo, L., Nevitt, M., Lynch, J., McCulloch, C., Link, T.: Obesity increases the prevalence and severity of focal knee abnormalities diagnosed using 3T MRI in middle-aged subjects - data from the osteoarthritis initiative. Skelet. Radiol. **41**(6), 633–641 (2012)
9. Marr, D., Poggio, T., Ullman, S.: Vision: A Computational Investigation into the Human Representation and Processing of Visual Information. The MIT Press, Cambridge (2010)
10. Schneider, E., Nevitt, M., McCulloch, C., Cicuttini, F., Duryea, J., Eckstein, F., Tamez-Pena, J.: Equivalence and precision of knee cartilage morphometry between different segmentation teams, cartilage regions, and MR acquisitions. Osteoarthr. Cartil. **20**(8), 869–879 (2012)
11. Śmieja, M., Tabor, J.: Image segmentation with use of cross-entropy clustering. In: Proceedings of the 8th International Conference on Computer Recognition Systems CORES 2013, pp. 403–409. Springer International Publishing (2013)

12. Sonka, M., Hlavac, V., Boyle, R.: Image Processing, Analysis, and Machine Vision. Cengage Learning, New York (2014)
13. Stehling, C., Liebl, H., Krug, R., Lane, N., Nevitt, M., Lynch, J., McCulloch, C., Link, T.: Patellar cartilage: T2 values and morphologic abnormalities at 3.0-T MR imaging in relation to physical activity in asymptomatic subjects from the osteoarthritis initiative. Radiology **254**(2), 509–520 (2010)
14. Tabor, J., Misztal, K.: Detection of elliptical shapes via cross-entropy clustering. In: Sanches, J.M., Micó, L., Cardoso, J.S. (eds.) Pattern Recognition and Image Analysis, pp. 656–663. Springer, Berlin (2013)
15. Tabor, J., Spurek, P.: Cross-entropy clustering. Pattern Recognit. **47**(9), 3046–3059 (2014)
16. Tomczyk, A., Szczepaniak, P.S., Pryczek, M.: Cognitive hierarchical active partitions in distributed analysis of medical images. J. Ambient Intell. Hum. Comput. **4**(3), 357–367 (2013)
17. Urish, K., Williams, A., Durkin, J., Chu, C.: Registration of magnetic resonance image series for knee articular cartilage analysis: data from the osteoarthritis initiative. Cartilage **4**(1), 20–27 (2013)

Text Detection in Document Images by Machine Learning Algorithms

Darko Zelenika, Janez Povh and Bernard Ženko

Abstract In the proposed paper, we consider a problem of text detection in document images. This problem plays an important role in OCR systems and is a challenging task. In the first step of our proposed text detection approach, we use a self-adjusting bottom-up segmentation algorithm to segment a document image into a set of connected components (CCs). The segmentation algorithm is based on the Sobel edge detection method. In the second step, CCs are described in terms of 27 features and a machine learning algorithm is then used to classify the CCs as text or nontext. For testing the approach, we have collected a dataset (ASTRoID), which contains 500 images of text blocks and 500 images of nontext blocks. We empirically compare performance of the proposed text detection method when using seven different machine learning algorithms.

Keywords Text detection · Document segmentation · Text/nontext classification · Machine learning

1 Introduction

In today's digital age, a lot of useful information is present as text in the form of digital document images such as invoices, business letters, web pages, etc. In order to effectively recognize this text with Optical Character Recognition (OCR)

D. Zelenika (✉) · J. Povh
Laboratory of Data Technologies, Faculty of Information Studies,
Ulica Talcev 3, 8000 Novo Mesto, Slovenia
e-mail: darko.zelenika@fis.unm.si

J. Povh
e-mail: janez.povh@fis.unm.si

B. Ženko
Jožef Stefan Institute, Department of Knowledge Technologies,
Jamova Cesta 39, 1000 Ljubljana, Slovenia
e-mail: bernard.zenko@ijs.si

R. Burduk et al. (eds.), *Proceedings of the 9th International Conference on Computer Recognition Systems CORES 2015*, Advances in Intelligent Systems and Computing 403, DOI 10.1007/978-3-319-26227-7_16

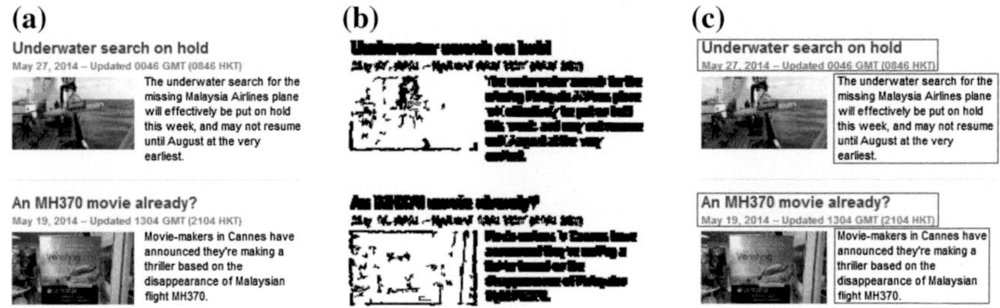

Fig. 1 **a** Image document, **b** segmented image and **c** detected text

technology, location of the text must be detected first. The first step of text detection in document images involves segmentation, which is followed by classification. Document segmentation is a task which splits a document image into blocks of interest, as shown in Fig. 1a where each connected component (CC) of black pixels represents one block. Blocks of interest usually appear in two forms: text and nontext. In this paper, we are mainly interested in text blocks, so our goal is to identify them and separate from nontext blocks (see Fig. 1c). It is important to understand that document segmentation and classification (identification) of segmented blocks can hardly be separated and are often treated together as "(physical) layout analysis" [1]. These tasks continue to be very challenging, especially for documents which have multicolored and complex background. Text in such documents may be of different sizes, orientations, colors, etc. Most of the currently available document segmentation algorithms require some predefined parameters due to different font sizes, layouts, and document image resolutions. Hence, robust and efficient techniques for document segmentation are required.

1.1 Text Detection Method

The purpose of this paper is to introduce and report results of a new text detection method. Our method consists of document segmentation and classification algorithms. The segmentation algorithm is based on bottom-up approach, which is obtained with edge detection methods. We created a custom dataset of text and nontext image blocks, called ASTRoID. From each image block we extracted 27 features, which are used by a machine learning algorithm to separate text from nontext image blocks; we test seven different machine learning algorithms. This work can be seen as an important step toward a parameter-free document segmentation and accurate text detection in complex text documents.

1.2 Document Segmentation Techniques

Document segmentation techniques are traditionally classified into three categories: top-down, bottom-up, and hybrid approaches. The top-down approach starts the segmentation from bigger blocks and repetitively segments the document image into smaller blocks until the document image is segmented into the smallest possible blocks. Top-down algorithms are usually fast but they tend to fail in segmenting documents with very complex layouts [2]; typical examples are X-Y Cuts [2–4], White streams [5], Run-length smearing [6], and other algorithms based on projection profile method [7, 8]. Kruatrachue et. al [4] used X-Y Cuts to build a fast segmentation method, but it is not well suited for complex documents because it is based on binarization. The bottom-up approach is the opposite of the top-down approach. It starts from the smallest segments (characters) and then joins them into bigger and bigger blocks (words, paragraphs, etc.). Algorithms based on bottom-up approach are flexible and robust but often slow because of time-consuming operations, some of these algorithms are based on the analysis of CCs [9–11] and some on morphological operations [12, 13]. An approach that does not fit into a top-down or bottom-up strategy or uses the combination of both is called a hybrid approach [14, 15]. Hybrid approaches often try to combine the speed of top-down approaches and the robustness of bottom-up approaches. In [15] authors used an adaptation of Scale Invariant Feature Transform (SIFT) approach for text character spotting in graphical documents. Their method uses a combination of bottom-up and top-down approaches to separate and locate text characters.

1.3 Classification

The task of classification algorithm is to classify the results of segmentation algorithm. Classification highly depends on the quality of segmentation algorithm. In [2] authors proposed a method to extract illustrations from digitized historical documents by using Support Vector Machine (SVM). Priyadharshini and Vijaya in [6] proposed a document block classification approach, which classifies the document blocks as text, image, drawing, and table. In their approach, a genetic programming-based classifier is used to classify document blocks. In order to detect text and nontext blocks, authors in [5, 9–11] extracted features from CCs in document images and classified them with machine learning algorithms. In this paper, the proposed text detection method uses the similar approach.

1.4 Contribution

The main contributions of this paper are: (1) a self-adjusting segmentation algorithm for finding text CCs that is independent of the image resolution and font size, (2) a new set of features which describe differences between text and nontext image blocks based on information about their shape and context, (3) a custom benchmark dataset ASTRoID of text and nontext image blocks, and (4) demonstration of performance of seven machine learning algorithms for separation between text and nontext image blocks. The rest of the paper is organized as follows. In section two, we introduce document segmentation algorithm, ASTRoID dataset and classification algorithm. The performance of our approach for five different machine learning algorithms is presented in section three. Obtained results are discussed in section four. Finally, section five concludes the paper.

2 Materials and Methods

Here we describe our text detection method, which performs two tasks: document image segmentation and classification. The text detection method extracts features from the segmented blocks (results of the segmentation algorithm), which are then classified with a machine learning algorithm as either text or nontext blocks.

2.1 Segmentation

The segmentation algorithm described in this paper follows the bottom-up strategy. The algorithm segments document into small CCs, which are constructed with a combination of Sobel edge detection and dilation method [1, 16]. The proposed document image segmentation algorithm is composed of three parts: (1) finding an optimal rectangular kernel, (2) edge detection and (3) extraction of standalone document image objects. Before the segmentation process begins, the document image needs to be converted to grayscale (Fig. 2a). The segmentation algorithm receives the grayscale image and outputs the binary image. The first part of the segmentation algorithm tries to find an optimal rectangular kernel, which is then used by the other two parts of the algorithm. The optimal rectangular kernel highly depends on the height of the dominant text in the document image, which we call height of the main text ($h_{mainText}$). In order to find the height of the main text, we applied the Sobel edge detection algorithm over the grayscale document image (Fig. 2b). The height of the main text is obtained from the heights of CCs, which are extracted from the binary image (Fig. 2b). After the height is obtained, we calculate the number of columns of optimal rectangular kernel based on the following *equation*:

$$N_{col} = \frac{h_{mainText}}{4} \tag{1}$$

The result of Eq. 1 is rounded and the optimal rectangular kernel is obtained (Eq. 2), which is a matrix of size $2XN_{col}$ containing only $1's$, which is going to be used to better emphasize textual objects on the document image.

$$K_{opt} = \begin{bmatrix} 1 & 1 & \dots & 1 \\ 1 & 1 & \dots & 1 \end{bmatrix} \tag{2}$$

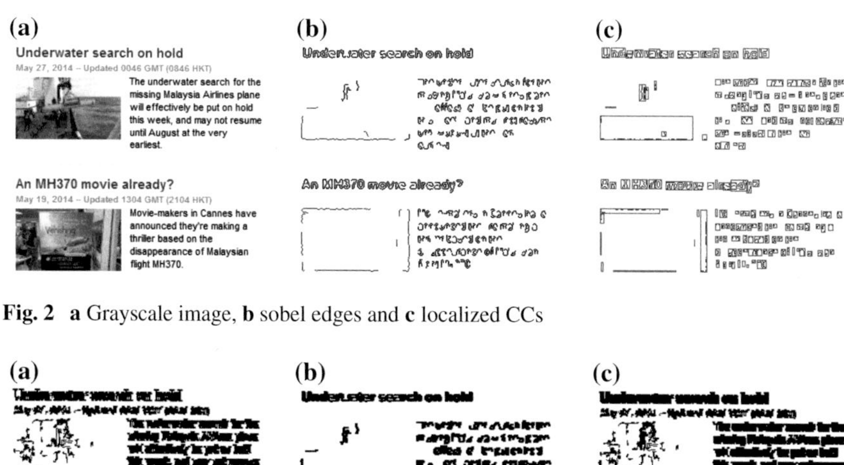

Fig. 2 **a** Grayscale image, **b** sobel edges and **c** localized CCs

Fig. 3 Segmentation steps, **a** 2nd part, **b** 3rd part and **c** final part

In the second part of the segmentation algorithm we apply different Sobel kernels (vertical, horizontal and diagonal) on the grayscale document image. Text blocks are composed of vertical, horizontal, and diagonal edges, and accordingly, in this part of the segmentation algorithm we used only a combination of vertical and diagonal Sobel kernels of different orientations to better emphasize text blocks. The result binary images of different Sobel kernels are dilated by the optimal rectangular kernel and combined into one image by logical *AND* and *OR* operations, as shown in Fig. 3a. In the third part of the segmentation algorithm, we again use the image with detected Sobel edges from Fig. 2b. We localize all CCs (red rectangles in Fig. 2c) on this image and keep only those CCs that intersect with two or less than two other CCs (due to characters such as "B" and "8") and we call them standalone CCs. In such a way most of CCs which do not belong to text blocks are removed. The result image of the third part of segmentation algorithm is binary image with standalone CCs, which is dilated by the optimal rectangular kernel (see Fig. 3b). The final part of the segmentation algorithm is to combine the obtained binary images from the second and third parts of the algorithm with logical OR operation (see Fig. 3c). In such a way, we segmented the document image into blocks that can either be text blocks or nontext blocks.

2.2 *Text Versus Nontext Classification*

2.2.1 ASTRoID Dataset

We created our custom dataset of text and nontext image blocks, which we called ASTRoID, in order to evaluate our method. We were choosing at random articles from ten web portals (one article per portal), took screenshots of full web pages and saved them as PNG (Portable Network Graphics) image files. These ten web portals were: abcnews.go.com, cnn.com, nationalgeographic.com, pcmag.com, telegraph.co.uk, racunalniske-novice.com, radio1.si, slovenskenovice.si, bljesak.info and dnevnik.hr, they include five portals in English language, three in Slovenian language, and two in Croatian language. We used the proposed segmentation algorithm over all ten article images to extract all CCs and save them as PNG image files. For our purposes, we manually chose 500 of image blocks which contain plain text of different size, length, color, font style, etc., and 500 image blocks of different size which do not contain text. It is important to state that our dataset has 150 image blocks of text that contain only one or two characters, which some avoid to use in order to get better classification results. The ASTRoID dataset is available to download at: http://dk.fis.unm.si/ASTRoID.zip.

2.2.2 Feature Extraction

In document images, most of the text blocks are uniformly structured and have a regular shape. On the other hand, nontext blocks have a lot of variability in shape, i.e., mostly they have an irregular shape. But only shape information is not enough to classify text from nontext blocks, we also need to take into account the information on the context of these blocks. Therefore, we need to create a set of features that describe the context and are able to improve the accuracy of distinguishing text from nontext blocks. In this paper, we used features similar to the ones proposed in [5, 6, 9, 10]. We took different approach to calculate some of the proposed features and introduced a new feature "color density". In our approach, we extract features from multiple images of the same segmented block (Fig. 4), which are obtained by different methods, unlike to the approaches found in the literature where features are extracted from usually one binary image. Before the actual feature extraction each segmented block is resized to 100 pixels in height while maintaining the width to height aspect ratio. In our approach we used the following features: number of CCs [5], aspect ratio [5, 6, 9], foreground density [5, 6, 9, 10], color density, standard deviation of the heights and widths of CCs [6, 9, 10], and standard deviation of the lengths of horizontal and vertical runs [6]. Most of the features are extracted from the different binary images (Fig. 4b–h), which are obtained with the following methods: skeletonization (Fig. 4d) [18], horizontal Sobel kernel (Fig. 4e), vertical Sobel kernel (Fig. 4f), diagonal Sobel kernels (Fig. 4g), and Canny edge detection method (Fig. 4h) [16, 19]. Number of CCs is computed after binarization of grayscale image by using

Otsu's [17] thresholding algorithm (Fig. 4b). Aspect ratio (A_r in Eq. 3) is defined as the ratio of a block's width-to-height if height is greater than width or height-to-width if width is greater than height.

$$A_r = \frac{w}{h} \; or \; A_r = \frac{h}{w} \tag{3}$$

The feature foreground density (D_f in Eq. 4) is defined as the ratio of the number of the foreground (black) pixels to the total number of pixels in the binary image, and is calculated two times, i.e., once from each of the following binary figures: Fig. 4b, h.

$$D_f = \frac{N_f}{N} \tag{4}$$

In order to determine the feature color density, we first extract the most frequent colors in the color image (Fig. 4a). The color density (D_c in Eq. 5) is defined as the ratio of the number of extracted colors to the total number of colors in the color image. We created binary image based on the location (coordinates) of extracted colors (Fig. 4c), which will be used for extraction of other features, by filling coordinates of extracted colors in new binary image with white pixel (background) and all the remaining coordinates with black pixel (foreground).

$$D_c = \frac{N_{extractedC}}{N_{totalC}} \tag{5}$$

Based on the extracted CCs from the binary images, we use heights and widths of each CC to calculate the features standard deviation of the heights and widths of CCs. Standard deviation of the heights of CCs is calculated eight times, i.e., once for each of the following figures: Fig. 4b–d, f–h, and for the Sobel vertical lines which are extracted from the Fig. 4c, h. Standard deviation of the widths of CCs is calculated two times, i.e., once for each of the following figures: Fig. 4e and for the Sobel horizontal lines which are extracted from the Fig. 4h. Also the binary images are used to calculate the features standard deviation of the lengths of the vertical and

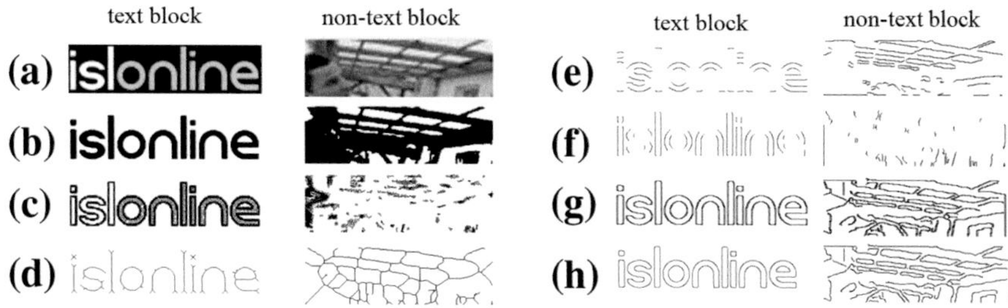

Fig. 4 **a** Color image, **b** binary image, **c** image of extracted colors, **d** skeleton image, **e** horizontal Sobel, **f** vertical Sobel, **g** diagonal Sobel and **h** canny

horizontal runs of black (foreground) pixels. The extraction of lengths of horizontal runs can be explained by using the following matrix [00110011110011100] where 0 is white and 1 black pixel. The lengths of horizontal runs of black pixels are 2, 4, and 3. Standard deviation of the vertical runs is calculated nine times, i.e., once for each of the following figures: Fig. 4b–d, f–h, and for the Sobel vertical lines which are extracted from the Fig. 4c, g, h. Standard deviation of the horizontal runs is calculated three times, i.e., once for each of the following figures: Fig. 4b, e, and for the Sobel horizontal lines which are extracted from the Fig. 4h. And finally, by using features described above each image block (text or nontext) is represented with the feature vector which consists of 27 features in total.

2.2.3 Classification with Machine Learning

As stated above, based on the proposed segmentation algorithm we created ASTRoID dataset which contains 500 images of text blocks and 500 images of nontext blocks. We extracted 27 features (described in previous section) from each image block of the dataset and appointed class label (text or nontext) to them. In this way we created a dataset, which we used for classification with seven popular machine learning algorithms that are frequently used in practical applications and typically give good results, in order to evaluate our choice of our features, and to find out which machine learning algorithm works best with our text detection method. We used: Naïve Bayes, C4.5 (decision tree), k-Nearest Neighbors (k-NN), Random Forest, Linear Support Vector Machine (SVM), Polynomial SVM, and Radial SVM. The accuracy of each of the above-mentioned algorithms is estimated by tenfold cross-validation. We tuned the parameters of some of the machine learning algorithms with an internal cross-validation. We used the implementations of machine learning algorithms in the WEKA data mining suite [20]. Naïve Bayes and C4.5 (decision tree) algorithms are used with default parameters while parameter k for the k-NN algorithm is tuned to 1 and parameter number of trees for the Random forest algorithm is tuned to 100. For SVMs, we normalized data by tuning normalize parameter. The parameter C for Linear SVM is tuned to 15. The parameters C, *degree*, *gamma* and *coefficient* for Polynomial SVM are tuned to 20, 3, 1, and 1, respectively. The parameters C and *gamma* for Radial SVM are tuned to 20 and 1, respectively.

3 Results

The classification results of all machine learning algorithms are shown in Table 1. The accuracy of all machine learning algorithms is higher than 90 % which suggests that the choice of our features for text/nontext differentiation is appropriate. The classification results show that Random forest and SVMs perform best for our text detection method. The machine learning algorithm that has the highest accuracy 98.2 % is Radial SVM.

Table 1 Classification results

Classifier	Naïve Bayes	k-NN	C4.5 - decision tree	Random forest	Lin. SVM	Poly. SVM	Rad. SVM
Accuracy (%)	90.1	93.6	94.3	97	97	97.3	98.2
Precision of text blocks	0.860	0.911	0.937	0.976	0.964	0.961	0.978
Precision of nontext blocks	0.953	0.964	0.949	0.964	0.976	0.986	0.986
Recall of text blocks	0.958	0.966	0.950	0.964	0.976	0.986	0.986
Recall of nontext blocks	0.844	0.906	0.936	0.976	0.964	0.960	0.978

4 Discussion

The results obtained by the proposed text detection method are promising. The chosen set of features, by which machine learning algorithms can separate text from nontext image blocks with good accuracy seems appropriate. Other authors who worked on similar problems also obtained comparable classification results. In [11] authors used SVM and classified text from nontext blocks with the accuracy of 96.62 %. In [6] authors obtained 97.5 % classification accuracy by using genetic programming to classify the document blocks as text, image, drawing, and table. In [9] authors used Multilayer Perceptron to classify text from nontext blocks and obtained 97.25 % of accuracy. The only disadvantage of approaches in [6, 9, 11] is that they fail to detect text on documents with complex layout, due to their segmentation algorithm. The advantage of our segmentation algorithm is that it self-adjusts to the document image regardless the image resolution and font size, it does not need any input parameters and it is able to segment documents with complex layout. The only disadvantage is that it fails to detect (segment) some text blocks of very light text color, and also some text blocks with very complex (with a lot of details) background and very decorated text strings. In the future, we plan to improve our method in order to avoid current disadvantages by using Canny together with the Sobel edge detection method. We also plan to increase the size of our ASTRoID dataset and test our text detection method on other datasets and compare it with other methods.

5 Conclusion

In this paper, we presented a text detection method, which consists of document segmentation and feature extraction algorithms. The proposed segmentation algorithm is based on the bottom-up strategy of analysis and segmentation is done by using the Sobel edge detection method. We created ASTRoID dataset of images, which consists of 500 image blocks of text and 500 image blocks of nontext. It is important to state that the dataset has 150 text blocks that contain either one or two characters, which some avoid to use to get better classification results. In order to classify text from nontext blocks, we used seven machine learning algorithms. Before classification, we first extracted features from image blocks, which are based on information about their shape and context. We used 27 different features in order to differentiate text from nontext regions. The accuracy of all machine learning algorithms is higher than 90 % which suggests that the choice of our features is appropriate. The classification results show that Random forest and SVMs are the best choices for our text detection method. SVM with radial kernel has the highest accuracy 98.2 %.

Acknowledgments The presented work was supported by Creative Core FISNM-3330-13-500033 'Simulations' project funded by the European Union, The European Regional Development Fund. The operation is carried out within the framework of the Operational Programme for Strengthening Regional Development Potentials for the period 2007–2013, Development Priority 1: Competitiveness and research excellence, Priority Guideline 1.1: Improving the competitive skills and research excellence.

References

1. Kise, K.: Page Segmentation Techniques in Document Analysis. Handbook of Document Image Processing and Recognition, pp. 135–175. Springer, London (2014)
2. Coppi, D., Grana, C., Cucchiara, R.: Illustrations segmentation in digitized documents using local correlation features. In: 10th Italian Research Conference on Digital Libraries, vol. 38, pp. 76–83. Procedia Computer Science, Padua (2014)
3. Shafait, F., Keysers, D., Breuel, T.: Performance evaluation and benchmarking of six-page segmentation algorithms. In: IEEE Transactions on Pattern Analysis and Machine Intelligence, pp. 941–954. IEEE Press (2008)
4. Kruatrachue, B., Moongfangklang, N., Siriboon, K.: Fast document segmentation using contour and X-Y cut technique. In: The Third World Enformatika Conference, WEC vol. 5, pp. 27–29. Turkey (2005)
5. Barlas, P., Kasar, T., Adams, S., Chatelain, C., Paquet, T.: A typed and handwritten text block segmentation system for heterogeneous and complex documents. In: 11th IAPR International Workshop on Document Analysis Systems, pp. 46–50, IEEE Press, Tours (2014)
6. Priyadharshini, N., Vijaya, M.S.: Genetic programming for document segmentation and region classification using discipulus. Int. J. Adv. Res. Artif. Intell. **2**, 15–22 (2013)
7. Priyanka, N., Pal, S., Mandal, R.: Line and word segmentation approach for printed documents. Int. J. Comput. Appl. **1**, 30–36 (2010)
8. Vikas, J.D., Vijay, H.M.: Devnagari document segmentation using histogram approach. Int. J. Comput. Sci. Eng. Inf. Tech. **1**, 46–53 (2011)
9. Bukhari, S.S., Azawi, M.A., Shafait, F., Breuel, T.M.: Document image segmentation using discriminative learning over connected components. In: 9th IAPR International Workshop on Document Analysis Systems, pp. 183–190. Boston (2010)
10. Bukhari, S.S., Asi, A., Breuel, T.M., El-Sana, J.: Layout analysis for arabic historical document images using machine learning. In: International Conference on Frontiers in Handwriting Recognition, pp. 639–644 (2012)
11. Zagoris, K., Chatzichristofis, S.A., Papamarkos, N.: Text Localization using standard deviation analysis of structure elements and support vector machines. EURASIP J. Adv. Sign. Process. **47**, 1–2 (2011)

12. Bukhari, S.S., Shafait, F., Breuel, T.M.: Improved document image segmentation algorithm using multiresolution morphology. In: 18th Document Recognition and Retrieval Conference, pp. 1–10. San Jose (2011)
13. Sumathi, C.P., Priya, N.: A combined edge-based text region extraction from document images. Int. J. Adv. Res. Comput. Sci. Softw. Eng. **3**, 827–835 (2013)
14. Kundu, M.K., Dhar, S., Banerjee, M.: A new approach for segmentation of image and text in natural and commercial color document. In: Proceedings of International Conference on Communication, Devices and Intelligent Systems, pp. 85–88. IEEE Press, India (2012)
15. Roy, P.P., Pal, U., Lladós, J.: Touching text character localization in graphical documents using SIFT. In: Proceedings of the 8th International Conference on Graphics Recognition: Achievements, Challenges, and Evolution, pp. 199–211. Springer, France (2010)
16. Vasuki, S., Ganesan, L.: Performance measure for edge based color image segmentation in color spaces. In: Proceedings of the International Conference on Emerging Technologies in Intelligent System and Control: Exploring, Exposing, and Experiencing the Emerging Technologies, pp. 621–626. Allied Publishers, Coimbatore (2005)
17. Otsu, N.: A threshold selection method from gray-level histograms. IEEE Trans. Syst. Man Cybern. **9**, 62–66 (1979)
18. Basilis, G.G.: Imaging Techniques in Document Analysis Processes. Handbook of Document Image Processing and Recognition. Springer, London (2014)
19. Burger, W., Burge, M.J.: Principles of Digital Image Processing. Springer, London (2009)
20. WEKA (Open source, Data Mining software in Java), University of Waikato, New Zealand. http://www.cs.waikato.ac.nz/ml/weka

Blind Source Separation for Improved Load Forecasting on Individual Household Level

Krzysztof Gajowniczek, Tomasz Ząbkowski and Ryszard Szupiluk

Abstract This paper presents the improved method for 24 h ahead load forecasting applied to individual household data from a smart metering system. In this approach we decompose a set of individual forecasts into basis latent components with destructive or constructive impact on the prediction. The main research problem in such model aggregation is the proper identification of destructive components that can be treated as some noise factors. To assess the randomness of signals and thus their similarity to the noise, we used a new variability measure that helps to compare decomposed signals with some typical noise models. The experiments performed on individual household electricity consumption data with blind separation algorithms contributed to forecasts improvements.

1 Introduction

Smart metering systems are expected to improve the way in which information about the electricity we use is collected and communicated [1, 2]. The primary goal is to encourage users to use less electricity through being better informed about their consumption patterns. Forecasting the usage provides individual customers with the means to link current usage behavior with future costs. Therefore, customers may benefit from forecasting solutions through greater understanding of their own energy consumption and future projections, allowing them to better manage costs of their usage. Load forecasting on the individual household level is a challenging task due to

---

K. Gajowniczek (✉)
Department of Informatics, Warsaw University of Life Sciences, Warsaw, Poland
e-mail: krzysztof_gajowniczek@sggw.pl

T. Ząbkowski
Warsaw School of Economics, Warsaw, Poland
e-mail: tomasz_zabkowski@sggw.pl

R. Szupiluk
Systems Research Institute, Polish Academy of Sciences, Warsaw, Poland
e-mail: rszupi@sgh.waw.pl

R. Burduk et al. (eds.), *Proceedings of the 9th International Conference on Computer Recognition Systems CORES 2015*, Advances in Intelligent Systems and Computing 403, DOI 10.1007/978-3-319-26227-7_17

the high volatility which is a result of many dynamic processes such as the operational characteristics of devices, user behaviors, economic factors, time of the day, day of the week, holidays, weather conditions, geographic patterns, and random effects. For this reason, time series forecasting methods are not effective in highly volatile data [3]. This paper presents a different approach. The individual household data were first grouped into segments of similar usage characteristics. Then, one neural network model to forecast a 24 h ahead electricity usage at each household was constructed. Next, the individual forecasts for each household were integrated and decomposed using blind source separation approach (BSS). The BSS system decomposes the original forecasts into a set of independent components. Some of these basic latent components represent essential information and some represent noise. Therefore, identification and elimination of noise should improve the final forecasts. Thanks to such processing the accuracy of individual forecasts may be significantly improved.

2 The General System for BSS Aggregation

The proposed method for load forecasting integrates different prognoses into one forecasting system. The results of each individual household prognosis generated by the model for the period used in training is delivered to the separation (BSS) system. The number of inputs to BSS is equal to the number N of the applied prognosis. The BSS system decomposes the original set of forecasts (signals) of length p, forming the matrix $\mathbf{X} \in \mathbb{R}^{N \times p}$ (p is the number of prognosis hours used in learning), into independent components using the matrix $\mathbf{W} \in \mathbb{R}^{N \times N}$. The independent component signals, generated by BSS, form the matrix $\mathbf{Y}$ of N rows and p columns. This transformation is described by $\mathbf{Y} = \mathbf{WX}$. Each row of matrix $\mathbf{Y}$ represents the independent component signals. Some of these signals represent essential information and some represent noise. Reconstructing the original time series back into real prognosis, on the basis of essential independent components only, will provide the prognosis without the noise which is possibly of better quality. Unfortunately, we do not know in advance which of the components are noise and which represents useful information. We solved the problem by assessing the randomness of signals and thus their similarity to the noise with new variability measures that help to compare the decomposed signals with typical noise models. In the data analysis practice, random signal or noise is observed when the present values give no precise information about the future values. The most popular example of a noise model is the white noise [4]. It is very convenient if the analyzed model or data includes white noise. In practice, the situation is more complex. There are colored noises with internal dependencies or mixtures of the random noises and deterministic signals. We expect that random signal should not include any predictable patterns (internal dependencies, correlations, trends) and should not be smooth. Therefore, for signals with a temporal structure we propose the following measures. Let us consider signal y with temporal structure and observations indexed by $k = 1, \ldots, N$. The variability (and thus unpredictability of the signal) might be measured with the following formula [5]:

$$Q(y) = \frac{\frac{1}{N-1}\sum_{k=2}^{N} y(k) - y(k-1)}{\rho(\max(y) - \min(y))}, \qquad (1)$$

where p symbol means unit indicator and it is introduced to avoid dividing by zero. The possible values of measure (1) range from 1 to 0. The measure has a simple interpretation: it is maximal when the changes in each step are equal to range (maximal change), and is minimal when data are constant. In both cases the signal is totally predictable, but between these marginal states the signal is random. After noise identification, reconstruction on the basis of essential components of the original data matrix $\mathbf{X}$ is done using the inverse operation $\hat{\mathbf{X}} = \mathbf{X}^{-1}\hat{\mathbf{Y}}$ where $\hat{\mathbf{X}}$ denotes the reconstructed time series matrix and $\hat{\mathbf{Y}}$ is the independent component matrix formed from the original matrix $\mathbf{Y}$ by omitting some row or rows. In recovering the signals we substitute the rejected components (appropriate rows of $\mathbf{Y}$) by zeros. The noise component removed and the resulting best prediction on the learning data is assumed to be the final solution. There are numerous BSS decomposition algorithms that can be effectively used for basis component estimation. In this paper, for numerical experiments we applied independent component analysis (ICA), which is a statistical tool that allows us to decompose an observed variable into independent components [6–9]. After ICA decomposition we obtain signals (variables) without any linear and nonlinear statistical dependencies.

3 Smart Metering Data

The data were obtained from Pecan Street Inc. via the WikiEnergy project [10]. The investigated households were located in Austin, Texas, USA. The dataset contains data from 61 homes, in which the household aggregate power demand are monitored at 1 h intervals over 7 months from December 2012 to July 2013. From these data we extracted six households which fitted into the segment of similar usage characteristics. There were 5075 observations available in total. The usage characteristics of one randomly chosen week at the analyzed households are shown in Fig. 1. In the research, we focused on forecasting the electricity usage of a particular household for 24 h ahead. In order to forecast the load we constructed a feature vector with attributes as presented in Table 1. These 51 attributes were empirically derived and they are the result of previous studies [11].

The individual, the average, the minimum, the maximum, the range loads information, and load of each hours of previous week were obtained from the hourly load time series. Each day was divided into five periods, namely morning, noon, afternoon, evening, and night. Seasonal indicator was derived for a given latitude of the investigated area. Moreover, holiday indicator was prepared to mark the holidays in USA.

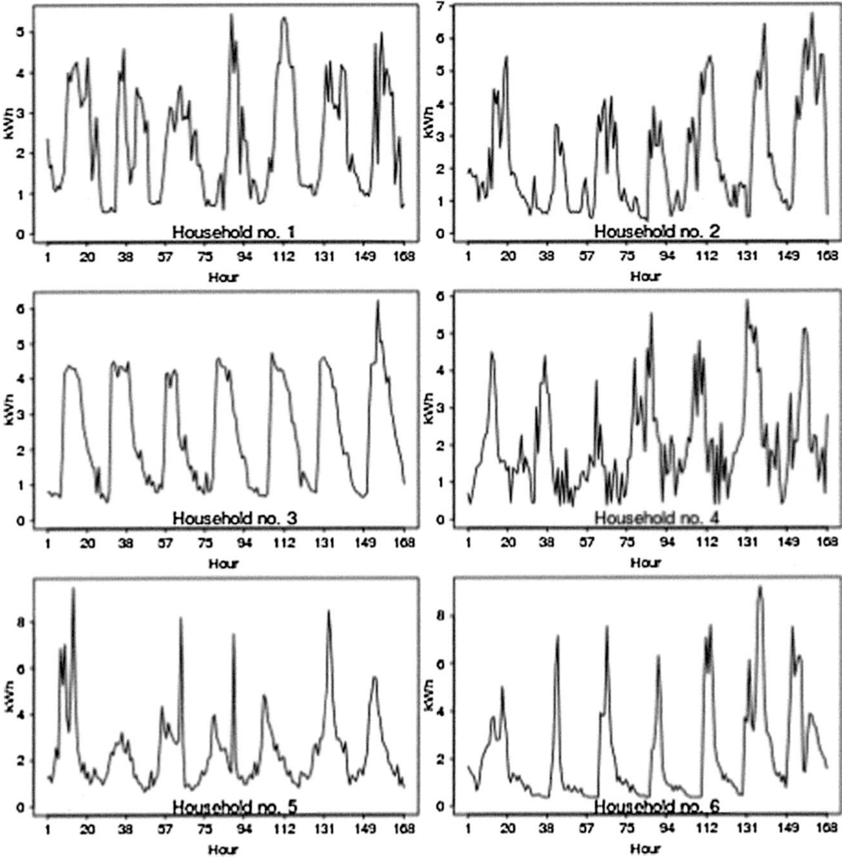

Fig. 1 Electricity usage characteristics at analyzed households

Table 1 Feature vector describing previous electricity usage

Attribute no.	Description	Formula
1–24	Load of previous 24 h	$L_{hi}, L_{h-1}, L_{h-24}$
25–28	Average load of previous 3, 6, 12, 24 h	$\frac{1}{i}\sum L_{hi}, i = 3, 6, 12, 24$
29–32	Maximum load of previous 3, 6, 12, 24 h	$\max\{L_{hi}\}, i = 3, 6, 12, 24$
33–36	Minimum load of previous 3, 6, 12, 24 h	$\min\{L_{hi}\}, i = 3, 6, 12, 24$
37–40	Range of load of previous 3, 6, 12, 24 h	$range\{L_{hi}\}, i = 3, 6, 12, 24$
41–47	Load of each hours of previous 7 days	$L_{hdi}, L_{hd-1}, L_{hd-7}$
48	Day of the week	D_w
49	Part of the day	P_d
50	Holiday indicator	H
51	Season indicator	S

4 The Experiments Design

4.1 The General Idea of the System

The individual household data were first grouped into segments of similar usage characteristics using Wards method [12] for clustering. Then, one neural network model to forecast a 24 h ahead electricity usage was constructed. The model was able to provide the forecast for each of the households. Next, the individual forecasts for each household were integrated and decomposed using blind source separation approach (BSS). The BSS system decomposes the original forecasts into set of independent components. Next, the components were classified into essential ones (constructive) and these were similar to noise according to the measure proposed in formula (1). We expect the identification and elimination of noise should improve the final forecasts. Figure 2 shows the graphical illustration of the proposed system.

4.2 Accuracy Measures

To assess the model performance for forecasting, we used two measures: MAPE (Mean Absolute Percentage Error) and resistant MAPE error. The MAPE error is defined as follows:

$$MAPE = \frac{1}{N}\sum_{i=1}^{N}\left|\frac{L_i - P_i}{L_i}\right|, \tag{2}$$

where L_i is the observed load in hour i, and P_i is the forecasted load in hour i. According to [13], any summary measure of error must meet five basic criteria: measurement validity, reliability, ease of interpretation, clarity of presentation, and

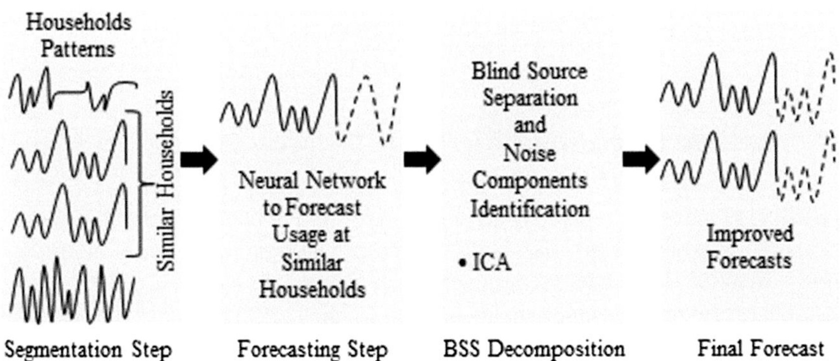

Fig. 2 The general scheme of the proposed solution

support of statistical evaluation. In attempt to meet these criteria, as the summary measure of forecast a MAPE (mean absolute percentage error) error is most often used. However, it does not meet the validity criterion due to the fact that the distribution of the absolute percentage errors is usually skewed to the right, with the presence of outlier values. In these cases, MAPE can be highly overinfluenced by some very bad instances and can overshadow quite good forecasts. In this paper, we also use an alternative index, called resistant MAPE or r-MAPE based on the calculation of the Huber M-estimator, which helps to overcome the aforementioned limitation [14]. An M-estimator for the location parameter μ using maximum likelihood (ML)-estimator is defined as a solution θ to

$$\sum_{i=1}^{N} \rho\left(\frac{\left|\frac{L_i-P_i}{L_i}\right| - \theta}{\sigma}\right) = \min_{\theta}, \tag{3}$$

or

$$\sum_{i=1}^{N} \varphi\left(\frac{\left|\frac{L_i-P_i}{L_i}\right| - \theta}{\sigma}\right) = 0, \tag{4}$$

where $\varphi = \rho'$ and σ is the scale parameter. For a given positive constant k, the Huber [15] estimator is defined by the following function in φ (4):

$$\varphi(x) = \begin{cases} k & \text{dla } x > k \\ x & \text{dla } -k \leq x \leq k \\ -k & \text{dla } x < k \end{cases} \tag{5}$$

where k is a tuning constant determining the degree of robustness set at 2.0. The above function is known as metric Winsorizing and brings in extreme observations to $\mu \pm k$. In reality σ is not known, thus a MAD robust estimator was used:

$$MAD = median(|x_i - median(x_i)|). \tag{6}$$

4.3 Model to Forecast the Electricity Usage

The calculations were prepared in an R software [16]. For each household in cluster, one model was built using three-layer backpropagation neural network algorithm. As loss function, least squares estimators were chosen. In the most general terms, least squares estimation is minimizing the sum of squared deviations of the observed values for the dependent variables from those forecasted by the model. Technically, the least squares estimator is obtained by minimizing the SOS (sum of squares) function:

$$SOS = \sum_{i=1}^{N} (L_i - P_i)^2, \tag{7}$$

where L_i is the observed load in hour i, and P_i is the forecasted load in hour i. Before estimating the artificial neural network, we randomly selected two samples. The training set was used to estimate the model, while the testing set was used to validate the model for better generalization. The calibration sample included 90 % of the observations (188 days) and the test sample included 10 % of the observations (23 days). In the experiment we tried several neural network structures for each hour to get the best result. The number of neurons in the hidden layer was proposed as a result of numerical procedure [17]. We started neural network learning with a small number of hidden units and then successively we increased the number of neurons until no significant improvement in terms of models performance was observed. As a result we used a neural network that consists of one hidden layer. The input layer consisted of 51 perceptrons (equal to the number of independent variables), hidden layer consisted of 20 perceptrons, and finally, the output layer consisted of one perceptron. All the perceptrons were activated by logistic function. For training neural networks we used the BFGS (Broyden–Fletcher–Goldfarb–Shanno) algorithm [18], which belongs to the broad family of quasi-Newton optimization methods. This method performs significantly better than traditional algorithms such as gradient descent, but it is more memory and computationally demanding.

5 The Results of BSS Improvement

In this section we present the effectiveness of the presented aggregation method for individual household data forecasting. The blind source separation system decomposes the streams of N input signals into N independent components. The basic assumption is that the input signals are the mixtures of some unknown basic original sources that are to be recovered by the separation algorithm. Based on the Q values calculated for each component we decided to substitute the third component by zeros. Next, the reconstruction taking into account only the essential components of the original data matrix was performed by using the inverse operation to decomposition. As a result, for each household we obtained statistically significant improved forecasts with lower error rates in comparison to primary models, as presented in Table 2. In general, the ICA decomposition improved the quality of forecasts as against the primary model. The highest improvement rate in terms of MAPE was observed for household No. 6. The error was reduced by nearly 25 % (0.9778 vs. 0.7372), but this was due to some coincidence rather than to regularity. We expect that the actual attainable level of relative improvement rate is somewhere between 6 and 8 % as observed for four households: No. 2, No. 3, No. 4 and No. 5. Finally, the behavior of the household No. 1 was least predictive and resulted in an improvement

Table 2 The results of electricity load forecasting on testing dataset

Household	Primary model		ICA decomposition		MAPE relative improvement (%)	r-MAPE relative improvement (%)
	MAPE	r-MAPE	MAPE	r-MAPE		
No. 1	0.4271	0.3311	0.4213	0.3556	1.4	−6.8
No. 2	0.5432	0.4203	0.5096	0.4147	6.2	1.3
No. 3	0.5035	0.4312	0.4676	0.4305	7.1	0.1
No. 4	0.7023	0.5010	0.6479	0.4848	7.7	3.3
No. 5	0.4778	0.3984	0.4412	0.3819	7.7	4.3
No. 6	0.9778	0.5473	0.7372	0.5091	24.6	7.4

rate of 1.4 %. In the case of r-MAPE, improvement was observed for the five households, namely the highest improvement rate was observed for household No. 6 once again. For households No. 5 and No. 4 the relative improvement rate was equal to 4.3 and 3.3 %, respectively. For household No. 3 the forecasting accuracy remained almost at the same level (slight improvement of 0.1 %) and 1.3 % improvement rate was observed for household No.2. Lastly, as previously the behavior of household No. 1 was the least predictive and resulted in deterioration rate of 6.8 %. The general conclusions of the experiment are twofold. First, the proposed decompositions applied for individual time series with household electricity usage manage to improve the forecasts in terms of MAPE for all of the analyzed households. Second,

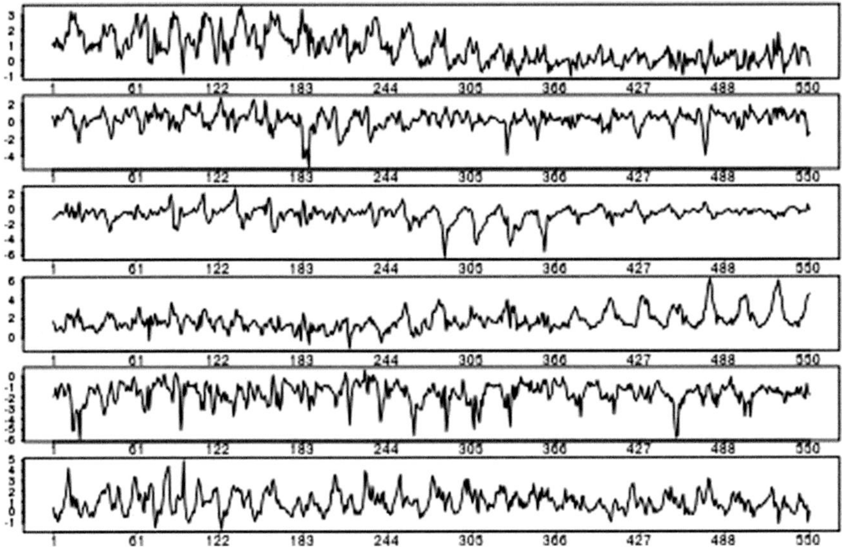

Fig. 3 The components after ICA decomposition

when r-MAPE was considered the forecasts for five out of six households improved. Additionally, we observed that the improvement rate for r-MAPE was influenced by the k parameter which brings in extreme observations to $\mu \pm k$ in accordance with (5). The values of $k < 2$ resulted in slight (inconsiderable) improvements while for $k \geq 2$ the results were notable (Fig. 3).

6 Conclusions

This paper has presented the BSS approach for 24 h ahead electricity demand forecasting applied to the individual household data available from the smart metering system. The proposed methodology aimed to eliminate common destructive components from model forecasts and helped to improve the final forecast accuracy due to combining only essential constructive components common to all the models. The general conclusions of the experiment are twofold. First, the proposed decompositions applied for individual time series with household electricity usage manage to improve the forecasts for all of the analyzed households. In some cases, it was possible to obtain nearly 25 % reduction of MAPE and 7.4 % reduction of r-MAPE. However, the realistic and attainable improvement rate is between 6 and 8% for MAPE as observed for four out of six households. Second, the method is universal, flexible, and applicable at any number of individual forecasts. It helps to improve the forecast accuracy and enhances generalization abilities due to combining the constructive components common to all the models. The investigation of common negative components can provide information about improper factors in forecasting or data acquisition process, since the destructive components can be present due to inadequate optimizations methods, poor quality learning data, or improper variables selection. Analysis of these questions may result in additional research tasks addressed for the future research.

Acknowledgments The study is cofounded by the European Union from resources of the European Social Fund. Project PO KL âInformation technologies: Research and their interdisciplinary applicationsâ, Agreement UDA-POKL.04.01.01-00-051/10-00.

References

1. Anda, M., Temmen, J.: Smart metering for residential energy efficiency: the use of community based social marketing for behavioural change and smart grid introduction. Renew. Energy **67**, 119–127 (2014)
2. Carroll, J., Lyons, S., Denny, E.: Reducing household electricity demand through smart metering: the role of improved information about energy saving. Energy Econ. **45**, 234–243 (2014)
3. Box, G.E.P., Jenkins, G.M.: Time Series Analysis. Forecasting and Control. Prentice-Hall, Englewood Cliffs (1994)
4. Hamilton, J.D.: Time Series Analysis. Princeton University Press, Princeton (1994)

5. Szupiluk, R., Ząbkowski, T.: In: Rutkowski, et al. (eds.) Signal Randomness Measure for BSS Ensemble Predictors. Lecture Notes in Computer Science, vol. 8468, pp. 570–578. Springer, Berlin (2014)
6. Cichocki, A., Amari, S.: Adaptive Blind Signal and Image Processing. Wiley, Chichester (2002)
7. Comon, P., Jutten, C.: Handbook of Blind Source Separation: Independent Component Analysis and Applications. Academic Press, Boston (2010)
8. Hyvarinen, A., Karhunen, J., Oja, E.: Independent Component Analysis. Wiley, New York (2001)
9. Szupiluk, R., Cichocki, A.: Blind signal separation using second order statistics. Proc. SPETO **2001**, 485–488 (2001)
10. Dataport. https://dataport.pecanstreet.org/ (2014)
11. Gajowniczek, K., Ząbkowski, T.: Short term electricity forecasting using individual smart meter data. Procedia Comput. Sci. **35**, 589–597 (2014)
12. Ward, J.H.: Hierarchical grouping to optimize an objective function. J. Am. Stat. Assoc. **58**(301), 236–244 (1963)
13. National Research Council: Estimating Population and Income for Small Places. National Academy Press, Washington (1980)
14. Moreno, J.J.M., Pol, A.P., Abad, A.S., Blasco, B.C.: Using the R-MAPE index as a resistant measure of forecast accuracy. Psicothema **25**(4), 500–506 (2013)
15. Huber, P.J.: Robust Statistics. Wiley, New York (1981)
16. R Development Core Team R: A language and environment for statistical computing. R Foundation for Statistical Computing, Vienna, Austria. ISBN 3-900051-07-0, http://www.R-project.org/ (2011)
17. Kuhn, M.: Building predictive models in r using the caret package. J. Stat. Softw. **28**(5), 1–26 (2008)
18. Dai, Y.D.: Convergence properties of the BFGS algorithm. SIAM J. Opt. **13**(3), 693–701 (2002)

Hierarchical Gaussian Mixture Model with Objects Attached to Terminal and Non-terminal Dendrogram Nodes

Łukasz P. Olech and Mariusz Paradowski

Abstract A hierarchical clustering algorithm based on Gaussian mixture model is presented. The key difference to regular hierarchical mixture models is the ability to store objects in both terminal and nonterminal nodes. Upper levels of the hierarchy contain sparsely distributed objects, while lower levels contain densely represented ones. As it was shown by experiments, this ability helps in noise detection (modeling). Furthermore, compared to regular hierarchical mixture model, the presented method generates more compact dendrograms with higher quality measured by adopted F-measure.

Keywords Background model · Outliers detection · Noise modeling · Hierarchical clustering · Hierarchic Gaussian mixture model

1 Introduction

This paper addresses the topic of *hierarchical data clustering* which is a countertype to *flat clustering*. Flat clustering approaches generate groups without structural connections between them. Hierarchical clustering algorithms generate groups and arrange them in a tree-structured manner. In such a tree structure, (known as a *dendrogram*) all child clusters are attached to their parent cluster. Clusters without any further children are called *terminal nodes* or *tree leaves*. Clusters with attached child clusters are called *nonterminal* or *internal nodes*. Hierarchical clustering algorithms can be divided into two categories depending on the objects attachment to the generated groups. The first category represents methods that attach objects only to *terminal nodes*, and *nonterminal* nodes of the hierarchy remain empty. These kind of methods are the majority of hierarchical data clustering methods. It is possible to fill an internal node with objects, by gathering all objects belonging to its child nodes.

Ł.P. Olech (✉) · M. Paradowski
Department of Computational Intelligence, Wroclaw University of Technology, Wroclaw, Poland
e-mail: lukasz.olech@pwr.edu.pl

R. Burduk et al. (eds.), *Proceedings of the 9th International Conference on Computer Recognition Systems CORES 2015*, Advances in Intelligent Systems and Computing 403, DOI 10.1007/978-3-319-26227-7_18

The second category represents methods attaching objects both to *internal nodes* and *tree leaves*. All tree leaves need to have at least one attached object. Internal nodes can have attached objects or remain empty. The *key difference* is that if an object is attached to the internal node, it is *not attached* to any of its child nodes. Methods belonging to this category are the minority. The presented research addresses this category. This paper is organized as follows. In following subsections we give the necessary background of clustering problems. The proposed method is introduced in the second section. The third section presents the experimental results and comparison with regular hierarchical Gaussian mixture model. Finally, the fourth section summarises this paper.

1.1 Hierarchical Approaches to Clustering

One of the earliest approaches of hierarchical clustering is *hierarchical agglomerative clustering* (HAC) [17]. HAC creates a dendrogram with all objects attached to its leaves. At each level of the hierarchy, two groups are merged. As a result, the created structure is an *unbalanced binary tree*. Various merging schemes are available, e.g., *Ward criterion* [22], *single-link* [19], or *complete-link* [5]. Both binary and nonbinary hierarchies can be constructed using various extensions of the *k-means* algorithm [10, 11]. Usage of hierarchical k-means leads to two major consequences comparing to flat k-means. First, the clustering process is *much faster* because the number of groups in a tree path is much lower. This is especially important if the number of clusters and the volume of data are high. Second, the overall quality of clustering tends to be *worse*, because cluster centers are not optimized simultaneously. One of the key problems of k-means clustering (both flat and hierarchical) is estimation of the number of clusters. There are many attempts to address this issue, e.g., *x-means* algorithm [18]. Hierarchical clustering using a probabilistic approach is also possible, e.g., [14]. The milestone in probabilistic clustering was the formulation of the *expectation maximization* (EM) algorithm [6]. Hierarchical setup of mixture models can be trained using modified EM [4]. One of the most common choices for mixture components is multivariate normal distribution.

1.2 Clustering in the Presence of Noise

Yet another important issue is clustering of data in the presence of *noise* or *outliers*. There are two common solutions to this problem. The first solution consists of two stages, e.g., [2]. In the initial stage data is filtered in order to detect and remove outliers. Then in the second stage clusterization is performed only on the accepted data. The second solution is to directly incorporate the noise model into the clustering process. Usually, the type or distribution of noise or outliers is not known. Various assumptions regarding these distributions have to be made. Exemplary, DBSCAN [7], and OPTICS [1] clustering algorithms assume a minimum density of the meaningful data. In probabilistic clustering, noise can be directly modeled by appropriate mixture components, e.g., [3, 9].

1.3 Problem Formulation, Motivation and Contribution

Probabilistic approach to clustering can be formulated using the parametric model. The key issue is the formulation of an appropriate *probability density function* (PDF). There are several forms of the probabilistic density function. *Gaussian mixture model* is one of the most prominent [9]. Let the *Gaussian mixture model* G with n mixture components be defined as

$$G(w, \mu, \Sigma) = \sum_{i=1}^{n} w_i N(\mu_i, \Sigma_i), \quad w_i \in \langle 0, 1 \rangle, \quad \sum_{i=1}^{n} w_i = 1, \tag{1}$$

and $N(\mu, \Sigma)$ represents the multivariate normal distribution, $w = [w_1, \ldots, w_n]$, $\mu = [\mu_1, \ldots, \mu_n]$, $\Sigma = [\Sigma_1, \ldots, \Sigma_n]$. In such case clustering problem becomes a probability density function estimation problem, where PDF parameters maximize *likelihood* $\mathcal{L}$

$$\langle w^*, \mu^*, \Sigma^* \rangle = \arg \max_{\langle w, \mu, \Sigma \rangle} \mathcal{L}(w, \mu, \Sigma | x_1, \ldots, x_m), \tag{2}$$

where $x_1, \ldots, x_m$ are the data vectors. This is typically solved by the EM algorithm, but other methods are also available, e.g., [8, 21, 23]. Gaussian mixture model fits to data distributed among several clusters, but does not model *outliers* [9]. Data not fitting to the assumed distribution can be interpreted in several ways: *noise*, *measurement errors*, or *sparser representation of meaningful objects*. Statistical modeling of noisy data requires making assumptions on the noise distribution. The data distribution is usually combined with noise distribution, e.g., [16]. In the presented approach we follow the third interpretation of the not fitting data, i.e., sparser representation of meaningful objects. We do not want to reject the data, but we want to model it on some level of the generated hierarchy. Data bound to the parent clusters should have lesser density comparing to the data bound to the child clusters. In the paper we show a simple approach to adapt hierarchical Gaussian mixture model to handle objects attached to any node in the tree. Similar to noise modeling [3, 9], we add an additional mixture component to the mixture model. But unlike these approaches, we do not estimate it but *directly take it from the higher level of the hierarchy*. As a consequence, parameters of the adapted mixture model are estimated in an identical manner as for the classic mixture model. They can be estimated by using EM or any other appropriate approach.

2 Proposed Approach

The proposed approach is an extension of a hierarchical setup of Gaussian mixture models. At each level of the hierarchy an additional mixture component, called *background component*, is introduced. This component is responsible for capturing outliers at a given level. Unlike all other mixture components, it is not estimated, but directly inherited from the higher level of the hierarchy. Root level also has this additional component. Its parameters are estimated (by definition) from all available data.

2.1 Formal Model of the Hierarchy

Let us define the model of the hierarchy in a *recursive* way. Any *parent node* has all its *child nodes*. A tree node T generated from a data set X is defined as

$$T(X) : \langle n, G_B, B \subseteq X, [T_1(X_1), \ldots, T_n(X_n)]\rangle, \tag{3}$$

where

$$B \cup \bigcup_{i=1}^{n} X_i = X, \quad \forall_{i\in[1,n]} B \cap X_i = \emptyset, \quad \forall_{i,j\in[i,n]} i \neq j \Rightarrow X_i \cap X_j = \emptyset \tag{4}$$

and n is the maximum number of child nodes (and mixture components), G_B is the *Gaussian mixture model with background component* $N(\mu_B, \Sigma_B)$:

$$\begin{aligned} G_B(\alpha, w, \mu_B, \mu, \Sigma_B, \Sigma) &= \alpha N(\mu_B, \Sigma_B) + (1-\alpha)G(w, \mu, \Sigma) \\ &= \alpha N(\mu_B, \Sigma_B) + \sum_{i=1}^{n}(1-\alpha)w_i N(\mu_i, \Sigma_i), \end{aligned} \tag{5}$$

$$\alpha \in \langle 0, 1\rangle, \quad \mu_B = E[X], \quad \Sigma_B = Var[X], \tag{6}$$

$B \subseteq X$ is the data subset attached to the node T, related to background mixture component $N(\mu_B, \Sigma_B)$, $T_1, \ldots, T_n$ are child nodes or *void*. Mixture component G_B and set B are representing the data that remain in tree node T. They are the key difference when comparing to classic hierarchical clustering methods.

2.2 Hierarchy Generation

Cluster hierarchy generation is done in a recursive way. First, the top level is generated and its parameters are estimated. Later, child levels are added sequentially in *breadth-first* manner. For each level the process terminates if a stop criterion is reached. This process is similar to the one used in hierarchical k-means approach [20]. It allows a dynamic generation of the hierarchical structure. As shown in the formal model, each level of the hierarchy contains only a subset of the data. The top level starts with all the data. Expectation maximization method is used to estimate the Gaussian mixture model. Because the proposed method is iterative, stochastic, and strongly depended on cluster initialization, several cluster reinitializations should be performed. Thus the number of cluster reinitialization R and number of EM iterations N are the parameters. Clusters initialization is based on choosing random n distinct points from the data and set them as initial centers μ of new clusters. Initial covariances Σ of that clusters are the same as parent cluster covariance. Full covariance matrices are used. When covariance matrix is non-invertible, regularization is introduced. Mixing coefficients (see Eq. 6) are initialized as equal values

$$\alpha = \frac{1}{n+1}, \quad (1-\alpha)w_i = \frac{1}{n+1}. \tag{7}$$

The denominator takes into account n newly created clusters and a background cluster. The data are distributed to all mixture components, according to data probability assignments. A single data instance is assigned to the mixture component with the highest probability of generating that instance. As a result, some mixture components, including the background component, may remain empty. After initialization, the EM algorithm works through N iterations, changing initial values of μ, Σ, w and α. After performing R reinitializations, a solution with the largest likelihood is chosen (see Eq. 2) as the final one. All mixture components with assigned data instances generate child nodes. The above process repeats for every generated node. In case a mixture component does not receive any data, it also does not generate a child node. The child nodes generation process is terminated when a stop criterion is reached. There are two stop criteria and each of them terminates the method. The first stop criterion is connected with the content of current leaf nodes. The clustering process proceeds only on those leaf nodes that contain at least k different data samples. The algorithm terminates when there are no leaf nodes to split or all data is assigned to background mixture component B. The second criterion occurs when provided W overall number of nodes was created.

Table 1 Original (without additional noise) UCI dataset statistics

Dataset name	Instances	Attributes	Classes
Iris	150	4	3
Wine	178	13	3
Glass identification	214	9	6
Image segmentation	2100	19	7

3 Experimental Verification

Experimental verification of the proposed approach consists of two parts. In the first part we give illustrative examples to demonstrate the idea behind the method. Manually prepared toy datasets are used for visualization purposes. In the second part we test the proposed approach on a set of benchmark datasets from UCI repository [13]. We choose well-known *iris, wine, glass identification* and *image segmentation* datasets varying in number of classes, attributes, and instances, as shown in Table 1. Since the mentioned datasets do not contain any noise points we added them manually. Noise points are uniformly distributed among original points. In each dataset, the number of noise points is equal to the half of the number of original points. The proposed approach is compared to a standard hierarchical setup of Gaussian mixture model. In order to compare the obtained results on the benchmark datasets we use a metric based on *F-measure* [12]. It takes a class attribute into consideration and yield a grouping quality by considering the whole dendrogram, not only a chosen level. This makes the measure adequate for hierarchical methods. *F-measure* is calculated for each generated group B with respect to each class C

$$P(X_i, C_c) = \frac{N_{ic}}{|X_i|}, \quad R(X_i, C_c) = \frac{N_{ic}}{N_{C_c}}, \tag{8}$$

$$F_{ic} = \frac{2P(X_i, C_c)R(X_i, C_c)}{P(X_i, C_c) + R(X_i, C_c)}, \tag{9}$$

where F_{ic}—*F-measure* for *ith* group and *cth* class, $P(X_i, C_c)$—precision and $R(X_i, C_c)$—recall, for *ith* group with respect to *cth class*, N_{ic} is the number of objects from *cth* class which are within *ith* group, N_{C_c} is the number of object from *cth* class in the entire tree, and $|X_i|$ is the number of objects that are within *ith* cluster. Noise points are not regarded as an additional class, and they are only counted in each $|X_i|$. Given the above definitions, F-measure for a chosen class C_c is defined as the maximum value of the measure over all nodes of the tree

$$F(C_c) = \max_i F_{ic}. \tag{10}$$

Finally, it is averaged over all classes giving F-Measure for whole hierarchy

$$F = \frac{1}{N} \sum_{c=1}^{|C|} N_{C_c} F(C_c), \tag{11}$$

where $|C|$ is number of classes used in dataset, N is the total number of objects (Excluding noise points) and N_{C_c} is the number of data objects of class c. Proposed evaluation criterion has the ability to explore hierarchy structure, which is a key point in the proposed method. F maximum value is 1 and minimum is 0. Better hierarchies have higher F values.

3.1 Manually Generated Data with Noise—an Illustration

All results presented in this section are two dimensional toy examples. Their sole purpose is to illustrate the behavior of the proposed method. The following examples are presented:

1. three groups with a large central group and a small amount of noise (*LC*),
2. small circular data clusters with a small amount of noise (*LN*),
3. small circular data clusters with a large amount of noise (*HN*).

Both the data and clustering results for the toy datasets are shown in Fig. 1. The method has some ability to capture less dense data. This data is attached to the intermediate nodes of the hierarchy. The additional background model component captures these instances. In consequence they are automatically bound to the node related to the background component. At the same time, densely distributed data is moved to the bottom of the hierarchy. This can be observed (to some extent) at all presented test cases.

3.2 UCI Benchmark Datasets

The second part of the experiments addresses the clustering of the UCI benchmark datasets. Instances of all processed datasets have both feature vectors and class assignment. Feature vectors without class information are used in the clustering process. Available class assignment is used in the evaluation process. Two methods are compared: (1) the proposed Gaussian mixture model with outlier modeling and (2) the classic Gaussian mixture model. The first method is denoted as B and the second as G. Both methods are trained using the same expectation–maximization routine. Hierarchies of both models are constructed in the same manner. Two quality estimates are shown: (1) log-likelihood values to address data fitting to the distribution, (2) F-measure values to check if the generated groups are meaningful. Performed experiments consider mentioned quality estimators when W parameter varies between 2 and 10. In all conducted experiments we set n parameter as a constant equal to 2.

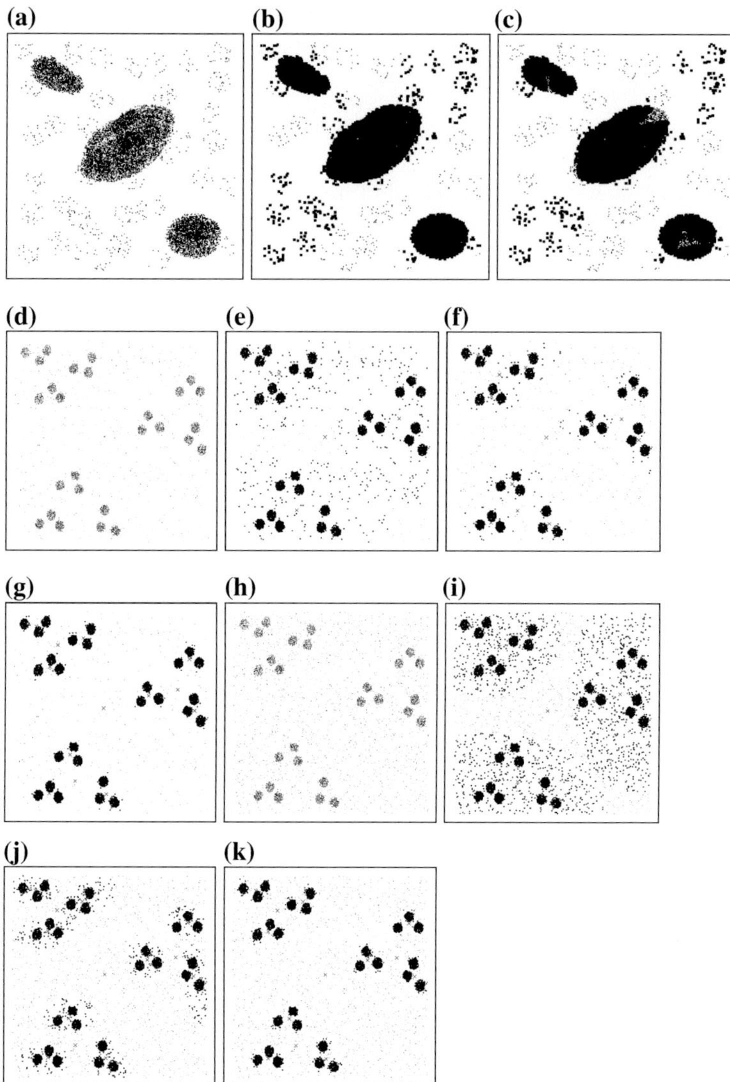

Fig. 1 Two-dimensional toy datasets with a various amount of noise. First column shows the data points. Second, third, and fourth columns show the clustering results at different levels of the hierarchy. Data attached to proposed background model are shown as *small pixels*, and data attached to the mixture model are shown as *large pixels*. **a** Large center (LC). **b** LC, level 1. **c** LC, level 2. **d** Low noise (LN). **e** LN, level 1. **f** LN, level 2. **g** LN, level 3. **h** High noise (HN). **i** HN, level 1. **j** HN, level 2. **k** HN, level 3

Table 2 Comparison of the proposed model (μ_B, σ_B) with the reference to Gaussian mixture model (μ_G, σ_G)

Dataset name	No of nodes (W)	Log-likelihood				F-measure				Significance test	
		μ_B	σ_B	μ_G	σ_G	μ_B	σ_B	μ_G	σ_G	U	Winner
Iris	2	247	7	107	0	**0.59**	0.045	0.36	0.000	0	B
	3	359	0	288	0	**0.75**	0.002	0.62	0.000	0	B
	4	359	1	355	14	**0.75**	0.003	0.62	0.000	0	B
	5	490	24	429	37	**0.77**	0.013	0.67	0.061	0	B
	6	483	27	503	53	**0.77**	0.081	0.69	0.062	830	B
	7	480	28	584	5	**0.77**	0.058	0.75	0.001	1385	B
	8	601	29	623	23	**0.79**	0.044	0.74	0.006	697	B
	9	629	24	642	29	**0.80**	0.046	**0.80**	0.071	3760	B
	10	614	19	680	43	**0.80**	0.046	0.78	0.070	2679	B
Wine	2	366	1	307	0	**0.40**	0.002	0.37	0.000	0	B
	3	402	2	392	1	**0.42**	0.005	0.41	0.005	2033	B
	4	403	2	445	21	**0.42**	0.003	0.41	0.005	1539	B
	5	466	8	478	20	**0.42**	0.005	**0.42**	0.017	5298	–
	6	466	11	529	8	0.42	0.005	**0.43**	0.016	6188	G
	7	470	8	561	8	0.42	0.004	**0.44**	0.005	10000	G
	8	527	19	588	23	0.42	0.010	**0.44**	0.006	9372	G
	9	524	20	599	20	0.42	0.008	**0.44**	0.008	9766	G
	10	531	20	617	31	0.42	0.008	**0.44**	0.010	9753	G
Glass	2	809	1	338	0	**0.41**	0.000	0.29	0.000	–	B
	3	1121	1	1121	0	**0.40**	0.001	**0.40**	0.000	4950	B
	4	1275	28	1227	15	**0.45**	0.045	0.40	0.000	600	B
	5	1299	48	1274	36	**0.46**	0.044	0.45	0.050	5084	–
	6	1314	49	1372	19	0.44	0.043	**0.46**	0.054	5537	–
	7	1469	23	1432	5	**0.50**	0.008	**0.50**	0.027	8715	G
	8	1477	19	1465	15	0.49	0.022	**0.50**	0.038	6476	G
	9	1495	5	1501	19	0.49	0.010	**0.51**	0.015	9032	G
	10	1532	14	1533	16	**0.50**	0.013	**0.50**	0.022	7632	G
Segmentation	2	−6605	11	−6678	0	**0.29**	0.012	0.28	0.000	–	–
	3	−6466	3	−6481	1	**0.50**	0.000	**0.50**	0.001	495	B
	4	−6466	3	−5413	154	**0.50**	0.000	**0.50**	0.000	600	–
	5	−5348	76	−5406	268	0.50	0.001	**0.52**	0.042	5084	–
	6	−5643	1001	−4524	72	0.49	0.017	**0.54**	0.042	5537	–
	7	−5470	710	−4483	115	0.50	0.014	**0.56**	0.038	8715	G
	8	−5257	715	−4388	366	0.50	0.019	**0.56**	0.038	6476	G
	9	−5342	763	−4253	359	0.50	0.026	**0.57**	0.049	9032	G
	10	−5257	850	−3959	278	0.50	0.018	**0.58**	0.050	7632	G

Both log-likelihood values are F-measure values are shown. Higher F-measure values are marked in bold

First of all, we found the best parameters configuration (N and R) for each method per single dataset instance and W value. Then, because of stochastic nature of both methods, we have performed 100 trials for each of dataset and W parameter value, calculating mean value μ and sample standard deviation σ. Moreover we conducted the *Wilcoxon rank sum test* [15] on calculated F-measure in order to show the statistic significance of the obtained results. Statistic value U is calculated with alpha level (α) equal to 0.05. Null hypotheses H_0 are equality of population distributions and alternative hypotheses H_A may vary (depending on the corresponding F-measure μ values). When F-measure mean values μ were different, then we performed a one-tailed test whereas equal means results in two-tailed. Achieved results are shown in Table 2. In that table the *winner* column shows whether there is statistical evidence to reject the null hypothesis and assume an alternative one. Experiments results in Table 2 show that the proposed background component improves the quality of generated dendrograms, when considering data class labels. This is especially visible when maximum number of nodes W is less than 5. Our method, though has the ability of creating compact dendrograms with better quality than the method without the background component, which is desired, because shorter trees have better generalization abilities. Moreover, considering the *iris* dataset, the background component helps obtaining higher F-measure in all cases, comparing to regular hierarchical Gaussian mixture model. Proposed method reaches statistically higher average F-measure results in 16 cases whereas the regular method wins only 13 times. There have been seven draws.

4 Summary

A hierarchical grouping method is presented. It has the ability to attach objects both to terminal and nonterminal nodes. It is an extension of the classic Gaussian mixture model. The mixture is extended with an additional component responsible for outlier modeling. Parameters of this mixture component are not estimated, but directly inherited from higher levels of the hierarchy. Conducted experiments show that the proposed modification allows to treat part of the data as sparser representation of meaningful objects. Though upper levels of hierarchy consist of sparsely distributes data, this can be used in noise or outliers modeling. Comparison between regular hierarchic GMM and hierarchic GMM with proposed modification shows that the background component helps to improve the quality of short hierarchies in real datasets with random noise.

References

1. Ankerst, M., Breunig, M.M., Kriegel, H., Sander, J.: OPTICS: ordering points to identify the clustering structure. In: Proceedings of ACM SIGMOD International Conference on Management of Data. pp. 49–60 (1999)
2. Byers, S., Raftery, A.E.: Nearest-neighbor clutter removal for estimating features in spatial point processes. J. Am. Stat. Assoc. **93**(442), 577–584 (1998)
3. Campbell, J.G., Fraley, C., Murtagh, F., Raftery, A.E.: Linear flaw detection in woven textiles using model-based clustering. Pattern Recognit. Lett. **18**(14), 1539–1548 (1997)

4. Carneiro, G., Chan, A.B., Moreno, P.J., Vasconcelos, N.: Supervised learning of semantic classes for image annotation and retrieval. IEEE Trans. Pattern Anal. Mach. Intell. **29**(3), 394–410 (2007)
5. Defays, D.: An efficient algorithm for a complete link method. Comput. J. **20**(4), 364–366 (1977)
6. Dempster, A.P., Laird, N.M., Rubin, D.B.: Maximum likelihood from incomplete data via the EM algorithm. J. R. Stat. Soc. Ser. B (Methodol.) **39**(1), 1–38 (1977)
7. Ester, M., Kriegel, H., Sander, J., Xu, X.: A density-based algorithm for discovering clusters in large spatial databases with noise. In: Proceedings of the 2nd International Conference on Knowledge Discovery and Data Mining (KDD). pp. 226–231 (1996)
8. Figueiredo, M.A.T., Jain, A.K.: Unsupervised learning of finite mixture models. IEEE Trans. Pattern Anal. Mach. Intell. **24**(3), 381–396 (2002)
9. Fraley, C., Raftery, A.E.: Model-based clustering, discriminant analysis, and density estimation. J. Am. Stat. Assoc. **97**(458), 611–631 (2002)
10. Hartigan, J.A., Wong, M.A.: Algorithm as 136: a k-means clustering algorithm. J. R. Stat. Soc. Ser. C **28**(1), 100–108 (1979)
11. Jain, A.K.: Data clustering: 50 years beyond k-means. Pattern Recognit. Lett. **31**(8), 651–666 (2010)
12. Larsen, B., Aone, C.: Fast and effective text mining using linear-time document clustering. In: Proceedings of the 5th ACM SIGKDD International Conference on Knowledge Discovery and Data Mining. pp. 16–22 (1999)
13. Lichman, M.: UCI machine learning repository (2013). http://archive.ics.uci.edu/ml
14. Liu, M., Chang, E., Dai, B.: Hierarchical gaussian mixture model for speaker verification. In: 7th International Conference on Spoken Language Processing (2002)
15. Mann, H.B., Whitney, D.R.: On a test of whether one of two random variables is stochastically larger than the other. Ann. Math. Stat. **18**(1), 50–60 (1947)
16. Minka, T.P.: Expectation propagation for approximate bayesian inference. In: Proceedings of the 17th Conference in Uncertainty in Artificial Intelligence. pp. 362–369 (2001)
17. Murtagh, F.: A survey of recent advances in hierarchical clustering algorithms. Comput. J. **26**(4), 354–359 (1983)
18. Pelleg, D., Moore, A.W.: X-means: Extending k-means with efficient estimation of the number of clusters. In: Proceedings of the 17th International Conference on Machine Learning. pp. 727–734 (2000)
19. Sibson, R.: SLINK: an optimally efficient algorithm for the single-link cluster method. Comput. J. **16**(1), 30–34 (1973)
20. Steinbach, M., Karypis, G., Kumar, V.: A comparison of document clustering techniques. Proceedings of Workshop on Text Mining, 6th ACM SIGKDD International Conference on Data Mining (KDD'00). pp. 109–110 (2000)
21. Verbeek, J.J., Vlassis, N.A., Kröse, B.J.A.: Efficient greedy learning of gaussian mixture models. Neural Comput. **15**(2), 469–485 (2003)
22. Ward, J.H.: Hierarchical grouping to optimize an objective function. J. Am. Stat. Assoc. **58**(301), 236–244 (1963)
23. Zivkovic, Z., van der Heijden, F.: Recursive unsupervised learning of finite mixture models. IEEE Trans. Pattern Anal. Mach. Intell. **26**(5), 651–656 (2004)

Real-Valued ACS Classifier System: A Preliminary Study

Olgierd Unold and Marcin Mianowski

Abstract A new model of learning classifier system is introduced to explore continuous-valued environment. The approach applies the real-valued anticipatory classifier system (rACS). In order to handle real-valued inputs effectively, the ternary representation has been replaced by an approach where the interval of real numbers is represented by a natural number. The rACS model has been tested on the 1D linear corridor and the 2D continuous gridworld environments. We show that modified ACS can evolve compact populations of classifiers which represent the optimal solution to the continuous problem.

Keywords Learning classifier system · Anticipatory classifier system · Continuous-valued input

1 Introduction

Learning classifier systems (LCSs) are a class of an evolutionary approach, in which knowledge about an environment is represented by a population of rule sets. Their task is to explore the most minimal set of the most general rules (called classifiers) that cover the solution space. The first learning classifier system was created by Holland [13], shortly after he proposed genetic algorithms (GAs) [12]. The rule-based LCSs are similar to immune systems [23]; it has been also shown that learning classifier systems are equivalent to neural networks [9]. Anticipatory classifier system (ACS) is a relatively new classifier system among different types of LCSs [18]. Introduced by Stolzmann [16], ACS differs from the other LCSs in its classifier. Instead of

O. Unold (✉) · M. Mianowski
Department of Computer Engineering, Faculty of Electronics, Wroclaw University of Technology, Wyb. Wyspianskiego 27, 50-370 Wroclaw, Poland
e-mail: olgierd.unold@pwr.edu.pl
URL: http://olgierd.unold.staff.iiar.pwr.edu.pl/

M. Mianowski
e-mail: djwmmianek@gmail.com

R. Burduk et al. (eds.), *Proceedings of the 9th International Conference on Computer Recognition Systems CORES 2015*, Advances in Intelligent Systems and Computing 403, DOI 10.1007/978-3-319-26227-7_19

(*condition*) $\longrightarrow$ (*action*), ACS manipulates (*condition*) (*action*) $\longrightarrow$ (*effect*) classifiers, where the last part represents the anticipated effect of the *action*. ACS, as many implementations of LCS, uses a ternary representation to encode the environmental state that a classifier matches. Such a representation is not sufficiently capacious to express most real-world problems. In this paper we introduce for the fist time an extension to ACS that allows the representation of continuous-valued inputs. The paper is organized as follows. In the next section, we give some background about real-valued LCSs. Section 3 contains a short introduction to the ACS. Then in Sect. 4, we introduce the real-valued anticipatory learning classifier system. In Sect. 5, we present the experimental study with rACS over real-valued environments, inaccessible to binary ACS. The last section summarizes the paper and makes plans for the future.

2 Related Work

Many real-world problems are not conveniently expressed using the ternary representation typically used by LCSs (true, false, and the "don't care" symbol). To overcome this limitation, Wilson [21] introduced a real-valued XCS classifier system for problems which can be defined by a vector of bounded continuous real-coded variables. In [22] min–max interval predicates were used. Stone and Bull [17] proposed unordered-bound interval predicates to be used in the XCS model. A ternary XCS model has been also extended to min-percentage representation [8], and ellipsoids and hyper-ellipsoids, respectively [2] and [6]. It is worth noting that not just XCS has been extended to deal with continuous inputs. Cielecki and Unold described real-valued extension to the grammar-based classifier system GCS [7], which explores a population of context-free grammar rules [19]. To our knowledge, there is not, as yet, any real-valued anticipatory classifier system.

3 Anticipatory Classifier System (ACS) in a Nutshell

Although Stolzmann is considered to be an inventor of the anticipatory classifier system [16], Riolo [15] was the first to publish LCS with anticipation (called CFSC2). ACS proposed by Stolzmann was later extended by Butz to ACS2 [3], and later to XACS [4]. The other line of research was YACS [10] and MACS [11], both models described by Gérard et al. Learning phase in ACS comprises three different algorithms: Anticipatory learning process (ALP), Reinforcement learning (RL), and Genetic algorithm (GA), run respectively. ACS, like the other LCS models, includes a reinforcement learning module which uses the well-known Q-learning idea [20]. It changes the parameters of classifiers by using the reward received from the environment. The main idea of ACS is that the system tries to predict changes of state in environment and learn from anticipations. In contrast to the standard LCS, where the classifiers consist of two parts: ⟨*condition*⟩ and ⟨*action*⟩, ACS's classifier has been extended by ⟨*effect*⟩ part. This part of a classifier represents the "anticipation" of the considered action, i.e., the expected state of an environment after the ⟨*action*⟩ has been taken. ACS2 model introduced yet another part of a classifier, ⟨*mark*⟩, in which the model records the states in which the classifier did not work correctly before. It follows that the classifier of ACS (more precisely ACS2) consists of four parts: ⟨*condition*⟩ ⟨*action*⟩ $\longrightarrow$ ⟨*effect*⟩⟨*mark*⟩. The ⟨*effect*⟩ part of a classifier, as well as a ⟨*condition*⟩ part, can contain a "don't care" symbol (denoted as "#"). This symbol placed in a ⟨*condition*⟩ will match whatever the value of the corresponding state attribute (i.e., true or false). The don't care symbol in ⟨*effect*⟩ denotes the attribute which does not change given that a certain action is fired in a given situation. This attribute of the saved state will remain unchanged at the next time step (in this case this symbol is called pass-through). The parts ⟨*effect*⟩ and ⟨*mark*⟩ are subject to the ALP. Central to ACS, the ALP algorithm is a new learning process in LCS framework that learns from anticipations. ALP makes use of ⟨*effect*⟩ and ⟨*mark*⟩ parts of classifiers to generate new classifiers. To present the behavior of ALP, ACS' single learning trial will be presented. At first, a new environmental state is perceived from environment $\sigma(t)$. Then, a match set $[M](t)$ is generated from the population of classifiers. Match set is a set of classifiers which condition parts matches of the current state. If this is not the first step in a trial, a learning phase takes place. After that, an action is selected from $[M](t)$ and an action set is formed $[A](t)$. An action is executed next and the system gets a reward from the environment and continues with the next step in the trial. Before finishing each trial, an anticipatory learning phase takes place. ALP works with an action set which was formed at the previous state of the environment. It compares the ⟨*effect*⟩ part of each classifier with the changes between the previous and current state. Two cases may occur in this comparison: *expected* and *unexpected*. The first one is a case in which the classifier predicted changes correctly. If it had not predicted incorrectly before (i.e., ⟨*mark*⟩ part does not contain any information), its quality is increased. Otherwise, a more specialized classifier is generated (using a result of a comparison between the actual state and the list of states taken from ⟨*mark*⟩ part). If the classifier did not predict correctly, an unexpected case occurs. In this case the quality of the classifier is decreased and if it is possible, a new classifier is generated. The ⟨*mark*⟩ part of the classifier is extended by an actual state. If action set does not contain a classifier that anticipated changes correctly, a new classifier which will be able to do so is generated. The goal of the last learning step—genetic algorithm—is to generalize conditional part of classifiers. It works only in the action set $[A](t-1)$ which was formed in a previous state. The first step is a determination if GA should be executed. If the condition of execution is fulfilled, two accurate, over-specified offsprings are selected. Then a mutation and crossover of these offsprings occur with a certain probability. If the size of the action set is too big, classifiers with noticeably lower quality than the others are deleted from the action set in order to make space for newly generated ones. For further reading and more detailed description of ACS and its parameters, readers are referred to [3, 4].

4 Real-Valued ACS (rACS)

All known implementations of ACS are based on the ternary representation, which is unnatural for continuous-valued inputs when using real-world data. To overcome this problem, we propose a new approach to the alphabet used by an environment and an ACS system. In the proposed representation, the alleles (so-called attributes or features) represent no longer the values 0, 1, and a special character #, but only a continuous numerical range. This range is limited by a closed interval [0, 1]. Originally we adopted the bounded continuous real-coded variables representation introduced by Wilson [21]. In this representation each allele consists of a pair of real numbers which limit its range, see Fig. 1. Preliminary research and analysis of the obtained results showed however a simpler idea. Explored environments have a regular structure, which means the intervals are evenly distributed throughout the investigated range. For that reason the interval representation can be simplified. The interval can be represented by a natural number, and the change of the range to a different one causes selection of a different number of natural numbers (most often shifting to neighboring intervals), see Fig. 2. In consequence, experiments are 30 % faster and occupy exactly 4x less memory. In a described environment the number of 0 represents the lowest interval, which starts at 0 and ends at a value equal to environment resolution. The upper bound also depends on the resolution and is equal to the number of intervals. Finally, positions taken as the value of an allele depends on the implementation of the environment, which forwards to the system which states can be achieved as a result of fire by the system action. A symbol '#' was replaced by the value of -1 because the natural numbers are reserved for specific intervals, and their number depends on the discretization of the continuous environment. rACS differs from ACS only at the input interface, and hence at the structure of the classifier. The other modules of the model are not to be changed. It is worth noting that extending ACS model to rACS changes the meaning of the "don't care" symbol. In the classic ACS, symbol "#" represents the two possible values, i.e., true (1) and false (0). "Don't care" symbol in the rACS model within the discretization described

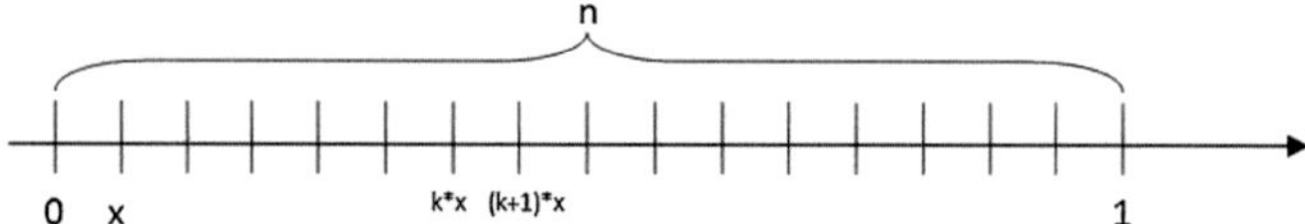

Fig. 1 Preliminary interval-bounded representation of allele in rACS

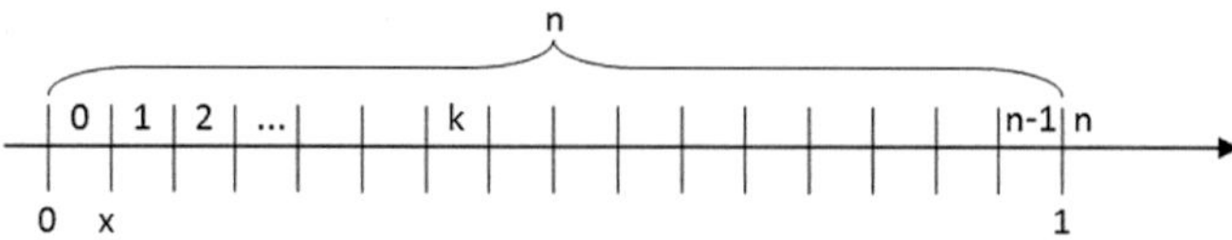

Fig. 2 Representation of allele in rACS as a natural number of partition

above is supposed to cover all possible values, and this means that it has to cover all possible intervals (represented by natural numbers). It can be concluded that this symbol can have a greater impact on learning than in ACS model. It can be seen also as a representation of a specific attribute, which does not affect action firing in all states matching the other attributes.

5 Experiments with rACS

We applied rACS to two classes of real-valued multi-step environments: the 1D linear corridor and the 2D continuous gridworld environments. These continuous environments were introduced in [1], and used to test XCSF system [14]. In each of these environments a state represents the coordinates of location, and each allele represents a separate axis in Cartesian coordinates. The goal of the system is to find placed somewhere food. In contrast to the standard experimental design used in multi-step experiments [14], rACS does not switch between learning and test (testing) problems, but for the first half of working time it performs only the learning phase, whereas for the rest of the time only testing phase. In the learning phase, the system selects actions randomly from those represented in the match set, and in the testing phase, the system always selects the action with the highest prediction. Note that this approach gives a real and deterministic system working time. When rACS solves the problem correctly, reaching the food position, it receives a constant reward equal to 1000. Otherwise it receives a constant reward equal to 0. The performance is computed as the average number of steps needed to reach a goal during the last 200 test problems. All results are averaged over 10 experiments. For every experiment, rACS was tested with the following settings: actionSet = 10, inadequacyThreshold = 0.05, reliabilityTreshold = 0.9, experienceTreshold = 50, learningRate = 0.03, discountFactor = 0.97, mutationRate = 0.04, crossoverProbability = 0.8, and geneticAlgorithmTreshold = 50.

5.1 The Continuous Linear Corridor

The aim of this experiment was to test the rACS system in the very simple, but continuous environment. In a 1D corridor, denoted Corr(s), the space is divided into n states describing the movement within the interval [0, 1]. There are two possible actions: *left* corresponding to move of "$-s$" and **right** corresponding to move of "$+s$" (step-size s equals $s = 1/n$); if the system tries to move below position 0, its position is set to 0. The system starts anywhere within a range [0, 1), and reaches the goal when the position of the system is greater or equal than 1. The average number of steps to goal is computed as $(s+1)/2s$ [14]. Note that smaller the step-size s, larger the environment. In the first experiments we apply rACS to a 1D linear corridor with a step-size of 0.05, Corr(0.05), and 0.025, Corr(0.025). These settings of the size of environment were tested for the XCSF system in [14]. The Figs. 3 and 4 show the performance of rACS for testing phase for $s = 0.05$ and $s = 0.025$, respectively. In both cases rACS reached the optimal performance. Figure 5 reports the performance of rACS in Corr(0.01), where the environment is larger with respect to the actions. Note that the XCSF system has not been tested for this step-size. In this case rACS also found an optimal solution.

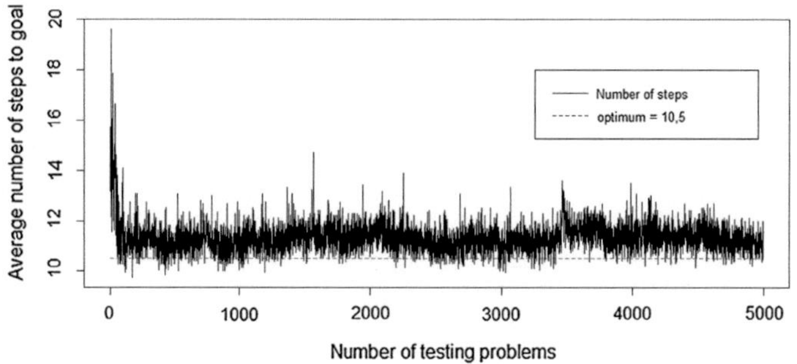

Fig. 3 rACS in corr(0.05)

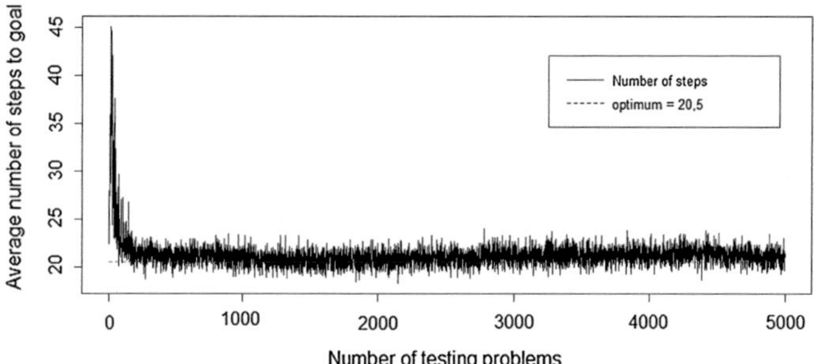

Fig. 4 rACS in corr(0.025)

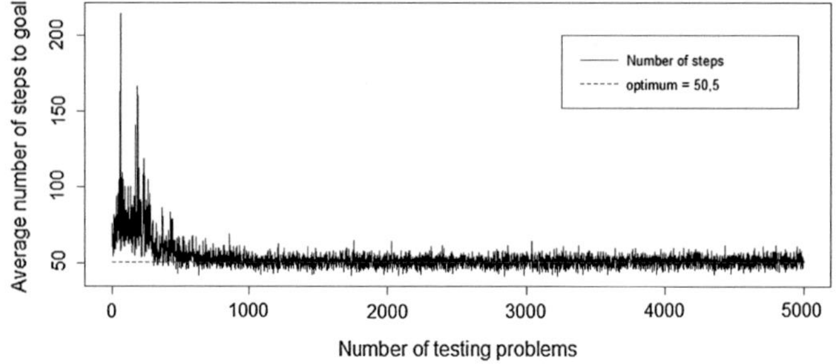

Fig. 5 rACS in corr(0.01)

5.2 The 2D Continuous Gridworlds

The second tested multi-step environment was 2D empty continuous Gridworld Grid(s). The every state is described by a pair of real-valued coordinates over $[0, 1]^2$ board. The food is located in the point 1, 1. There are four possible actions: *left*, *right*, *up*, and *down*. The exploring area is restricted to the domain $[0, 1]^2$, so if performed action exceeds one of the coordinate, the coordinate is assigned to the grid border. According to [14], for the step-size s the average number of steps to reach the aim is computed as $(s + 1)/s$. Figures 6 and 7 show the performance of rACS in Grid(0.05) and Grid(0.04). The proposed system deals with these environments very well, finding the optimal solution in both cases. We extended the tested grids in [14] by checking the larger border in Grid(0.025). The obtained results are presented in Fig. 8. Also in this case rACS reached full optimality. What is worth noting is [14] reports for XCSF system in Grid(0.05) on average more than 900 classifiers in the final population, whereas the final population of rACS in this environment contains less than 80 classifiers.

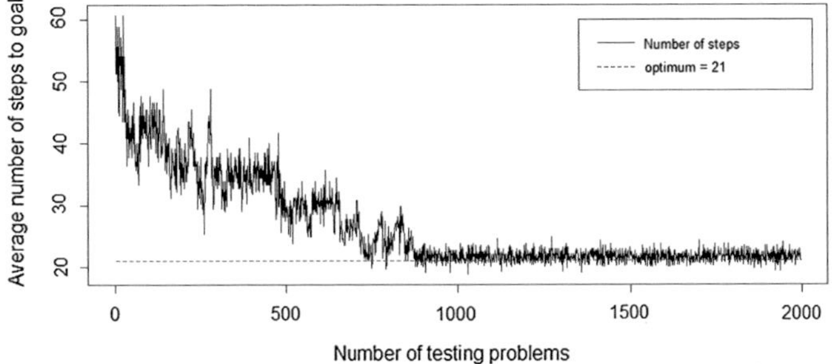

Fig. 6 rACS in grid(0.05)

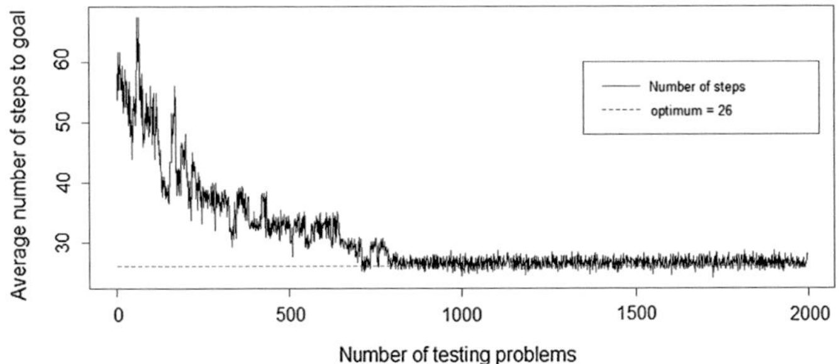

Fig. 7 rACS in grid(0.04)

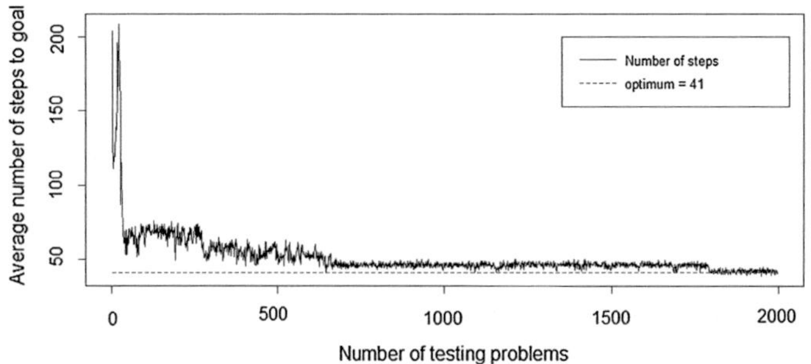

Fig. 8 rACS in grid(0.025)

6 Conclusion and Future Works

We have introduced the new model of learning classifier system called real-valued anticipatory classifier system. To the best of our knowledge, it is the first time that continuous-valued input was applied to a classifier system using anticipation learning. We have applied rACS to multi-step problems involving continuous inputs, i.e., the 1D linear corridor and the 2D continuous gridworld. The tested environments were even larger than in the [14], where the XCSF system coped with the same continuous problems. In all studied cases rACS can converge fast to an optimal solution, also producing a compact final population of classifiers. It is worth noting that during the experiments the (*mark*) part of the classifiers was used only in the case of crossing the border of the examined domain. It seems that this feature of the real-valued anticipatory classifier system finds particular application in dealing with obstacles placed in the explored environment. Our future plans include, among others, tuning the rACS system and applying it to the no-empty continuous 2D and 3D gridworlds.

References

1. Boyan, J.A., Moore, A.W.: Generalization in reinforcement learning: Safely approximating the value function. In: Tesauro, G., Touretzky, D.S., Leen, T.K. (eds.) Advances in Neural Information Processing Systems 7, pp. 369–376. The MIT Press, Cambridge (1995)
2. Butz, M.: Kernel-based, ellipsoidal conditions in the real-valued XCS classifier system. In: Proceedings of the 2005 conference on Genetic and evolutionary computation, pp. 1835–1842 (2005)
3. Butz, M., Stolzmann, W.: Advances in learning classifier systems. An Algorithmic Description of ACS2, pp. 211–229. Springer, Berlin (2002)
4. Butz, M.V., Goldberg, D.E.: Generalized state values in an anticipatory learning classifier system. In: Butz, Martin V., Sigaud, Olivier, Gérard, Pierre (eds.) Anticipatory Behavior in Adaptive Learning Systems. Lecture Notes in Computer Science (Lecture Notes in Artificial Intelligence), vol. 2684, pp. 282–301. Springer, Heidelberg (2003)

5. Butz, M.V., Goldberg, D.E., Stolzmann, W.: Introducing a genetic generalization pressure to the Anticipatory Classifier Systems part I: theoretical approach. In: Proceedings of the 2000 Genetic and Evolutionary Computation Conference (GECCO 2000), 34–41 (2000)
6. Butz, M., Lanzi, P., Wilson, S.W.: Hyper-ellipsoidal conditions in XCS: rotation, linear approximation, and solution structure. In: Proceedings of the 8th Annual Conference on Genetic and Evolutionary Computation, pp. 1457–1464, ACM, USA (2006)
7. Cielecki, L., Unold, O.: Real-valued GCS classifier system. Int. J. App. Math. Comput. Sci. **17**(4), 539–547 (2007)
8. Dam, H., Abbass, H., Lokan, C.: Be real! XCS with continuous-valued inputs. In: Proceedings of the 2005 Workshops on Genetic and Evolutionary Computation, pp. 85–87. ACM, USA (2005)
9. Farmer, J.D.: A Rosetta stone for connectionism. In: CNLS'89: Proceedings of the Ninth Annual International Conference of the Center for Nonlinear Studies on Self-organizing, Collective, and Cooperative Phenomena in Natural and Artificial Computing Networks on Emergent computation, pp. 153–187 North-Holland Publishing Co., The Netherlands (1990)
10. Gérard, P., Stolzmann, W., Sigaud, O.: YACS: a new learning classifier system with anticipation. J. Soft Comput.: Spec. Issue Learn. Classif. Syst. **6**(3–4), 216–228 (2002)
11. Gérard, P., Meyer, J.A., Sigaud, O.: Combining latent learning with dynamic programming in MACS. Eur. J. Oper. Res. **160**, 614–637 (2005)
12. Holland, J.H.: Adaptation in Natural and Artificial System. University of Michigan Press, Ann Arbor (1975)
13. Holland, J.H.: Adaptation. In: Rosen, R.F. (ed.) Progress in Theoretical Biology, pp. 263–293. Plenum Press, New York (1976)
14. Lanzi, P.L., Loiacono, D., Wilson, S.W., Goldberg, D.E.: XCS with Computed Prediction in Continuous Multistep Environments. IlliGAL Report No. 2005018, (2005)
15. Riolo, R.L.: Lookahead planning and latent learning in a Classifier System. In: Meyer, J.A., Wilson, S.W. (eds.) From animals to animats: Proceedings of the First International Conference on Simulation of Adaptative Behavior, pp. 316–326. MIT Press, Cambridge (1991)
16. Stolzmann, W.: Anticipatory classifier systems. In: Koza, J., et al. (eds.) Proceedings of the 1998 Genetic and Evolutionary Computation Conference, pp. 658–664. Morgan Kaufmann Publishers Inc, San Francisco (1998)
17. Stone, C., Bull, L.: For real! XCS with continuous-valued inputs. Evolut. Comput. **11**(3), 299–336 (2003)
18. Urbanowicz, R.J., Moore, J.H.: Learning classifier systems: a complete introduction, review, and roadmap, J. Artif. Evol. Appl. (2009)
19. Unold, O.: Context-free grammar induction with grammar-based classifier system. Arch. Control Sci. **15**(LI)(4), 681–690 (2005)
20. Watkins, C.J.: Learning from Delayed Rewards, Doctoral Disseration. King's College, Cambridge (1989)
21. Wilson S.W.: Get real! XCS with Continuous-Valued Inputs. In: Lanzi Pier Luca, Stolzmann, W., Wilson, S.W (eds.) IWLCS 1999, vol. 1813, LNCS (LNAI) Springer, Heidelberg, pp. 209–222 (2000)
22. Wilson, S.W.: Mining Oblique Data with XCS. In: Revised Papers from the Third International Workshop on Advances in Learning Classifier Systems, pp. 158–176 (2000)
23. Vargas, P.A., de Castro, L.N., Zuben, F.J.V.: Mapping Artificial Immune Systems into Learning Classifier Systems. In: Lanzi, P.L., Stolzmann, W., Wilson, S.W. (eds.) Lecture Notes in ComputerScience, vol. 2661, pp. 163–186, Springer, Berlin (2002)

Random Forest Active Learning for Retinal Image Segmentation

Borja Ayerdi and Manuel Graña

Abstract Computer-assisted detection and segmentation of blood vessels in retinal images of pathological subjects is difficult problem due to the great variability of the images. In this paper we propose an interactive image segmentation system using active learning which will allow quick volume segmentation requiring minimal intervention of the human operator. The advantage of this approach is that it can cope with large variability in images with minimal effort. The collection of image features used for this approach is simple statistics and undirected morphological operators computed on the green component of the image. Image segmentation is produced by classification by a random forest (RF) classifier. An initial RF classifier is built from seed set of labeled points. The human operator is presented with the most uncertain unlabeled voxels to select some of them for inclusion in the training set, retraining the RF classifier. We apply this approach to a well-known benchmarking dataset achieving results comparable to the state of the art in the literature.

1 Introduction

Ocular fundus image assessment is a diagnostic tool of vascular and non vascular pathology. Retinal vasculature inspection may reveal hypertension, diabetes, arteriosclerosis, cardiovascular disease and stroke. The major challenges for the retinal vessel segmentation are: (1) the presence of lesions which may be misdetected as blood vessels; (2) low contrast around thinner vessels; and (3) multiple scales of vessel size. Retinal image segmentation has been tackled from many points of view:

B. Ayerdi (✉) · M. Graña
Department of CCIA, Computer Intelligence Group, UPV/EHU, San Sebastian, Spain
e-mail: ayerdi.borja@gmail.com

M. Graña
ENGINE Centre, Wrocław University of Technology, Wybrzeże Wyspiańskiego 27, 50-370 Wrocław, Poland
e-mail: manuel.grana@ehu.es

R. Burduk et al. (eds.), *Proceedings of the 9th International Conference on Computer Recognition Systems CORES 2015*, Advances in Intelligent Systems and Computing 403, DOI 10.1007/978-3-319-26227-7_20

- as supervised classification problem, where classifiers are trained to discriminate vessel from non-vessel pixels,
- by unsupervised methods based on a priori assumptions on vessel properties in the image which are appropriately detected by specific filters, either matched filters or edge detection oriented filters.
- multi-scale approaches combining several scale filters into a single map for decision.
- many approaches end up with some kind of post-processing or cleaning of the result image in order to remove false positives. Often these processes are heuristics for removing isolated detections.

In our approach, we compute several simple local spatial features over the green band of the image, which contains most of the contrast information. The size of the local window is related to the detection scale. These features are then presented to a classifier to obtain image segmentation. The classifier is built following an active learning approach, where the most uncertain unlabeled pixels are presented to a human operator which labels them for addition to the training dataset and classifier retraining. Section 2 contains a review of methods for retinal vessel segmentation. Section 3 contains the description of the computational methods for segmentation. Section 4 contains the experimental results. Section 5 comments on the conclusions and future work.

2 State of the Art

Though some taxonomy of methods can be outlined, attending to the main claim by the authors, most of the approaches combine several procedures. Here we summarize some of the procedures found in the literature as an example. A matched filter is an image pattern of the expected appearance of some image structure. The response by the matched filters can then be processed by morphological processes, such as thinning, to obtain estimations of the vessel branches which are combined heuristically. [13]. Kande et al. [15] compute a collection of responses by rotated matched filters which are then aggregated as features for fuzzy clustering to obtain the thresholding needed for detection. A connected components filter is applied to remove isolated detections. Ng et al. [23] computed a bank of second derivative Gaussian filters at different scales. The inverse model tries to find the maximum likelihood estimates of the parameters of the vessels in the image, including a noise term which is estimated from the image. Post-processing is done on the basis of the estimation that the result can be obtained by pure noise sampling. Morphological processes, such as multi-directional top-hat filters, have been also been combined with centerline detection in order to obtain a map of vessel detections in [10]. Final cleaning removes isolated non-vessel pixels. The approach of supervised classification needs some features extracted from the spatial information in the image. The AdaBoost classifier in [16] uses a battery of 41 features that include gaussian based vesselness and ridgeness features, 2D Gabor features and curvature measures. The support vector machine in [31] uses the residuals of wavelet and curvelet multi-scale transforms after a thresholding as the classification features. The result of the classification is subject to thinning and line tracking to find the vessel network structure.

3 Methods

3.1 Random Forest Classifiers

The random forests (RF) algorithm is a classifier [6] that encompasses bagging [5] and random decision forests [2, 12], being used in a variety of applications [4]. RF became popular due to its simplicity of training and tuning while offering a similar performance to boosting. Consider a RF collection of tree predictors, i.e., a RF is a large collection of decorrelated decision trees

$$h(\mathbf{x}; \psi_t), t = 1, \ldots, T,$$

where $\mathbf{x}$ is a d-dimensional random sample of random vector X, ψ_t are independent identically distributed random vectors whose nature depends on their use in the tree construction, and each tree casts a unit vote for the most popular class of input $\mathbf{x}$. Given a dataset of N samples, a bootstrapped training dataset is used to grow tree $h(\mathbf{x}; \psi_t)$ by recursively selecting a random subset of data dimensions $\hat{d}$ such that $\hat{d} \ll d$ and picking the best split of each node based on these variables. Unlike conventional decision trees, pruning is not required. The independent, identically distributed random vectors ψ_t determine the random dimension selection. The trained RF can be used for classification of a new input $\mathbf{x}$ by majority vote among the class prediction of the RF trees $C_u(x)$. The critical parameters of the RF classifier for the experiments reported below are set as follows. The number of trees in the forest should be sufficiently large to ensure that each input class receives a number of predictions: we set it to 100. The number of variables randomly sampled at each split node is $\hat{d} = 5$.

3.2 Active Learning

The performances of supervised algorithms strongly depend on the information gain provided by the data used to train the classifier. This makes the construction of the training set a cumbersome task requiring extensive manual analysis of the image. This is typically done by visual inspection of the scene and successive labeling of each sample. Consequently, the training set is highly redundant and training phase of the model is significantly slowed down. Besides, noisy pixels may interfere with the class statistics, which may lead to poor classification performances and/or overfitting. For these reasons, a training set should also be kept as small as possible and focused on those pixels effectively improving the performance of the model. Therefore a desirable training set must be constructed in a smart way, meaning it must represent correctly the class boundaries by sampling discriminative pixels. In the machine learning literature this approach to sampling is known as active learning. Active learning [9, 29] focuses on the interaction between the user and the classifier. In the context of classifier- based image segmentation [32], the system returns the user the pixels whose classification outcome is most uncertain. After accurate labeling by the user, pixels are included into the training set in order to retrain the classifier [11]. The classification model is optimized on well-chosen difficult examples, maximizing its generalization capabilities. Let $X = \{\mathbf{x}_i, y_i\}_{i=1}^{l}$ be training set labeled samples, with $\mathbf{x}_i \in \mathbb{R}^d$ and $y_i \in \{1, \ldots, N\}$. Let be $U = \{\mathbf{x}_i\}_{i=l+1}^{l+u} \in \mathbb{R}^d$ the *pool of candidates*, with $u \gg l$, corresponding to the set of unlabeled pixels to be classified. In a given iteration ϵ, the active learning algorithm selects from the pool U^ϵ the q candidates that will at the same time maximize the gain in performance and reduce the uncertainty of the classification model if added to the current training set X^ϵ. The selected samples $S^\epsilon = \{\mathbf{x}_m\}_{m=1}^{q} \subset U$ are labeled with labels $\{y_m\}_{m=1}^{q}$ by an oracle, which can be a human operator in interactive segmentation, or the ground truth. Finally, the set S^ϵ is added to the current training set ($X^{\epsilon+1} = X^\epsilon \cup S^\epsilon$) and removed from the pool of candidates ($U^{\epsilon+1} = U^\epsilon \backslash S^\epsilon$). The process is iterated until a stopping criterion is met. Algorithm 4 summarizes the active learning process. We apply a committee approach to the estimation of the unlabeled sample uncertainty [29] From the output of the RF we obtain T labels $\left\{y_i^k\right\}_{k=1}^{T}$ for each candidate sample $\mathbf{x}_i \in U$. The standard deviation σ^k of the distribution of the class predictions is used as heuristic measure of the uncertainty of its classification. The *standard deviation query-by-bagging* heuristic selection of samples to be added to the train set is stated as follows:

Algorithm 4 Active learning general algorithm

repeat
  Train a model with current training set X^ϵ
  for *each* candidate in U^ϵ **do**
    Evaluate a user-defined heuristic
  end for
  Rank the candidates in U^ϵ according to the score of the heuristic
  Select the q most interesting pixels $S^\epsilon = \{\mathrm{x}_k\}_{k=1}^{q}$
  The system assigns a label to the selected pixels $S^\epsilon = \{\mathrm{x}_k, y_k\}_{k=1}$
  Add the batch to the training set $X^{\epsilon+1} = X^\epsilon \cup S^\epsilon$
  Remove the batch from the pool of candidates $U^{\epsilon+1} = U^\epsilon \backslash S^\epsilon$
  $\epsilon = \epsilon + 1$
until accuracy > 0.99

$$\hat{\mathrm{x}}^{\mathrm{SDQB}} = \arg\max_{\mathrm{x}_i \in U} \{\sigma^k(\mathrm{x}_i)\}$$

Standard deviation is a natural multi-class heuristic measure of classification uncertainty. A candidate sample for which all the classifiers in the committee agree has a zero prediction standard deviation, thus its inclusion in the training set does not bring additional information. On the contrary, a candidate with maximum disagreement between the classifiers results in maximum standard deviation, and its inclusion will be highly beneficial.

3.3 Active Learning for Image Segmentation

The goal is to classify image pixels into at least two classes, the interested region and the background [32]. Image intensity is not a discriminant value, because often many unrelated regions have similar pixel intensity values. Therefore, a feature vector is computed for each pixel location using information extracted from its neighboring pixels. This information comes from the result of linear and/or nonlinear filtering performed on the pixel neighborhood. In this paper the features are: the voxel intensity, the mean, variance, maximum, and minimum of the voxel neighborhood, for different values of the neighborhood radius $(1,2,4...2^n)$. The definition of these features increases the data dimensionality and the complexity of the classifiers built on them.

4 Experiments

Datasets. We have performed computational experiments over 40 images of the DRIVE[1] dataset to test the proposed active learning based image classification approach. We are using 20 image for training set and the remaining 20 ones for testing purposes. *Segmentation problem.* We are looking for the segmentation of the blood vessels. Therefore, we deal with a two-class problem. *Validation.* The performance measure results of the experiments are the classification accuracy, overall error, specificity, and sensibility. We provide summary comparison of results from state-of-the-art approaches in the literature. Apart from that we are also including some testing images visual results for a visual validation. Table 1 shows the comparative results of our approach with state-of-the-art algorithms. It can be appreciated that specificity and accuracy results are close to the best in the literature. Also, sensitivity is comparable to some of them. In order to have a fair comparison, the reader must take into account that there is no post-processing or cleaning step in our algorithm, that the number of features (24) is lower than many other approaches, and the

[1] http://www.isi.uu.nl/Research/Databases/DRIVE/.

Table 1 Comparison of summary results of our approach with approaches reported in the literature

Method	Sensitivity	Specificity	Accuracy
Human rater	0.7763	0.9723	0.9470
Proposed method	0.6499	0.9803	0.9501
Abramoff et al. [24]	0.7145	–	0.9416
Staal et at. [28]	–	–	0.9442
Soares et al. [27]	–	–	0.9466
Ricci and Perfetti [25]	–	–	0.9563
Lupascu et al. [16]	0.72	–	0.9597
Xu and Luo [31]	0.7760	–	0.9328
You et al. [33]	0.7410	0.9751	0.9434
Marin et al. [17]	0.7067	0.9801	0.9452
Ng et al. [23]	0.7000	0.9530	–
Kande et at. [15]	–	–	0.8911
Salem et al. [26]	0.8215	0.9750	–
Chaudhuri et al. [7]	–	–	0.8773
Hoover et al. [13]	0.6751	0.9567	0.9267
Xiaoji and Mojon [14]	–	–	0.9212
Al-Rawi et al. [1]	–	–	0.9535
Zhanget al. [35]	0.7120	0.9724	0.9382
Cinsdikici and Aydin [8]	–	–	0.9293
Zana and Klein [34]	0.6971	–	0.9377
Mendonca and Campilho [21]	0.7344	0.9764	0.9452
M.M Fraz et al. [10]	0.7152	0.9769	0.9430
Miri and Mahloojifar [22]	0.7352	0.9795	0.9458
Martinez-Perez et al. [18]	0.6389	–	0.9181
Martinez-Perez et al. [19]	0.7246	0.9655	0.9344
Perez et al. [20]	0.6600	0.9612	0.9220
Anzalone et al. [3]	–	–	0.9419
Vlachos and Dermatas [30]	0.747	0.955	0.929

computation of these features is very straightforward, much more than the ones reported in the literature. Regarding the number of pixel samples used, in our approach we used much lesser than the other approaches in the literature due to the active learning strategy.
Visual results

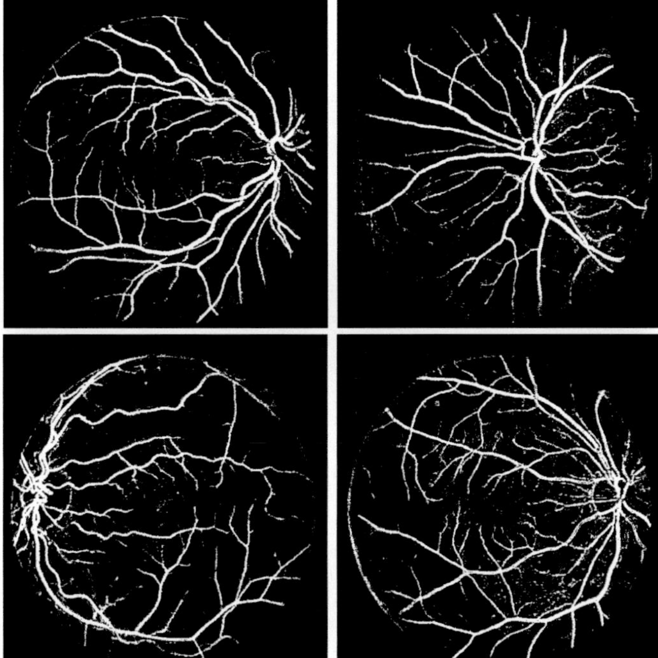

Fig. 1 Segmentation result

5 Conclusion and Future Works

In this paper we propose an active learning approach for training RF classifiers for the segmentation of the blood vessels in retinal images. The approach is very efficient in time, requiring 10 % of the training samples used by other approaches, while the image features are very simple to compute and much more efficient that those reported in the literature. Results have been provided without a cleaning or post-processing step, which is often used to increase sensitivity. Future work may include centerline and other geometrical informations for the enhanced process (Fig. 1).

Acknowledgments This research has been partially funded by Basque Government grant IT874-13 for the research group. Manuel Graña was supported by EC under FP7, Coordination and Support Action, Grant Agreement Number 316097, ENGINE European Research Centre of Network Intelligence for Innovation Enhancement.

References

1. Al-Rawi, M., Qutaishat, M., Arrar, M.: An improved matched filter for blood vessel detection of digital retinal images. Comput. Biol. Med. **37**(2), 262–267 (2007). http://dx.doi.org/10.1016/j.compbiomed.2006.03.003
2. Amit, Y., Geman, D.: Shape quantization and recognition with randomized trees. Neural comput. **9**(7), 1545–1588 (1997)
3. Anzalone, A., Bizzarri, F., Parodi, M., Storace, M.: A modular supervised algorithm for vessel segmentation in red-free retinal images. Comput. Bio. Med. **38**(8), 913–922 (2008). http://dblp.uni-trier.de/db/journals/cbm/cbm38.html#AnzaloneBPS08
4. Barandiaran, I., Paloc, C., Graña, M.: Real-time optical markerless tracking for augmented reality applications. J. R.-Time Image Process. **5**, 129–138 (2010)
5. Breiman, L.: Bagging predictors. Mach. Learn. **24**(2), 123–140 (1996)
6. Breiman, L.: Random forests. Mach. Learn. **45**(1), 5–32 (2001)
7. Chaudhuri, S., Chatterjee, S., Katz, N., Nelson, M., Goldbaum, M.: Detection of blood vessels in retinal images using two-dimensional matched filters. IEEE Trans. Med. Imaging **8**(3), 263–269 (1989). http://dx.doi.org/10.1109/42.34715
8. Cinsdikici, M.G., Aydin, D.: Detection of blood vessels in ophthalmoscope images using mf/ant (matched filter/ant colony) algorithm. Comput. Methods Programs Biomed. **96**(2), 85–95 (2009). http://dblp.uni-trier.de/db/journals/cmpb/cmpb96.html#CinsdikiciA09
9. Cohn, D., Atlas, L., Ladner, R.: Improving generalization with active learning. Mach. Learn. **15**, 201–221 (1994). doi:10.1007/BF00993277
10. Fraz, M.M., Barman, S., Remagnino, P., Hoppe, A., Basit, A., Uyyanonvara, B., Rudnicka, A.R., Owen, C.G.: An approach to localize the retinal blood vessels using bit planes and centerline detection. Comput. Methods Programs Biomed. **108**(2), 600–616 (2012). http://dblp.uni-trier.de/db/journals/cmpb/cmpb108.html#FrazBRHBURO12
11. Geremia, E., Menze, B., Clatz, O., Konukoglu, E., Criminisi, A., Ayache, N.: Spatial decision forests for MS lesion segmentation in multi-channel MR images. MedicalImage Computing and Computer-Assisted Interventio—MICCAI 2010. Lecture Notes in Computer Science, vol. 6361, pp. 111–118. Springer, Heidelberg (2010)
12. Ho, T.: The random subspace method for constructing decision forests. IEEE Trans. Patt. Anal. Mach. Intell. **20**(8), 832–844 (1998)
13. Hoover, A., Kouznetsova, V., Goldbaum, M.: Locating blood vessels in retinal images by piecewise threshold probing of a matched filter response. IEEE Trans. Med. Imaging **19**, 203–210 (2000)
14. Jiang, X., Mojon, D.: Adaptive local thresholding by verification-based multithreshold probing with application to vessel detection in retinal images. IEEE Trans. Pattern Anal. Mach. Intell. **25**(1), 131–137 (2003). http://dx.doi.org/10.1109/TPAMI.2003.1159954
15. Kande, G.B., Subbaiah, P.V., Tirumala, S.S.: Unsupervised fuzzy based vessel segmentation in pathological digital fundus images. J. Med. Syst. **34**(5), 849–858 (2010). http://dblp.uni-trier.de/db/journals/jms/jms34.html#KandeST10
16. Lupascu, C.A., Tegolo, D., Trucco, E.: Fabc: retinal vessel segmentation using adaboost. IEEE Trans. Inf. Tech. Biomed. **14**(5), 1267–1274 (2010). http://dblp.uni-trier.de/db/journals/titb/titb14.html#LupascuTT10
17. Marin, D., Aquino, A., Gegundez-Arias, M.E., Bravo, J.M.: A new supervised method for blood vessel segmentation in retinal images by using gray-level and moment invariants-based features. IEEE Trans. Med. Imaging **30**(1), 146–158 (2011). http://dblp.uni-trier.de/db/journals/tmi/tmi30.html#MarinAGB11
18. Martínez-Pérez, M.E., Hughes, A.D., Stanton, A.V., Thom, S.A., Bharath, A.A., Parker, K.H.: Retinal blood vessel segmentation by means of scale-space analysis and region growing. In: Taylor, C., Colchester, A (eds.) Medical Image Computing and Computer-Assisted Intervention–MICCAI'99. Lecture Notes in Computer Science, vol. 1679, pp. 90–97. Springer, Heidelberg (1999). http://dblp.uni-trier.de/db/conf/miccai/miccai1999.html#Martinez-PerezHSTBP99

19. Martinez-Perez, M.E., Hughes, A.D., Thom, S.A., Bharath, A.A., Parker, K.H.: Segmentation of blood vessels from red-free and fluorescein retinal images. Med. Image Anal. **11**(1), 47–61 (2007). http://dblp.uni-trier.de/db/journals/mia/mia11.html#Martinez-PerezHTBP07
20. Martinez-Perez, M., Hughes, A., Thom, S., Parker, K.: Improvement of a retinal blood vessel segmentation method using the insight segmentation and registration toolkit (itk). IEEE Comput. Soc. 892–895 (2007). http://dx.doi.org/10.1109/IEMBS.2007.4352434
21. Mendonca, A.M., Campilho, A.C.: Segmentation of retinal blood vessels by combining the detection of centerlines and morphological reconstruction. IEEE Trans. Med. Imaging **25**(9), 1200–1213 (2006). http://dblp.uni-trier.de/db/journals/tmi/tmi25.html#MendoncaC06
22. Miri, M.S., Far, A.M.: Retinal image analysis using curvelet transform and multistructure elements morphology by reconstruction. IEEE Trans. Biomed. Eng. **58**(5), 1183–1192 (2011). http://dblp.uni-trier.de/db/journals/tbe/tbe58.html#MiriM11
23. Ng, J., Clay, S.T., Barman, S.A., Fielder, A.R., Moseley, M.J., Parker, K.H., Paterson, C.: Maximum likelihood estimation of vessel parameters from scale space analysis. Image Vision Comput. **28**(1), 55–63 (2010). http://dx.doi.org/10.1016/j.imavis.2009.04.019
24. Niemeijer, M., Staal, J., van Ginneken, B., Loog, M., Abramoff, M.: Comparative study of retinal vessel segmentation methods on a new publicly available database. In: Fitzpatrick, J.M., Sonka, M. (eds.) SPIE Med. Imaging, vol. 5370, pp. 648–656. SPIE, USA (2004)
25. Ricci, E., Perfetti, R.: Retinal blood vessel segmentation using line operators and support vector classification. IEEE Trans. Med. Imaging **26**(10), 1357–1365 (2007)
26. Salem, S.A., Salem, N.M., Nandi, A.K.: Segmentation of retinal blood vessels using a novel clustering algorithm (racal) with a partial supervision strategy. Med. Biol. Eng. Comput. **45**(3), 261–273 (2007). http://dblp.uni-trier.de/db/journals/mbec/mbec45.html#SalemSN07
27. Soares, J.V.B., Le, J.J.G., Cesar, R.M., Jelinek, H.F., Cree, M.J., Member, S.: Retinal vessel segmentation using the 2-d gabor wavelet and supervised classification. IEEE Trans. Med. Imaging **25**, 1214–1222 (2006)
28. Staal, J., Abramoff, M., Niemeijer, M., Viergever, M., van Ginneken, B.: Ridge-based vessel segmentation in color images of the retina. IEEE Trans. Med. Imaging **23**(4), 501–509 (2004)
29. Tuia, D., Pasolli, E., Emery, W.: Using active learning to adapt remote sensing image classifiers. Remote Sensing of Environment (2011). http://linkinghub.elsevier.com/retrieve/pii/S0034425711001507
30. Vlachos, M., Dermatas, E.: Multi-scale retinal vessel segmentation using line tracking. Comput. Med. Imaging Graph. **34**(3), 213–227 (2010). http://dblp.uni-trier.de/db/journals/cmig/cmig34.html#VlachosD10
31. Xu, L., Luo, S.: A novel method for blood vessel detection from retinal images. BioMed. Eng. OnLine **9**(1), 14 (2010). http://www.biomedical-engineering-online.com/content/9/1/14
32. Yaqub, M., Javaid, M., Cooper, C., Noble, J.: Improving the Classification Accuracy of the Classic RF Method by Intelligent Feature Selection and Weighted Voting of Trees with Application to Medical Image Segmentation. In: Suzuki, Kenji, Wang, Fei, Shen, Dinggang, Yan, Pingkun (eds.) Machine Learning in Medical Imaging. Lecture Notes in Computer Science, vol. 7009, pp. 184–192. Springer, Heidelberg (2011)
33. You, X., Peng, Q., Yuan, Y., Cheung, Y.m., Lei, J.: Segmentation of retinal blood vessels using the radial projection and semi-supervised approach. Patt. Recognit. **44**(10–11), 2314–2324 (2011). http://dx.doi.org/10.1016/j.patcog.2011.01.007
34. Zana, F., Klein, J.C.: Segmentation of vessel-like patterns using mathematical morphology and curvature evaluation. IEEE Trans. Image Process. **10**(7), 1010–1019 (2001). http://dblp.uni-trier.de/db/journals/tip/tip10.html#ZanaK01
35. Zhang, B., Zhang, L., Zhang, L., Karray, F.: Retinal vessel extraction by matched filter with first-order derivative of gaussian. Comput. Biol. Med. **40**(4), 438–445 (2010). http://dx.doi.org/10.1016/j.compbiomed.2010.02.008

A Comparison of Differential Evolution and Genetic Algorithms for the Column Subset Selection Problem

Pavel Krömer and Jan Platoš

Abstract The column subset selection problem is a well-known complex optimization problem that has a number of appealing real-world applications including network and data sampling, dimension reduction, and feature selection. There are a number of traditional deterministic and randomized heuristic algorithms for this problem. Recently, it has been tackled by a variety of bio-inspired and evolutionary methods. In this work, differential evolution, a popular and successful real-parameter optimization algorithm, adapted for fixed-length subset selection, is used to find solutions to the column subset selection problem. Its results are compared to a recent genetic algorithm designed for the same purpose.

Keywords Differential evolution · Genetic algorithms · Column subset selection

1 Introduction

This work investigates the ability of a straightforward variant of differential evolution (DE) to solve a well-known combin atorial optimization problem, the column subset selection problem (CSSP). The considered approach is purely meta-heuristic and exploits the ability of the DE to find good CSSP solutions without domain-specific knowledge. Such strictly meta-heuristic strategy allows a straightforward re-use of this approach for a wider class of fixed-length subset selection problems that includes, for example, network sampling [17], feature selection [23], and the p-median problem [18]. The DE is in this work compared to another entirely meta-heuristic method based on genetic algorithms (GA) on a number of real-world and randomly generated CSSP instances. The GA-based algorithm for CSSP used in this

P. Krömer (✉) · J. Platoš
IT4Innovations & Department of Computer Science, VŠB Technical University of Ostrava, Ostrava, Czech Republic
e-mail: pavel.kromer@vsb.cz

J. Platoš
e-mail: jan.platos@vsb.cz

R. Burduk et al. (eds.), *Proceedings of the 9th International Conference on Computer Recognition Systems CORES 2015*, Advances in Intelligent Systems and Computing 403, DOI 10.1007/978-3-319-26227-7_21

study as a baseline meta-heuristic method has been introduced recently in [17]. It uses a novel solution representation and modified genetic operators. The algorithm was developed for network sampling and later used to address the p-median problem [18] and the CSSP [20] with encouraging results. Another recent study has demonstrated that a plain DE performs on the p-median problem equally well [16]. This work takes a similar approach and investigates the ability of a simple differential evolution to solve the CSSP. The quality of obtained results is compared to the GA first used for the CSSP in [20]. Such side-by-side comparison of purely meta-heuristic nature inspired methods for CSSP was not available prior to this study.

2 Column Subset Selection Problem

The CSSP is a combinatorial optimization problem known from linear algebra and computer science [4].

Definition 1 (*Column subset selection problem*) Given the matrix $A \in R^{m \times n}$ and positive integer $k < n$, pick k columns of A forming a new matrix $C \in R^{m \times k}$ such that the residual

$$\|A - P_C A\|_\xi \tag{1}$$

is minimized over all possible $\binom{n}{k}$ column choices for the matrix C. Here, $P_C = CC^\dagger$ denotes the projection onto the k-dimensional space spanned by the columns of C, $\xi = \{2, F\}$ denotes the spectral norm and Frobenius norm, respectively [4], and $C^\dagger$ stands for the Moore–Penrose pseudoinverse [14] of C.

Informally, the goal of the CSSP is to find a subset of exactly k columns selected from all n columns of A with a minimum error. The problem is appealing due to the variety of its applications and challenging because of its complexity and expected NP-hardness. A recent study shows that the CSSP is unique games (UG) hard [6], which, under the acceptance of the unique games conjecture, confirms its NP-hardness. The CSSP was in the past solved by a number of deterministic and non-deterministic methods. Čivril and Magdon-Ismail [7] proposed a procedure based on a sparse approximation of the singular value decomposition. The algorithm utilized a combination of a greedy solution to the generalization of the sparse approximation problem and an existence result on the possibility of sparse approximation. Boutsidis et al. [4] presented a hybrid two-stage algorithm that combines a randomized and deterministic steps. In the randomized phase, the algorithm randomly selects $O(k \log k)$ columns according to the probability distribution, which depends on the information in the top-k right singular subspace of matrix A. Another study [2] describes a different set of algorithms that approximate the CSSP in both, a randomized and a deterministic way. Farahat et al. [11, 12] proposed a novel greedy algorithm for large-scale data matrices which uses a recursive function to compute the reconstruction error and a massively parallel computational strategy based on the MapReduce paradigm. The algorithm uses a random projection of data columns and then solves the generalized CSSP on each machine participating in the distributed computation. The authors also presented a fast greedy algorithm for generalized CSSP [13]. Couvreur and Bresler [8] investigated the optimality of the greedy approach to the generalized CSSP and described a simple implementation based on a QR downdating scheme using Givens rotations. Balzano et al. [3] studied the CSSP with special attention to data with missing values. The authors performed several simulations and compared their results to traditional deterministic algorithms for CSSP. Deshpande and Rademacher [9] presented a randomized algorithm with greater efficiency than in [4]. Recently, Santana and Canuto [10] solved a slightly different problem of the selection of a subset of features for ensemble systems. The authors used bio-inspired algorithms including ant colony optimization, particle swarm optimization, and genetic algorithms to solve the problem and the experiments showed that all three algorithms were able to solve that problem very efficiently. A genetic algorithm for CSSP is due to Krömer et al. [20]. The authors used a new GA for fixed-length subset selection based on ordering of alleles in the chromosomes and modified genetic operators employed to find good CSSP solutions. The quality of obtained results was demonstrated on several randomly generated CSSP instances. A hybrid bio-inspired ensemble method for feature selection and classification was developed by Tan et al. [23]. The authors used a modified micro-GA to form an ensemble optimizer for selecting a small number of input features and to optimize a neural network used for the classification task.

3 Differential Evolution

The DE is a population-based optimizer that evolves a population of real encoded vectors representing the solutions to given problem. It was introduced by Storn and Price in 1995 [22] and quickly became a popular alternative to the more traditional types of evolutionary algorithms. The algorithm evolves a population of candidate solutions by an iterative modification of candidate solutions by the application of the differential mutation and crossover [20]. In each iteration, so-called trial vectors are created from current population by the differential mutation and further modified by various types of crossover operator. At the end, the trial vectors compete with existing candidate solutions for survival in the population. The DE starts with an initial population of N real-valued vectors. The vectors are initialized with real values either randomly or so, that they are evenly spread over the problem space. The latter initialization leads to better results of the optimization [20]. During the optimization, the DE generates new vectors that are scaled perturbations of existing population vectors. The algorithm perturbs selected base vectors with the scaled difference of two (or more) other population vectors in order to produce the trial vectors. The trial vectors compete with members of the current population with the same index called the target vectors. If a trial vector represents a better solution than the corresponding target vector, it takes its place in the population [20]. The basic operations of the classic DE can be summarized using the following formulas [20]: the random initialization of the ith vector with N parameters is defined by

$$x_j^i = urand(b_j^L, b_j^U), \ \ j \in \{1, \ldots, N\} \tag{2}$$

where b_j^L is the lower bound of jth parameter, b_j^U is the upper bound of jth parameter and $urand(a, b)$ is a function generating a uniform random number from the range (a, b). A simple form of standard differential mutation is given by

$$\boldsymbol{v}^i = \boldsymbol{v}^{r1} + F(\boldsymbol{v}^{r2} - \boldsymbol{v}^{r3}) \tag{3}$$

where F is the scaling factor and $\boldsymbol{v}^{r1}$, $\boldsymbol{v}^{r2}$ and $\boldsymbol{v}^{r3}$ are three random vectors from the population. The vector $\boldsymbol{v}^{r1}$ is the base vector, $\boldsymbol{v}^{r2}$ and $\boldsymbol{v}^{r3}$ are the difference vectors, and the ith vector in the population, $\boldsymbol{v}^i$ is the trial vector. It is required that $i \neq r1 \neq r2 \neq r3$. The uniform (binomial) crossover that combines the target vector, $\boldsymbol{x}^i$, with the trial vector, $\boldsymbol{v}^i$, is given by

$$v_j^i = \begin{cases} v_j^i & \text{if}(urand(0, 1) < C) \text{ or } j = j_{\text{rand}} \\ x_j^i, & \text{otherwise} \end{cases} \tag{4}$$

for each $j \in \{1, \ldots, N\}$. The random index j_{rand} is in the above selected randomly as $j_{\text{rand}} = urand(1, N)$. The uniform crossover replaces the parameters in $\boldsymbol{v}^i$ by the parameters from the target vector $\boldsymbol{x}^i$ with probability $1 - C$.

4 Experiments

A series of computational experiments with several randomly generated and real-world data sets was conducted in order to assess the ability of a simple DE to solve the CSSP. A recent genetic algorithm for column subset selection [20] was used to solve the same CSSP instances to provide a comparison with another purely meta-heuristic evolutionary approach.

4.1 Simple DE for CSSP

The DE considered in this work uses a simple real-valued representation of candidate solutions. Each solution is represented by a real-valued vector, $\boldsymbol{c}$, of the size k. The vector is decoded into a set of column indices, K, $|K| = k$, using the following simple strategy: every candidate vector coordinate, c_i, is truncated and added to K. If $j = trunc(c_i)$ already exists in K, next available column index that is not included in K yet, is added to the set. The fitness function is the only problem-specific part of the investigated algorithm. It is based on the general definition of CSSP from Eq. (1) and uses Frobenius norm [15],

$$||A||_F = \sqrt{\sum_{i=1}^{m}\sum_{j=1}^{n}|a_{ij}|^2}, \quad (5)$$

to compute the matrix residual defined by chromosome $\boldsymbol{c}$

$$fit(\boldsymbol{c}) = ||A - CC^{\dagger}A||_F, \quad C = [a_{ij}], \ i = \{0, \ldots, m\}, \ j = \{c_1, \ldots, c_k\}. \quad (6)$$

The evaluation of fitness function is a computationally costly procedure that involves calculation of $C^{\dagger}$ and three matrix multiplications. In this work, it was offloaded to a GPU to accelerate the computation. However, it must be noted that an efficient data-parallel implementation is not the focus of this work.

4.2 Genetic Algorithm for CSSP

A novel GA for fixed-length subset selection was introduced in [17] and used for p-median problem and CSSP in [18] and [20], respectively. It uses a compact chromosome encoding and modified genetic operators to facilitate efficient evolution of fixed-length subsets. A subset of k objects (out of n) is in this GA represented as an ordered set of k indices, $\boldsymbol{c} = (c_1, \ldots c_k)$, $\forall(i, j) \in \{0, \ldots k\} : i < j \implies c_i < c_j$. To avoid creation of invalid individuals, the GA uses modified mutation and crossover operators that maintain the ordering of indices within the chromosomes. *Order-preserving mutation* replaces the ith gene, c_i, in a chromosome, $\boldsymbol{c}$, with a random value drawn from the interval defined by its left and right neighbour as defined in Eq. (7)

$$mut(c_i) = \begin{cases} urand^*(0, c_{i+1}), & \text{if } i = 0 \\ urand(c_{i-1}, c_{i+1}), & \text{if } i \in (0, N-1) \\ urand(c_{i-1}, N), & \text{if } i = N-1 \end{cases}, \quad (7)$$

where $i \in \{0, \ldots, N\}$ and $urand(a, b)$ selects a uniform pseudo-random integer from the interval (a, b) (whereas $urand^*(a, b)$ selects a uniform pseudo-random integer from the interval $[a, b)$). Such mutation operator guarantees that the ordering of the indices within the chromosome remains valid through the evolution. *Order-preserving crossover* is an extension of the traditional one-point crossover operator [1]. It selects a random position, i, in parent chromosomes $\mathbf{c1}$ and $\mathbf{c2}$, which is inspected for suitability to be a crossover point. A position, i, is suitable for crossover it the condition $c1_i < c2_{i+1} \wedge c2_i < c1_{i+1}$ holds. If it does not hold for i, the remaining positions in the chromosomes are searched for a suitable crossover point.

4.3 Algorithms and Settings

Both DE and GA employed in this work were implemented from scratch in C++ and used to find solutions of several CSSP instances. The DE was */DE/rand/1* with scaling factor $F = 0.1$, adaptive crossover rate $C = \frac{1}{\text{max\_fitness}}$, and population size 100. The genetic algorithm was a steady-state GA with generation gap 2 (parents were instantly replaced by superior offspring chromosomes), population 100, order-preserving crossover with crossover probability $c = 0.9$, and order-preserving mutation with probability $m = 0.4$. It used a combination of roulette-wheel selection and elitist parent selection. The GA was used to minimize the objective function. The variants of used algorithms as well as their parameters were selected on the basis of previous experience and initial trials. Because of the stochastic nature of both, DE and GA, all experiments were executed 30 times independently and presented results are averages over the 30 runs.

4.4 Test Data

A number of randomly generated CSSP instances and problem instances generated from real-world data sets were used to evaluate the ability of the investigated DE to search for good column subsets from different matrices with various properties. Subsets of 5, 15, and 25 columns were selected from every test CSSP instance. Four artificial data sets named R50, R100, R150, and R200 were generated randomly. All values of these matrices, r_{ij}, were drawn from the interval [0, 10] at random. The matrices represent dense and highly irregular problem instances with different dimensions. The execution of DE and GA was terminated after 50,000 fitness function evaluations when looking for solutions of these random instances. The first real-world data set used in presented experiments is the Yale collection of 165 faces of 15 individuals [5]. Each picture has a resolution of 32×32 pixels. The collection contains 11 images per subject, one per different facial expression or configuration. All images were merged into a data matrix with 165 rows and 1024 columns termed *yale*. Three remaining real-world data sets were taken from the Feature Selection Challenge which has been created for NIPS 2003 Workshop on Feature Extraction. The *arcene*, *madelon*, and *dexter* data sets were selected as realistic CSSP instances with different properties. The execution of DE and GA on the real-world instances was terminated after 10,000 fitness function evaluations.

The properties of all used CSSP instances are summarized in Table 1 and the matrices are visualized in Figs. 1 and 2. It can be seen that the random instances are irregular and noisy while the real-world instances are much more structured. However, some of them (i.e. *dexter*) are sparse with only few non-zero elements (shown as darker pixels) and therefore yield a rather flat fitness landscape. It can be also noted that the range of values in these matrices is much wider than the range of values in random CSSP instances.

Table 1 Data set properties

Name	Rows (m)	Features (n)	Name	Rows (m)	Features (n)
R50	50	50	*Yale*	165	1,024
R100	100	100	*Arcene*	100	10,000
R150	150	150	*Madelon*	2,000	500
R200	200	200	*Dexter*	20,000	300

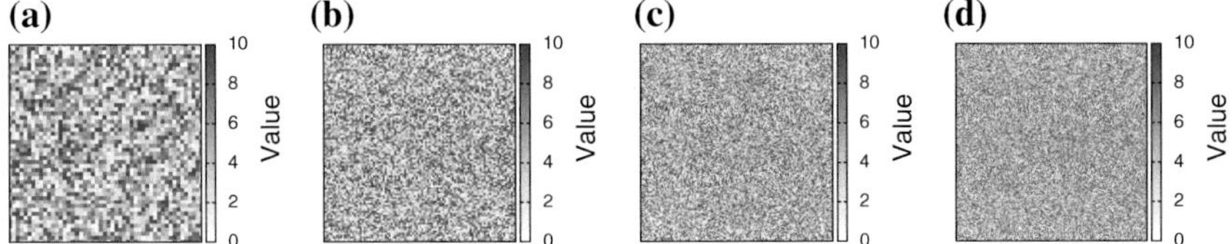

Fig. 1 Randomly generated CSSP instances. **a** R50. **b** R100. **c** R150. **d** R200

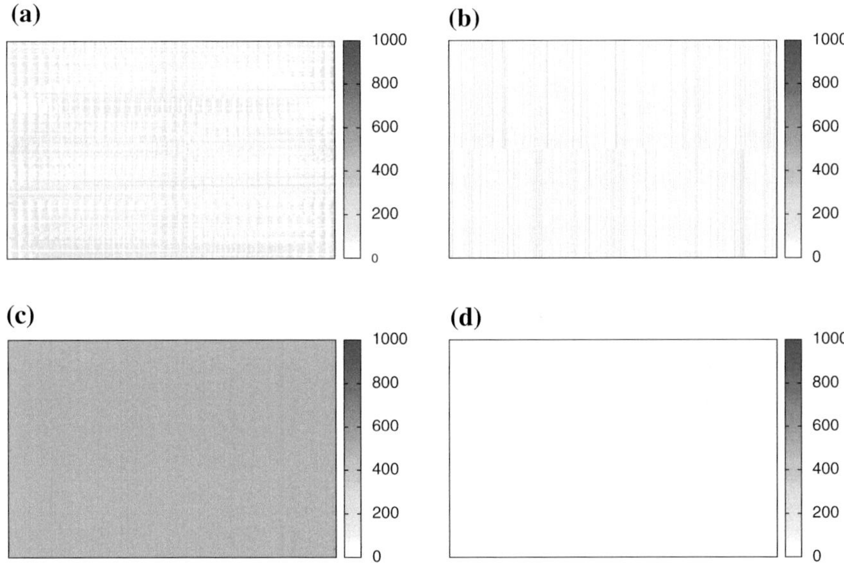

Fig. 2 Real-world data sets as CSSP instances. **a** Yale. **b** Arcene. **c** Madelon. **d** Dexter

4.5 Results

The results of evolutionary search for column subsets by DE and GA are summarized in Tables 2 and 3, respectively. Table 2 shows that the DE was for the randomly generated CSSP instances able to find, on average, slightly better column subsets. The average best solution, found by independent DE runs, was better than the average best solution found by the GA for all problem instances and all values of k with the

Table 2 Average minimum and mean fitness of CSSP solutions found by DE and GA for randomly generated problem instances

Instance	k	DE		GA	
		Minimum fit.	Mean fit.	Minimum fit.	Mean fit.
R50	5	*133.0452*	*133.0475*	133.197	133.7344
	15	*97.80042*	*98.09294*	97.8785	98.76577
	25	*63.82975*	*64.32986*	64.6106	65.69874
R100	5	*281.6881*	*281.9820*	281.961	283.3103
	15	*244.2703*	*244.6569*	244.569	245.4339
	25	*210.9884*	*211.6895*	211.003	212.0016
R150	5	*435.5153*	*436.8175*	437.257	439.5902
	15	*393.1180*	*393.6290*	393.566	394.1902
	25	*356.1570*	*356.5939*	356.322	356.9892
R200	5	*590.7519*	*592.2415*	592.218	594.2606
	15	*539.5831*	*540.8179*	540.240	541.0546
	25	505.6092	*506.1364*	*505.250*	506.2188

Table 3 Average minimum and mean fitness of CSSP solutions found by DE and GA for real-world problem instances

Instance	k	DE		GA	
		Minimum fit.	Mean fit.	Minimum fit.	Mean fit.
yale	5	15577.6738	*15672.4795*	*15563.6000*	15709.4100
	15	12167.6268	12244.3520	*12128.8000*	*12229.1500*
	25	10413.0359	10467.8728	*10292.4000*	*10426.1266*
arcene	5	40825.1781	41422.5747	*40605.8000*	*41008.8052*
	15	29952.5296	30198.9125	*29597.0000*	*29890.2304*
	25	25088.1984	25269.0355	*24686.1000*	*24919.5666*
madelon	5	25894.9217	26013.1776	*25892.2000*	*26009.5550*
	15	*24996.3896*	*25008.0498*	25000.9000	25025.2923
	25	*24201.8284*	*24224.3746*	24224.7000	24253.1400
dexter	5	20381.8832	20992.2281	*20150.0000*	*20468.7909*
	15	*19526.0790*	*19971.7958*	19722.3000	20003.5666
	25	19275.3857	19684.5465	*18967.7000*	*19373.2777*

exception of R200 and $k = 25$. The average mean fitness of final populations found by the DE was in all cases better than the average mean fitness of final populations found by the GA. Different results were obtained when solving CSSP instances created from real-world data sets. As shown in Table 3, the GA was in this case more successful than the DE. The average minimum fitness, found by independent DE runs, was better than the average minimum fitness of solutions found by the GA in just 3 out of 12 test cases (*madelon* with $k = 15$ and $k = 25$ and *dexter* with $k = 15$).

The average mean fitness of final populations found by the DE was better in four cases (*madelon* with $k = 15$ and $k = 25$, *dexter* with $k = 15$, and *yale* with $k = 5$). The GA has found better CSSP solutions in all other test cases. This behaviour suggests that the used GA might be a better choice for more structured CSSP instances. However, the difference between the fitness of solutions found by DE and GA was again rather modest, especially considering the dimension and structure of these instances.

5 Conclusions

A simple variant of differential evolution was in this work used to search for optimal column subsets. The algorithm was implemented without any type of local search and its results were compared to another pure meta-heuristic method based on genetic algorithms. Two sets of CSSP instances were in this work used to assess the ability of the DE to find good CSSP solutions. Although the DE obtained slightly better results for randomly generated problem instances and the GA performed better for CSSP instances generated from real-world data, the results of both algorithms were generally comparable. The results suggest that even a simple DE can be used as a valid evolutionary method for meta-heuristic search for CSSP solutions. This is an encouraging result because DE is, due to its simplicity, suitable for efficient high-performance implementation, e.g. massively parallel environments [19].

Acknowledgments This work was supported by the IT4Innovations Centre of Excellence project (CZ.1.05/1.1.00/02.0070), funded by the European Regional Development Fund and the national budget of the Czech Republic via the Research and Development for Innovations Operational Programme and by Project SP2015/146 of the Student Grant System, VŠB - Technical University of Ostrava.

References

1. Affenzeller, M., Winkler, S., Wagner, S., Beham, A.: Genetic Algorithms and Genetic Programming: Modern Concepts and Practical Applications. Chapman & Hall/CRC, Boca Raton (2009)
2. Avron, H., Boutsidis, C.: Faster subset selection for matrices and applications. SIAM J. Matrix Anal. Appl. **34**(4), 1464–1499 (2013)
3. Balzano, L., Nowak, R., Bajwa, W.U.: Column subset selection with missing data. In: NIPS workshop on low-rank methods for large-scale machine learning (2010)
4. Boutsidis, C., Mahoney, M.W., Drineas, P.: An improved approximation algorithm for the column subset selection problem. In: Proceedings of the Twentieth Annual ACM-SIAM Symposium on Discrete Algorithms, pp. 968–977. SODA '09. Society for Industrial and Applied Mathematics, Philadelphia (2009)
5. Cai, D., He, X., Han, J.: Spectral regression for efficient regularized subspace learning. In: Proceedings of the International Confeence on Computer Vision (ICCV'07) (2007)
6. Çivril, A.: Column subset selection problem is UG-hard. J. Comput. Syst. Sci. **80**(4), 849–859 (2014)

7. Çivril, A., Magdon-Ismail, M.: Column subset selection via sparse approximation of SVD. Theoret. Comput. Sci. **421**, 1–14 (2012)
8. Couvreur, C., Bresler, Y.: On the optimality of the backward greedy algorithm for the subset selection problem. SIAM J. Matrix Anal. Appl. **21**(3), 797–808 (2000)
9. Deshpande, A., Rademacher, L.: Efficient volume sampling for row/column subset selection. In: Proceedings of the 2010 IEEE 51st Annual Symposim on Foundations of Computer Science, pp. 329–338. FOCS '10. IEEE Computer Society, Washington (2010)
10. dos S. Santana, L.E.A., de Paula Canuto, A.M.: Filter-based optimization techniques for selection of feature subsets in ensemble systems. Expert Syst. Appl. **41**(4, Part 2), 1622–1631 (2014)
11. Farahat, A.K., Elgohary, A., Ghodsi, A., Kamel, M.S.: Distributed column subset selection on mapreduce. In: 2013 IEEE 13th International Conference on Data Mining (ICDM), pp. 171–180, December 2013
12. Farahat, A.K., Elgohary, A., Ghodsi, A., Kamel, M.S.: Greedy column subset selection for large-scale data sets. CoRR abs/1312.6838 (2013)
13. Farahat, A.K., Ghodsi, A., Kamel, M.S.: A fast greedy algorithm for generalized column subset selection. CoRR abs/1312.6820 (2013)
14. Friedberg, S.: Linear Algebra, 4th edn. Prentice-Hall Of India Pvt Limited, New Delhi (2003)
15. Golub, G., Van Loan, C.: Matrix Computations. Johns Hopkins Studies in the Mathematical Sciences, 3rd edn. Johns Hopkins University Press, Baltimore (1996)
16. Kromer, P., Platos, J.: Solving the p-median problem by a simple differential evolution. In: 2014 IEEE International Conference on Systems, Man and Cybernetics (SMC), pp. 3503–3507, October 2014
17. Kromer, P., Platos, J.: Genetic algorithm for sampling from scale-free data and networks. In: Proceedings of the 2014 Conference on Genetic and Evolutionary Computation, pp. 793–800. GECCO '14. ACM, New York (2014)
18. Krömer, P., Platoš, J.: New genetic algorithm for the p-median problem. In: Pan, J.-S., Snasel, V., Corchado, E.S., Abraham, A., Wang, S.-L. (eds.) Intelligent Data Analysis and Its Applications, Volume II. Advances in Intelligent Systems and Computing, vol. 298, pp. 35–44. Springer, Heidelberg (2014)
19. Kromer, P., Snasel, V., Platos, J., Abraham, A.: Many-threaded implementation of differential evolution for the cuda platform. In: Proceedings of the 13th Annual Conference on Genetic and Evolutionary Computation, pp. 1595–1602. GECCO '11. ACM, New York (2011)
20. Kromer, P., Platos, J., Snasel, V.: Genetic algorithm for the column subset selection problem. In: 2014 Eighth International Conference on Complex, Intelligent and Software Intensive Systems (CISIS), pp. 16–22, July 2014
21. Price, K.V., Storn, R.M., Lampinen, J.A.: Differential Evolution A Practical Approach to Global Optimization. Natural Computing Series. Springer, Berlin (2005)
22. Storn, R., Price, K.: Differential evolution- a simple and efficient adaptive scheme for global optimization over continuous spaces. Technical report (1995)
23. Tan, C.J., Lim, C.P., Cheah, Y.: A multi-objective evolutionary algorithm-based ensemble optimizer for feature selection and classification with neural network models. Neurocomputing **125**, 217–228 (2014)

Experiments on Data Classification Using Relative Entropy

Michal Vašinek and Jan Platoš

Abstract Data classification is one of the basic tasks in data mining. In this paper, we propose a new classifier based on relative entropy, where data to particular class assignment is made by the majority good guess criteria. The presented approach is intended to be used when relations between datasets and assignment classes are rather complex, nonlinear, or with logical inconsistencies; because such datasets can be too complex to be classified by ordinary methods of decision trees or by the tools of logical analysis. The relative entropy evaluation of associative rules can be simple to interpret and offers better comprehensibility in comparison to decision trees and artificial neural networks.

Keywords Category · Data classification · Data mining · Relative entropy · Kullback–Leibler

1 Introduction

Data classification and data compression shares several common concepts, first of all they both try to reduce provided data into some smaller unit. In this sense, data classification can be considered as a lossy compression, but in data classification ability to recover former data from resulting class is not our ambition, we are perfectly confident with reduction and recovery is not needed. There are several basic categories of classification algorithms, there are algorithms based on decision trees, learning set of rules, neural networks, naive Bayesian classifiers, instance-based learning, and support vector machines, a review of algorithms can be found in Kotsiantis [1]. Classification algorithms presented in this paper belong to the learning set of rules class, extensive overview of the learning set of rules class of algorithms is provided

M. Vašinek (✉) · J. Platoš
FEECS, Department of Computer Science, VŠB-Technical University of Ostrava, 17.listopadu, 708 33 Ostrava, Poruba, Czech Republic
e-mail: michal.vasinek@vsb.cz

J. Platoš
e-mail: jan.platos@vsb.cz

R. Burduk et al. (eds.), *Proceedings of the 9th International Conference on Computer Recognition Systems CORES 2015*, Advances in Intelligent Systems and Computing 403, DOI 10.1007/978-3-319-26227-7_22

in [2]. Experiments made in this paper use several concepts from Information Theory [3, 4], especially concept of entropy and relative entropy. The role of entropy in data classification was already studied and several algorithms reducing entropy of training dataset like ID3 [5] and PRISM [6] were developed. The main idea covered in this paper is to use relative entropy to evaluate rules, such evaluation can then be used to sort rules and consequently to select first n of them for classification purposes. Class of classifiers presented in this paper is in the present state able to distinguish only between two classes, so our presented results will deal only with binary classification. We compared our results with the work of Thabtah [7] and Li [8]. The rest of the paper is organized as follows. Section 2 contains description of entropy, relative entropy, and introduces basics of these concepts. Section 3 describes rules, their types, and how can be rules evaluated by relative entropy. Section 4 describes the proposed classifier. Section 5 contains discussion and presents results achieved on the selected datasets. Last Sect. 6 concludes the paper and discusses the future experiments.

2 Entropy

Entropy is the key concept of Information Theory, but the term itself has many interpretations, statisticians would say it is uncertainty in random variable, Information Theory scientist would say it is the amount of information, data compression scientist would say it is the average number of bits needed to describe symbols and we can continue with physicists and so on; in this paper, we will follow Fano's [9] interpretation of entropy, as in his point of view the entropy is an average number of binary questions that we would put in infinitely many trials to distinguish between different events. When probabilities of classes are given, we can compute entropy by Shannon's equation:

$$H = -\sum_{x} p(x) \log p(x) \tag{1}$$

Entropy is always nonnegative and is zero only when one item x_i has probability $p(x_i) = 1$, since when probability of some event is equal to one, then we do not need to put any questions about incoming event, because we know exactly what the event is. All logarithms in this paper are based two. For the given set of events, the entropy is maximal when distribution of events is uniform: $p(x_i) = p(x_j)$ for all indices i, j of events x_i, x_j in $p(x)$.

2.1 Relative Entropy

Using entropy, we get an amount of information respective number of binary questions about single probability distribution. In Information Theory, the concept of relative entropy $D(P||Q)$ is used to measure distance between two different probability distributions $p(x)$ and $q(x)$.

$$D(P||Q) = \sum_{x \in \Sigma} p(x) \log \frac{p(x)}{q(x)} \tag{2}$$

Relative entropy defined by (2) measures distance in bits, respectively; in extra binary questions, we have to put if instead of proper distribution $p(x)$ use other distribution $q(x)$. Relative entropy in (2) is nonnegative but it is not a metric function, because symmetry condition fails: $D(P||Q) \neq D(Q||P)$. In the former paper of Kullback and Leibler [10], authors derived symmetric measure later called by their names as Kullback–Leibler divergence D_{KL}:

$$D_{KL} = D(P||Q) + D(Q||P) \tag{3}$$

When computing rules, we analyze individual terms in summations of (2) and (3), the individual term:

$$D(x) = p(x) \log \frac{p(x)}{q(x)} \tag{4}$$

uncovers several properties about single shared event x from the two distributions p, q when they are compared. Since probabilities are defined on closed interval $p(x) \in < 0; 1 >$, then $D(x)$ is positive when $\log \frac{p(x)}{q(x)} > 0$ and so must hold that $p(x) > p(y)$. When both distributions are equal $p(x) = q(x)$ for all x then each term $\log \frac{p(x)}{p(x)} = 0$ and also $D(P||Q) = D_{KL}(P||Q) = 0$. If we view entropy as an average number of questions to differentiate between classes, then relative entropy can be interpreted as the increase of an average number of question, we have to put, if instead of distribution P distribution Q is used. We hope that concepts of entropy and relative entropy can be more comprehensible for interpretation than, for example, decision tree structure or weight given by artificial neural network.

2.2 Zeros and Infinities in Relative Entropy

By definition $0 \log 0 = 0$, even when $\log 0$ is undefined, the factor in front of logarithm will force the term to be zero, but the case when $\log x/0$ is present then it is interpreted as infinity, because $\lim_{x \to 0^+} \log 1/x = \infty$. Zeros and infinities in relative entropy bring several problems, infinities in comparison of a relative entropy of different probability distributions causes their incomparability. In computational implementation, division by zero problem appears. For comparison purposes, the error into computation of relative entropy is introduced, suppose that in the training data set, the frequency of some particular event x from class c_1 is $f_1(x) > 0$ and for class c_2 is $f_2(x) = 0$, then we set the zero frequency to be equal to one: $f_2(x) = 1$.

2.3 Relative Entropy of Multiple Attributes

When we need to compute relative entropy over several attributes $a_1, a_2, \ldots, a_n$, we simply substitute $p(x_i)$ for its joint form $p(x_1, x_2, \ldots, x_n)$:

$$D(P||Q) = p(x_1, x_2, \ldots, x_n) \log \frac{p(x_1, x_2, \ldots, x_n)}{q(x_1, x_2, \ldots, x_n)} \quad (5)$$

2.4 Example

Suppose a simple dataset given in Table 1 consisting of two attributes and two classes into which we would like to assign individual records. Each class-attribute combination has associated joint probability vector $p(c_i, a_i)$, in our example case, classes c_1 and c_2 have for attribute a_1 corresponding vectors:

$$p(c_1, a_1) = (p(c_1, x_1), p(c_1, y_1)) = (1, 0) \quad (6)$$

and

$$p(c_2, a_1) = (p(c_2, x_1), p(c_1, y_1)) = (0, 1) \quad (7)$$

Probabilities are computed only from records belonging to particular class. When the relative entropy is computed over attribute's a_1 probability vectors, we get infinities since: $D(P(c_1, a_1)||Q(c_2, a_1)) = 1 \log \frac{1}{0} + 0 \log \frac{0}{1} = \infty + 0 = \infty$, the same value is achieved when the measuring set is P: $D(Q(c_2, a_1)||P(c_1, a_1)) = \infty$. When attribute a_2 is considered then its corresponding joint probability vectors are: $p(c_1, a_2) = (1, 0)$ and for the second class $p(c_2, a_2) = (0.5, 0.5)$. In this case, relative entropies will be $D(P(c_1, a_2)||Q(c_2, a_2)) = 1 \log \frac{1}{0.5} + 0 \log \frac{0}{0.5} = 1 + 0 = 1$. To get a better understanding of the topic, consider a following situation, let the classifier knowledge of the incoming event be equal to $q = (0.5, 0.5)$, but the real distribution of events is given by $p = (1, 0)$, in this situation classifier is forced to put one question to reveal the value of attribute, but if classifier would knew the real case, the vector p, then the classifier would not reveal any information about event at all.

Table 1 Description of example dataset

Class (c_i)	Attribute - 1 (a_1)	Attribute - 2 (a_2)
c_1	x_1	x_2
c_1	x_1	x_2
c_2	y_1	x_2
c_2	y_1	y_2

3 Rules

Let $r = \{a_1 = x_1, \ldots, a_n = x_n\}$ is a rule, where a_i is attribute and x_i is a particular value of attribute a_i. The number of different attributes in rule r is a length of the rule. Let R be a set of rules r_i over data source S. In the present section, we will describe several classes of rules we distinguish:

- correct rules,
- mostly correct rules,
- neutral rules,
- incorrect rules.

Definition 1 Correct rule is a rule, when applied on training data set, which makes only good predictions.

In PRISM algorithm, the author proposed a method that works only with rules of 'correct' class, rules of this class make only good predictions over training data. From relative entropy perspective, every time some rule is classified as 'correct' then its single relative entropy (4) is infinite.

Definition 2 The mostly correct rule is a rule, that when applied on training data set, makes majority of predictions correct.

The mostly correct rules have single relative entropy positive. In our experiments, we deal primarily with rules, which are correct over majority number of training samples. This class contains as a subclass 'correct' rules class.

Definition 3 Neutral rule is a rule, that when applied has equal number of correct and incorrect predictions.

The neutral rule has its corresponding single relative entropy equal to zero. The last class of rules is a class of incorrect rules. The incorrect rule has the single relative entropy negative. We do not use neutral and incorrect rules in our experiments, since these rules contribute mainly to misprediction.

Definition 4 Incorrect rule is a rule, that when applied, makes majority number of predictions incorrect.

3.1 Relative Entropy of Rules

Suppose again individual summation terms from Eq. (3) for some event x_i and suppose that $p(x_i) > q(x_i)$, then there are exactly two individual relative entropies that can be computed $D(P = x_i||Q = x_i)$ and $D(Q = x_i||P = x_i)$, meanwhile the former is positive the latter is negative and because the relative entropy is nonnegative function $D(P = x_i||Q = x_i) > -D(Q = x_i||P = x_i)$. In our experiments we considered two ways of rules comparison:

- the ratio between relative entropies,
- and the sum of relative entropies.

In the case when the ratio between relative entropies is applied, when nominator and denominator are evaluated, and the whole equation is simplified we realize that the ratio between relative entropies is exactly the negative ratio between probabilities:

$$r = \frac{p(x_i)\log\frac{p(x_i)}{q(x_i)}}{q(x_i)\log\frac{q(x_i)}{p({}_ix)}} = \frac{p(x_i)\log\frac{p(x_i)}{q(x_i)}}{-q(x_i)\log\frac{p(x_i)}{q(x_i)}} = -\frac{p(x_i)}{q(x_i)} \tag{8}$$

When rules are being sorted, the absolute value of (8) is taken. The ratio is dimensionless parameter, meanwhile in the second case, when the sum of relative entropies is applied, then we are using one term of Kullback–Leibler divergence from Eq. (3) and the unit of measure is a bit (binary question):

$$s = p(x_i)\log\frac{p(x_i)}{q(x_i)} + q(x_i)\log\frac{q(x_i)}{p(x_i)} = (p(x_i) - q(x_i))\log\frac{p(x_i)}{q(x_i)} \tag{9}$$

Our philosophy is that every rule in binary classification is a rule that must predict at least neutrally, rules that are not neutral are always positive when interpreted as classifying one of the classes.

4 Basic Principles of the Classifier

The most important concept in the present paper is the concept of relative entropy and its applicability to data classification. This section describes training and test phases of the classification algorithm. The training data are prepared in the following way:

1. Prepare a set of all accessible rules of length n.
2. For each rule from the step 1, compute single relative entropies by Eq. (8) resp. (9).
3. Sort rules by values of relative entropies from step 2.

When the training data were processed and n-best rules were produced, we can apply these rules on particular record from the test dataset in the following way:

1. Select the currently best rule.
2. Check if the record satisfies the rule, i.e., the record has exact pairs of attribute-values like the rule. If the record do not satisfies the rule, then go to step 4.
3. Since each rule classifies particular class, then if the rule is present we add one to counter of corresponding class.
4. Select the next best rule, if there is one, and repeat step 2, otherwise go to step 5.

5. If the sum of predictions of one class is higher than the sum of predictions of the second class, then the record is classified as a class with higher prediction counter, otherwise the record is not classified.

Finally, all predictions of records from the test dataset are merged together and if the predicted class is equal to the class corresponding to the test record, then the classification of the record is considered to be correct. In the step 5, classifier makes the prediction by majority voting, when there is more rules in one class then in the other one, then classifiers selects as a prediction the one with more rules (voters are rules).

5 Results and Discussion

Results were produced on datasets from UCI, since in present time our experiments have been prepared for prediction of binary classes from categorical data only, there is a limited number of datasets available to evaluate. We compared our approach with other algorithms dealing with classification by learning set of rules. Accuracies were achieved by performing tenfold cross-validation. In Table 2, the results that were achieved in [7] are summarized. Our experiments were setup to compare two characteristics, the comparison of n-best rules selection by relative entropies based on (8) and (9), and evaluation of the case when all mostly correct rules are applied. We did not setup weights to rules so far as the intention of this paper is an initial study and we focus on basic properties before we introduce more complex classification system. Accuracies of predictions are summarized in Tables 3, 4, and 5 based on two comparison criteria: classifier D-Ratio is a classifier that sorts rules by the ratio between the relative entropies, meanwhile D-Sum is a classifier based on a single Kullback–Leibler relative entropy. Both classifiers are evaluated in two scenarios, in the first scenario, rules consist of only one attribute-value pair(Attrs-1) and in the second scenario, rules consist of two attribute-value pairs(Attrs-2). In the Breast dataset, classifier was able to achieve accuracy comparable of other classifiers. One attribute sized rules performed better than in the case when two attributes were used. The best result on the dataset was achieved by CBA algorithm. To examine if classifier is able to deal with data that are logically structured, Tic Tac Toe—

Table 2 Description and results on datasets from UCI

Dataset	Size	Attr. no.	Accuracy (%)				No. of rules			
			C4.5	RIPPER	CBA	RMR	C4.5	Ripper	CBA	RMR
Breast	699	9	94.66	95.42	**98.84**	95.92	14	6	45	60
Tic-Tac	958	9	83.71	96.97	**100.00**	**100.00**	95	9	25	26
Votes	435	16	88.27	87.35	86.91	**88.70**	4	4	40	84

Comparison of different algorithms: C4.5 [11], RIPPER [12], CBA [13] and RMR [7]. Results from [7]

Table 3 UCI dataset—Breast

No. of rules (algorithm)	D-Ratio		D-Sum	
	Attrs-1	Attrs-2	Attrs-1	Attrs-2
6 (RIPPER)	93.21	92.30	91.6	92.48
14 (C4.5)	94.75	92.64	93.38	91.65
45 (CBA)	**96.50**	93.81	95.87	91.80
60 (RMR)	96.38	93.90	96.21	92.13
500	96.39	**95.85**	**96.48**	**95.77**
All	96.41	93.45	96.48	93.36

Comparison of prediction accuracy (%) for number of rules achieved by different algorithms with the n-best rules derived by the presented classifier. Attrs-N denotes the length of rules (number of attributes in a rule) used by the classifier

Table 4 UCI dataset—Tic Tac Toe—Endgame

No. of rules (algorithm)	D-Ratio		D-Sum	
	Attrs-1	Attrs-2	Attrs-1	Attrs-2
9 (RIPPER)	58.08	51.57	55.94	52.23
25(26) (CBA,RMR)	60.58	61.08	59.96	60.16
95 (C4.5)	60.81	64.66	**61.84**	**65.14**
500	60.45	**65.11**	61.66	64.04
All	**61.14**	65.04	61.54	64.20

The prediction accuracy (%) in the case of Tic Tac Toe—Endgame dataset is very weak

Table 5 UCI dataset—Votes

No. of rules (algorithm)	D-Ratio		D-Sum	
	Attrs-1	Attrs-2	Attrs-1	Attrs-2
4 (C4.5 and RIPPER)	**88.18**	**91.84**	**91.95**	**92.29**
40 (CBA)	87.70	87.19	87.99	89.88
84 (RMR)	88.02	89.65	87.77	88.48
500	87.72	89.13	88.00	89.38
All	87.93	88.25	88.37	88.18

Endgame dataset was used and the results are summarized in Table 4. There are several reasons why classifier is unable to deal with a logically structured data, but the main reason is that the classifier uses mostly correct rules that misclassifies many records and in comparison with classifiers that are building the least set of correct rules cannot succeed. The prediction problems can be solved when we permit only rules of correct class and selects three or more attributes, such rules always leads to good prediction no matter of which subset of training data was used, because these rules are logically correct and they would mispredict only in cases when provided test dataset is logically inconsistent. The last case examined in the experiment was a UCI Votes dataset, in this particular case the accuracies of predictions achieved using four-best rules two attribute classifier were better in comparison with other algorithms. In comparison with techniques that constructs the least size set of rules, we prefer to build as large set as possible and discriminate rules afterward. Let s_l is the size of the set of all rules of length l, then the presented technique selection by n-best rules will discriminate $l - n$ rules. In the presented results, we saw that usage of all rules does not lead to as good prediction as in cases with less rules, so in the future work we will consider application of weights to rules and we will try to make the all (resp. nearly all) rules prediction more accurate.

6 Conclusion

In this paper, we proposed and experimentally examined classification of categorical data using comparison of rules based on their relative entropies and selecting n-best of them. The experiments showed that the classifier has ability to classify data and even on one dataset and particular setup of classifier it was able to exceed accuracies achieved by other algorithms, but it should be also mentioned that the classifier is unable to distribute logically based datasets correctly. In the future work, we would like to prepare a version of classifier that would be able to decide between more than two classes as well as to allow the classifier to process continuous data, as that would allow us to make many more experiments and comparisons.

Acknowledgments This work was supported by the SGS in VSB—Technical University of Ostrava, Czech Republic, under the grant No. SP2015/146.

References

1. Kotsiantis, S.B.: Supervised machine learning: a review of classification techniques. Informatica **31**, 249–268 (2007)
2. Fürnkranz, J., Flach, P.A.: ROC 'n' rule learning—towards a better understanding of covering rules. Mach. Learn. **58**, 39–77 (2005)
3. Shannon, C.E.: A mathematical theory of communication. Bell Syst. Tech. J. **27**, 379–423, 623–656 (1948)
4. Cover, T.M.: Elements of Information Theory. Wiley-Interscience, New York (1991)
5. Quinlan, J.R.: Learning efficient classification procedures and their application to chess endgames. Machine Learning: An Artificial Inteligence Approach, pp. 463–482. Palo Alto, Tioga (1983)
6. Cendrowska, J.: PRISM: an algorithm for inducing modular rules. Int. J. Man-Mach. Stud. **27**, 349–370 (1987)
7. Thabtah, F.A., Cowling, P.I.: A greedy classification algorithm based on association rule. Appl. Soft Comput. **7**, 1102–1111 (2007)
8. Li, J., Wong, L.: Using rules to analyse bio-medical data: a comparison between C4.5 and PCL. Adv. Web-Age Inf. Manag. **4**, 254–265 (2003)
9. Fano, R.M.: Transmission of Information. A Statistical Theory of Communications. M.I.T. Press, New York (1961)

10. Kullback, S., Leibler, R.A.: On information and sufficiency. Ann. Math. Stat. **22**, 79–86 (1951)
11. Quinlan, J.R.: C4.5: Programs for Machine Learning. Morgan Kaufmann Publishers Inc., San Francisco (1993)
12. Cohen, W.: Fast effective rule induction. In: Proceedings of ICML-95, pp. 115–123 (1995)
13. Liu, B., Hsu, W., Ma, Y.: Integrating classification and association rule mining. In: Proceedings of the KDD, pp. 80–86. New York (1998)

Object Recognition Based on Comparative Similarity Assessment

Juliusz L. Kulikowski

Abstract In the paper a concept of object recognition based on their similarity assessment in case of nonhomogenous qualitative and quantitative objects' features is presented. Moreover, it is assumed that the features' intensity values are not given directly but by their pairwise comparative assessment. This corresponds to an intuitive, on human experience-based assessment of the objects' properties. The proposed object recognition method is based on reference sets divided into credibility layers, according to a relative logical model and conceptual classes of similarity. This concept is illustrated by an example of a conceptual class of "irregular" objects, the "irregularity" being intuitively assessed. The method is presented in the form of an algorithm.

Keywords Object recognition · Relative logic · Similarity assessment · Conceptual similarity classes

1 Introduction

Most of object (pattern) recognition methods are based on a paradigm of *similarity*. An observed object ω is declared belonging to a class A if it is found to be "similar" to other objects of this class. However, the concepts of similarity may be different. They may be considered as a sort of geometric *proximity* to other objects in the given class [1], a *probabilistic* or *membership measure* of some class of objects [2], relative position in a linear semiordered vector space [3], product of a *formal grammar* or other "*mechanism*" generating similar objects [4], level of *acceptance* by a *neural*

J.L. Kulikowski (✉)
Nalecz Institute of Biocybernetics and Biomedical Engineering PAS,
4 Ks. Trojdena str., 02-109 Warsaw, Poland
e-mail: juliusz.kulikowski@ibib.waw.pl

R. Burduk et al. (eds.), *Proceedings of the 9th International Conference on Computer Recognition Systems CORES 2015*, Advances in Intelligent Systems and Computing 403, DOI 10.1007/978-3-319-26227-7_23

network set to distinguish specific object parameters [5], etc. The probabilistic- or formal grammar-based recognition methods need a relatively high level of knowledge about the properties of the classes of objects to be recognized. Other methods are oriented toward collection of primary information in the form of learning sets of objects used to evaluate the "proximities" of objects (e.g., in the *k-NN* [1] or in *SVM* [6] recognition algorithms) or to set the weight coefficients of a classifier (in artificial neural networks). In such case the learning sets are assumed to consist of the elements (representatives of real objects) whose classification is a priori exactly fixed. In this paper a step further is done. Instead of strongly fixed learning sets $S_A, S_B, \ldots$, etc., where $A, B, \ldots$, etc., are the names of the classes of objects the existence of which is assumed, a relation of soft similarity between exemplary elements is assumed. Such approach is based on an attempt to simulate the objects recognition mechanisms observed in the nature. The paper is organized as follows. In Sect. 2 basic concepts of a soft approach to object recognition are described. Section 3 presents the proposed object recognition method in detail. Section 4 contains concluding remarks.

2 Basic Concepts

There will be considered basic statements of the form "$\omega \in A$" which in general can be read as "ω *belongs to* A", "ω *is of type* A" or "ω *has the property* A". It is assumed that ω can be exactly distinguished from any other element of a real or abstract world, while A denotes a concept of a community of all elements such that the statement "$\omega \in A$" can be logically assessed. In the case if A denotes a *property* (an adjective) rather than a *collection* of objects, the notation "$\omega \propto A$" ("ω *is* A") will be used instead of "$\omega \in A$" ("ω *belongs to* A"). If $A \subseteq \Omega$ denotes a subset of Cartesian product of a family of nonempty sets $\Omega = C^{(p)} \times C^{(q)} \times \cdots \times C^{(r)}$ then in the sense of classical set theory [7] A can be interpreted as a multivariable *relation* described on Ω and for any $\omega \in \Omega$ an expression "$\omega \in A$" can be read as a statement: "ω *satisfies the relation* A". It is assumed that the predicate algebra holds and its products (composite statements) are subjected to the below-defined *logical equivalence* and *precedence* rules.

Definition 1 Logical equivalence ($\approx$, read: is as credible as) of statements is a relation satisfying the following conditions:

1. Reflexivity: each statement s is logically equivalent to itself, $s \approx s$;
2. Symmetry: for any statements s', s'' if $s' \approx s''$ then $s'' \approx s'$;
3. Transitivity: for any statements s', s'', s''' if $s' \approx s''$ and $s'' \approx s'''$ then $s' \approx s'''$.

Definition 2 Logical precedence ($\precsim$, read: is at most so credible as) of statements is a relation satisfying the following conditions:

1. For any two statements s', s'' both $s' \precsim s''$ and $s'' \precsim s'$ hold if and only if $s' \approx s''$;
2. For any statements s', s'', s''' if $s' \precsim s''$ and $s'' \precsim s'''$ then $s' \precsim s'''$.

The logical equivalence and precedence rules are introduced in order to replace the classical binary (*true, false*) logical scale by relative assessment of the credibility of statements [8]. We denote by $s' \, ?s''$ a situation when neither $s' \preccurlyeq s''$ nor $s'' \preccurlyeq s'$ takes place. For our purposes, the following assumptions are introduced:

Assumption 1 If $s' \preccurlyeq s''$ then:

$$1.\ (s' \vee s'') \approx s''; \qquad 2.\ (s' \wedge s'') \approx s'.$$

Assumption 2 If $s' \preccurlyeq s''$ then $s' \preccurlyeq (\neg s')$.

The predicate algebra operations together with the Definitions 1 and 2 and the Assumptions 1 and 2 constitute a *relative logical model* (*RLM*) over any given set S of basic statements describing a selected application area. The *RLM* does not use the notion of "*fully true*" or "*fully false*" statements; the statements are less or more *credible* with respect to the other ones [9]:

Definition 3 A statement s will be called:

(a) $\Re s$ (read: "rather s") if $\neg s \preccurlyeq s$;
(b) ambivalent if $(\neg s) \approx s$.

On the basis of the above-given concepts a notion playing a basic role in object recognition can be formulated.

Definition 4 Let Ω be a set of elements ω_p called objects and S_A denote a set of statements:

$$s^A(\omega_p):\ \text{“}\omega_p \propto A\text{”}$$

such that $\Re s_A(\omega_p)$ holds for any ω_p. Then a pair $(S_A,\ \sigma_A)$ where σ_A is a logical precedence relation described on $S_A \times S_A$, will be called A-conceptual class (A-*cc*).

There is a basic difference between a conceptual class where for a given A the credibility of statements "$\omega_p \propto A$" are pairwise relatively assessed and a "classical" similarity class where the statements "$\omega_p \in A$" may be only "*true*" or "*false*". The following example may illustrate why the notion of conceptual classes may be useful. Assume that Ω is a set of radiological images in which some specific types of tumors are presented. The type and form of tumors by medical experts may be characterized as *ameloblastic, cellular, colloidal, desmoid, fibroid, medullary, sebaceous, solid, stellate,* etc. However, no strong mathematical definitions or models of the above-mentioned types of tumors are given. Therefore, the similarity classes of images should be considered as rough [10] or fuzzy sets [11] rather than as the classical ones. The *conceptual classes* come also from the attempt to use a soft approach to similarity classes definition and, if possible, to remove from considerations numerical membership measures. In general, any similarity class can be described empirically—using a subset of reference objects representing the exemplary cases of object-to-class matching.

Definition 5 Let, for a certain set of objects Ω, it will be given a defined on it A-conceptual class (S_A, σ_A). Then, for a finite subset $Q \subset \Omega$, the corresponding subset of statements $S_{A/Q} \subset S_A$ and a subrelation of $\sigma{*}_{A/Q} \subseteq S_{A/Q} \times S_{A/Q}$ consisting of selected pairs $\left[s_A(\omega_i),\ s_A(\omega_j)\right] \in \sigma_A$, the pair

$$\Sigma_{A/Q} = \left(S_{A/Q},\ \sigma{*}_{A/Q}\right), \tag{1}$$

will be called a reference set of the A-cc based on the subset Q.

The so-defined reference sets are thus a sort of conceptual classes; they do not consist of objects but of statements concerning some objects. In practice it may happen that certain pairs of objects $\omega_i,\ \omega_j \in Q$ in $\Sigma_{A/Q}$ remain incomparable $[s_A(\omega_i)?s_A(\omega_j)]$; such reference sets may be called *incomplete*. Any reference set $\Sigma_{A/Q}$ consists of statements that can be subjected to a partition into layers in the below-described way.

Definition 6 Let it will be given a reference set $\Sigma_{A/Q}$. If it is given a partition of $S_{A/Q}$:

$$S_{A/Q} = L^{(1)} \cup L^{(2)} \cup \cdots \cup L^{(k)} \tag{2}$$

into subsets $L^{(\kappa)}$, $\kappa = 1, 2, \ldots, k$, called credibility layers, such that:

(a) the credibility layers are pairwise disjoint;
(b) all statements $s_A(\omega_i), s_A(\omega_j)$, etc., belonging to a given $L^{(\kappa)}$ satisfy the condition $s_A(\omega_i) \approx s_A(\omega_j)$;
(c) if any $s_A(\omega_i) \propto L^{(\kappa)}$, $s_A(\omega_j) \propto L^{(\lambda)}$, and $\kappa > \lambda$ then $s_A(\omega_j) \preccurlyeq s_A(\omega_i)$,

then it will be called that a credibility gradation has been imposed on $S_{A/Q}$.

Any credibility gradation imposed on the reference set induces a corresponding partition of the subset Q of objects into subsets $Q^{(\kappa)}$, $\kappa = 1, 2, \ldots, k$, such that $\omega_i \propto Q^{(\kappa)}$ if and only if $s_A(\omega_i) \propto L^{(\kappa)}$.

Definition 7 For given credibility gradation imposed on $S_{A/Q}$ and its partition into certainty layers $L^{(\kappa)}$, $\kappa = 1, 2, \ldots, k$, an induced by it partition of Q into subsets $Q^{(\kappa)}$, $\kappa = 1, 2, \ldots, k$, will be called credibility gradation induced in Q.

Any credibility gradation induced in Q can be presented and illustrated by a directed graph:

$$G_{A/Q} = [Q,\ \Lambda,\ \varrho_{A/Q}] \tag{3}$$

where, in this context, Q is the set of nodes, $\Lambda = \{\rightarrow,\ \leftrightarrow\}$ is the set of types of edges and $\varrho_{A/Q}$ is a function assigning unidirectional ($\rightarrow$) edges to the pairs of nodes corresponding to the relation $s_A(\omega_i) \preccurlyeq s_A(\omega_j)$ (and not the reverse) and bidirectional ($\leftrightarrow$) edges to the pairs of nodes corresponding to the relation $s_A(\omega_i) \approx s_A(\omega_j)$.

Fig. 1 A set Q of exemplary geometrical shapes of various regularity

Example 1 Let Ω denote a set of geometrical shapes of some objects potentially observable in a sort of experiment. We would like to describe a conceptual class of *irregular* objects, the *irregularity* (a property shortly denoted by A) being not a strongly defined concept. Let $Q \subset \Omega$ be a subset of geometrical shapes shown in Fig. 1. It will be constructed a reference set $\Sigma_{A/Q}$ of *irregularity conceptual class* based on Q (see Definition 5). For this purpose, some pairs of objects in Q should be taken into consideration and the credibility of the statements $s_A(\omega_p)$: "ω_p *is irregular*" should be assessed by comparison in pairs. First, ω_3 should be removed from consideration because a circle can be assessed as a *rather* regular shaper. For the rest of objects ω_p the property $\Re s_A(\omega_p)$ can be stated. Then, looking at the shapes the following relationships among them can be established:

$$s_A(\omega_2) \approx s_A(\omega_{11}) \approx s_A(\omega_{12}); \quad s_A(\omega_1) \approx s_A(\omega_6) \approx s_A(\omega_7) \approx s_A(\omega_{10});$$
$$s_A(\omega_8) \approx s_A(\omega_9); \quad s_A(\omega_5) \preccurlyeq s_A(\omega_9); \quad s_A(\omega_8) \preccurlyeq s_A(\omega_6);$$
$$s_A(\omega_1) \preccurlyeq s_A(\omega_4); \quad s_A(\omega_4) \preccurlyeq s_A(\omega_2).$$

This makes possible a partition of the statements into credibility layers:

$$\begin{aligned} L^{(1)} &= \{s_A(\omega_2),\ s_A(\omega_{11}),\ s_A(\omega_{12})\}, \\ L^{(2)} &= \{s_A(\omega_4)\}, \\ L^{(3)} &= \{s_A(\omega_1),\ s_A(\omega_6),\ s_A(\omega_7),\ s_A(\omega_{10})\}, \\ L^{(4)} &= \{s_A(\omega_8),\ s_A(\omega_9)\}, \\ L^{(5)} &= \{s_A(\omega_5)\}. \end{aligned}$$

A graph $G_{A/Q}$ illustrating the given credibility gradation is shown in Fig. 2. The nodes correspond to the ω_p such that $\Re s_A(\omega_p)$ holds; for the sake of simplicity, the

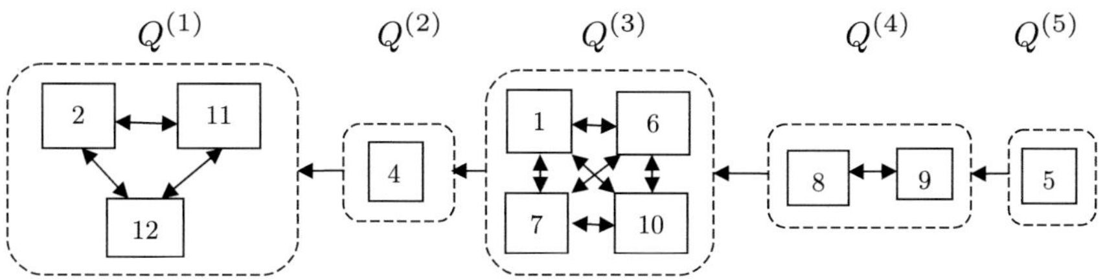

Fig. 2 Graph $G_{A/Q}$ illustrating a credibility gradation imposed on the subset $S_{A/Q}$ of statements concerning irregularity of forms presented in Fig. 1

nodes have been labeled by the indexes p of the objects. The credibility gradation plays a role analogous to this of membership functions in fuzzy sets theory [11]. However, the numerical membership values of the elements of a fuzzy set A in this case have been replaced by different credibility layers characterizing the level of objects matching the conceptual similarity class.

3 Object Recognition

The *object recognition* problem in the *RLM* terms can be formulated as follows:

- It is given a set Ω of objects, it is assumed that an *A-cc* has been defined and is given by a reference set $\Sigma_{A/Q}$ based on a subset $Q \subset \Omega$.
- A credibility gradation has been imposed on $\Sigma_{A/Q}$.
- Let $\xi \in \Omega \backslash Q$ be given as a new-observed object.
- Find the credibility level of the statement $s_A(\xi)$ related to the credibility gradation imposed on $\Sigma_{A/Q}$.

Unlike typical pattern recognition problem [1], in this formulation the concept of a similarity class A is replaced by an *A-cc* approximated by a reference set $\Sigma_{A/Q}$ and instead of asking about belonging/not belonging of a new-observed object ξ to A its relative similarity to the elements ω described by the statements $s_A(\omega)$ in the reference set $\Sigma_{A/Q}$ is asked about and assessed. The question arises, how to assess the similarity between the new-observed object ξ and the objects of reference set, as needed to assign to $s_A(\xi)$ a logically justified credibility layer. The credibility gradation of a reference set $\Sigma_{A/Q}$ is usually based on experts' experience or intuition. On the other hand, the similarity between ξ and other objects $\omega \in Q$ can be assessed only on the basis of selected measurable or observable objects' features.

3.1 Assessing Comparative Similarity Between Objects

It is assumed that Ω can be developed into a Cartesian product of a finite family of sets $X^{(\kappa)}$ of quantitative or qualitative *features* describing the objects:

$$\Omega = X^{(1)} \times X^{(2)} \times \cdots \times X^{(n)} \times \cdots \times X^{(N)} \tag{4}$$

The nature of features may be different, however, it is assumed that each set $X^{(n)}$ is, independently on the other ones, linearly ordered so that for any $\xi_\alpha^{(n)}, \xi_\beta^{(n)} \in X^{(n)}$ (α, β being indexes specifying the values) exactly one of the following situations takes place:

$$\xi_\alpha^{(n)} \prec \xi_\beta^{(n)}, \quad \xi_\beta^{(n)} \prec \xi_\alpha^{(n)} \text{ or } \xi_\alpha^{(n)} \sim \xi_\beta^{(n)},$$

symbol $\prec$ being read as "*is lower than*" and $\sim$ as "*is equivalent to*". The relation $\xi_\alpha^{(n)} \prec \xi_\beta^{(n)}$ can also be presented in an equivalent form $\xi_\beta^{(n)} \succ \xi_\alpha^{(n)}$ (read as "*is higher than*"). It is assumed that if $\xi_\alpha^{(n)} = \xi_\beta^{(n)}$ then $\xi_\alpha^{(n)} \sim \xi_\beta^{(n)}$; however, the reverse is not assumed because the relation $\sim$ holds also for qualitative features for which strong mathematical equality may be meaningless. The objects will be thus presented by *pseudovectors*, i.e., strings of features' values, generally, of various formal nature:

$$\omega_i = \left(\xi_i^{(1)},\ \xi_i^{(2)}, \ldots \xi_i^{(n)}, \ldots \xi_i^{(N)}\right). \tag{5}$$

The pairs $(\xi_i^{(n)},\ \xi_j^{(n)})$ of respective features' values of any two objects can be compared in the sense of $\prec$ relation and the results can be presented in the form of a binary vector:

$$\Delta_{ij} = \left(\delta_i^{(1)},\ \delta_{ij}^{(2)}, \ldots, \delta_{ij}^{(N)}\right) \tag{6}$$

where for any $n = 1, 2, \ldots, N$:

$$\delta_{ij}^{(n)} = \begin{cases} 1 & \text{if } \xi_i^{(n)} \prec \xi_j^{(n)} \text{ or } \xi_i^{(n)} \succ \xi_j^{(n)}, \\ 0 & \text{if } \xi_i^{(n)} \sim \xi_j^{(n)}. \end{cases} \tag{7}$$

The vectors Δ_{ij} averaged over their components:

$$H(\omega_i, \omega_j) \equiv H_{ij} = \frac{1}{N}\sum_{n=1}^{N} \delta_{ij}^{(n)} \tag{8}$$

can be considered as a sort of Hamming *distance* between the vectors ω_i, ω_j. Two other notions based on the notion of distance between the vectors will be used below.

Definition 8 Let C be a set of pseudovectors ω. The average

$$d(C) = \frac{1}{K_C}\sum_{i \neq j} H_{ij} \tag{9}$$

where K_C denotes the number of pairs of different unordered pairs of pseudovectors in C, will be called the spread of the set C.

Example 2 Let the objects presented in Fig. 1 be additionally characterized by the following features:

- $\xi^{(1)}$: *total surface*;
- $\xi^{(2)}$: *length of contour line*;
- $\xi^{(3)}$: *sharpness of contour line*;
- $\xi^{(4)}$: *exhibiting inner holes or cracks*;
- $\xi^{(5)}$: *sophisticated form*;
- $\xi^{(6)}$: *asymmetry*.

None of the (arbitrarily chosen) features directly indicates on object's *irregularity*. However, it may be guessed that *irregularity* is in some sense influenced by the above-proposed set of features. In order to justify this guess *logical compactness* of the credibility layers should be proven. For this purpose, Hamming distances [12] between the pairs of pseudovectors corresponding to the objects and the spreads of the credibility layers should be calculated and analyzed. The results are presented below, by Tables 1, 2, and 3 (the layers $L^{(2)}$ and $L^{(5)}$ as consisting of single elements have the spreads equal 0 by definition). The spreads of the certainty layers, according to Eq. (7), are: $d^{(1)} = 0.277, d^{(2)} = 0, d^{(3)} = 0.47, d^{(4)} = 0.17, d^{(5)} = 0$. A spread less than 0.5 shows that the given certainty layer consists of vectors whose components in most cases are mutually equivalent. If they are removed, the remaining features $\xi^{(2)}, \xi^{(3)}, \xi^{(4)}$, and $\xi^{(5)}$ reduce the spread to $d^{(3)} = 0.33$. Low spread of a layer shows that its elements have been correctly chosen for representation of the *A-cc* on the given credibility level. In Example 2, taking into account the data given in Table 1, we obtain $G^{(3)} = 0.277$. However, for the layer $L^{(2)}$ containing a single element $s_A(\omega_4)$ we obtain $G^{(2)} = 0$. Therefore, low spreads are adequate quality characteristics only of multi-element credibility layers.

Table 1 Comparison of objects' features in layer $L^{(1)}$

	$\xi^{(1)}$	$\xi^{(2)}$	$\xi^{(3)}$	$\xi^{(4)}$	$\xi^{(5)}$	$\xi^{(6)}$	H_{ij}
$\Delta_{2\,11}$	*1*	*0*	*0*	*0*	*0*	*0*	0.17
$\Delta_{2\,12}$	*1*	*1*	*0*	*0*	*0*	*0*	0.33
$\Delta_{11\,12}$	*0*	*1*	*0*	*0*	*0*	*1*	0.33

Table 2 Comparison of objects' features in layer $L^{(3)}$

	$\xi^{(1)}$	$\xi^{(2)}$	$\xi^{(3)}$	$\xi^{(4)}$	$\xi^{(5)}$	$\xi^{(6)}$	H_{ij}
$\Delta_{1\,6}$	*1*	*1*	*0*	*0*	*0*	*0*	0.33
$\Delta_{1\,7}$	*0*	*0*	*0*	*0*	*1*	*1*	0.33
$\Delta_{1\,10}$	*1*	*0*	*0*	*0*	*0*	*1*	0.33
$\Delta_{6\,7}$	*1*	*1*	*1*	*0*	*1*	*1*	0.83
$\Delta_{6\,10}$	*0*	*0*	*0*	*0*	*0*	*1*	0.17
$\Delta_{7\,10}$	*1*	*1*	*1*	*0*	*1*	*1*	0.83

Table 3 Comparison of objects' features in layer $L^{(4)}$

	$\xi^{(1)}$	$\xi^{(2)}$	$\xi^{(3)}$	$\xi^{(4)}$	$\xi^{(5)}$	$\xi^{(6)}$	H_{ij}
$\Delta_{8\,9}$	*0*	*1*	*0*	*0*	*0*	*0*	0.17

Definition 9 For given set C of pseudovectors and a pseudovector $\xi, \xi \notin C$, the ratio

$$\eta(\xi/C) = \frac{1 + d(C)}{1 + d(C \cup \{\xi\})} \tag{10}$$

will be called a ξ/C matching factor.

The matching factor takes maximal value $\eta^{(\kappa)}(\xi/C) = 1$ if all vectors ω in C as well as vector ξ are equivalent.

3.2 *Object Recognition Algorithm*

Object recognition can be based on the notions given above by the Definitions 4 and 5 in particular case, when the subsets $Q^{(\kappa)}$ are induced by credibility gradation imposed on the reference set $S_{A/Q}$. Let ξ be a pseudovector of features of a new-observed object.

Definition 10 It will be said that credibility level of the statement $s_A(\xi)$ *matches* the layer $L^{(\kappa)}$ in $S_{A/Q}$ if

$$\kappa = \arg\max_{\{\mu\}}[\eta(\xi/Q^{(\mu)})], \qquad \mu = 1, 2, \ldots, k. \tag{11}$$

On the basis of above-defined notions an algorithm of object recognition can be proposed. For a given set Ω of objects, assumed A-*cc*, a subset of exemplary objects $Q \subset \Omega$ given in the form of pseudovectors (Eqs. (4) and (5)), a reference set $S_{A/Q}$ imposed on it credibility gradation (Eq. (2)) inducing a credibility gradation in Q:

Algorithm

Object matching to credibility layer
input: reference set $\Sigma_{A/Q}$, empty subset $Q^{(0)}$, family of subsets $[Q^{(\kappa)}]$, $\kappa \in [1, 2, \ldots, k]$, induced in Q by credibility gradation imposed on $S_{A/Q}$, pseudovector $\xi \in \Omega\ Q$; **output:** indices $\mu \in [1, 2, \ldots, k]$ maximizing $\eta(\xi/Q^{(\mu)})$;

1. **initialize** $\mu := 0$, $\eta(\xi/Q) := 0$;
2. **for** $\kappa = 1$ **step** 1 **to** k **do**
3.   **for** all pairs (ω_i, ω_j), (ω_i, ξ), $\omega_i, \omega_j \in Q^{(\kappa)}$, **do**
4.   assess Δ_{ij}, $\Delta_{i\xi}$
5.   calculate $d(Q^{(\kappa)})$, $d(Q^{(\kappa)} \cup \{\xi\})$, $\eta\{\xi/Q^{(\kappa)}$
6.   **if** $\eta(\xi/Q^{(\kappa)}) > \eta(\xi/Q^{(0)})$ **then**
7. $\eta(\xi/Q^{(0)}) := \eta(\xi/Q^{(\kappa)})$, $\mu := \kappa$
8.   **goto** 3
9.   **else goto** 3
10. **return** μ

In realization of point 4 of the Algorithm a sort of human–system interaction is assumed. This can be expanded as follows:

Sub-algorithm

Assessing distance components between pseudovectors **input:** integers I,N, pairs of pseudovectors $\omega_i,\ \omega_j \in Q^{(\kappa)} \subseteq \Omega = X^{(1)} \times X^{(2)} \times \cdots \times X^{(n)} \times \cdots \times X^{(N)}$, $i = 1, 2, \ldots, I$, $j = i+1,\ i+2, \ldots, I$, $X^{(n)}$ for $n = 1, 2, \ldots N$ being linearly ordered sets of objects' features; **output:** vectors $\Delta_{ij} = \delta_{ij}^{(1)},\ \delta_{ij}^{(2)}, \ldots, \delta_{ij}^{(N)})$; **for** $i = 1$ **step** 1 **to** I **do for** $j = i + 1$ **step** 1 **to** I **do for** $n = 1$ **step** 1 **to** N **do if** $\xi_i^{(n)} \sim \xi_j^{(n)}$ **then** $\delta_i^{(n)} j := 0$ **else** $\delta_{ij}^{(n)} := 1$ **return** δ_{ij}

4 Conclusions

Any object or pattern recognition method is based on a concept of similarity. In general, similarity can be defined by a numerical similarity measure, e.g., based on an objects' distance measure. Alas, a large class of real objects cannot be presented by elements of a metric space. In the paper a general concept of object recognition based on comparative assessment of objects similarity has been formally described. The concept is based on a notion of conceptual class of similarity which can be considered as an extension of the fuzzy set concept. For the purpose of object recognition the conceptual class is represented by a reference set divided according to the indications of experts into credibility layers. Evaluation of the similarity of a new-observed object to the reference objects is also based on assessment of relative levels of objects' features of both qualitative and/or qualitative type. Such approach to object recognition can be implemented in computer system. Two problems need to be explained in connection with the above-proposed method. First, are the formal properties of conceptual classes, possible algebraic operations on such classes, their relation to fuzzy set, and classical set theory. Second, is the problem of convergence of conceptual classes to classical sets when the number of their elements is increasing. Answering the last questions will be possible when the method is implemented in several exemplary application areas.

References

1. Duda, R.O., Hart, P.E., Stork, D.G.: Pattern Classification, 2nd edn. Wiley, Chichester (2001)
2. Devijver, P., Kittler, J.: Pattern Recognition: A Statistical Approach. Prentice Hall, London (1982)
3. Kulikowski J.L., Przytulska M.: Pattern recognition based on similarity in linear semi-ordered spaces. In: Hybrid Artificial Intelligent Systems, Part I. LNAI, vol. 6678, pp. 22–29. Springer, Heidelberg (2011)
4. Fu, K.S.: Syntactic Methods in Pattern Recognition. Academic Press, New York (1974)
5. Ripley, B.: Pattern Recognition and Neural Networks. Cambridge University Press, Cambridge (1996)
6. Abe, S.: Support Vector Machines for Pattern Classification. Springer, New York (2005)
7. Rudeanu, S.: Sets and Ordered Structures. Bentham Science Publishers, New York (2012)

8. Vessel H.A.: About topological logic (in Russian). In: Neklassičeskaya logika. Izd. Nauka, Moscow (1970)
9. Kulikowski J.L.: Decision making in a modified version of topological logic. In: Proceedings of Seminar on Nonconventional Problems of Optimization, Part I. Prace IBS PAN No 134, Warsaw (1986)
10. Pawlak, Z.: Rough Sets—Theoretical Aspects of Reasoning About Data. Kluwer Academic Publishers, Boston (1991)
11. Bezdek, J.C.: Pattern Recognition with Fuzzy Objective Function Algorithms. Plenum Press, New York (1982)
12. Hamming, R.W.: Error detecting and error correcting codes. Bell Syst. Tech. J. **29**, 147–160 (1950)

An Efficiency K-Means Data Clustering in Cotton Textile Imports

Dragan Simić, Vasa Svirčević, Siniša Sremac, Vladimir Ilin and Svetlana Simić

Abstract Data clustering is a technique of finding similar characteristics among the data sets which are always hidden in nature, and dividing them into groups. The major factor influencing cluster validation is choosing the optimal number of clusters. A novel random algorithm for estimating the optimal number of clusters is introduced here. The efficiency hybrid random algorithm for good k and modified classical k-means data clustering method in cotton textile imports country clustering and ranking is described and implemented on real-world data set. The original real-world U.S. cotton textile and apparel imports data set is taken under view in this research.

Keywords Data clustering · Cluster · k-means algorithm · Random algorithm

1 Introduction

Clustering is deemed to be one of the most difficult and challenging problems in machine learning, particularly due to its unsupervised nature. The unsupervised nature of the problem implies that its structural characteristics are not known, except

D. Simić (✉) · S. Sremac · V. Ilin
Faculty of Technical Sciences, University of Novi Sad,
Trg Dositeja Obradovića 6, 21000 Novi Sad, Serbia
e-mail: dsimic@eunet.rs

S. Sremac
e-mail: sremacs@uns.ac.rs

V. Ilin
e-mail: v.ilin@uns.ac.rs

V. Svirčević
Lames Ltd., Jarački put bb., 22000 Sremska Mitrovica, Serbia
e-mail: vasasv@hotmail.com

S. Simić
Faculty of Medicine, University of Novi Sad, Hajduk Veljkova 1-9,
21000 Novi Sad, Serbia
e-mail: drdragansimic@gmail.com

R. Burduk et al. (eds.), *Proceedings of the 9th International Conference on Computer Recognition Systems CORES 2015*, Advances in Intelligent Systems and Computing 403, DOI 10.1007/978-3-319-26227-7_24

if there is some sort of domain knowledge available in advance. Clustering is a synonym for the decomposition of a set of entities into 'natural groups'. Clustering is the process of assigning data objects into a set of disjoint groups called clusters so that objects in each cluster are more similar to each other than objects from different clusters. The goal of clustering is to group similar objects in one cluster and dissimilar objects in different clusters. Data clustering is a technique of finding similar characteristics among the data sets which are always hidden in nature and dividing them into groups, called clusters. Different data clustering algorithms exhibit different results, since they are very sensitive to the characteristics of original data set especially noise and dimension. Originally, clustering was introduced to the data mining research as the unsupervised classification of patterns into groups [16]. Clustering techniques offer several advantages over manual grouping process. First, a clustering algorithm can apply a specified objective criterion consistently to form the groups. Second, a clustering algorithm can form the groups in a fraction of time required by manual grouping, particularly if long list of descriptors or features is associated with each object. The speed, reliability, and consistency of clustering algorithm in organizing data constitute an overwhelming reason to use it. Probably, the most frequently used clustering algorithm is the classic k-means with applications in any real life domain. The k-means clustering is characterized by nonoverlapping, clearly separated ('crisp') clusters with bivalent memberships: an object either belongs to or does not belong to a cluster. The U.S. cotton textile and apparel import, real-world dataset investigated in this research, is originally taken from TRADE DATA U.S. Textiles and Apparel Imports [15]. However, this research uses data sets for 200 countries, and only for years 1990, 2000, and 2010. Some data are eliminated because countries no longer exist or because data are incomplete. The rest of the paper is organized in the following way: Sect. 2 provides some approaches of data clustering techniques and related work. Section 3 presents k-means clustering method, *Random Algorithm for good* k, and data collection of the original data set. An application of hybrid *Random Algorithm for good* k and modified k-means classical method experimental result, clustering, and ranking of the thirty countries presented in Sect. 4. Finally, Sect. 5 gives concluding remarks.

2 Clustering and Related Work

Organizing data into sensible groupings is one of the most fundamental models of understanding and learning. Cluster analysis is the formal study of algorithms and methods for grouping or classifying objects. An object is described either by a set of measurements or by relationships between the object and other objects. Cluster analysis does not use category labels that tag objects with prior identifications. The absence of category labels distinguishes cluster analysis from discriminant analysis, pattern recognition, and decision analysis. Clustering differs from classification in which there is no target variable for clustering. Instead, clustering algorithms seek to segment the entire data set into relatively homogeneous subgroups or clusters [19].

In practice, classifying objects according to the perceived similarities is the basis for majority scientific classifications. The objective of clustering is simply to find a convenient and valid data organization, not to establish rules for separating future data into categories. Clustering algorithms are geared toward finding structure in the data. A cluster is composed of a number of similar objects collected or grouped together. The crucial problem in identifying clusters in data is to specify proximity and how to measure it. As to be expected, the notion of proximity is problem dependent [10]. General references regarding data clustering are presented in [13, 25]. A very good presentation of contemporary data mining clustering techniques can be found in the textbook [12]. There is a close relationship between clustering techniques and many other disciplines. Clustering has always been used in statistics and science. The classic introduction to pattern recognition framework is given in [7]. Typical applications include speech and character recognitions. Machine learning clustering algorithms were applied to image segmentation and computer vision [17]. Statistical approaches to pattern recognition are discussed in [6, 9]. Clustering can be viewed as a density estimation problem. This is the subject of traditional multivariate statistical estimation [23]. Clustering is also widely used for data compression in image processing, which is also known as vector quantization [11]. Data fitting in numerical analysis provides still another venue in data modeling [5]. Such clustering is characterized by large data sets with many attributes of different types. Clustering in data mining was brought to life by intense developments in information retrieval and text mining [26], spatial data base applications, for example, GIS or astronomical data [8], sequence and heterogeneous data analysis [4], Web applications [14], DNA analysis [3, 28], and computational biology [22]. The applicability of clustering is manifold, ranging from market segmentation [2] and image processing [16] through document categorization and web mining [20]. An application field that has been shown to be particularly promising for clustering techniques is bioinformatics [29]. Indeed, the importance of clustering gene-expression data measured with the aid of microarray and other related technologies has grown fast and persistently over the recent years [18].

3 Data Clustering

Data clustering is a technique of partitioning the data set without known prior information. It finds its use in most of the applications where unsupervised learning occurs. A wide range of clustering algorithms is available in the market for grouping low-dimensional data and data of higher dimensions. The different kinds of clustering algorithms when used for varying data sets produce different kinds of results based on the initial input parameters, environment conditions, and the data set nature. In such a scenario, since there are no predefined classes or groups known in clustering process, finding an appropriate metric for measuring, if found cluster configuration, number of clusters, cluster shapes, etc., is acceptable or not, has always been an issue. Clustering is the process of assigning data objects into a set of disjoint groups called clusters so that objects in each cluster are more similar to each other than objects from different clusters. Let $\{x(q) : q = 1, \ldots, Q\}$ be a set of Q feature vectors. Each feature vector $x(q) = (x_1(q), \ldots, x_N(q))$ has N components with weights $w(q) = (w_1(q), \ldots, w_N(q))$ and distances metrics $D(q) = (d_1(q), \ldots, d_N(q))$. The process of clustering is to assign the Q feature vectors into K clusters $\{c(k) : k = 1, \ldots, K\}$ usually by the minimum distance assignment principle. Choosing the representation of cluster centers is crucial to clustering. Feature vectors that are farther away from the cluster center should not have as much weight as those that are close. These more distant feature vectors are outliers usually caused by errors in one or more measurements or a deviation in the processes that formed the object. Most clustering methods assume some sort of structure or model for the cluster (spherical, elliptical). Thus, they find cluster of that type, regardless of whether they really are present in the data or not.

3.1 K-Means Clustering Method

The wide popularity of k-means algorithm is well deserved. It is simple, straightforward, and based on the firm foundation of analysis of variances. The k-means algorithm also suffers from all the usual suspects:

- The result strongly depends on the initial guess of centroids (or assignments);
- Computed local optimum is known to be a far cry from the global one;
- It is not obvious what is a good k to use;
- The process is sensitive with respect to outliers;
- The algorithm lacks scalability;
- Only numerical attributes are covered;
- Resulting clusters can be unbalanced.

A large area of research in clustering has focused on improving the clustering process such that the clusters are not dependent on the initial identification of cluster representation. The major factor which influences cluster validation is the internal cluster validity measure of choosing the optimal number of clusters. In this research, classical MATLAB k-means clustering algorithm with the squared Euclidean distance measure and the k-means ++ algorithm for cluster center initialization is improved. This structure is very strict, with proximity defined as Euclidean distance given by

$$\delta_{ij} = \sqrt{\sum_k \left(x_{ik} - x_{jk}\right)^2}, \tag{1}$$

where x_{ik} is the kth element in the ith observation. In k-means case a cluster is represented by its centroid, which is a mean, usually weighted average, of points within a cluster. This works conveniently only with numerical attributes and can be negatively affected by a single outlier. On the other hand, centroids have the advantage of clear geometric and statistical meaning. The classical MATLAB k-means function does not support higher number of clusters, than four, for usage data set. K-means function does not support five clusters. The number of clusters strongly depends on the initial guess of centroids, and in many experimental cases after some iterations one centroid does not support any points. There were the reasons to improve original k-means function to support at least one point in one centroid. On the other side, this research introduces a novel random algorithm to suggest number of clusters—*good k*.

3.2 Data Collection

The U.S. cotton textile and apparel import, real-world dataset investigated in this research, is originally taken from TRADE DATA U.S. Textiles and Apparel Imports [15]. The entire data set presented in U.S. cotton textile and apparel imports originally contains cotton textile and apparel import data for the period between 1988 and 2012 and includes 238 countries. However, this research uses data set for 200 countries and only for years 1990, 2000, and 2010. Some data are eliminated because countries no longer exist or the data are incomplete, but data set used here is quite sufficient to show changes in world's cotton production and production changes in countries that export cotton textile to the USA.

3.3 Random Algorithm for Good k

A major challenge in data clustering is the estimation of the optimal number of clusters. There are many methods where some researches were suggested. Some of them are as follows: Rule of thumb for number of clusters; Information criterion approach—Akaike's information criterion (AIC) [1]; Bayesian inference criterion (BIC) [24]; Choosing k using the silhouette [21]; and Gap statistics [27]. *Random algorithm for good k* is introduced in this research. It is done in the following way: a random subset of data points is selected and used as input to the random algorithm for clustering. Then, the cluster centers as mean/median are computed for each cluster group. It is important to mention that the number of clusters is less or equal ($\leq$) with subset data points. Usually, the data points in subset data make less than 5 % of all objects in data set. Finally, for each instance that was not selected in the subset, its distance to each of the centroids is simply computed and assigned it to the closest one.

4 Experimental Results

The experimental results can be presented in two steps. First, experimental results from *Random algorithm for good k* will be presented. This algorithm estimates the optimal number of clusters. According to the nature of algorithm, it is repeated 200 times and experimental results for optimal number of clusters are summarized and presented in Table 1. Ten data points are used in subset which is exactly 5 % of 200 objects in data set. The expected number of clusters is less or equal to 10. On the other side, *Rule of thumb for number of clusters* is defined as $k \approx \sqrt{n/2}$ where n is the number of data points and k is the number of clusters. Considering that data set contains $n = 200$ data points, it can be assumed that the optimal number of clusters is $k = 10$ (Figs. 1 and 2).

Table 2 shows experimental results of clusters, number of countries in clusters, and cluster value in (thousands of pounds) for years 1990, 2000, 2010, and for the period 1990–2010. Selected data set in this research consists of U.S. cotton textile imports for years 1990, 2000, and 2010, for 200 countries, and 8 clusters are used. Every country is clustered depending on the value of U.S. cotton textile imports, for every year, respectively—1990, 2000, 2010. After that, in fourth measure, every country is clustered depending on the value of U.S. cotton textile imports for all

Table 1 *Random algorithm for good k*—frequency of number of clusters

Number of clusters	Frequency
10	27
9	54
8	60
7	35
6	19
5	3
4	2
Experiments	200
Suggestion clusters	**8.09**

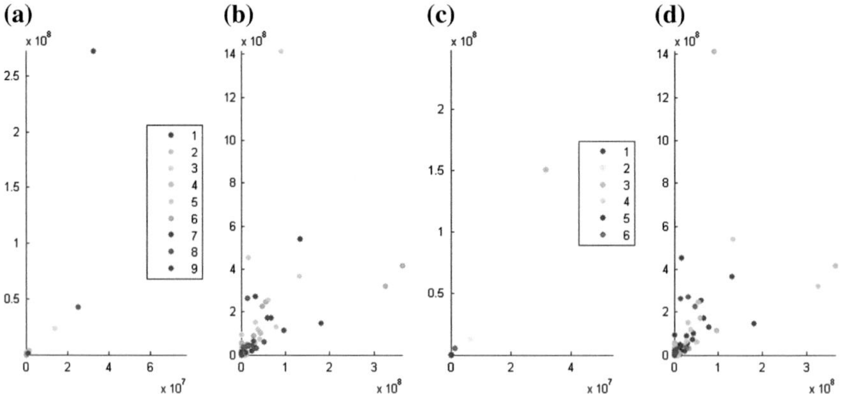

Fig. 1 Scatter plots data visualization *Random algorithm for good k*: **a** $k = 9$ number of clusters for subset; **b** $k = 9$ number of clusters for data set; **c** $k = 6$ number of clusters for subset; **d** $k = 6$ number of clusters for data set

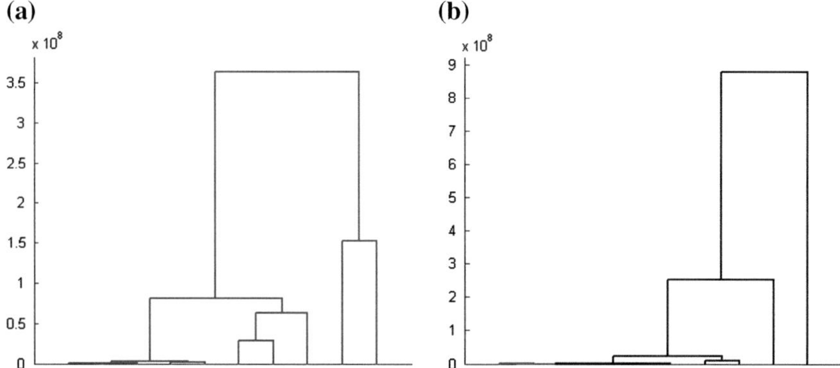

Fig. 2 Tree-structured graph—dendrogram for data visualization *Random algorithm for good k*: **a** $k = 9$ number of clusters for subset; **b** $k = 6$ number of clusters for subset

three years. The fourth measure shows the influence of U.S. cotton textile imports in period of 20 years. Clustering and ranking countries presents the influence of a specific country in U.S. cotton textile imports. Experimental results of clustering and ranking thirty of two hundred countries in eight clusters for four measurements of U.S. cotton textile imports are presented in Table 3. Experimental results show that it is possible to cluster and rank countries in economy and business. Presented results also show trend of cotton production in the world. It can be seen that some countries grow much more than the others, such as China and Cambodia. Recently, China in 2010 produces 10 times more than in 1990, but in 2010 Cambodia produces cotton 22 thousand times more than in 1990. Cambodia was positioned in the eighth cluster, as 113th producer in the world in 1990, but due to increase in cotton production in sixth cluster in 2000, it is positioned in the third cluster in 2010. Also, it can be seen that the number of high producers in 2010 is lower than in 1990. In 1990, in first five clusters there were 26 countries but in 2010, in the same five clusters, there were

Table 2 *Random algorithm for good k*—frequency of number of clusters

Cluster	Year 1990		Year 2000		Year 2010		Year 1990–2010	
	Country	Value	Country	Value	Country	Value	Country	Value
1	2	343,431	1	1,416,516	1	2,147,483	1	975,664
2	3	147,247	3	471,273	4	701,748	1	711,597
3	4	82,174	2	344,492	5	341,827	4	392,918
4	7	50,885	5	252,627	1	178,477	4	174,993
5	10	29,678	4	160,917	6	115,357	4	144,748
6	14	12,841	6	107,317	9	52,544	8	89,635
7	17	4,752	17	46,216	15	17,995	6	57,675
8	143	262	162	2,751	159	386	172	3,110

Table 3 Experimental results of clustering and ranking 30 of 200 countries in 8 clusters for 1990, 2000, and 2010 of U.S. cotton textile imports (thousands of pounds) (**first cluster bold**, second cluster underline, *third cluster italic*)

	Country	1990		2000		2010		1990–2010	
		Cotton	Clu	Cotton	Clu	Cotton	Clu	Cotton	Clu
1	China	**362,211**	**1**	417,299	2	**3,310,700**	**1**	**1,363,403**	**1**
2	Hong Kong	**324,650**	**1**	*321,282*	*3*	16,630	7	220,854	5
3	China (Taiwan)	180,525	2	147,089	5	40,881	6	122,832	7
4	Pakistan	131,029	2	541,966	2	925,986	2	*532,994*	*3*
5	India	130,189	2	*367,701*	*3*	788,252	2	*428,714*	*3*
6	South Korea	*94,847*	*3*	112,650	6	85,823	5	97,773	6
7	Mexico	*89,710*	*3*	**1,416,516**	**1**	628,563	2	711,597	2
8	Philippines	*77,208*	*3*	130,464	6	71,522	6	93,065	6
9	Indonesia	*66,930*	*3*	172,558	5	*380,960*	*3*	206,816	4
10	Dominican Rep.	60,720	4	253,419	4	70,711	6	128,283	5
11	Thailand	58,393	4	173,370	5	133,693	5	121,819	6
12	Bangladesh	53,975	4	247,243	4	628,189	2	*309,802*	*3*
13	Brazil	51,565	4	60,455	7	33,025	7	48,348	7
14	Turkey	48,101	4	226,583	4	69,778	6	114,820	5
15	Sri Lanka	42,525	4	100,885	6	89,768	5	77,726	6
16	Malaysia	40,917	4	71,831	7	36,603	6	49,784	7
17	Costa Rica	36,937	5	118,369	6	14,140	7	56,482	8
18	Singapore	33,078	5	29,730	7	1,018	8	21,275	8
19	Canada	32,389	5	272,652	4	40,065	6	115,035	5
20	Guatemala	31,622	5	150,651	5	117,633	5	99,969	6
21	Portugal	29,646	5	57,091	7	19,492	7	35,410	7
22	Egypt	28,572	5	88,829	6	130,668	5	82,690	6
23	Jamaica	28,132	5	34,093	7	27	8	20,751	8
24	Macao	27,928	5	65,942	7	5,716	8	33,196	7
25	Colombia	24,792	5	42,758	7	27,600	7	31,717	8
26	Japan	23,679	5	16,592	8	13,875	7	18,048	8
27	Peru	19,201	6	44,458	7	38,298	6	33,986	8
28	Honduras	16,429	6	454,554	2	*429,508*	*3*	*300,163*	*3*
29	Israel	15,601	6	42,143	7	17,266	7	25,003	8
30	El Salvador	14,261	6	263,241	4	*284,013*	*3*	187,172	4
…	…	…	…	…	…	…	…	…	…
113	Cambodia	12	8	92,702	6	*272,825*	*3*	121,846	5

only 14 countries, which means that big countries produced even more, while the smaller countries produced less. Also, it is interesting to mention that 159 countries in cluster 8 in 2010 produced as much as one country in cluster 3.

5 Conclusion and Future Work

The proposed hybrid *Random algorithm for good k* and modified k-means classical method is described and implemented on real-world data set. Random algorithm for good k estimates the optimal number of clusters. Modified k-means classical method is implemented in clustering and ranking countries in U.S. cotton textile import. This research is quite sufficient to show changes in world's cotton production and production changes in countries that export cotton textile to the USA. In future work, ensemble algorithm for estimating optimal number of clusters will be implemented, and the attempts will be made to improve the existing k-means algorithm to get less than two objects—data points, in one cluster.

References

1. Akaike, H.: A new look at statistical model identification. IEEE Trans. Autom. Control **19**(6), 716–723 (1974)
2. Bigus, J.P.: Data Mining with Neural Networks. McGraw-Hill, New York (1996)
3. Ben-Dor, A., Yakhini, Z.: Clustering gene expression patterns. In: Proceedings of the 3rd Annual International Conference on Computational Molecular Biology (RECOMB 99), pp. 11–14, Lyon (1999)
4. Cadez, I., Smyth, P., Mannila, H.: Probabilistic modeling of transactional data with applications to profiling, visualization, and prediction. In: Proceedings of the 7th ACM SIGKDD, pp. 37-46, San Francisco (2001)
5. Daniel, C., Wood, F.C.: Fitting Equations to Data: Computer Analysis of Multifactor Data. Wiley, New York (1980)
6. Dempster, A., Laird, N., Rubin, D.: Maximum likelihood from incomplete data via the EM algorithm. J. R. Stat. Soc. Ser. B **39**(1), 1–38 (1977)
7. Duda, R., Hart, P.: Pattern Classification and Scene Analysis. Wiley, New York (1973)
8. Ester, M., Frommlet, A., Kriegel, H.P., Sander, J.: Spatial data mining: database primitives, algorithms and efficient DBMS support. Data Min. Knowl. Discov. **4**(2–3), 193–216 (2000)
9. Fukunaga, K.: Introduction to Statistical Pattern Recognition. Academic Press, San Diego (1990)
10. Gaertler, M.: Clustering. In: Brandes, U., Erlebach, T. (eds.) Network Analysis: Methodological Foundations. LNCS, vol. 3418, pp. 178–215 (2004)
11. Gersho, A., Gray, R.M.: Vector Quantization and Signal Compression. Communications and information theory. Kluwer Academic Publishers, Norwell (1992)
12. Han, J., Kamber, M.: Data Mining. Morgan Kaufmann Publishers, San Francisco (2001)
13. Hartigan, J.: Clustering Algorithms. Wiley, New York (1975)
14. Heer, J., Chi, E.: Identification of Web user traffic composition using multimodal clustering and information scent. 1st SIAM ICDM, Workshop on Web Mining, pp. 51–58, Chicago (2001)
15. http://otexa.trade.gov/Msrcat.htm. Accessed 29 April 2015

16. Jain, A.K., Dubes, R.C.: Algorithms for Clustering Data. Prentice Hall, Upper Saddle River (1988)
17. Jain, A.K., Murty, N., Flynn, P.J.: Data clustering: a review. ACM Comput. Surv. **31**(3), 264–323 (1999)
18. Jiang, D., Tang, C., Zhang, A.: Cluster analysis for gene expression data: a survey. IEEE Trans. Knowl. Data Eng. **16**(11), 1370–1386 (2004)
19. Larose, D.T.: Discovering Knowledge in Data: An Introduction to Data Mining. Wiley, New York (2005)
20. Mecca, G., Raunich, S., Pappalardo, A.: A new algorithm for clustering search results. Data Knowl. Eng. **62**(3), 504–522 (2007)
21. Milligan, G.W., Cooper, M.C.: An examination of procedures for determining the number of clusters in a data set. Psychometrika **50**(2), 159–179 (1985)
22. Piórkowski, A., Gronkowska-Serafin, J.: Towards precise segmentation of corneal endothelial cells. Lect. Notes Comput. Sci. **9043**, 240–249 (2015)
23. Scott, D.W.: Multivariate Density Estimation. Wiley, New York (1992)
24. Schwarz, G.: Estimating the dimension of a model. Ann. Stat. **6**(2), 461–464 (1978)
25. Spath, H.: Cluster Analysis Algorithms. Ellis Horwood, Chichester (1980)
26. Steinbach, M., Karypis, G., Kumar, V: A comparison of document clustering techniques. In: 6th ACM SIGKDD, World Text Mining Conference, Boston (2000)
27. Tibshirani, R., Walther, G., Hastie, T.: Estimating the number of clusters in a data set via the gap statistic. J. R. Stat. Soc. **63**(2), 411–423 (2001)
28. Tibshirani, R., Hastie, T., Eisen, M., Ross, D., Botstein, D., Brown, P.: Clustering methods for the analysis of DNA microarray data. Department of Statistics, Stanford University, Stanford, Technical Report, http://statweb.stanford.edu/tibs/ftp/jcgs.ps (2015). Accessed 29 April 2015
29. Valafar, F.: Pattern recognition techniques in microarray data analysis: a survey. Ann. N. Y. Acad. Sci. **980**, 41–64 (2002)

Discriminant Function Selection in Binary Classification Task

Robert Burduk

Abstract The ensemble selection is one of the important problems in building multiple classifier systems (MCSs). This paper presents dynamic ensemble selection based on the analysis of discriminant functions. The idea of the selection is presented on the basis of binary classification tasks. The paper presents two approaches: one takes into account the normalization of the discrimination functions, and in the second approach, normalization is not performed. The reported results based on the data sets form the UCI repository show that the proposed ensemble selection is a promising method for the development of MCSs.

Keywords Ensemble selection · Multiple classifier system · Binary classification task

1 Introduction

The pattern recognition task is one of the trends in research on machine learning [1]. The classification task can be accomplished by a single classifier or by a team of classifiers. In the literature, the use of multiple classifiers for a decision problem is known as MCSs or an ensemble of classifiers [4, 11, 27]. These methods are popular for their ability to fuse together multiple classification outputs for better accuracy of classification. The building of MCSs consists of three phases: generation, selection, and integration [3]. The final decision which is made in the third phase uses the prediction of the selected classifiers. The output of an individual classifier can be divided into three types [17].

R. Burduk (✉)
Department of Systems and Computer Networks, Wroclaw University of Technology, Wybrzeze Wyspianskiego 27, 50-370 Wroclaw, Poland
e-mail: robert.burduk@pwr.edu.pl

R. Burduk et al. (eds.), *Proceedings of the 9th International Conference on Computer Recognition Systems CORES 2015*, Advances in Intelligent Systems and Computing 403, DOI 10.1007/978-3-319-26227-7_25

- The abstract level—classifier ψ assigns the unique label j to a given input x.
- The rank level—in this case for each input x, each classifier produces an integer rank array. Each element within this array corresponds to one of the defined class labels. The array is usually sorted with the label at the top being the first choice.
- The measurement level—the output of a classifier is represented by a measurement value that addresses the degree of assigning the class label to the given output x. An example of such a representation of the output is a posteriori probability returned by Bayes classifier.

Following these three types of outputs of the base classifier, various problems of combination function (third phase) of classifier outputs are considered. The problems studied in [19, 25] belong to the abstract level. The combining outputs for the rank level are presented in [13] and problems studied in [16, 18] belong to the last level. The selection of classifiers is one of the important problems in the creation of EoC [15, 24]. This task is related to the choice of a set of classifiers from all the available pool of classifiers. Formally, if we choose one classifier, it is a classifier selection process. But if we choose a subset of classifiers from the pool, then it is an ensemble selection. In this work, these terms will be used interchangeably. Here you can distinguish between the static or dynamic selection [22, 26]. In the static classifier selection one set of classifiers is selected to create EoC. This EoC is used in the classification of all the objects from the testing set. The main problem in this case is to find a pertinent objective function for selecting the classifiers. One of the best objective functions for the abstract level of classifier outputs is the simple majority voting error [23]. In the dynamic classifier selection for each unknown sample a specific subset of classifiers is selected [5]. It means that we are selecting different EoCs for different objects from the testing set. In this type of classifier selection, the classifier is chosen and assigned to the sample based on different features [28] or different decision regions [7, 14]. In this work we will consider the dynamic ensemble selection. In detail, we propose the new selection method based on the analysis of the discriminant functions in the contents of the binary classification. The paper presents two approaches: one takes into account the normalization of the discrimination functions. In the second approach, normalization is not performed. The text is organized as follows: in Sect. 2 the ensemble of classifiers and combination functions based on the sum method are presented. Section 3 contains the new method for the dynamic ensemble selection based on the analysis of the discriminant functions. Section 4 includes the description of research experiments comparing the suggested algorithms with base classifiers. Finally, the discussion and conclusions from the experiments are presented.

2 Ensemble of Classifiers

Let us consider the binary classification task. It means that we have two class labels $M = \{1, 2\}$. Each pattern is characterized by a feature vector X. The recognition algorithm maps the feature space X to the set of class labels M according to the general formula:

$$\Psi : X \rightarrow M. \tag{1}$$

Let us assume that K different classifiers $\Psi_1, \Psi_2, \ldots, \Psi_K$ are available to solve the classification task. In MCSs these classifiers are called base classifiers. In the binary classification task K is assumed to be an odd number. As a result, for all the classifiers' actions, their K responses are obtained. The output information from all K component classifiers is applied to make the ultimate decision of MCSs. This decision is made based on the predictions of all the base classifiers. One of the possible methods for integrating the output of the base classifier is the sum rule. In this method the score of MCSs is based on the application of the following sums:

$$s_i(x) = \sum_{k=1}^{K} \hat{p}_k(i|x), \qquad i \in M, \tag{2}$$

where $\hat{p}_k(i|x)$ is an estimate of the discrimination functions for class label i returned by classifier k. The final decision of the MCSs is made following the maximum rule:

$$\Psi_S(x) = \arg\max_i s_i(x). \tag{3}$$

In the presented method (3) the discrimination functions obtained from the individual classifiers take an equal part in building MCSs. This is the simplest situation in which we do not need additional information about the testing process of the base classifiers except for the models of these classifiers. One of the possible methods in which weights of the base classifier are used is presented in [2].

2.1 Selection of Discrimination Functions

The classifier selection phase is often criticized due to usually observed high computational cost [3]. Bearing this in mind, we propose the dynamic selection of discrimination functions, which has minor computing requirements. It uses the assumption that the more credible the classifier, the larger the differences in the discrimination function. In the discussed binary classification task this means that during the selection we compare the discriminant functions $|\hat{p}_k(1|x) - \hat{p}_k(2|x)| < \alpha$. The parameter α determines the size of the difference in the discriminant functions. The values are derived from the interval $< \alpha \in [0, 1)$. For value $\alpha = 0$ the selection process does not occur. The proposed selection process is performed before the calculation of the coefficients $s_i(x)$. The final decision is made according to the sum rule, and MCSs with the selection is labelled as $\Psi^{\alpha}_{S_{ON}}$. In MSCs model labeled as $\Psi^{\alpha}_{S_N}$, in addition the normalization of the discrimination functions is carried out. Normalization is performed for each label class i according to the rule:

$$\widehat{p'}_k(i|x) = \frac{\hat{p}_k(i|x) - \min(\hat{p}_1(i|x), \ldots, \hat{p}_k(i|x))}{\max(\hat{p}_1(i|x), \ldots, \hat{p}_k(i|x)) - \min(\hat{p}_1(i|x), \ldots, \hat{p}_k(i|x))}, \quad k \in K. \tag{4}$$

The aim of the experiments presented in the next section was among other things, to compare two proposed methods of selection.

3 Experimental Studies

A series of experiments were carried out to illustrate the quality of classifications. The aim of the experiments was to compare the proposed selection method algorithms with the base classifiers and ensemble classifiers based on the majority voting and sum methods. In the experiential research 11 benchmark data sets were used. Two of them were generated randomly—they are the so-called Banana and Highleyman sets. The other nine benchmark data sets come from the UCI repository [10]. All the data sets concern the binary classification task. The numbers of attributes, examples, and ration in the classes are presented in Table 1. The studies did not include the impact of the feature selection process on the quality of classifications. Therefore, the feature selection process [12, 21] was not performed. The results are obtained via tenfold-cross-validation method. In experiment 7, base classifiers were used. Two of them work according to $k - NN$ rule where the k parameter is equal to 5 or 7. Two base classifiers are used as decision trees algorithms, with the number of branches denoted as 2 and the depth of the precision tree having at most 6 levels. In the decision-making nodes the Gini index or entropy is used. One of the classifiers uses a combination of the decision trees' base models. It is Gradient boosting algorithm. The other two classifiers use Support Vector Machines models. One of them uses Least Squares SVM method and the second Decomposed Quadratic Programming

Table 1 Description of data sets selected for the experiments

Data set	Example	Attribute	Ration (0/1)
Banana	400	2	1.0
Blood	748	5	3.2
Breast cancer wisconsin	699	10	1.9
Highleyman	400	2	1.0
Ionosphere	351	34	1.8
Indian liver patient	583	10	0.4
Mammographic mass	961	6	1.2
Parkinson	197	23	0.3
Pima Indians diabetes	768	8	1.9
Sonar (mines versus rocks)	208	60	0.9
Statlog	690	14	0.8

Table 2 Classification accuracy and mean rank positions for the proposed selection algorithm with normalization produced by the Friedman test

Data set	Ψ_{S_N} with $\alpha =$							
	0	0.1	0.2	0.3	0.4	0.5	0.6	0.7
Banana	0.975	0.975	0.975	0.975	0.977	0.977	0.977	0.977
Blood	0.657	0.660	0.655	0.655	0.660	0.665	0.657	0.647
Cancer	0.947	0.945	0.945	0.947	0.947	0.942	0.942	0.940
Higle	0.955	0.955	0.955	0.952	0.952	0.952	0.952	0.952
Ion	0.967	0.970	0.972	0.970	0.960	0.957	0.962	0.955
Liver	0.602	0.597	0.595	0.622	0.647	0.667	0.660	0.662
Mam	0.625	0.620	0.615	0.632	0.637	0.640	0.632	0.632
Park	0.977	0.977	0.987	0.980	0.985	0.985	0.980	0.982
Pima	0.572	0.575	0.572	0.582	0.592	0.577	0.582	0.577
Sonar	0.980	0.980	0.980	0.980	0.980	0.980	0.980	0.980
Statlog	0.912	0.912	0.915	0.915	0.915	0.920	0.922	0.910
Mean rank	4.36	4.36	4.09	3.36	2.45	2.72	3.18	4.63

method. The experiments were performed in an SAS Enterprise Miner environment. Table 2 shows the results of the classification for the proposed classifier selection with normalization of the posteriori probability functions. Additionally, the mean ranks obtained by the Friedman test were presented. The values show that the best value of the parameter α is 0.4 for that classifier selection method. Table 3 shows the results

Table 3 Classification accuracy and mean rank positions for the proposed selection algorithm without normalization produced by the Friedman test

Data set	$\Psi_{S_{ON}}$ with $\alpha =$							
	0	0.1	0.2	0.3	0.4	0.5	0.6	0.7
Banana	0.977	0.977	0.977	0.977	0.977	0.977	0.977	0.977
Blood	0.672	0.672	0.660	0.605	0.647	0.660	0.662	0.572
Cancer	0.942	0.945	0.945	0.945	0.940	0.940	0.937	0.935
Higle	0.952	0.952	0.950	0.952	0.952	0.957	0.957	0.940
Ion	0.962	0.962	0.965	0.962	0.957	0.960	0.950	0.960
Liver	0.675	0.675	0.655	0.662	0.665	0.665	0.665	0.617
Mam	0.642	0.637	0.632	0.632	0.632	0.635	0.625	0.625
Park	0.987	0.990	0.990	0.992	0.990	0.987	0.987	0.982
Pima	0.600	0.602	0.597	0.605	0.595	0.587	0.585	0.580
Sonar	0.987	0.987	0.985	0.985	0.985	0.980	0.980	0.977
Statlog	0.910	0.910	0.907	0.907	0.910	0.912	0.910	0.907
Mean rank	2.18	1.63	3.63	3.09	3.81	3.63	4.54	6.81

Table 4 Classification accuracy and mean rank positions for the base classifiers ($\Psi_1, \ldots, \Psi_7$) and MCSs algorithms produced by the Friedman test

Data set	Ψ_1	Ψ_2	Ψ_3	Ψ_4	Ψ_5	Ψ_6	Ψ_7	Ψ_{MV}	Ψ_{Sum}	$\Psi_{S_N}^{0.4}$	$\Psi_{S_{ON}}^{0.1}$
Banana	0.130	0.130	0.027	0.027	0.030	0.007	0.012	0.022	0.022	0.022	0.022
Blood	0.445	0.422	0.377	0.377	0.372	0.350	0.372	0.327	0.327	0.340	0.327
Cancer	0.067	0.065	0.077	0.077	0.057	0.060	0.067	0.060	0.057	0.052	0.055
Higle	0.167	0.170	0.050	0.050	0.072	0.047	0.052	0.045	0.047	0.047	0.047
Ion	0.060	0.057	0.062	0.062	0.037	0.110	0.127	0.040	0.037	0.040	0.037
Liver	0.417	0.417	0.412	0.412	0.337	0.315	0.335	0.402	0.325	0.352	0.325
Mam	0.485	0.485	0.377	0.377	0.372	0.350	0.380	0.365	0.357	0.362	0.362
Park	0.052	0.060	0.015	0.015	0.012	0.052	0.057	0.027	0.012	0.015	0.010
Pima	0.425	0.422	0.435	0.435	0.405	0.377	0.390	0.415	0.400	0.407	0.397
Sonar	0.052	0.042	0.047	0.047	0.010	0.070	0.072	0.015	0.012	0.020	0.012
Statlog	0.100	0.105	0.090	0.090	0.067	0.157	0.155	0.092	0.090	0.085	0.090
Mean rank	9.09	8.90	7.09	7.09	4.36	5.00	7.36	4.63	2.27	3.36	2.09

for the case where there is no normalization of the posteriori probability functions. For this case the optimal value of the parameter α is 0.1. Classifiers with the selected values of parameter α were compared with the base classifiers and ensemble methods based on the majority voting and sum methods. The results of classification with the mean ranks obtained by the Friedman test are presented in Table 4. To compare the results the post-hoc Nemenyi test was used [24]. The critical difference for this test at $p = 0.05$ is equal to $CD = 4.51$. Since the difference between Ψ_{Sum} and the proposed algorithms $\Psi_{S_N}^{0.4}$, $\Psi_{S_{ON}}^{0.1}$ is already smaller than 4.51, we can conclude that the post-hoc test is not powerful enough to detect any significant differences between these algorithms. However, the proposed selection of the posteriori probability functions algorithm achieved the best result, which is the lowest average rank (Table 3).

4 Conclusion

This paper discusses the classifier selection for the binary classification task. The proposal in the paper process concerns the selection of the posteriori probability functions. The paper presents two approaches. In one of them normalization of the posteriori probability functions is carried out. In the second case under consideration, normalization is not performed. The distributed computing approaches enable the efficient and parallel processing of the complicated data analysis task, also in the context of classification systems with multiple classifiers [20]. The methods of classifier selection presented in the paper show the ability to work in a parallel and distributed environment. Parallel processing provides the possibility to speed up the selection of the posteriori probability functions, whose results are needed to make the decision by the classifier ensemble. Additionally, the proposed approach can be applied in various practical tasks involving multiple elementary classification tasks [6, 8, 9]. In the paper several experiments were carried out on the data sets available from the UCI repository and on the synthetical data sets. The aim of the experiments was to compare the proposed selection method algorithms with the base classifiers and ensemble classifiers based on the majority voting and sum methods. For the proposed selection method, we obtained improvement of the classification quality measured by average values from the Friedman test. However, the difference in average ranks is too small to detect statistically significant differences between the proposed selection method and the ensemble method based on the sum rule.

Acknowledgments This work was supported by the Polish National Science Center under the grant no. DEC-2013/09/B/ST6/02264 and by the statutory funds of the Department of Systems and Computer Networks, Wroclaw University of Technology.

References

1. Bishop, C.M.: Pattern Recognition and Machine Learning (Information Science and Statistics). Springer, New York (2006)
2. Burduk, R.: Classifier fusion with interval-valued weights. Pattern Recognit. Lett. **34**(14), 1623–1629 (2013)
3. Britto, A.S., Sabourin, R., Oliveira, L.E.S.: Dynamic selection of classifiers—a comprehensive review. Pattern Recognit. **47**(11), 3665–3680 (2014)
4. Cyganek, B.: One-class support vector ensembles for image segmentation and classification. J. Math. Imaging Vis. **42**(2–3), 103–117 (2012)
5. Cavalin, P.R., Sabourin, R., Suen, C.Y.: Dynamic selection approaches for multiple classifier systems. Neural Comput. Appl. **22**(3–4), 673–688 (2013)
6. Cyganek, B., Woźniak, M.: Vehicle Logo Recognition with an Ensemble of Classifiers. Lecture Notes in Computer Science, vol. 8398, pp. 117–126. Springer, Berlin (2014)
7. Didaci, L., Giacinto, G., Roli, F., Marcialis, G.L.: A study on the performances of dynamic classifier selection based on local accuracy estimation. Pattern Recognit. **38**, 2188–2191 (2005)
8. Frejlichowski, D.: An Algorithm for the Automatic Analysis of Characters Located on Car License Plates. Lecture Notes in Computer Science, vol. 7950, pp. 774–781. Springer, Berlin (2013)
9. Forczmański, P., Łabędź, P.: Recognition of Occluded Faces Based on Multi-subspace Classification. Lecture Notes in Computer Science, vol. 8104, pp. 148–157. Springer, Berlin (2013)
10. Frank, A., Asuncion, A.: UCI Machine Learning Repository (2010)
11. Giacinto, G., Roli, F.: An approach to the automatic design of multiple classifier systems. Pattern Recognit. Lett. **22**, 25–33 (2001)
12. Guyon, I., Elisseeff, A.: An introduction to variable and feature selection. J. Mach. Learn. Res. **3**, 1157–1182 (2003)
13. Ho, T.K., Hull, J.J., Srihari, S.N.: Decision combination in multiple classifier systems. IEEE Trans. Pattern Anal. Mach. Intell. **16**(1), 66–75 (1994)
14. Jackowski, K., Woźniak, M.: Method of classifier selection using the genetic approach. Expert Syst. **27**(2), 114–128 (2010)
15. Jackowski, K., Krawczyk, B., Woźniak, M.: Improved adaptive splitting and selection: the hybrid training method of a classifier based on a feature space partitioning. Int. J. Neural Syst. **24**(3) (2014)
16. Kuncheva, L.I.: A theoretical study on six classifier fusion strategies. IEEE Trans. Pattern Anal. Mach. Intell. **24**(2), 281–286 (2002)
17. Kuncheva, L.I.: Combining Pattern Classifiers: Methods and Algorithms. Wiley, Hoboken (2004)
18. Kittler, J., Alkoot, F.M.: Sum versus vote fusion in multiple classifier systems. IEEE Trans. Pattern Anal. Mach. Intell. **25**(1), 110–115 (2003)
19. Lam, L., Suen, C.Y.: Application of majority voting to pattern recognition: an analysis of its behavior and performance. IEEE Trans. Syst., Man, Cybern., Part A **27**(5), 553–568 (1997)
20. Przewoźniczek, M., Walkowiak, K., Woźniak, M.: Optimizing distributed computing systems for k-nearest neighbours classifiers-evolutionary approach. Logic J. IGPL **19**(2), 357–372 (2010)
21. Rejer, I.: Genetic Algorithms in EEG Feature Selection for the Classification of Movements of the Left and Right Hand. In: Proceedings of the 8th International Conference on Computer Recognition Systems CORES 2013. Advances in Intelligent Systems and Computing, vol. 226, pp. 581–590. Springer, Heidelberg (2013)
22. Ranawana, R., Palade, V.: Multi-classifier systems: review and a roadmap for developers. Int. J. Hybrid Intell. Syst. **3**(1), 35–61 (2006)
23. Ruta, D., Gabrys, B.: Classifier selection for majority voting. Inf. Fusion **6**(1), 63–81 (2005)
24. Smętek, M., Trawiński, B.: Selection of heterogeneous fuzzy model ensembles using self-adaptive genetic algorithms. New Gener. Comput. **29**(3), 309–327 (2011)

25. Suen, C.Y., Legault, R., Nadal, C.P., Cheriet, M., Lam, L.: Building a new generation of handwriting recognition systems. Pattern Recognit. Lett. **14**(4), 303–315 (1993)
26. Trawiński, K., Cordon, O., Quirin, A.: A study on the use of multiobjective genetic algorithms for classifier selection in furia-based fuzzy multiclassifiers. Int. J. Comput. Intell. Syst. **5**(2), 231–253 (2012)
27. Ulas, A., Semerci, M., Yildiz, O.T., Alpaydin, E.: Incremental construction of classifier and discriminant ensembles. Inf. Sci. **179**(9), 1298–1318 (2009). Apr
28. Woloszyński, T., Kurzyński, M.: A probabilistic model of classifier competence for dynamic ensemble selection. Pattern Recognit. **44**(10–11), 2656–2668 (2011)

Comparison of Multi-label and Multi-perspective Classifiers in Multi-task Pattern Recognition Problems

Edward Puchała and Krzysztof Reisner

Abstract This paper deals with the comparison of two different approaches for multi-task pattern recognition problem—multi-label and multi-perspective. The experiment performed measured the hamming loss and mean accuracy of both classifiers, to judge which of these two better fit to this kind of problem.

1 Introduction

In many practical cases, like medical diagnosis, the pattern recognition problem can be, and should be described as a set of tasks. This means that the classifier must be able to deliver not a single class number, but a vector whose components represent the results of classification for a particular task. Of course in real life, the decision of the previous task can affect the results of the next one; in this paper we will omit such dependencies between tasks and focus on the similarities and differences between multi-label and multi-perspective classifiers [1].

In the paper two different approaches for multi-task recognition pattern problem—multi-label and multi- perspective are presented. These two methods are analogous enough to compare them without any special conditions.

2 Definitions

This section contains a short description and the basic conception of the multi-task pattern recognition method. Also, two approaches to problem solving are discussed.

E. Puchała (✉) · K. Reisner
Wroclaw University of Technology, Department of Systems and Computer Networks, Wyb. Wyspianskiego 27, 50-370 Wroclaw, Poland
e-mail: edward.puchala@pwr.edu.pl

R. Burduk et al. (eds.), *Proceedings of the 9th International Conference on Computer Recognition Systems CORES 2015*, Advances in Intelligent Systems and Computing 403, DOI 10.1007/978-3-319-26227-7_26

2.1 Multi-task Pattern Recognition Problem

Let us consider the N-task pattern recognition problem. This means that we have a set of tasks $T = \{1, 2, \ldots, N\}$, where each task has a defined set of class numbers. If we consider kth task we have set of classes $j_k \in M_k$ with numbers $j_k = \{1, 2, \ldots, M_k\}$. In this problem recognized objects are described using the compound features vector $X = \{x_1, x_2, \ldots, x_N\}$ and vector of classes $J = \{j_1, j_2, \ldots, j_N\}$ [6, 7].

2.2 Multi-label Classifier

Multi-label classifiers allow to assign more than one class number to the recognized object. The main concern about this classification method is that the length of vector of assigned class numbers can be different in each recognition task. This could lead to an inconclusive decision (e.g., vector contains more than one class number for task). Existing methods of multi-label classification can be grouped into two categories:

- problem transformation methods and
- algorithm adaptation methods [3].

We focused on problem transformation methods, because this could lead us to a more reliable comparison of multi-label and multi-perspective approaches due to the same implementation of base classifier that could be used in both cases. There are several methods of problem transformation which allow to perform multi-label classification efficiently [3]. We decided to choose the disassemble problem to a set of binary decisions—we check if class number is applicable for the recognized object or not. Then we combine all positive results in one result vector. An example of this transformation is shown below where in Table 1 is presented example vectors of class numbers. Table 2 presents sets that are used to make binary decisions [5]. This is a simple but effective way to apply the multi-label classification method as a multi-task problem. However, there is one limitation—class numbers across all tasks must be unique.

Let us assume that we have n-tasks pattern recognition problem. For kth task we have available class numbers $j_k \in M_k = \{1, 2, \ldots, M_k\}$. This means that class numbers across tasks are not unique. To provide uniqueness of every class number in the set of labels (whole class numbers in problem) we have to perform translation from n sets of class numbers to one:

Table 1 Representation of example objects with assigned classes

	Class1	Class2	Class3	Class4
Object1	x		x	
Object2		x		x
Object3	x			x

Table 2 Results of problem transformation into the set of binary decisions

	Decision1	Decision2	Decision3	Decision4
	Class1	Class2	Class3	Class4
Object1	1	−1	1	−1
Object2	−1	1	−1	1
Object3	1	−1	−1	1

$$T(M_1, M_2, \ldots, M_k) \rightarrow L \tag{1}$$

The translation used in this paper is quite simple. We add class numbers one after another to a new set L and assign the lowest available number in set L. This translation can be present as

$$L = M_1' \cup M_2' \cup \ldots \cup M_k' \cup \ldots \cup M_n' \tag{2}$$

where:

$$M_k' = \begin{cases} \{1, 2, \ldots, M_k\} & \text{for } k = 1 \\ \{1 + M_{k-1}', 2 + M_{k-1}', \ldots, M_k + M_{k-1}'\} & \text{for } k > 1 \end{cases} \tag{3}$$

2.3 Multi-perspective Classifier

The multi-perspective classifier was designed to resolve the pattern recognition problem, where an object is recognized from different perspectives—as in medical diagnostic processes. Perspectives differ between each other in the count of class numbers and their interpretations.

There are generally two approaches to multi-perspective classification methods: composed (direct) and decomposed. In this paper we focus on composed approach, which is simpler and omits dependencies between tasks as in multi-label classifier. Figure 1 represents the idea of a multi-perspective classifier in composed approach [1, 4].

Analyzing Fig. 1, it is easy to find similarities to the multi-label classifier. In fact, if we limit the number of predicted class numbers in a multi-label classifier to

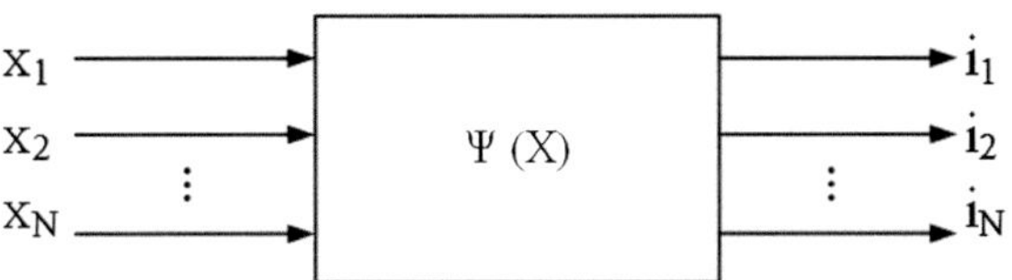

Fig. 1 Scheme of multi-perspective composed classifier

the dimension of problem we will get a multi-perspective classifier, but with one exception—the class numbers will still not be related to a particular task. Although we introduce class numbers transformation described by Eq. 1, we can relate each decision to a specified task, but we cannot preserve a situation when classifier assigns object to two class numbers from the same task. This is one of the most important differences between the two approaches.

3 Experiment

Here, we describe how we prepare and perform the experiment that allows us to compare the two chosen approaches. First, we need to create a data set that allows us to perform experiment. We have to generate a data set because we do not have real-life data, which is suitable for multi-task pattern recognition.

3.1 Preliminaries

In the experiment we used 3-task recognition problem. Vectors of features are generated randomly according to normal distribution with given parameters. All features are correlated to each other and this correlation describes the covariation matrix, which is presented in Table 3.

We also decided to bind the class number with chosen features and those class numbers to tasks. This hierarchy is shown in Table 4. Then we assign randomly the parameters of normal distribution to each feature in class number scope. The last step was to choose the layout of class numbers in each task. Due to the nature of multi-task pattern recognition we decide on that first task by a binary decision, where class number **1** means that recognition is done in the first step (e.g. in medical utilization first task determines if patient is healthy or not. If it is then other tasks which may represent diagnostic or treatment process should be omitted). For uniformity sake, we decided to assign class numbers: **3** and **6** to object **1**, which represents the end of decision making in the first step case.

Table 3 Covariation matrix

	x_1	x_2	x_3	x_4	x_5
x_1	6.135	−0.187	0.222	−0.271	−0.819
x_2	−0.187	6.863	0.114	1.713	0.901
x_3	0.222	0.114	2.622	0.039	0.172
x_4	−0.271	1.713	0.039	8.322	0.268
x_5	−0.819	0.901	0.172	0.268	4.757

Table 4 Dependencies between class number and feature

	Class number	Binded feature
Task1	1,2	x_1,x_2
Task2	3,4,5	x_3,x_4
Task3	6,7,8,9,10	x_5

Table 5 Used means in order to class number for random sample generation according to normal distribution

Class number	Feature	Mean
1	x_1	1
	x_2	11
2	x_1	18
	x_2	10
3	x_3	16
	x_4	9
4	x_3	2
	x_4	4
5	x_3	14
	x_4	19
6	x_5	5
7	x_5	16
8	x_5	7
9	x_5	8
10	x_5	1

Table 6 Generated objects used in experiment

	Assigned class for task 1	Assigned class for task 2	Assigned class for task 3
Object 1	1	3	6
Object 2	2	5	7
Object 3	2	5	8
Object 4	2	4	10
Object 5	2	4	9

The procedure described above gave us a data set of random values with normal distribution. It also provides correlation between features and class numbers. The results are presented in Table 5. It shows parameters dependency between class number and mean of normal distribution. Please notice that each set x_k is disjoined to each other. Table 6 shows the generated objects definition to classifications. We decide to use the same number of objects—100. This means that the total number of objects in the data sets is always 500.

3.2 Design of Experiment

Experiment was designed based on some common rules. First of all we decided to use Hamming loss and mean accuracy, which is described by Eqs. 4 and 5, respectively, as the statistics that allow us to compare two classifiers (multi-label and multi-perspective).

$$HammingLoss(j,i) = \frac{1}{N}\sum_{k=1}^{N}\frac{|j_k \Delta i_k|}{|L|} \tag{4}$$

$$MeanAccuracy(j,i) = \frac{1}{N}\sum_{k=1}^{N}\frac{|j_k \cap i_k|}{|j_k|} \tag{5}$$

where i is the set of vectors which contains predicated class numbers of recognized object by classifier and j is a set of vectors which contains correct class numbers for object to recognize. L is the set of available class numbers and the Δ operator means the symmetric difference between sets. It is like the xor operator but is applicable to sets.

Also, we provide a percentage of inconclusive classifications, but only for multi-label classifier because this statistic for multi-perspective classifier will always be 0 %. The experiment is repeated 25 times and for each repetition the data are generated randomly according to normal distributions with defined parameters. In each iteration we use 3-folded cross-validation protocol [2]. This means that in each iteration we have 332 training objects and 168 testing objects.

One of the reasons that we are used these specific types of classifiers with restriction, like method of multi-label classifier design (a problem transformation approach), is that we use the same implementation of base classifiers. This provides results without any concerns about quality of implementation in both cases. We decided to use, as a base classifier, decision trees and neural networks, which are both implemented in R language. The first one—neural network has a simple structure—we use a single hidden layer. In the results of experiments we set down the number of nodes as 5. The size of te output layer depends on the class numbers' set count. The maximum number of epochs was set to 100 [8]. For decision tree classifier we decided to use a conditional inference tree [9].

3.3 Results

The tables show the results of experiment described above. Table 7 represents the data extracted from experiment with decision trees and Table 8 the neural networks. Both tables present the Mean Accuracy for each task in a multi-task problem. There is also a column called 'Overall', which presents the accuracy for the whole problem or, in other words, how many assigned sets of class numbers are exactly the same as testing.

Table 7 Results for decision trees

	Mean accuracy					
	Task1	Task2	Task3	Overall	Hamming loss	Inconclusive (%)
Multi-perspective	0.997	0.994	0.953	0.952	0.006	0
Multi-label	0.998	0.975	0.766	0.757	0.031	2

Table 8 Results for neural networks

	Mean accuracy					
	Task1	Task2	Task3	Overall	Hamming loss	Inconclusive (%)
Multi-perspective	0.995	0.985	0.897	0.880	0.0108	0
Multi-label	0.997	0.966	0.696	0.672	0.0450	4

Table 9 Values of standard derivation of statistics in case of decision trees

	Standard derivation of mean accuracy	Standard derivation of hamming loss
Multi-perspective	1.151	0.001
Multi-label	1.380	0.007

Table 10 Values of standard derivation of statistics in case of neural networks

	Standard derivation of mean accuracy	Standard derivation of hamming loss
Multi-perspective	6.185	0.006
Multi-label	9.103	0.012

Additionally, we measured the standard derivation of overall Mean Accuracy and Hamming Loss. Values of standard derivation for Mean Accuracy and Hamming loss in case of decision trees are presented in Table 9 and for neural network in Table 10.

3.4 Results and Discussion

The differences between both combined classifiers is not significant, but it clearly points out that a multi-perspective classifier gives better Mean Accuracy (especially in case of Task 3). Also, it provides a smaller Hamming Loss, which means that the statistically multi-perspective classifier should give more reliable results. Also, the standard derivation points that the multi-perspective classifier has more consistent results.

What is interesting is the fact that both classifiers has worse results of Mean Accuracy for Task 3 than the others. This could be due to the characteristics of data set, because as we can see in Table 4 in Task 3 we have more available class numbers than in others. Also, only one feature is assigned to this task.

4 Conclusions

The superiority of the multi-perspective classifier in composed (direct) version over the muli-label (decision trees, neural network) demonstrates the effectiveness of this concept in such multitask classification problems for which decomposition is necessary from the functional or computational point of view (e.g. in medical diagnosis). A direct approach to multi-perspective recognition algorithms gives better results than the decomposed approach, because such algorithms take into consideration correlation between individual classification problems.

For future consideration, we should investigate the influence of the number of available class numbers in particular and features correlated to this task on classification quality, because the results presented in this paper suggest that this could be a significant factor.

References

1. Puchała, E.: Podstawy komputerowego rozpoznawania wielozadaniowego—algorytmy i ich zastosowania w diagnostyce medycznej Politechnika Wroc^{3}awska, Wroc^{3}aw (1993)
2. Kuncheva, L.: Combining Pattern Classifiers Methods and Algorithms. Wiley, New Jersey (2014)
3. Tsoumakas, G., Katakis, I.: Multi-Label Classification: An Overview. Department of Informatics, Aristotle University of Thessaloniki, Greece (2007)
4. Kurzynski, M.W., Puchala, E.: Algorithms of the multiperspective recognition. In: Proceedings of the 11th International Conference on Pattern Recognition, Hague (1992)
5. Bromuria, S., Zufferey, D., Hennebert, J., Schumacher, M.: Multi-label classification of chronically ill patients with bag of words and supervised dimensionality reduction algorithms. J. Biomed. Inf. **51**, 165–175 (2014)
6. Kurzynski, M.W., Puchala, E., Blinowska, A.: A branch-and-bound algorithm for optimization of multiperspective classifier pattern recognition. In: Proceedings of the 12th IAPR International Conference on Computer Vision and Image Processing—Conference B, **2**:235–239 (1994)
7. Duda, R., Hart, P.: Pattern Classification and Scene Analysis. Wiley, New York (1973)
8. Ripley, B., Venables, W.: Feed-Forward Neural Networks and Multinomial Log-Linear Models. http://cran.r-project.org/web/packages/nnet/index.html
9. Hothorn, T., Hornik, K., Strobl, C, Zeileis, A.: A Laboratory for Recursive Partytioning. http://cran.r-project.org/web/packages/party/index.html

New Data Level Approach for Imbalanced Data Classification Improvement

Katarzyna Borowska and Magdalena Topczewska

Abstract The article concerns the problem of imbalanced data classification. The algorithm improving a standard SMOTE method has been proposed and tested. It is a synergy of the existing approaches and was designed to be more versatile than other similar solutions. To measure the distance between objects, the Euclidean or the HVDM metrics were applied, depending on the number of nominal attributes in a data set.

Keywords Class imbalance · Oversampling · Classification

1 Introduction

One of the area of data analysis experts interest is infrequent phenomenon, like rare diseases, failure detection, etc., that can be a source of valuable knowledge. Data consisting of few examples belonging to one class and simultaneously a large number of observations from the second class are called *imbalanced data*. Thus, in the two-class problem, the distribution of observations can be characterized by large disproportion—we have considerably less objects belonging to the *positive* or *minority* class than to the *negative* or *majority* class. To determine a level of differentiation of class disproportion, the *imbalance ratio* (*IR*) measure is specified [1]. It is calculated as the quotient of the negative and positive class cardinalities. In real data sets the values of IR can be at the level of 100:1 or 1000:1; however, high values are not a source of complications for the analysts. The problem arises from the assumption of data distribution. Typically, the distribution of objects in a data set is predicted to be uniform and costs of misclassification are expected to be equal for all classes, to which objects belong, but in fact, the most relevant is the cost of the minority data misclassification. This cost substantially outweighs conse-

K. Borowska · M. Topczewska (✉)
Faculty of Computer Science, Bialystok University of Technology,
Bialystok, Poland
e-mail: m.topczewska@pb.edu.pl

R. Burduk et al. (eds.), *Proceedings of the 9th International Conference on Computer Recognition Systems CORES 2015*, Advances in Intelligent Systems and Computing 403, DOI 10.1007/978-3-319-26227-7_27

quences of errors while majority data are misclassified, for instance, diagnosing a cancer patient as healthy is much more serious mistake than recognizing the healthy as the sick. Since many real-life domains suffer from the class imbalance problem, it has emerged as one of the challenges in data mining community [1]. Describing the causes of deficiency, two aspects can be mentioned: relative and absolute imbalances [2, 3]. In the relative imbalance case, objects are not rare in an absolute sense, but compared to the negative class. In the absolute imbalance case, the cardinality of the positive class is insufficient to perform decision process correctly and successfully. The imbalance of the class distributions can cause a poor performance when standard classifiers are used. The complex distribution and concepts such as overlapping of classes, small disjuncts, or noise can be additional nuisances. Three main methodologies have been developed to overcome negative effects of difference between positive and negative class cardinalities: data level techniques, algorithm level approaches, and cost-sensitive methods [1, 4]. The first set of approaches is associated with an inclusion of a preliminary step to the analysis. In this step cardinalities of classes are modified to reduce the imbalance consequences. In the second set of techniques, not data set but the existing algorithms are adjusted to maintain a special attention on the minority class objects. The last set combines data level and algorithm level modifications. Data level approaches are the most versatile and consist of three fundamental groups of methods: undersampling (when a subset of the original data set is created), oversampling (when new objects are created), and hybrid methods. Widely used and well-known method of this type is the SMOTE algorithm (*Synthetic minority oversampling technique*) [5]. The main idea of this algorithm is to create new minority class objects along the line segments between each positive class object and any of the k-nearest neighbors. New instances are generated by selecting randomly an appropriate number of the k-nearest neighbors of a positive class example and creating a combination of features describing each of them and sample under consideration. The number of neighbors involved in oversampling depends on the number of required minority objects. Although the SMOTE was successfully applied to many domains, it is not deprived of some drawbacks. There are numerous specific data sets that make this technique insufficient. Especially, in medical diagnosis, even a small depletion of the classifier performance can lead to the catastrophic effects. In this article a novel approach is proposed as a modification of a standard SMOTE algorithm. It is a hybrid technique of three algorithms, two of which have been described in [6].

2 Algorithm

The advantages of data preprocessing methods such as SMOTE were proven in various real-world applications. These kinds of algorithms are designed to have larger decision regions of minority class points built by the classifier. Certain complex issues may cause the SMOTE method not as effective as it is expected. The fact that the differences between minority and majority class examples are not obvious is considered as the main source of the problem. The standard classifier tends to generalize, whereas peculiar information about the minority class is usually ignored. This unique information is indeed crucial in imbalanced data classification process. The following difficulties mainly impede the proper recognition of the minority class instances: overlapping, small disjuncts, and noise [7]. These undesirable factors lead to the growth of data complexity and therefore degradation of the standard classifier performance. In order to address this important issue, the versatile improved SMOTE (VIS) algorithm is proposed in this paper.

2.1 Description

The novel algorithm comprises a synergy of the existing approaches and provides a brand new way of dealing with the imbalanced data problem. The VIS technique was designed to be more versatile than other similar solutions. It is a flexible tool which automatically adapts to the specificity of the imbalanced data problem. The first phase of the processing is described in Algorithm 1.

At the beginning of Algorithm 1, loaded data are analyzed to define which metric should be applied. Standard SMOTE algorithm does not include the impact of the distance function to the performance of the abused k-nearest neighbor method. In this novel technique the characteristics of the attributes determines whether the Euclidean distance is used or the HVDM [8]. More than half nominal attributes indicates that the HVDM metric should be applied. Otherwise, the Euclidean distance is calculated. The idea presented in [8] was the inspiration for developing this approach. The authors assumed that the HVDM effectiveness should be closely related to the number of nominal attributes. On the other hand, the simplicity and the good computational complexity of the Euclidean metric make this solution the most suitable for the data described with the majority of linear attributes. The next step harnesses the kNN algorithm to obtain the distance between each minority example and all other instances from both classes. The value of k-nearest neighbors is a parameter, which can be modified in order to tune the algorithm's performance. The mentioned calculations are essential in recognizing the local characteristics of the examples. This is very important part of the algorithm since the further processing depends on it. The idea of dividing data into specific groups and oversampling each subconcept, respectively, occurs in the literature [9–11] as the effective, groundbreaking solution. However, the methods of detecting these within-class subconcepts differentiate among various works. Likewise, a new alternative approach is presented in this paper. The minority objects are divided into the following groups: NOISE (all of the k-nearest neighbors represent the majority class); DANGER (half or more than half of the k-nearest neighbors belong to the majority class); and SAFE (more than half of the k-nearest neighbors represent the same class as the example under consideration (namely the minority class). This mechanism of the minority data division includes the location of each example in the feature space. It enables to customize the strategy of the oversampling regarding the belongingness to the groups. Specifying examples

---

Algorithm 1 VIS (S, M, k)

Require: Number of all instances S; Number of minority class samples M; Number of nearest neighbors k;
metrics: keeps the name of used evaluation metric
SampleMinority: array for the original minority class samples
Synthetic: array for the new examples
nominal: stores the number of nominal attributes
continuous: stores the number of linear attributes
label: array for examples labels
safe, *danger*, *noise*: each of them keeps the number of the minority class objects belonging to the *SAFE*, *DANGER* and *NOISE* group, respectively
mode: defines the mode of the algorithm
Verify the number of two kinds of attributes: nominal and linear. Save the number of nominal attributes in *nominal* variable and the number of linear attributes in *continuous* variable. Obtain value of the *metrics* variable by comparing values saved in *nominal* and *continuous* variables.
for $i \leftarrow 0$ **to**M **do**
  Calculate the distance between minority class examples and all other examples using kNN method with measure written in *metrics* variable.
  Indexes of k-nearest neighbors write in *nnarray* array.
  $label[i] := LabelMinorityData(nnarray, i, k)$
end for
Calculate the number of objects with the same label for the following groups: SAFE, DANGER and NOISE. Save these numbers in *safe*, *danger*, *noise* variables.
if $safe == 0$ **then**
  $mode := noSAFE$
else
  if $danger \geqslant 30\,\% M$ **then**
    $mode := HighComplexity$
  else
    $mode := LowComplexity$
  end if
end if
Calculate the needed number of minority class examples to create. The result save in N variable.
for $i \leftarrow 0$ **to**M **do**
  if $label \neq NOISE$ **then**
    Run the kNN algorithm for the object i using distance measure saved in *metrics* variable, indexes save in *nnarray* array
    $Populate(N, i, nnarray, label[i], mode)$
  end if
end for

---

that are surrounded only by the majority class instances as the NOISE helps to avoid the serious difficulties in learning process [12]. The relatively homogeneous areas are assumed to be composed of the SAFE objects. Examples placed in the area surrounding class boundaries are labeled as DANGER, which results in the appropriate consequences in the next phase. These borderline instances determine the boundaries between classes. Therefore, a general purpose of this paper is to verify the impact of DANGER examples on the classification process. The main process, responsible for generating new data, strictly relates to the assigned groups. The data oversampling techniques necessitates the number of the minority class examples to be generated.

Algorithm 2 PopulateVIS (N, i, nnarray, label[i], mode)

Require: Number of examples to create N; Indexes of k-nearest neighbors *nnarray*; Index of the example under consideration i; Label of ith example; Mode of the algorithm *mode*

```
if mode == NoSAFE then
  Create N examples on the interval connecting ith DANGER object and half the distance to the neighbor with the index drawn from nnarray
end if
if mode == HighComplexity then
  if label[i] = DANGER then
    Create one example at a shorter distance to the ith object than to its nearest neighbor with nnarray[0] index
  else
    if label[i] = SAFE then
      Create N examples based on a combination of features of the ith object and one of the nearest neighbors with index drawn from nnarray
    end if
  end if
else
  if mode == LowComplexity then
    if label[i] = SAFE then
      Create one example based on a combination of features ith object and its nearest neighbor with nnarray[0] index
    else
      if label[i] = DANGER then
        Create N examples. Each should be at a shorter distance to ith object than to its nearest neighbors with index drawn from nnarray
      end if
    end if
  end if
end if
```

In the proposed VIS algorithm this number is obtained automatically. The minority instances are created to even the amount of objects from both classes. In the last step of this preliminary phase the distances between samples representing only the minority class are calculated. For this computation the kNN method is applied again. Once these calculations are finished, the process of generating new synthetic samples can begin.

2.2 *Modes*

There are three modes of the main phase. The first one, "HighComplexity," represents the case when at least 30 % of the minority class instances are the borderline ones. According to the published studies [11], the occurrence of the 30 % and more of the DANGER objects indicates the significant complexity of the problem, see Fig. 1b. Generating a lot of the minority synthetic samples in these regions may lead to the undesirable overlapping effect. This issue was presented in [12]. We cannot assume that there are no majority instances between two borderline minority class objects. If this problem occurred, creating new example along the line segment joining the object under consideration and its neighbor may cause too large similarity between a new sample and the majority instance placed nearby. In the extreme case synthetic sample and the adjacent negative class example can be identical. The more objects are created in the complex area surrounding class boundaries, the more probable the generation of wrong minority samples is. To avoid this problem, the reduced number of new objects is created in the ambiguous areas. As a result, the number of DANGER objects is doubled. Each individual minority sample is created closer to the example under consideration, along the line segment between half of the distance from the current DANGER object and one of its k-nearest neighbors. On the other hand, a plenty of new data is created in homogeneous regions. As SAFE objects located in these areas are relatively far from the majority examples, they should be considered as the data which has specific properties of the minority class. The SAFE instances are assumed to be representatives of the minority class. Hence, the density of minority class examples is especially increased in these regions. Less complex problems (Fig. 1a) are processed in "LowComplexity" mode. In this approach the borderline data are amplified. Since objects recognized as DANGER comprise less than 30 % of all minority class examples, they may be dominated in the learning process by the majority class neighbors. Due to this fact, the plenty of synthetic instances is generated in the area surrounding the class boundaries. The number of SAFE objects is only doubled. It should be sufficient to define properly decision rules regarding these examples. New SAFE samples are created along the line segment between the object under consideration and its nearest neighbor. If analyzed sample is DANGER, new data are generated at the line segments between half of the distance from the current example and some of the k-nearest neighbors. The NOISE examples are omitted in described oversampling technique. However, the augmented number

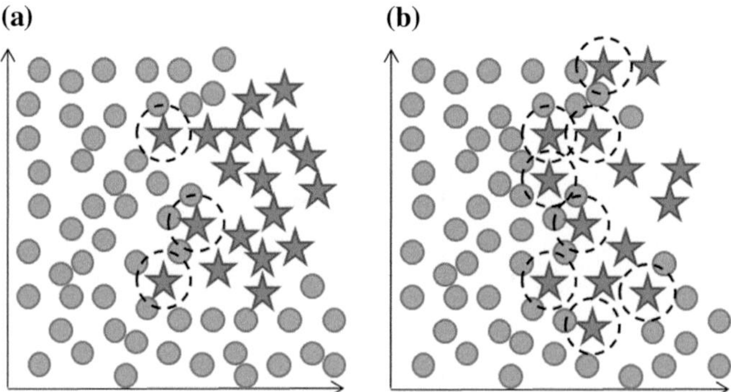

Fig. 1 Example of **a** low complexity, **b** high complexity

of these objects (more than 30 %) causes the increase of the method's sensitivity to the noise. Hence, the distance-weighted nearest neighbor algorithm is applied. Greater weights are given to closer neighbors. Although samples considered as NOISE do not take part in processing, they can contribute to the generation of wrong synthetic samples. In the proposed solution, the impact of these objects to the newly created objects is strongly restricted. The third mode is named "NoSAFE." It represents the case, when there are no SAFE samples in the minority class data. In this approach, instead of increasing density of relatively homogeneous regions (as it is normally performed in "HighComplexity" mode), all new examples are generated in the area surrounding class boundaries. It prevents the situation, when not enough synthetic objects are created.

3 Experiments

Two experiments have been performed to test the new method as the improvement of the SMOTE algorithm. To compare the results of classification five measurements were calculated: accuracy (Q); sensitivity—percent of positive class objects classified correctly (TP_{rate}); specificity—percent of negative class objects classified correctly (TN_{rate}); ratio of weighted sensitivity and precision significance ($F - measure$); and area under the curve (AUC).

3.1 Experiment 1

In the first experiment the artificial data set, containing only 25 objects, has been used to present the performance of new methods. These data are characterized by a moderate IR value. The size of the majority class four times exceeds the number of objects in the minority class, thus the analyzed data set may be specified as implicitly imbalanced, because the size of the minority class is extremely small. The obtained results of classification are shown in Table 1. In Fig. 2 the majority class objects are marked as circles, while the minority class objects are squares. Newly generated objects are marked as diamonds. After preprocessing step a new IR value equals 1, thus the imbalance of the class cardinalities was compensated. The best results were obtained for the number of neighbors equaled 3 for both—SMOTE and VIS—methods. The accuracy increased from 87.50 % for the SMOTE to 95 % for the VIS technique. Remaining all parameters were also higher for the VIS technique apart from sensitivity equaled for both methods (90 %): the rate of true negatives (100 %), the $F - measure$ (0.95), and the area under the curve (the level 95 %). The same highest results have been gained for the number of neighbors 5.

Table 1 Classification results for `artifficialData` ($IR = 4$): Q—accuracy, TP_{rate}—rate of true positives, TN_{rate}—rate of true negatives, AUC—area under the curve, nIR—new value of IR

Method	k	Q	TP_{rate}	TN_{rate}	F-measure	AUC	nIR
SMOTE	1	80.00	80.00	80.00	0.80	80.00	1.00
VIS	1	80.00	90.00	70.00	0.82	80.00	1.00
SMOTE	3	87.50	90.00	85.00	0.88	87.50	1.00
VIS	3	95.00	90.00	100.00	0.95	95.00	1.00
SMOTE	5	80.00	80.00	80.00	0.80	80.00	1.00
VIS	5	90.00	85.00	95.00	0.89	90.00	1.00
SMOTE	7	80.00	80.00	80.00	0.80	80.00	1.00
VIS	7	75.00	65.00	85.00	0.72	75.00	1.00

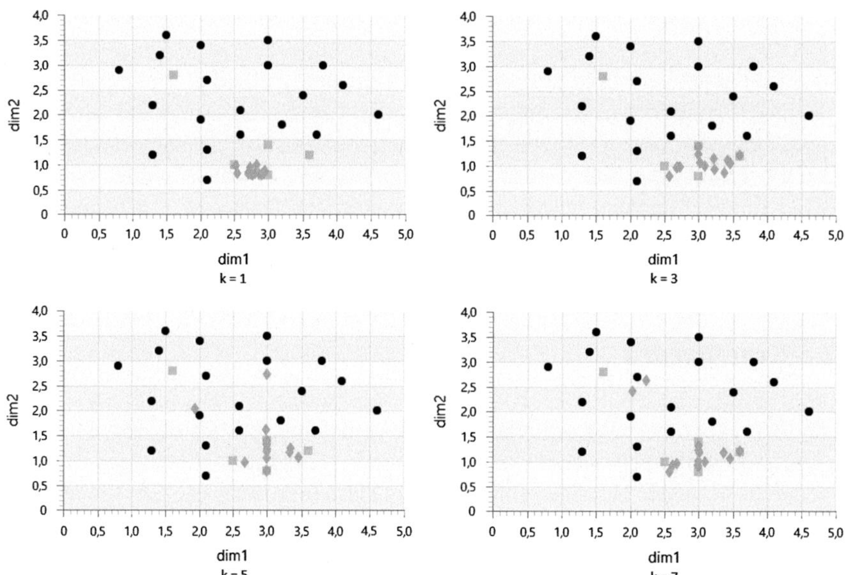

Fig. 2 Distributions after VIS preprocessing

Hereby, the assumption that the correctness of the classification depends largely on the complexity of the data distribution was confirmed. Placement of objects from the minority class in a homogeneous area is one of the main success factors in creating the correct model.

3.2 Experiment 2

In the second experiment the data sets from the UCI (*University of California at Irvine Repository*) [13] were performed. The characteristics of the selected data sets are presented in Table 2. Ten data sets were chosen and then preprocessed using the VIS method. Detailed results are presented in Table 3 with a comparison to the SMOTE algorithm results and the situation when no preprocessing was performed before classification process (noPRE). Additionally, the mean values of parameters, calculated for the VIS method and odd number of neighbors from 1 to 15, are included. The improvement for most parameters has been obtained for both methods that generate new objects comparing to no preprocessing approach (in both cases Q: $p < 0.008$; TP: $p = 0.005$; F: $p = 0.005$; AUC: $p < 0.002$). The t test or Wilcoxon signed-rank test for paired data was applied. Analyzing the SMOTE and the VIS techniques, for six out of ten data sets, the improvement has been achieved for the accuracy ($p = 0.055$), $F - measure$ ($p = 0.052$), and the area under the curve in favor of the VIS method ($p = 0.055$). For two data sets (breast cancer, lymphography) these parameters remained at the same level. For four data sets the sensitivity was higher for the VIS method, for four - the SMOTE technique gave better results, and for two data sets results obtained by two methods were equal. As expected, the specificity decreased in almost all data sets.

Table 2 Characteristics of the data sets

Data set	Number of objects	Number of attributes (numeric; symbolic)	Missing data	*IR*
abalone9-18	731	8 (7;1)	No	16.60
blood transfusion	748	4 (4;0)	No	3.20
breast cancer	286	9 (0;9)	Yes	2.36
german credit	1000	20 (7;13)	No	2.33
haberman	306	3 (2;1)	No	2.78
hepatitis	155	19 (6;13)	Yes	3.84
lymphography	148	18 (3;15)	No	23.67
spectf	267	44 (44;0)	No	3.89
thoracic surgery	470	17 (3;14)	No	5.71
vowel0	988	13 (13;0)	No	9.98

Table 3 Classification results for the selected UCI data sets: Q—accuracy, TP_{rate}—rate of true positives, TN_{rate}—rate of true negatives, F—F measure, AUC—area under the curve, nIR—new value of IR

Method	k	Q	TP_{rate}	TN_{rate}	F	AUC	k	Q	TP_{rate}	TN_{rate}	F	AUC
abalone9-18							blood transfusion					
noPRE	–	94.12	35.71	97.68	0.41	66.70	–	76.07	33.71	**89.30**	0.40	61.50
SMOTE	3	95.72	**97.10**	94.34	0.96	95.72	15	78.16	73.68	82.63	0.77	78.16
VIS	3	**97.31**	96.23	**98.40**	**0.97**	**97.31**	3	**81.49**	**79.82**	83.16	**0.81**	**81.49**
VIS_{avg}	–	96.43	96.23	96.63	0.96	96.43	–	78.95	77.30	80.60	0.79	78.95
breast cancer							german credit					
noPRE	–	69.58	41.18	**81.59**	0.45	61.38	–	69.60	47.33	79.14	0.48	63.24
SMOTE	7	**74.63**	**78.11**	71.14	**0.75**	**74.63**	7	**79.21**	78.00	**80.43**	**0.79**	**79.21**
VIS	3	**74.63**	76.62	72.64	**0.75**	74.63	3	79.07	**78.86**	79.29	**0.79**	79.07
VIS_{avg}	–	72.19	73.83	70.55	0.73	72.19	–	78.77	79.46	78.09	0.79	78.77
haberman							hepatitis					
noPRE	–	70.26	25.93	**86.22**	0.32	56.07	–	85.81	53.13	**94.31**	0.61	73.72
SMOTE	15	70.00	71.11	68.89	0.70	70.00	3	90.65	91.06	90.24	0.91	90.65
VIS	3	**76.44**	**79.11**	73.78	**0.77**	**76.44**	11	**91.46**	**93.50**	89.43	**0.92**	**91.46**
VIS_{avg}	–	74.35	75.29	73.42	0.75	74.35	–	89.84	91.36	88.29	0.90	89.84

(continued)

Table 3 (continued)

Method	k	Q	TP_{rate}	TN_{rate}	F	AUC	k	Q	TP_{rate}	TN_{rate}	F	AUC
`lymphography`							`spectf`					
noPRE	–	97.97	50.00	**100.00**	0.67	75.00	–	77.32	29.09	89.72	0.34	59.41
SMOTE	3	**99.30**	**98.59**	**100.00**	**0.99**	**99.30**	3	89.02	**91.12**	86.92	0.89	89.02
VIS	3	**99.30**	**98.59**	**100.00**	**0.99**	**99.30**	5	**89.49**	**91.12**	**87.85**	**0.90**	**89.49**
VIS_{avg}	–	99.30	98.59	100.00	0.99	99.30	–	88.41	91.31	85.51	0.89	88.41
`thoracic surgery`							`vowel0`					
noPRE	–	78.94	10.00	91.00	0.12	50.50	–	99.49	96.67	**99.78**	0.97	98.22
SMOTE	3	86.75	**87.00**	86.50	0.87	86.75	3	**99.89**	**100.00**	**99.78**	**1.00**	**99.89**
VIS	3	**89.00**	85.95	**92.75**	**0.89**	**89.00**	3	99.78	99.78	**99.78**	**1.00**	99.78
VIS_{avg}	–	87.38	87.15	87.60	0.87	87.38	–	99.69	99.76	99.62	1.00	99.69

4 Conclusions

Imbalance in data distributions is one of the challenges in data analysis or exploration. Growing interest in this topic is related to the prevalence of this phenomenon and the ability to collect great amount of information. Many aspects of everyday life can be described by data sets with distorted distributions, like in medical diagnostics. Thus, it is crucial to ensure that the systems for decision support operate correctly especially for singular objects belonging to the positive class. To achieve a high accuracy of classifiers for minority class, datum is not an easy task and many approaches have been created to address the problem. The versatile improved SMOTE (VIS) algorithm was proposed in this paper. It is a synergy of the existing approaches and provides a brand new way of dealing with the imbalanced data problem. The VIS technique was designed to be more versatile than other similar solutions.

References

1. Galar, M., Fernandez, A., Barrenechea, E., Bustince, H., Herrera, F.: A review on ensembles for the class imbalance problem: bagging-, boosting-, and hybrid-based approaches. IEEE Trans. Syst. Man, Cybern. Part C: Appl. Rev. **42**(4), 463–484 (2012)
2. He, H., Garcia, E.A.: Learning from imbalanced data. IEEE Trans. Knowl. Data Eng. **21**(9), 1263–1284 (2009)
3. Weiss, G.M.: Mining with rarity: a unifying framework. SIGKDD Explor. Newsl **6**(1), 7–19 (2004)
4. Sun, Y., Kamela, M.S., Wongb, A.K.C., Wangc, Y.: Cost-sensitive boosting for classification of imbalanced data. Pattern Recognit. **40**(12), 3358–3378 (2007)
5. Chawla, N.V., Bowyer, K.W., Hall, L.O., Kegelmeyer, W.P.: SMOTE: synthetic minority over-sampling technique. J. Artif. Int. Res. **16**(1), 321–357 (2002)
6. Borowska, K., Topczewska, M.: Data preprocessing in the classification of the imbalanced data. Adv. Comput. Sci. Res. **11**, 31–46 (2014)
7. Taeho, J., Japkowicz, N.: Class imbalances versus small disjuncts. SIGKDD Explor. Newsl. **6**(1), 40–49 (2004)
8. Wilson, D.R., Martinez, T.R.: Improved heterogeneous distance functions. J. Artif. Int. Res. **1**, 1–34 (1997)
9. Han H., Wang W.Y., Mao B.H.: Borderline-SMOTE: a new over-sampling method in imbalanced data sets learning. In: Proceedings of the 2005 International Conference on Advances in Intelligent Computing—vol. Part I, pp. 878–887 (2005)
10. Hu S., Liang Y., Ma L., He Y.: MSMOTE: improving classification performance when training data is imbalanced. In: Proceedings of the 2009 Second International Workshop on Computer Science and Engineering—vol. 02, pp. 13–17 (2009)
11. Napierała, Krystyna, Stefanowski, Jerzy, Wilk, Szymon: Learning from Imbalanced Data in Presence of Noisy and Borderline Examples. In: Proc. of the 7th InternationalConference on Rough Sets and Current Trends in Computing, pp. 158–167 (2010)
12. Barua, S., Islam, M., Murase, K.: A novel synthetic minority oversampling technique for imbalanced data set learning. In: Proceedings of the 18th InternationalConference on Neural Information Processing—vol. Part II, pp. 735–744 (2011)
13. UC Irvine Machine Learning Repository, http://archive.ics.uci.edu/ml/ (2004). Accessed 20 May 2014

Automatic Syllable Repetition Detection in Continuous Speech Based on Linear Prediction Coefficients

Adam Kobus, Wiesława Kuniszyk-Jóźkowiak and Ireneusz Codello

Abstract The goal of this paper is to present a syllable repetition detection method based on linear prediction coefficients obtained by the Levinson–Durbin method. The algorithm wrought by the authors of this paper is based on the linear prediction spectrum. At first the utterance is automatically split into continuous fragments that correspond with syllables. Next, for each of them the formant maps are being obtained. After dimension reduction by the K-means method they are being compared. The algorithm was verified based on 56 continuous utterances of 14 stutterers. They contain fluent parts, as well as syllable repetitions on Polish phonemes. The classifying success reached 90 % of sensitivity with 75–80% precision.

1 Introduction

The occurrence and similarity of stuttering can be observed in many languages. This disorder can be divided into several types of dysfluencies: prolongations, blocks, interjections, repetitions of syllables or even parts of words [15]. This paper deals with the syllable and fragment repetitions in continuous speech. The analysis of the possibility of automatic dysfluency detection in continuous speech is a very important topic. Stuttering people have to deal with not only the interpersonal communication problems, but also with the more and more spreading voice actuation. Additionally, this analysis increases the knowledge of the nature of stuttering and helps to prepare even better stuttering therapies.

A. Kobus (✉) · I. Codello
Institute of Computer Science, Marie Curie-Skłodowska University,
Pl. M. Curie-Skłodowskiej 1, 20-031 Lublin, Poland
e-mail: adam.kobus@poczta.umcs.lublin.pl; kobus.adam@gmail.com

W. Kuniszyk-Jóźkowiak
Faculty of Physical Education and Sport in Biała Podlaska, Józef Piłsudski University of Physical Education in Warsaw, ul. Akademicka 2, 21-500 Biała Podlaska, Poland

R. Burduk et al. (eds.), *Proceedings of the 9th International Conference on Computer Recognition Systems CORES 2015*, Advances in Intelligent Systems and Computing 403, DOI 10.1007/978-3-319-26227-7_28

1.1 Related Work

In the research on the automatic dysfluency detection, the initial listening selection of speech fragments into fluent and dysfluent parts is made [1, 5, 7, 8, 16, 19–21, 24–29]. This research aims at detecting dysfluent fragments continuous utterances. First dysfluency detection trials without the initial dysfluent words extraction were undertaken in the works of Howell [9–11], afterwards in the research of Suszyński [23], Wiśniewski [30, 31], Codello [2–4] and Kobus [13, 15]. Those works have as their input the four-second-long recordings initially classified as fluent or dysfluent. The comparison of differences between the results of various analyses of the fluent and dysfluent samples allows to notice the significant disparities between these two types of samples. Thanks to this fact, the prolongation and phoneme repetitions' detection gives good results. The syllable repetition detection is a more complex issue. Still, it was addressed in the Suszyński's research [25], based on the worse of Hiroshima [7] and Codello [4]. For this aim Suszyński performed a comparison of spectrograms. He applied the correlation of the one-third octave spectrums in the connection with the procedures for the time–amplitude file structure analysis to detect the repeating fragments. The classification was made by exceeding the obtained correlation coefficient of the two fragments. This analysis allows to detect and localise syllable repetitions with 70 % efficiency. Codello's research [4] was based on the application of CWT in the Bark scale. The results were split into vectors and compared by the correlation method. This analysis reached 80 % of efficiency. For the speech dysfluency analysis the authors programmed the "Dabar" application. Beside the implementation of the prolongation [15] and block [13] detection algorithms the possibility of syllable and word fragment detection was also analysed. Amongst the many functionalities of the application are the following: dimension reduction by the Kohonen networks, evaluation of linear prediction and PARCOR coefficients, the spectrum evaluation on the basis of LPC and the creation of the elliptic model of the vocal tract [14]. Previous works allow to ascertain that the representation of the speech signal by means of linear prediction coefficients serves perfectly for detecting plosive repetition [13] and phoneme prolongation [15]. For this research, the splitting of utterances into syllables and/or speech fragments was implemented, together with the K-means centres obtaining method.

2 Methodology

The goal of this research is the analysis of the possibility of automatic syllable and speech parts' repetition detection using linear prediction method. In each utterance, the places where the repetition of syllable or speech fragment occurs were marked.

2.1 Speech Data

For the analysis, the several seconds long utterances were used. All 56 files of the total length of 3 min 47 s are in *WAV* format. Fourteen speakers were recorded—nine men and five women. Men were 11–25 years old (30 recordings), women were 13–24 years old (26 recordings). All recordings contain 262 pairs of syllables, 117 of which were dysfluent. Materials were recorded with Creative Wave Studio using the SoundBlaster card with 22050 Hz frequency and 16 bits per sample.

2.2 Preprocessing

The independent recordings were split into non-overlapping frames with 512 samples. Each of the frames was multiplied by the Hanna window function (1). Based on the previous research, it allows to achieve frequency spectrums with the best characteristic of those obtained by the linear prediction method [13].

$$w(n) = 0.5\left(1 - \cos\left(\frac{2\pi n}{N-1}\right)\right), \tag{1}$$

where $N = 512$ is the frame size.

2.3 Feature Extraction

2.3.1 Linear Prediction

The most basic characteristic of the linear prediction method is that it stores the knowledge of the speech signal change in a vector of a few coefficients. Consecutive speech signal samples vary to a small degree [18], thus the approximation of the next sample can be evaluated from the p previous samples using the proper linear prediction coefficients α. This idea is expressed by the following equation:

$$\tilde{s}(m) = \sum_{k=1}^{p} \alpha_k s(m-k) \tag{2}$$

where $\tilde{s}(m)$ is the mth value of the predicted voice sample, $s(m)$ is the mth value of the input speech sample, p is a prediction order and α_k are the obtained linear prediction coefficients. Basing on the linear prediction coefficients, the continuous frequency spectrum with an arbitrary number of stripes may be evaluated. The evaluation is expressed by Eq. (3) derived from the definition of the LP (Eq. 2):

$$H(f_i) = \frac{G}{1 - \sum_{k=1}^{p} \alpha_k \left(\cos\left(k\frac{f_i}{f}2\pi\right) - i\sin\left(k\frac{f_i}{f}2\pi\right)\right)} \tag{3}$$

where f is the chosen number of the spectrum stripes, f_i is the number of the chosen spectrum stripe, where $0 \leq f_i \leq f$, and G is the gain factor.

2.3.2 Analysis

In this algorithm the linear prediction coefficients were evaluated using the Levinson–Durbin method [18] with G gains from the obtained frames. The number of coefficients for each of them was 15, which is sufficient for good signal characteristics [22]. Next, the frequency spectrum was evaluated from Eq. (3) for each of the frames. In the order to obtain a precise spectrum, the number of stripes was set to 300. The evaluated coefficient vectors were used as the input vectors in the method of splitting speech into fragments Sect. 2.4. They were also the basis of the values parameterising the speech for further analysis Sect. 2.6.

2.4 *Segmentation*

The algorithm of splitting the speech into fragments was implemented for the purpose of this research. The algorithm is based on the frequency spectrum obtained from the linear prediction coefficients. From the obtained spectrums, the average value of the powers of all the m spectrum stripes values was evaluated for each frame. On the basis of the obtained values, the threshold was defined. It was evaluated as the average of the two primary values from the vector increased by the 3 db.

2.5 *Pairing*

Each fragment of the examined utterance was compared with the succeeding fragment. If the succeeding fragment was longer than the preceding one, it was cut into the length of the first fragment. This is due to the fact that the succeeding fragment may also contain the further, fluent part of the utterance and the repeated fragment is in its initial phase.

2.6 *Formants Extraction*

For each of the frames the formants are defined [6, 17, 22]. They are determined by those maxima F from the local maximum points of the frequency spectrum which fulfil the bandwidth condition [12, 22]:

$$B_j = -(f_j/\pi)\ln(r_0) < B \tag{4}$$

where B_j is the 3 db bandwidth, j is the ordinal number of the maximum from the chosen frame, f_j is the frequency of the jth spectrum maximum and r_0 are the roots of the polynomial $A(z) = 1 - \sum_{k=1}^{p} \alpha_k z^{-k}$, where $z = e^{i2\pi f_j/f_s}$, f_s is the sampling frequency. B is the maximum bandwidth permissible for the formant. For the purpose of this analysis $B = 1000$ Hz. The points have three coordinates: frequency, amplitude and time. For each frame several formants were obtained. As a result of that, a fragment is represented by a set of P points. Two of those dimensions, the frequency and the amplitude, are the basis for the K-means algorithm. The main goal of this method is to reduce m characteristic points from the P set to k characteristic centres for the whole fragment.

2.7 *Clusterisation*

The main rule of the K-means algorithm is to gradually adjust the midpoints (centres) so that they would divide the space in the best possible way into groups of points concentrated around each other.

2.7.1 The Normalisation of Space of Characteristic Points

The normalisation of space of characteristic points consists in scaling each of the dimensions with respect to the average value and the variance of the value of a set of points, that is, on the basis of the Mahalanobis distance:

$$\overline{p_j} = \frac{p_j - \mu}{\sigma} \tag{5}$$

where $\overline{p_j}$ is a normalised point in the R^N space, p_j is a point in the R^N space, μ is a point of average values for p_j points and δ is a vector of variances for these points.

2.7.2 Input Data

The P set of the normalised points p_j is given, where $0 < j < m$, m is the number of the P set elements, as well as the vector $\mathbf{k}$ of the normalised initial centres c_i, where $0 < i < k$ and k is the length of the $\mathbf{k}$ vector. Both the set of points and the vector of the centres should be normalised. In order to measure the distance between two points in N-dimensional space the space normalisation is needed for the comparison of distances. The vector $\mathbf{k}$ of the centres should be also initialised by the values of the coordinates.

2.7.3 Centres Initialisation

The random initialisation of the centres is often applied; however, thanks to the knowledge of the structure of the set of points, the initialisation based on this knowledge may be applied. The uniform distribution method was used for the purpose of this research. The basic data for the initialisation are points with two dimensions: amplitude and frequency. The input points are sorted in the order of frequency. Next, k_p points distributed uniformly in frequency are chosen. If there are only few of such points, these points are being duplicated up to the k number so that they would not affect the result of measurement of the distances between the vectors.

2.8 *Distance Measuring*

For the analysis of the distances between the vectors **k** and consequently between the points, a few distance measures were applied.

2.8.1 Metrics

For measuring the distance between the points two metrics were applied.

Euclidean Metric

This metric was applied in the measurement of the distance between points in the K-means algorithm, as well as in the classification for measurement of the distance between the tested pair and the threshold between the fluent and dysfluent pairs (**MEDIAN**, **AVERAGE**).

Mahalanobis Metric

This metric was applied in the classification for examining the distance between the tested pair and the group of fluent and dysfluent pairs (**MAHALANOBIS**).

2.8.2 Distance Between Vectors

After the K-means analysis, each fragment is represented by a vector of centres. They are combined into pairs and the distance between them in two dimensions is being evaluated. The evaluation consists in obtaining the sum of minimal Euclidean distances between the centres of the succeeding and preceding fragment.

2.9 *Classification*

The obtained distances were used for the classification of 262 pairs of fragments into dysfluent (fragment repetitions) and fluent ones. The previously classified pairs were randomly divided into three parts. Two of them were used as a training part (172 pairs) and one as a testing part (90 pairs). Three methods were used for classification:

AVERAGE the minimal distance from the threshold in the Euclidean metric—the threshold was obtained experimentally as $n + (f - n)/5$, where n is an arithmetic average of the distances between contiguous recurring fragments and f is an arithmetic average of the distances between contiguous nonrecurring fragments (fluent speech).

MEDIAN analogous to the above mentioned, but the average distance was substituted with a median.

MAHALANOBIS the minimal distance from the group in a Mahalanobis metric.

For the evaluation of the average and the median, the training part was used.

3 Results and Discussion

The results refer to the testing part of extracted pairs. The result figures contain a number of parameters related to the quality assessment of the classification.

1. Sensitivity—$100 * TP/(TP + FN)$—where TP—true positive—is a number of correctly classified dysfluencies, FN—false negative—a number of not recognised dysfluencies
2. Precision—$100 * TP/(TP + FP)$—where TP is a number of non-fluent pairs with recognised non-fluency, FP—false positive—a number of fluent pairs classified as non-fluent
3. K—a number of the centres in a representation vector
4. Categorisation method Sect. 2.9: AVERAGE, MEDIAN, MAHALANOBIS (Figs. 1, 2 and 3).

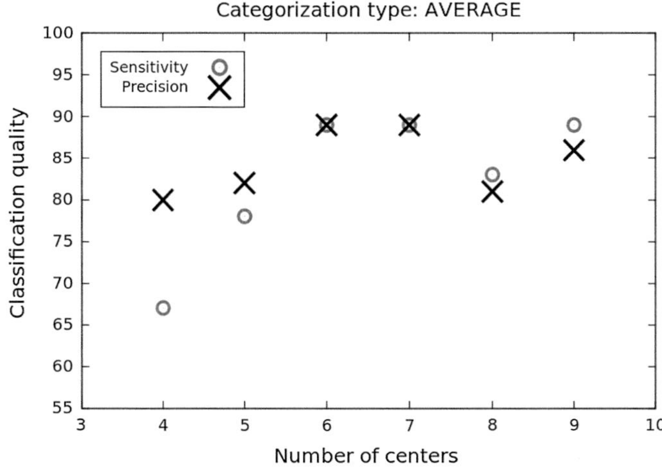

Fig. 1 Classification quality percentage graph for the best precision and sensitivity in relation to the number of centres for the AVERAGE type of categorisation

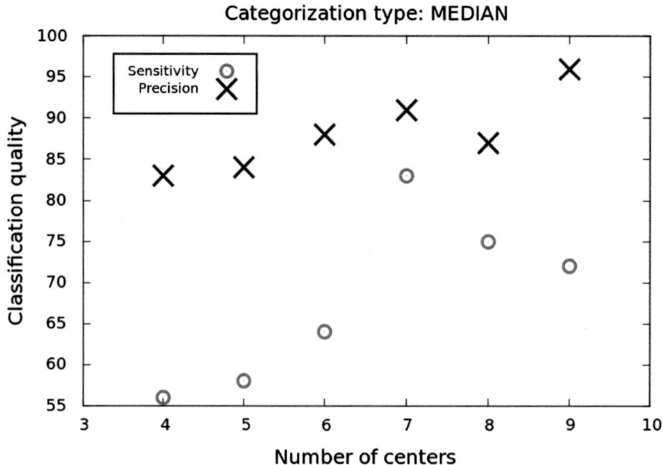

Fig. 2 Classification quality percentage graph for the best precision and sensitivity in relation to the number of centres for the MEDIAN type of categorisation

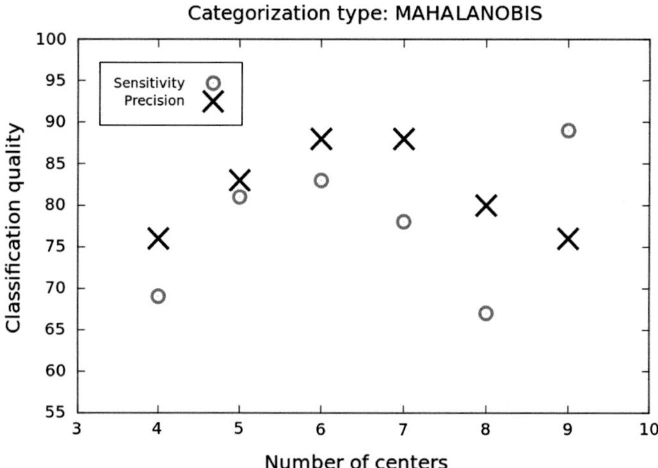

Fig. 3 Classification quality percentage graph for the best precision and sensitivity in relation to the number of centres for the MAHALANOBIS type of categorisation

4 Conclusion

The results of the research proved the effectiveness of the syllable and fragment repetition detection using the proposed method. Regarding the number K of the centres, the best results were reached for $5 \leq K \leq 7$. The results for $8 \leq K \leq 9$ were good and for $K = 4$, a little bit worse. It may result from the number of formants which represent the centres. If K is too small, some of the formants may not be duly represented, e.g. in the case of two or three phonemes. If K is too big, the representation is good but the excess of points has an influence on the distortion of the results for the lower number of the phonemes. Fortunately, the differences were not significant. The second aspect which was examined was categorisation. It was observed that regardless of its type, over 80 % of sensitivity was achieved. The best results were obtained using the AVERAGE categorisation. The MAHALANOBIS and MEDIAN categorisations were a little bit worse but similar to each other. The results of the analysis lead to the conclusion that detecting speech fragment repetitions, using the K-means method as a dimension reducer, the distance analysis as a classifier and the linear prediction coefficients as the input values, is possible and reaches 89 % precision with 89 % sensitivity. The applied method of automatic speech segmentation into syllables and fragments allows to determine the places where the syllable, phoneme or fragment repetition occurs with an 89 % efficiency. The algorithm described in this paper, equipped with a database of recordings with predefined dysfluencies' occurrences, may also be applied in a real-time detection of repetitions.

References

1. Chia Ai, O., Hariharan, M., Yaacob, S., Chee, L.S.: Classification of speech dysfluencies with MFCC and LPCC features. Expert Syst. Appl. **39**, 2157–2165 (2012)
2. Codello, I., Kuniszyk-Jóźkowiak, W., Smołka, E., Kobus, A.: Disordered sound repetition recognition in continuous speech using CWT and Kohonen network. J. Med. Inform. Technol. **17**, 123–130 (2011)
3. Codello, I., Kuniszyk-Jóźkowiak, W., Smołka, E., Kobus, A.: Automatic prolongation recognition in disordered speech using CWT and Kohonen network. J. Med. Inform. Technol. **20**, 137–144 (2012)
4. Codello, I., Kuniszyk-Jóźkowiak, W., Smołka, E., Kobus A.: Automatic disordered syllables repetition recognition in continuous speech using CWT and correlation. In: Proceedings of the 8th International Conference on Computer RecognitionSystems CORES 2013, Advances in Intelligent Systems and Computing, vol. 226, pp 865–874 (2013)
5. Czyżewski, A., Kaczmarek, A., Kostek, B.: Intelligent processing of stuttered speech. J. Intell. Inf. Syst. **21**(2), 143–171 (2003)
6. Halberstam, B., Raphael, L.J.: Vowel normalization: the role of fundamental frequency and upper formants. J. Phon. **32**, 423–434 (2004)
7. Hiroshima, S.M.: A spectrographic analysis of speech disfluencies: characteristics of sound/syllable repetitions in stutterers and nonstutterers, In: Proceedings 24th IALP Congress Amsterdam, pp. 712–714, Amsterdam (1999)
8. Geetha, Y.V., Pratibha, K., Ashok, R., Ravindra, S.K.: Classification of childhood disfluencies using neural networks. J. Fluen. Disord. **25**(2), 99–117 (2000)
9. Howell, P., Sackin, S.: Automatic recognition of repetitions and prolongations in stuttered speech. In: Proceedings of the First World Congress on Fluency Disorders, pp. 1–4 (1995)
10. Howell, P., Sackin, S., Glenn, K.: Development of a two-stage procedure for the automatic recognition of dysfluencies in the speech of children who stutter: I. Psychometric procedures appropriate for selection of training material for lexical dysfluency classifiers. J. Speech, Lang. Hear. Res. **40**(5), 1073–1084 (1997)
11. Howell, P., Sackin, S., Glenn, K.: Development of a two-stage procedure for the automatic recognition of dysfluencies in the speech of children who stutter: II. ANN recognition of

repetitions and prolongations with supplied word segment markers. J. Speech, Lang. Hear. Res. **40**(5), 1085–1096 (1997)
12. Kim, C., Seo, K., Sung, W.: A robust formant extraction algorithm combining spectral peak picking and root polishing. EURASIP J. Appl. Signal Process. **2006**, 1–16 (2006)
13. Kobus, A., Kuniszyk-Jóźkowiak, W., Smołka, E., Codello, I.: Speech nonfluency detection and classification based on linear prediction coefficients and neural networks. J. Med. Inform. Technol. **15**, 135–144 (2010)
14. Kobus, A., Kuniszyk-Jóźkowiak, W., Smołka, E., Suszyński, W., Codello, I.: A new elliptical model of the vocal tract. J. Med. Inform. Technol. **17**, 131–140 (2011)
15. Kobus, A., Kuniszyk-Jóźkowiak W., Smołka, E., Codello, I., Suszyński W.: The prolongation-type speech non-fluency detection based on the linear prediction coefficients and the neural networks. In: Proceedings of the 8th InternationalConference on Computer Recognition Systems CORES 2013, Advances inIntelligent Systems and Computing, vol. 226, pp. 885–894 (2013)
16. Kuniszyk-Jóźkowiak, W., Suszyński, W., Smołka, E., Dzieńkowski, M.: Automatic recognition and measurement of durations of fricative prolongations in the speech of persons who stutter. Speech Lang. Technol. **8** (2004). (in polish)
17. Millhouse, T., Clermont, F., Davis, P.: Exploring the importance of formant bandwidths in the production of the singer's formant. In: Proceedings of the 9th Australian International Conference on Speech Science & Technology, Melbourne, pp. 373–378 (2002)
18. Rabiner, L.R., Schafer, R.W.: Theory and Applications of Digital Speech Processing Chap. 9. Pearson Higher Education Inc (2011)
19. Ravikumar, K., Reddy, B., Rajagopal, R., Nagaraj, H.: Automatic detection of syllable repetition in read speech for objective assessment of stuttered disfluencies. In: Proceedings of world academy science, engineering and technology, pp. 270–273 (2008)
20. Ravikumar, K.M., Rajagopal, R., Nagaraj, H.C.: An approach for objective assessment of stuttered speech using MFCC features. ICGST Int. J. Digit. Signal Process., DSP **9**(1), 19–24 (2009)
21. Ravikumar, K.M., Ganesan, S.: Comparison of multidimensional MFCC feature vectors for objective assessment of stuttered disfluencies. Int. J. Adv. Netw. Appl. **02**(05), 854–860 (2011)
22. Snell, R.C., Milinazzo, F.: Formant location from LPC analysis data. IEEE Trans. Speech Audio Process. **1**(2), 129–134 (1993)
23. Suszyński, W.: Automatic detection of speech non-fluencies. In: 50th Opened Acoustic Seminar, pp. 386–390 (2003). (in polish)
24. Suszyński, W., Kuniszyk-Jóźkowiak, W., Smołka, E., Dzieńkowski, M.: Automatic recognition of nasals prolongations in the speech of persons who stutter. Structures—Waves—Human Health, pp. 175–184 (2003)
25. Suszyński, W.: Computer analysis and speech dyspluency recognition. Doctoral dissertation, Politechnika Śląska, Gliwice (2005). (in polish)
26. Szczurowska, I., Kuniszyk-Jóźkowiak, W., Smołka, E.: The application of Kohonen and multilayer perceptron networks in the speech nonfluency analysis. Arch. Acoust. **31**(4), 205–210 (2006)
27. Szczurowska, I., Kuniszyk-Jóźkowiak, W., Smołka, E.: Speech nonfluency detection using Kohonen networks. Neural Comput. Appl. **18**, 677–687 (2009)
28. Świetlicka, I., Kuniszyk-Jóźkowiak, W., Smołka, E.: Artificial neural networks in the disabled speech analysis. Computer Recognition Systems 3, vol. 57/2009, Springer, Heidelberg, pp. 347–354 (2009)
29. Tian-Swee, T., Helbin, L., Ariff, A.K., Chee-Ming, T., Salleh, S.H.: Application of malay speech technology in malay speech therapy assistance tools. In: International Conference on Intelligent and Advanced Systems, pp. 330–334 (2007)
30. Wiśniewski, M., Kuniszyk-Jóźkowiak, W., Smołka, E., Suszyński, W.: Automatic detection of disorders in a continuous speech with the hidden markov models approach computer recognition systems 2, Vol. 45/2008, pp. 445–453. Springer, Berlin (2007)
31. Wiśniewski, M., Kuniszyk-Jóźkowiak, W.: Automatic detection and classification of phoneme repetitions using HTK toolkit. J. Med. Inform. Technol. **17**, 141–148 (2011)

Part II
Biometrics

Chain Code-Based Local Descriptor for Face Recognition

Paweł Karczmarek, Adam Kiersztyn, Witold Pedrycz and Przemysław Rutka

Abstract Local descriptors have been one of the most intensively examined mechanisms of image analysis. In this paper, we propose a new chain code-based local descriptor. Unlike many other descriptors existing in the literature, this descriptor is based on string values, which are obtained when starting from a particular point of the image and searching for extrema in a given neighborhood and memorizing a path being traversed through the consequent pixels of the image. We demonstrate that this approach is efficient and helps us preserve both local and global properties of the object. To compare the words we apply the Levenshtein distance. Moreover, four similarity measures (correlation, histogram intersection, chi-square, and Hellinger) are used to compare the histograms of words in the process of classification.

Keywords Local descriptors · Chain code · Similarity measures · Levenshtein distance

---

P. Karczmarek (✉) · A. Kiersztyn · P. Rutka
Institute of Mathematics and Computer Science, The John Paul II Catholic University of Lublin, ul. Konstantynów 1H, 20-708 Lublin, Poland
e-mail: pawelk@kul.pl

A. Kiersztyn
e-mail: kierat@kul.pl

P. Rutka
e-mail: rootus@kul.pl

W. Pedrycz
Department of Electrical and Computer Engineering, University of Alberta, Edmonton, AB T6R 2V4, Canada
e-mail: wpedrycz@ualberta.ca

W. Pedrycz
Department of Electrical and Computer Engineering, Faculty of Engineering, King Abdulaziz University, 21589 Jeddah, Saudi Arabia

W. Pedrycz
Systems Research Institute, Polish Academy of Sciences, Warsaw, Poland

R. Burduk et al. (eds.), *Proceedings of the 9th International Conference on Computer Recognition Systems CORES 2015*, Advances in Intelligent Systems and Computing 403, DOI 10.1007/978-3-319-26227-7_29

1 Introduction

Since we have stepped into the digital era of modern technologies, a lot of elements of our new computerized life demand various interactions with the machines. To make them more human-centric and useful, the machines have to be endowed with vision capabilities. This, in particular, is the main reason why the face recognition techniques have emerged so intensively in recent times, becoming one of the most popular areas of computer science. Now, it is still an open domain of intensive research, bringing innovative applications [33]. The generic machine facial recognition is, in fact, a problem of automated identification/verification of a 3D object (real person) in 2D images (still or video) of a scene based on a stored database of known faces. Its general solution involves detection of face in a scene, appropriate effective feature extraction from this face, recognition by classification of these features, and returning the identity of a recognized person or confirmation/rejection of claimed identity [33]. The feature extraction and classification constitute the crucial part of any face recognition system. Roughly speaking, there are: global and local approaches, to be utilized to perform these tasks [6, 33]. The global methods operate on the appearance of whole face, e.g., [4, 28], whereas the local methods analyze the local features of face, based on some characteristics of so-called fiducial points or neighborhoods of pixels, e.g., [1, 6, 11, 29]. One of the most important approaches of the latter kind is an approach based on the local descriptors. Here, the most impressive seems to be the LBP operator which labels the pixels of an image by thresholding the 3×3-neighborhood of each pixel with the central value and returns the result as a concatenation of such binary comparisons. At the end, this binary value is converted into a decimal equivalent. Ahonen et al. [1] inspired by the works [21, 22] proposed a new approach in face recognition. In their idea, the face image is first divided into small parts (regions) from which the local binary pattern (LBP) components are elicited and concatenated into a histogram of features. The textures of the facial regions are locally encrypted by the LBP operator while the overall structure of the face is regained by assembly of the feature histogram. This method has been further studied and improved. For example, in [15] proposed was a novel representation, called multiscale block LBP (MB-LBP). In MB-LBP, the estimation is done based on average values of block subregions, as an alternative of single pixels. Each subregion is a square block of contiguous pixels. The whole filter is composed of nine blocks. In addition, it is worth to mention the center-symmetric local binary patterns (CSLBP) [11] and extended version of local binary patterns (ELBP) [25]. Moreover, noteworthy are combinations of LBP and Gabor filters. Contrary to the mainstream approaches based on statistical learning [32] devoted to conceptualize a new nonstatistics-based face representation approach, local Gabor binary pattern histogram sequence (LGBPHS) is an exposition approach based on multiresolution spatial histogram combining local density with the spatial information and multi-scale Gabor filters. Similarly, in [27] it was shown that combining Gabor wavelets and LBP gives massively better performance than either of them alone. Another interesting approach is a combination of Markov models and LBP proposed in

[24, 30], etc. Finally, one of the most promising descriptors is Full Ranking proposed in [9]. The authors used the full ranking of a set of pixels as a local descriptor and nearest neighbor classifier using chi-square distance for visual recognition in the Bag-of-Visual-Words (BoVW) paradigm [26]. Comprehensive overviews of local descriptors or their comparisons can be found in [5, 6, 18]. One of the important techniques used in the field of computer vision is a chain code. It was initially applied to represent shapes in digital images [10], by coding consecutive directions of movement along a boundary of the object in the form of ordered sequence of symbols. It is now common to use either 4 or 8 directional chain codes, with both numeral and literal labeling of directions. The chain code technique has become successful since it turned out to possess several significant advantages. In particular, it preserves information of any shape feature and provides a lossless data reduction [8]. It has been also found useful in the process of recognizing objects in images [16]. The main purpose of this paper is to present a new type of local descriptor, namely chain code-based local descriptor (CCBLD). It is constructed by using the simplified chain codes obtained by finding the consequent local extrema of grayscale image. Instead of traditionally used in local descriptors theory binary or decimal values, the string values describing the chain codes and the Levenshtein distance for their comparison are used. Furthermore, our objective is to apply the so-called Bag-of-Visual-Words (BoVW) paradigm stating that each local descriptor is assigned to the visual word in a dictionary, ultimately in a histogram [9, 26]. We are also interested in an examination of its efficiency in the process of face recognition using well-known public databases such as the one coming from AT&T. One of the most important features is the distance function applied to compare the words. In our tests, we use the well-known Levenshtein distance (edit distance), which can be defined as the minimal number of deletions, insertions, and replacements needed to make two strings equal, see [14, 19]. Similarly, the appropriate choice of a measure of similarity for comparing the histograms can crucially affect the recognition rate, see, e.g., [5, 7, 20, 23, 31]. Therefore, in the ensuing series of experiments, we use the following measures to express the distance between two vectors $\mathbf{x} = (x_1, \ldots, x_n) \in \mathbb{R}^n$, $\mathbf{y} = (y_1, \ldots, y_n) \in \mathbb{R}^n$:

1. Correlation $d(\mathbf{x}, \mathbf{y}) = 1 - \dfrac{\sum\limits_{i=1}^{n}\left(x_i - \frac{1}{n}\sum\limits_{j=1}^{n} x_j\right)\left(y_i - \frac{1}{n}\sum\limits_{j=1}^{n} y_j\right)}{\sqrt{\sum\limits_{i=1}^{n}\left(x_i - \frac{1}{n}\sum\limits_{j=1}^{n} x_j\right)^2 \sum\limits_{i=1}^{n}\left(y_i - \frac{1}{n}\sum\limits_{j=1}^{n} y_j\right)^2}}$;
2. Histogram intersection $d(\mathbf{x}, \mathbf{y}) = 1 - \dfrac{\sum\limits_{i=1}^{n} \min(x_i, y_i)}{\sum\limits_{i=1}^{n} \max(x_i, y_i)}$;
3. χ^2-statistics $d(\mathbf{x}, \mathbf{y}) = \sum\limits_{i=1}^{n} \frac{(x_i - y_i)^2}{x_i + y_i}$;
4. Hellinger (or Bhattacharyya coefficient) $d(\mathbf{x}, \mathbf{y}) = \sqrt{1 - \dfrac{\sum\limits_{i=1}^{n} \sqrt{x_i y_i}}{\sqrt{\sum\limits_{i=1}^{n} x_i \sum\limits_{i=1}^{n} y_i}}}$.

Therefore, one of the goals of this study is to determine which of these commonly used similarity measures are the most appropriate for the comparison of the histograms of words obtained in the process of the descriptor construction. The paper is organized as follows. Section 2 is devoted to the CCBLD local descriptor. In Sect. 3, we present the analysis of its application in the facial features description in the face recognition process while conclusions are presented in Sect. 4.

2 Chain Code-Based Local Descriptor (CCBLD)

A new local descriptor is based on the local path which may be reached when we start from an arbitrary point (pixel) of an image. It is worth noting that instead of the commonly used way of descriptors construction involving decimal or binary values, we apply a string (or character) of values of the local path. However, the main difference between our approach and the descriptors published so far in the literature is that the description of the neighborhood of a given pixel depends on the properties of the path which beginning is placed at the pixel and the end may be in any place of the image. In other words, the values taken to obtain the result of the description are not taken from the same sources, e.g., the circular or rectangular neighborhood. It means that in our approach the neighborhood depends on the local dynamic of the changes in an image. The full process of path finding and forming its description can be outlined as follows. Let us consider a starting point $I\,(x, y)$ and assume that we have to move in the image according to the following rule. If there exists a point having a higher grayscale value than the starting point $I\,(x, y)$ and if it belongs to the set of neighboring points, viz. $I_L\,(x, y) = I\,(x-1, y)\,,\ I_R\,(x, y) = I\,(x+1, y)\,,\ I_U\,(x, y) = I\,(x, y-1)\,,\ I_D\,(x, y) = I\,(x, y+1)$, then we can start the description from one of the four letters "*L*", "*R*", "*U*", and "*D*", corresponding to the way we move (namely, left, right, up, down). Obviously, if there is no maximum in such neighborhood of the original point, the description returns an empty string. If there are two or more maxima, then we can choose one of the circumstances. Here it is important that the rule of the choice must be the same in all such cases. Next, we start from the point, which was found as the maximum at the previous step and repeat the process. It finishes when there is no maximum fulfilling the above-mentioned criteria. We choose such a manner of finding the next elements of a path since finding the consecutive extremal points in the nearest neighborhood seems to express and bring out the small shifts and details of the image object feature which may stay illumination invariant, for instance. This kind of using characters is a very popular way of lines and shapes description in image analysis (see, e.g., [17]). Similarly, we build similar word using the "l", "r", "d", and "u", characters when looking for the consecutive minima starting from the original pixel $I\,(x, y)$. Both string values (i.e., the minima and maxima paths) are created at the same time and they are complementary. Clearly, the process always terminates since the grayscale is bounded by 0 and 255. An illustrative example of this scanning process is presented in Fig. 1. As the potential number of all the strings built in the above-mentioned

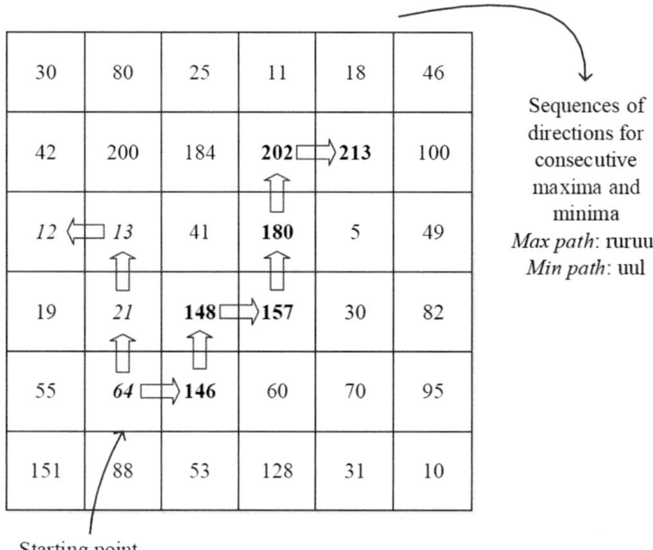

Fig. 1 The process of finding the maxima and minima of the consecutive neighborhoods starting from a given point and the result of the process. The values of the pixels forming the path of minima are shown in *italics* while the maximal values are shown in *boldface*

way may be large, we have to constrain their number in a dictionary. We can do this by building a dictionary of selected words and, at the level of particular images description, using the most similar ones. This approach to create the dictionary was efficiently applied in [9] for numerical descriptor called Full Ranking according to the so-called Bag-of-Visual-Words paradigm [26]. To improve the efficiency of the descriptor, one can divide an image into smaller parts and build separate dictionaries for each region. Of course, these dictionaries might be similar. However, in practice, they are quite different with some possible overlap.

3 Performance of the Descriptor

We evaluate the performance of the proposed descriptor through a series of experiments using the well-known AT&T (formerly ORL) image dataset [2]. This database contains 400 two-dimensional images of faces of 40 people (10 per each person) with various pose, illumination, and emotion. For the purpose of testing of the algorithm, all the images coming from the dataset have been preprocessed. The faces are detected, scaled, cropped, and the histogram equalization is made for each image. The results are the 90×94 images of the faces with no “external” features such as hair or ears. To fully address the verification process for the descriptor, we divided our tests into two stages. At the first stage, we apply the leave-one-out strategy for the examination of the classifier. We established the size of the dictionary to be 500. Similar number of words was proposed in [9]. Moreover, we observed that if the number of words and, in consequence, the length of the histograms of words from the dictionary and relative to a given image is substantially higher, then these histograms contain relevantly higher number of zero values. All the images are divided into the 25 equal subregions. This means, that to describe such subregion one can use just 20 words. The words in the dictionary are collected randomly in the following way. For each of 25 subregions, there is one image randomly selected from the dataset and one starting pixel from the corresponding area of this image. For such pixel, the string descriptor is formed according to the rules described in Sect. 2. The result is built with two words: one describing the consecutive minima and second—the maxima. Using some formal notation, the dictionary is a vector of the form:

$$D = \left[w_{1,1}^{+}, w_{1,1}^{-}, \ldots, w_{1,10}^{+}, w_{1,10}^{-}, \ldots, w_{ij}^{+}, w_{ij}^{-}, \ldots, w_{25,1}^{+}, w_{25,1}^{-}, \ldots, w_{25,10}^{+}, w_{25,10}^{-}\right],$$

where w_{ij}^{+}, $i = 1, \ldots, 25$, $j = 1, \ldots, 10$, stands for the word constructed starting from the jth randomly chosen pixel from the randomly chosen image from the ith subregion of this face image by using the successive maxima. Similarly, w_{ij}^{-} denotes the word built from the successive minima starting from this point. Hence, if we describe any pixel of a chosen image using the words of D and if we assume that this pixel belongs to the ith subregion, then its "maximal" path is matched to one of the words $w_{i,1}^{+}, \ldots, w_{i,10}^{+}$ and the "minimal" is one of the words $w_{i,1}^{-}, \ldots, w_{i,10}^{-}$. This way of dictionary building is similar to the technique presented in [9]. Next, all the images from the dataset are described using the histograms of words collected in the dictionary. In particular, each pixel is described separately and there is found the nearest word present in the dictionary, but only in the set of words allowed to describe a given subregion. Moreover, to express the path of the successive maxima, one can use the words formed from the "maximal" paths. The same procedure applies to the formation of the "minimal" paths. In this manner, for each image we obtain the histogram containing 500 bins. The above procedure was repeated 100 times and the average results are presented in Fig. 2. Our descriptor produces very good results at the level of 97.5–98.5 %, particularly when the histogram intersection and Hellinger measures of similarity between histograms of words are used. Slightly worse results have been obtained for the χ^2-statistics distance. At the second stage of the experiments, we split the set of images into training and testing sets by taking 5 images of each person for each of the sets. The dictionary of 500 words was created using the training images only (again, done in a random manner). The average results of 100 times repeated recognition process are presented in Fig. 3. The results are about 5 % worse than in case of the leave-one-out strategy. However, the relationships between the measures are relatively similar, i.e., in both cases the Hellinger measure follows the histogram intersection, correlation, and chi-square measure. The standard deviations obtained during the experimental studies and corresponding to both types of tests are listed in Table 1. As one can see from the results of the experiments, our

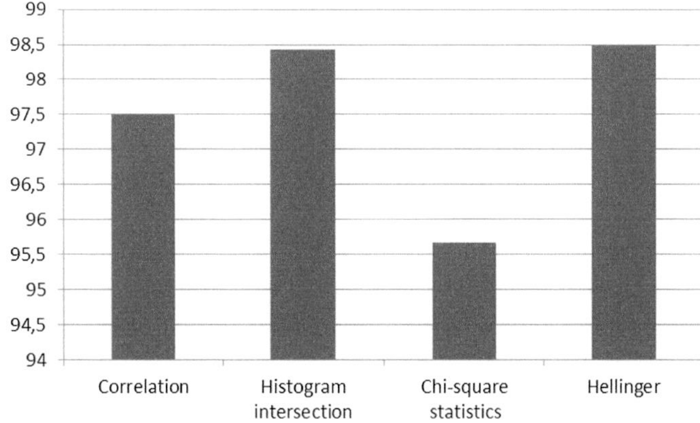

Fig. 2 Recognition rates (%) obtained for the proposed method using the leave-one-out strategy following the four measures of histogram similarity

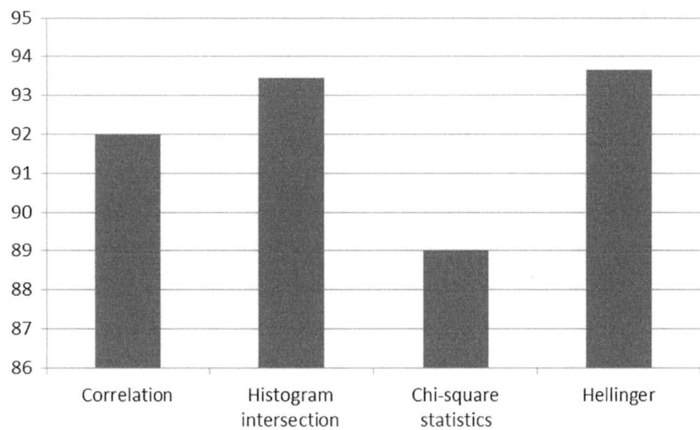

Fig. 3 Recognition rates (%) obtained for the chain code-based local descriptor using in case of 50 % of images included to the training set following the four measures of histogram similarity

approach is comparable with other state-of-the-art methods presented in the literature. For instance, the well-known eigenfaces [28] and Fisherfaces [4] with similar settings of experiments (5 images per person being in the training set and 5 in the testing set) give the recognition rates 75.35 % and 93.7 %, respectively. Moreover, according to [1] the modified LBP called uniform Local Binary Pattern with 16 points on a circle of radius 2 is reported to give accuracy at the level 98 %.

Table 1 Standard deviations obtained in the numerical evaluation of the proposed approach (leave-one-out strategy and division of the dataset into two equal sets)

Similarity measure	Leave-one-out strategy	Training/testing sets (50/50)
Correlation	0.46	2.07
Histogram intersection	0.37	1.94
Chi-square statistics	0.69	2.47
Hellinger	0.33	2.03

4 Conclusions and Future Work

In this study, we have introduced a new type of the local descriptor (chain code-based local descriptor) and in a series of experiments we showed its potential applicability to the facial recognition process. The results were obtained by running two validation strategies using four similarity measures for histogram (correlation, chi-square, histogram intersection, and Hellinger). The descriptor reached the recognition rate at about 98.5 % in the leave-one-out strategy. Therefore, we can assume that the next modifications and examinations of such kinds of descriptor which are based on string may give more satisfactory results. The future work may contain the optimization of the dictionary length, application of other than Levensthein distance functions for comparing the strings (e.g., Hamming distance or Levenshtein distance modifications, see [3]). Finally, our next investigations will include the construction of other string-based descriptors, assigning the weights for the particular face areas according to their saliency in the process of recognition, according to the conceptions presented in e.g., [13] or [12], etc.

Acknowledgments The authors are supported by the National Science Centre, Poland (grant no. 2014/13/D/ST6/03244). Support from the Canada Research Chair (CRC) program and Natural Sciences and Engineering Research Council of Canada (NSERC) is gratefully acknowledged (W. Pedrycz).

References

1. Ahonen T., Hadid A., Pietikäinen M.: Face recognition with local binary patterns. In: Proceedings of the 8th European Conference on Computer Vision, pp. 469–481 (2004)
2. AT&T Database of Faces, URL:http://www.cl.cam.ac.uk/research/dtg/attarchive/facedatabase.html
3. Bartyzel, K.: Invariant Levenshtein distance for comparison of Brownian strings. J. Appl. Comput. **18**, 7–17 (2010)
4. Belhumeur, P.N., Hespanha, J.P., Kriegman, D.J.: Eigenfaces vs. fisherfaces: recognition using class specific linear projection. IEEE Trans. Pattern Anal. Mach. Intell. **19**, 711–720 (1997)
5. Bereta, M., Karczmarek, P., Pedrycz, W., Reformat, M.: Local descriptors in application to the aging problem in face recognition. Pattern Recogn. **46**, 2634–2646 (2013)

6. Bereta, M., Pedrycz, W., Reformat, M.: Local descriptors and similarity measures for frontal face recognition: a comparative analysis. J. Vis. Commun. Image R. **24**, 1213–1231 (2013)
7. Bharkad, S.D., Kokare, M.: Performance evaluation of distance metrics: application to fingerprint recognition. Int. J. Pattern. Recogn. **25**, 777–806 (2011)
8. Bribiesca, E.: A new chain code. Pattern Recogn. **32**, 235–251 (1999)
9. Chan, C.H., Yan, F., Kittler, J., Mikolajczyk, K.: Full ranking as local descriptor for visual recognition: a comparison of distance metrics on Sn. Pattern Recogn. **48**, 1328–1336 (2015)
10. Freeman, H.: On the encoding of arbitrary geometric configurations. IRE T. Electron. **10**, 260–268 (1961)
11. Heikkilä, M., Pietikäinen, M., Schmid, C.: Description of interest regions with local binary patterns. Pattern Recogn. **42**, 425–436 (2009)
12. Karczmarek, P., Pedrycz, W., Reformat, M., Akhoundi, E.: A study in facial regions saliency: a fuzzy measure approach. Soft Comput. **18**, 379–391 (2014)
13. Kwak, K.-C., Pedrycz, W.: Face recognition: a study in information fusion using fuzzy integral. Pattern Recogn. Lett. **26**, 719–733 (2005)
14. Levenshtein, V.I.: Binary codes with correction for deletions and insertions of the symbol 1. Probl. Peredachi Inf. **1**, 12–25 (1965)
15. Liao, S., Zhu, X., Lei, Z., Zhang, L., Li, S.: Learning multi-scale block local binary patterns for face recognition. In: Lee, S.-W., Li, S.Z. (eds.) Advances in Biometrics. Lecture Notes in Computer Science, vol. 4642, pp. 828–837 (2007)
16. McKee, J.W., Aggarwal, J.K.: Computer recognition of partial views of curved objects. IEEE T. Comput. **26**, 790–800 (1977)
17. Mehtre, B.M., Kankanhalli, M.S., Lee, W.F.: Shape measures for content based image retrieval: a comparison. Inform. Process. Manag. **33**, 319–337 (1997)
18. Mikolajczyk, K., Schmid, C.: A performance evaluation of local descriptors. IEEE Trans. Pattern Anal. Mach. Intell. **27**, 1615–1630 (2005)
19. Navarro, G.: A guided tour to approximate string matching. ACM Comput. Surv. **33**, 31–88 (2001)
20. Naveena, C., Manjunath Aradhya, V.N., Niranjan, S.K.: The study of different similarity measure techniques in recognition of handwritten characters. In: Proceedings of the International Conference on Advances in Computing, Communications and Informatics (ICACCI-2012), ACM, pp. 781–787 (2012)
21. Ojala, T., Pietikäinen, M., Harwood, D.: A comparative study of texture measures with classification based on feature distributions. Pattern Recog. **29**, 51–59 (1996)
22. Ojala, T., Pietikäinen, M., Mäenpää, T.: Multiresolution gray-scale and rotation invariant texture classification with local binary patterns. IEEE Trans. Pattern Anal. Mach. Intell. **24**, 971–987 (2002)
23. Perlibakas, V.: Distance measures for PCA-based face recognition. Pattern Recogn. Lett. **25**, 711–724 (2004)
24. Phan-Ngoc, P.-T., Jo, K.-H.: Color-based face detection using combination of modified local binary patterns and embedded hidden Markov models. SICE-ICASE, pp. 5595–5603 (2006)
25. Rodriguez, Y., Marcel, S.: Face authentication using adapted local binary pattern histograms. Computer Vision—ECCV 2006. Lecture Notes in Computer Science, vol. 3954, pp. 321–332 (2006)
26. Sivic, J., Zisserman, A.: Video Google: a text retrieval approach to object matching in videos. computer vision, 2003. In: Ninth IEEE International Conference on Proceedings, vol 2, pp. 1470–1477 (2003)
27. Tan, X., Triggs, B.: Fusing gabor and LBP feature sets for kernel-based face recognition. In: Zhou, S.K., Zhao, W., Tang, X., Gong, S. (Eds.) Analysis and Modeling of Face and Gestures, vol. 4778, pp. 235–249 (2007)
28. Turk, M., Pentland, A.: Eigenfaces for recognition. J. Cogn. Neurosci. **3**, 71–86 (1991)
29. Wiskott, L., Fellous, J.-M., Krüger, N., von der Malsburg, C.: Face recognition by elastic bunch graph matching. IEEE Trans. Pattern Anal. Mach. Intell. **19**, 775–779 (1997)

30. Wu, W., Li, J., Wang, T., Zhang, Y.: Markov chain local binary pattern and its application to video concept detection. In: Proceedings of 15th ICIP 2008, pp. 2524–2527 (2008)
31. Xue, Y., Tong, C.S., Zhang, W.: Survey of distance measures for NMF-based face recognition. In: Wang, Y., Cheung, Y., Liu H. (Eds.): CIS 2006, LNAI 4456, pp. 1039–1049 (2007)
32. Zhang, W., Shan, S., Gao, W., Chen, X.: Local gabor binary pattern histogram sequence (LGBPHS): a novel non-statistical model for face representation and recognition. In: Proceedings of ICCV'05, vol. 1, pp. 786–791 (2005)
33. Zhao, W., Chellappa, R., Phillips, P.J., Rosenfeld, A.: Face recognition: a literature survey. ACM Comput. Surv. **35**, 399–458 (2003)

Face Recognition Method with Two-Dimensional HMM

Janusz Bobulski

Abstract This paper presents an automatic face recognition system, which bases on two-dimensional hidden Markov models. The traditional HMM uses one-dimensional data vectors, which is a drawback in the case of 2D image processing, as part of the information is lost during conversion. The paper presents the full ergodic 2D-HMM and uses it to identify faces. The experimental results demonstrate that the system, basing on two-dimensional hidden Markov models, is able to achieving an average recognition rate of 94 %.

Keywords 2D hidden Markov model · Face recognition · Image processing

1 Introduction

Automatic identification of a person based on face image has been of great interest for a long time, because it is the most natural and the most commonly used by people for verification of identity. Human recognition technology based on facial image is noninvasive, noncontact, and the most natural of all methods of human identification. The increase in interest in this technology is due to the increased safety requirements, as well as the possibility of using it in practice.

The most popular method of face identification is Principal Component Analysis (PCA) [1], which bases on that proposed by Turk and Pentland [2] Eigenfaces and employed by Karhunen–Loeve Transform (KLT) [3]. PCA is an unsupervised learning method that treats samples of different classes in the same way. Fisherfaces proposed by Belhumeour and Hespanha [4] is a supervised learning method using the category information associated with each sample to extract the most discriminatory features. Other popular methods use Wavelet Transform [5], Hidden Markov Models [6], or characteristic points [7]. Previous methods which based on HMM processed

J. Bobulski (✉)
Institute of Computer and Information Science, Czestochowa University of Technology, Dabrowskiego Street 73, 42-200 Czestochowa, Poland
e-mail: januszb@icis.pcz.pl

R. Burduk et al. (eds.), *Proceedings of the 9th International Conference on Computer Recognition Systems CORES 2015*, Advances in Intelligent Systems and Computing 403, DOI 10.1007/978-3-319-26227-7_30

one-dimensional data. This is not a problem in applications such as speech recognition, because feature vectors are of only one dimension. 1D HMM is unpractical in image processing, because the images are two-dimensional. When we convert an image from 2D to 1D, we lose some information. So, if we process two-dimensional data, we should apply two-dimensional HMM, and the same 2D HMM should work with 2D data. One of the solutions is the pseudo 2D HMM [8, 9], which is an extension of the classic 1D HMM. There are superstates hiding linear one-dimensional hidden Markov models. So, we have 1D model with 1D data in practice. Article [10] presents an analytic solution and proof of correctness two-dimensional HMM, which is similar to MRF [11, 12], and works with one-dimensional data. Additionally, it can be applied only for left–right type of HMM. This paper presents a real solution for the 2D problem in HMM. There is shown true 2D HMM which processes 2D data. Similar to 1D HMM, the most important thing for 2D HMMs is also to solve two basic problems, namely probability evolution and parameters estimation. Moreover, the presented algorithms are regarding ergodic models rather than of type "left-right" [10]. In this paper we focus on the face recognition method from single digital images with two-dimensional hidden Markov models (2D-HMM) based on feature extraction with wavelets.

2 Proposed Method

2.1 *Preprocessing Procedure*

The preprocessing steps aim to reduce the effects of noise, address intensity inhomogeneities, and perform global intensity level correction and are applied prior to segmentation. These are based on existing techniques and are only presented here for completeness, and are not discussed in detail. The ideal output of processing is to obtain face images which have normalized intensity, equal size, and containing the whole face in a vertical pose. Moreover, this procedure should also eliminate the effect of illumination and lighting. The database [13] used in the experiment provides the majority of these conditions. The preprocessing procedure of our system performs the following steps in converting image to a normalized face image for feature extraction: (1) locate and crop the face region using a rectangle according to face shape; (2) select face area; (3) scale image in that way so that the distance between the inner corners of the eyes is equal to 120 pixels; (4) histogram equalization. The points' coordinates of the inner corners of the eyes are obtained from the database. The effect of the preprocessing procedure is shown in Fig. 1.

2.2 *Features Extraction*

One of the parts personâs identification systems is features extraction, and this process is very important because effectiveness of system depend of it. The features

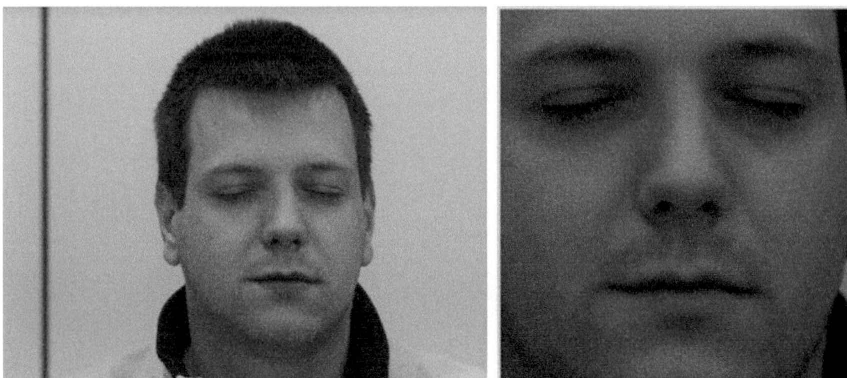

Fig. 1 The effect of preprocessing procedure (image from [13])

extraction has to get out information from a signal (image), which will be the base for person identification. The separation of useful information from the face is important, because this data will be of use for identification and should clearly describe the face. One of the popular techniques for features extraction is the Wavelet Transform (WT) [5]. One major advantage afforded by wavelets is the ability to perform local analysis—that is, to analyze a localized area of a larger signal. In wavelet analysis, we often speak about approximations and details. The approximations are the high-scale, low-frequency components of the signal. The details are the low-scale, high-frequency components. Using 2D WT, the face image is decomposed into four subimages via high-pass and low-pass filtering. The image is decomposed along column direction into subimages to high-pass frequency band H and low-pass frequency band L. Assuming that the input image is a matrix of $m \times n$ pixels, the resulting subimages become $m/2 \times n$ matrices. In the second step, the images H and L are decomposed along row vector direction and, respectively, produce the high and low frequency band HH and HL for H, and LH and LL for L. The four output images become the matrices of $m/2 \times n/2$ pixels. Low frequency subimage LL possesses high energy, and is the smallest copy of the original images. The remaining subimages LH, HL, and HH, respectively, extract the changing components in horizontal, vertical, and diagonal directions. An important aspect of features extraction with WT is the suitable choice of wavelet function [14]. The choice should adapt the shape of wavelet to the individual case and take into consideration the properties of the signal or image. A bad choice of wavelet will cause problems of analysis and identification processed signal. Experiment was conducted in order to point the best wavelet function. The best result was achieved with function db 10 from among the accessible functions.

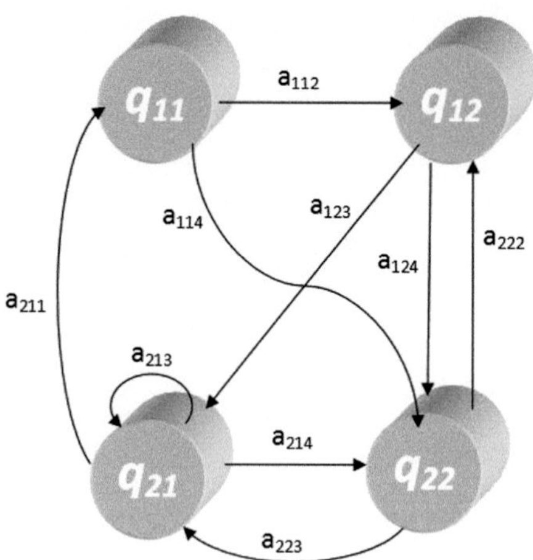

Fig. 2 Two-dimensional ergodic HMM

2.3 Face Recognition with 2D HMM

Two-dimensional hidden Markov models (2D HMM) are a development of one-dimensional HMM and provide a reasonable method of two-dimensional data modeling, e.g. images. In [10] is shown the definition and the correctness proof of the 2DHMM concept. The presented solutions to these problems were carried out for input data, which is transformed to one-dimensional vectors. This causes a loss on the part of the information contained in the input signal, which may be useful in recognition systems. For patterns that are images, their structure, that is, the relative position of pixels is important for the information communicated. Therefore, the presented solutions are not fully satisfactory. As a result of the analysis of this problem, it was decided to develop a solution for the use of two-dimensional input data in two-dimensional Markov models. The statistical parameters of the 2D model are shown in Figs. 2 and 3.

- The number of states of the model N^2.
- The number of data streams $k_1 \times k_2 = K$.
- The number of symbols M.
- The transition probabilities of the underlying Markov chain, $A = \{a_{ijl}\}$, $1 \leq i, j \leq N$, $1 \leq l \leq N^2$, where a_{ij} is the probability of transition from state ij to state l.
- The observation probabilities, $B = \{b_{ijm})\}$, $1 \leq i, j \leq N$, $1 \leq m \leq M$, which represents the probability of generate the mth symbol in the ijth state.
- The initial probability, $\Pi = \{\pi_{ijk}\}$, $1 \leq i, j \leq N$, $1 \leq k \leq K$.
- Observation sequence $O = \{o_t\}$, $1 \leq t \leq T$, o_t is square matrix simply observation of size $k_1 \times k_2 = K$.

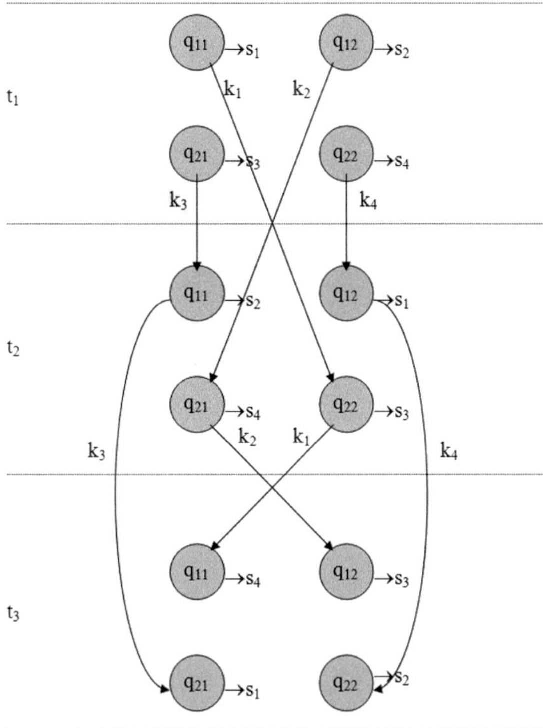

Fig. 3 The idea of 2DHMM

In the proposed structure of the 2D HMM, $K = N^2$ processes operate in parallel. They share the matrix B probabilities of observation generation and are dependent on each other. This feature distinguishes this solution from pseudo 2DHMM proposing solutions. There are two fundamental problems of interest that must be solved for HMM to be useful in face recognition applications. These problems are given below.

1. Given observation $O = (o_1, o_2, \ldots, o_T)$ and model $\lambda = (A, B, \Pi)$, efficiently compute $P(O|\lambda)$.
2. Given observation $O = (o_1, o_2, \ldots, o_T)$, estimate model parameters $\lambda = (A, B, \Pi)$ that maximize $P(O|\lambda)$.

2.3.1 Solution to Problem 1

The modified forward algorithm

- Define forward variable $\alpha_t(i, j, k)$ as:

$$\alpha_t(i, j, k) = P(o_1, o_2, \text{âŚ}, o_t, q_t = ij|\lambda) \tag{1}$$

- $\alpha_t(i, j, k)$ is the probability of observing the partial sequence $(o_1, o_2, âŚ, o_t)$ such that the state q_t is i, j for each kth stream of data
- Induction

1. Initialization:

$$\alpha_1(i, j, k) = \pi_{ijk} b_{ij}(o_1) \tag{2}$$

2. Induction:

$$\alpha_{t+1}(i, j, k) = \left[\sum_{l=1}^{N} \alpha_t(i, j, k) a_{ijl} \right] b_{ij}(o_{t+1}) \tag{3}$$

3. Termination:

$$P(O|\lambda) = \sum_{t=1}^{T} \sum_{k=1}^{K} \alpha_T(i, j, k) \tag{4}$$

2.3.2 Solution to Problem 2

The modified parameters re-estimation algorithm

- Define $\xi(i, j, l)$ as the probability of being in state ij at time t and in state l at time $t + 1$ for each kth stream of data

$$\begin{aligned} \xi_t(i, j, l) &= \frac{\alpha_t(i, j, k) a_{ijl} b_{ij}(o_{t+1}) \beta_{t+1}(i, j, k)}{P(O|\lambda)} \\ &= \frac{\alpha_t(i, j, k) a_{ij} b_{ij}(o_{t+1}) \beta_{t+1}(i, j, k)}{\sum_{k=1}^{K} \sum_{l=1}^{N^2} \alpha_t(i, j, k) a_{ijl} b_{ij}(o_{t+1}) \beta_{t+1}(i, j, k)} \end{aligned} \tag{5}$$

- Define $\gamma(i, j)$ as the probability of being in state i, j at time t, given observation sequence.

$$\gamma_t(i, j) = \sum_{l=1}^{N^2} \xi_t(i, j, l) \tag{6}$$

- $\sum_{t=1}^{T} \gamma_t(i, j)$ is the expected number of times state i, j is visited
- $\sum_{t=1}^{T-1} \xi_t(i, j, l)$ is the expected number of transitions from state ij to l

Update rules:

- $\bar{\pi_i} jk$ = expected frequency in state i, j at time $(t = 1) = \gamma_1(i, j)$

- $\bar{a}_{ij}$ = (expected number of transition from state i, j to state l)/(expected number of transitions from state i, j):

$$\bar{a}_{ijl} = \frac{\sum_t \xi_t(i, j, l)}{\sum_t \gamma_t(i, j)} \tag{7}$$

- $\bar{b_{ij}}(k)$ = (expected number of times in state j and observing symbol k)/(expected number of times in state j):

$$\bar{b}_{ij}(k) = \frac{\sum_{t, o_t = k} \gamma_t(i, j)}{\sum_t \gamma_t(i, j)} \tag{8}$$

3 Experiment

The experiment uses the image database *UMB-DB*. The University of Milano Bicocca 3D face database is a collection of multimodal (3D + 2D colour images) facial acquisitions. There are 1473 recorded images of 143 subjects (98 male, 45 female). The images show the faces in variable conditions of lighting, rotation, and size [13]. The database is available to universities and research centers interested in face recognition. We chose 30 persons in order to verify the method, and for each individual we chose three images for learning and three for testing. The size of each image is 640 × 480 pixels, with 24 bits color. During preprocessing process the image is converted to 256 gray levels per pixel and resized to 220 × 220 pixels after normalization selected area of face. The features extraction technique chosen was 2D wavelet transform of second level, and as wavelet function db 10. The size of features vector was 48400 elements. The 2D HMM implemented with parameters $N = 4$; $N^2 = 16$; $K = 16$; $M = 25$. The parameters were chosen experimentally and the values obtained were the best results (see Table 1). Table 2 gives the results of experiment with another database FERET. The experiment on this database was conducted to compare our method with other popular methods.

Table 1 The results of the experimental parameters selection of HMM

Number of states N^2	Number of symbols O	Recognition rate (%)
9	10	50
9	20	80
16	10	66
16	20	90
16	25	94
16	50	74
25	25	90
25	50	72

Table 2 Comparison of recognition rate [15]

Method	Databases	Number of images	Recognition rate (%)
PCA	AR-Faces	100	70
LDA	AR-Faces	100	88
ICA	FERET	200	89
NN	FERET	200	99
PCA	UMB-DB	90	94
1D HMM	UMB-DB	90	90
2D HMM	UMB-DB	90	93
2D HMM	FERET	200	96

4 Conclusion

A novel face recognition method is proposed in this paper. A new concept of two-dimensional hidden Markov models working with two-dimensional data is presented. We show solutions of principal problems for ergodic 2D HMM, which may be applied for 2D data. The presented method allows for faster face processing and recognition because they do not have to change the two-dimensional input data in the image form to a one-dimensional data, and thus we do not lose the information contained in the image. The obtained results are satisfactory in comparison to other methods and the proposed method may be the alternative solution to others. The average recognition rate of the method is 94 %, which is better than the classical one-dimensional HMM. Experiments confirmed the validity of the concept of two-dimensional hidden Markov models. In addition, 2D-HMM is a fast method because it needs only about 0.7 s to compare two faces.

References

1. Kirby, M., Sirovich, L.: Application of the Karhunen-Loeve procedure for the characterization of human faces. IEEE Trans. Pattern Anal. Mach. Intell. **12**(1), 103–108 (1990)
2. Turk, M., Pentland, A.: Eigenfaces for recognition. J. Cogn. Neuro-sci. **3**, 71–86 (1991)
3. Duda, R.O., Hart, P.E., Stork, D.G.: Pattern Classification. Wiley, New York (2001)
4. Belhumeour, P.N., Hespanha, J.P., Kriegman, D.J.: Eigenfaces vs. fisherfaces: recognition using class specific linear projection. IEEE Trans. Pattern Anal. Mach. Intell. **19**, 711–720 (1997)
5. Garcia, C., Zikos, G., Tziritas, G.: Wavelet pocket analysis for face recognition. Image Vis. Comput. **18**, 289–297 (2000)
6. Samaria, F., Young, S.: HMM-based architecture for face identification. Image Vis. Comput. **12**(8), 537–543 (1994)
7. Kubanek, M.: Automatic Methods for Determining the Characteristic Points in Face Image. Lecture Notes in Artificial Intelligence, vol. 6114, Part I, pp. 523–530. Springer, Berlin (2010)
8. Eickeler, S., Mller, S., Rigoll, G.: High performance face recognition using pseudo 2-D hidden Markov models. In: European Control Conference. http://citeseer.ist.psu.edu (1999)

9. Bevilacqua, V., Cariello, L., Carro, G., Daleno, D., Mastronardi, G.: A face recognition system based on Pseudo 2D HMM applied to neural network coefficients. Soft Comput. **12**(7), 615–621 (2008)
10. Yujian, L.: An analytic solution for estimating two-dimensional hidden Markov models. Appli. Math. Comput. **185**, 810–822 (2007)
11. Li, J., Najmi, A., Gray, R.M.: Image classification by a two dimensional hidden Markov model. IEEE Trans. Signal Process. **48**, 517–533 (2000)
12. Kindermann, R., Snell, J.L.: Markov Random Fields and Their Applications. American Mathematical Society, Providence (1980)
13. Colombo, A., Cusano, C., Schettini, R.: UMB-DB: a database of partially occluded 3D faces. In: Proceedings of the ICCV 2011 Workshops, pp. 2113–2119 (2011)
14. Antoniadis, A., Oppenheim, G.: Wavelets and Statistics. Lecture Notes in Statistics, vol. 103. Springer, New York (1995)
15. Abate, A.F., Nappi, M., Riccio, D., Sabatino, G.: 2D and 3D face recognition: a survey. Pattern Recognit. Lett. **28**(14), 1885–1906 (2007)

Shape-Based Eye Blinking Detection and Analysis

Zeyd Boukhers, Tomasz Jarzyński, Florian Schmidt, Oliver Tiebe and Marcin Grzegorzek

Abstract Methods for automated eye blinking analysis can be applied to support people with certain disabilities in interaction with technical systems, to analyse human deceptive behaviour, in driver fatigue assessment, etc. In this paper we introduce a robust shape-based algorithm for automatic eye blinking detection in video sequences. First, all video frames are classified separately into those showing an open and those corresponding to a closed eye. Second, these classification results are cleverly combined for blinking detection so that the influence of single misclassified frames gets compensated almost completely. In addition to that, we present our investigations on the user behaviour in terms of eye blinking frequency in two different everyday life situations. The most relevant scientific contributions of this paper are (1) the introduction of a new and robust feature extraction technique for the representation of images displaying eyes, (2) a smart fusion scheme improving the results for single-frame classification and (3) the compensation of wrong classification results for single frames providing an almost perfect eye blinking detection rate.

Keywords Eye blinking · Human behaviour analysis · Support vector machines

Z. Boukhers (✉) · T. Jarzyński · F. Schmidt · O. Tiebe · M. Grzegorzek
Pattern Recognition Group, University of Siegen, Siegen, Germany
e-mail: zeyd.boukhers@uni-siegen.de
URL: http://www.pr.informatik.uni-siegen.de

T. Jarzyński
e-mail: jarzynski.tomasz@gmail.com
URL: http://www.pr.informatik.uni-siegen.de

F. Schmidt
e-mail: florian.schmidt@uni-siegen.de
URL: http://www.pr.informatik.uni-siegen.de

O. Tiebe
e-mail: oliver.tiebe@student.uni-siegen.de
URL: http://www.pr.informatik.uni-siegen.de

M. Grzegorzek
e-mail: marcin.grzegorzek@uni-siegen.de
URL: http://www.pr.informatik.uni-siegen.de

R. Burduk et al. (eds.), *Proceedings of the 9th International Conference on Computer Recognition Systems CORES 2015*, Advances in Intelligent Systems and Computing 403, DOI 10.1007/978-3-319-26227-7_31

1 Introduction

Research has shown that eye blinking decreases when cognitive demand increases, but during gaps of cognitive demand a flurry of blinks occur [7]. This might be a proper source of information for a number of applications, e.g. an automatic lie detector. Liars, who invent an alibi about the target period, experience a more cognitive demand when recalling the target period deceptively than when recalling the baseline periods truthfully [7]. Obviously, if any kind of eye blinking analysis is required to be performed and understandable automatically [17], a robust and method for automatic detection of eye blinking in video data is necessary.

After a state-of-the-art study (Sect. 2), we propose a robust technique for shape-based eye blinking detection and exemplarily use it for human behaviour analysis. For this, we first analyse a video frame by frame. After a necessary image preprocessing step, we identify the region of interest including the area of eyes in each image and perform the eye segmentation (Sect. 3). Subsequently, we compute a three-dimensional feature vector distinguishing images with closed eyes from those with an open eye (Sect. 4). Later, we classify each frame into one of the mentioned categories using the support vector machine (Sect. 5). Although the image classification itself is not perfect (classification rate of about 96 %), the eye blink detection software works almost perfectly (eye blink detection rate of more than 99 %). The reason for this is that we apply a smart information fusion method and combine the classification results for single frames so that wrong results become compensated (Sect. 6). Although our work mainly focuses on eye blinking detection, its final purpose is interpreting human behaviour [16]. Therefore, we additionally analyse the user behaviour in terms of eye blinking frequency in two different situations which provides interesting conclusions (Sect. 7).

The most relevant scientific contributions of this paper clearly distinguishing our approach from the state of the art are (1) the introduction of a new and robust feature extraction technique for the representation of images displaying eyes, (2) a clever fusion scheme improving the results for single-frame classification and (3) the compensation of wrong classification results for single frames providing the almost perfect eye blink detection rate.

2 Related Work

Until now, various algorithms have been proposed for eye blinking detection. In recent work, Minkow et al. [10] compared different features for automatic eye blinking detection with an application to analysis of deceptive behaviour. They use features such as raw image intensities, the magnitude of the responses of Gabor filters, histograms of oriented gradients (HOGs) and optical flow. The classification itself is performed by a support vector machine. In [5] a vision-based eye blinking monitoring system for human–computer interfacing is proposed. In [14] a novel approach to the eye movement analysis using a high-speed camera is described. It combines spatial and temporal derivatives. The pixels of each frame are divided into two groups according to the direction and magnitude of the hybrid gradient vectors and the distance between their centres of gravity is used for the determination of the eye movement characteristics. Choi et al. [2] propose an eye and eye blink detection technique that uses AdaBoost learning and grouping. The algorithm detects eyes by applying the eye detector to eye candidate regions of a face. To eliminate outliers, eye candidates are grouped before. In [8] an eye blinking detection technique based on multiple Gabor response waves is proposed. Here, eye blinking is considered as an evidence of aliveness to exclude fakes, mainly two-dimensional photographs, to spoof face recognition systems. Panning et al. [12] propose a colour-based approach for eye blinking detection in image sequences. The algorithm is also able to detect eye lid movements. In [15] a real-time eye blinking detection algorithm that uses local ternary patterns and SVM is proposed. Its application is the interaction with smart devices like tablet PCs, etc. In [6] an algorithm using SIFT features to detect eye blinks in low contrast images under near-infrared illumination is described. Danisman et al. [3] introduce an approach to get drowsiness levels from eye blinks and alert car drivers in case of getting too drowsy. The work by Park et al. [13] goes in the same direction. They propose a method which can detect eye blinks under various illumination conditions and assess the drowsiness level.

Inspired by some of these methods, we propose an eye blinking detection technique outperforming them and describe additional investigations towards human behaviour analysis in this paper.

3 Eye Segmentation

The eye segmentation process consists of the steps depicted in Fig. 1 and described below.

Face Localisation: The face localisation is performed with the algorithm proposed by [18]. After localising the face, we use only the red image channel from the RGB colour system for further processing. The reason for this is that the iris, the eyebrow and the eyelash have usually a high response in the red channel, while pixel values corresponding to skin are rather small in this channel.

Region of Interest: In this step, two rectangular regions of interest ($\boldsymbol{R}_{\text{left}}$, $\boldsymbol{R}_{\text{right}}$) corresponding to the left and to the right eye are determined based on the face region $\boldsymbol{I}$ computed in the last step. We consciously make the regions large enough so that

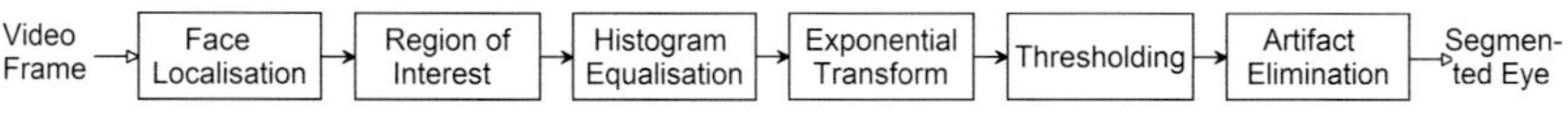

Fig. 1 Processing steps for eye segmentation in video frames

they also include eyebrows. The reason for this is the diversity of people being subject to our experiments.

Histogram Equalisation: The goal of this step is the contrast enhancement in the region of interest $\boldsymbol{R}_{\Phi\times\Lambda} = (r(\phi, \lambda))$. For this, we use histogram equalisation given by

$$\begin{aligned} \boldsymbol{s} &= (s_1, \ldots, s_{256})^{\mathrm{T}} && \text{with } \boldsymbol{s} = \tfrac{\boldsymbol{h}(\boldsymbol{R})}{\Phi\Lambda} \\ \boldsymbol{v} &= (v_1, \ldots, v_{256})^{\mathrm{T}} && \text{with } v_j = \operatorname{floor}\left(256 \sum_{i=1}^{j} s_i\right), \\ \boldsymbol{Q}_{\Phi\times\Lambda} &= (q(\phi, \lambda)) && \text{with } q(\phi, \lambda) = v_{r(\phi,\lambda)+1} \end{aligned} \tag{1}$$

where $\boldsymbol{h}(\boldsymbol{R})$ represents the histogram of $\boldsymbol{R}$ with 256 bins. Φ and Λ correspond to the height and to the width of $\boldsymbol{R}$, respectively.

Exponential Transform: In order to improve the results of thresholding conducted in the next step, we increase the difference among the intensity of pixels in $\boldsymbol{Q}$ according to the method presented in [11]. For this, the exponential transform of each pixel in $\boldsymbol{Q}$ is computed as follows:

$$\boldsymbol{E}_{\Phi\times\Lambda} = (e(\phi, \lambda)) \quad \text{with} \quad e(\phi, \lambda) = \exp(q(\phi, \lambda)) \quad . \tag{2}$$

Thresholding: In this step, we binarise the image $\boldsymbol{Q}$ by thresholding the values in the matrix $\boldsymbol{E}$ as follows:

$$\boldsymbol{T}_{\Phi\times\Lambda} = (t(\phi, \lambda)); \quad t(\phi, \lambda) = \begin{cases} 0, & \text{if } e(\phi, \lambda) < \varepsilon \\ 255, & \text{otherwise} \end{cases}, \tag{3}$$

where ε is the average value of $\boldsymbol{E}$ elements.

Artefact Elimination: The thresholded image $\boldsymbol{T}$ usually contains not only the desired elements (eyebrow and eyelash for a closed eye; eyebrow and iris for an open eye), but also irrelevant parts and artefacts like hair, eye dark circles or unusual lighting effects that are eliminated in this step. In addition to this, the eyebrow is also eliminated in this step, since it is not needed for further processing. The relevant image segments (the iris and the eyelash) are supposed to be in the centre of the image $\boldsymbol{T}$. The elimination method proposed in this paper selects the nearest bloc of pixels to the centre of $\boldsymbol{T}$. Exemplary results are shown in Figs. 2e and 3e.

4 Feature Extraction

For each result of the eye segmentation (Sect. 3), a three-dimensional feature vector $\boldsymbol{c} = (c_1, c_2, c_3)^{\mathrm{T}}$ is extracted. Its first element c_1 is determined based on the outcome of a circle detection in the area of the segmented eye (circle detected $\rightarrow c_1 = 1$, circle not detected $\rightarrow c_1 = 0$). It is computed using a circle detection method based on the Hough Transform introduced in [1] with the sensitivity equal to 0.89, the minimum radius of 10 pixels, and the maximum radius of 25. The second ele-

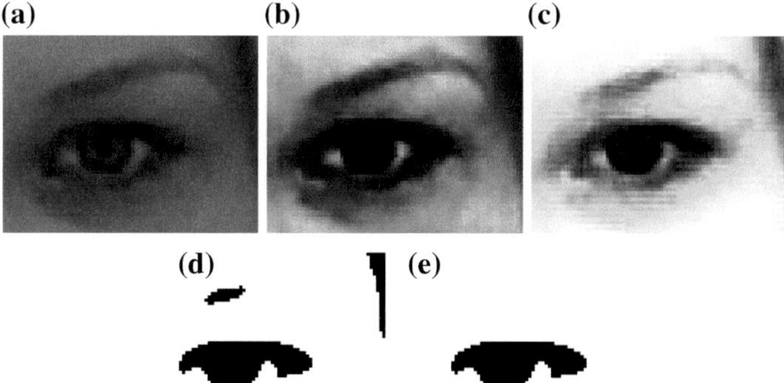

Fig. 2 Open eye segmentation on an example image. **a** Original ROI. **b** Histogram equalisation. **c** Red channel. **d** Thresholding. **e** Artefact elimination

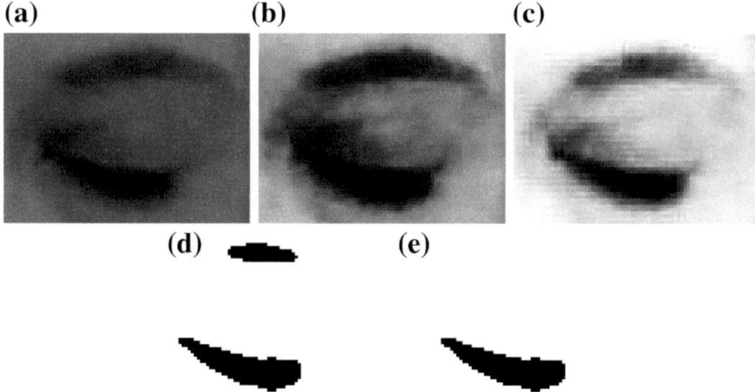

Fig. 3 Closed eye segmentation on an example image. **a** Original ROI. **b** Histogram equalisation. **c** Red channel. **d** Thresholding. **e** Artefact elimination

ment c_2 equals the ratio of width to height of the minimum bounding box enclosing the segmented eye. The third element c_3 is the size of the segmented eye in relation to the overall bounding box size. In order to equalise the influence of each element of the feature vector $\boldsymbol{c}$ to the whole process, we linearly transform their values into the range [0, 1].

To evaluate the discriminative properties of the feature space, we applied the Fisher linear discriminant analysis [9] and achieved the following results: $\lambda'(c_1) = 0.095$, $\lambda'(c_2) = 0.785$ and $\lambda'(c_3) = 0.372$ for the different dimensions of the feature space, respectively. The value $\lambda'(c_i)$ expresses the overall discrimination power for the feature space dimension c_i and is calculated as the ratio of the determinant of intra-class covariance matrix to the determinant of the total covariance matrix. The values of $\lambda'(c_i)$ belong to the range [0, 1], whereas 0 corresponds to perfect discriminative properties and 1 denotes no discrimination.

According to the analysis of the feature space described above, the first feature c_1 possesses the strongest discriminative power for distinguishing open and closed

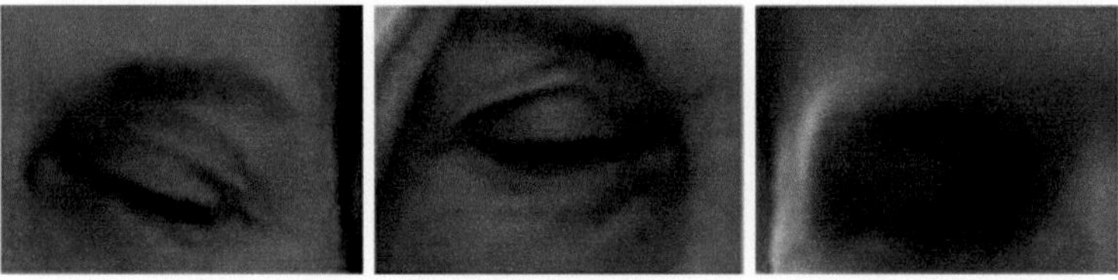

Fig. 4 Example images with wrong iris detection

eyes. This is due to the iris visible in images of an open eye that, however, can only very rarely be found in closed eye images. Unfortunately, the circle detection algorithm sometimes delivers wrong results, especially under drastically changing light conditions, while considering the diversity of ethnic groups, and due to varying people's age. Examples of such images can be seen in Fig. 4.

5 Eye Blinking Detection

For single frame classification, we consider two classes (ω_1 and ω_2) which represent a closed and an open eye, respectively. In this work a KNN and a linear SVM are used to arrange the training sets into these two classes. We conclude from [4] that the left and right eyes are blinking synchronously. For this reason we fuse the classifier outcomes for both eyes. To do so, we first calculate the Euclidean distances $d_{i,\text{left}}$ and $d_{i,\text{right}}$ of the left eye and the right eye image to the learnt SVM hyperplane $g(x)$. The distance is given by the orthogonal projection and its value is positive when the sample belongs to ω_1 and negative in the contrary case. The final classifier outcome for all frames in a video is stored in the vector $\boldsymbol{b} = (b_1, \ldots, b_i, \ldots, b_n)^{\mathrm{T}}$. Its elements are determined as follows:

$$b_i = \begin{cases} \max(d_{i,\text{left}}, d_{i,\text{right}}), & \text{if } d_{i,\text{left}} + d_{i,\text{right}} > 0 \\ 0, & \text{otherwise} \end{cases} \tag{4}$$

This vector may of course contain wrong single-frame classification results. However, if the number of misclassifications is small, the eye blinking detection can still be performed with an almost perfect result. For this, we process the vector $\boldsymbol{b}$ by a low pass filter, so that single wrong results corresponding to big changes in the signal can be filtered out

$$\boldsymbol{b} \longrightarrow \boldsymbol{b}^* = (b_1^*, \ldots, b_n^*)^{\mathrm{T}} \quad \text{with}$$

$$b_i^* = \begin{cases} \frac{b_{i-1}+b_{i+1}}{2}, & \text{if} \begin{array}{l} \text{sign}(b_i) \neq \text{sign}(b_{i-1}) \\ \text{sign}(b_i) \neq \text{sign}(b_{i+1}) \end{array} \\ \frac{b_{i-1}+b_i+b_{i+1}}{3}, & \text{otherwise.} \end{cases} \tag{5}$$

Table 1 Single-frame classification rates

	$\omega_1 \rightarrow$ closed eye (%)	$\omega_2 \rightarrow$ open eye (%)
SVM	92.50	98.75
KNN	77.50	80.00

Table 2 Blinking detection accuracy

	Number	Percentage (%)
Full blink detected	311	99.36
Full blink not detected	2	0.64
Half blink detected	261	80.31
Half blink not detected	64	19.69

6 Experiments and Results

To evaluate the robustness of this methodology, we carried out an experiment using a high-speed camera which delivers videos with 100 frames per second and the resolution 1280 × 720 pixels. We recorded 21 videos (60 s long each) of people from diverse ethnic groups. So, we have 6,000 frames per video and 126,000 frames in total. randomly selected 240 images per class. The single-frame classification accuracy is shown in Table 1. Due to this result, we have used the SVM outcomes in the further process of eye blinking detection. Evaluating eye blinking detection we distinguished between three states: full blink, half blink, no blink. The accuracy of detection can be seen in Table 2.

6.1 Eye Blinking Frequency Analysis

In our second experiment we studied the human blinking behaviour in two different situations. For this, faces of 50 probands have been recorded as they sequentially watched two kinds of videos (action and calm) on a laptop in two different cognitive states. In the first cognitive state, the probands did not know the objective of our test, in the second one, we informed them about it. So, altogether we consider four different situations: (1) watching an action video without the background knowledge (AV-KNOW), (2) watching a calm video without the background knowledge (CV-KNOW), (3) watching an action video with background knowledge (AV+KNOW) and (4) watching a calm video with background knowledge (CV+KNOW). For all these situations, we statistically analysed the blinking frequency behaviour. The results can be seen in Figure 5.

It can be clearly observed that the blinking behaviour of users watching calm videos is significantly different as for the case of action videos. However, no significant differ-

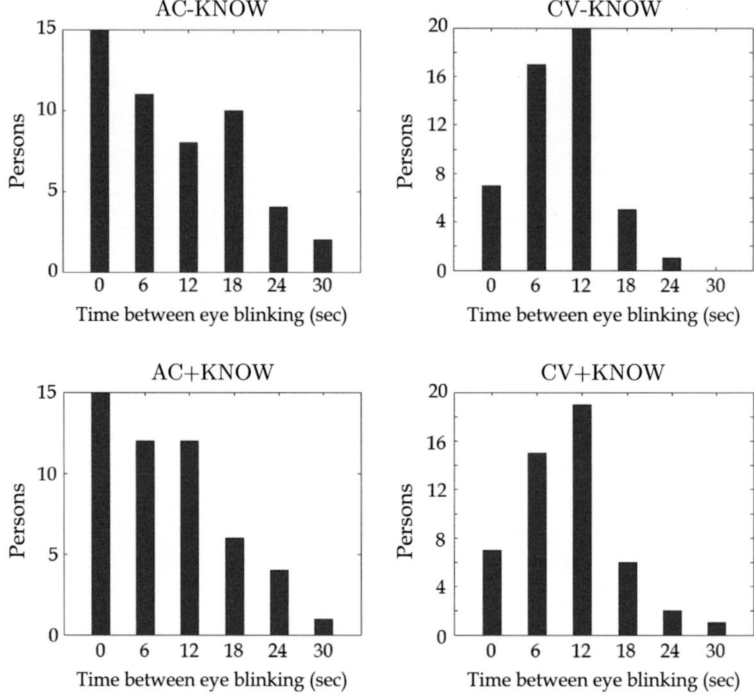

Fig. 5 Histograms of distances between two subsequent eye blinks measured in time (seconds) in four different cognitive user states. The histogram bin "0" corresponds to videos with one blink at maximum, i.e. the computation of the time distance was not possible. *AC-KNOW* watching an action video without the background knowledge; *CV-KNOW* watching a calm video without the background knowledge *AC+KNOW* watching an action video with background knowledge; *CV+KNOW* watching a calm video with background knowledge

ence in blinking behaviour could be observed between people knowing the objective of our experiments (eye blinking analysis) and those without this knowledge.

7 Conclusion

In this paper we introduced and evaluated a method for automated shape-based eye blinking detection. We also analysed eye blinking frequency in different cognitive stage. The impressive performance of our eye blinking detection technique (over 99 %) could be achieved thanks to the new feature space introduced in this paper possessing a lot of discriminative power and the clever fusion method compensating single-frame misclassifications. The eye blinking frequency analysis has shown that eye blinking behaviour is highly dependent on user's reaction to the videos. In the future we will apply our eye blinking detection algorithm for psychological studies that will support a framework for human behavioural biometry.

References

1. Atherton, T., Kerbyson, D.: Size invariant circle detection. Image Vis. Comput. **17**(1), 795–803 (1999). http://www.sciencedirect.com/science/article/pii/S0262885698001607
2. Choi, I., Han, S., Kim, D.: Eye detection and eye blink detection using adaboost learning and grouping. In: 2011 Proceedings of the 20th International Conference on Computer Communications and Networks (ICCCN), pp. 1–4, July 2011
3. Danisman, T., Bilasco, I., Djeraba, C., Ihaddadene, N.: Drowsy driver detection system using eye blink patterns. In: 2010 International Conference on Machine and Web Intelligence (ICMWI), pp. 230–233 (2010)
4. Fukuda, K.: Analysis of eyeblink activity during discriminative tasks. Percept. Mot. Skills **79**, 1599–1608 (1994)
5. Krolak, A., Strumillo, P.: Vision-based eye blink monitoring system for human-computer interfacing. In: 2008 Conference on Human System Interactions, pp. 994–998, May 2008
6. Lalonde, M., Byrns, D., Gagnon, L., Teasdale, N., Laurendeau, D.: Real-time eye blink detection with gpu-based sift tracking. In: Fourth Canadian Conference on Computer and Robot Vision, 2007, CRV '07, pp. 481–487 (2007)
7. Leal, S., Vrij, A.: Blinking during and after lying. J. Nonverbal Behav. **32**(4), 187–194 (2008)
8. Li, J.W.: Eye blink detection based on multiple gabor response waves. In: International Conference on Machine Learning and Cybernetics, vol. 5, pp. 2852–2856, July 2008
9. Lindeman, R.H., Merenda, P.F., Gold, R.: Robust real-time face detection (1980)
10. Minkov, K., Zafeiriou, S., Pantic, M.: A comparison of different features for automatic eye blinking detection with an application to analysis of deceptive behavior. In: 2012 5th International Symposium on Communications Control and Signal Processing (ISCCSP), pp. 1–4, May 2012
11. Nixon, M., Aguado, A.S.: Robust real-time face detection, vol. 2 (2007)
12. Panning, A., Al-Hamadi, A., Michaelis, B.: A color based approach for eye blink detection in image sequences. In: 2011 IEEE International Conference on Signal and Image Processing Applications (ICSIPA), pp. 40–45 (2011)
13. Park, I., Ahn, J.H., Byun, H.: Efficient measurement of eye blinking under various illumination conditions for drowsiness detection systems. In: 18th International Conference on Pattern Recognition, ICPR 2006, vol. 1, pp. 383–386 (2006)
14. Radlak, K., Smolka, B.: A novel approach to the eye movement analysis using a high speed camera. In: 2012 2nd International Conference on Advances in Computational Tools for Engineering Applications (ACTEA), pp. 145–150, December 2012
15. Ryu, J.B., Yang, H.S., Seo, Y.H.: Real time eye blinking detection using local ternary pattern and SVM. In: 2013 Eighth International Conference on Broadband and Wireless Computing, Communication and Applications (BWCCA), pp. 598–601 (2013)
16. Shirahama, K., Grzegorzek, M.: Towards large-scale multimedia retrieval enriched by knowledge about human interpretation -retrospective survey. Multimedia Tools and Applications (2014)
17. Tadeusiewicz, R., Ogiela, M.R.: Why automatic understanding? In: Bartlomiej, B., Dzielinski, A., Iwanowski, M., Bernerdete, R. (eds.) Adaptive and Natural Computing. Lecture Notes on Computer Science, pp. 477–491. Springer, Heidelberg (2007). http://www.springer.com
18. Viola, P., Jones, M.: Robust real-time face detection. In: Proceedings of the Eighth IEEE International Conference on Computer Vision, ICCV 2001, vol. 2, pp. 747–747 (2001)

Lip Print Pattern Extraction Using Top-Hat Transform

Lukasz Smacki, Jacek Luczak and Zygmunt Wrobel

Abstract Lip print examination is a very difficult and complex task even for modern forensic departments. Computer systems that will assist a crime scene expert in identification of this kind of evidence are very desired in the forensic science community. Unfortunately, such softwares do not exist as methods of automatic lip print identification are still insufficiently developed. This paper presents an original method of lip print pattern extraction that can be used as a preprocessing stage in the lip print identification process. Research shows that the proposed method increased lip print identification accuracy for all tested template matching algorithms. After further improvements, our method can be used as a base for creating a personal identification system based on lip prints.

Keywords Cheiloscopy · Lip print · Pattern extraction · Top-hat transform · Forensic science

1 Introduction

1.1 Lip Prints in Criminal Identification

Lip prints are the impressions of human lips left on objects such as glasses, cups, drink containers, cutlery, food or cigarettes. In 1932, Edmond Locard, the French criminologist, first recommended the use of lip prints for criminal identification [1]. The first serious study of human lips as a means of personal identification was

---

L. Smacki (✉) · J. Luczak · Z. Wrobel
Department of Biomedical Computer Systems, University of Silesia,
Bedzinska 39, 41200 Sosnowiec, Poland
e-mail: lukasz.smacki@us.edu.pl

J. Luczak
e-mail: jacek7.17@gmail.com

Z. Wrobel
e-mail: zygmunt.wrobel@us.edu.pl

R. Burduk et al. (eds.), *Proceedings of the 9th International Conference on Computer Recognition Systems CORES 2015*, Advances in Intelligent Systems and Computing 403, DOI 10.1007/978-3-319-26227-7_32

Fig. 1 Lip prints obtained using different latent print powders. **a** Fluorescent *green*. **b** Oxide *red*. **c** Magnetic *black*

presented in 1970s by two Japanese scientists Suzuki and Tsuchihasi and concluded "lip print has its own individual characteristics" and "can be used as one of the techniques for identification" [2, 3]. Later research [4–10] proved that lip prints are determined by one's genotype, and are therefore unique and stable throughout the life of a human being. In 1999 the Illinois Appellate Court accepted the fact that "lip print identification is generally acceptable within the forensic science community as a means of positive identification" and "the F.B.I. and the Illinois State Police consider that lip prints are unique like fingerprints and are a positive means of identification" [11]. If properly lifted and examined, lip prints left at a crime scene can contain useful data leading to class and individual identification [1, 12]. Cheiloscopy, a forensic investigation technique that deals with identification of humans based on lips traces, has already been successfully used as evidence in lawsuits [4, 5, 13, 14]. However, the use of lip prints in criminal cases is still limited because the credibility of lip prints has not been firmly established in some countries. This follows naturally from the fact that this domain is still relatively new, being explored and developed only over the recent years. Cheiloscopy can be also used for gender [15] and affinity [16] recognition as well as postmortem [17] identification. Figure 1 shows lip print images recovered using different latent print powders.

1.2 Existing Solutions

There are several approaches to lips recognition reported in the literature such as texture, geometry and motion based. In texture-based approaches, type and direction of lip grooves are analyzed [18, 19] or lip regions are directly compared using SURF and SIFT algorithms [20]. Geometry-based approaches analyze outline [21], dimensions [22] or characteristic points [23] of lips. In motion-based approaches, motion vectors are computed to represent the lip movement during speaking [24, 25]. Fusions of the above approaches also exist in the literature [26]. Unfortunately, the above approaches require high-quality photos or videos of human lips. Therefore it is impossible to apply them in criminal identification based on latent lip prints. Computer-based lip print recognition has not been extensively researched so far, although there are some published approaches that deal with this topic. In 2010 Smacki et al. [27–29] the first computer lip print recognition method. It extracted line segments from a lip print image using Hough Transform and compared them using similarity measure based on Euclidean distance. Two years later, Bhattacharjee et al. [30] presented a lip print recognition approach based on statistical analysis of average direction of groves in a lip print image. In 2013 Wrobel et al. [31] published a method of lip print recognition based on Generalized Hough Transform where, similar to [27], line segments were used as characteristic features. Unfortunately, all of the above approaches require high-quality full lip print (containing both upper and lower lip) so they are not suitable for real-life criminal identification where usually partial and blurred lip prints are analyzed.

1.3 Purpose of the Research

The main purpose of this paper is to present a new lip print pattern extraction method which can be used as a preprocessing stage in lip print identification process. Lip print pattern is a system of lines present on a lip print. Criminal identification using lip prints is based on characteristic features that are part of a lip print pattern (e.g., bifurcations, bridges, dots). Extraction of a lip print pattern (ipso facto emphasize of lip grooves and removal of background distractions) can improve lip print identification accuracy for both human- and computer-based lip print examination.

2 Lip Print Pattern Extraction

The presented lip print pattern extraction method continues our previous research [27–29, 32] but it is a completely new approach that meets the criteria of real-life criminal identification. It can be used to extract pattern from any kind of lip prints found at a crime scene, containing full lips, single lip, or just a fragment of a lip. The proposed approach consists of two main stages: segmentation and pattern extraction. Please see Fig. 2 for a detailed diagram. The input of the proposed method is a lip print image in PNG file format (Fig. 2a). The procedure to acquire such image consists of several steps that are normally performed by a forensic expert. First, the backing card containing lip print from a crime scene or collected from a suspect is digitalized using scanner or digital camera. Then, the image is aligned horizontally, resized to a standard size (e.g., 300 dpi), cropped to remove unnecessary objects (e.g., photo markers) and saved as a PNG file. Any graphics software can be used for such purpose (e.g., Gimp, Photoshop). We use the PNG file format because it offers small file size and high-quality image thanks to lossless compression. Our algorithm is set up to 300 dpi by default but it can adapt to different dpi values. Output of the proposed method is a PNG file that contains binary lip print pattern image (Fig. 2k).

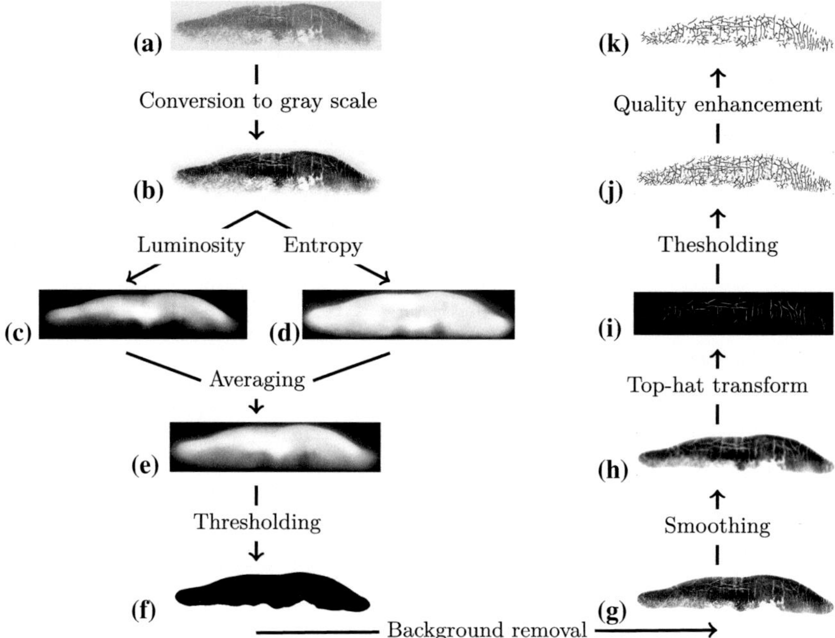

Fig. 2 Diagram of lip print pattern extraction

2.1 Segmentation

First step of the segmentation stage is grayscale conversion. It standardizes lip print images acquired using different methods (e.g., fluorescent, oxide, magnetic or regular latent print powder). We convert input RGB image I (Fig. 2a) to gray scale image G (Fig. 2b) by eliminating the hue and saturation information while retaining the luminance. RGB values are converted to grayscale values using standard NTSC conversion formula [33] that calculates the effective luminance of a pixel: $0.2989R + 0.5870G + 0.1140B$. For the standardized lip print image, we obtain luminosity (light intensity) and entropy (disorder intensity) images. On average background has higher luminosity and lower entropy then lip print area. To generate the luminosity image, we apply median filtering [34] to negative of an input lip print image G. Each output pixel contains the median value in the m-by-n (m = n = 51 for dpi = 300) neighborhood around the corresponding pixel in the input image. The optimal size of the neighborhood was determined experimentally. Figure 2c shows the output image L. Each pixel of the entropy image E is generated calculating entropy value of a neighborhood around the corresponding pixel in the input image G. Experiments showed that the neighborhood of 9-by-9 is the most optimal for 300 dpi image. Then we apply blurring using median filter described above. Blurring gets rid of the noise and prepares the image for further processing. The output image E is showed in

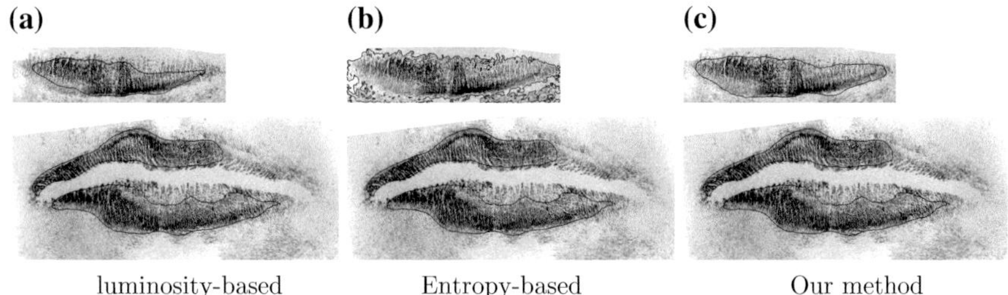

Fig. 3 Output of lip print area segmentation

Fig. 2d. Luminosity and entropy images are then merged into one image A. This procedure is performed by averaging the corresponding pixels in both of the images. First, we perform normalization [35] of the images L and E which improves image contrast by stretching the range of intensity values it contains to span the full range of pixel values. Then, we average the corresponding pixel values of both of the images producing a gray scale image A (Fig. 2e). The next step is thresholding. First, our algorithm replaces all pixels in the input image A with luminance greater than a given threshold with the value 1 (white) and replaces all other pixels with the value 0 (black) generating a binary image. We used Otsu's method [36] that calculates the threshold automatically to minimize the interclass variance of the black and white pixels. Then, we remove small insignificant objects from the image using connected-component labeling [37] and object area estimation [38]. These functions remove foreground objects that are smaller in area than a given threshold that was established experimentally at 90,000 pixels (for 300 dpi). Figure 2f shows the output image T. Finally, we perform background removal to delete unnecessary background around the original lip print area. First, the background is converted to white color by sum of the original image G and the binary image T. Then, we crop the resulting image by removing white space around the lip area creating an output image B containing white background (Fig. 2g). Figure 3 shows sample lip print areas segmented by our method (c) and basic threshold methods: luminosity- (a) and entropy-based (b). As you can see, our approach is the most accurate for both full and partial lip prints. The luminosity-based method gives poor results for lip prints of unequal contrast distribution while entropy-based method is sensitive to background noise.

2.2 *Pattern Extraction*

The first step of pattern extraction is smoothing. This is an original algorithm developed by our team. Smoothing improves image quality by making the lines forming lip print pattern smoother. First, we create structuring elements that contain lines at all possible angles. We obtained the best results in 300 dpi images for structuring elements of size 9-by-9 pixels. For the selected size, we generate 16 structuring ele-

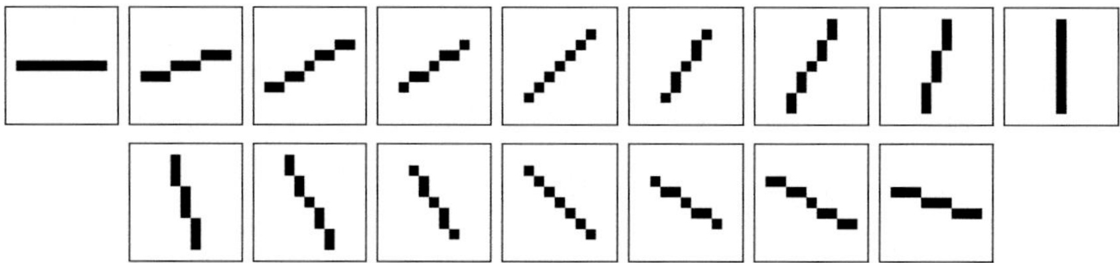

Fig. 4 Structuring elements used for smoothing

ments (Fig. 4). Then, we perform image filtering by applying structuring elements to all pixels in the input image. For every pixel, the average value of neighborhood pixels defined by each structuring element is calculated. The maximum value obtained replaces the original pixel value. This way we get a new image S where lines creating lip print pattern are smooth while keeping sharpness of their edges (Fig. 2h). After smoothing we perform morphological top-hat transform [39] which extracts lip print pattern by making it contrast to the background color. Top-hat computes the morphological opening of the image S and then subtracts the result from the image S. In our implementation, top-hat transform uses four flat linear structuring elements that are symmetric with respect to the neighborhood center. Elements are 3 pixels long (for 300 dpi) and are rotated: 0°, 45°, 90°and 135° (Fig. 5). Optimal sizes and shapes of the structuring elements were obtained experimentally. Output image H of the top-hat transform is showed in Fig. 2i. The next step of our method is thresholding. It converts the image H generated by top-hat transform to binary image T for further processing. For thresholding, we use the algorithm described in Sect. 2.1. As a result, we get a binary image containing black lip print pattern on white background (Fig. 2j). In the final step, we perform quality enhancement that improves the quality of image T generated by the thresholding function. First, we fix edges of the lines forming lip print pattern using morphological spur operation [40] and morphological holes operation [40]. Then, we remove small insignificant objects that incorporate noise in the lip print pattern image. For this purpose, we use small objects removal procedure described in Sect. 2.1. Output of the quality enhancement algorithm was shown in Fig. 2k. Figure 6 shows sample lip prints (a) and their pattern extracted using our previous method (b) [32] and the method presented in this paper (c). As you can see, the new approach is more accurate. Lip print pattern from the previous method contains noise and continuity of lines is disturbed.

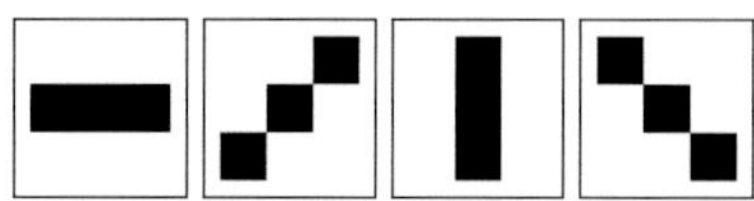

Fig. 5 Structuring elements used for top-hat transform

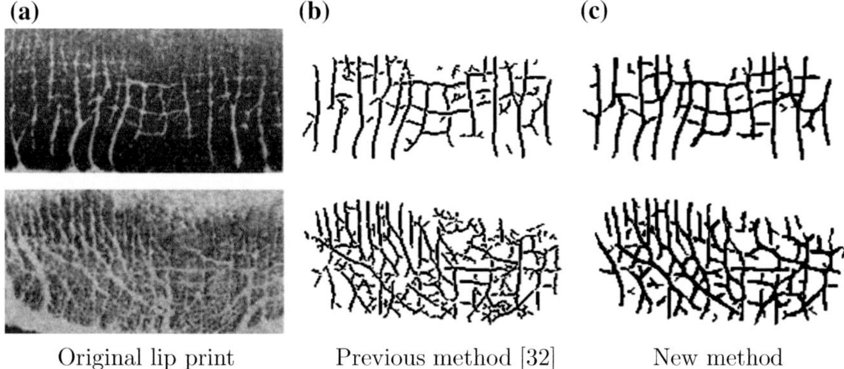

Fig. 6 Output of pattern extraction

3 Results

3.1 Design of the Experiment

The experiments were designed to verify if the proposed lip print pattern extraction method increases lip print identification accuracy comparing to the previous method [32] . The second aim was to determine if lip print identification based on lip print patterns obtains higher accuracy comparing to lip print identification based on original lip prints. The study was conducted on a data set of 300 lip prints (100 test images and 200 template images) taken from 50 individuals. Both types of lip prints were acquired with tools and methods used by crime scene examinators. The test images contained lower lip similar to lip prints usually found at crime scenes. The template images contained both upper and lower lip similar to lip prints taken from crime suspects. This way we obtained more real-life test environment comparing to [32] where both test and template images contained full lips. Test images were compared "round robin" with template images in close-set identification mode. Similarity score was calculated using two template matching algorithms available in the OpenCV library: CV_TM_CCORR_NORMED and CV_TM_CCOEFF_NORMED [41]. The above algorithms were chosen because they work for both RGB and binary images and include image normalization. We did not use the algorithm from [32] as it was designed for binary images containing full lips. As a measure of accuracy, we used the probability of identification calculated for different ranks. Tests were conducted on Apple MacBook Pro with 2 GHz Intel Core i7 processor and 8 GB 1600 MHz DDR memory.

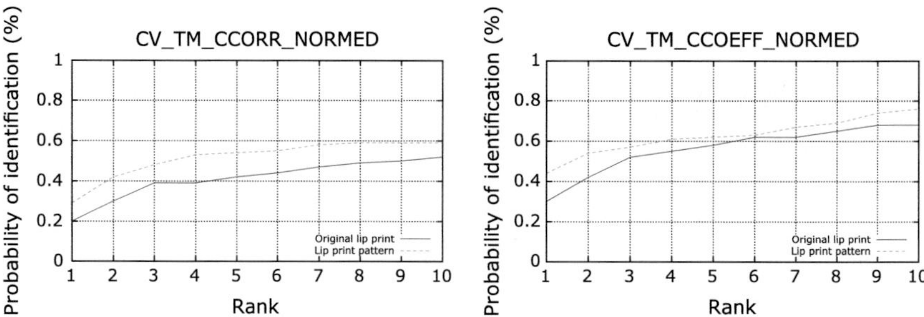

Fig. 7 Accuracy of identification based on original lip prints and patterns

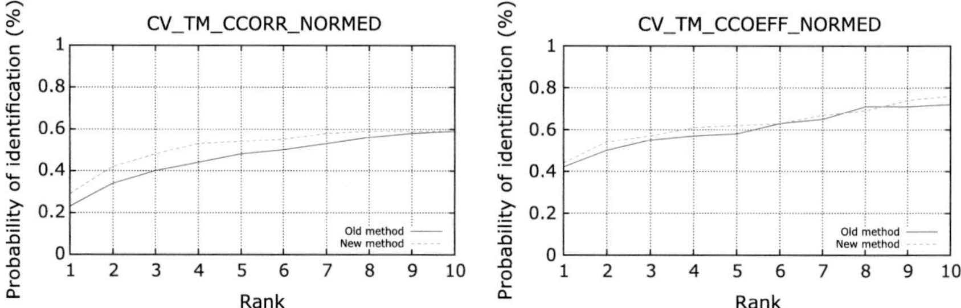

Fig. 8 Identification accuracy for the previous and new approach

3.2 Main Findings

The following charts show probability of identification for different ranks. Rank 1 is the probability that owner of the test lip print is on the first place in the ranking list, rank 2—on the 2nd place, etc. Every chart shows data obtained for one of two aforementioned template matching methods. From the charts in Fig. 7 we can see that for both template matching methods identification based on lip print patterns gives better results comparing to identification based on original lip prints. From the charts in Fig. 8 we can see that for both template matching algorithms, the proposed lip print pattern extraction method improves identification accuracy comparing to our previous approach [32].

4 Conclusions

The results acknowledge our expectations that lip print identification based on lip print pattern gives better results than based on original lip prints. This was confirmed using all tested template matching methods. Although further tests are needed to verify if this dependence is true, other template matching methods. We observed improvement in lip print identification accuracy for the proposed approach comparing to the lip print pattern extraction algorithm used in [32]. The presented algorithm is not perfect yet but the results show that our research is moving in the right direction. After further improvements, the proposed lip print pattern extraction method will have a potential to be a base for creating a lip print identification system for real-life criminal identification purposes. Future work will include further improvement of the lip print pattern extraction algorithms to obtain even more accurate lip print pattern. We will also start working on a lip print identification method based on the approach presented in this paper.

Acknowledgments Lukasz Smacki is a scholarship holder of the DoktoRIS project cofunded by the European Union under the European Social Fund. Jacek Luczak is supported by the Forszt project cofunded by the European Union under the European Social Fund.

References

1. Siegel, J., Saukko, P., Knupfer, G.: Encyclopedia of Forensic Science, pp. 358–362. Academic Press, London (2000)
2. Suzuki, K., Tsuchihashi, Y.: Personal identification by means of lip prints. J. Forensic Med. **17**, 52–57 (1970)
3. Tsuchihashi, Y.: Studies on personal identification by means of lip prints. Forensic Sci. **3**, 233–248 (1974)
4. Kasprzak, J.: Possibilities of cheiloscopy. Forensic Sci. Int. **46**, 145–151 (1990)
5. Kasprzak, J.: Cheiloscopy, Forensic Science International, pp. 358–362 (2000)
6. Hirth, L., Gottsche, H., Goedde, H.W.: Lip prints—variability and genetics. Natl Libr. Med. **30**, 47–62 (1975)
7. Sivapathasundharam, B., Prakash, P.A., Sivakumar, G.: Lip prints (Cheiloscopy). Indian J. Dental Res. **12**, 234 (2001)
8. Coward, R.C.: The stability of lip pattern characteristics over time. J. Forensic Odontostomatol. **25**(2), 40–56 (2007)
9. Eldomiaty, M.A., Anwar, R.I., Algaidi, S.A.: Stability of lip-print patterns: A longitudinal study of Saudi females. J. Forensic Legal Med. **22**, 154–158 (2014)
10. Popa, M.F., Stefanescu, C., Corici, P.D.: Medical-legal identification methods with the aid of cheiloscopy. Roman. J. Legal Med. **21**, 215–218 (2013)
11. The People of the State of Illinois, Plaintiff-Appellee, v. Lavelle L. Davis, Defendant-Appellant, http://www.leagle.com
12. Newton, M.: The Encyclopedia of Crime Scene Investigation, p. 42 (2008)
13. Castello, A., et al.: Luminous lip-prints as criminal evidence. Forensic Sci. Int. **155**, 185–187 (2005)
14. Sharma, P., Saxwna, S., Rathod, V.: Comparative reliability of cheiloscopy and palatoscopy in human identification. Indian J. Dental Res. **20**(4), 453–457 (2009)
15. Gupta, S., Gupta, K., Gupta, O.P.: A study of morphological patterns of lip prints in relation to gender of North Indian population. J. Oral Biol. Craniofac. Res. **1**(1), 12–16 (2011)
16. Eldomiaty, M.A., Al-gaidi, S.A., Elayat, A.A., Safwat, M.D.E., Galal, S.A.: Morphological patterns of lip prints in Saudi Arabia at Almadinah Almonawarah province. Forensic Sci. Int. **200**, 179.e1–179.e9 (2010)
17. Utsuno, H., et al.: Preliminary study of post mortem identification using lip prints. Forensic Sci. Int. **149**(2–3), 129–132 (2005)
18. Kim, J.O., et al.: Lip print recognition for security systems by multi resolution architecture. Future Gener. Comput. Syst. **20**, 295–301 (2004)

19. Sharma, P., Deo, S., Venkateshan, S., Vaish, A.: Lip print recognition for security systems: an up-coming biometric solution. In: Tsihrintzis George, A., Virvou, M., Jain Lakhmi, C., Howlett Robert, J. (eds.) IIMSS 2011, vol. 11, pp. 347–359. SIST Springer, Heidelberg (2011)
20. Bakshi, S., Raman, R., Sa, P.K.: Lip pattern recognition based on local feature extraction. In: India Conference (INDICON), pp. 1–4 (2011)
21. Gomez, E., Travieso, C.M., Briceno, J.C., Ferrer, M.A.: Biometric identification system by lip shape. In: Security Technology, pp. 39–42 (2002)
22. Choras, M.: The lip as a biometric. Pattern Anal. Appl. **13**(1), 105–112 (2010)
23. Reshmi, M.P., Karthick, V.J.: Biometric identification system using lips. Int. J. Sci. Res. (IJSR) **2**(4), 304–307 (2013)
24. Faraj, M., Bigun, J.: Audio-visual person authentication using lip-motion from orientation maps. Pattern Recognit. Lett. **28**(11), 1368–1382 (2007)
25. Cetingul, H.E., Erzin, E., Yemez, Y., Tekalp, A.M.: Multimodal speaker/speech recognition using lip motion, lip texture and audio. Signal Process. **86**(12), 3549–3558 (2006)
26. Wanga, S.L., Liewb, W.C.: Physiological and behavioral lip biometrics: A comprehensive study of their discriminative power. Pattern Recognit. **45**(9), 3328–3335 (2012)
27. Smacki, L., Porwik, P., Tomaszycki, K., Kwarcinska, S.: Lip print recognition using the Hough transform. J. Med. Inf. Technol. **14**, 31–38 (2010)
28. Smacki, L., Wrobel, K.: Lip print recognition based on mean differences similarity measure. In: Burduk, R., Kurzyński, M., Woźniak, M., Żołnierek, A. (eds.) Computer Recognition Systems 4, vol. 95, pp. 41–49. AISC Springer, Heidelberg (2011)
29. Smacki, L., Wrobel, K., Porwik, P.: Lip print recognition based on the DTW algorithm. In: Proceedings of the Third World Congress on Nature and Biologically Inspired Computing, Salamanca, Spain, pp. 601–606 (2011)
30. Bhattacharjee, S., Arunkumar, S., Bandyopadhyay, S.K.: Personal identification from lip-print features using a statistical model. Int. J. Comput. Appl. **55**(13), 30–34 (2012)
31. Wrobel, K., Doroz, R., Palys, M.: A method of lip print recognition based on sections comparison. In: International Conference on Biometrics and Kansei Engineering (ICBAKE 2013), Tokyo, Japan, pp. 47–52 (2013)
32. Smacki, L.: Latent lip print identification using fast normalized cross-correlation. In: International Conference on Biometrics and Kansei Engineering (ICBAKE 2013), Tokyo, Japan, pp. 189–192 (2013)
33. Ware, C.: Information Visualization: Perception for Design (2004)
34. Huang, T., Yang, G., Tang, G.: A fast two-dimensional median filtering algorithm. Acoust. Speech Signal Process. **27**(1), 13–18 (1979)
35. Jain, A.: Fundamentals of Digital Image Processing, Chap. 7, p. 235 (1989)
36. Otsu, N.: A threshold selection method from gray-level histograms. IEEE Trans. Syst. Man Cybern. **9**(1), 62–66 (1979)
37. Shapiro, L., Stockman, G.: Computer Vision (2002)
38. Pratt, W.K.: Digital Image Processing (1991)
39. Dougherty, E.: An Introduction to Morphological Image Processing (1992)
40. Haralick, T.M., Shapiro, L.G.: Computer and Robot Vision, vol. 1 (1992)
41. Template Matching - OpenCV 2.4.8.0 documentation, http://docs.opencv.org

Effective Lip Prints Preprocessing and Matching Methods

Krzysztof Wrobel, Piotr Porwik and Rafal Doroz

Abstract This paper presents a method of recognition of human lips. It can be treated as a new kind of biometric measure. During image preprocessing, the features are extracted from the lip print image. In the same step, image is denoised and normalized and ROI is determined. In the next stage, the normalized cross-correlation method was applied to estimation of the biometric parameters EER, FAR, and FRR. Investigations were conducted on 120 lip print images. These images come from University of Silesia public repository http://biometrics.us.edu.pl. The results obtained are very promising and suggest that the proposed recognition method can be introduced into professional forensic identification systems.

Keywords Biometrics · Lip print · Image preprocessing · Normalized cross-correlation

1 Introduction

Today, many anatomical or behavioral traits can be used in verification or identification of person. These traits can be revealed in specialized chemical laboratories only. Chemical markers allow also to analyzed blood, semen, etc. Other kind of

K. Wrobel (✉) · P. Porwik · R. Doroz
Institute of Computer Science, University of Silesia, ul. Bedzinska 39, 41-200 Sosnowiec, Poland
e-mail: krzysztof.wrobel@us.edu.pl
URL: http://zsk.tech.us.edu.pl

P. Porwik
e-mail: piotr.porwik@us.edu.pl
URL: http://biometrics.us.edu.pl

R. Doroz
e-mail: rafal.doroz@us.edu.pl

R. Burduk et al. (eds.), *Proceedings of the 9th International Conference on Computer Recognition Systems CORES 2015*, Advances in Intelligent Systems and Computing 403, DOI 10.1007/978-3-319-26227-7_33

traits can be observed by using the image processing technique—for example, pistol bullets with their ballistic traits or human faces can be recognized during specialized image processing. In this paper, only visible imprints will be considered. To develop latent or faintly visible imprints, chemical or luminescent methods can be applied. If features are properly captured and denoised, then features contain useful data leading to personal identification. Lip prints are extremely flexible, therefore capturing them is often incomplete, of low quality, or deformed. Additionally, lip prints are left on different kind of surfaces. Therefore in practice, capturing and reliable analysis of such traits is difficult. The forensic human lip prints investigation is named cheiloscopy [4, 10], where furrows (lines, bifurcations bridges, dots, crossings, and other marks) on the red part of human lips are searched. Lip print images are left on many materials such as cups, glasses, or even envelopes. According to some commentators, lip-based biometric systems, based on image analysis techniques, cannot be used due to troubles with extraction of topological features from the images [10]. From these opinions, only geometrical features, such as lip shape for example, can be taken into account and the comparison of the lip images can only be conducted by means of the shapes measurements. This is not true. In this paper, it will be shown that important lip print features can be directly extracted from images. It should be noticed that also other biometrics are intensively developed [5–7]. Even though lip print analysis is not substantially developed, this new form of identification is slowly becoming accepted and being introduced into practice all over the world. Nowadays, technique of lip print analysis is used for human identification, for example during postmortem identification [11]. Additionally, the popularity of lip print sensors is not yet high and investigations in this area are still being developed. It should be mentioned that a literature review shows that there is small number of works in which lip print technology (acquisition, automatic capture, analysis, and recognition) have been described and practically utilized. From characteristics of lip imprints, it follows that the analysis of such images can bring extensive benefits in many domains. Lip prints can, in the first place, be photographed and their shape parameters and different dimensions measured and compared. From color images, the red part of the lips can be extracted via a color space separation and geometric lip parameters can then be measured. In practice, lip prints are often captured fragmentarily, in this case geometrical-based lip print recognition is impossible [12]. In the solution preferred by forensic laboratories, lip imprints are transferred to special materials (white paper, a special cream, and a magnetic powder). Here, in the first step, the lip print is rendered onto a durable surface using a magnetic fingerprint powder (a different type of powder can be used depending on the surface type). To record lip prints using the magna brush method, the person should impress lips against a smooth nonporous surface. The image thus developed, is then converted into a digital image by a scanner. It should be noticed that related works are difficult to find because proposed approach was initialized by the authors.

2 Database Preparing

Investigations presented in this paper were carried out on the basis of the lip print images repository. This repository contains 120 lip print images. All images were captured in the resolution of 300 dpi. Captured lip prints come from different individuals, adult both men and women. All images were captured by means of the same technique. For the proper preservation, lip imprints should be carefully captured and then copied to appropriate materials by means of a special cream and a magnetic powder. This procedure allows for the fixation of an image. The process of capturing an image of the lips starts from one corner of the mouth and ends at the opposite site of the mouth, all details of the lips being thus copied onto a paper surface. In successful prints, the lines on the lips must be recognizable, not smudged, neither too light nor too dark, and the entire upper and lower lip must be visible. In the next step, this trace needs to be digitalized because a digital lip print preparation is necessary for the subsequent techniques. Details of this technique can be found in [12, 13]. Repository is publicly available with free access. Resources of repository are maintained by Computer Systems Department, University of Silesia, Katowice, Poland: http://biometrics.us.edu.pl.

3 Lip Print Image Preprocessing

Digital image processing can automatically detect furrows, grooves, and wrinkles like patterns. In the literature, this process is often referred to as ridge–valley structure extraction [2]. Many types of images contain similar structures (e.g., medical images of veins and vessels, images of fingerprints, etc.) and the common goal is to extract or detect these patterns in order to separate them from the rest of the image. An overview of different pattern detection techniques can be found in [10]. These include different types of thresholding, the use of multi-oriented filters (such as Gabor filters), and a plethora of morphological transforms. Lip imprints have often many artifacts: fragments of skin, hairs, and others, all of which generate unwanted noise. For this reason, the lip prints image should be improved and all of its artifacts be removed. Lip prints are extremely flexible, so they may vary in appearance according to the pressure, direction, and method used in making the print. From this point of view, selection and extraction of important lip print features is a challenge.

3.1 Detection of the Lip Area

In the first stage, lip furrows and wrinkles should be extracted from the image background. This task can be performed in the binarization process. Binarization threshold is automatically established by well-known iterative Ridler–Calvard algorithm [8].

This algorithm gives a good binarization results for images with low contrast [14]. The Ridler–Calvard algorithm automatically analyzes average intensities of the background and foreground pixels as long as binarization threshold will be stable. Lip print area should be extracted from the lip print image. It is an important region of interest—ROI. Contours of this area are disclosed in the erosion and binarization procedures. After binarization all black pixels form the ROI—lip print area. Figure 1 presents three images. The first of them depicts initial image, the second shows Ridlers–Calvards binarization results, whereas the third presents image after erosion. Image (Fig. 1c) is still noised and contains many artifacts, which should be removed. These artifacts can be wrongly classified as the lip print areas. Figure 2 presents example magnified area of the artifacts of the binarized image from Fig. 1c. All these artifacts form so-called pixel regions. Regions are nonoverlapped clusters of pixels and consist of black color pixels only. All pixels inside of a given region have to share at least one boundary pixel, except single pixels which are automatically treated as separate regions. Image regions can be automatically indicated in computer programs. For example, in the MATLAB environment or OpenCV library in C++ there are functions which can determine regions in an analyzed image. In the next step, number of black pixels inside of the every region should be fixed. It allows to build classifier which will remove regions, which do not belong to the lip print image. Size of the region is an important factor. Removing very large regions of pixels will destroy structure of original lip prints, while removing very small regions lead to many artifacts that will be still visible. In this paper, we propose a new method

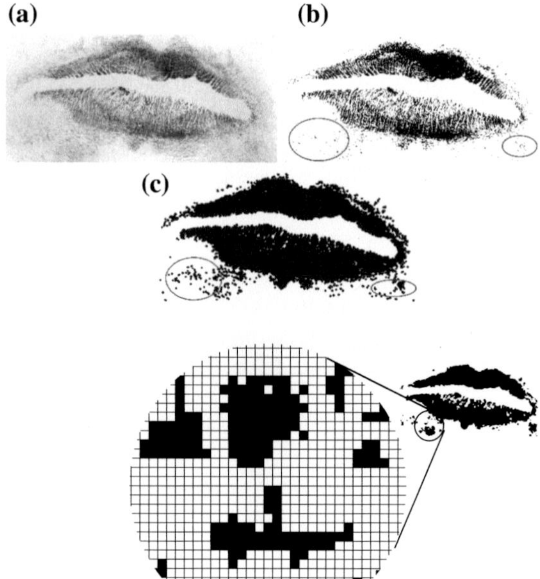

Fig. 1 Originally captured lip print image (**a**), the same lip print with artifacts after binarization (**b**) and image after erosion (**c**)

Fig. 2 Separated regions of pixels belonging to background of the lip print image

of region size determination. Let image domain be defined as $G = [0, 1] \times [0, 1]$. On the other hand, image is discretizing in the $X \times Y$ pixels grid. Hence, domain of image I consists of numbers belonging to the set $\mathbb{N}$, transformed by some function $F : G \rightarrow I$. More precisely the grid is a quantum rectangle of size $1/X \times 1/Y$. From this moment, image domain is defined as a grid of pixels, I has dimension of $X \times Y$, and p denotes pixel in the grid point. Size of the region R_i is defined as follows:

$$Size(R_i) = \iint_{R_i \in I} R_i \, dxdy \triangleq \sum_{p \in R_i} p, \tag{1}$$

where: $R_i \in \{R_1, \ldots, R_n\}$ is a given pixel-based region, and n is the number of detected regions. The number of regions is automatically established. For simplification, instead of pixel-based region, the name "region" will be used. Regions $R_i, i = 1, \ldots, n$ which not belonging to the lip print area should be removed. It can be done by appropriate inspection of the regions. Let histogram H count the regions with the number of pixels belonging to the defined ranges (so-called bins). Let the number of regions in the jth bin be denoted as $H(j)$. Let histogram H consist of m bins. It means that histogram H shows $H(j), j = 1, \ldots, m$ independent values. It is depicted in Fig. 3. For this assumption, we can determine the minimal number of pixels in the region as

$$T = \arg\min_j \{H(j) : j = 1, \ldots, m\} \cdot \frac{\max\{Size(R_i) : i = 1, \ldots, n\}}{m}. \tag{2}$$

During regions inspection, the value T is always checked. If number of pixels which belongs to the given region R_i is less than T (what means that $Size(R_i) < T$), then the region R_i is completely removed from the binarized lip print image. After removing all inappropriate regions of pixels, denoised lip print image is formed. It depicts Fig. 4.

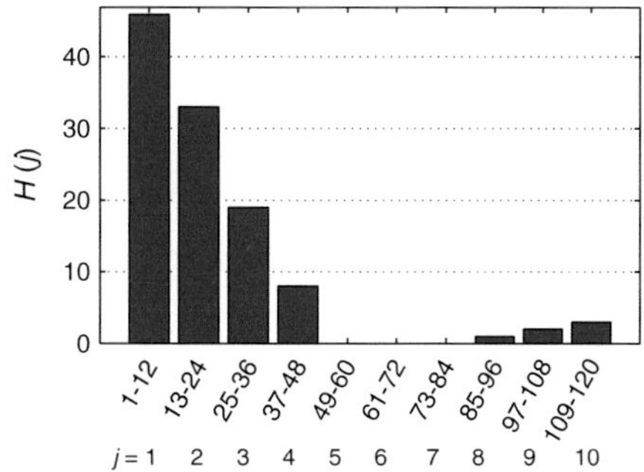

Fig. 3 Number of regions in the form of histograms

Fig. 4 Lip print image after removing all regions which do not belong to the lip print area—the ROI area

4 Lip Image Analysis Inside of the ROI

In previous sections we had to deal with preliminary preparation of the image, where different image processing techniques were applied. It allowed determining the ROI on the lip print image. From the ROI, all artifacts and noise are removed or at least significantly reduced. The ROI includes important individual lip print features, which have to be isolated and enhanced. Further, on the basis of isolated features, individuals will be recognized in the biometric approach. Isolation of distinctive features on the surface of lip requires using another image processing techniques.

4.1 Lip Features Enhancement and Improvement

In the first stage, lip print image (Fig. 1a) inside of ROI (Fig. 4) is enhanced. This process aims to improve the quality level of the furrows and wrinkles forming the lip pattern. The image ROI enhancement is realized as a convolution of ROI with special designed filters. Dimension of these filters should be always selected with respect to source image resolution. The filters enhance the image but together with enhancement some additional artifacts will be also created. It is a disadvantageous phenomenon. For this reason, in the comparative studies, the optimal dimension of the filters was established. All filters have the same dimension of $N \times N$ pixels. The value N is always an odd number. The best isolation of furrows, wrinkles, and grooves in the ROI provides filters depicted on Fig. 5. Here, only part filters are presented. Elements of each filter consist of binary values, 0 or 1. The white squares (elements) of the filter have the value of 0. Other elements have the binary value of 1.

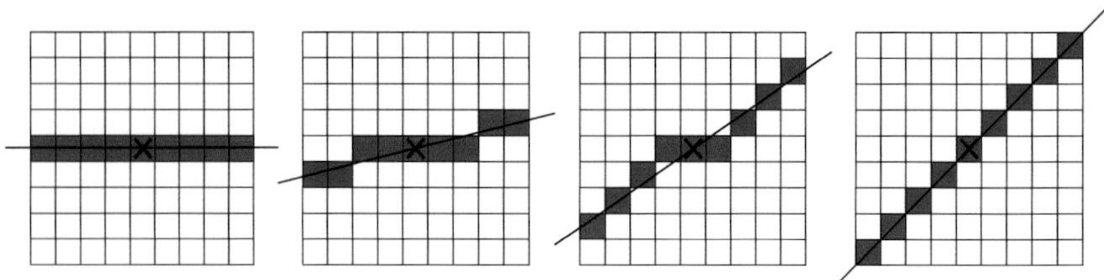

Fig. 5 Filters used in the lip print image enhancement procedure with central points of filter

Active elements of the filter reflect the most important directions of furrows, gloves, and wrinkles on the human lip surface. Figure 5 presents example of four optimal filters of dimension 9×9. It should be noted that only small part of these filters have been presented. In practice, $L = 24$ various filters have been formed. Within the ROI (Fig. 4), the filters traverse the image from its upper left corner, being applied to each pixels of the lip image. It should be noted that elements of the mask are not placed outside the source lip print image I. To calculate the value of the central point of the mask, here marked with a cross, the values of the pixels of the original image I covered by the mask are convolved with the mask's elements. Let filters be denoted as S^J, where $J = 1, \ldots, L$ and $S^J(a, b) = \{0, 1\}$. Then, for each position of the filter S^J (see Fig. 5), the value of its central pixel can be calculated as

$$p^J(x, y) = \sum_{a=1}^{N} \sum_{b=1}^{N} I(x + a - k, y + b - k) \cdot S^J(a, b), \quad x, y \in \text{ROI}, \tag{3}$$

where $k = \lceil \frac{N}{2} \rceil$, $N = 9$. Area of the filter S^J must always be located entirely inside the image I, hence the procedure (3) begins at the point $x = y = \lceil \frac{N}{2} \rceil + 1$ of the image, hence:

$$I(x, y) = round(N^{-1} \cdot \max\{p^J(x, y) : x, y \in \text{ROI}\}). \tag{4}$$

The convolution procedure (4) is repeated for each pixel of the image I over the range $x \in [\lceil \frac{N}{2} \rceil + 1, W - \lceil \frac{N}{2} \rceil + 1]$ and $y \in [\lceil \frac{N}{2} \rceil + 1, H - \lceil \frac{N}{2} \rceil + 1]$, where W and H represent dimension in pixels of the ROI.

4.2 Lip Image Features Extraction Inside of the ROI

In the previous section, the most important lip print features (furrows, wrinkles, and grooves) were enhanced in the ROI of original image. Unfortunately, apart from lip features, some details are also disclosed. These details are part of the features but their detailed analysis is not needed, because the main lip patterns have already been disclosed. Unessential details in the ROI will be removed in the two steps procedure. In the first step, Sauvola's binarization algorithm is performed [9]. After binarization the ROI includes well visible characteristic of furrows and wrinkles but in this area there are some pixels clusters which are artifacts. It presents Fig. 6a. These clusters have to be localized and then removed. It is done in the second step. To achieve this goal, the procedure described in the Sect. 3.1 is applied again. From the formulas (1) and (2) inappropriate pools (regions) of pixels are removed from the ROI. Finally, lip print image area with principal lip features is formed, what presents Fig. 6b.

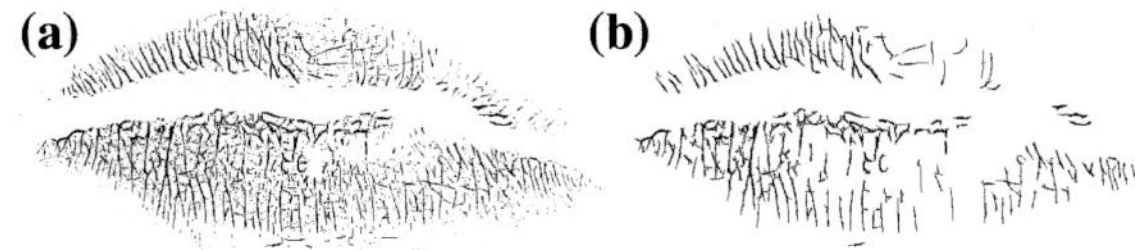

Fig. 6 The ROI after Sauvola's binarization (**a**). The same ROI after removing inappropriate pools of pixels (**b**)

5 Similarity Determination

Investigations presented in this paper were carried out on the basis of the lip print images repository. This repository contains 120 lip print images captured from 30 individuals (4 imprints per one person). All images were captured by means of the same technique. From the every image the fragment of $P \times P \in$ ROI pixels has been extracted. Additionally, these fragments have been randomly rotated in the range $\pm 360°$. In the images comparison, the normalized cross-correlation (NCC) method has been applied [1], where matching function has been calculated as

$$\gamma(u, v) = \frac{\sum_{x,y} [I(x, y) - \bar{I}_{u,v}][g(x - u), g(y - v) - \bar{g}]}{\left\{\sum_{x,y} [I(x, y) - \bar{I}_{u,v}]^2 \sum_{x,y} [g(x - u), g(y - v) - \bar{g}]^2\right\}^{0.5}}, \quad (5)$$

where: I is a template print image with biometric features (see Fig. 6b), g is a sub-image consists of $P \times P$ pixels extracted from the template image I. Finally, NCC value γ as total similarity between two images I and g is computed as

$$\gamma = \max\{\gamma(u, v) : u = 1, \ldots, X,\ v = 1, \ldots, Y\}. \quad (6)$$

For each point (u, v), value of the function γ is stored in the matrix M. Similarity sim between images I and g is then calculated from various formulas:

$$sim(I_k, g_i) = \{MAX(M) \vee MIN(M) \vee AVG(M)\}, \quad (7)$$

where:

$$w_{\max} = \max\{M[a, b] : b = 1, \ldots, 4\},\ a = 1, \ldots, 30, \quad (8)$$

$$MAX(M) = \max(w_{\max} : a = 1, \ldots, 30\}, \quad (9)$$

$$w_{\min} = \min\{M[a, b] : b = 1, \ldots, 4\},\ a = 1, \ldots, 30, \quad (10)$$

$$MIN(M) = \max(w_{\min} : a = 1, \ldots, 30\}, \quad (11)$$

$$w_{avg} = avg\{M[a, b] : b = 1, \ldots, 4\},\ a = 1, \ldots, 30, \quad (12)$$

$$AVG(M) = \max(w_{avg} : a = 1, \ldots, 30\}, \tag{13}$$

and $k = 1, \ldots, 120$, $i = 1, \ldots, 120$. As mentioned above, the sub-images was randomly rotated, therefore images I were also rotated in range of $\pm 360°$ with step of $\pm 1°$. It allows to match the sub-images to the images I. Images of lip prints are stored in the database. The database contains lip print images of 30 individuals, 4 imprints per individual. For every sub-image g of size $P \times P$, similarities $sim(I_k, g_i)$ are computed and stored in the matrix M.

6 The Results Obtained

For the values $MAX(M)$, $MIN(M)$, and $AVG(M)$ the equal error rate (EER) was established. Figure 7a presents various EER values for different size of sub-images g. From investigation follows that optimal size of sub-images g is 90×90. This image size will be used in the following experiments. During investigations carried out, the FAR and FRR curves for different recognition threshold levels were checked. Dependencies between FAR and FRR factor for different threshold level presents Fig. 7b. Table 1 summarizes the results and recognition factors obtained for optimal size 90×90 pixels of sub-images g, what follows from the observation of the Fig. 7a.

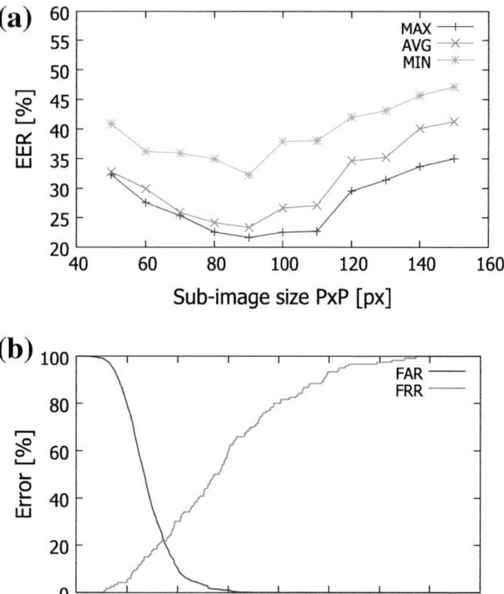

Fig. 7 The EER values in dependency on size of sub-image g (**a**). FAR/FRR curves for different recognition level (**b**)

Table 1 Performance of the lip print recognition system

True Positive (TP)	108
True Negative (TN)	3132
False Positive (FP)	12
False Negative (FN)	232
EER (%)	21.63
Sensitivity (SE) (%)	31
Specificity (SP) (%)	99
Accuracy (ACC) (%)	93

7 Conclusions

The investigations performed demonstrate the validity of our results. Our studies make it clear that if lip features are appropriately captured, the imprints may become an important tool of biometrics. As was demonstrated, the method developed here is similar to that used in the case of fingerprints. For this reason, no special equipment is needed for such investigations. It is well known that as the database of biometric features increases, then the recognition performance generally decreases [3]. In our approach, recognition performance is stable for up to 120 database records, but in the future, additional investigations with larger databases will also be programmed. The lip biometrics proposed in this paper could also be used as an additional method to enhance the effectiveness of other well-known biometrics through the implementation of multimodal systems, including fusion or hybrid systems.

References

1. Briechle, K., Hanebeck, U.D.: Template matching using fast normalized cross-correlation. In: Proceedings of SPIE, Aero-Sense Symposium, vol. 4387. Orlando, Florida (2001)
2. Dougherty, E.: An Introduction to Morphological Image Processing (1992)
3. Johnson, A.Y., Sun, J., Bobick, A.F.: Using similarity scores from a small gallery to estimate recognition performance for larger galleries. In: IEEE International Workshop on analysis and modeling of faces and gestures, AMFG2003, pp. 100–103 (2003)
4. Kasprzak, J., Leczynska, B.: Cheiloscopy. Human Identification on the Basis of a Lip Trace (in Polish) (2001)
5. Kudlacik, P., Porwik, P.: A new approach to signature recognition using the fuzzy method. Pattern Anal. Appl. **17**(3), 451–463 (2014)
6. Porwik, P., Doroz, R., Orczyk, T.: The k-NN classifier and self-adaptive Hotelling data reduction technique in handwritten signatures recognition. Pattern Anal. Appl. 1–19. doi:10.1007/s10044-014-0419-1
7. Porwik, P., Orczyk, T.: DTW and voting-based lip print recognition system. In: Cortesi, A., Chaki, N., Saeed, K., Wierzchoń, S. (eds.) Computer Information Systems and Industrial Management. LNCS, vol. 7564, pp. 191–202. Springer, Heidelberg (2012)
8. Ridler, T.W., Calvard, S.: Picture thresholding using an iterative selection method. IEEE Trans. Syst. Man Cybern. SMC-8, 630–632 (1978)

9. Sauvola, J., Pietikainen, M.: Adaptive document image binarization. Pattern Recognit. **33**, 225–236 (2000)
10. Siegel, J., et al.: Encyclopedia of Forensic Science, pp. 358–362 (2000)
11. Utsuno, H., et al.: Preliminary study of post mortem identification using lip prints. Forensic Sci. Int. **149**(23), 129–132 (2005)
12. Wrobel, K., Doroz, R., Palys, M.: A method of lip print recognition based on sections comparison. In: Proceedings of 2013 IEEE International Conference on Biometrics and Kansei Engineering (ICBAKE 2013), pp. 47–52. Tokyo, Japan (2013)
13. Wrobel, K., Doroz, R., Palys, M.: Lip print recognition method using bifurcations analysis. In: Nguyen, N.T., Trawiński, B., Kosala, R. (eds.) Intelligent Information and Database Systems. LNCS, vol. 9012, pp. 72–81. Springer, Heidelberg (2015)
14. Yang, H., Kot, A.C., Jiang, X.: Binarization of low-quality barcode images captured by mobile phones using local window of adaptive location and size. IEEE Trans. Image Process. **21**(1), 418–425 (2012)

Local Texture Pattern Selection for Efficient Face Recognition and Tracking

Maciej Smiatacz and Jacek Rumiński

Abstract This paper describes the research aimed at finding the optimal configuration of the face recognition algorithm based on local texture descriptors (binary and ternary patterns). Since the identification module was supposed to be a part of the face tracking system developed for interactive wearable computer, proper feature selection, allowing for real-time operation, became particularly important. Our experiments showed that it is unfeasible to reduce the computational complexity of the process by choosing discriminant regions of interest on the basis of the training set. The application of simulated annealing, however, to the selection of the most discriminant LTP codes provided satisfactory results.

Keywords Face recognition · Feature selection · Local binary patterns

1 Introduction

Face recognition is a general term for the wide range of problems that involve some kind of face image classification. In fact, face *verification* (that we deal with when a user claims some identity that must be confirmed by the system in order to grant the access to some protected resource) requires substantially different approach than the typical multi-class recognition (where the goal of the algorithm is to pick one of the known class labels). Face tracking is yet another scenario, in which temporal

This work was funded in part by NCBiR, FWF, SNSF, ANR, and FNR in the framework of the ERA-NET CHIST-ERA II, project *eGlasses—The interactive eyeglasses for mobile, perceptual computing*, and by statutory funds of the Faculty of Electronics, Telecommunications and Informatics, Gdańsk University of Technology.

M. Smiatacz (✉) · J. Rumiński
Faculty of Electronics, Telecommunications and Informatics,
Gdańsk University of Technology, Gdańsk, Poland
e-mail: maciej.smiatacz@pg.gda.pl

J. Rumiński
e-mail: jacek.ruminski@pg.gda.pl

R. Burduk et al. (eds.), *Proceedings of the 9th International Conference on Computer Recognition Systems CORES 2015*, Advances in Intelligent Systems and Computing 403, DOI 10.1007/978-3-319-26227-7_34

information plays particularly important role. In the previous work [1], we showed that the most popular methods of face recognition are hardly useful in the context of face verification. This time we change the perspective: our goal is to recognize and track faces of a limited number of people in a live video stream coming from a head-mounted camera. This is a part of a larger project, aimed at developing the *eGlasses*, a highly interactive wearable computer equipped with computer vision mechanisms [2, 3]. Such a device could be used, for example, by the medical staff at hospitals: by taking a look at a patient, the physician would be able to immediately access the data (presented on the see-through display) of the recognized person. In the rest of the paper, we will not address the issues of face tracking, concentrating instead on the problem of real-time identification of a relatively small number of persons. Based on our previous experience [4], we decided to use the local texture descriptors as face features and the simplest and nearest neighbor algorithm as the classifier. This straightforward configuration was chosen due to the limited computational resources of our mobile device; additionally, application of a more sophisticated classification method would make little sense in this case, because of the assumption that during the enrollment process only about 1–3 template images of each patient should be collected. In this circumstances, the simplification of the feature extraction process has become our main task. Since the local texture descriptors come in many versions, most of which are highly parameterized, finding the best option has also been particularly important. In the following sections, we will describe the concept of local binary/ternary patterns and known attempts to optimize their performance. Then we will present our preliminary experiments related to the LBP/LTP parameter selection, followed by the report on the application of simulated annealing and principal component analysis to the local texture feature extraction.

2 Background

Local binary pattern operator was proposed in [5]. It considers a small (3×3 in the basic case) neighborhood of each pixel, and performs the binarization in this area, using the intensity of the central point as the threshold value. The original 8-bit form (taking the values of eight neighbors into account) was defined as

$$LBP(x_c, y_c) = \sum_{n=0}^{7} 2^n s(i_n - i_c), \tag{1}$$

where (x_c, y_c) denotes the coordinates of the central point, i_c and i_n represent, respectively, the intensity of the central pixel and the intensity of its nth neighbor, while the $s(u)$ function returns 1 when $u \geq 0$ or 0 otherwise. Obviously, the original neighborhood can be extended: if the interpolation is applied, any number N of i_n values may be extracted from a circle of any radius r. The basic 3×3 window can be treated as the case where $N = 8$, $r = 1$ and the interpolation is switched off. The image is usually described by a set of histograms of LBP codes created for particular regions, which may be either uniformly distributed or located around some characteristic points. The so-called uniform codes, i.e., binary sequences containing at most one 0-1 and at most one 1-0 transition are said to be especially important, as they correspond to the characteristic structures such as edges or corners. Typically, the LBP histogram comprises 58 bins for uniform codes and one for all nonuniform ones. Having computed two sets of histograms for two images **A** and **B**, we can measure their similarity using the following formula

$$\chi^2(\mathbf{A}, \mathbf{B}) = \sum_{i,j} \frac{(a_{i,j} - b_{i,j})^2}{a_{i,j} + b_{i,j}}, \tag{2}$$

where $a_{i,j}$ denotes the value of the jth bin in the ith histogram describing image **A**. By introducing another parameter, t, and the modified $s(u)$ function,

$$s'(i_n, i_c, t) = \begin{cases} 1 & i_n \geq i_c + t \\ 0 & |i_n - i_c| < t \\ -1 & i_n \leq i_c - t \end{cases}, \tag{3}$$

local ternary patterns [6] can be defined. Here, the t parameter introduces some level of tolerance to the binarization process. Usually, each LTP code is converted into two LBP codes with the help of the following operations: (a) $-1 \rightarrow 0$, (b) $1 \rightarrow 0$, $-1 \rightarrow 1$. Consequently, the two "channels" are created, so that the LTP histogram is twice as big as the LBP version. The initial success of the LBP concept has led to the introduction of several other local texture descriptors. We ignore them, however, in this study, following the results of the recent comparison [7], indicating that in the case of face recognition, the differences between them are negligible. Still, our goal is to find the most efficient way to calculate the most discriminant set of features. Such attempts has been made before; the authors of [8], for example, instead of extracting the neighbors from the circle of a given radius, proposed to learn the irregular sampling pattern corresponding to arbitrary pixels within a small distance from the center. They used hill climbing algorithm and the Fisher class separability criterion, achieving some improvements over the state of the art algorithms. Authors of [9] suggested that the optimal LBP structure should be task-dependent and proposed a new method to learn discriminative LBP patterns, based on maximal joint mutual information criterion and binary quadratic programming. They were using, however, large numbers of training images—around 100 samples per class. In [10] the linear discriminant analysis was employed to learn the dominant patterns and to determine the optimal neighborhood sampling strategy that best differentiates face images. It remains unclear, however, how the within-class scatter matrices were constructed, i.e., how many training images per person were used. In the light of our experiments described below, the reported improvements seem highly disputable. The efforts most similar to ours were presented in [11], though in this case they were aimed at facial expression recognition, the task considerably different from

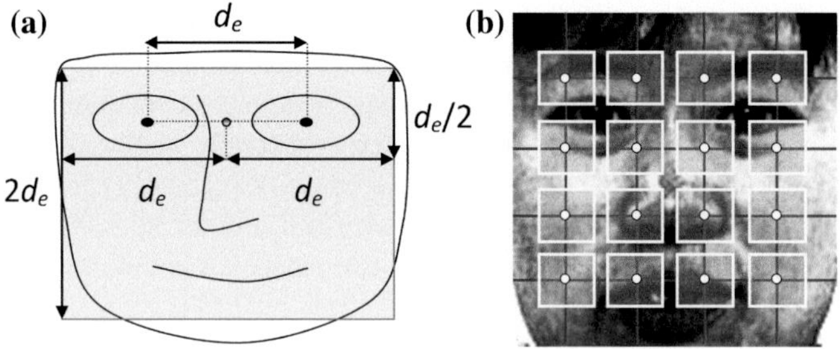

Fig. 1 **a** Image normalization based on eye center detection, **b** the regular grid for LBP/LTP histogram calculations

face identification. The authors used AdaBoost approach, training binary classifiers in the one-against-rest manner. They managed to determine the most discriminative set of LBP histogram bins together with the optimal collection of regions for which the histograms were constructed.

3 Preliminary Experiments

The first series of experiments was aimed at comparing different variants of LBP and LTP operators in the context of face recognition, meeting the following requirements: (1) the training set size is limited to 1 image per person, (2) the algorithm is fast enough to facilitate the real-time tracking on a wearable device. Out of the many data sets available [12], the CMU-PIE database [13] was chosen as a source of face

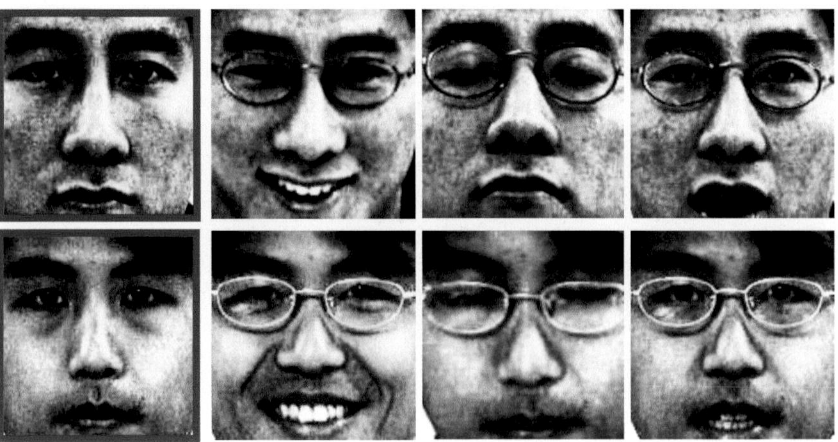

Fig. 2 The gallery images (framed) and selected testing samples from two classes

images. The training samples came from the *gallery* folder containing portraits of 68 persons. Since the local texture operators are not illumination and pose invariant and the current configuration of our system does not include any additional modules that could handle those variations properly, the testing set contained approximately frontal images captured under constant illumination conditions (1208 images taken by cameras no 07, 09, and 27 during the *expression* and *talking* sessions). The initial normalization consisted of histogram equalization and cropping based on the automatic detection of eye centers and the calculation of interocular distance d_e (Fig. 1a). The LBP/LTP histograms were extracted around the regularly distributed nodes (Fig. 1b) and then concatenated. The initial test (using basic configuration) showed that by changing the face image resolution from 32×32, through 64×64 to 128×128 the error rate can be reduced from 98.2 and 62.7 to 6.4 %, respectively. Therefore, we used the 128×128 images throughout the rest of experiments. Despite the very limited variations of lighting and pose, the differences between gallery and testing images were often significant (Fig. 2). During the first experiment only the basic 8-bit LBP codes were used, the resolution of the grid, however, was changing together with the window size. The results (Table 1) show that, in general, the error rate decreases when the image description becomes more detailed, i.e., when we use more nodes and smaller windows. On the other hand, increasing the grid resolution over 8×8 leads to marginal improvements, achieved at the cost of unacceptable computation times. Results obtained at lower resolutions seem to confirm the observation known from the literature, according to which the nonuniform codes should be ignored. This is not so obvious, however, when the grid gets more dense. Due to the lack of space we will present only the short summary of the rest of our experiments. All the tests were carried out using a PC equipped with Intel Core i7 920 2.67 MHz processor. (1) When the interpolation is applied, i.e., when the truly circular neighborhood is taken into account, results are better even in the case of the simplest (8-bit) LBP algorithm. Additionally, the nonuniform codes seem to be carrying some important information. For the 11×11 grid the error rate with all nonuniform codes represented by just one bin was equal to 5.88 %, while the introduction of the full-sized histogram reduced the error to 4.97 %. Interpolation, however, takes time (26 instead of 19 ms to calculate the 7139 features). (2) Nonuniform codes become even more important when the number of features increases (neighborhood contains 12 or 16 elements). Interestingly, 16-bit codes provide worse results. Moreover, increasing the size of histograms slows the algorithm radically: for the 16-bit case and 8×8 grid the full feature vector contains 4,194,304 elements and the identification of one face takes 4500 ms. (3) LTP performs better than LBP and it is usually unfavorable to neglect nonuniform LTP codes. For the 8-bit case with 8×8 grid and $t = 5$ the error rate was equal to 3.81 % without nonuniform codes, and 3.39 % with nonuniform codes included. The 12-bit variant of LTP provided even better results (3.06 % for 11×11 grid). (4) Although the rise of t value in (2) lowers the sensitivity of the algorithm, it seems that the choice of t is not critical. For the 12-bit LTP and 11×11 grid the best results were obtained for $t = 7$ (2.82 %). To sum up, we decided to concentrate on 12-bit LTP codes, calculated with interpolation for the 8×8 or 11×11 grid. Moreover, the preliminary experiments showed that the nonuniform codes should not be

Table 1 The results of experiments involving basic 8-bit LBP codes

Grid size	Window size	Nonuniform codes	Error rate (%)	Calculation time (ms/frame)	Vector length
2 × 2	63 × 63	all in one bin	31.46	6	236
		no	28.23	6	232
		yes	32.53	7	1024
4 × 4	31 × 31	all in one bin	10.51	6	944
		no	7.45	6	928
		yes	11.76	11	4096
8 × 8	15 × 15	all in one bin	6.37	11	3776
		no	6.37	11	3712
		yes	7.12	30	16,384
11 × 11	11 × 11	all in one bin	6.13	19	7139
		no	6.37	19	7018
		yes	6.04	49	30,976
14 × 14	9 × 9	all in one bin	5.96	30	11,564
		no	5.96	30	11,368
		yes	5.63	75	50,176

completely discarded, although some kind of feature selection would be helpful in the case of full-sized histograms (e.g., the 11 × 11 grid generates 991,232 elements and increases the processing time to 1092 ms).

4 Feature Selection

In the context of the face tracking application for wearable computer, the real-time operation of the recognition algorithm is essential. In our case, two approaches to the computational complexity reduction might be taken. (1) Selection of the useful LTP *codes*. If we were able to find a reasonably small subset of histogram bins (including those corresponding to nonuniform codes) the feature vectors could become considerably shorter (991,232 elements mentioned above is way too much). However, in this manner only the decision-making stage and some auxiliary operations are simplified. This is because all the histogram values are always calculated—we can decide not to process some of them but this does not influence the main loop of the algorithm that checks the neighborhoods of subsequent pixels. Still, our experiments show that the selection of codes is worth the effort: in the case of 11 × 11 grid and histograms containing 14,278 bins (only uniform 12-bit LTP codes) one frame was processed in 48 ms, while for full-sized histograms (61,952 bins) the processing time reached 97 ms—although in both situations all the codes had to be calculated. (2)

Selection of *regions* for which the LTP histograms are extracted. Potentially, this is the best way to reduce the computational complexity of the method. If we could identify just a few regions of interest, the main loop of the program would have to inspect much smaller number of neighborhoods. Both of the above approaches require some criterion driving the optimization process. Since we initially assumed that only one training (gallery) image per person would be available, the only option was to measure the dissimilarity of those images after the feature selection. To this end, the metrics defined by (2) was adopted. We decided to base our feature selection algorithm on the well-known simulated annealing (SA) procedure, according to which one can start with a random solution (feature subset in this case) and then accept new candidates with the probability defined as

$$P(E_C, E_N, T) = 1/\left(1 + \exp((E_N - E_C)/T)\right), \tag{4}$$

where E_C and is the so-called energy of the current solution, E_N is the energy of the new candidate, and T denotes the parameter (called a temperature), the value of which decreases throughout the process. Since the algorithm tries to minimize the energy of the solution, we can use the following definitions of the energy, based on the maximization of the distance between gallery images:

$$\text{a)}\quad E = \left(\frac{1}{M}\sum_{i=0}^{M}\sum_{j\neq i}\chi^2(\mathbf{I}_i, \mathbf{I}_j)\right)^{-1} \qquad \text{b)}\quad E = \left(\min_{i,j,i\neq j}\chi^2(\mathbf{I}_i, \mathbf{I}_j)\right)^{-1}, \tag{5}$$

where $\mathbf{I}_i$ is the ith sample from the training set containing M images. New candidate solutions are generated by changing one element of the feature subset at the time—i.e., one parameterized region of interest or one histogram bin. For some of the experiments, our training set was expanded so that each class was represented by three images instead of one. This allowed us to replace the χ^2 distances in (5) with the following class separability metrics

$$J_{p,q} = \frac{\chi^2\left(\boldsymbol{\mu}_p, \boldsymbol{\mu}_q\right)}{\sigma^2_{\chi p} + \sigma^2_{\chi q}}, \tag{6}$$

where $p, q = 1 \ldots L$ (L is the number of classes), $\boldsymbol{\mu}_p$ denotes the mean of class no. p, and $\sigma^2_{\chi p}$ is the variance of χ^2 distances in this class. Apart from the above feature *selection* methods one could also apply some feature *extraction* algorithm, such as the widely used principal component analysis (PCA), based on the calculation of eigenvectors and eigenvalues of the scatter matrix describing the whole training set [1]. In our case, however, one should not expect much from this approach: although the final feature vectors passed to the classification module may be much shorter than the original ones, all the full-sized histograms still have to be calculated, and the transformation itself (i.e., multiplying histograms by the matrix of eigenvectors) also takes time. Still, we decided to use the PCA as a reference method in our experiments.

5 LTP Optimization Experiments

The goal of the first part of our experiments (see Table 2) was to check if it is possible to select a relatively small set of image regions, for which the LTP histograms could be calculated efficiently without compromising the recognition accuracy. The solution containing 4320 features based on 12-bit LTP uniform codes calculated (with interpolation) over 4×4 grid was chosen as a reference point. In this case the SA algorithm was allowed to construct similarly sized feature vectors using a large number of options—it could pick up any window from 4×4, 8×8, 11×11 or 14×14 grid and then calculate 8-bit or 12-bit LTP histogram inside that region (with or without nonuniform codes), using t values ranging from 3 to 27. As we can see, the SA algorithm was unable to create a better feature vector than the reference one **although** the energy values were much lower for the solutions provided by SA. This means that the measures defined by (5) and (6) are irrelevant, especially in the case of so small size of the training set: better separation of the training samples does not guarantee lower recognition rates. We have also noticed that SA tended to select windows located near the **border** of the image, not around the prominent face features such as eyes. This behavior was logical since the selected regions usually covered the most discriminant and stable parts of the images: fragments of hair or elements of the background. The haircut remained unchanged across all the sessions represented in CMU-PIE database, while the background elements implicitly coded the shape differences between the faces. This led us to the conclusion that, despite the cropping illustrated in Fig. 2, during the first experiment the *classification was based on features related to the characteristics of the dataset, not the intrinsic face features*. Consequently, we changed the normalization procedure by reducing the width of the face region by 10 % and its height by 7.5 %. As expected, this operation generally increased the error rates but made us more confident that the algorithm was actually recognizing faces, not artifacts. Without feature selection, the best results for the reasonably sized (7552 element) feature vectors were obtained in the case of 8-bit LTP (with uniform codes only) and the 8×8 grid. This configuration achieved the error rate of 8.53 %. The best version of the SA-based algorithm still provided worse results for the same vector length, i.e., 9.02 %. Once again, the selected areas tended to concentrate along the borders of the image. This confirmed our supposition that in the context of texture-based face recognition it is impossible to select the most

Table 2 The results of experiments involving SA-based region selection

Method of feature vector construction	Error rate (%)
12-bit LTP, 4×4 grid (31×31 windows), uniform codes only, $t = 5$	4.14
SA, energy defined by (5a)	11.34
SA, energy defined by (5b)	7.04
SA, energy defined by (5a) with (6)—3 training images per class	11.84
SA, energy defined by (5b) with (6)—3 training images per class	6.3

Table 3 The results of experiments involving SA-based LTP code selection

Method of feature vector construction	Vector length	Error rate (%)
No selection, 118 uniform codes only	7552	8.53
80 bins selected out of 512, incl. nonuniform codes	5120	7.29
59 bins selected out of 512, incl. nonuniform codes	3776	7.37
30 bins selected out of 512, incl. nonuniform codes	1920	7.65

discriminant face regions with the help of the small-sized training set. Intuitively, we know that the fragments containing eyes or mouth carry important information about a face, but at the same time the texture around these features varies considerably while talking, blinking or changing the expression. This is why the algorithm looks for some more stable features, which, unfortunately, are often not present outside of the training set. The next part of our experiments was aimed at checking the possibility of selecting the discriminant LTP codes, not regions (see Table 3). This time the training set contained only one gallery image per person. The grid size was fixed at 8×8 (15×15 window size), 8-bit LTP was used. Initial tests showed that the most stable results were obtained when the energy of the solution was calculated as a weighted sum of the two criterions—the contribution of (5a) and (5b) was set to 0.25 and 0.75, respectively. The error rates shown in Table 3 were averaged over five cycles of training and testing on the same dataset. In this experiment, the relation between the energy (criterion value) and the error rate was noticeable. This means that the codes that provided better separation of the training samples were also better at discriminating between testing images, and the measure that we created can be used for feature selection. Thanks to the application of the simulated annealing we are now able to operate on feature vectors which are almost four times shorter than the original ones (1920 instead of 7552 elements) and still achieve better results (the error rate of 7.65 instead of 8.53 %). The last stage of experiments was related to the application of the standard method of feature extraction, i.e., the principal component analysis. Two versions of PCA were implemented: global one that calculated eigenvectors of the covariance matrix created from feature vectors describing whole images, and the region-based PCA, which calculated separate covariance matrices for the subsequent regions defined by the grid. The resultant error rates were poor (10.27 and 9.02 %, respectively), and the processing time became unacceptable (in the case of region-based PCA the classification of one image took almost 2 s).

6 Conclusions

Our experiments showed that in the case of face recognition based on local texture descriptors and a small training set the problem of automatic selection of the most discriminant regions of interest is ill posed. However, the methods such as simulated annealing can be successfully applied to select a small set of LTP codes that provide error rates which are lower than those obtained for full-sized LTP histograms. The introduction of the feature selection mechanism described in the paper to our face recognition system designed for wearable computer (*eGlasses*), allowed us to improve the performance of the face tracking module by reducing the computational complexity of the classification algorithm.

References

1. Smiatacz, M.: Eigenfaces, Fisherfaces, Laplacianfaces, Marginfaces—how to face the face verification task. In: Burduk, R., Jackowski, K., Kurzynski, M., Wozniak, M., Zolnierek, A. (eds.) CORES 2013. Advances in Intelligent Systems and Computing, vol. 226, pp. 187–196. Springer Int., Heidelberg (2013)
2. Ruminski, J., Bujnowski, A., Wtorek, J., Andrushevich, A., Biallas, M., Kistler, R.: Interactions with recognized objects. In: 7th International Conference on Human System Interactions (HSI), pp. 101–105. IEEE eXplore (2014)
3. Czuszynski, K., Ruminski, J.: Interaction with medical data using QR-codes. In: 7th International Conference on Human System Interactions (HSI), pp. 182–187. IEEE eXplore (2014)
4. Smiatacz M.: Face recognition: shape versus texture. In: Choraś, R.S. (ed.) IP&C 2014. Advances in Intelligent Systems and Computing, vol. 313, pp. 211–218. Springer Int. (2015)
5. Ojala, T., Pietikainen, M., Harwood, D.: A comparative study of texture measures with classification based on featured distributions. Pattern Recognit. **29**, 51–59 (1996)
6. Tan, X., Triggs, B.: Enhanced local texture feature sets for face recognition under difficult lighting conditions. IEEE Trans. Image Process. **19**, 1635–1650 (2011)
7. Bereta, M., Karczmarek, P., Pedrycz, W., Reformat, M.: Local descriptors in application to the aging problem in face recognition. Pattern Recognit. **46**, 2634–2646 (2013)
8. Maturana, D., Mery, D., Soto, A.: Learning discriminative local binary patterns for face recognition. In: IEEE International Conference on Automatic Face & Gesture Recognition, pp. 470–475 (2011)
9. Ren, J., Jiang, X., Yuan, J., Wang, G.: Optimizing LBP structure for visual recognition using binary quadratic programming. IEEE Signal Process. Lett. **21**, 1346–1350 (2014)
10. Lei, Z., Pietikainen, M., Li, S.Z.: Learning discriminant face descriptor. IEEE Trans. Pattern Anal. Mach. Intell. **36**, 289–302 (2014)
11. Shan, C., Gritti, T.: Learning discriminative lbp-histogram bins for facial expression recognition. In: Proceedings of British Machine Vision Conference, pp. 1–10 (2008)
12. Forczmanski, P., Furman, M.: Comparative analysis of benchmark datasets for face recognition algorithms verification. In: Bolc, L., Tadeusiewicz, R., Chmielewski, L.J., Wojciechowski, K. (eds.) ICCVG 2012. LNCS, vol. 7594, pp. 354–362. Springer, Heidelberg (2012)
13. Sim, T., Baker, S., Bsat, M.: The CMU pose, illumination, and expression (PIE) database. In: Proceedings of 5th International Conference on Automatic Face and Gesture Recognition (2002)

Part III
Data Stream Classification and Big Data Analytics

Online Extreme Entropy Machines for Streams Classification and Active Learning

Wojciech Marian Czarnecki and Jacek Tabor

Abstract When dealing with large evolving datasets one needs machine learning models able to adapt to the growing number of information. In particular, stream classification is a research topic where classifiers need an ability to rapidly change their solutions and behave stably after many changes in training set structure. In this paper we show how recently proposed Extreme Entropy Machine can be trained in an online fashion supporting not only adding/removing points to/from the model but even changing the size of the internal representation on demand. In particular we show how one can build a well-conditioned covariance estimator in an online scenario. All these operations are guaranteed to converge to the optimal solutions given by their offline counterparts.

Keywords Online learning · Extreme entropy machine · Stream classification · Covariance estimation

1 Introduction

Classic classification models are developed to perform batch-based learning where one is given a training set of pairs $\{(\mathbf{x}_i, t_i)\}_{i=1}^N$ and is supposed to build a concept of mapping $\mathbf{x}$ to t. There are many successful techniques including Support Vector Machines, Random Forests [3] or Extreme Learning Machines [10]. These three models achieved lots of attention due to the recent comprehensive evaluation of

W.M. Czarnecki (✉) · J. Tabor
Faculty of Mathematics and Computer Science, Jagiellonian University, Krakow, Poland
email: wojciech.czarnecki@uj.edu.pl
URL: http://gmum.net

J. Tabor
email: jacek.tabor@uj.edu.pl
URL: http://gmum.net

R. Burduk et al. (eds.), *Proceedings of the 9th International Conference on Computer Recognition Systems CORES 2015*, Advances in Intelligent Systems and Computing 403, DOI 10.1007/978-3-319-26227-7_35

multiple classification tools [8] showing their applicability in wide range of problems. However, for large, real life datasets it is preferable to be able to train model in an online fashion. Many models have their online versions, but many of them only roughly approximate batch equivalent or might even not converge at all. In this paper, we show online version of Extreme Entropy Machines [5], recently proposed robust, information theoretic, nonlinear classifier. We do not only show that it can be efficiently trained in an incremental fashion but is also able to forget any point and it is guaranteed to return the exact same model as its batch counterpart (up to numerical errors) which is a very rare feature for online learning methods. Most of the existing state-of-the-art online training only roughly approximate the batch solution [9, 13]. Paper is structured as follows. First, we briefly introduce Extreme Entropy Machines, then we focus on optimal covariance estimation under general asymptotics. Next, we provide algorithms for online training of both optimal estimator as well as EEM classifier. We finish with empirical evaluation on 10 datasets and discussion.

2 Extreme Entropy Machine

Extreme Learning Machines (ELM [10]) is a recent machine learning model which uses single feed forward learning network without tuning the hidden layer's weights. This concept is present in machine learning society for quite a long time, spanning from approximation techniques [7], through random projections techniques to methods based on random subspaces and beyond [3, 5]. As the result, ELM can be trained very fast and the resulting implementation is trivial. Furthermore, they have closed-form solution of the training procedure. Recently, we proposed Extreme Entropy Machines (EEM [5]) which combines ELM's random, nonlinear projections with Renyi's quadratic cross entropy optimization used in Multithreshold Entropy Linear Classifier [6] to form an information theoretical, robust, nonlinear classifier.

Optimization problem: **Extreme Entropy Machine**

$$\begin{aligned}
\underset{\beta}{\text{minimize}} \quad & \boldsymbol{\beta}^T \boldsymbol{\Sigma}^+ \boldsymbol{\beta} + \boldsymbol{\beta}^T \boldsymbol{\Sigma}^- \boldsymbol{\beta} \\
\text{subject to} \quad & \boldsymbol{\beta}^T (\mathbf{m}^+ - \mathbf{m}^-) = 2 \\
\text{where} \quad & \boldsymbol{\Sigma}^\pm = \text{cov}_\dagger(\mathbf{H}^\pm) \\
& \mathbf{m}^\pm = \text{mean}(\mathbf{H}^\pm) \\
& \mathbf{H}^\pm = \varphi(\mathbf{X}^\pm)
\end{aligned}$$

It appears [5] that up to scaling factor the solution is given by

$$\boldsymbol{\beta} = (\boldsymbol{\Sigma}^+ + \boldsymbol{\Sigma}^-)^{-1}(\mathbf{m}^+ - \mathbf{m}^-), \tag{1}$$

where $\mathbf{\Sigma}^{\pm}$ is a covariance estimator for each class projection through some random φ. The classification of the new point can be done using

$$cl(\mathbf{x}) = \operatorname*{argmax}_{t \in \{+,-\}} \mathcal{N}(\boldsymbol{\beta}^T \mathbf{m}^t, \boldsymbol{\beta}^T \mathbf{\Sigma}^t \boldsymbol{\beta})[\boldsymbol{\beta}^T \varphi(\mathbf{x})].$$

There appears this nontrivial question as how to estimate this covariance from small sample size and in the highly (possibly infinite) dimensional Hilbert space. One possible definition of the optimal covariance estimator $\mathbf{\Sigma}_{\dagger}$ is the one minimizing the L^2 norm of the difference between estimator and true covariance $\mathbf{\Sigma}_{\mathbf{H}}$ of distribution from which $\mathbf{H}$ has been sampled, formally

$$\mathbf{\Sigma}_{\dagger} = \min_{\mathbf{\Sigma}} \mathbb{E}[\|\mathbf{\Sigma} - \mathbf{\Sigma}_{\mathbf{H}}\|^2]. \tag{2}$$

This problem is highly non trivial for two main reasons, first $\mathbf{\Sigma}_{\mathbf{H}}$ is unknown, so the above problem seems unsolvable. Second, $\mathbf{H} \in \mathbb{R}^{N \times h}$ is relatively small and embedded in highly dimensional space, in other words $N < h$.

2.1 *Optimal Closed-Form Covariance Estimator*

Ledoit and Wolf have showed [14] that under general asymptotics, where both number of samples and number of dimensions grow to infinity (with an assumption that their ratio is constant) there is a closed-form equation for shrinkage coefficient for the estimator of the form

$$\text{cov}_{\dagger}(\mathbf{H}) = \alpha_1 \text{cov}(\mathbf{H}) + \alpha_2 \mathbf{I},$$

which solves the optimization problem 2. Let us assume that $\mathbf{H} \in \mathbb{R}^{N \times h}$ is centered and that its empirical covariance $\mathbf{\Sigma}_{\text{emp}} = \text{cov}(\mathbf{H})$ and[1] $\mathbf{\Sigma}_{\text{sqr}} = (\mathbf{H}^2)^T(\mathbf{H}^2)$. Then

$$\alpha_1 = 1 - \alpha \text{ and } \alpha_2 = \alpha\mu,$$

where

$$\mu = \tfrac{1}{h}\text{tr}(\mathbf{\Sigma}_{\text{emp}}) \text{ and } \alpha = \tfrac{\text{sum}(\mathbf{\Sigma}_{\text{sqr}} - (\mathbf{\Sigma}_{\text{emp}})^2)}{N\|\mathbf{\Sigma}_{\text{emp}} - \mu\mathbf{I}\|^2}.$$

As a result we get a closed-form covariance estimator which is well-conditioned. Recall that in Eq. 1 we need to invert the sum of two such estimators, but as each of them is positive definite also their sum is, as a result the inversion is always possible.[2]

---

[1] A^2 denotes element-wise squaring of A.

[2] Up to numerical errors.

2.2 Stream Learning of EEM

The above equations and model are designed to perform offline learning, where all the training instances are available at the initialization time. However in real life applications it is often not a case, one can have a stream of data [12] where new samples appear nearly in real time [11], an active learning loop [16], where samples are labeled by some oracle during the training procedure or one needs to adapt the model to changing environment (so called concept drift) [2]. For all these applications, one needs more adaptive strategies which should approximate the batch solution while trained one sample at a time. In order to create online EEM we need to be able to compute LW-shrinkage in the incremental fashion. For this reason we will store the estimates of $\mathbf{\Sigma}_{\text{emp}}$ and $\mathbf{\Sigma}_{\text{sqr}}$ for each class, and update it after each new point. This leads to the online formulas for Ledoit-Wolf covariance estimator which to authors best knowledge have not yet been provided nor used in any previous work. As our aim is to provide EEM with capabilities of online and window-based stream learning we need two operations:

- ADD POINT, which adds information about the given point into machine's memory,
- REMOVE POINT, which removes it from machine's structures.

We expect these operations to perform in $\mathcal{O}(h^2)$ (linear in terms of the size of the covariance estimators). Algorithm 1 shows implementation of these operations. Using basic algebra one can show that resulting estimator is equivalent to the one obtained in batch mode. Both are based on the update of the empirical covariance estimator (see $\mathcal{N}$- UPDATE operations) and update of the Ledoit Wolf shrinkage coefficient (see LW- UPDATE). As one can see the most expensive is the update of the empirical covariance matrix (line 9), which modifies h^2 elements and further shrinkage update which requires evaluation of code in line 3, which is also quadratic in terms of h. Notice that these operations complexity do not involve the size of the window nor size of the training samples. Consequently, if we assume models hyperparameters constant than these operations have complexity $\mathcal{O}(1)$ which makes them optimal solutions. Once we get new covariance estimators, our linear operator $\boldsymbol{\beta}$ is no longer a solution for our optimization problem 1 so we need to update it. There are two basic ways of performing this operation. One could simply solve Eq. 1 by inverting the covariance matrix which takes $\mathcal{O}(h^3)$ time. However, as $\boldsymbol{\beta}$ did not change much, iterative methods of solving systems of linear equations could be of use. We are facing here a problem of solving

$$(\mathbf{\Sigma}'^{+} + \mathbf{\Sigma}'^{-})\boldsymbol{\beta}' = \mathbf{m}'^{+} - \mathbf{m}'^{-}, \tag{3}$$

given $\boldsymbol{\beta}$ which solves

$$(\mathbf{\Sigma}^{+} + \mathbf{\Sigma}^{-})\boldsymbol{\beta} = \mathbf{m}^{+} - \mathbf{m}^{-}, \tag{4}$$

Algorithm 1 Stream learning operations, updating machine's memory

LW- UPDATE $(\mathbf{\Sigma}_{\text{emp}}, \mathbf{\Sigma}_{\text{sqr}}, N)$
  $\mu \leftarrow \frac{1}{h}\text{tr}(\mathbf{\Sigma}_{\text{emp}})$
  $\alpha \leftarrow (\text{sum}(\mathbf{\Sigma}_{\text{sqr}} - (\mathbf{\Sigma}_{\text{emp}})^2))/(N\|\mathbf{\Sigma}_{\text{emp}} - \mu\mathbf{I}\|^2)$
  return $\alpha\mu\mathbf{I} + (1-\alpha)\mathbf{\Sigma}_{\text{emp}}$

$\mathcal{N}$-UPDATE $(\mathbf{h}, \mathbf{\Sigma}, \mathbf{m}, N, \pm)$
  $\bar{\mathbf{m}} \leftarrow \frac{1}{N\pm 1}(N\mathbf{m} \pm \mathbf{h})$ {compute new mean}
  $\mathbf{\Sigma} \leftarrow \frac{1}{N\pm 1}((\mathbf{\Sigma} + \mathbf{m}\mathbf{m}^T)N \pm \mathbf{h}\mathbf{h}^T) - \bar{\mathbf{m}}\bar{\mathbf{m}}^T$ {update covariance estimate}
  return $\mathbf{\Sigma}, \bar{\mathbf{m}}, N \pm 1$

ADD POINT $(\mathbf{x}, t)$
  $\mathbf{h} \leftarrow \varphi(\mathbf{x})$ {project new point to Hilbert's space}
  $\bar{\mathbf{m}}^t \leftarrow \mathbf{m}^t_{\text{emp}}$
  $\mathbf{\Sigma}^t_{\text{emp}}, \mathbf{m}^t_{\text{emp}}, N^t_{\text{emp}} \leftarrow \mathcal{N}$- UPDATE $(\mathbf{h}, \mathbf{\Sigma}^t_{\text{emp}}, \mathbf{m}^t_{\text{emp}}, N^t_{\text{emp}}, +)$
  $\mathbf{\Sigma}^t_{\text{sqr}}, \mathbf{m}^t_{\text{sqr}}, N^t_{\text{sqr}} \leftarrow \mathcal{N}$- UPDATE $((\mathbf{h} - \bar{\mathbf{m}}^t)^2, \mathbf{\Sigma}^t_{\text{sqr}}, \mathbf{m}^t_{\text{sqr}}, N^t_{\text{sqr}}, +)$
  $\mathbf{\Sigma}^t \leftarrow$ LW- UPDATE$(\mathbf{\Sigma}^t_{\text{emp}}, \mathbf{\Sigma}^t_{\text{sqr}}, N^t_{\text{emp}})$
  $\mathbf{M}^t$.enqueue$(\mathbf{x})$ {add point to machine's memory}
  $\mathbf{i}$.enqueue(t) {add label to the memory index}

REMOVE POINT
  $t \leftarrow \mathbf{i}$.dequeue() {pop label from memory index}
  $\mathbf{h} \leftarrow \varphi(\mathbf{M}^t$.dequeue()) {project corresponding point from memory}
  $\mathbf{\Sigma}^t_{\text{emp}}, \mathbf{m}^t_{\text{emp}}, N^t_{\text{emp}} \leftarrow \mathcal{N}$- UPDATE $(\mathbf{h}, \mathbf{\Sigma}^t_{\text{emp}}, \mathbf{m}^t_{\text{emp}}, N^t_{\text{emp}}, -)$
  $\mathbf{\Sigma}^t_{\text{sqr}}, \mathbf{m}^t_{\text{sqr}}, N^t_{\text{sqr}} \leftarrow \mathcal{N}$- UPDATE $((\mathbf{h} - \bar{\mathbf{m}}^t)^2 + \bar{\mathbf{m}}^t, \mathbf{\Sigma}^t_{\text{sqr}}, \mathbf{m}^t_{\text{sqr}}, N^t_{\text{sqr}}, -)$
  $\mathbf{\Sigma}^t \leftarrow$ LW- UPDATE $(\mathbf{\Sigma}^t_{\text{emp}}, \mathbf{\Sigma}^t_{\text{sqr}}, N^t_{\text{emp}})$

knowing that $\mathbf{\Sigma}^{\pm} \approx \mathbf{\Sigma}'^{\pm}, \mathbf{m}^{\pm} \approx \mathbf{m}'^{\pm}$. As a result we can efficiently use conjugate gradients method [4] on Eq. 3 starting from the solution of Eq. 4. Given that we run K iterations, the complexity of such tuning is $\mathcal{O}(Kh^2)$ (obviously we choose $K < h$). As the number of hidden neurons (projections) is one of the main hyperparameters of the EEM model it would be desirable to be able to change its value during the online training procedure. First, if we train for a long time it might be reasonable to increase the size of the feature space so it is not too small for the new information present in the data. In the same time due to some memory limitations or decreasing sliding window size one might prefer to reduce the size of the feature space to prevent overfitting. Algorithm 2 shows methods which are able to add new projection as well as remove the old one. They both require estimators update and $\boldsymbol{\beta}$ tuning, which as shown before, can be done in $\mathcal{O}(Kh^2)$. The only additional cost comes from the fact that one needs to project each point in the window through the new projection, which is linear in terms of number of such points. As a result we are equipped with the robust, nonlinear classifier which can be trained in an iterative fashion, using online version of the optimal covariance estimator. We can add and remove points on demand as well as add and remove projections (size of the feature space), each

Algorithm 2 Memory size adapting, updating machine's representation

EXTEND MEAN AND COVARIANCE $(\mathbf{m}, \bar{\mathbf{m}}, \boldsymbol{\Sigma}, \mathbf{h}, \mathbf{H}, N)$

  return $[\mathbf{m}_1 \ldots \mathbf{m}_h\ \bar{\mathbf{m}}'], \begin{bmatrix} \boldsymbol{\Sigma} & \frac{1}{N}\mathbf{H}^T\mathbf{h} \\ \frac{1}{N}\bar{\mathbf{h}}^T\mathbf{H} & \frac{1}{N}\mathbf{h}^T\mathbf{h} \end{bmatrix}$

ADD PROJECTION $(\mathcal{G})$

  $\bar{\mathrm{G}} \sim \mathcal{G}$ {draw a random function from $\mathcal{G}$ family}

  $\varphi \leftarrow \lambda\mathbf{x} : [\mathrm{G}_1(\mathbf{x}) \ \ldots \ \mathrm{G}_h(\mathbf{x})\ \bar{\mathrm{G}}(\mathbf{x})]$ {update projection and Hilbert space}

  $\mathbf{H}_{\pm} \leftarrow \varphi(\mathbf{M}_{\pm}) - \mathbf{m}^{\pm}_{\mathrm{emp}}$ {project memory to the new space}

  $\mathbf{h}_{\pm} \leftarrow \bar{\mathrm{G}}(\mathbf{M}_{\pm})$ {get the projection through new function}

  $\bar{\mathbf{m}}_{\pm} \leftarrow \frac{1}{N_{\pm}} \sum_{a \in \mathbf{h}_{\pm}} a$ {compute mean in new dimension}

  $\mathbf{m}^{\pm}_{\mathrm{emp}}, \boldsymbol{\Sigma}^{\pm}_{\mathrm{emp}} \leftarrow$ EXTEND MEAN AND COVARIANCE $(\mathbf{m}^{\pm}_{\mathrm{emp}}, \boldsymbol{\Sigma}^{\pm}_{\mathrm{emp}}, \mathbf{h}_{\pm}, \mathbf{H}_{\pm}, N_{\pm})$

  $\mathbf{H}'_{\pm} \leftarrow (\mathbf{H}_{\pm})^2$ {get the projection through new function}

  $\mathbf{h}'_{\pm} \leftarrow (\mathbf{h}_{\pm} - \mathbf{m}^{\pm}_{\mathrm{emp}})^2$ {get the projection through new function}

  $\bar{\mathbf{m}}'_{\pm} \leftarrow \frac{1}{N_{\pm}} \sum_{a' \in \mathbf{h}'_{\pm}} a'$ {compute mean in new dimension}

  $\mathbf{m}^{\pm}_{\mathrm{sqr}}, \boldsymbol{\Sigma}^{\pm}_{\mathrm{sqr}} \leftarrow$ EXTEND MEAN AND COVARIANCE $(\mathbf{m}^{\pm}_{\mathrm{sqr}}, \boldsymbol{\Sigma}^{\pm}_{\mathrm{sqr}}, \mathbf{h}'_{\pm}, \mathbf{H}'_{\pm}, N_{\pm})$

  $h \leftarrow h + 1$ {increase dimensionality}

  $\boldsymbol{\Sigma}_{\pm} \leftarrow$ LW- UPDATE $(\boldsymbol{\Sigma}^{\pm}_{\mathrm{emp}}, \boldsymbol{\Sigma}^{\pm}_{\mathrm{sqr}}, N^{\pm}_{\mathrm{emp}})$

REDUCE MEAN AND COVARIANCE $(\mathbf{m}, \boldsymbol{\Sigma})$

  return $[\mathbf{m}_2 \ \ldots \ \mathbf{m}_h], [\boldsymbol{\Sigma}_{ij}]_{i \neq 1, j \neq 1}$ {remove dimension of covariance and mean}

REMOVE PROJECTION

  $\varphi \leftarrow \lambda\mathbf{x} : [\mathrm{G}_2(\mathbf{x}) \ \ldots \mathrm{G}_h(\mathbf{x})]$ {restrict projection to $h - 1$ dimensions}

  $\mathbf{m}^{\pm}_{\mathrm{emp}}, \boldsymbol{\Sigma}^{\pm}_{\mathrm{emp}} \leftarrow$ REDUCE MEAN AND COVARIANCE $(\mathbf{m}^{\pm}_{\mathrm{emp}}, \boldsymbol{\Sigma}^{\pm}_{\mathrm{emp}})$

  $\mathbf{m}^{\pm}_{\mathrm{sqr}}, \boldsymbol{\Sigma}^{\pm}_{\mathrm{sqr}} \leftarrow$ REDUCE MEAN AND COVARIANCE $(\mathbf{m}^{\pm}_{\mathrm{sqr}}, \boldsymbol{\Sigma}^{\pm}_{\mathrm{sqr}})$

  $\boldsymbol{\beta} \leftarrow [\boldsymbol{\beta}_2 \ \ldots \ \boldsymbol{\beta}_h]$

  $h \leftarrow h - 1$ {decrease dimensionality}

  $\boldsymbol{\Sigma}_{\pm} \leftarrow$ LW- UPDATE$(\boldsymbol{\Sigma}^{\pm}_{\mathrm{emp}}, \boldsymbol{\Sigma}^{\pm}_{\mathrm{sqr}}, N^{\pm}_{\mathrm{emp}})$

operation guaranteed to run in $\mathcal{O}(Kh^2)$ complexity (besides adding new projection requiring $\mathcal{O}(k + Kh^2)$ operations where k is window size). Table 1 sums up the exact cost of each operation with the list of updated internal OnlineEEM structures.

Table 1 List of all algorithms used for stream learning of EEM

Operation	Complexity	Updated structures	Algorithm	When to run
Add point $(\mathbf{x}, t)$	$\mathcal{O}(h^2)$	$\boldsymbol{\Sigma}^t_*, \mathbf{m}^t_*, N^t_*, \mathbf{M}^t, \mathbf{i}$	1	Each new point
Remove point $(\mathbf{x}, t)$	$\mathcal{O}(h^2)$	$\boldsymbol{\Sigma}^t_*, \mathbf{m}^t_*, N^t_*, \mathbf{M}^t, \mathbf{i}$	1	Each new point
Add projection	$\mathcal{O}(kh + h^2)$	$\boldsymbol{\Sigma}^{\pm}_*, \mathbf{m}^{\pm}_*, \varphi, h$	2	Window size increases
Remove projection	$\mathcal{O}(h^2)$	$\boldsymbol{\Sigma}^{\pm}_*, \mathbf{m}^{\pm}_*, \varphi, h$	2	Window size decreases
Retrain [exact]	$\mathcal{O}(h^3)$	$\boldsymbol{\beta}, \mathcal{N}_{\pm}$	[5]	Once each h new points
Retrain [approx.]	$\mathcal{O}(Kh^2)$	$\boldsymbol{\beta}, \mathcal{N}_{\pm}$	[4]	Each new point
Stream learning	$\mathcal{O}(Nh^2)$		1, 2, [4, 5]	
Batch learning	$\mathcal{O}(Nh^2)$		[5]	

3 Evaluation

Performed evaluation is based on 10 well known datasets from UCI repository [1] including in particular iris dataset reduced to binary problem (called iris-reduced). All experiments were performed using code written in python with use of scikit-learn [15]. We implemented online version of Weighted Extreme Learning Machines [17] as this is the most similar model to the proposed online Extreme Entropy Machines and to authors best knowledge there is no stream-based ELM nor truly online WELM (as adding new samples changes the weights which should be used in weighted least squares solution). In order to give ELM valid classes priors we informed it about exact priors before an experiment. This way WELM scores are overestimated, assuming that we do not know these prior values. EEM on the other hand is a balanced model thus does not need such information. We also compare with Mondrian Forest (MF [13]), which is a recently proposed online version of tree bagging (Random Forest, one of the best classifiers in the wide range of problems [8]), using that model's authors implementation.[3] In order to train MF in the window scenario we did retrain it at each iteration on the corresponding subset of examples. Each extreme method was using exact retraining (matrix inversion $\mathcal{O}(h^3)$) for the first 10 iterations and after that the approximate conjugate gradient solution. In order to measure quality of the learning process we focus on the area under the BAC[4] curve as a function of iteration number (denoted AUC-BAC). All results are mean values over 10-fold cross validation with randomized ordering. Each model has fitted hyperparameters, for WELM and EEM it is the number of hidden neurons ($h = 10, 20, 50, 100$) and for Mondrian Forest it is the number of trees ($h = 20, 50, 100, 200$) and lifetime parameter ($\lambda = \infty, 1000, 100, 10$). We consider separately four activation functions, namely sigmoid $\varphi(\langle w, \mathbf{x}\rangle - b) = (1 + \exp(-\langle w, \mathbf{x}\rangle + b))^{-1}$, RBF $\varphi(b\|\mathbf{x} - w\|^2) = \exp(-b\|\mathbf{x} - w\|^2)$, ReLU $\varphi(\langle w, \mathbf{x}\rangle - b) = \max\{0, \langle w, \mathbf{x}\rangle - b\}$ and leaky ReLU which is equal to ReLU for positive signals, and equal to $0.01 \cdot (\langle w, \mathbf{x}\rangle - b)$ for $\langle w, \mathbf{x}\rangle - b < 0$. We did not fit any other hyperparameters (as for example, additional regularization in WELM[5] and EEM) as in the stream/online scenario time is an important aspect. We wanted to test the efficiency of these models in the fastest, black-box scenarios. We first focus on the online learning scenario, here we start with an empty model and iteratively add points from each dataset. Top part of Table 2 summarizes the obtained AUC-BAC after the whole experiment. One can easily notice that in most experiments Online EEM behaves better than Online WELM. It is worth noting that these results are comparable (and often much better) than those obtained by Mondrian Forest. This not only confirms that online training of EEM is effective

[3] https://github.com/balajiln/mondrianforest.

[4] $\text{BAC} = \frac{1}{2}\left(\frac{\text{TP}}{\text{TP+FN}} + \frac{\text{TN}}{\text{TN+FP}}\right)$.

[5] We use Moore–Penrose pseudoinverse solution.

Table 2 AUC-BAC in online and stream scenarios using conjugate gradients

Online	EEM	WELM	EEM	WELM	EEM	WELM	EEM	WELM	MF
	leaky ReLU		RBF		ReLU		sigmoid		
Australian	**520**	483	**531**	491	**523**	483	**514**	510	521
Breast-cancer	**604**	522	**600**	595	**604**	539	**605**	**605**	603
Crashes	**369**	336	**398**	315	**367**	329	**363**	334	255
Diabetes	**541**	472	**537**	532	**540**	330	539	**548**	512
Fourclass	**758**	669	**737**	726	**758**	670	636	**731**	750
Heart	195	**197**	193	**203**	196	**197**	**210**	205	202
Ionosphere	**289**	278	215	**228**	**289**	277	**276**	249	286
Iris-reduced	**89**	88	**89**	88	**89**	88	88	**89**	85
Liver-disorders	176	**186**	**182**	**182**	176	**177**	178	**185**	**189**
Sonar	**147**	137	**122**	110	**148**	94	**135**	104	126
Stream									
Australian	**524**	447	**520**	488	**524**	454	**528**	514	522
Breast-cancer	**591**	524	**598**	572	**591**	527	**593**	587	584
Crashes	**331**	287	**305**	287	**331**	290	**350**	316	254
Diabetes	**495**	441	**499**	471	**495**	315	**498**	481	474
Fourclass	**733**	600	687	**733**	**734**	603	571	**738**	714
Heart	176	**183**	**180**	176	175	**183**	187	**189**	185
Ionosphere	**268**	253	210	**227**	**268**	255	**264**	235	**278**
Iris-reduced	**87**	86	**87**	86	**87**	86	**87**	**87**	86
Liver-disorders	**200**	193	183	**192**	**200**	165	189	**200**	200
Sonar	127	**130**	**113**	111	**127**	94	**129**	96	**141**

and stable but also that model itself is well suited for these kinds of problems. It appears that EEM gains biggest advantage when using strongest activation functions such as ReLU and leaky ReLU, while achieves significantly worse results using sigmoid. It might be valuable to further investigate this phenomenon, our current hypothesis is that the image of the projection through sigmoid activation function is less Gaussian than while using ReLU. One more interesting thing is an observation that although EEM performs better, WELM constructed simpler models (selected smaller number of hidden neurons). It is probably a consequence of regularizing impact of our covariance estimation which makes use of higher dimensional spaces possible. As opposed to our model WELM had to choose small number of hidden neurons in order to deal with overfitting. To further investigate this scenario we plot AUC scores for each iteration of the australian dataset in Fig. 1. One can notice that better results of EEM are the consequence of rapid learning of the classified concepts in first 50 iterations. Interestingly it seems that both EEM and ELM struggled to model ionosphere dataset using RBF activation function. Remaining projections behaved much better but one can notice that ELM appears to be much less stable. This is probably a consequence of the fact that ELM uses internally $\mathbf{H}^T\mathbf{H}$ equation which (up to removing the mean) is the empirical covariance of the whole dataset as opposed to well-conditioned covariance estimator of EEM. Second experiment involved training our models in stream fashion using sliding window technique of size 100. It simply means that for first 100 iterations models were trained in the online fashion and then in each iteration new point was added and the one leaving the sliding window was removed. Results, summarized in bottom part of Table 2, are very similar to those obtained for online scenario. One can easily notice that EEM achieves significantly better results than both WELM and MF in most datasets. In two cases (sonar and ionoshpere) it appears that using MF is a much better solution. We recall that MF has been retrained on each window so its results are overestimated [13], but this phenomenon should be further investigated as for online scenario MF did not achieve better scores. We did also perform tests checking whether Ledoit–Wolf estimator is truly needed. Results obtained using empirical covariance estimation were about 20–30 % worse, which supports our claim of importance of applying this regularized estimator and its proposed online version. Use of gradient descent instead of conjugate gradients lead to similar results, however number of required iterations to converge was about 10× bigger. Consequently, conjugate gradients also seems an important element of the model. At the same time, using the exact retraining did not lead to any improvements, however for big datasets we still suggest to run it once each h iterations (as it does not increase overall complexity and ensures stability).

4 Conclusions

We showed how to build an optimal covariance estimator in an online fashion and apply it to an online (and stream) version of Extreme Entropy Machines. Resulting model is mathematically equivalent to EEM trained in the batch setting (rare feature

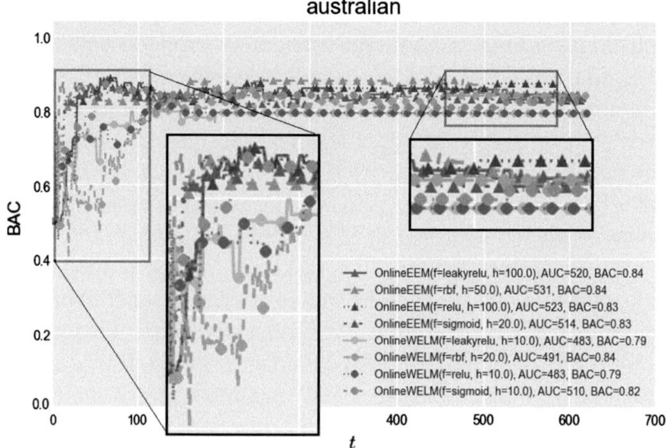

Fig. 1 Plot of BAC in each iteration during online learning

in online learning methods [9]). Further empirical evaluation shows that it usually achieves significantly better results (it learns faster) than online Weighted Extreme Learning Machines as well as Mondrian Forest. Resulting model is well adapted for unbalanced problems in particular it can easily work with changing classes' ratios.

Acknowledgments The work of the first author was partially founded by National Science Centre Poland grant no. 2013/09/N/ST6/03015, while the work of the second one by National Science Centre Poland grant no. 2014/13/B/ST6/01792.

References

1. Bache, K., Lichman, M.: UCI machine learning repository. http://archive.ics.uci.edu/ml (2013)
2. Bartocha, K., Podolak, I.T.: Classifier ensembles for virtual concept drift-the DEnboost algorithm. In: Corchado, E., Kurzyński, M., Woźniak, M. (eds.) Hybrid Artificial Intelligent Systems, pp. 164–171. Springer, Berlin (2011)
3. Breiman, L.: Random forests. Mach. Learn. **45**(1), 5–32 (2001)
4. Chong, E.K., Zak, S.H.: An Introduction to Optimization, vol. 76. Wiley, New York (2013)
5. Czarnecki, W.M., Tabor, J.: Extreme entropy machines: robust information theoretic classification. Pattern Anal. Appl. (2015). doi:10.1007/s10044-015-0497-8
6. Czarnecki, W.M., Tabor, J.: Multithreshold entropy linear classifier: theory and applications. Expert Syst. Appl. **42**, 5591–5606 (2015)
7. Drineas, P., Mahoney, M.W.: On the nyström method for approximating a gram matrix for improved kernel-based learning. J. Mach. Learn. Res. **6**, 2153–2175 (2005)
8. Fernández-Delgado, M., Cernadas, E., Barro, S., Amorim, D.: Do we need hundreds of classifiers to solve real world classification problems? J. Mach. Learn. Res. **15**(1), 3133–3181 (2014)
9. Gaber, M.M., Zaslavsky, A., Krishnaswamy, S.: Mining data streams: a review. ACM Sigmod Rec. **34**(2), 18–26 (2005)

10. Huang, G.B., Zhu, Q.Y., Siew, C.K.: Extreme learning machine: a new learning scheme of feedforward neural networks. In: Proceedings. 2004 IEEE International Joint Conference on Neural Networks, vol. 2, pp. 985–990. IEEE (2004)
11. Kosina, P., Gama, J.: Very fast decision rules for classification in data streams. Data Min. Know. Discov. **29**(1), 168–202 (2015)
12. Krawczyk, B., Stefanowski, J., Wozniak, M.: Data stream classification and big data analytics. Neurocomputing **150**, 238–239 (2015)
13. Lakshminarayanan, B., Roy, D.M., Teh, Y.W.: Mondrian forests: efficient online random forests. In: Advances in Neural Information Processing Systems, pp. 3140–3148 (2014)
14. Ledoit, O., Wolf, M.: A well-conditioned estimator for large-dimensional covariance matrices. J. Multivar. Anal. **88**(2), 365–411 (2004)
15. Pedregosa, F., Varoquaux, G., Gramfort, A., Michel, V., Thirion, B., Grisel, O., Blondel, M., Prettenhofer, P., Weiss, R., Dubourg, V., et al.: Scikit-learn: machine learning in python. J. Mach. Learn. Res. **12**, 2825–2830 (2011)
16. Settles, B.: Active learning. Synth. Lect. Artif. Intel. Mach. Learn. **6**(1), 1–114 (2012)
17. Zong, W., Huang, G.B., Chen, Y.: Weighted extreme learning machine for imbalance learning. Neurocomputing **101**, 229–242 (2013)

A Context-Driven Data Weighting Approach for Handling Concept Drift in Classification

Lida Barakat

Abstract Adapting classification models to concept drift is one of the main challenges associated with applying these models in dynamic environments. In particular, the learned concept is not static and may change over time under the influence of varying conditions (i.e. varying context). Unlike existing approaches where only the most recent data are considered for adapting the model, we propose incorporating *context awareness* into the adaptation process. The goal is to utilise knowledge of relevant context variables to facilitate the selection of more relevant training data. Specifically, we propose to weight each training example based on the degree of similarity with the current context. To detect such similarity, we utilise two approaches: a simple difference between the context variable values and a distribution-based distance metric. The experimental analyses show that such explicit context utilisation results in a more effective data selection strategy and enables to produce more accurate predictions.

Keywords Concept drift · Context awareness · Data weighting

1 Introduction

The continuous flow of information in many domains requires the ability of the classification model to incorporate new data so that it can update its concept description (i.e., the underlying function between the variable to be predicted and the respective input data) and make accurate predictions. One of the main challenges associated with such learning is that, in the real world, the learned concept is not static and may change over time (a phenomenon referred to as *concept drift*) under the influence

L. Barakat (✉)
Department of Management Science, Lancaster University, Lancaster, UK
e-mail: l.barakat@lancaster.ac.uk

R. Burduk et al. (eds.), *Proceedings of the 9th International Conference on Computer Recognition Systems CORES 2015*, Advances in Intelligent Systems and Computing 403, DOI 10.1007/978-3-319-26227-7_36

of varying conditions (i.e., varying context). Such contextual changes, for example, may correspond to changes in economic conditions for financial prediction, the time of the year for weather prediction, etc. In the presence of concept changes, the previously seen data examples (input–output pairs) may not remain relevant. Hence, it is essential for the classification model not only to accommodate new information, but also to have some data selection mechanism in order to be able to adapt to changes. This task, however, is far from trivial since drift can occur at any time, and its type (gradual, abrupt, etc.) cannot be known in advance. This poses a challenge on how to select *relevant* and *sufficient* training data to adapt the model. Many adaptive learning approaches have been proposed in the literature to address concept drift. These approaches can be mainly classified in terms of how adaptation is achieved into change detection-based adaptation and continuous adaptation [16]. The former involves monitoring the performance of the classifier over time to systematically identify when (and how) it should be rebuilt (e.g., [2, 4, 6, 19, 20]). The latter approach, on the other hand, assumes continuous drift, and updates the classification model on data arrival without attempting to identify when drift occurs (e.g., [9, 11–14]). In both approaches, time is considered as a measure for training example relevance, with recent examples being considered the most relevant, while older ones being forgotten eventually. In particular, the classifier is (re-)estimated using either a window of the latest observed examples (e.g., [4]) or a time-based example weighting function (e.g., [11]). However, such time-based approach may neglect more relevant older examples (in the case of reappearing concepts), and may capture data irrelevant for the current concept.

In response, we propose utilising knowledge of relevant contextual information to facilitate the selection of more relevant training data for the classification model. The motivation is that context changes are usually considered the main reason behind concept drift [7, 17, 19]. Specifically, we propose to *weight* each training example based on the degree of similarity with the current context. To detect such weights, we *learn* the similarity among context values based on how these values discriminate between occurring concepts. To validate the approach, we utilise Naive Bayes classifier with the proposed context-aware weighting scheme, and test it under different conditions of changes and concept recurrence. The experimental analyses, on simulated and benchmark datasets, show that such explicit utilisation of contextual information enables to produce more accurate prediction results. The rest of the paper is organised as follows. The proposed context-driven example weighting approach is introduced in Sect. 2. Evaluation setup and results are reported in Sect. 3, while Sect. 4 concludes the paper.

2 Context-Driven Example Weighting Model

In classification problems, the predictive model at time step t is normally derived based on the previously observed labeled examples $\{(x_i, y_i)\}_{i=1}^{t}$, where $x_i = (x_1, x_2, \dots, x_m)$ represents the value vector of input variables at time step i, and y_i

is the target class label. In the presence of concept drift, the examples may not remain relevant, and hence the contribution of each example in estimating the classification model should be governed by its relevance for the current situation (the current concept). In this work, we propose to identify such relevance by weighting the training examples based on the degree of similarity between the context under which the training example is collected and the current context. Formally, the proposed context-aware example weighting model can be represented as a tuple $(C, w, vrel)$, where: $C = \{C_1, C_2, \dots, C_p\}$, is the set of relevant context variables (i.e. those affecting the concept of interest, e.g. for electricity price prediction, $C = \{$season, day of the week$\}$); $w(u_i)$, is an example weighting function, which associates each training example, $u_i = (x_i, y_i)$, with a weighting factor reflecting its relevance for the current situation; and finally $vrel(C_j, v_1, v_2)$, is a value relevance function, which determines the similarity degree between two different values, v_1 and v_2, of a context variable, $C_j \in C$. For example, for $C_j =$ season: $v_1, v_2 \in \{$winter, spring, summer, autumn$\}$. The value relevance function governs the calculation of $w(u_i)$, as detailed in Sect. 2.1. Assuming knowledge of relevant context variables (when unknown a priori, these variables could be identified according to the approach proposed in [1]), our goal is to define the example weighting function and the associated value relevance function. More details on these functions are presented next.

2.1 Example Weighting Function

As stated earlier, our goal is to associate each training example, $u_i = (x_i, y_i)$, with a weighting factor that reflects its relevance for the current concept. The proposed weighting scheme is based on measuring the degree of context similarity. That is, data examples collected under conditions more similar to the current context (the context of the example to be classified) should receive higher weights, and the weights should decrease correspondingly as this similarity decreases. Formally, our example weighting function, $w(u_i) \in [0, 1]$ (which we henceforth refer to as w_i, for simplicity) can be defined as follows:

$$w_i = 1 - d(ctx_i, ctx_t) \tag{1}$$

where: $ctx_i = (c_1^i, c_2^i, \dots, c_p^i)$ is the context instance under which example u_i is collected (i.e. the value vector of context variables $(C_1, C_2, \dots, C_p)$ at time step i); $ctx_t = (c_1^t, c_2^t, \dots, c_p^t)$ is the context instance at the current time step t; and $d(.,.) \in [0, 1]$ is a distance measure between two context instances, given as

$$d(ctx_i, ctx_t) = \sum_{C_j \in C} (1 - vrel(C_j, c_j^i, c_j^t)) \tag{2}$$

where $vrel$ is the value relevance function measuring the similarity between values c^i_j and c^t_j of context variable C_j (this function is discussed next).

2.2 Value Relevance Function

Given a context variable $C_j \in C$, we distinguish two possible approaches for detecting the similarity between two values v_1 and v_2 of C_j, i.e., $vrel(C_j, v_1, v_2)$: a simple direct difference approach, and a learning-based approach utilising conceptual similarity, as outlined below.

2.2.1 Simple Approach

The simplest realisation of function $vrel(C_j, v_1, v_2)$ is to utilise the direct difference between v_1 and v_2, as follows (where $min(C_j)$ and $max(C_j)$ are the minimum and maximum values, respectively, of variable C_j)

$$vrel(C_j, v_1, v_2) = \begin{cases} 1 - \frac{|v_1 - v_2|}{max(C_j) - min(C_j)} & \textbf{if } C_j \text{ is numerical} \\ 1 & \textbf{if } C_j \text{ is categorical } \textbf{and } v_1 = v_2 \\ 0 & \textbf{if } C_j \text{ is categorical } \textbf{and } v_1 \neq v_2 \end{cases} \quad (3)$$

2.2.2 Learning-Based Approach

The above realisation of function $vrel$ may not necessarily reflect the value similarity of interest in our approach. This is because we are interested in identifying two values of variable C_j as similar if their associated underlying concepts are the same, regardless of the actual difference between these values. Hence, an alternative realisation of the value relevance function would be to *learn* the relevance among context values from historical data, based on how these values discriminate between occurring concepts. In probabilistic terms, two different values v_1 and v_2 of context variable C_j should still be considered similar if the two probability distributions, $P(Y, X|\ v_1)$ and $P(Y, X|\ v_2)$, do not differ much under those values. Here, Y and X are the target variable and the vector of input variables, respectively. In order to compare distributions $P(Y, X|\ v_1)$ and $P(Y, X|\ v_2)$, we utilise an information theoretic measure suitable for this purpose, namely the Entropy Absolute Difference (EAD) [16]. It detects changes by measuring the dispersions of the differences between two distributions P and Q

$$H(P(Y, X)||Q(Y, X)) = -\sum_y \sum_x |p(y, x) - q(y, x)| log_2 |p(y, x) - q(y, x)| \quad (4)$$

Here, $p(y, x)$ and $q(y, x)$ corresponds to $p(y, x| v_1)$ and $p(y, x| v_2)$, respectively, in our problem. In this measure, smaller values indicate smaller differences between the distributions, and equal zero if the two distributions are identical.

3 Evaluation

In this section, we conduct empirical evaluation of the proposed context-aware example weighting approach, focusing on its influence on classification results. We assume an incremental learning scenario where the classification model is (re-)estimated on the arrival of each new example (consisting of the input value vector, the target class label, and the associated context). Further details regarding the experimental setup and results are presented next.

3.1 Hypotheses

Our aim is to test the following hypotheses. *Hypothesis 1*. The proposed context-driven approach utilising the learning-based value relevance function outperforms that utilising the direct-difference-based value relevance function. *Hypothesis 2*. The proposed context-driven approach outperforms (or at least performs as well as) other data selection strategies, including window-based and time-based-weighting approaches (see Sect. 3.3), in settings with no recurring concepts, while outperforming these strategies in settings with recurring concepts.

3.2 Datasets

We utilise two artificial datasets to study the behavior of the proposed approach under different settings (under different types of concept drift and recurring concepts). In addition, we utilise two publically available real-world datasets that have been widely used for evaluating concept drift handling systems (both exhibit concept drift and contextual characteristics, which make them suitable for the purpose of our evaluation). These datasets are described below.

3.2.1 Artificial Datasets

STAGGER Concepts [15]: This dataset exhibits abrupt concept drift, and is defined by three categorical input variables: size, colour, and shape. Each variable has three possible values: $size \in$ {small, medium, large}, colour $\in$ {red, green, blue}, and shape $\in$ {square, circular, triangular}. The target concept switches between a sequence of the following three descriptions: Concept 1 (size = small $\wedge$ colour = red); Concept 2 (colour = green $\wedge$ shape = circular); and Concept 3 (size = medium $\vee$ large). *Rotating Hyperplane* [18]: This dataset exhibits gradual concept drift, and is defined by equation: $\sum_{k=1}^{m} a_k x_k = a_0$, where a_k and x_k are the weight and value of input variable X_k, respectively, m is the input vector's dimensionality, and a_0 is a threshold value distinguishing two classes for a given concept. The value of a_0 is determined such that $a_0 = \frac{1}{2} \sum_{k=1}^{m} a_k$. That is, the hyperplane divides the space of the examples into two parts of approximately the same size, where examples satisfying $\sum_{k=1}^{m} a_k x_k \leq a_0$ could be labelled as positive, and as negative otherwise. The drift is generated by continuously changing the weights a_k of the input variables during each concept. Specifically, at each time step, a_k is updated by quantity $\frac{\delta}{N}$, where δ is the magnitude of the change during each concept, and N is the duration of the concept. For the purpose of our analysis, we generate 3 uniformly distributed input variables over the range $x_k \in [0, 1]$. The weights, a_k, are randomly initialised between 0 and 1, i.e. $a_k \in [0, 1]$, and then made to change during each concept by $\delta = 1.5$. For both datasets, we simulate context by introducing an additional variable, $C \in \{v_1, v_2, v_3, v_4, v_5, v_6\}$, where each value is associated with a certain concept. Specifically, $C = v_1 \vee v_2$ under Concept 1, $C = v_3 \vee v_4$ under Concept 2, and $C = v_5 \vee v_6$ under Concept 3. This context variable is exposed to 5 % noise (i.e. has a 5 % probability to take on values not associated with the current concept).

3.2.2 Real-World Datasets

Electricity Market Dataset (*Elec*) [8]: This real-world dataset consists of 45312 records concerning electricity price changes, obtained from the Australian New South Wales Electricity Market, between May 1996 and December 1998. Each record represents a period of 30 min, with a binary class label and four categorical input variables. The class label refers to the price change (up/down) with respect to a moving average of the last 24 h. As context, we utilise day of the week (7 categorical values) and season (four categorical values) as relevant context variables [1]. Furthermore, due to the dependencies among the labels in this dataset [21] (which cause even the simplest classifier predicting the next label based only on the previously seen label to achieve very high accuracy), we subsample the data to break such dependencies [20]. Specifically, we take every twentieth observation at regular intervals, resulting in a total of 2265 records utilised for classifier evaluation. *Emailing List Dataset* (*Elist*) [10]: This real-world dataset contains 1500 examples on how to filter email messages into either interesting or junk according to user preferences. Each record consists of 913 Boolean input variables representing words frequently appearing in the body of the message, with a binary class label. Two recurring concepts are exhibited: Concept 1 (the user is only interested in medical messages), and Concept 2 (the user changes their interest to space and baseball). Such interest change is associated with the context variable Location. Changes in concept are made to occur at every 300 time steps.

3.3 Classification Strategies

Throughout the presentation of our experimental results, we refer to the following classification strategies. *Expanding Window*: The classification model is re-estimated based on all the examples observed so far, assigning equal weights to all examples. *Context as Input*: The classification model is similar to the *Expanding Window*, but with considering context variables as additional input variables. *Sliding Window*: The classification model is re-estimated based on a fixed window of the latest observed examples (assigning equal weights to all examples). *Dynamic Window*: The classification model is re-estimated based on a dynamic window of the latest observed examples. The window size is determined according to the change detection approach proposed in [4]. *Time-based Weighting*: The classification model is re-estimated utilising a time-based weighting scheme inspired by the approach proposed in [11], such that the weight of each example decreases with time. *Context-based Weighting*: The classification model is re-estimated utilising the proposed context-based weighting scheme. The classification model adopted in our experiments is Naive Bayes classifier due to its ability to learn incrementally, its computational efficiency, and the ability to produce accurate results despite its simplicity [3]. Note, however, that the proposed context-based weighting approach is not tailored to a particular classifier, but should generally be applicable with a number of classification models supporting learning from weighted examples.

3.4 Performance Measures

We are interested in measuring the predictive accuracy (*AC*) of the classification strategy in the presence of concept changes, computed as the ratio between the number of correctly classified examples and the total number of classified examples. In addition, to account for the class imbalance problem, we utilise the Kappa Statistic (*KS*). The values of the Kappa Statistic range between 0 and 1, where 0 indicates that the achieved accuracy is completely random [5].

3.5 Results

In the first set of experiments, we assess *Hypothesis 1* (see Sect. 3.1). For this purpose, we compare the performance of *Context-based Weighting* under the alternative variations proposed for the value relevance function: the simple direct difference approach (*Simple Difference*) and the learning-based approach utilising EAD (*EAD-based*). The experiments are conducted on the Stagger and Hyperplane datasets with the artificial context variable. We generate 1200 examples and 3 concepts, with each concept occurrence lasting for 200 examples following concept sequence 1-2-3-1-2-3,

Table 1 Comparison of different realisations of the value relevance function (standard deviations are indicated in parentheses)

Context-based weighting	Stagger		Hyperplane	
	AC	KS	AC	KS
Simple difference	0.86 (0.01)	0.71 (0.03)	0.78 (0.02)	0.57 (0.04)
EAD-based	0.89 (0.01)	0.77 (0.03)	0.84 (0.02)	0.67 (0.04)

with added 5 % label noise (i.e., 5 % of the examples in each concept have incorrect class labels). The relevance among context values (for *EAD-based*) is first learned separately in an offline mode (the first 600 examples), and then incorporated during classification into the adaptation process (the following 600 examples). Table 1 reports the corresponding results averaged over 30 runs. The advantage of using *EAD-based* over *Simple Difference* is evident in both datasets, with the former outperforming the latter in terms of both *AC* (accuracy) and *KS* (Kappa Statistic). To further analyse the results, Fig. 1 studies the accuracy evolution over time (as a moving average of the last 20 observations) of both approaches after a new concept (Concept 2 in the figure) is encountered. As can be seen, the *Simple Difference* suffers from an accuracy drop during the period of concept stability (in particular, after time step 300, when the context variable switches from value v_3 to v_4 while still being under Concept 2). This is because the appearance of a new value for the context variable is interpreted as an indication of a new concept, thus neglecting relevant examples prior to this value change (i.e., examples under context value v_3). In contrast, *EAD-based* approach recognises the new value of the context variable as a relevant one (belonging to the same concept), avoiding such degradation in the predictive accuracy. In the second set of experiments, we assess *Hypothesis 2* (see Sect. 3.1). For this purpose, we compare the classification strategies of Sect. 3.3, considering *EAD-based* implementation for *Context-based Weighting*. Table 2 reports the cor-

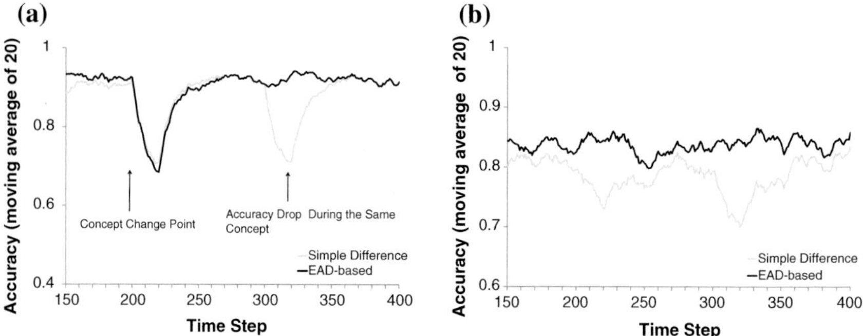

Fig. 1 Comparison of different realisations of the value relevance function over time. **a** Stagger. **b** Hyperplane

Table 2 Comparison of different classifiers for Stagger and Hyperplane datasets (standard deviations are indicated in parentheses)

Classification strategy	Stagger		Hyperplane	
	AC	KS	AC	KS
Expanding window	0.57 (0.02)	0.15 (0.03)	0.83 (0.03)	0.65 (0.05)
Context as input	0.70 (0.02)	0.41 (0.04)	0.82 (0.02)	0.64 (0.05)
Sliding window	0.68 (0.01)	0.37 (0.02)	0.84 (0.01)	0.67 (0.03)
Dynamic window	0.81 (0.09)	0.62 (0.19)	0.82 (0.02)	0.64 (0.04)
Time-based weighting	0.59 (0.01)	0.19 (0.03)	0.83 (0.02)	0.66 (0.04)
Context-based weighting	0.91 (0.01)	0.81 (0.01)	0.82 (0.02)	0.64 (0.04)

responding results (averaged over 30 runs) for Stagger and Hyperplane datasets, with concept sequence settings being the same as previously. As can be seen, in the case of abrupt changes (Stagger dataset), *Context-based Weighting* outperforms all the other considered strategies (including *Context as Input*). Further analysis of this case is provided in Fig. 2, where we depict the evolution of predictive accuracy over time, distinguishing the cases of new and recurring concepts. In particular, when a new concept is encountered (Fig. 2a), all the considered strategies suffer from performance degradation. However, *Context-based Weighting* achieves the fastest recovery due to the utilisation of contextual evidence, which enables capturing only relevant observations for learning the new concept. In contrast, time-based strategies suffer from slower reaction to changes since the effect of older irrelevant observations takes longer to be forgotten. Furthermore, when encountering a recurring concept (Fig. 2b), and unlike other strategies, *Context-based Weighting* always maintains high accuracy without experiencing performance degradation after a change point, due to its ability to utilise old (but relevant again) data. In the case of gradual changes (Hyperplane dataset in Table 2), where the examples remain relevant for longer periods (as opposed to abrupt changes), the proposed approach approximates the performance of recency-based approaches (the best performing in such a case).

Similar observations hold for the real-world datasets, with the results being reported in Table 3. For these datasets, we utilise the McNemar's test [5] to evaluate the statistical significance of the performance differences obtained.

4 Conclusions and Future Work

The paper presented a context-based approach for adapting classification models to concept changes. In particular, each example is weighted according to the degree of similarity with the current context. The experimental analysis demonstrated that the proposed approach enables faster recovery after change points and helps avoiding performance degradation in the case of recurring concepts. Future work involves

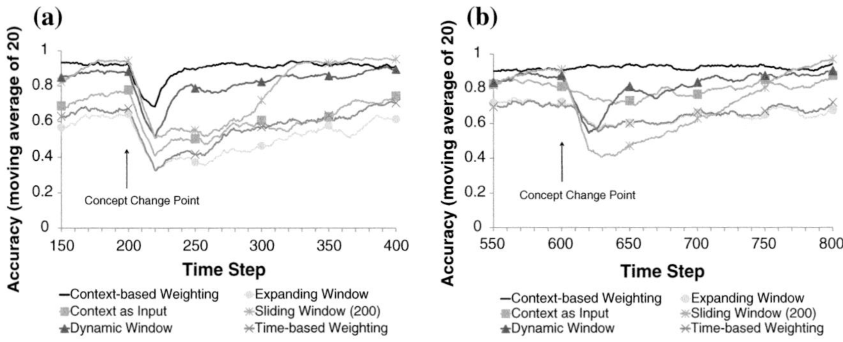

Fig. 2 Accuracy evolution of different classifiers for Stagger dataset. **a** Encountering new concept. **b** Encountering recurring concept

Table 3 Comparison of different classifiers for Elist and Elec datasets

Classification strategy	Elist		Elec	
	AC	KS	AC	KS
Expanding window	0.62*	0.23	0.63*	0.26
Context as input	0.69*	0.38	0.64*	0.28
Sliding window	0.62*	0.24	0.64*	0.26
Dynamic window	0.70*	0.39	0.66	0.29
Time-based weighting	0.62*	0.24	0.64*	0.27
Context-based weighting	0.81	0.61	0.67	0.34

* Significant difference according to McNemar Test (0.05 significance level)

testing the proposed approach with other classification models, and utilising the value relevance function proposed for drift detection with ensemble learning.

References

1. Barakat, L.: Context identification and exploitation in dynamic data mining—an application to classifying electricity price changes. In: Bouchachia, A. (ed.) Adaptive and Intelligent Systems. Lecture Notes in Computer Science, vol. 8779, pp. 80–89. Springer, Heidelberg (2014)
2. Du, L., Song, Q., Jia, X.: Detecting concept drift: an information entropy based method using an adaptive sliding window. Intell. Data Anal. **18**, 337–364 (2014)
3. Frank, E., Hall, M., Pfahringer, B.: Locally weighted naive bayes. In: Proceedings of the 19th Conference on Uncertainty in Artificial Intelligence, pp. 249–256 (2003)
4. Gama, J., Medas, P., Castillo, G., Rodrigues, P.: Learning with drift detection. In: Proceedings of the 17th Brazilian Symposium on Artificial Intelligence, LNAI 3171, pp. 286–295 (2004)
5. Gama, J., Bifet, A., Pechenizkiy, M., Bouchachia, A.: A survey on concept drift adaptation. ACM Comput. Surv. **46**(4), 1–37 (2014)
6. Garcia, M.B., del Campo-Avila, J., Fidalgo, R., Bifet, A., Gavalda, R., Morales-Bueno, R.: Early drift detection method. In: ECML PKDD 2006 Workshop on Knowledge Discovery from Data Streams (2006)

7. Harries, M., Sammut, C., Horn, K.: Extracting hidden context. Mach. Learn. **32**(2), 101–126 (1998)
8. Harries, M.: Splice-2 Comparative Evaluation: Electricity Pricing. Technical report, University of New South Wales (1999)
9. Hulten, G., Spencer, L., Domingos, P.: Mining time-changing data streams. In: Proceedings of the 7th ACM SIGKDD International Conference on Knowledge Discovery and Data Mining, pp. 97–106 (2001)
10. Katakis, I., Tsoumakas, G., Vlahavas, I.: Tracking recurring contexts using ensemble classifiers: an application to email filtering. Knowl. Inf. Syst. **22**(3), 371–391 (2010)
11. Koychev, I.: Gradual forgetting for adaptation to concept drift. In: Proceedings of ECAI Workshop Current Issues in Spatio-Temporal Reasoning, pp. 101–106 (2000)
12. Kubat, M.: Floating approximation in time-varying knowledge bases. Pattern Recognit. Lett. **10**(4), 223–227 (1989)
13. Pavlidis, N.G., Tasoulis, D.K., Adams, N.M., Hand, D.J.: λ-Perceptron: an adaptive classifier for data streams. Pattern Recognit. **44**(1), 78–96 (2011)
14. Pavlidis, N.G., Tasoulis, D.K., Adams, N.M., Hand, D.J.: Adaptive consumer credit classification. J. Oper. Res. Soc. **63**(12), 1645–1654 (2012)
15. Schlimmer, J., Granger, R.: Incremental learning from noisy data. Mach. Learn. **1**(3), 317–354 (1986)
16. Sebastiao, R., Gama, J.: Change detection in learning histograms from data streams. In: Proceedings of the Portuguese Conference on Artificial Intelligence. LNAI 4874, pp. 112–123 (2007)
17. Tsymbal, A.: The Problem of Concept Drift: Definitions and Related Work. Computer Science Department, Trinity College Dublin (2004)
18. Wang, H., Fan, W., Yu, P. S., Han, J.: Mining concept-drifting data streams using ensemble classifiers. In: Proceedings of the 9th ACM SIGKDD International Conference on Knowledge Discovery and Data Mining, pp. 226–235 (2003)
19. Widmer, G., Kubat, M.: Learning in the presence of concept drift and hidden contexts. Mach. Learn. **23**(1), 69–101 (1996)
20. Zliobaite, I., Kuncheva, L.: Determining the training window for small sample size classification with concept drift. In: Proceedings of the IEEE International Conference on Data Mining Workshop, pp. 447–452 (2009)
21. Zliobaite, I.: How Good is the Electricity Benchmark for Evaluating Concept Drift Adaptation. CoRR, abs/1301-3524 (2013)

Ontology Learning from Graph-Stream Representation of Complex Process

Radosław Z. Ziembiński

Abstract Societies around the world faced arrival of smart technologies in the last decade. Often interconnected, intelligent devices form new entity called Internet of Things (IoT). Mounted to commodities they are versatile tools for collecting various sorts of data about our behavior. Related applications require novel knowledge exploration methods handling large amount of observations containing complex data. Therefore, this paper introduces graph-stream structure as a capable tool for the complex process description. Further, it delivers a method for graph-stream processing making possible extraction of the compact ontological description of the recorded process. Introduced method uses novel online clustering algorithm and was verified experimentally on synthetic data sets.

Keywords Ontology learning · Clustering · Data streams

1 Introduction

Internet of Things relies on knowledge explored from data produced at enormous rate by e.g., sensors, logs, registers, AV devices. Information collected from multiple sources constitute data streams. Until now, researches have developed a significant number of methods handling them. However, in majority of considered problems the data stream was defined as a single pipeline. It is a strong simplification of information. Especially, if it is collected in observations of process containing many active subprocesses. It may lower the processing quality by random blending of data

R.Z. Ziembiński (✉)
Faculty of Computing, Poznan University of Technology, Poznan, Poland
e-mail: radoslaw.ziembinski@cs.put.poznan.pl
URL: http://www.etacar.put.poznan.pl/radoslaw.ziembinski

R. Burduk et al. (eds.), *Proceedings of the 9th International Conference on Computer Recognition Systems CORES 2015*, Advances in Intelligent Systems and Computing 403, DOI 10.1007/978-3-319-26227-7_37

generated simultaneously by different subprocesses. Therefore, the approach proposed here introduces more complex representation of data. It is a graph-stream that is a data structure capable to store information about the complex process embracing a population of more of less dependent subprocesses. The structure natively separates data generated by parallel threads and allows expressing causal relationships [4]. In consequence, many problems and errors associated to elements order, possible in single pipeline approaches, may be avoided. This paper will introduce only one example method for extraction of information from graph-stream structure. Here, data in the graph-stream will be processed by online clustering algorithm to obtain compact description of the observed process (its "ontology"). Retrieved description would contain information about frequent subprocesses enriched in causal dependencies between them. Obtained model is familiar to Bayesian network and can be utilized in semi-supervised machine learning. Added value of this paper can be considered in:

- Novel data representation reducing problems with ordering errors and preserving causal relationships between subprocesses.
- Ontology learning algorithm delivering compact description of the process. It finds causal relations and similarities between frequent subprocesses.
- A modified pyramid clustering algorithm used for the purpose of the ontology learning in the online environment.

This paper runs as follows. It begins from related works coverage introducing to this research area. Afterwards, it introduces the definition of the graph-stream structure and the ontology learning problem. Finally, it experimentally evaluates the implementation of the algorithm on artificial data sets.

2 Related Works

Many developments in considered research area have focused on problems formulated as extensions of ones applicable previously to stationary data sets. As such, they often inherits a lot of assumptions (and design limitations altogether) related to processed data from stationary counterparts e.g., clustering and classification problems that were adapted to online applications like in [6, 7, 12]. Such approach leads to methods where consecutiveness of "things" is often considered narrowly. In a consequence, above methods are adaptable to recent changes in data distribution (the concept drift). However, causal relationships present in historical data are exploited poorly [10]. Efforts of other researchers focused on data mining in streams. They delivered online algorithms for frequent item sets, sequences, subtrees and graphs mining, e.g., in [1, 11]. Patterns found by these methods represent structural and time-depended knowledge about the process. Moreover, these methods are relatively fast and reliable but are highly sensitive on order of elements (when structural mining plays role). They work well for nominal data with simply defined identity relation. Further, many available methods have been proposed for multi-dimensional

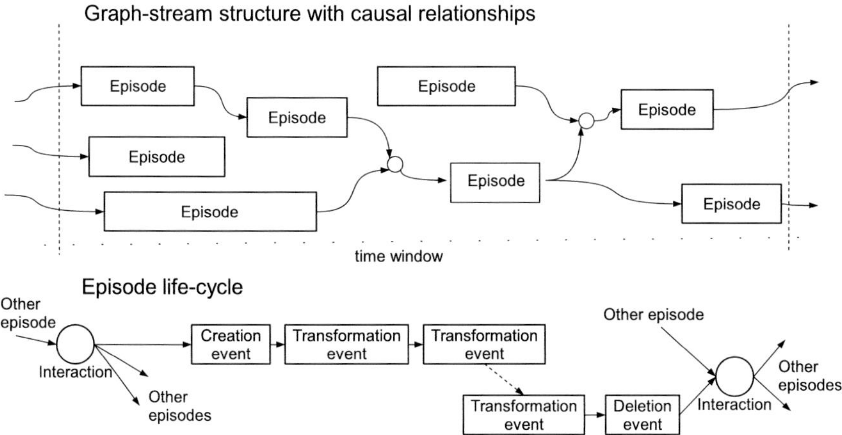

Fig. 1 Graph-stream structure and episode life-cycle

time-series analysis, e.g., like motifs finding algorithms. They can handle data streams coming directly from sensors making the processing less prone on errors related to simplification [5]. For them, a fixed number of dimensions is assumed (separate single dimensional streams). Such assumption delimits the flexibility of this representation. Moreover, considered single dimensional streams are synchronized and do not carry information allowing for easy identification and separation of interplaying subprocesses. Therefore, structural information about relations between subprocesses is buried in low level description of the process and its retrieval may be nontrivial [3]. A solution proposed in this paper contributes to above state of art extending ideas from [8]. It introduces a graph-stream structure designed to preserve data from complex interacting subprocesses. The graph-stream is a network (DAG) laid along time axis and storing information about subprocesses in episodes connected by causal relationships. These connections are managed by interactions events. It is capable to store separate motifs found in time-series revealing different subprocesses of the observed process. The number of subprocesses described by parallel episodes is not limited in this representation and they are synchronized only by interaction events and causal relations. Thus, such representation allows for storing information even if some uncertainty about exact events time exists. The proposed data structure schema is presented on Fig. 1. This data structure can be a source of inspiration for some new methods of information processing. This paper puts attention on the ontology learning. Such learning focuses on the compact description extraction which should explain the observables (in our case the complex process). Looking from the data stream processing perspective, it involves finding concepts and different sorts of relations between them. Among them the causality is also considered as the ontology building block e.g., in [4]. Ontology learning algorithms often employs clustering algorithm for finding prototypes (definitions) of concepts. However, clustering results strongly depends on similarity function used for objects. The complex makeup of episodes hardly can be compared by a measure. Thus, nonmetric clustering algorithm should be employed to retrieve adequate result. This problem was considered in [9] for iterative algorithms revealing some caveats. The solution is nonobvious for nonmetric similarity functions since they not obey triangle inequality axiom. That paper concludes that concept and terms prototypes should come from clusters that are inherently very similar. Thus, there was proposed a heuristic iterative method for finding cliques in similarity graph as the best quality approach. It outperforms hierarchical and partitioning clustering since formers insufficiently check similarities between all pairs of objects in found clusters (see [2]). Unfortunately, computational costs of cliques finding problem grows exponentially to the task size if we seek an exact solution. From the perspective of ontology learning this paper also delivers some new results. First of all, it introduces online clustering heuristic method to find approximation of cliques defining concepts. It performs its job within limited per iteration deadline. Additionally, it delivers a method for probabilistic analysis of causal dependencies between subprocesses. Then, as the result it delivers graphical model that may be used in the process development forecasting or subprocesses (semi-supervised) classification.

3 Problem Definition

3.1 Graph-Stream Data Structure

Let's define a heterogeneous set of attributes A carrying information gathered by sensors observing process P. Then, *node* $n_{p_{type},i}(p_{ts}, A_i) \in N$ is an inseparable unit of information collected at a given moment (at event). It contains attributes values $A_i \in A$ and is associated with two properties: event's type p_{type} and occurrence timestamp p_{ts}. The type property is used to distinguish different sorts of nodes. Following types of nodes are recognized:

- *Interaction node* $n_{E,i}(p_{ts}, A_i) \in N$ describes event where at least two subprocesses interact at the same moment exchanging information. The interaction involves destruction of subprocesses and creation of some new descendants.
- *State node* $n_{S,i}(p_{ts}, A_i) \in N$ is a node containing all values of attributes describing particular observed subprocess. It delivers complete data about subprocess state and its boundary, e.g., just after its creation, deletion or interaction.
- *Transformation node* $n_{T,i}(p_{ts}, A_i) \in N$ describes action performed inside subprocess. The nature of transformation event is intrinsic to subprocess and it does not involve interactions with other ones.

Directed edges $e_{D,i,j}(n_i, n_j) \in E$ constitute connections. They connect nodes $n_i \in N$ and $n_j \in N$ and represents causality relation between nodes (graphically denoted using $\rightarrow$). From above, a directed cycle is a sequence of nodes $n_1 \rightarrow n_2, \ldots, n_{m-1} \rightarrow n_m \in N$ collected along a path made from directed edges where $n_1 \equiv n_m$. Following constrains are imposed on connections:

- Interaction node can be connected by directed edges only to state nodes. Creation event node can connect to interaction node by ingoing edge while deletion event node by outgoing one.
- Transformation node connects only to state nodes or causally adjacent transformation nodes (from the same episode). It binds by one edge in each direction forming ordered sequence—the body of episode. Creation event node can connect to transformation node by outgoing edge while deletion event node by ingoing one.

A directed acyclic graph that meets above connections constraints is called the **graph-stream** $G_{ST}(N, E)$. The description provided by graph-stream resembles Gantt charts or Feynman diagrams. Its concept is illustrated on Fig. 1. It is assumed that number of parallel episodes is much lower than a length of path in the graph-stream that lies along time axis. Thus, the graph-stream structure usually has to be accessible through sliding window for the purpose of the processing.

3.2 Ontology Learning

Ontology is a data structure that describes observed (usually encapsulated and limited) "world". In the considered case, it refers to behavior of process and its sub-processes. Usually, it is denoted as a graph where nodes represent concepts and edges relations between them. It usually assumes existence of many kinds of relations allowing for, e.g., modeling compositions, hierarchical or causal ordering of concepts. Ontology learning is a process of its construction from observed data. Concepts are usually found by unsupervised learning methods, e.g., clustering of information about objects. Hence, clusters are considered as classes of events described by their most suitable representatives (prototypes) and their attributes. In our case the considered objects are episodes separated on state nodes and transformation bodies. In the proposed model, concepts $c_m \in C$ are derived from clusters of state nodes. They contain associations of attributes that belong to clusters' prototypes (objects exemplifying clusters). Concept's c_m support is equal to the cluster size and should be greater than threshold Θ_C to consider cluster as reliable candidate for the concept. Different relations can be found by clustering of the graph-stream. First of all, hierarchical structure of concepts can be derived during nodes clustering from clusters overlapping analysis. It is natural feature for clique-based and hierarchical clustering algorithms. Specializations of general concept may be related to subsets inclusion in one bigger set. Hierarchy relations form a set denoted R_H, where each $r_H(c_k, c_l) \in R_H$ connects more general concept c_k to specific one c_l. Further, non-taxonomic relations can be derived from episodes bodies. This kind of relation can transform one concept (state) into another by an application of operations sequence. Functional relation is directed and identified by a cluster of similar sequences of transformations found by their clustering. Their set is denoted as R_T where $r_T(c_k, c_l) \in R_T$ represents functional relationships transforming c_k into c_l. Finally, another kind of relation can be obtained from interaction nodes and their bindings to state nodes. It tells about causal dependencies between episodes involving exchange of information between them. Clusters of similar state nodes determine where set of entry concepts (delete states) interact together to produce a set of resulting concepts (creation states). Interaction is unitary operation in proposed model. However complex operations within interaction may form separate intermediating episode. Symbol R_I denotes such kind of relations where $r_i(C_k, C_l) \in R_I$ is a relationship between entry nodes $C_k \subseteq C$ and resulting nodes $C_l \subseteq C$. A multi-graph $O(C, R_H \cup R_T \cup R_I)$ is considered as **an ontology describing the process** P during a period from t_1 to t_2 if all its concepts are supported. Thus, $\forall c_m \in C : \theta_C(c_m) \geqslant \Theta_C$ in the period $[t_1, t_2]$. Above model may be updated if relations and attributes values distributions in analyzed graph-stream would evolve. In experimental sections the algorithmic solution would focus on R_T and R_I relations.

3.3 Similarity Evaluation

A pair of nodes n_i, n_j is compared using following similarity function $\sigma(n_i, n_j) = \sigma_v(n_i, n_j)$. Similarity between sets of attributes A_i, A_j values related to nodes is calculated as follows:

$$\sigma_v(n_i, n_j) = \left(\prod_{a \in (A_i \cap A_j)} \exp\left(\frac{-\alpha_a \cdot \left| v_i(a) - v_j(a) \right|}{\left| vmax(a) - vmin(a) \right|} \right) \right)^{|A_i \cap A_j|} \quad (1)$$

Symbol $v_i(a)$ represents value of attribute $a \in A$ for node n_i. Pair $vmax(a)$, $vmin(a)$ represents bounds on range of possible attribute's values. Parameter α_a is used as weight. A similarity of occurrence timestamps is calculated in relation to the longer episode. It is used for transformation episodes. Then is can be obtained by equation $\sigma(n_i, n_j) = \sqrt{\sigma_t(n_i, n_j) \cdot \sigma_v(n_i, n_j)}$, where $\sigma_t(n_i, n_j) = exp(|t_R(n_i) - t_R(n_j)| \cdot \beta)$. Parameter β is used as weight. The function $t_R(n_i)$ represents relative occurrence time of node n_i. It is calculated as normalized time of event occurrence in the respect to episode timestamps related to creation and deletion events. All similarity measures return values from 0 to 1, where value equal to 1 means identity. A pair of episodes is compared in a more complex manner. Similarities for pairs of respectable state nodes are evaluated using above functions regardless their timestamps. Then, there are calculated similarities between all pairs of transformation nodes from both episodes considering timestamps. From, all extracted pairs there are selected nonoverlapping ones with higher similarities. Then these values for pairs are aggregated using arithmetic mean. Finally, similarity between episodes may be calculated as mean of similarities between state nodes and aggregated similarity of transformation nodes. However, for the purpose of ontology learning all these similarities are considered separately.

4 Experimental Evaluation

4.1 Algorithmic Solution

The proposed solution is and extension of pyramid clustering algorithm used to find cliques of similar objects. It is iterative algorithm with constrained upper processing complexity applicable for the online processing. The algorithm does not require that the similarity function has to obey metric axioms. It works in cycles connected to object's insertion or removal.

Algorithm 9 Clustering algorithm SimCliq

Data: Inserted(removed) object o, set of objects S, clusters population P, similarity matrix M, layer maximum size L_{max}
Result: Update clustering result R

```
// object insertion with replacement
if S.isFull() then
| P.eraseAllClustersContaining(S.getOldestObject())
end
S.add(o)      P.populateFirstLayer(s, o)      // browses layers ascending to
   candidates sizes
foreach L ∈ increasingSizes(P) do
|  L.trim(L_max)
|  G = L.mergeNewCandidateClusters()
|  G.evaluateInternalSimilarity(M)
|  P.addNewCandidates(G)
end
// finds set of candidates with the highest similarities
R = P.selectCandidates()
R_old.propagateIdentifiers(R); R_old = R;
```

Algorithm stores similarities between objects in matrix M. The implementation uses multi-layered and fixed-size population Q of candidate clusters. Each layer $linL$ stores some candidate clusters of a given size sorted according to their internal similarity. Integral similarity is calculated from:

$$\sigma_I M(c) = exp\left(\frac{T_{simsum}(c)}{T_{count}(c)} - 1\right)/exp\left(\frac{D_{simsum}(c)}{D_{count}(c)}\right) \tag{2}$$

Let M_s be a sub-matrix M containing similarities between objects of cluster c and all of them in M (in fetched rows). Then, $T_{simsum}(c)$ is sum of similarities in M_s and $T_{count}(c)$ contains their number. While, $D_{simsum}(c)$ is a sum of differences of similarities between cells in all distinct pairs of rows in M_s and $D_{count}(c)$ is their number. The function returns higher value if a number of similar cells is larger and differences between pairs of objects are lower. The population is browsed from smaller clusters to larger ones. New candidates are created by merging pairs of smaller ones found on the currently processed layer. Newly created clusters are evaluated (using internal similarity measure) and inserted to the proper layer dependent on their sizes. Then cleaning procedure swaps out clusters of the same size with similar content (objects) but poorer quality measure. They are removed from the population. The layers trimming procedure delivers stability to the processing time since binomial explosion of the population size is suppressed. Removal of an object from the clustered set due to replacement or aging, enforces removal of all candidate clusters from layers that contains it. On the other hand, addition of a new object requires recalculation of internal similarity measure for all clusters in all layers that would contain it. The alteration propagates from bottom to top layers and involves candidates reordering and trimming. After the phase of candidates generation, the selection of clusters occurs for the output set. Follows-up procedure browses the population from the layer containing the clusters with higher internal similarities. Usually they can be found in the middle layers. It extracts nonoverlapping ones (or overlapping to parametrized threshold). Then it goes to following layer and continues search for subsequent nonoverlapping smaller clusters. It ends when none can be found meeting similarity loss condition defined by a quality threshold. Finally, found clusters are matched against the result of the previous cycle. The best fit matching procedure relates current clusters to previous ones and finds pairs with the highest similarities (using Jaccard coef.). This relation is used to propagate cluster identifiers for the next generation of updated clusters. However, at this step once assigned cluster identifier cannot be used once again. The complexity of algorithm is $O((L_{max} \cdot log(L_{max}))^2)$. M size has to be set adequately to window size and L_{max} impacts on clustering accuracy.

4.2 Performance Results

Experiments evaluating algorithm's performance have been conducted on artificial data stream. The generated graph-stream contains a fixed number of distinct episodes (4 types) occurring simultaneously and repeating in unlimited cycles. Each episode contained 8 events described by 4 attributes each. Results for different clustering algorithm setting are provided on Fig. 2. Processing costs of pyramid clustering heuristics strongly depends on the population size. Even, if this dependency is not exponential as for the clique finding algorithms still strong increase of costs is observable. However, the impact of the matrix size M is relatively low in respect to the population layer size L_{max}. Visible wobbling on plots is caused by three separate instances of the clustering algorithms. They are used for isolated processing of state nodes types and transformation bodies.

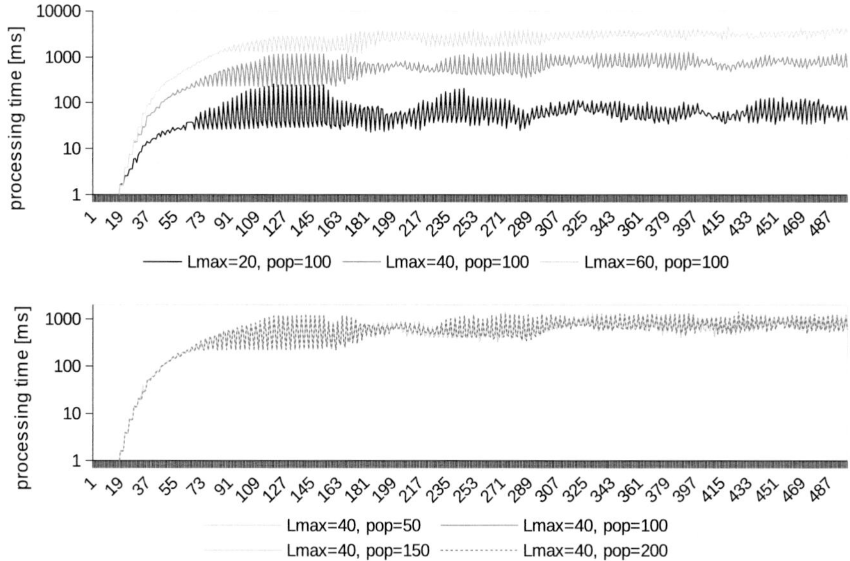

Fig. 2 The performance of the ontology updating procedure

4.3 Quality Evaluation

Quality is measured according to labels assigned to episodes by the generator (episodes types). For each episode, the ontology learning algorithm assigns concept id (cluster identifier) to the creation node, deletion node and transformation body. The quality is measured as the proportion of majority class (concept id assigned to given type of episode in most cases) to number of all identifiers assigned to episode of a particular type. Relatively low quality values in some cases are connected to drift of identifiers assigned to clusters. The process of id passing between cycles sometimes favors new identifier over existing ones lowering the result. It is a consequence of using heuristics that removes some information about clustered objects during cycle. Moreover, extensions of the population or layers sizes do not result in the better quality. It can happen because more candidate clusters not always carry important information from the perspective of the result's quality (Table 1).

Table 1 Quality for the ontology creation algorithm after 200 cycles (Adjusted Rand Index) and retrieved mode

Parameters	Creation nodes	Deletion nodes	Transf. bodies	Retrieved model
$L_{size} = 20, Pop = 20$	0.5700	0.5700	0.4228	
$L_{size} = 60, Pop = 20$	0.5700	0.5700	0.7809	
$L_{size} = 20, Pop = 50$	0.8540	0.8540	0.6097	
$L_{size} = 60, Pop = 50$	0.8540	0.8540	0.8323	
$L_{size} = 20, Pop = 100$	0.8540	0.8540	0.6379	
$L_{size} = 40, Pop = 100$	0.8540	0.8540	0.8323	
$L_{size} = 60, Pop = 100$	0.8540	0.8540	0.8323	

5 Conclusions

This paper discussed how graph-stream structure can be used to build an ontology explaining the complex process. Undoubtedly, following work on algorithms can led to more accurate and faster solutions. Particularly improvements in phases of candidate clusters creation and selection can lead to better ones. Obtained results shown that even complex algorithms can be used in the context of the data stream if the window size is delimited. Problem concerned in this paper can have practical applications in data streams analysis. Some of them have been discussed e.g., [10].

This paper is a result of the project financed by National Science Centre in Poland grant no. DEC-2011/03/D/ST6/01621.

References

1. Aggarwal, C.C., Li, Y., Yu, P.S., Jin, R.: On dense pattern mining in graph streams. In: Proceedings of VLDB Endowment, vol. 3, pp. 975–984 (2010)
2. Baraty, S., Simovici, D., Zara, C.: Foundations of intelligent systems. The Impact of Triangular Inequality Violations on Medoid-Based Clustering. LNCS. Springer, Heidelberg (2011)
3. Böttcher, M., Höppner, F., Spiliopoulou, M.: On exploiting the power of time in data mining. ACM SIGKDD Explor. Newsl. **10**, 3–11 (2008)
4. Griffiths, T.L., Tenenbaum, J.B.: Theory-based causal induction. Psychol. Rev. **116**, 661–716 (2009)
5. Khaleghi, B., Khamis, A., Karray, F.O., Razavi, S.N.: Multisensor data fusion: a review of the state-of-the-art. Inf. Fusion **14**, 28–44 (2013)
6. Kranen, P., Kremer, H., Jansen, T., Seidl, T., Bifet, A., Holmes, G., Pfahringer, B., Read, J.: Stream data mining using the MOA framework. In: Proceedings of the 17th International Conference on Database Systems for Advanced Applications—Volume Part II, pp. 309–313, Springer, Heidelberg (2012)
7. Krawczyk, B., Stefanowski, J., Wozniak, M.: Data stream classification and big data analytics. Neurocomputing **150**, 238–239 (2015)

8. Lauritzen, S.L., Richardson, T.S.: Chain graph models and their causal interpretations. J. R. Stat. Soc.: Ser. B (Stat. Methodol.) **64**, 321–348 (2002). Blackwell Publishers
9. Liu, J., Zhang, Q., Wang, W., McMillan, L., Prins, J.: PoClustering: lossless clustering of dissimilarity data. In: Proceedings of the Seventh SIAM International Conference on Data Mining, pp. 557–562. Minneapolis, Minnesota, USA, 26–28 April 2007
10. Marinazzo, D., Pellicoro, M., Stramaglia, S.: Causal information approach to partial conditioning in multivariate data sets. Comput. Math. Meth. Med. **17** (2012)
11. Rassi, C., Plantevit, M.: Mining multidimensional sequential patterns over data streams. In: Song, IY., Eder, J., Nguyen, TM. (eds.) Data Warehousing and Knowledge Discovery. Lecture Notes in Computer Science, vol. 5182, pp. 263–272. Springer, Heidelberg (2008)
12. Wan, L., Ng, W.K., Dang, X.H., Yu, P.S., Zhang, K.: Density-based clustering of data streams at multiple resolutions, ACM Trans. Knowl. Discov. Data, ACM, 3, 14:1–14:28 (2009)

On Properties of Undersampling Bagging and Its Extensions for Imbalanced Data

Jerzy Stefanowski

Abstract Undersampling bagging ensembles specialized for class imbalanced data are considered. Particular attention is paid to Roughly Balanced Bagging, as it leads to better classification performance than other extensions of bagging. We experimentally analyze its properties with respect to bootstrap construction, deciding on the number of component classifiers, their diversity, and ability to deal with the most difficult types of the minority examples. We also discuss further extensions of undersampling bagging, where the data difficulty factors influence sampling examples into bootstraps.

1 Introduction

Learning classifiers still reveal difficulties if data are class imbalanced. The standard classifiers are biased toward the majority classes and instances of the minority class tend to be misclassified. The methods for improving classifiers learned from imbalanced data are categorized in *data level* and *algorithm level* ones [4]. Methods from the first category are based on preprocessing, and they transform the original data distribution into more appropriate one for learning classifiers. The other category involves modifications of the learning algorithm, its classification strategy, or adaptation to the cost-sensitive framework [5]. *Ensembles* are also quite often considered. However, as the standard ensembles do not sufficiently recognize the minority class, new extensions have been introduced. They either employ preprocessing methods before learning component classifiers or embed the cost-sensitive framework in the ensemble learning process; see their review in [3]. Most of these ensembles are based on known strategies from bagging, boosting, or random forests. Although the ensemble classifiers are recognized as a remedy to imbalanced problems, there is still a lack of a wider study of their properties. Up to now, only few comprehensive studies were

J. Stefanowski (✉)
Institute of Computing Science, Poznań University of Technology,
60-965 Poznań, Poland
e-mail: jerzy.stefanowski@cs.put.poznan.pl

R. Burduk et al. (eds.), *Proceedings of the 9th International Conference on Computer Recognition Systems CORES 2015*, Advances in Intelligent Systems and Computing 403, DOI 10.1007/978-3-319-26227-7_38

carried out in different experimental frameworks [3, 8]. The main conclusion from [3] is that simple versions of undersampling or oversampling combined with bagging work better than more complex solutions. The results of [8] show that extensions of bagging significantly outperform boosting ones. Finally, the results of our previous comparative study [1] showed that Roughly Balanced Bagging (RBBag)—a variant of undersampling extensions of bagging—[6] achieved the best results and was significantly better than other oversampling extensions of bagging. However, all these studies did not attempt to more deeply analyze properties of the best undersampling bagging—namely Roughly Balanced Bagging. This paper is an experimental study which attempts to examine (1) the most influential aspects of constructing this kind of ensemble and its main properties (with respect to bootstrap construction, deciding on the number of component classifiers, their diversity, methods for aggregating predictions); (2) directions for its further extension and improvements.

2 Undersampling Extensions of Bagging

The original Breiman's bagging ensemble is based on the *bootstrap* aggregation, where the training set for each classifier is constructed by random uniformly sampling (with replacement) instances from the original training set. The component classifiers are induced by the same learning algorithm from the bootstrap samples and their predictions form the final decision with the equal weight majority voting. As bootstrap samples are still biased toward the majority class, many authors proposed to apply preprocessing techniques, which change the balance between classes in bootstraps.

In *Underbagging* approaches the number of the majority class examples in each bootstrap sample is randomly reduced. In simplest versions, the entire minority class is just copied and combined with randomly chosen subsets of the majority class to exactly balance cardinalities between classes. The *Roughly Balanced Bagging* (RBBag) [6] results from the critique of these approaches. Instead of fixing the constant sample size, it equalizes the sampling probability of each class. For each of T iterations the size of the majority class in the bootstrap sample (S_{maj}) is determined probabilistically according to the negative binominal distribution. Then, $N_{\min}$ examples are drawn from the minority class and S_{maj} examples are drawn from the entire majority class using bootstrap sampling as in the standard bagging (with or without replacement). The class distribution inside the bootstrap samples maybe slightly imbalanced and varies over iterations. According to [6] this approach is more consistent with the nature of the original bagging and performs better than simple underbagging. Another way to transform bootstrap samples includes oversampling the minority class before training classifiers. In this way the number of minority examples is increased in each bootstrap sample while the majority class is not reduced as in underbagging. This idea is realized with different oversampling techniques, either plan random replication of the minority examples or generating synthetic ones by SMOTE method, see their description in [3].

3 Experimental Studying of Roughly Balanced Bagging

The aims of the first part of experiments are to study the basic properties of constructing Roughly Balanced Bagging, which have not been sufficiently studied in the literature. More precisely we want to examine: (1) Using different learning algorithms to built component classifiers; (2) The influence of the number of component classifiers on the final performance of the ensemble; (3) The role of diversity of component classifiers inside Roughly Balanced Bagging.

We conduct our analysis on 15 real-world imbalanced data sets representing different domains, sizes, and imbalance ratio. Most of data sets come from the UCI repository and have been used in other works on class imbalance. One data set—`abdominal`—comes from our medical applications. For data sets with more than two classes, we chose the smallest one as a minority class and combined other classes into one majority class. Due to the space limits we skip the table with detailed characteristics and refer the reader to [2]. The performance of all ensembles is measured using: *sensitivity* of the minority class (the minority class accuracy), its *specificity* (an accuracy of recognizing majority classes), their aggregation to the *geometric mean* (G-mean). We have chosen these point measures, instead of AUC or other curve-based characteristics, as the majority of applied learning algorithms produce deterministic outputs. For the definitions of these measures, see, e.g., [5, 7]. These measures are estimated with the stratified ten fold cross-validation repeated three times to reduce the variance. All implementations are done for the WEKA framework.[1]

3.1 Using Different Learning Algorithms

While looking into the related works the reader notices that Roughly Balanced Bagging as well as other undersampling extensions of bagging are usually constructed with decision tress. Component classifiers are learned with C4.5 tree learning algorithm (J4.8 in case WEKA), which uses standard parameters except disabling pruning. In this study we check whether classification performance of Roughly Balanced Bagging may depend on using other learning algorithms. Besides J4.8 unpruned tree we considered Naive Bayes tree, rule algorithms—Ripper and PART, Naive Bayes classifiers and SVM (SMO) available in WEKA. The RBBag ensemble is constructed with different numbers (30, 50, and 70) of component classifiers. All considered evaluation measures do not indicate significant differences for using these algorithms. We skip the precise results and summary the average ranks according to Friedman test (which has not rejected the null hypothesis on equal performance of all versions of RBBag). Average ranks in Friedman test (the smaller, the better) are the

[1]We are grateful to our Master students Lukasz Idkowiak and Mateusz Lango for their help in implementing and testing these algorithms.

following: SVM 4.1; RIPPER 4.12; NBTree 4.4; J4.8tree/PART 4.5; NB 4.8. Furthermore, we verified (with the paired ranked Wilcoxon test) that in all versions, RBBag was significantly better than its standard bagging equivalent.

3.2 The Influence of the Number of Component Classifiers

In the literature, authors usually applied 30, 50, or 70 component classifiers inside Roughly Balanced Bagging. As in the previous experimental studies RBBag was a winner over other extensions of bagging for all these numbers, see, e.g., our studies as [1, 2], we have decided to examine more systemically other sizes of this ensemble. We stayed with constructing the component classifiers with J4.8 unpruned trees, and for each data set we constructed the same splits into learning and testing sets (inside the cross-validation). Then, we constructed a series of Roughly Balanced Bagging ensembles increasing one by one the number of component classifiers since 5 trees up to 50 ones. For each of these classifiers we estimated the evaluation measures. Due to page limits, in Table 1 we present results of G-mean for selected sizes of component classifiers only. We have observed that for all considered data sets increasing the number of component classifiers improves the evaluation measures up to the certain size of the ensemble. Then, the performance stays at a stable value for most cases. For some data sets it slightly varies around the certain level—however, we can still notice a stable constant reference line without any increasing or decreasing trend. A more surprising observation is that the RBBag ensemble achieves this good performance for a relatively small number of component classifiers. For most data sets, the stabilization of G-mean is observed approx. between 10 and 15 trees. In case of the sensitivity measure (accuracy of correct recognition of the minority class) we have noticed quite similar tendencies in performance. Although we can say that RBBag needs few more trees (approx 15–20 trees) to obtain the best value of the sensitivity. To sum up, these experimental results lead to conclusions that the best performance of Roughly Balanced Bagging comes from rather a small number of component classifiers. It is interesting observation, as this ensemble may require a heavy undersampling. For some large data sets (e.g., containing around 1000 examples) with the high imbalance ratio, the minority class may contain 50–60 examples, so the number of majority examples sampled into each bootstrap is also quite small. One could expect that due to such strong changes inside distributions in these bootstrap samples, their variance will be high, and undersampling bagging should reduce it by applying many components. However, the experimental results have showed that it is not a case. Therefore, we have decided to study more precisely the performance of single component classifiers and their diversity—see the next subsection.

Table 1 G-mean [%] of RBBag for a different number of component classifiers

Data set	No. of component classifiers						
	5	10	15	20	30	40	50
abdominal pain	79.10	80.0	80.11	80.09	80.16	80.06	80.10
balance scale	45.70	52.30	53.42	53.12	53.36	52.71	52.90
car	95.21	95.29	96.49	97.09	96.80	96.68	96.9
cleveland	64.02	68.03	71.77	72.14	71.6	70.75	71.3
cmc	62.21	64.74	65.95	64.86	64.81	64.73	64.33
credit-g	64.87	66.68	67.75	66.89	66.48	66.94	66.89
ecoli	84.41	86.38	87.42	88.31	87.52	86.74	87.0
haberman	61.22	62.02	63.51	63.92	62.24	62.65	62.8
hepatitis	71.32	73.92	75.47	76.16	76.23	76.19	75.33
ionosphere	88.51	90.88	90.30	90.47	90.25	90.76	89.95
new thyroid	95.35	95.18	97.36	97.15	96.73	97.02	96.8
solar flareF	81.67	82.04	84.07	84.91	84.80	83.13	83.8
transfusion	62.45	64.33	66.96	65.83	66.39	65.98	64.56
vehicle	93.32	45.13	95.34	94.61	94.77	95.19	94.91
yeast-ME2	81.22	82.59	84.41	83.70	84.37	83.35	80.86

3.3 *Diversity of Component Classifiers*

It is often claimed that performance of ensembles is decided by two factors: accuracy of an individual component classifier and diversity among all these classifiers [9]. Diversity is the degree to which component classifiers make different decisions on one problem. In particular, if they do not make the same wrong decision for a given classified instance, it allows the voting strategy to still produce the correct final prediction of the ensemble. Although, such an intuition behind constructing diverse component classifiers is present in many solutions, research on defining, so called, diversity measures and exploiting them to analyze the performance of ensembles is still an open research direction, see, e.g., a discussion on "good" and "bad" diversity in L. Kuncheva's book [10]. Furthermore, it is still not clear that how diversity affects classification performance especially on minority classes. The only work on ensembles dedicated for imbalance data has been done by S.Wang and X.Yao, who empirically studied diversity of specialized oversampling ensembles. Their conclusions do not provide a clear message as to increasing diversity among component classifiers. They have noticed that larger diversity caused better recognition for the minority class, however, at the cost of deteriorating the majority classes. They also concluded that tendency of F-measure and G-mean are decided by classifier accuracy and diversity together and the best F-measure and G-mean values appeared with rather medium accuracy and medium diversity [15]. However, nobody analyzed diversity of undersampling extensions of bagging as Roughly Balanced Bagging. Similarly to [15], we have selected Q-statistic as a basic diversity measure. For two classifiers C_i and C_k, Q-statistic is defined as [10]:

$$Q_{i,k} = \frac{N(11) \cdot N(00) - N(10) \cdot N(01)}{N(11) \cdot N(00) + N(10) \cdot N(01)} \tag{1}$$

where $N(ab)$ is the number of training instances for which classifier C_i gives result a and classifier C_k gives result b (It is supposed that the result here is equal to 1 if an instance is classified correctly and 0 if it is misclassified). The averaged Q-statistics is calculated over all pairs of M classifiers:

$$Q_{avr} = \frac{2}{M(M-1)} \sum_{i=1}^{M-1} \sum_{k=i+1}^{M} Q_{i,k} \tag{2}$$

The value of Q is normalized in the range -1 and 1. For independent classifiers Q value is 0. It has a high positive value if classifiers tend to recognize the same instances correctly, and will be negative if they commit errors on different instances. The larger the value is, the less diverse classifiers are [10]. We calculated Q-statistics for predictions in both classes and also for the minority class (denoted as Q(min)). Roughly Balanced Bagging was constructed with J4.8 tree classifiers and different number of components (15, 20, 30, 40, and 50). Additionally, we calculated the global Q for overbagging, SMOTEbagging to have a reference to earlier experiments from [15]. In Table 2 we present values of Q-statistics for ensembles constructed with 30 component trees. One can easily notice that Q values for Roughly Balanced Bagging are relatively high. For nearly all data sets they are least 0.6 (even around 0.8 for some data sets). The only exceptions are `new thyroid` (Q = 0.5 and Q(min) = 0.05) and `german credit` data (Q = 0.4 and Q(min) = 0.27). The small diversity concerns all class predictions (Q) and minority class ones—Q(min). Moreover, we noticed that the number of component classifiers in RBBag do not influence values of Q-statistics. They are very stable for all considered sizes of RBBag. Comparing RBBag to oversampling extensions of bagging, we can conclude that SMOTEbagging is more diversified—which is consistent with motivations of its authors [15]. However, both simple oversampling bagging and SMOTEbagging are less accurate than any undersampling extension of bagging. Previous comparative studies on a wider collection of imbalanced data, see [1], showed that Roughly Balanced Bagging significantly outperformed both these ensembles with respect to G-mean, F-measure, and Sensitivity of the minority class. Therefore, the high accuracy of Roughly Balanced Bagging is not directly related to its higher diversity. We have even analyzed predictions of particular pairs of classifiers and noticed that they quite often make the same decisions (most often correct ones). Note that the part of equation (1)—$N(10) \cdot N(01)$ (referring to different decisions) is relatively small comparing to $N(11)$. This could be also noticed by analyzing special diagrams of average pairwise error versus Q values. In Fig. 1 we show results of tested errors

Table 2 Q-statistics for RBBag and two extensions of overbagging

Data set	SMOTEBag	OverBag	RBBag-Q	RBBag-Qmin
abdominal pain	0.76	0.81	0.79	0.71
balance scale	0.49	0.73	0.70	0.5
car	0.9	0.95	0.91	0.80
cleveland	0.42	0.53	0.57	0.24
cmc	0.57	0.74	0.59	0.46
credit-g	0.38	0.48	0.49	0.34
ecoli	0.61	0.78	0.82	0.29
haberman	0.52	0.68	0.67	0.76
hepatitis	0.39	0.47	0.56	0.17
ionosphere	0.38	0.43	0.57	0.39
new thyroid	0.15	0.38	0.46	0.14
solar flareF	0.91	0.95	0.92	0.61
transfusion	0.71	0.86	0.83	0.79
vehicle	0.67	0.71	0.86	0.17
yeast-ME2	0.79	0.84	0.87	0.37

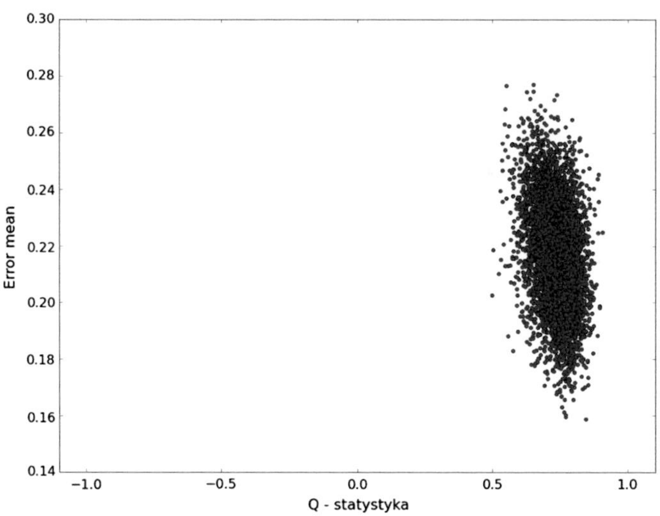

Fig. 1 Plot of an average error of the component classifiers versus $Q_{i,j}$ statistics for abalone data.

for RBBag built for `abalone` data set. The cloud of points referring to pairs of component classifiers shows that most of them are quite accurate (with small error values) independently of their small diversity (relatively high values of Q_{ij} statistics).

4 Impact of Imbalanced Data Difficulty Factors

Many studies showed that the degradation of classification performance is linked to the difficult characteristics of the minority class distribution, see, e.g., [11, 12, 16]. Recall that Napierala and Stefanowski proposed in [13] to model data difficulty factors concerning distributions of the minority class by identifying types of examples. Authors differentiate between safe and unsafe examples. *Safe examples* are ones located in the homogeneous regions populated by examples from one class only. Other examples are unsafe and more difficult for learning. Unsafe examples are categorized into *borderline* (placed close to the decision boundary between classes), *rare cases* (isolated groups of few examples located deeper inside the opposite class), or *outliers*. Following the method introduced in [13, 14] we assign types of examples using information about class labels in their k-nearest local neighborhood. Repeating conclusions from previous experimental studies [13, 14] we claim that most of data sets considered in this paper contain rather a small number of safe examples. The exceptions are two data sets composed of many safe examples: `new thyroid`, and `car`. On the other hand, there are data sets such as `cleveland`, `balance scale`, or `solar flareF`, which do not contain any safe examples but many outliers and rare cases. We have already observed in [2] that nearly all bagging extensions achieve similar high performance for safe data sets. The best improvements of all evaluation measures for RBBag are observed for the unsafe data sets. For instance, consider `cleveland` (no safe examples, nearly 50 % of outliers) where RBBag gives 72 % G-mean comparing to overbagging with 22.7 %. Similar highest improvements of RBBag over other bagging extension occur `balance scale`, `yeast-ME2`, `ecoli`, `haberman`, or `solar flare`.

To study more deeply the nature of RBBag we perform the same neighborhood analysis and labeling types of minority examples inside their bootstraps. For each bootstrap sample we label types of minority examples basing on class labels of the k-nearest neighbors. Then, we average results from all bootstraps. Due to page limits we cannot show precise results for all data sets. Selected ones are presented in Table 3. We notice that RBBag has changed types of the minority class distributions into safer ones inside their bootstraps. For many data sets which originally contain quite high numbers of rare cases or outliers, the undersampled bootstrap contain more safe examples. For instance, consider the very difficult `balance scale` data set, where RBBag creates bootstrap samples with at most 4 % outliers and 7.5 % rare cases while moving the rest of examples into safe and borderline ones. Similar data type shift could be observed for: `yeast-ME2` (originally 5 % safe examples, now over 70 %), `solar flareF`, `ecoli`, `ionosphere`, `hepatitis`, `cleveland`. We can hypothesize that this unexpected characteristics of undersampling may support constructing quite accurate component classifiers.

Table 3 Distributions of types of minority examples [in%] inside bootstrap samples of standard bagging and Roughly Balanced Bagging ensembles

Data set	Ensemble	Safe	Border	Rare	Outlier
adominal pain	bagging	61.39	23.76	6.93	7.92
	RBBag	72.60	19.15	5.11	3.15
balance scale	bagging	0.0	0.0	8.16	91.4
	RBBag	39.68	59.02	0.05	1.25
cleveland	bagging	0.0	45.71	8.57	45.71
	RBBag	42.86	53.33	0.44	3.37
ecoli	bagging	28.7	54.29	2.86	14.29
	RBBag	85.33	11.75	0.00	2.92
hepatitis	bagging	18.75	62.5	6.25	12.5
	RBBag	67.01	26.60	1.25	5.14
ionosphere	bagging	44.44	30.95	11.9	12.70
	RBBag	51.98	31.10	6.60	10.32
solar flareF	bagging	2.33	41.86	16.28	39.53
	RBBag	70.52	21.37	2.69	5.43
yast-ME2	bagging	5.88	47.06	7.84	39.22
	RBBag	64.31	34.29	0.22	1.18

5 Future Directions and Final Remarks

Let us first summarize results of the experiments. We can conclude that the choice of the learning algorithm to construct component classifiers inside the ensemble does not influence too much classification results of Roughly Balanced Bagging. The more interesting observation results from next experiments clearly showed that the number of component classifiers is not necessarily too high. In series of experiments we observed that the performance measures reach the stable level after generating approximately 15 classifiers. This could be a motivation for further research in at least two directions: (1) constructing an ensemble in an iterative way, starting from a small-sized ensemble and adding a new component by controlling the evaluation measure on the validation set; (2) incremental constructing relatively small online undersampling ensembles for evolving and imbalanced data streams—this is a particularly new and still open research challenge. Calculations of Q diversity measure for pairs of component classifiers clearly show that Roughly Balanced Bagging is less diversified than other variants of bagging. What is even more interesting—quite often pairs of classifiers both make the same, but correct, decisions. We observed that Roughly Balanced Bagging is more accurate while improving classification of imbalanced data than other ensembles which construct more diversified component classifiers (as SMOTEbagging and oversampling bagging). This opens several questions as to better studying the role of diversity and its trade off with accuracy in case of imbalanced data. Perhaps, following observations from [15] we should not focus too much on high diversity. Moreover, we think that current diversity measures may not sufficiently well capture characteristics of minority versus majority class predictions. Studying local accuracies for types of classified examples shows that Roughly Balanced Bagging can improve performance for unsafe distributions of the minority class. Comparing to original data distribution the sampling procedure of Roughly Balanced Bagging changes the nature of examples from the minority class—by analyzing the class labels inside the neighborhood we claim that the minority examples become safer (easier) for learning component classifiers. This part of experiments refers also to studying data difficulty factors for new versions of undersampling bagging, called Nearest Balanced Bagging [2], which is based on changing weights of sampled examples with respect to the results of analyzing the class labels inside the neighborhood of these examples. The preliminary results on ongoing research on this extended bagging show that modifications of sampling examples into bootstraps are more promising than modifications of aggregation techniques—as changing the simple majority voting inside RBBag into different versions of weighted voting does not significantly improve the prediction abilities.

Acknowledgments The paper was partially funded by the Polish National Science Center Grant No. DEC-2013/11/B/ST6/00963.

References

1. Blaszczynski, J., Stefanowski, J., Idkowiak L.: Extending bagging for imbalanced data. In: Proceedings of the 8th CORES 2013. Springer Series on Advances in Intelligent Systems and Computing, vol. 226, pp. 269–278 (2013)
2. Blaszczynski, J., Stefanowski, J.: Neighbourhood sampling in bagging for imbalanced data. Neurocomputing **150**-Part B, 529–542 (2015)
3. Galar, M., Fernandez, A., Barrenechea, E., Bustince, H., Herrera, F.: A review on ensembles for the class imbalance problem: bagging-, boosting-, and hybrid-based approaches. IEEE Trans. Syst. Man Cybern. Part C: Appl. Rev. **99**, 1–22 (2011)
4. He, H., Garcia, E.: Learning from imbalanced data. IEEE Trans. Data Knowl. Eng. **21**(9), 1263–1284 (2009)
5. He, H., Ma, Y. (eds.): IEEE Imbalanced Learning. Foundations, Algorithms and Applications. Wiley, New York (2013)
6. Hido, S., Kashima, H.: Roughly balanced bagging for imbalance data. Stat. Anal. Data Min. **2**(5–6), 412–426 (2009)
7. Japkowicz, N., Mohak, S.: Evaluating Learning Algorithms: A Classification Perspective. Cambridge University Press, Cambridge (2011)
8. Khoshgoftaar, T., Van Hulse, J., Napolitano, A.: Comparing boosting and bagging techniques with noisy and imbalanced data. IEEE Trans. Syst. Man Cybern.-Part A **41**(3), 552–568 (2011)
9. Krawczyk, N., Woźniak, M.: Analysis of diversity assurance methods for combined classifiers. In: Choraś, R.S. (ed.) Image Processing and Communications Challenges 4. Advances in Intelligent Systems and Computing, vol. 184, pp. 177–184. Springer, Heidelberg (2013)
10. Kuncheva, L.: Combining Pattern Classifiers: Methods and Algorithms, 2nd edn. Wiley, New York (2014)
11. Lopez, V., Fernandez, A., Garcia, S., Palade, V., Herrera, F.: An insight into classification with imbalanced data: empirical results and current trends on using data intrinsic characteristics. Inf. Sci. **257**, 113–141 (2014)

12. Napierała, K., Stefanowski, J., Wilk, S.: Learning from imbalanced data in presence of noisy and borderline examples. In: Szczuka, M., Kryszkiewicz, M., Ramanna, S., Jensen, R., Hu, Q. (eds.) Rough Sets and Current Trends in Computing. Lecture Notes in Computer Science, vol. 6086, pp. 158–167. Springer, Heidelberg (2010)
13. Napierala, K., Stefanowski, J.: Identification of different types of minority class examples in imbalanced data. In: Corchado, E., Snášel, V., Abraham, A., Woźniak, M., Graña, M., Cho, S.-B. (eds.) Hybrid Artificial Intelligent Systems. Lecture Notes in Computer Science, vol. 7209, pp. 139–150. Springer, Heidelberg (2012)
14. Napierala, K., Stefanowski, J.: Types of minority class examples and their influence on learning classifiers from imbalanced data. J. Intell. Inf. Syst. (accepted) (2015). doi:10.1007/s10844-015-0368-1
15. Wang, S., Yao, T.: Diversity analysis on imbalanced data sets by using ensemble models. In Proc. IEEE Symp. Comput. Intell. Data Min. pp. 324–331 (2009)
16. Weiss, G.M.: Mining with rarity: a unifying framework. ACM SIGKDD Explor. Newsl. **6**(1), 7–19 (2004)

Part IV
Image Processing and Computer Vision

Object Tracking Using the Parametric Active Contour Model in Video Streams

Marcin Ciecholewski

Abstract This article proposes a new approach which helps to prevent the formation of self-crossings and loops in the parametric active contour model while tracking moving objects in video streams. The presented solutions mean that the process of tracking is stable in all subsequent video sequences. On the other hand, self-crossings of nodes and the resultant contour loops happen very often in the basic model and ruin the tracking process. The Gaussian filter was used to eliminate noise from video streams.

1 Introduction

Active contour models are widely used for tracking and segmenting. Generally, active contour models are divided into two classes: parametric and region-based. In parametric models, the parametric form of the curve is defined and the fitting of the contour to the analysed shape in the digital image consists in minimising the total contour energy [3, 6, 8, 10]. The advantage of curve parameterisation is that the contour moves fast, so this approach dominates in the literature on the uses of active contours for tracking moving objects in video streams [1]. Unfortunately, the use of parametric models can be problematic for disturbed images because contour can loop (Fig. 1), which destabilises the tracking process. The literature does describe some solutions that make it possible to avoid the creation of false loops of the moving contour during subsequent iterations. Article [12] presents a solution in which self-crossings of active contour points are eliminated using calculated angular values between pairs of adjacent nodes, and also using four-connected line interpolation. Publication [9], in turn, demonstrates an approach in which it is possible to remove tangles only from a contour which is inflating, i.e. increasing its surface area, in segmenting blood vessels and the heart in single MRI images of the latter. In region-based models, it is not

M. Ciecholewski (✉)
Chair of Optimization and Control, Faculty of Mathematics and Computer Science, Jagiellonian University, ul. Łojasiewicza 6, 30-348 Krakow, Poland
e-mail: marcin.ciecholewski@uj.edu.pl

R. Burduk et al. (eds.), *Proceedings of the 9th International Conference on Computer Recognition Systems CORES 2015*, Advances in Intelligent Systems and Computing 403, DOI 10.1007/978-3-319-26227-7_39

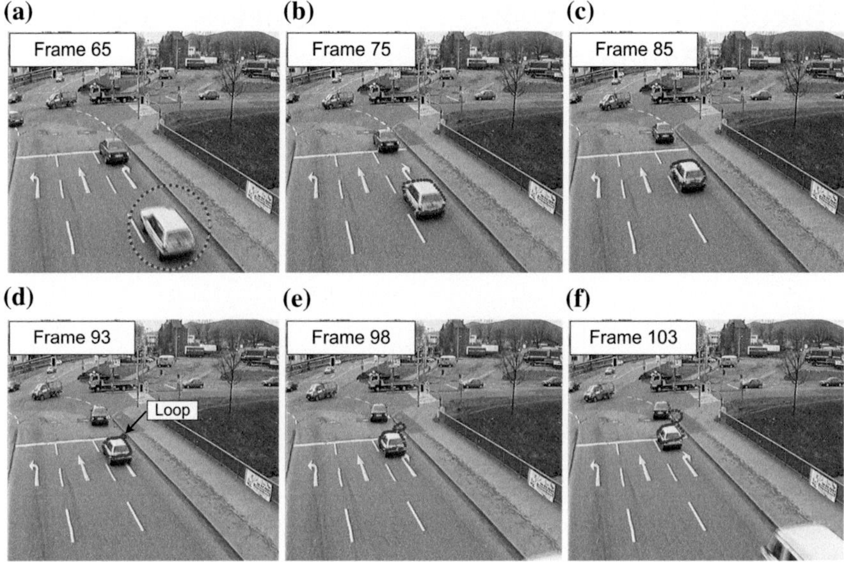

Fig. 1 A "false" loop forming in a parametric model based on the video file "rheinhafen.mpg"—source [7]. **a** Contour initialisation. **b**, **c** Video frames number 75 and 85, the active contour enables the moving car to be tracked. **d**–**f** Video frames number 93, 98 and 103, the contour loops and further tracking becomes impossible

necessary to eliminate self-crossings and loops, what is more the initial location of the contour has much less influence on the way the analysed shape is approximated in subsequent iterations. Unfortunately, the contour moves slower in region-based models than it does in parametric ones [12]. Region-based active contour models represent a certain optimal energy, calculated taking into account statistical data from sub-regions, based on which the model fits the analysed image best [4, 11, 13]. This study introduces new solutions to prevent active contour looping during moving object tracking in video sequences, based on calculated distances between adjacent nodes and the simultaneous control of angular values of subsequent node pairs. This solution was tested for various video streams in which moving cars and pedestrians were tracked. The contents of this article are arranged as follows. Section 2 describes the parametric active contour model. Section 3 presents the computer implementation of at tracking algorithm for video sequences, and Sect. 4—selected experiments conducted. The last section is the summary.

2 Parametric Active Contour Model

In a continuous domain, an active contour is defined using the parametric curve $v(s) = (x(s), y(s))$ whereas $s \in [0, 1]$. In order to allow the model to fit the analysed shape in the image, the energy components of this model are linked so as to minimise the total contour energy. The contour energy ξ depends both on its current shape and the function of image brightness $I(x, y)$. The total contour energy can be presented using the following formula:

$$\xi(v) = \alpha(v) + \beta(v) \tag{1}$$

whereas the internal energy which models the contour shape is presented as follows:

$$\alpha(v) = \int_0^1 \left(w_1(s)|\frac{\delta v}{\delta s}|^2 + w_2(s)|\frac{\delta^2 v}{\delta s^2}|^2 \right) ds \tag{2}$$

where w_1 and w_2 are certain weight functions. The external energy, i.e. the effect of the image, is expressed by the following relationship:

$$\beta(v) = \int_0^1 P(v(s))ds \tag{3}$$

where $P(x, y)$ is a function with minimums in the points of the image in which the gradient is the steepest:

$$P(x, y) = -c||\nabla(G_\sigma * I(x, y))|| \tag{4}$$

The convolution $G_\sigma * I(x, y)$ represents the digital image after Gaussian filtration with the parameter σ. According to the calculus of variations theory, the contour $v(s)$ which minimises the energy ξ fulfils the Euler–Lagrange equation:

$$-\frac{\delta}{\delta s}\left(w_1 \frac{\delta v}{\delta s} \right) + \frac{\delta^2}{\delta s^2}\left(w_2 \frac{\delta^2 v}{\delta s^2} \right) + \nabla P(v(s)) = 0 \tag{5}$$

Let contour v have the following form: $v(s, t) = (x(s, t), y(s, t))$ whereas $v(., t)$ represents the contour at the moment $t \in [0, \infty]$. The contour v equation then has the following form:

$$\mu(s)\frac{\delta^2 v}{\delta t^2} + \gamma(s)\frac{\delta v}{\delta t} - \frac{\delta}{\delta s}\left(w_1 \frac{\delta v}{\delta s} \right) + \frac{\delta^2}{\delta s^2}\left(w_2 \frac{\delta^2 v}{\delta s^2} \right) + \nabla P(v(s, t)) = 0 \tag{6}$$

where $\mu(s)$ represents the mass, and $\gamma(s)$ the damping coefficient. In the discrete domain, the active contour is shown as a set of N nodes $v_i(t) = (x_i(t), y_i(t))$ for $i = 1, 2, \ldots, N$. Hence the active contour equation (6) can be presented in the iterative form:

$$v_i(t+1) = v_i(t) - \frac{1}{\gamma}\left(\alpha F_i^{tensile}(t) + \beta F_i^{flexural}(t) - F_i^{external}(t) - F_i^{inflation}(t) \right) \tag{7}$$

Equation (7) accounts for the fact that the masses of nodes are equal to 0 and a certain constant value dumping the movement of nodes is assumed. $F_i^{inflation}(t) = F(I_s(x_i, y_i))n_i(t)$ is the inflation force acting on the node marked with index i at iteration t in a direction normal to the contour, determined using the vector $n_i(t)$. This force makes it possible to stretch or expand the contour in the direction of the edge being approximated using the following function:

$$F(x, y) = \begin{cases} +1 & \text{if } I(x, y) \geq T \\ -1 & \text{if } I(x, y) < T \end{cases} \tag{8}$$

where T is the defined brightness threshold. $F_i^{external}(t)$, in turn, is the external force based on image data, defined as

$$F_i^{external}(t) = \nabla P(x_i(t), y_i(t)) \tag{9}$$

whereas function P is defined in Eq. (4). $F_i^{tensile}(t)$, the tensile force, is the tension which prevents expansion:

$$F_i^{tensile}(t) = 2v_i(t) - v_{i-1}(t) - v_{i+1}(t) \tag{10}$$

while $F_i^{flexural}(t)$, the flexural force, expresses the bending force which prevents flexing:

$$F_i^{flexural}(t) = 2F_i^{tensile}(t) - F_{i-1}^{tensile}(t) + F_{i+1}^{tensile}(t) \tag{11}$$

3 Computer Implementation of the Tracking Algorithm in Video Sequences

In order to simplify the computer implementation, Eq. (7) can be written as follows:

$$v_{i+1}(t) = v_i(t) - \left(\hat{\alpha}F_i^{tensile}(t) + \hat{\beta}F_i^{flexural}(t) - \eta F_i^{external}(t) - \tau_0 F_i^{inflation}(t)\right) \tag{12}$$

what is more, the inflation force need not be dampened for subsequent sequences, so it can be established that $\tau_0 = 1$ and the inflation force is constant for each nodal point. In addition, a certain value of the brightness threshold T must be assumed for the determined edges, and also a direction of the inflation force has to be determined (a positive direction means the contour is expanding to approximate the edges of the analysed object, otherwise this will be a contracting contour), and it is also necessary to assume a certain constant number of nodes N needed to approximate the moving object. The user also sets the minimum and the maximum value of the angles between adjacent nodes (represented by $\theta_{min}, \theta_{max}$), and also the minimum and maximum

Algorithm 1 The tracking algorithm for video sequences.

Input: V - a set of N nodes; T, $\hat{\alpha}$, $\hat{\beta}$, η, D_{min}, D_{max}, θ_{min}, θ_{max} – set by the user
Output: V - a set of N nodes
for $i = 1 \rightarrow N$ **do**
  $n \leftarrow \text{NormalVector}(v_{i+1} - v_{i-1})$
  if $I(v_i) < T$ **then**
    $n \leftarrow -n$ {vector reversal}
  end if
  $d \leftarrow ||v_i - v_{i-1}||$
  $\theta \leftarrow \angle(v_{i-1} - v_i, v_{i+1} - v_i)$
  if $(d > D_{max}$ **and** $\theta > \theta_{min})$ **or**$(d > D_{min}$ **and** $\theta > \theta_{max})$ **then**
    Calculate $MaxDistance$ {the greatest distance between subsequent nodal points}
    Remove node v_i
    Add node $v_i = \frac{\hat{v}+\tilde{v}}{2}$ {where $\|\hat{v} - \tilde{v}\| = MaxDistance$}
  else if $d < D_{min}$ **then**
    Calculate $MaxDistance$
    Remove node v_i
    Add node $v_i = \frac{\hat{v}+\tilde{v}}{2}$
  end if
  Calculate $F_i^{tensile}$
  Calculate $F_i^{flexural}$
  Calculate the external force $F_i^{external}$ and image gradient values for point v_i
  Compute node coordinates using equation v_i (12)
end for

distance between nodes (represented by D_{min}, D_{max}). As Algorithm 1 includes a control of the value of these for parameters, it is possible to prevent the formation of self-crossings and loops of the active contour. If the node breaches the permissible angular values (θ_{min}, θ_{max}) and distances (D_{min}, D_{max}), then according to Algorithm 1, this node is placed between a pair of adjacent nodes for which the distance is the greatest in the contour for the given video sequence. Examples of topological changes of the contour (e.g. merging and splitting operations) are presented in the previous research described in the publication [5]. Publication [5] contains examples of the algorithm behaviour in the case of exceeding fixed angle values and distances between adjacent nodes, and examples of preventing the formation of self-crossings and loops are also given. Publication [5] presents an algorithm which allows extracting gall bladder shape from ultrasonographic (USG) images and its modified form can also be used for tracking moving objects in video streams, which is the subject of the current research.

4 Selected Experimental Results

An example of Algorithm 1 operation based on a video sequence is presented in Figs. 2 and 3. After the initiation—Fig. 2a—the contour contracted, zooming in on the car which was then tracked in subsequent iterations and in all video frames. When

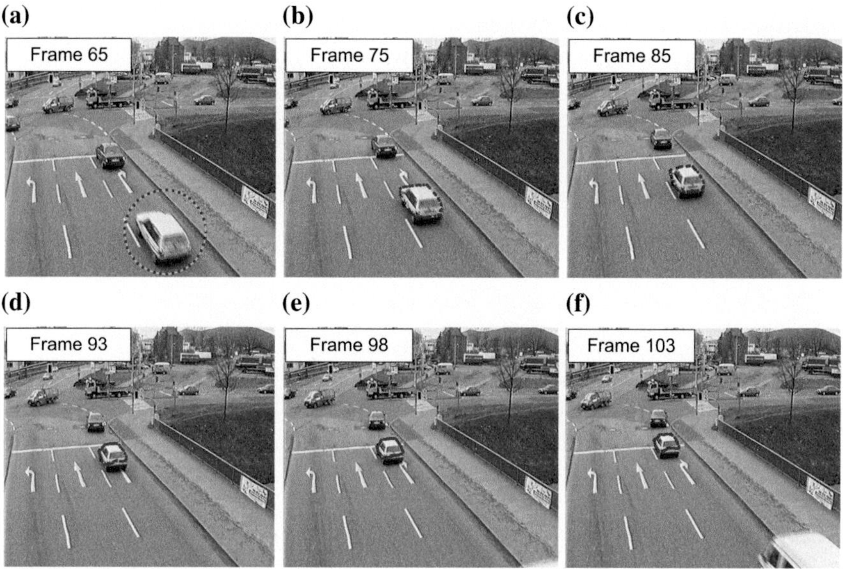

Fig. 2 Using Algorithm 1 to track a moving car in video sequences based on an example video file "rheinhafen.mpg"—source [7]. Values of parameters for the experiment conducted are presented in Table 1. **a** Contour initiation. **b–f** Frames number 75, 85, 93, 98, 103, the contour allows the moving car to be tracked

the basic version of active contour was implemented, the moving car tracking was prohibited by a loop that formed, as shown by the examples in Fig. 1d–f. According to Algorithm 1, at every iteration new coordinates of nodes are determined (the V set), and their location is updated in the next frame of the video so as to adjust the nodes to the current (shifted) location of the tracked object using Eq. (12). Equation (12) is updated based on forces $F_i^{tensile}$, $F_i^{flexural}$, $F_i^{external}$ calculated at every iteration, and also using the values of parameters T, $\hat{\alpha}$, $\hat{\beta}$, η, D_{min}, D_{max}, θ_{min}, θ_{max} which do not change for every iteration. Algorithm 1 was implemented in the C++ language using the OpenCV library [2]. Table 1 contains the parameters of the active contour for an example experiment presented in Fig. 2. If video sequences contain noise and the edges of the moving objects are not clear cut, the contour can also destabilise, like in the case of self-crossings. This is why, before the active contour is applied, a Gaussian filtration with a specified sigma value (given in Table 1) experimentally selected for the analysed case should be carried out to remove the noise found in video sequences. The 'resistance' of the algorithm to interference or its susceptibility to destabilisation depending on the intensity of interference will form the subject of subsequent research. Figure 3 presents experiments in which the parameters of the brightness threshold T are incorrectly selected and in which no Gaussian filtration was carried out before the contour was initialised. Values of parameters for the experiments conducted and shown in Fig. 3 are presented in Table 2. In examples from Fig. 3b–f, setting an excessive value of brightness T caused the contour to

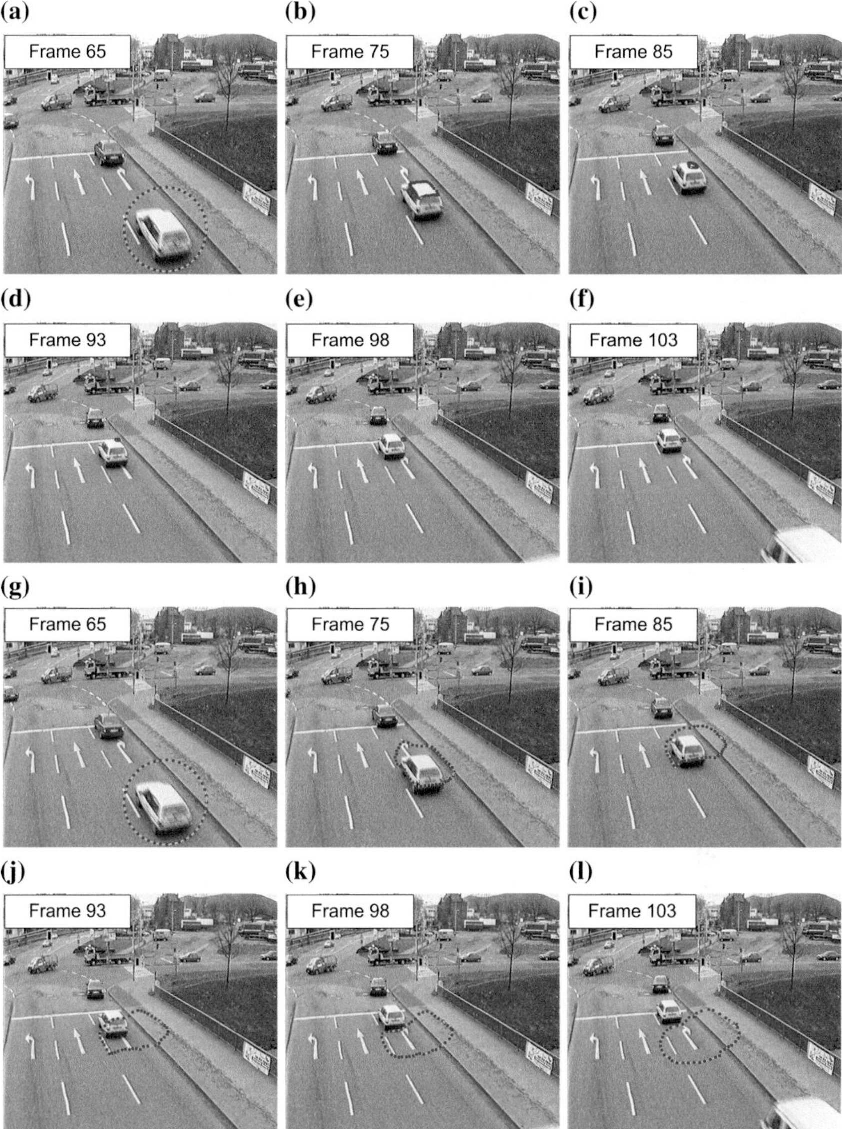

Fig. 3 Examples of an incorrectly set brightness threshold T and no Gaussian filtration (σ) in Algorithm 1. Values of parameters for the presented examples are given in Table 2. The experiments were carried out on the video file "rheinhafen.mpg"—source [7]. **a** Contour initialisation. **b–f** At subsequent iterations the contour clearly deflates and as a result this prohibits the moving car from being tracked. **g** Contour initialisation. **h–l** The contour has been inflated too much and the tracking has failed

Table 1 Values of parameters for the parametric method of active contour for examples from Fig. 2

Parameter name	N	$\hat{\alpha}$	$\hat{\beta}$	η	T	D_{min}	D_{max}	θ_{min}	θ_{max}	σ
Parameter value	41	1	0.5	40	0.75	3 pixels	8 pixels	21°	30°	3

Table 2 Values of parameters for the parametric method of active contour for examples from Fig. 3

Parameter name	N	$\hat{\alpha}$	$\hat{\beta}$	η	T	D_{min}	D_{max}	θ_{min}	θ_{max}	σ
Parameter value	41	1	0.5	40	0.85	3 pixels	8 pixels	21°	30°	–
List of parameters for examples from Fig. 3a–f										
Parameter name	N	$\hat{\alpha}$	$\hat{\beta}$	η	T	D_{min}	D_{max}	θ_{min}	θ_{max}	σ
Parameter value	41	1	0.5	40	0.6	3 pixels	8 pixels	21°	30°	–
List of parameters for examples from Fig. 3g–l										

deflate too much and it became impossible to track the moving car. Figure 3h–l, in turn, present examples of tracking destabilisation when the brightness threshold T is set at too low a value. Here, too, the Gaussian filtration was not executed.

5 Conclusions

This article proposes new solutions which make it possible to control the location of nodes in the parametric active contour model and thus prevent the formation of node self-crossings and loops during subsequent iterations when moving objects are tracked in video streams. The solutions applied ensure a stable behaviour of the active contour at all iterations, in comparison to the basic version, but the number of parameters which have to be initiated in the active contour has increased and the position of the initial contour also must be determined. Then, the active contour fits the edges of the analysed object and makes it possible to track it in subsequent frames. Due to the frequent disturbances of video streams, they have to be preprocessed using a Gaussian filter. It is planned to further develop and test this method both on video streams and single digital images. Because the method presented in this paper is only at the prototype stage, in further research it will be worthwhile to compare it to other methods of tracking moving objects in video streams.

References

1. Blake, A., Isard, M.: Active Contours: The Application of Techniques From Graphics, Vision, Control Theory and Statistics to Visual Tracking of Shapes in Motion. Springer, New York (1998)
2. Bradski, G.: The OpenCV library. Dr. Dobb's J. Softw. Tools **25**(11), 120–123 (2000)
3. Caselles, V., Kimmel, R., Sapiro, G.: Geodesic active contours. Int. J. Comput. Vis. **22**, 61–79 (1997)

4. Chan, T.F., Vese, L.A.: Active contours without edges. IEEE Trans. Image Process. **10**, 266–277 (2001)
5. Ciecholewski, M., Chocholowicz, J.: Gallbladder shape extraction from ultrasound images using active contour models. Comput. Biol. Med. **43**(12), 2238–2255 (2013)
6. Cohen, L.: On active countor models and balloons. Comput. Vis. Graph. Image Process. **53**, 211–218 (1991)
7. Image Sequence Server. Available on-line: http://i21www.ira.uka.de/image_sequences/#rheinhafen
8. Kass, M., Witkin, A., Terauzopoulos, D.: Snakes: active contour models. Int. J. Comput. Vis. **1**(4), 321–331 (1988)
9. Makowski, P., Sørensen, T.S., Therkildsen, S.V., Materka, A., Stødkilde-Jørgensen, H., Pedersen, E.M.: Two-phase active contour method for semiautomatic segmentation of the heart and blood vessels from MRI images for 3D visualization. Comput. Med. Imaging Graph. **26**(1), 9–17 (2002)
10. McInerney, T., Terzopoulos, D.: Topologically adaptable snakes. In: Proceedings of the International Conference on Computer Vision, pp. 840–845 (1995)
11. Mumford, D., Shah, J.: Optimal approximations of piecewise smooth functions and associated variational problems. Commun. Pure Appl. Math. **42**, 577–685 (1989)
12. Nakhmani, A., Tannenbaum, A.: Self-crossing detection and location for parametric active contours. IEEE Trans. Image Process. **21**, 3150–3156 (2012)
13. Yezzi, A.J., Tsai, A., Willsky, A.S.: A fully global approach to image segmentation via coupled curve evolution equations. J. Vis. Commun. Image Represent. **13**, 195–216 (2002)

Vision Diagnostics of Power Transmission Lines: Approach to Recognition of Insulators

Angelika Wronkowicz

Abstract Due to requirements related to the maintenance of power transmission lines, it is necessary to diagnose their condition regularly. Among approaches being nowadays applied fundamental technical diagnostic methods are the vision inspections, often performed with use of cameras. The aim of this paper is to develop a method of automated recognition of insulators in images for the purposes of further computer analysis of their technical condition. Application of image segmentation by the statistical region merging method lets separate objects visible in images of very composed backgrounds. In order to recognize the insulators the template matching by an improved brute force method was used. The author proposes an approach which makes use of the fact that insulators are longitudinal and the problem with scaling and rotation variability is solved. The proposed algorithm can be applied to automatic recognition of insulators as well as any oblong elements in images.

Keywords Image segmentation · Image recognition · Template matching · Power transmission lines · Insulators

1 Introduction

Power transmission lines are ones of the technical structures which are fundamental to our daily life. Requirements for them, which are high efficiency, reliability and safety, lead to a need of performing their regular inspections. Regular monitoring of power lines is very important since they are permanently exposed to influence of environment as well as vandalism or thefts. The most common damage that may occur in the power lines structure are mechanical failures or collapses of pylons, cracks or degradation of insulators (respectively, for the glass or polymer ones), corrosion and fatigue (of pylons, foundations or elements of insulators such as cores, pins, nuts),

---

A. Wronkowicz (✉)
Institute of Fundamentals of Machinery Design, Silesian University of Technology,
Konarskiego Street 18A, 44-100 Gliwice, Poland
e-mail: angelika.wronkowicz@polsl.pl

R. Burduk et al. (eds.), *Proceedings of the 9th International Conference on Computer Recognition Systems CORES 2015*, Advances in Intelligent Systems and Computing 403, DOI 10.1007/978-3-319-26227-7_40

breaks of electrical conductors, as well as appearance of so-called corona discharges. Examples of diagnostic methods of such structures are the vision-based ones (often with use of mobile devices with a regular or infrared camera) [1–3], procedures connected with electrical parameters measurement [4, 5] as well as vibration and acoustic measurements [6, 7]. The vision-based methods are very important since it is highly needed to notice defects or potential failures in their earliest possible stage in order to avoid damage and thus its serious consequences. Damage of power lines, or even adjacent vegetation, leads to disruption in energy supply and may result in big financial losses or even disasters, like fires. Difficulties in carrying vision inspections result mainly from a difficult access to the inspected object, its extensiveness and large dimensions as well as danger arising from operation under high or very high voltage. Recently noted a big interest in performing inspections with use of unmanned aerial vehicles (UAVs), remotely controlled or autonomous, also called drones, together with a camera (or various types of cameras simultaneously). UAVs have found a very significant application in a military industry [8] as well as commercial applications, for instance for inspecting bridges [9], power lines, windmills [10], buildings [11] or for the aim of acquisition of mapping data [12]. Because in case of power lines a number of data collected during such inspections is huge, certain problems with its analysis occur. The analysis of such a big amount of data by human experts is problematic, long lasting and can be unreliable due to eye strain. Therefore, it is needed to aid such procedures and develop a method for automated detection of defects or any other irregularities of diagnosed structures basing on acquired images during vision inspections. Numerous studies on power lines elements detection have been developed. A big part of them is connected with identifying electrical conductors by detecting linear objects in images [13–16]. The literature analysis indicated that detection of the other elements of power lines has not been deeply investigated to-date. The examples, where the failures detection was investigated are the studies on inspecting the damage of insulators [17] as well as breakers [18]. The author of this paper works on developing a mobile vision system for automated reasoning about power transmission lines condition [19]. In this paper, the author would like to present an approach to automated recognition of insulators in images. There must be a problem taken into consideration that the acquired images have variable and diverse backgrounds, thus the method must be universal for different scenarios.

2 Theoretical Background

The first step before object recognition is usually the image segmentation which is used in order to divide or separate objects visible in an image. This is one of the most difficult steps in image processing since its accuracy extremely influences on consecutive steps of pattern recognition. Image segmentation is usually performed based on a pixel colour, intensity or texture. The most common methods used for image segmentation are the amplitude segmentation, clustering segmentation, boundary-based segmentation, region-based segmentation and texture segmentation. The amplitude segmentation methods are based on thresholding of luminance or colour components of an image. The clustering techniques consist in grouping a set of objects based on similarity of their features. By the boundary-based method particular regions can be separated by edges that can be detected using the Canny, Prewitt, Sobel, Roberts, Laplacian of Gaussian or zero-cross method. The examples of region-based methods are region growing, region splitting and merging or watershed segmentation and they are based on assumption that pixels in neighbouring regions have similar values of colour and intensity. The texture segmentation consists in identifying regions basing on their texture [20, 21]. Since the considered problem in this paper is very complicated mainly because of the complexity and variability of backgrounds that can be recorded during vision inspections of power transmission lines, a region-based segmentation method was decided to be applied because of its high accuracy. The separation of regions in the analyzed image, by their colour and intensity, and subjecting all of them to further analysis ensures that the region of interest will not be excluded from an image like it could happen using other methods of image segmentation. Object recognition methods rely on matching, learning or pattern recognition algorithms using feature-based or appearance-based techniques. The examples include statistical or syntactic pattern recognition, artificial neural networks, template or graph matching [22]. In this paper, the simple template matching by the brute force method using normalized cross-correlation was applied. In such a method, a template is compared to all regions in an analyzed image and if the match between a template and a region is close enough, this region is labelled as the template object. Because a template may have a different size and a position angle than a region of interest (ROI) some additional procedures should be performed in order to make such a method rotation- and scale-invariant. There have been numerous researches done to develop a translation-, rotation-, and/or scale-invariant techniques of template matching. Many researches have been reported where the authors performed the template matching after rotating a template by numerous values of angles and scaling by selected scale factors, e.g. in [23, 24]. Although such a classical method is effective, it is time-consuming. Examples of other approaches include a wavelet decomposition approach using the ring projection transform [25], moments-based approach [26] or FFT-based matching method [27]. The author of this paper would like to propose a simple rotation- and scale-invariant approach that can be used for recognizing longitudinal objects and ensures a great acceleration of the classical brute force algorithm.

3 Algorithm Description

The preliminary step of recognition of insulators is to first distinguish them among other details visible in an image. This step consists in image segmentation by the region merging method. The author decided to apply the statistical region merging (SRM) method proposed by the authors of [28], which is very effective and has brought the expected results. Afterwards, template matching was applied in order to

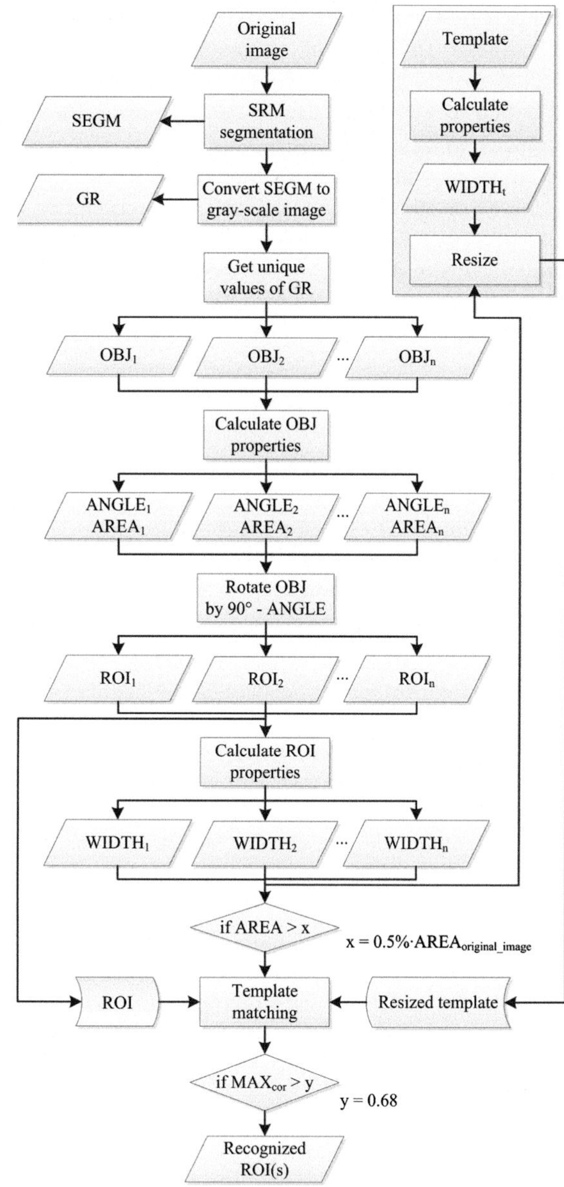

Fig. 1 The algorithm of automated recognition of insulators

recognize insulators among all objects extracted by image segmentation. The brute force method of template matching, using the normalized cross-correlation matrix function [29], was used. Because this method is not rotation- and scale-invariant, the author proposes introduction of additional steps in the algorithm. All the applied algorithms were implemented in Matlab® and its block diagram is presented in Fig. 1. The detailed description about particular steps is presented in the subsections below.

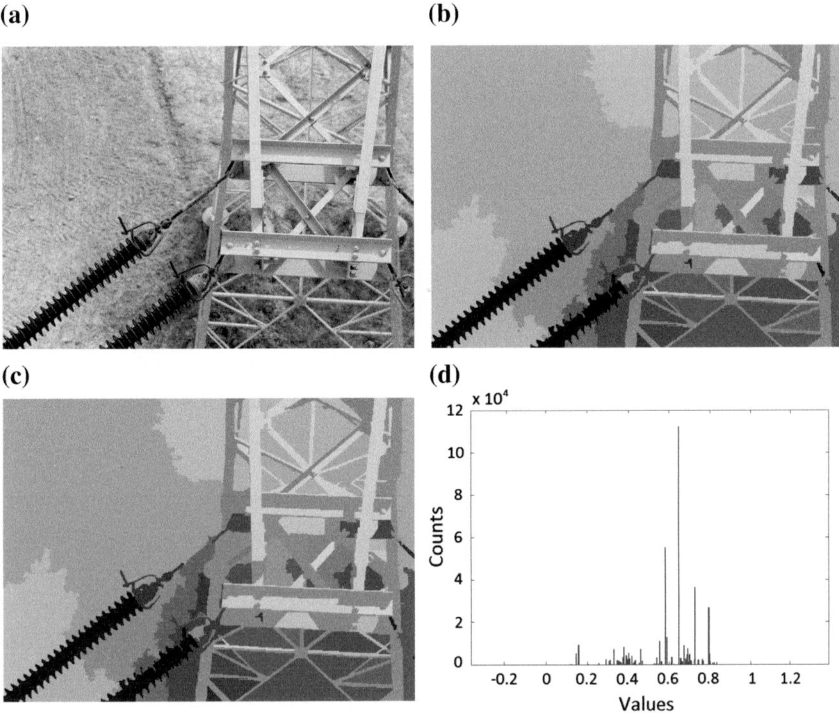

Fig. 2 **a** The original image (taken from [10] under permission), **b** the image after segmentation, **c** the grey-scale image, **d** the histogram of the grey-scale image

3.1 Image Segmentation

The original image (Fig. 2a), took by the Aibotix GmbH company [10] (use under permission) during vision inspection of power lines, was divided into separate regions by image segmentation using the SRM method. The results are presented in Fig. 2b. Then, this image was converted into a grey-scale one (Fig. 2c) in order to obtain exactly one value for each pixel occurring in the image. All unique values of the grey-scale image were collected (Fig. 2d) and each of them represents a single detected object or region in the image. Few examples of detected objects are presented in Fig. 3, where we can notice that Fig. 3a, f represent the insulators and the others show other parts of a power line or a background.

3.2 Template Matching

Certain preparatory steps before template matching were introduced in order to eliminate the problem with rotation and scaling variability of an object with respect to

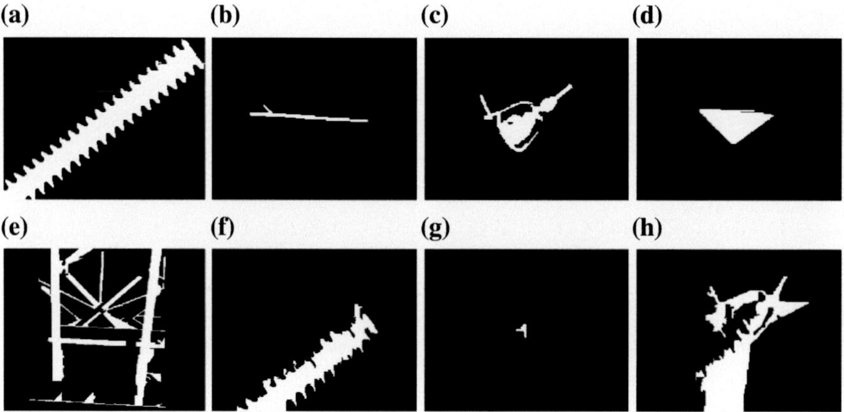

Fig. 3 The exemplary detected objects

a template. These steps were performed for each of detected regions, iteratively. The region properties, such as an orientation and a surface area were calculated. Because the ROIs in this case insulators are longitudinal, all detected regions were positioned to the vertical position by rotating them (see example in Fig. 4c) by 90° minus the calculated orientation (Fig. 4b). Then, a width of the object was calculated and the template (Fig. 4a), representing a segment of an insulator, was resized by the calculated ratio of its width with respect to the width of the analyzed object. Another procedure was added in order to eliminate very small detected regions from the matching procedure. Such exclusion is very beneficial since it accelerates the algorithm and reduces the risk of recognition of inappropriate regions, such as small parts of a field in the background which may have a similar shape to the template. With certain values of parameters settings, even a rectangular area could be identified as an insulator. Having the information about approximate distance from a camera to a power line during acquisition of images one can assume a safe number of pixels (surface area) which can be excluded from the template matching process. In the considered case, this threshold was estimated as a 0.5 % part of the whole image.

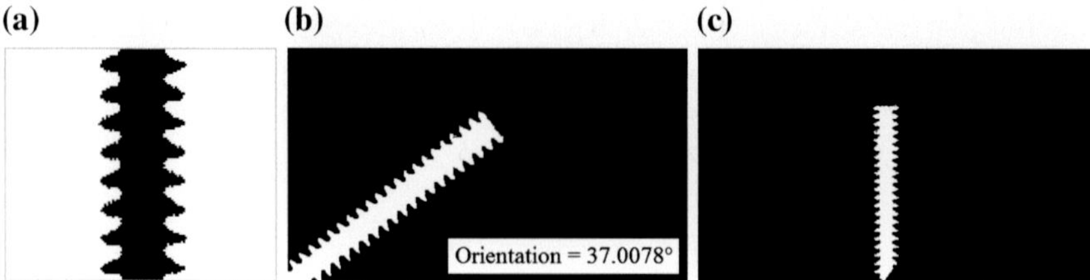

Fig. 4 **a** The template, **b** the exemplary ROI and its calculated orientation, **c** the ROI after rotation to vertical position

3.3 Final Results

The computation of the normalized cross-correlation of matrices with the template and the analyzed image returns the correlation values that consist in the range of $[-1, 1]$ since matching can be positive or negative depending on uniformity or inverse of colours values. Therefore, the absolute values (ABS) of correlation are taken into consideration, i.e. the closer this value to 1 the better the matching is. The coordinates of a point for the highest ABS correlation value correspond to coordinates of a specified point in a template. It means that in this position matching of the template to the element in the picture is the best. Knowing these coordinates and dimensions of the template one can precisely localize the recognized object in the analyzed image. In this paper, these locations were marked with a red rectangle (see Fig. 6). The results of recognition using the proposed algorithm for the first considered case is presented in Fig. 6a. The threshold of the ABS of the highest correlation value, above which an analyzed object is labelled as the ROI, was experimentally selected and set to 0.6. The highest ABS correlation value for the detected ROIs equalled 0.9279 (Fig. 5) and 0.6845, respectively, for the insulator on the left side and on the right side of the image. The correlation values for the rest of analyzed ROIs were much lower (approximately 0.2–0.4) and not higher than 0.58. In Fig. 5 one can observe one of the results presenting correlation coefficients between the rotated ROI (Fig. 4c) and the template (Fig. 4a) after scaling with respect to coordinates of the analyzed image (in pixels). The method was tested on several another figures using slightly different templates (with various number of ribs of an insulator). The examples are presented in Fig. 6b, where both insulators were recognized properly (with correlation value above 0.65) and in Fig. 6c, where only one insulator was

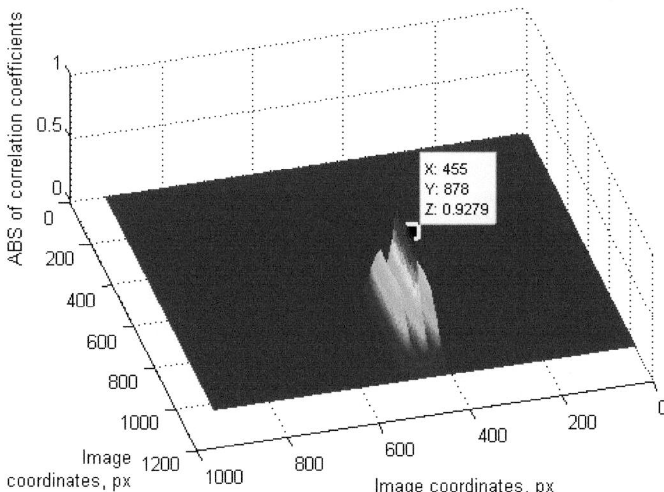

Fig. 5 The exemplary resulted ABS of correlation coefficients

Fig. 6 Examples of results of recognition of insulators (original images taken from [10] under permission)

recognized (with correlation value above 0.7). The omission of the second insulator was caused by its merging with a part of the first one (they are too close to each other in this image) during the segmentation process. Moreover, there is also one other object recognized improperly. Other tests were performed on images of poor quality and they did not bring good results since there were many undesirable recognized objects. One can notice that application of the proposed algorithm using the selected values of parameters in most cases has brought the expected results. However, the method is very sensitive to noise or any irregularities of shapes of analyzed objects, thus in order to eliminate recognition of inappropriate elements, the acceptable correlation value should be very high. On the other hand, setting this value at too high level would then lead to a risk of omitting the authentic ROIs, for example noised after the operation of image segmentation. The establishment of the acceptable correlation value is therefore a problematic matter in the considered problem. Furthermore, the quality of analyzed images should be sufficiently high for the purpose of proper image segmentation.

4 Conclusions

In the paper, the method of recognition of insulators in images acquired during vision inspections of power transmission lines is presented. For the purpose of separating the insulators from the image with a complex background the statistical region merging method was applied. The recognition of insulators among all detected objects after image segmentation was performed using the template matching by the brute force method using normalized cross-correlation. The author proposed simple techniques by which the algorithm can be significantly accelerated. The method is translation-, rotation- and scale-invariant and is useful for recognition of insulators in images, which was verified by the presented case studies. The presented approach can be used for the recognition purposes in vision diagnostics of elements of power transmission lines. The proposed techniques can be also used for other applications where image recognition of any longitudinal objects is needed.

References

1. Ahmad, J., Malik, A.S., Xia, L., Ashikin, N.: Vegetation Encroachment monitoring for transmission lines right-of-ways: a survey. Electr. Pow. Syst. Res. **95**, 339–352 (2013)
2. Bagavathiappan, S., Lahiri, B.B., Saravanan, T., Philip, J., Jayakumar, T.: Infrared thermography for condition monitoring—a review. Infrared Phys. Techn. **60**, 35–55 (2013)
3. Gonçalves, R.S., Carvalho, J.C.M.: Review and latest trends in mobile robots used on power transmission lines. Int. J. Adv. Robot. Syst **10**(408), 1–14 (2013)
4. Silva, K.M., Souza, B.A., Brito, N.S.D.: Fault detection and classification in transmission lines based on wavelet transform and ANN. IEEE T. Power Deliver. **21**(4), 2058–2063 (2006)
5. Bockarjova, M., Sauhats, A., Andersson, G.: Statistical algorithms for fault location on power transmission lines. In: IEEE Russia Power Tech 2005, pp. 1–7, St. Petersburg (2005)
6. Saadabad, N.A., Moradi, H., Vossoughi, G.: Semi-active control of forced oscillations in power transmission lines via optimum tuneable vibration absorbers: with review on linear dynamic aspects. Int. J. Mech. Sci. **87**, 163–178 (2014)
7. Tadeusiewicz, R., Wszołek, T., Izworski, A., Wszołek, W.: Recognition of defects in high voltage transmission lines using the acoustic signal of corona effect. In: Proceedings of the 2000 IEEE Signal Processing Society Workshop., Neural Networks for Signal Processing X (2)NSW, Sydney, pp. 869–875 (2000)
8. Current and future UAV military users and applications. Air & Space Europe **1**(5–6), 51–58 (1999)
9. Metni, N., Hamel, T.: A UAV for bridge inspection: visual servoing control law with orientation limits. Automat. Constr. **17**, 3–10 (2007)
10. Aibot X6 Multicopter for Mapping and Industry—Aibotix International, www.aibotix.com
11. Roca, D., Lagüela, S., Díaz-Vilariño, L., Armesto, J., Arias, P.: Low-cost aerial unit for outdoor inspection of building façades. Automat. Constr. **36**, 128–135 (2013)
12. Siebert, S., Teizer, J.: Mobile 3D mapping for surveying earthwork projects using an unmanned aerial vehicle (UAV) system. Automat. Constr. **41**, 1–14 (2014)
13. Li, Z., Liu, Y., Walker, R., Hayward, R., Zhang, J.: Towards automatic power line detection for a UAV surveillance system using pulse coupled neural filter and an improved hough transform. Mach. Vis. Appl. **21**(5), 677–686 (2010)
14. Rajeev, M.B., Adithya, V., Hrishikesh, S., Balamurali, P.: Detection of power-lines in complex natural surroundings. Comput. Sci. Inf. Technol. (CS & IT) **3**(9), 101–108 (2013)
15. Candamo, J., Kasturi, R., Goldgof, D., Sarkar, S.: Detection of thin lines using low-quality video from low-altitude aircraft in urban settings. IEEE T. Aero. Elec. Sys. **45**(3), 937–949 (2009)
16. Song, B., Li, X.: Power line detection from optical images. Neurocomputing **129**, 350–361 (2014)
17. Liu, G., Zhu, Z., Jie, X.: Orientation and damage inspection of insulators based on tchebichef moment invariants. In: IEEE 2008 International Conference on Neural Networks & Signal Processing, pp. 48–52, Zhenjiang, China (2008)
18. Shinohara, A.H., Santana, D.M.F., Oliveira, P.P.J.C., Silva, R.J.G., Magalhães, O.H., Silveira, C.G., Khoury, H.J., Wavrik, J.F.A.G., Branco, F.M.A.C., Leite, M.A., Galindo, T.C.: Defects detection in electrical insulators and breaker for high voltage by low cost computed radiography systems. In: International Symposium on Digital industrial Radiology and Computed Tomography. Lyon, France (2007)
19. Wronkowicz, A.: Concept of diagnostics of energy networks by means of vision system. Diagnostyka **15**(2), 13–18 (2014)
20. Pratt, W.K.: Digital Image Processing: PIKS Scientific Inside, 4th edn. Willey-Interscience, Hoboken (2007)
21. Gonzalez, R.C., Woods, R.E., Eddins, S.L.: Digital Image Processing using MATLAB, 2nd edn. Gatesmark Publishing, Knoxville (2009)
22. Sonka, M., Hlavac, V., Boyle, R.: Image Processing, Analysis, and Machine Vision, 4th edn. Cengage Learning, Stamford (2014)

23. Ventura, A.S., Borrego, J.Á., Solorza, S.: Adaptive nonlinear correlation with a binary mask invariant to rotation and scale. Opt. Commun. **339**, 185–193 (2015)
24. Torres-Méndez, L.A., Ruiz-Suárez, J.C., Sucar, L.E., Gómez, G.: Translation, rotation and scale-invariant object recognition. IEEE Trans. Systems, Man, and Cybernetics—part C: Applications and Reviews **30**(1), 125–130 (2000)
25. Tsai, D.-M., Chiang, C.-H.: Rotation-invariant pattern matching using wavelet decomposition. Pattern Recog. Lett. **23**(1–3), 191–201 (2002)
26. Liao, S.X., Pawlak, M.: On image analysis by moments. IEEE T. Pattern Anal. **18**(3), 254–266 (1996)
27. Reddy, B.S., Chatterji, B.N.: An FFT-based technique for translation, rotation, and scale-invariant image registration. IEEE Trans. Image Process. **5**(8), 1266–1271 (1996)
28. Nock, R., Nielsen, F.: Statistical region merging. IEEE T. Pattern. Anal. **26**(11), 1452–1458 (2004)
29. Lewis, J.P.: Fast normalized cross-correlation. Vis. Interface **10**(1), 120–123 (1995)

Evaluation of Touchless Typing Techniques with Hand Movement

Adam Nowosielski

Abstract Hands-free control of electronic devices gains increasing interest. The interaction based on the interpretation of hand gestures is convenient for users. However, it requires adequate techniques to capture user movement and appropriate onscreen interface. The hand movements in touchless graphical user interface are translated into the motion of a pointer on a display. The main question is how to convert hand gestures into casual and comfortable text entry. The paper focuses on the evaluation of text input techniques in a touchless interface. Well-known traditional solutions and some innovations for text input have been adapted to noncontact onscreen keyboard interface and subjected to examination. The examined solutions include: single hand and double hands QWERTY-based virtual keyboard, swipe text input, and the 8pen (The 8pen solution, http://www.8pen.com/ [1]) based technique.

Keywords Touchless typing · Virtual keyboard · Swipe typing · Gesture interaction · Human–computer interaction

1 Introduction

New touchless interfaces provide the user the ability to interact in a natural way with electronic devices. They are useful for increasing and enhancing the user accessibility. Touchless interaction enables users to operate without additional medium and intermediate equipment. This is of particular interest with large-display environments and intermittent or casual interaction. Touchless interaction with large displays is particularly suitable for the following scenarios [2]: public spaces, clinical environments, consumer electronics, brainstorming, and visualization. In public spaces like airports or shopping malls, people expect from interactive displays a rapid response to their spontaneous interaction. In surgical settings new image-guided procedures are

A. Nowosielski (✉)
Faculty of Computer Science and Information Technology, West Pomeranian University of Technology, Szczecin, Żołnierska 52, 71-210 Szczecin, Poland
e-mail: anowosielski@wi.zut.edu.pl

R. Burduk et al. (eds.), *Proceedings of the 9th International Conference on Computer Recognition Systems CORES 2015*, Advances in Intelligent Systems and Computing 403, DOI 10.1007/978-3-319-26227-7_41

introduced [3]. A surgeon through noncontact interaction can manipulate on screen without the assistance in sterile conditions. Wide field of touchless applications is observed in consumer electronics where smart television and games are main representatives. Collaborative brainstorming using touchless interfaces for wall-sized displays is another concept [4]. Other applications include [5, 6]: vision-based augmented reality, 3D CAD modeling, sign language recognition, robot control, etc. In all the above-mentioned examples the interaction is usually limited to some kind of manipulation on a display in graphical user interface by the hand movements. More opportunities of control via movements of the whole body is provided for games. In these solutions the user movements or hand gestures are interpreted and transformed to expected action. A somewhat neglected problem in contemporary research is touchless typing. In classic human–computer interaction text entry and mouse control constitute together an integral interface. In touchless environments text entry is not the most important element, however the possibility of writing using the same medium of communication is still necessary. The paper focuses on the problem of touchless text input with hand movement in a large-display environment. The rest of the paper is structured as follows. In Sect. 2 the interface concept for touchless typing is presented. In Sect. 3 single hand and double hands QWERTY-based interfaces are presented. Section 4 describes the solutions proposed for touchless swipe text input. The fourth approach, discussed in Sect. 3, is the adaptation of the 8pen [1] based technique. Evaluation and comparison of the interfaces are provided in Sect. 5. The paper ends with a summary.

2 Interface Concept

Natural user interface (NUI) enables the user to interact with electronic devices in a natural way. NUIs have to be imperceptible and should base on nature or natural elements [7]. These principles are guaranteed by the specifically designed hardware and natural actions. All the examples mentioned in the previous section are based on NUI's principles. The user movements or hand gestures are interpreted and transformed to specific action. The problem of text entry in these solutions is predominantly solved with voice recognition. This can become a problem in silence-required or high-background-noise environments. Although an alternative in the form of silent speech recognition exist [8] the possibility of casual writing using the same medium of communication is still necessary. Touchless typing with hand movement requires good hand gesture recognition solution. There is an abundance of work devoted to the problem of hand gesture recognition and the reader can find a good literature surveys in [5, 6, 9]. The main problem in gesture recognition is the hand modeling, since it has approximately 27 degrees of freedom [5]. Two types of gestures are available [5, 6]: temporal (dynamic) and spatial (static, with a certain shape arranged). There are two distinct categories in spatial domain [5]: 2D model (appearance based or view based) and 3D model. Researchers are actively engaged in the development of touchless interaction. However, the problem of touchless typing is neglected in

Fig. 1 The environment for large-display touchless text entry

the contemporary research. Text entry is not the most important element of touchless environment but the possibility for spontaneous short writing should be provided. In such context some well-known traditional solutions and some innovations for text input have been adapted to noncontact onscreen keyboard interface and reported further in the paper. It is assumed that the user interacts in the environment resembling the one depicted in Fig. 1. There have been varied approaches to handle gesture recognition and in addition to the theoretical aspects many different imaging and tracking devices or gadgets have been proposed [6]: gloves, body suits, marker-based optical tracking, and vision-based techniques. Most recent works in the NUI field take advantage of 3D depth sensors [10]. Proprietary solutions such as Kinect from Microsoft or Xtion PRO LIVE from ASUS are willingly used by researchers since in evaluation process they provide unified and coherent environment. The application programming interface for Kinect is available as a free download from the manufacturer's home page [7] and an example framework for rapid prototyping of touchless user interfaces using Kinect sensor is proposed in [10]. For the huge popularity and in order to ensure similar imaging conditions for the evaluation process, Kinect sensor has been chosen for tracking hand movements in the evaluated interfaces.

3 Single Hand and Double Hands QWERTY-Based Virtual Keyboard

The most advantageous feature of QWERTY keyboard is its ubiquity and known layout. Many approaches to change this standard resulted in a failure. The QWERTY layout is therefore the first choice in trials to noncontact writing. A single hand

QWERTY-based approach to touchless typing is presented in [11] where four directional movements are interpreted and converted to the process of character selection. This solution proved its worth in touchless input interface for the physically challenged people [12]. Gestures are performed there with selected part of the body (usually hand or head). The referred solution, however, turned out to be slow for spontaneous casual writing [11]. Since typing on QWERTY layout keyboard requires a physical process of keystroke, in current approach to touchless text entry we decided to imitate the process of typing. It consists of two stages now: a selection and the following confirmation. Hand movements in the 3D environment are mapped to hand-image pointer above the virtual keyboard on the screen—thus the potential character is selected. The push gesture corresponds to key pressing. To make the process more similar to keyboard typing a second hand indicator has been introduced and the possibility of double hands writing allowed. In sum, two interfaces have been developed. Both base on QWERTY layout. One of them is handled with one hand and the other with both hands.

4 Touchless Swipe Text Input

Swipe text input is a solution introduced to small-sized touchscreens. In these devices specific key areas are small. For the large fingers it renders the process of text entry cumbersome. To facilitate the whole process of entering the text, specific sliding techniques have been introduced. In swipe text entry, the user makes a path between successive letters of the word. The recognition system matches the possible letters and infers appropriate word. The user is liberated from the precise indication of a specific letter. Usually an approximate location on the path is sufficient. The original swipe input keyboard is the Swype [13]. Many swipe input keyboards for touchscreen devices similar in concept have emerged in the market with examples like: SwiftKey, SwipeIt, SlideIT, and many others. The idea of swipe text input appears to be interesting for touchless text entry. However, it must be noticed that in noncontact environment the user can move freely and is not limited only to the expected behavior. In such conditions a method defining the beginning of text entry should be provided. In the developed interface the clench fist gesture is used. To put it differently, the word entry process is carried out with the clenched fist. Opening or closing the hand indicates the start and the end of word entry. The clench fist gesture can be easily performed by the user. It is, for example, widely used in smart TV environment. The crucial part of developed interface is the approach to word matching. While the user moves the hand above the virtual keyboard the trajectory (mapped to 2D screen coordinates) and temporal characteristics (for a pairs of succeeding screen coordinates) are recorded. All characters located on the performed path are selected. Temporal characteristics are used to remove characters with the shortest time of focus. The resultant string is dictionary matched. For the task, the Hunspell [14] spell checker is engaged.

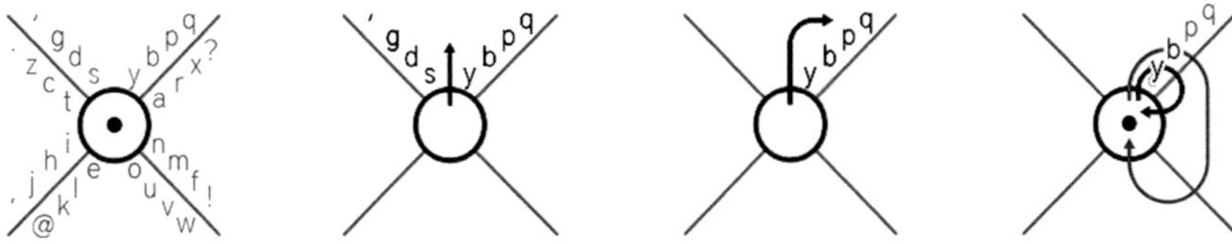

Fig. 2 Entering a letter on the 8pen keyboard [1]

5 The 8pen Adaptation to Touchless Typing

The 8pen text entry method is a proprietary solution that bases on own keyboard layout and stroke-based input. The keyboard is divided into four directional sectors (top, down, left, and right) with specifically arranged letters on the borders. The central part is the starting point. Movement to the directional sector limits potential letters to the certain group. The choice of intended letter is made by further circular movements through remaining sectors. The number of sectors crossed corresponds to the position of the letter on the border. Selection of the letter ends with the return to the center. The process is symbolically presented in Fig. 2. The described method performs well for touchscreens. The text entry begins with the touch on the center part of the keyboard. Otherwise, the user can freely move the finger above the screen. Adaptation of such approach to touchless environment requires some modifications. We have already proposed some adjustment with the clench fist gesture to distinguish the sequence of free hand movements from the sequence of text input in touchless environment [11]. This solution proved to work well. In fact, with average speed of 02:06 min it was faster form our previous approach to QWERTY-based touchless typing where average speed of 03:49 min was achieved [11]. For that reason, the adaptation of the 8pen technique to touchless environment presented in [11] will be used in current evaluation without any modifications.

6 Text Entry Methods Evaluation

Text entry methods evaluation in general is user-based. During the experiment the participant has the task of writing a text quickly and accurately [15]. At the same time, performance measures are gathered. Finally, the presented text and the transcribed text are compared. Ideally, both texts should be identical. However, the following errors occur [15–17]: substitutions, insertions, or deletions. Many error rate measures have been introduced. Their calculation is predominantly based on counting the number of primitive operations required to convert the incorrectly entered text phrase to the correct one. A good reference of the error rate measures can be found in [17, 18]. A crucial factor of text entry techniques assessment is the approach to error correction. There are three possibilities [18]: *none*, *recommended*, and *forced* error correction. In the *none* error correction condition, participants are prohibited to correct errors.

If the mistake is made, it is taken into account in error rate measure. The *error rate* (*ER*, calculated as the ratio of the number of incorrect characters in the resultant text to the length of text) and the *minimum string distance error rate* (*MSD ER*, consisting in the calculation of the minimum distance between two strings) are usually taken into consideration [18]. With the *forced* error correction condition the participant is forced to correct each error. The presented text and the transcribed text must be identical. Since the resultant text does not contain errors, measures like *Total ER* and *EKS ER* are used [18]. *Total error rate* (*Total ER*) is equivalent to the number of erroneous keystrokes, to the total number of correct, incorrect, and corrected characters [18]. *Erroneous keystroke error rate* (*EKS ER*), in turn, is the ratio of the total number of erroneous keystrokes (*EKS*) to the length of the presented text [18]. The criteria of error correction are somewhat mitigated in the *recommended* error correction condition where the participant is allowed to correct errors if he identifies them. The final text contains both corrected and uncorrected errors and in this case the *Total ER* is taken into account for error calculation. In our experiments we tested presented touchless text input techniques in forced error correction condition. All experiments were performed in front of the projector screen. There were 44 participants in the experiments and all of them were third-grade computer science students, aged mostly 22–23 years. All participants were good typist with fluent use of the QWERTY layout keyboard. Most of them have declared some experience with swipe typing on touch devices. A large part of them use this solution on a daily basis. Some of the participants have already have previous contact with the 8pen keyboard but none of them was an everyday user of that layout. All participants were divided into small groups. For each group the order of examined solution was randomized. After the introduction of each interface, the participants were invited to practice by writing some arbitrary sequences. Participants familiar with the given interface individually proceed to the proper experiment. All participants had the task to enter the same text quickly and accurately with the *forced* error correction condition. Text input language was Polish without diacritics. The study was conducted using phrases without numerical and any special characters. To ensure correct transcription of the resultant text in the *forced* approach, the participants were notified (by sound, as in [18]) when an inappropriate character was entered. To evaluate the effectiveness of text entry methods, the time of typing was measured. The results are provided in Fig. 3. The 8pen method in touchless environment proved to be the slowest with the average time of 02:08 min. This result is almost identical to our earlier study [11] where time of 02:06 min was obtained with 35 participants. It must be underlined here that all participants were novice to the 8pen layout and with some practice better results should be obtained. As regards touchless QWERTY typing and an older version of the interface in the previous experiments (reported in [11]) we acquired average of 03:49 min. With new interfaces with push gesture implementation it took an average of 01:14 min to write a sequence in both interfaces operated with one and both hands. The standard deviation was a bit greater in the first case (17 s) compared to two-hands interface (14 s). Interestingly, it would seem that both hands manipulation should ensure better typing speed, however it turned out that all gains were overcome by the difficulty of controlling two indicators. The best results however were achieved for the QWERTY swipe approach with the

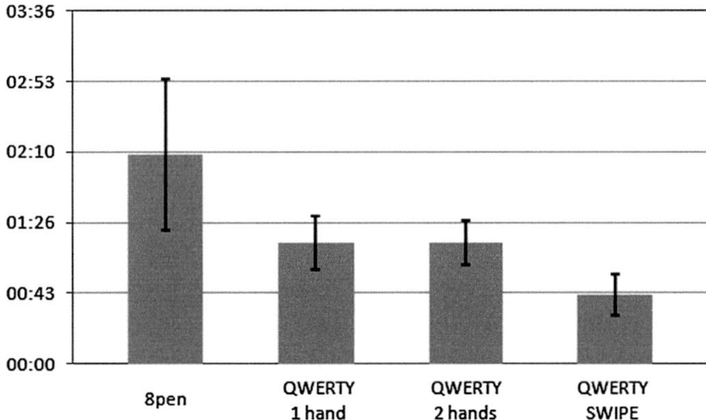

Fig. 3 Comparison of typing time (mean value) of the examined interfaces

time of 42 s (12 s of standard deviation). These results prove that known layout accompanied with a proper approach to the interface only together can provide the appropriate means to touchless typing. The second observation is that swipe typing technique can successfully be applied to touchless environment. Errors (especially in *forced* error condition) extend the process of text entry. Although good measures consider the cost of committing errors, it is cumbersome to compare swipe text typing techniques to single letter selection techniques. With the latter one and case of incorrect character selected there is a single error. How many errors are there when unintended word is wrongly recognized from traced path in swipe text entry? With the possibility to select other word from available suggestions it is possible to indicate the correct word with one choice (gesture). With one selection many characters are modified to the proper final transcription. With one correction many letters are modified. For that reason and the fact that short sentences were taken into consideration in the experiments, the simple measure of *no errors nor corrections* (NENC) was proposed. It indicates how many participants have not committed an error during the text entry experiments. Then NENC is a ratio of perfect sentence entry (without any corrections in the process of typing and with no errors in final transcription) to the total number of trials (performed by successive participants). It must be noticed that proposed measure is limited to short-length texts. The longer the text, the higher the probability of error. The results are provided in Fig. 4. For not-swipe text entry techniques, the *EKS ER* metric (*erroneous keystroke error rate*) was also calculated. For the 8pen, QWERTY 1 hand, QWERTY 2 hands, the following results have been accordingly achieved: 6.74, 2.44, and 3.41 %. Participants, new to the 8pen layout, made many errors. The EKS ER for both QWERTY interfaces was significantly smaller. Slightly higher values were observed for double hands operated QWERTY. Nevertheless, the average time of typing was equal in both cases (see Fig. 3), which leads to the conclusion that operating with one hand QWERTY touchless interface is slower but more resistant to mistakes. With the QWERTY swipe interface, the word is suggested after the path is performed. If the proposed word is

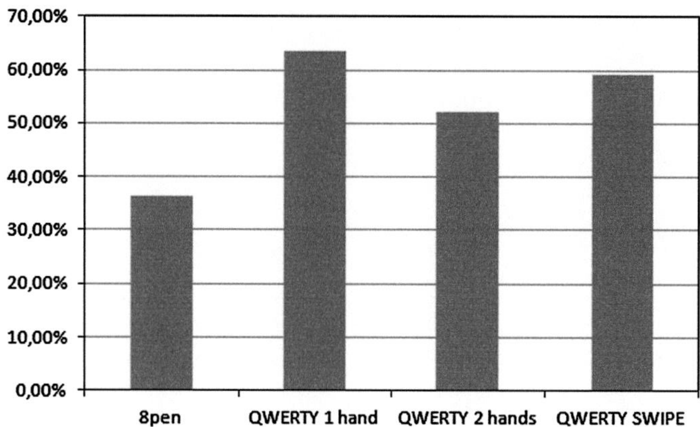

Fig. 4 The effectiveness of error-free typing with *no errors nor corrections* (NENC) measure

inappropriate then a selection of the expected one from other suggestions is needed. In that matter, wrong word suggestion and later correction in swipe typing may be equated to the error correction in single character selection writing. Both corrections are carried out after some action (appropriate for the given method). Replacing the erroneous keystroke in *EKS ER* metric by the erroneous word suggestion in swipe text typing, 2.19 % of errors were reported.

7 Conclusions

In the paper, the problem of touchless text input with hand movements has been addressed. Four solutions: one- and two-hand QWERTY-based virtual keyboard, swipe text input and the 8pen [1] based technique have been adopted to noncontact onscreen keyboard interface with typing intended to be performed in front of the projector screen. All solutions have utilized the proprietary Kinect sensor. In QWERTY-based techniques, the push action was utilized for character selection. In the 8pen approach, circular gestures and the state of hand (clenched or open) were used for operation. In swipe-based QWERTY input simple yet effective algorithm for traced path recognition was proposed. All solutions have been subjected to thorough examination conducted by 44 participants. Obtained results prove that known keyboard layout accompanied with proper approach to manipulation in the interface only together can provide the appropriate means to touchless typing. What is more, approaches known from haptic devices like swipe typing, can be transferred to noncontact environment with success.

References

1. The 8pen solution, http://www.8pen.com/
2. Chattopadhyay, D., Bolchini, D.: Understanding visual feedback in large-display touchless interactions: an exploratory study. Research report, IUPUI Scholar Works, Indiana University (2014)
3. Johnson, R., O'Hara, K., Sellen, A., Cousins, C., Criminisi, A.: Exploring the potential for touchless interaction in image-guided interventional radiology. In: ACM Conference on Computer-Human Interaction (CHI'11), pp. 3323–3332 (2011)
4. Dostal, J., Hinrichs, U., Kristensson, P.O., Quigley, A.: SpiderEyes: designing attention- and proximity-aware collaborative interfaces for wall-sized displays. In: Proceedings of the 19th international conference on Intelligent User Interfaces., IUI'14Haifa, Israel, pp. 143–152 (2014)
5. Sarkar, A.R., Sanyal, G., Majumder, S.: Hand gesture recognition systems: a survey. Int. J. Comput. Appl. **71**(15), 25–37 (2013)
6. Mitra, S., Acharya, T.: Gesture recognition: a survey. IEEE Trans. Sys. Man, Cybern. Part C: Appl. Rev. **37**(3), 311–324 (2007)
7. Giorio, C., Fascinari, M.: Kinect in Motion—Audio and Visual Tracking by Example Starting. Packt Publishing, Birmingham (2013)
8. Denby, B., Schultz, T., Honda, K., Hueber, T., Gilbert, J.M., Brumberg, J.S.: Silent speech interfaces. Speech Commun. **52**(4), 270–287 (2010)
9. Ibraheem, N.A., Khan, R.Z.: Survey on various gesture recognition technologies and techniques. Int. J. Comput. Appl. **50**(7), 38–44 (2012)
10. Placitelli, A.P.: Toward a framework for rapid prototyping of touchless user interfaces. In: 2012 Sixth International Conference on Complex, Intelligent, and Software Intensive Systems, pp. 539–543 (2012)
11. Nowosielski, A.: QWERTY- and 8pen- based touchless text input with hand movement. In: Chmielewski, L.J., Kozera, R., Shin, B.-S. (eds.) ICCVG 2014, vol. 8671, pp. 470–477. LNCS, Springer, Heidelberg (2014)
12. Nowosielski, A., Chodyła, Ł.: Touchless input interface for disabled. In: Advances in Intelligent Systems and Computing 226, Proceedings of the 8th International Conference on Computer Recognition Systems CORES 2013, pp. 701–709 (2013)
13. The Swype solution, http://www.swype.com/
14. The Hunspell solution, http://hunspell.sourceforge.net/
15. Costagliola, G., Fuccella, V., DiCapua, M.: Interpretation of strokes in radial menus: the case of the KeyScretch text entry method. J. Vis. Lang. Comput. **24**, 234–247 (2013)
16. MacKenzie, I.S., Soukoreff, R.W.: A character-level error analysis technique for evaluating text entry methods. In: Proceedings of the Second Nordic Conference on Human-Computer Interaction—NordiCHI 2002, pp. 241–244 (2002)
17. Soukoreff, R.W., MacKenzie, I.S.: Metrics for text entry research: an evaluation of MSD and KSPC, and a new unified error metric. In: Proceedings of the ACM Conference on Human Factors in Computing Systems—CHI 2003, pp. 113–120 (2003)
18. Arif, A.S., Stuerzlinger, W.: Analysis of text entry performance metrics. In: 2009 IEEE Toronto International Conference on Science and Technology for Humanity (TIC-STH), pp. 100–105 (2009)

A Hybrid Field of View Vision System for Efficient Robot Self-localization with QR Codes

Marta Rostkowska and Michał Topolski

Abstract This paper presents an outcome of experiments on a self-localization system for small mobile robots that involves the use of QR codes as artificial landmarks. The QR codes are employed for the double purpose of localization marks and data carriers for navigation-relevant informations. As we want to make our robots autonomous and independent from external cameras for localization, we investigate the use of an omnidirectional vision system. Then, we demonstrate that a hybrid solution consisting of an omnidirectional camera with a mirror and a classic front-view camera provides better localization results and application flexibility than either of these configurations applied alone. The hybrid vision system is inspired by the peripheral and foveal vision cooperation in animals. We demonstrate that the omnidirectional camera enables the robot to detect quickly landmark candidates and track the already recognized QR codes, while the front-view camera guided by the omnidirectional information enables precise measurements of the landmark position over extended distances and reading of the extra data carried by the QR code.

Keywords QR code · Self-localization · Mobile robotics · Omnidirectional camera · Front-view camera · Hybrid system

This research was funded by the Faculty of Electrical Engineering, Poznan University of Technology DS-MK/2015 grant.

M. Rostkowska (✉) · M. Topolski
Institute of Control and Information Engineering,
ul. Piotrowo 3a, 60-965 Poznań, Poland
e-mail: marta.a.rostkowska@doctorate.put.poznan.pl

M. Topolski
e-mail: michal.m.topolski@doctorate.put.poznan.pl

R. Burduk et al. (eds.), *Proceedings of the 9th International Conference on Computer Recognition Systems CORES 2015*, Advances in Intelligent Systems and Computing 403, DOI 10.1007/978-3-319-26227-7_42

1 Introduction

One of the most important functions of an autonomous mobile robot is the ability to estimate its own position with respect (w.r.t.) to the environment. For wheeled robots moving on a plane it is enough to compute the position and orientation $\mathbf{x}_R = [x_r\ y_r\ \theta_r]^T$. The robot can localize itself by matching some features of the environment observed at the given moment in time to known features in the predefined map. Unfortunately, it is tedious and time-consuming to obtain a map of the environment that is precise enough for self-localization, and it is not obvious which natural features are suitable for self-localization [1]. In practical applications of mobile robots these problems can be avoided to a large extent by employing artificial landmarks, which enhance the efficiency and robustness of self-localization either with a map known a priori [2], or in the simultaneous localization and mapping (SLAM) scenario [3]. While there are many possibilities to construct a passive marker that meets the basic requirements for a landmark in vision-based self-localization [4], an efficient solution investigated in our previous research is to employ the matrix barcodes commonly used to recognize packages and other goods in the logistics supply chains. Our tests revealed that among the available matrix barcodes the most suitable for navigation are the QR (quick response) codes. We tested the application of QR code-based landmarks in localization of small mobile robots in a multi-robot system, where the landmarks are attached to the robots and observed by an overhead camera, and on a single autonomous robot that has a forward-view camera attached [4]. Both solutions turned out to enable reliable robot localization in real-time, but both have some practical drawbacks. The overhead camera is a low-cost localization solution for a multi-robot team, but limits the autonomy of the robots and provides localization data only over a relatively small area. In contrast, the on-board front-view perspective camera enables the robot to perceive the QR codes autonomously, and makes the self-localization task independent from the possible communication problems. However, the QR codes are detectable and decodable only over a limited range of viewing configurations. Thus, the robot has to turn the front-mounted camera towards the landmark before it starts to recognize the code. Therefore, in this paper we demonstrate how the passive visual landmarks based on the QR codes may be used to create a simple and affordable self-localization solution utilizing a hybrid vision system, consisting of both the omnidirectional and the front-view camera. The omnidirectional part, using an upward-looking camera and a catadioptric mirror provides to the robot an analogy of the peripheral vision in animals: the ability to quickly detect interesting objects over a large field of view. Conversely, the front-view camera provides foveal vision: the ability to focus on details in a much narrower field of view. The cooperation of these two subsystems enables to track in real-time many QR code-based landmarks located in the environment.

2 Related Work

Nowadays, passive cameras are considered the most affordable and universal sensors for self-localization of mobile robots. They are widely used in research and practical applications. Unfortunately, visual navigation algorithms that employ natural salient features or dense, optical-flow-like approaches are sensitive to unpredictable changes in the environment and lighting conditions [5]. Moreover, they often require considerable computing power on-board of the mobile robot. Therefore, many practical self-localization methods employ artificial landmarks that enhance the environment by introducing easily detectable and recognizable visual features. Many approaches to the design of a passive landmark have been described in the literature. Simple geometric shapes can be quickly extracted from the images, particularly if they are enhanced by colors [3]. Another approach is to employ the idea of barcode [6]. The main advantage of this approach is that the barcodes may encode additional information, while the simple geometric landmarks usually contain only some form of identification number (ID). Two-dimensional matrix codes that proliferated to our everyday life together with their use in smartphones enable to easily fabricate even more information-rich landmarks, which are robust to partial occlusion or damages of the surface. Landmarks based on QR codes are unobtrusive—their size can be easily adapted to the requirements of particular task and environment. As they are monochromatic, they can be produced in a color matching the environment, partially blending into the surrounding. Moreover, the use of omnidirectional camera can decrease the number of landmarks required for robust self-localization in comparison to the common solution with a front-view camera. Recently, Figaj and Kasprzak [7] demonstrated the applicability of QR codes on a NAO small humanoid for navigation and object labeling. However, their solution requires the NAO to approach the landmark at a viewing angle smaller than 50° to recognize it. Another example of using the matrix QR codes in mobile robotics has been presented in [8], where the information-carrying capability of matrix codes was utilized as well. An application of QR codes to improve the robustness and quality of the probabilistic Monte Carlo self-localization algorithm in a dynamic environment was presented in [9], where the landmarks based on matrix codes were used together with the depth data from a RGB-D sensor. Hybrid field of view vision systems on mobile robots are rarely reported in the literature. Perhaps the reason is that they are complicated, requiring to mount two tightly coupled cameras on a robot. One example of a system that is similar to our approach and mimics the relationship between the peripheral vision and the foveal vision in humans is given by Menegatti and Pagello [10]. They investigate cooperation between an omnidirectional catadioptric camera and a typical perspective camera in the framework of a distributed vision system, with the RoboCup Soccer being one of its primary applications. However, only simple geometric and color features are considered in this system. Also Adorni et al. [11] describe the use of a combined peripheral/foveal vision system including an omnidirectional camera in the context of mobile robot navigation. Their system uses both cameras in a stereo vision setup and implements obstacle detection and avoidance, but not self-localization.

3 Self-localization

3.1 Landmarks

The base chosen to create landmarks was QR code, which is placed in the center of a landmark. The QR code is encompassed with a black frame, which is used for initial sorting and reducing the number of potential objects, which can be subjected to decoding and further processing. Landmarks are monochromatic (usually black-and-white), because they should be extremely low-cost and printable on any material, not only paper. An example of a QR code, in which value '1' has been encoded is shown in Fig. 1a. An example of prepared landmark, with added bounding black frame, is shown in Fig. 1b. The same code, decoded in real-life environment is presented in Fig. 1c.

3.2 Mobile Platform Setup and Self-localization Idea

The mobile robot with the hybrid field of view vision system has been shown in Fig. 2a, b. It consists of the small, differential-drive mobile platform (SanBot Mk II [12]), the front-view camera, and the omnidirectional camera. Figure 2c shows an idea of the self-localization method employing QR landmarks: a passive landmark orientation and distance to the robot can be calculated from a single image taken by the front-view camera. The information about the landmark position and orientation with the reference to a global coordinate system is encoded in the QR code of the landmark itself, so the robot doesn't need to keep a map of known landmarks in the memory. Therefore, if at least one landmark can be recognized and decoded, the position of the robot and its orientation can be computed. The omnidirectional camera enables the minimum-effort tracking of the already known (discovered) landmarks, as the robot moves across the environment. If any new potential landmark is found on the omnidirectional image, it is pointed at with the front-view camera and recognized.

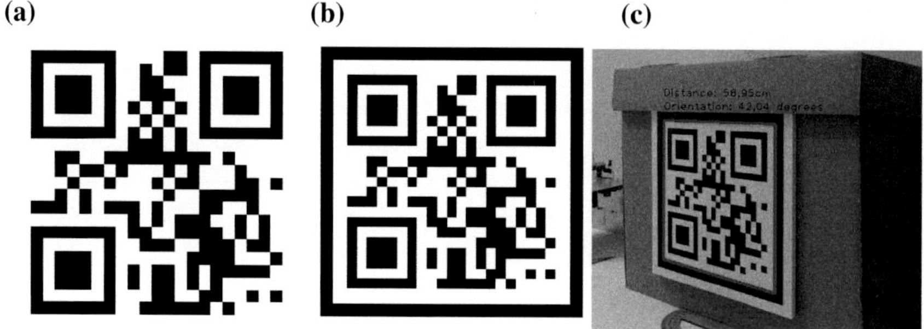

Fig. 1 QR code-based landmark used in research (encoded value: '1') **a** QR code, **b** landmark (QR code with frame); size: 16.5 × 16.5 cm, **c** application output

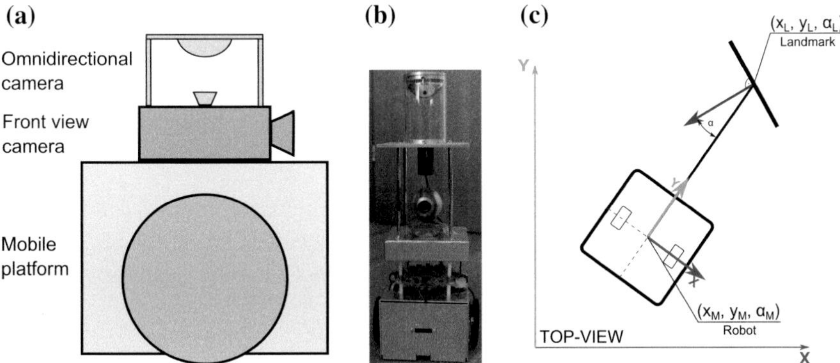

Fig. 2 **a**, **b** Mobile platform setup, **c** idea of self-localization utilizing landmarks

3.3 Vision Systems

Using only an omnidirectional camera or a front-view camera has important disadvantages. By combining both systems these drawbacks are mutually compensated. The front-view camera is the most natural approach, easiest to understand and implement. It provides a reasonable trade between useful range and camera's resolution and thus the computing power. However, in order to overwhelm the surrounding environment, the robot has to constantly scan the area changing its heading, which is inconvenient. On the other hand, while the omnidirectional camera is capable of inspecting the whole proximity, it requires a lot of computing power required to rectify the whole images acquired by the catadioptric system. To match these antagonisms, we implemented a solution, where both cameras are used: the omnidirectional to provide 360° view and tracking, thus providing robust self-localization and front view to enable proper landmarks decoding from an undistorted image.

3.4 Recognition and Localization of Landmarks

The program processing the images from the cameras has been created using C# programming language in Microsoft Visual Studio 2010. Finding the landmarks' contours has been programmed used Emgu CV [13], which is a C# port of OpenCV [14]. The decoding of extracted QR codes is done by MessagingToolkit [15]. Image processing begins with acquiring images from the camera, which are filtrated. Then, the frames cordoning QR code are looked for, by searching the quadrangles. It is assumed that landmarks are mounted on surfaces sufficiently rigid, so that they will not bend, deforming the squares into ellipses (the image used at this point is assumed to be not distorted). Later, it is assumed that among found quadrangles, some contain directly decodable QR codes, which means that their surfaces are parallel to the camera

sensor's surface. If such landmarks are found, the distance to them is calculated using Eq. (1).

$$Z_{xy} = \frac{f \cdot D_{xy}}{d_{xy}}, \tag{1}$$

where Z_{xy} is distance of object from the camera [mm], f is focal length [mm], D_{xy} is object's true dimension [mm], and d_{xy} is object's dimension on the image [px]. The angle between the camera optical axis and the landmark's normal vector is calculated using Eq. (2).

$$\alpha = \arctan\left(\frac{x}{Z_x}\right) \cdot \frac{180^\circ}{\pi} \tag{2}$$

where α is the angle between camera optical axis and the landmark's normal vector [°], Z_x is the distance from camera to the object, calculated using x-axis data [mm] and x is the distance between camera optical axis and landmark's normal vector, calculated in x-axis using Eq. (3):

$$x = \frac{x_p \cdot D}{d} \tag{3}$$

where x_p is the distance, calculated in x-axis, between center of the image (camera optical axis) and center of the landmark, calculated as center of the bounding frame [mm]. For front-view configuration, where landmarks' surfaces are not parallel to camera sensor's surface, the situation is much more difficult, leading to very complex mathematical problems. In order to properly calculate the landmark's position and rotation, the dependency between 3D points and the image plane must be found. This relation is described by equations and functions that can be found in [4].

4 Experimental Results

4.1 *Front-View Camera*

The front-view configuration is useful when working in partially unknown or changing environment. Figure 3a presents a view from robot's camera. The resolution is 960 × 720 pixels. Figure 3b presents results of decoding and localization, and Fig. 3c shows system's capability of recognition and decoding of landmarks at extreme conditions: 170 cm of distance and 55° of the viewing angle. Measurements have been conducted in the range from 50 to 170 cm, for landmarks' angles in correlation to camera sensor's plane in the range from −60° to +60°. These ranges have been chosen as the most common and useful in indoor environment. This limitation implies impossibility of recognizing or even tracking landmarks, that are located on the sides or on the back of the robot. Therefore, the front-view camera is useless in a situa-

Fig. 3 Results obtained in the front-view configuration **a** overview of configuration in room environment, **b** results as seen by robot's application, **c** extreme conditions

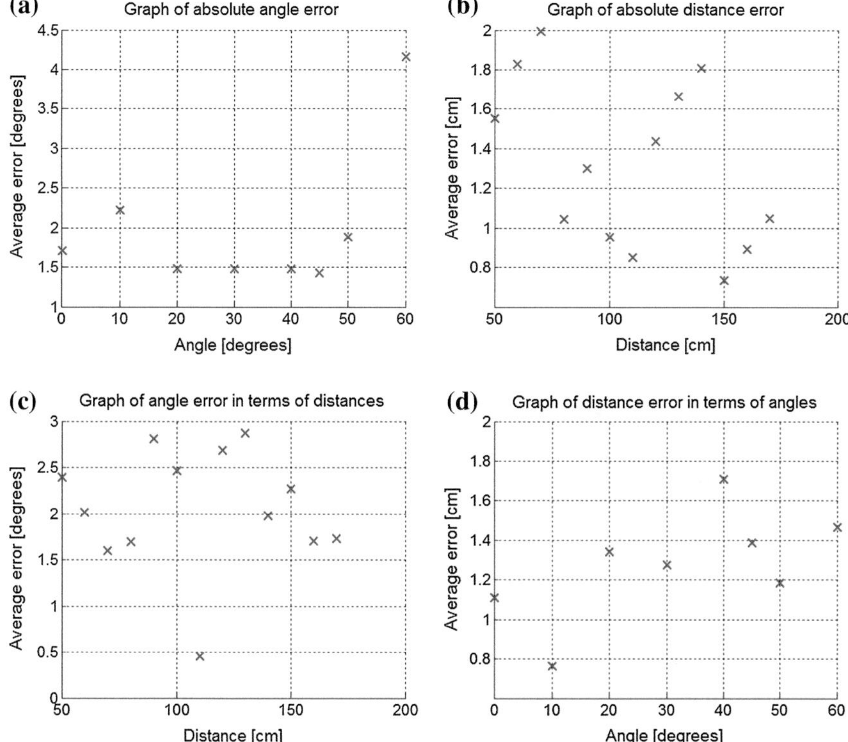

Fig. 4 Landmark localization precision for the front-view camera **a** average error of absolute angle measured in terms of different angles, **b** average error of absolute distance measured in terms of different distances, **c** average error of absolute angle measured in terms of different distances, **d** average error of absolute distance measured in terms of different angles

tion, when i.e. the mobile platform is moving across a corridor, where landmarks are mounted on its walls. The outcome of these experiments is shown in Fig. 4. Average precision of the measurements turned out to be 1.3 cm for distance and 2° for angle [4].

4.2 Omnidirectional Vision System

Omnidirectional camera handling turned out to be extremely difficult. It has proven to be ineffective in terms of computation power, as the highly distorted images required complicated transformations. Therefore, its use has been limited to the minimum, acquiring only the most essential information needed in order to successfully track the known landmarks and find the potential landmark candidates. The camera used in the omnidirectional vision system provides image with the resolution of 1280×720 pixels. Processing data from the omnidirectional camera is accomplished as follows:

1. Raw image is acquired from the camera.
2. Image is cropped and unwound (Fig. 5a).
3. Extreme parts of the image are duplicated on the opposing ends in order to obtain a continuous image in places where landmarks can appear (Fig. 5b).
4. Candidates for landmarks are found and marked on the image. The rotation angle needed to lock the front-view camera on each of them is computed.

Unfortunately, the used omnidirectional camera consists of a perspective view camera and parabolic mirror, which forecloses the decoding of QR codes directly from the unwrapped image, as the distortions are highly nonlinear [16, 17], and thus the image would require complex calibration and transformations, obviously complicating the computations and disabling the online tracking.

4.3 Hybrid System

Hybrid system combines results obtained from previously described systems, switching between cameras when necessary. Dual-vision system benefits from lowering the number of landmarks located in the environment. On the other hand, it also has forced a specific design of robot's chassis, that ensures a proper field of view for both

Fig. 5 Results from the omnidirectional camera **a** cropped and unwrapped image, **b** extended image

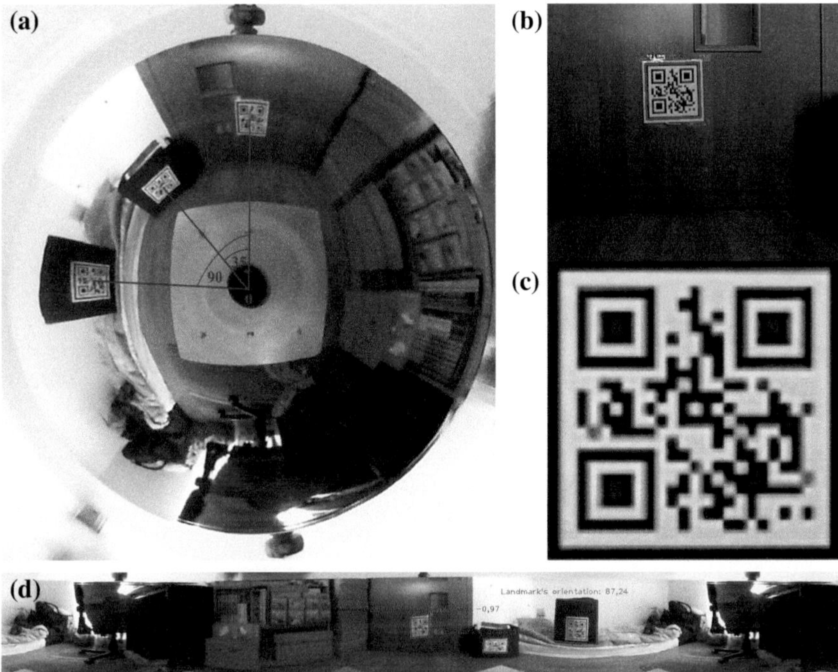

Fig. 6 **a** Image acquired from omnidirectional camera, **b** computed parameters of a landmark, **c** extracted QR code, **d** unwrapped and extended with marked candidates

cameras. Particularly, the omnidirectional camera had to be mounted on the top of the robot. Although the front-view camera is capable of recognizing landmarks that are visible at the angles up to $\pm 60°$, to ensure robustness the QR codes are decoded when they are visible at an angle at most $\pm 20°$. By this way, the robot is not forced to aim directly at the potential landmark, speeding up the recognition process. Figure 6a, d presents images acquired from the omnidirectional camera, where potential landmarks have been marked. Figure 6b shows an example landmark and its parameters found after positioning the front-view camera to the angle computed using the omnidirectional camera. Figure 6c shows QR code extracted from the landmark.

5 Conclusions

Presented research shows that online self-localization utilizing QR codes is possible, but is demanding in terms of computational power. The proposed hybrid approach consisting of two complementary cameras comes along as a compromise solution, enabling both online landmarks tracking and offline acquiring of complex data encoded in passive landmarks themselves. The system has proven to be robust to changing environment conditions, especially the varying light. In our further research the presented system will be extended by the ability to create semantic maps of the environment consisting of objects labeled with the QR codes.

References

1. Skrzypczyński, P.: Simultaneous localization and mapping: a feature-based probabilistic approach. Int. J. Appl. Math. Comput. Sci. **19**(4), 575–588 (2009)
2. Kasiński, A., Skrzypczyński, P.: Perception network for the team of indoor mobile robots. Concept. Architecture. Implementation. Eng. Appl. Artif. Intell. **14**(2), 125–137 (2001)
3. Bąaczyk, R., Kasiński, A.: Visual simultaneous localisation and map-building supported by structured landmarks. Int. J. Appl. Math. Comput. Sci. **20**(2), 281–293 (2010)
4. Rostkowska, M., Topolski, M.: On the application of QR codes for robust self-localization of mobile robots in various application scenarios. In: Szewczyk, R., et al. (eds.) Progress in Automation, Robotics and Measuring Techniques. AISC. Springer, Berlin (2015)
5. Lemaire, T., Berger, C., Jung, I.-K., Lacroix, S.: Vision-based SLAM: Stereo and monocular approaches. Int. J. Comput. Vis. **74**(3), 343–364 (2007)
6. Briggs, A., Scharstein, D., Braziunas, D., Dima, C., Wall, P.: Mobile robot navigation using self-similar landmarks. In: Proceedings of IEEE Interbnational Conference on Robotics and Automation, pp. 1428–1434. San Francisco (2000)
7. Figat, J., Kasprzak, W.: NAO-mark vs. QR-code recognition by NAO robot vision. In: Szewczyk, R., et al. (eds.) Progress in Automation, Robotics and Measuring Techniques, Vol. 2 Robotics. AISC, vol. 351, pp. 55–64. Springer, Heidelberg (2015)
8. Rahim, N.A., Ayob, M.N., Ismail, A.H., Jamil, S.J.: A comprehensive study of using 2D barcode for multi robot labelling and communication. Int. J. Adv. Sci. Eng. Inf. Technol. **2**(1), 80–84 (1998)
9. McCann, E., Medvedev, M., Brooks, D., Saenko, K.: Off the grid: self-contained landmarks for improved indoor probabilistic localization. In: Proceedings of IEEE International Conference on Technologies for Practical Robot Applications, pp. 1–6. Woburn (2013)
10. Menegatti, E., Pagello, E.: Cooperation between omnidirectional vision agents and perspective vision agents for mobile robots. In: Gini, M., et al. (eds.) Intelligent Autonomous Systems, vol. 7, pp. 231–135. IOS Press, Amsterdam (2002)
11. Adorni, G., Bolognini, L., Cagnoni, S., Mordonini, M.: A non-traditional omnidirectional vision system with stereo capabilities for autonomous robots. In: AIIA 2001: Advances in Artificial Intelligence. LNCS, vol. 2175, pp. 344–355. Springer, Berlin (2001)
12. Rostkowska, M., Topolski, M., Skrzypczyński, P.: A Modular Mobile Robot for Multi-Robot Applications. Pomiary Automatyka Robotyka **2**, 288–293 (2013)
13. EmguCV, http://www.emgu.com/wiki/index.php/Main_Page
14. OpenCV Documentation, http://docs.opencv.org
15. MessagingToolkit, http://platform.twit88.com
16. Bazin, J.: Catadioptric vision for robotic applications, Ph.D. Dissertation, Korea Advanced Institute of Science and Technology, Daejeon (2010)
17. Potúček, I.: Omni-directional image processing for human detection and tracking, Ph.D. Dissertation, Brno University of Technology, Brno (2006)

Morphologic-Statistical Approach to Detection of Lesions in Liver Tissue in Fish

Małgorzata Przytulska, Juliusz Kulikowski and Adam Jóźwik

Abstract The problem of light microscope images enhancement by filtering for recognition pathologic liver tissues in fish is considered in the paper. The problem follows from the necessity of monitoring the sea water pollutions caused by mercury compounds and their influence on living organisms. It is proposed to use image filtering based on morphological spectra to enhance visibility of liver lesions in the images in order to extract morphologic-statistical parameters useful in automatic tissues classification into *normal* and *pathologic* classes. It is shown that selected components of the 4th range morphologic spectra ($MS4$) are the most suitable to discriminate *normal* and *pathologic* liver tissues. The selected spectral components are characterized by their estimated mean values, standard deviations and kurtoses. The so-obtained morphologic-statistical parameters have been used to construct the learning sets for two types of image classifiers: based on the *nearest mean* and *k nearest neighbors rules*. It is shown that preliminary image filtering by morphological spectra-based filters improves spatial distribution of the recognized *normal* and *pathologic* objects in the parameter space.

Keywords Texture analysis · Morphological spectra · Lesion recognition

1 Introduction

Monitoring of the level of air and water pollutions and of their influence on the health-state of plants, animals and humans is a greatly important and topical problem. The process of propagation of pollutions runs slowly but its effects are deep

---

M. Przytulska (✉) · J. Kulikowski
Nalecz Institute of Biocybernetics and Biomedical Engineering PAS,
4 Ks. Trojdena Str., 02-109 Warsaw, Poland
e-mail: mprzytulska@ibib.waw.pl

J. Kulikowski
e-mail: jkulikowski@ibib.waw.pl

A. Jóźwik
Faculty of Physics and Applied Informatics, Department of Computer Science,
University of Łódź, 149/153 Pomorska Str., 90-236 Łódź, Poland

R. Burduk et al. (eds.), *Proceedings of the 9th International Conference on Computer Recognition Systems CORES 2015*, Advances in Intelligent Systems and Computing 403, DOI 10.1007/978-3-319-26227-7_43

and hardly reversible. That is why the methods of long-term pollutions monitoring should be sufficiently sensitive in order to prevent potential threat of life deterioration. On the other hand, they should be also accurate in order to reduce the costs of ineffectual, on wrong data based preventing actions. Computer methods may be an effective tool for reaching the above-mentioned goals. They, in particular, may help to analyze large numbers of biological specimens in order to detect and to evaluate the level of lesions caused by various types of pollutions in inner organs of living organisms. Typical example of dangerous water pollution are mercury (Hg) compounds dissolved in sea water, penetrating fish organisms, including those consumed by humans. The presence of Hg compounds in the organism is manifested by increased number of macrophages and necrotic liver cells. The aim of an image analyzing computer program consists in this case in detection of lesions and evaluation of their size, intensity etc. However, the measured parameters are charged by statistical errors. Therefore, at the next state of specimens examination it arises the problem of damaged (pathologic) tissues recognition and assigning to them a class of "abnormality". Final results of analysis strongly depend on the method of image preprocessing which should enhance the visibility of morphological structures to be analyzed. Second problem consists in selection of measurable parameters to construct an observation space in which the pattern recognition problem could be effectively solved. In the paper an approach to the above-mentioned problems solution based on a combination of selected morphological spectra components and some their statistical parameters calculation is presented. In our opinion, this approach in much larger class of lesions detection problems can be used. The paper is organized as follows. In Sect. 2 materials used in our experiments are described. Section 3 presents in details the proposed image processing and lesions recognition method. Section 4 contains the results of experiments. Conclusions are summarized in Sect. 5.

2 Materials

The below-described image enhancement and lesions recognition method has been tested on a collection of light microscope images of *yelloweye rockfish* liver tissue specimens. 169 color images in *RGB* have been provided by B.D. Barst [1, 2]. The images were of 1030 × 1300 pixels size. They have been provided in three groups of magnification: ×40 (55 images), ×200 (57 images) and ×40 (57 images). The subgroups of recognized images used as learning sets were much smaller and consisted of 6 normal and 8 pathologic images. Examples of typical images of normal and pathologic liver tissues are shown in Fig. 1. The lesions are visible in Fig. 1b as randomly distributed dark spots of various size and forms.

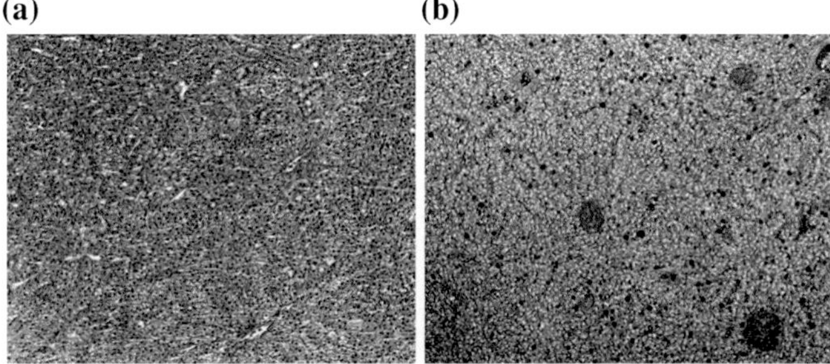

Fig. 1 Light microscope images of *yelloweye rockfish* liver tissue: **a** Normal, **b** damaged by mercury compounds

3 Methods

The image analysis method aimed at pollution level evaluation consists of three steps:

1. Selection of effective image enhancing filters.
2. Selection of parameters for recognition of pathologic tissues.
3. Recognition of pathologic tissues.
4. Numerical evaluation of the level of pathology.

In the paper only points 1, 2 and 3 are presented, point 4 is going to be described in a forthcoming paper.

3.1 Selection of Effective Image Enhancing Filters

Color images have been transformed into grey-level images. For finding effective method of image enhancement by filtering the results of comparative filters evaluation described in [3] were taken into consideration. Three filters sensible to granular morphological structures: *MS2EE*, Laplace 7×7 and Sobel [4, 5] were preliminarily chosen for further examinations. The *MS2EE* is a filter reinforcing the *SX* component of 2nd order morphological spectrum, described by the mask of weight coefficients imposed on the filtered image pixels shown in Fig. 2: The filtering (reinforcing)

Fig. 2 The mask of weights used in calculation of the *SX* component of *MS*2 spectrum

−1	1	−1	1
1	−1	1	−1
−1	1	−1	1
1	−1	1	−1

Table 1 Weight coefficients of *MS2EE* filter

MS2	SS	SV	SH	SX	VS	VV	VH	VX	HS	HV	HH	HX	XS	XV	XH	XX
w	1	1	1	2	1	1	1	1	1	1	1	1	1	1	1	1

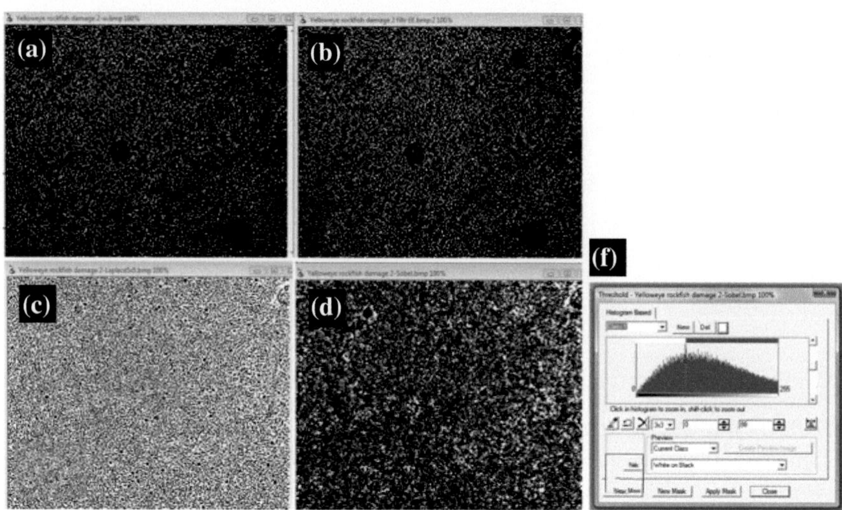

Fig. 3 Filtered and binarized liver tissue images: **a** Non-filtered image, **b** *MS2EE* filter, **c** Laplace 7×7 filter, **d** Sobel filter, **e** luminance histogram with indicated median used as a threshold level in all filtered images

weight coefficient of *MS2EE* filter are shown in Table 1. There is a difference between the ways of using the weight coefficients; in *MS2EE* the mask is imposed on the image horizontally and vertically with step 4 (without overlapping) while in Laplace and Sobel filters they are shifted by step 1. Filtered images have been binarized on a threshold level corresponding to the median of luminance histogram of original image. The results of filtering a typical (pathologic) liver tissue are shown in Fig. 3. Black spots visible in Fig. 3a being of particular interest are hardly visible in images filtered by Laplace and Sobel filters. However, the *MS2EE* filter improves mainly the visibility of small granular details. This suggests that larger morphological structures might be reinforced if the *SX* component is calculated not for single but for larger groups of pixels. This can be reached in one of two ways: (1) by analyzing images of lower magnitude or (2) by using higher level morphological spectra: *MS*3, *MS*4 etc. In order to evaluate the influence of *MS* range on discrimination of large morphological details there were calculated the following averaged over the sets of *normal* (6 cases) and in *pathologic* (8 cases) values of three basic types of *MS* components:

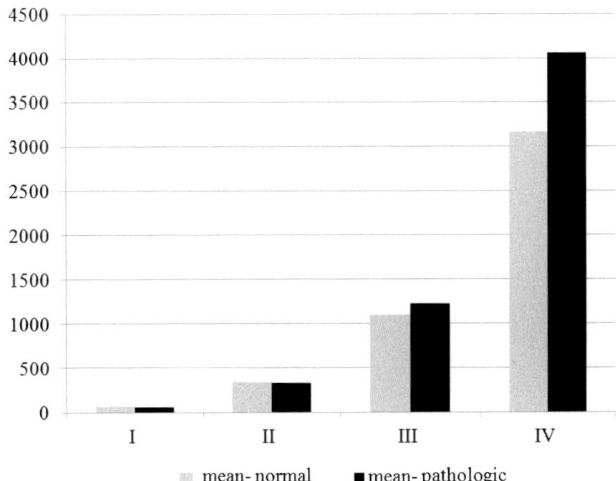

Fig. 4 Averaged selected *MS* components in images of normal (*grey*) and pathologic (*black*) liver tissues in fish

I For *MS*1: $Av(\overline{V}, \overline{H}, \overline{X})$,
II For *MS*2: $Av(\overline{VS}, \overline{HS}, \overline{XS})$,
III For *MS*3: $Av(\overline{VSS}, \overline{HSS}, \overline{XSS})$,
IV For *MS*4: $Av(\overline{VSSS}, \overline{HSSS}, \overline{XSSS})$,

The results are shown in Fig. 4. The grey (left) columns correspond to *normal*, the black (right) ones to *pathologic* tissues. Substantial differences between *normal* and *pathologic* data are visible in averaged values of *MS*3 and, particularly, *MS*4 components. The differences have been calculated for the series consisting of at least 6 images, each image containing 5230 basic windows of maximal size 16×16 pixels for 3 selected components. This gave 94,140 samples for calculation of each averaged value, which meant that the differences are statistically substantial and the given *MS*4 components can be preferred for recognition of pathologic liver tissues.

3.2 Selection of Parameters for Recognition of Pathologic Tissues

Each *MS* component of an analyzed image can be considered as a matrix of observed values of a multi-component random variable. It has been found in Sect. 3.1 that the random values in *MS*4 are different in the images of normal and pathologic tissues. This suggests that the differences with different intensity may be manifested by various statistical parameters of the corresponding probability distributions of the random matrices. In particular, three widely used standard statistical parameters: mean value, standard deviation and kurtosis [6] have been taken into consideration. The parameters have been calculated for all 256 *MS*4 components in normal and in

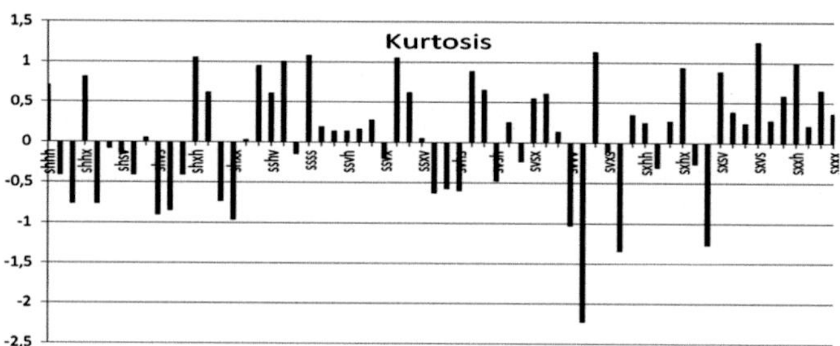

Fig. 5 Differences of kurtosis of $MS4$ series S^{***} components in images of normal and pathologic tissues

pathologic images and their differences have been taken into consideration. In Fig. 5 example of calculation of the differences of kurtosis in a series S^{***} of $MS4$ components is shown; analogous results have been calculated for mean value and standard deviation of the same series of $MS4$ components. In all cases the calculations were performed and averaged over 6 images of normal and 8 images of pathologic tissues. The differences of parameters were analyzed. For each statistical parameter 3 $MS4$ components corresponding to maximal absolute values of differences discriminating normal and pathologic tissues have been selected. The results of selection are shown in Fig. 6. The results once again have been analyzed in order to choose $MS4$ components whose statistical parameters are the most significant for discrimination of normal and pathologic tissues. As a result, the following components have been selected:

- due to high difference of *mean values*: *SSVS, SVSS, VSSS, HSSS, XSSS*;
- due to high difference of *standard deviations*: *SSVS, SVSS, VSSS, HSSS, XSSS*;
- due to high difference of *kurtoses: HVVS, HHSH, HHHS, XSVX, XHHS.*

The components *SSVS, SVSS, VSSS, HSSS* and *XSSS* discriminate tissues by more than one statistical parameter. On the other hand, recognition of pathologic tissues should be independent of rotation of the specimens. This leads to a condition of symmetry that should be imposed on the classifier: if it is sensible to a component containing symbol V in its notation, like $*V*$, then it should be equally sensible to the components $*H*$ and vice versa. Finally, the following series of 15 $MS4$ components has been selected for recognition of pathologic tissues:

SVSS, SVVS,	*VSSS, VSVS,*	*VHSS, VXSS,*	*XVSS, XSSS,*
SHSS, SHHS,	*HSSS, HSHS,*	*HVSS, HXSS,*	*XHSS.*

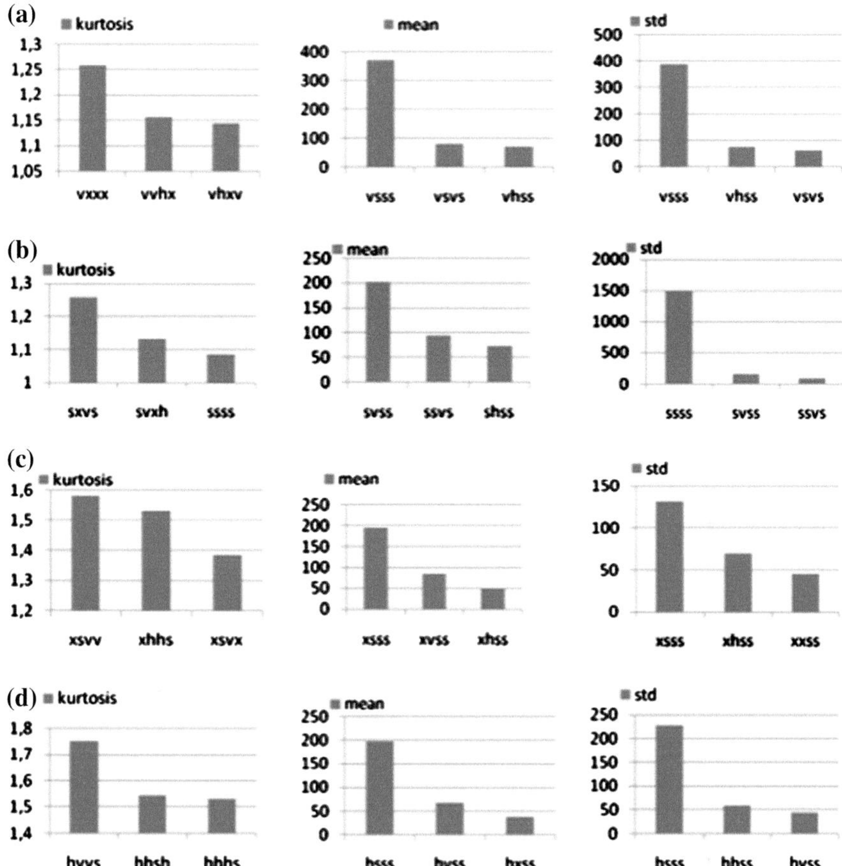

Fig. 6 The best "normal" and "pathologic" images discriminating morphologic-statistical parameters for: **a** S^{***} components, **b** V^{***} components, **c** H^{***} components, **d** X^{***} components

4 Experiments

For the selected *MS*4 components the corresponding morphologic-statistical parameters: *mean values, standard deviations* and *kurtoses* have been used to construct a 45-dimensional vector space representing the objects (images of specimens) to be classified into the *normal* and *pathologic* classes. For this purpose, two types of pattern recognition algorithms were used:

1. A simplified *NM* (*nearest mean*) algorithm based on choosing the lower distances between the classified objects and the mean vectors of the *normal* and *pathologic* learning sets.
2. A standard *k-NN* rule assigning the classified object to the same class as a majority of its *k* "nearest neighbors" in a given learning set. The value of *k* is being

established experimentally. The misclassification rates, for all possible number of nearest neighbors, are estimated by the leave one out method and the k offering the smallest error rate is finally chosen [7–9].

In both algorithms a standard Manhattan-distance was used. The experiments were performed on the sets of images of $\times 400$, $\times 200$ and $\times 40$ magnification (see Sect. 2). Figure 7 presents the results of classification of 55 images of $\times 400$ magnification by using the *NM* algorithm. The results are presented as spatial distribution of the recognized objects on the *mean* versus *kurtosis* plane of coordinates where *mean* and *kurtosis* in each classified object have been averaged over the above-selected 15 *MS*4 components. The mean vectors of the *normal* and *pathologic* learning sets have been denoted, respectively, by a rhombus and a square. The objects recognized as *normal* are marked by crosses while the *pathologic* ones are marked by triangles. Similar calculations have been performed for the subsets of images of $\times 200$ and $\times 40$ magnification. Next experiment was aimed at investigation of the influence of image filtering on separation of the *normal* and *pathologic* classes of objects. For this purpose, all images have been filtered by two types of *MS*4-based filters characterized by the following weight coefficients:

(a) Weights: $w = 1.5$ for the components *SVSS, SVVS, SHSS, SHHS, VSSS, VSVS, VHSS, VXSS, HSSS, HSHS, HVSS, HXSS, XSSS, XVSS, XHSS*; $w = 1$ for all other components;
(b) Weights: $w = 2.0$ for the components *SVSS, SVVS, SHSS, SHHS, VSSS, VSVS, VHSS, VXSS, HSSS, HSHS, HVSS, HXSS, XSSS, XVSS, XHSS*; $w = 1$ for all other components.

The results are shown in Fig. 8 (for type (a) filter) and in Fig. 9 (for type (b) filter). A comparison of Figs. 8 and 9 shows that image filtering caused better concentration of recognized objects near the mean points of *normal* and *pathologic* learning sets. Analogous calculations performed for the sets of images of $\times 200$ and $\times 40$ magnification led to similar results. Similar influence of filtering on image classification has

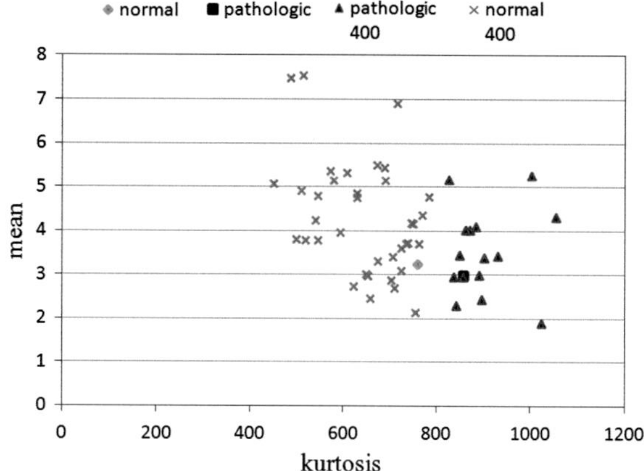

Fig. 7 Spatial distribution of 55 classified *normal* ($\times$) and *pathologic* (▲) liver tissue images in averaged *mean* versus *kurtosis* coordinates; image magnification $\times 400$

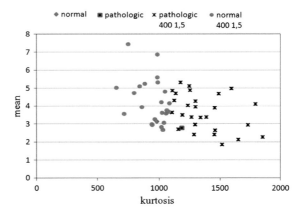

Fig. 8 Spatial distribution of 55 classified *normal* (×-) and *pathologic* (▲-) liver tissue images filtered by *MS*4 type (a) filter presented in averaged *mean* versus *kurtosis* coordinates; image magnification ×400

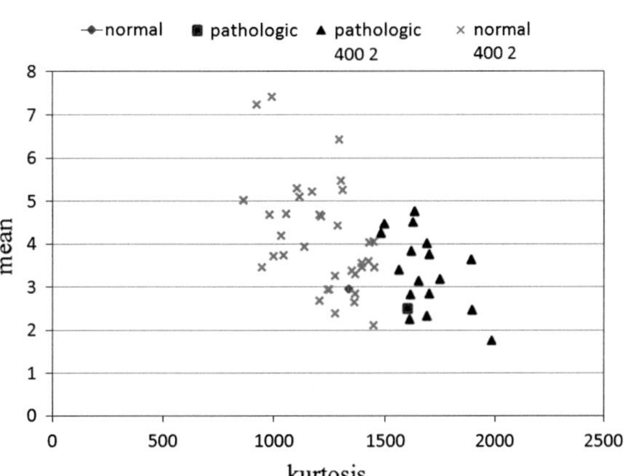

Fig. 9 Spatial distribution of 55 classified *normal* (×) and *pathologic* (▲) liver tissue images filtered by *MS*4 type (b) filter presented in averaged *mean* versus *kurtosis* coordinates; image magnification ×400

also been observed when *k-NN* algorithm was used. However, the results of classification reached by using *NM* and *k-NN* algorithms were slightly different. The results might also be different if Euclidean distance instead of Manhattan distance are used. This point was not analyzed in details because our aim was investigation of the influence of image enhancement by filtering on image classification independently of the classification method.

5 Conclusions

Monitoring of the processes of pollution expansion in natural environment needs, among many other tools, effective methods of recognition and evaluation of lesions caused by harmful chemical compounds in inner organs of living organisms. Morphological spectra can be used to improve the visibility of lesions in the images of liver tissue of fish affected by mercury compounds dissolved in the sea water. Selected components of the 4th range morphological spectra (*MS*4) are particularly useful in detection of morphological forms specific for this type of lesions. For classification of liver tissue images into the *normal* and *pathologic* groups estimated mean values, standard deviations and kurtoses of selected *MS*4 components should be used. The so-obtained morphologic-statistical parameters can be used to construct the *normal* and *pathologic* learning sets in a multi-dimensional parameter space and to design the corresponding classifiers for the liver tissue images. Preliminary image filtering consisting in reinforcement of the selected *MS*4 components improves the results of classification. The approach to lesions detection based on morphological spectra of the texture of biological tissues can be applied to larger class of applications.

Acknowledgments We would like to express our gratitude to Benjamin Daniel Barst, for providing images of liver tissues in fish for our experiments and to Diana Wierzbicka for her help in image filtering.

References

1. Barst, B.D.: Hepatotoxicity of mercury to fish. B.S. Thesis, University of North Texas, August 2010
2. Barst, B.D., Gevertz, A.K., Chumchal, M.M., et al.: Laser ablation ICP-MSCo-Localization of mercury and immune response in fish. Environ. Sci. Technol. **45**, 8982–8988 (2011)
3. Przytulska, M., Kulikowski, J.L., Wierzbicka, D.: Biomedical images enhancement based on the properties of morphological spectra. Biocybern. Biomed. Eng. **35**, 206–215 (2014)
4. Maitre, H.: Image Processing. Wiley, New York (2008)
5. González, R., Woods, R.: Digital Image Processing. Pearson/Prentice Hall, Upper Saddle River (2008)
6. Kulikowski, J., Przytulska, M., Wierzbicka, D.: Description of biomedical textures by statistical properties of their morphological spectra. Biocybern. Biomed. Eng. **30**(3), 19–34 (2009)
7. Fix, E., Hodges, J.L.: Discriminatory analysis: nonparametric discrimination small sample performance, project 21–49-004, report number 11, USAF School of Aviation Medicine, Randolph Field, Texas, pp. 280–322 (1952)
8. Dasarathy, B.V.: NN Pattern Classification Techniques, pp. 40–56. IEEE Computer Society Press, Washington (1991)
9. Devijver, P.A., Kittler, J.: Pattern Recognition: A Statistical Approach. Prentice Hall, London (1982)

Artificial Photoreceptors for Ensemble Classification of Hyperspectral Images

Paweł Ksieniewicz and Michał Woźniak

Abstract Data obtained by hyperspectral imaging gives us enough information to recreate the human vision, and also to extend it by a new methods to extract features coded in a light spectra. This work proposes a set of functions, based on abstraction of natural photoreceptors. The proposed method was employed as the feature extraction for the classification system based on combined approach and compared with other state-of-art methods on the basis of the selected benchmark images.

Keywords Artificial photoreceptors · Ensemble classification · Machine learning · Hyperspectral imaging

1 Introduction

Bare human perception of electromagnetic radiation reflected from observable objects is limited to only four[1] narrow quants of information spectrum. They are the single channels, which are composed together by a human brain bring its owner a chemical illusion called *color vision*. The phenomena of colors give human an ability to classify observed objects and detect their attributes. For example, an apple with dominant reflection from the red channel is more ripen than one with a supremacy of green. With progress of the civilization, information coded in colors become easy to learn, naturally non-interruptive language of signs as traffic lights or color-coded subway maps. Color can be interpreted as short vector, most often builded by three values. Its most popular representation for computers is RGB model, based on human perception on daylight, described by Svaetichin in 1956 [12]. Place of S M *and* L

[1]Three for color vision, and one for limited night vision.

P. Ksieniewicz (✉) · M. Woźniak
Department of Systems and Computer Networks, Faculty of Electronics, Wroclaw University of Technology, Wroclaw, Poland
e-mail: pawel.ksieniewicz@pwr.edu.pl

M. Woźniak
e-mail: michal.wozniak@pwr.edu.pl

R. Burduk et al. (eds.), *Proceedings of the 9th International Conference on Computer Recognition Systems CORES 2015*, Advances in Intelligent Systems and Computing 403, DOI 10.1007/978-3-319-26227-7_44

cone cells is taken there for channels of particular light impressions. But, as a matter of fact, red, green, and blue are nothing more than just the features, mixed into colors by a human brain. Moving to the last decade of XX century, we reached an area of rapid and intense research and development on idea of *hyperspectral* (HS) *imaging*. HS images are acquired by a remote sensors, sensitive on way much wider spectral range and with much denser channel spacing than human eye. The current industrial standard, AVIRIS spectrometer, captures images with 224 channels in range 0.4–2.5 μm. Imitating the natural photoreceptors is an enhancement of HS image visualization. The beginnings of HS imaging came with a collection of basic methods of generating false-color pictures. The simplest possible maps three bands from spectral signature into RGB model. Main idea of this approach it is to reduce the length of feature vector, to be describable by some of existing color models (RGB, HSL, HSV). Most popular standard used to reduce spectral depth of data is PCA[2] [1]. The first three principal components, capturing the most information from HS cube are mapped to the color channels. Due to the high influence of atmospheric effects on quality of HS images, many works attempts to balance S/N[3] to reduce impact of noise and enhance contrast of image [6]. Our work also proposes a method of easy noise filtering, which is necessary to make proposed processing possible. In 2005 [7], *Jacobson* and *Gupta* defined *the design goals* for the HS image visualization. Since that, during last decade, the problem of color display for HS images has received significant attention. Some works implement and review a series of supervised and unsupervised data transformation and classification algorithms for HS images purposes [5], where the main goal of other works is focused on redefinition of original design goals [3]. Often, instead of the fourth dimension—irrepresentable in a flat picture—the time or interactivity is used. Usage of PCA is also extended with linear programming, or popular technics like bilateral filtering [9]. Lately, there are many works in topic of linear [8] or optimization-based fusion [10].

2 Artificial Photoreceptors

Counterpart of color for HS imaging is called a spectral signature. High density of information spanned on wide range of spectrum results with a structure similar to continuos section of polynomial function. It emphasizes the interpretation of image not as a flat matrix of colors, but a bold cube of reflectance values. Retina of human is equipped with average close to 4.5 millions of cone cells [11]. Letting ourselves a huge simplification, we can look at it like an organic 4.5 Mpix CCD matrix. According to this comparison it is over twice as dense as 1080p—a current standard of HD Television. But it is also only a half of incoming standard of 4K. It is absolutely imageable amount of information, we process in our everyday RTV devices. From computational point of view, we can percept the matrix of cone cells of same type as function, processing a spectral cube of data into flat, monochrome image. Quick look

[2]Principal Components Analysis.

[3]Signal-to-noise ratio.

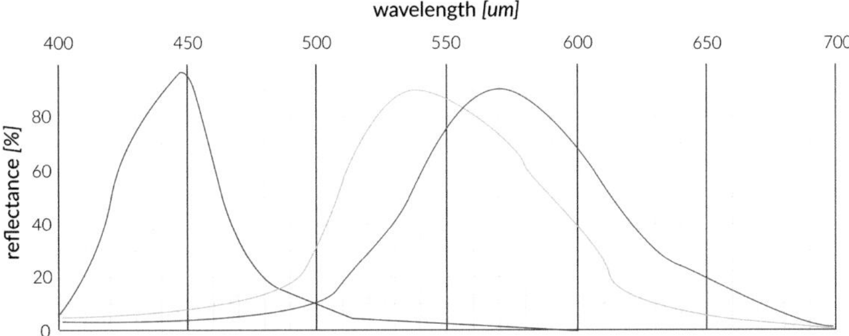

Fig. 1 Sensing spectra of human cone cells S, M, and L types

at sensing spectra of all types of human cones (Fig. 1) clarifies why it is a *function* and *not only a quant* of spectrum. According to this observation, we can imagine other functions generating color channel information. As long as they are not real cone cells, we can name them as *artificial photoreceptors* (APs). Being equipped with an spectral cube from HS sensor, we are not longer limited by theirs natural properties. APs can express literally anything we are able to calculate from a spectral signature.

3 Processing Procedure

Procedure will be presented according to Salinas A image, which, in form of false-color image, is presented in Fig. 2. By combining *channels* (outputs of APs) describing wider range than visible light into the RBG model, we can *translate* an HS image to human. Such method protects against lost of sensitivity of the naturally invisible spectrum covered by HS sensor. This work tries to prove this thesis, proposing a new, four-staged conversion method. Processing chain is presented in Fig. 3. Let us describe the successive stages of the proposed algorithm.

Fig. 2 Salinas A image in false-color visualization

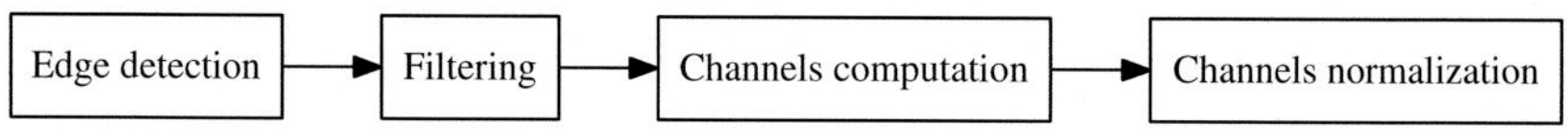

Fig. 3 Processing chain

3.1 Edge Detection

Detecting edges of regions on an image can be useful, because rejection of unsure, border-pixels could have positive influence on process of machine learning and contrast increasing by normalization accomplished by ignoring borders between regions of picture. Increase of contrast is significant and easy-observable for human eye, so it is not experimentally proven in this paper. Easiest possible method to detect borders of areas on a picture is to calculate an difference between minimal and maximal value in nearest neighborhood for its every pixel [4]. The result of this operation can be imagined as a map of dynamics of every pixels neighborhood (Fig. 4).

- For HS cube we are starting with three-dimensional area.
- We are building a bigger, four-dimensional image, which additional dimension is an image of a shift alongside layer by two-dimensional unit vector in every direction possible on a surface.
- It results with nine three-dimensional images joined into one four-dimensional matrix. We can decrease its dimensions into three by replacing last one with the value of a difference between its highest and lowest elements.
- It results with a new three-dimensional image of dynamics. A flat mask would be less complex in later computations, so we are flattering it alongside wavelength axis.

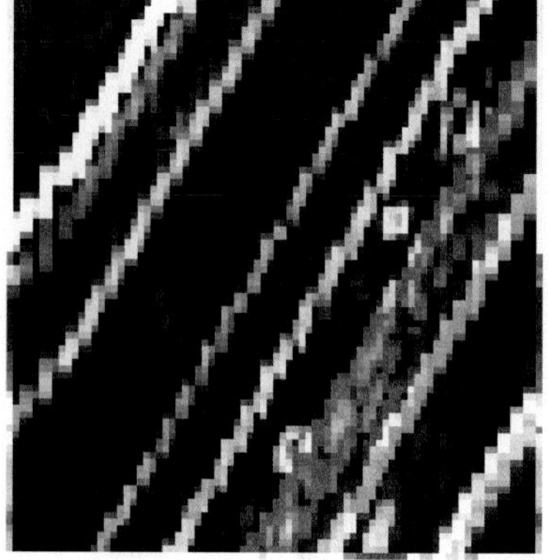

Fig. 4 The exemplary edge mask generated by the proposed method

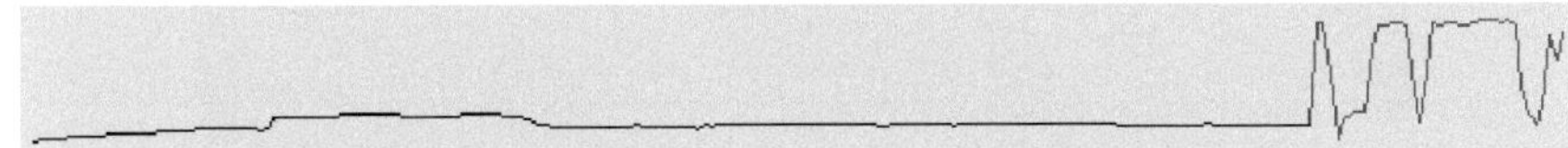

Fig. 5 Information and noise separation

3.2 Filtering

The method described in last section gives us an useful side effect. It is very easy to obtain a measurement of entropy ($\bar{\mathrm{H}}$) for every layer of image from it.

- Take the sum of all values (ρ) in every layer and divide it by amount of pixels per layer (ppl). It results with vector containing normalized value of entropy.

$$\bar{\mathrm{H}} = \frac{\sum \rho}{ppl} \tag{1}$$

- Making an assumption that every HS image have ranges with high amount of noises,[4] we can also assume that enough threshold for filtering most of them should be *mean value of entropy*.
- Unfortunately, this kind of filter is terrible in separating slopes of entropy changes. This lacking ability can be obtained by enhancing entropy vector by extra set of data, describing its dynamics ($D\bar{H}$). Compute it using the same method as for edge detection, but for an one-dimensional data. Calculate difference between current ($\bar{H}$) and neighbor value ($\bar{H}'$) on the entropy vector.

$$\mathrm{D}\bar{\mathrm{H}} = |\bar{H} - \bar{H}'| \tag{2}$$

- Generate an dynamics vector in the same way as entropy.
- Final filter is the **union** of mean entropy and mean dynamics filter.

Ability of separation informations from noises across available spectrum is presented in Fig. 5.

3.3 Channels Computation

A cleaned-up spectral signature can be used to generate its description using simple statistical methods. For example, it is impossible to obtain minimum or maximum from vector, when the noises in some ranges are way stronger than regular

[4]Noises in spectral signature comes from atmospheric effect and are an immanent part of every HS image coming from every HS sensor.

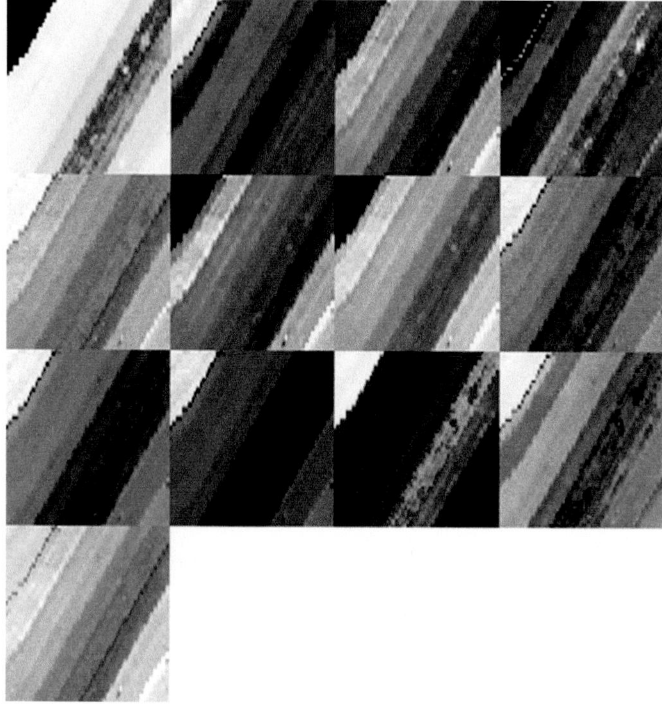

Fig. 6 Normalized channels for *Salinas A* dataset. From *top left* hue, saturation, brightness, red, green, blue, minimum, maximum, mean, median, mode, standard deviation, and variance

information. There was proposed 13 example APs, which can be easily and used in parallel for every signature vector. We can use functions for minimum, maximum, mean or median values and mode, standard deviation or variance, to enumerate only a few. Proposed collection is very simple and easy to calculate, even during the loading of HS cube to memory. Figure 6 shows all of them, calculated for *Salinas A* dataset.

3.4 Channels Normalization

All channels (outputs of APs) are normalized using the regular equation, but the values of minimal and maximal pixels of image are obtained by ignoring mask of the border-pixels calculated during the first stage of process. Difference between channel before and after normalization is presented in Fig. 7.

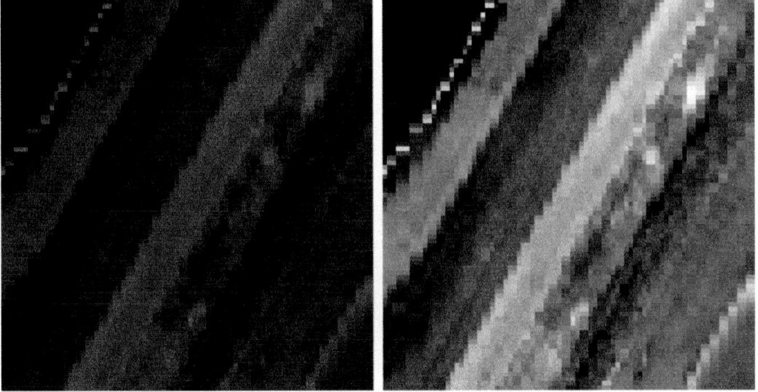

Fig. 7 The example channel (*left*) and its normalized version (*right*)

4 Experimental Evaluation and Results

Experiment aims to compare accuracy of classification proceeded on full-length signature and only on the features obtained during proposed algorithm. We want to know what is an information and accuracy loss of proposed method and which classification algorithms are the best fit for it. This part was processed in KNIME framework. All experiments were carried out using 5×2 cross-validation and presented as average accuracy. We have also used a statistical test to judge statistically significant differences between classificators. Choice was a 5×2 cv F Test [2], where algorithm proceeded for each fold. Evaluation and comparison between classification results was proceeded using four hyperspectral datasets, which are conventional benchmarks, according to the literature. Three of them were taken over agricultural terrains, by the AVIRIS (airborne visible infra-red imaging spectrometer) and the last one was acquired using the ROSIS (reactive optical system imaging spectrometer) over an urban scene. Datasets were provided with background label marks, assumed by expert as not important for analysis. Like in most articles, it was decided to eliminate those part from analysis. We decide to compare classifier ensembles based on popular individual classifiers:

- **k-NN**: k Nearest Neighbor,
- **NB**: Naive Bayes,
- **MLP**: Multiple Layer Perceptron with two hidden layers,
- **DT**: Decision Tree,
- **SVM**: Support Vector Machine.

To obtain, how signature reduced to set of channels preserves information, k-NN classifier used the full-length signature. As classifier ensemble, we selected the tree ensemble, available in KNIME software. Results of experiment for every dataset are presented in Fig. 8. Averaged results for all classifiers accuracies are presented in the Table 1. As we can observe, an classifier ensemble and k-NN win every competition with statistically proven superiority over rest of classifiers. The worst classifiers are

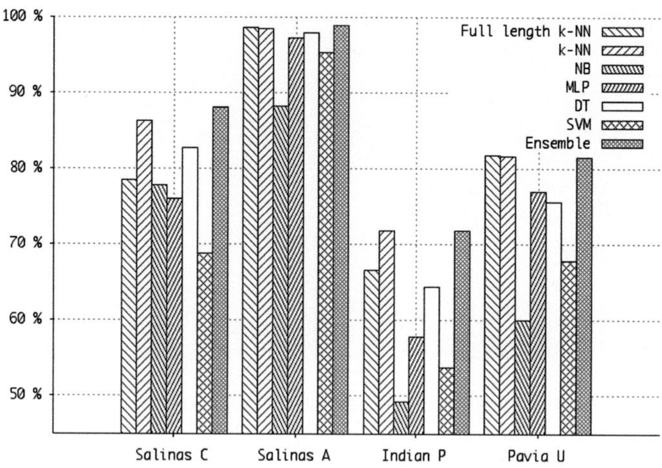

Fig. 8 Classification algorithms accuracy

Table 1 Results of the experiment with respect to accuracy [%], extended by information from 5×2 *cv F Test*. Values in table denote for how many datasets, algorithm from row is statistically better than algorithm from column

	F.l.	k-NN	NB	MLP	DT	SVM	Ens.
F.l.	–	0	4	0	0	4	0
k-NN	1	–	4	2	2	4	0
NB	0	0	–	0	0	0	0
MLP	0	0	4	–	0	4	0
DT	0	0	4	0	–	4	0
SVM	0	0	1	0	0	–	0
Ens.	1	0	4	2	2	4	–

Naive Bayes and SVM. We can assume that using classifier ensemble or k-NN with the features obtained by artificial photoreceptors are the best and most stable solution between compared. Comparison between full-length k-NN and ensemble tells us, that the proposed method is able to minimize the information loss, and in some cases overpower the full-length representation. It reduces the resources needed for computation with preserving the high accuracy of classification.

5 Conclusions

The classifier ensemble based on the spectral signature basic statistical parameters seems to be reliable, proven tool to classify regions on hyperspectral images. Also, the proposed simple method of HS dimensions reduction seems to be enough for classification purposes, and its simplicity allows us to make lightweight solutions for hyperspectral analysis. SVM, NB, DT and MLP classifiers are proven to be a bad choice for HS data. In future works, we can extend the proposed processing by methods of fusion or available methods of dimensions reduction to product new method of HS image visualization.

Acknowledgments The work was supported by the statutory funds of the Department of Systems and Computer Networks, Wroclaw University of Technology and by The Polish National Science Centre under the grant agreement no. DEC-2013/09/B/ST6/ 02264.

References

1. Agarwal, A., El-Ghazawi, T., El-Askary, H., Le-Moigne, J.: Efficient hierarchical-PCA dimension reduction for hyperspectral imagery. In: 2007 IEEE International Symposium on Signal Processing and Information Technology, pp. 353–356, December 2007
2. Alpaydin, E.: Combined 5 x 2 cv f test for comparing supervised classification learning algorithms. Neural Comput. **11**(8), 1885–1892 (1999)
3. Cui, M., Razdan, A., Hu, J., Wonka, P.: Interactive hyperspectral image visualization using convex optimization. IEEE Trans. Geosci. Remote Sens. **47**(6), 1673–1684 (2009)
4. Davies, E.R.: Machine Vision: Theory, Algorithms, Practicalities. Elsevier, Amsterdam (2004)
5. Du, Q., Raksuntorn, N., Cai, S., Moorhead, R.: Color display for hyperspectral imagery. IEEE Trans. Geosci. Remote Sens. **46**(6), 1858–1866 (2008)
6. Durand, J., Kerr, Y.: An improved decorrelation method for the efficient display of multispectral data. IEEE Trans. Geosci. Remote Sens. **27**(5), 611–619 (1989)
7. Jacobson, N., Gupta, M.: Design goals and solutions for display of hyperspectral images. IEEE Trans. Geosci. Remote Sens. **43**(11), 2684–2692 (2005)
8. Jacobson, N., Gupta, M., Cole, J.: Linear fusion of image sets for display. IEEE Trans. Geosci. Remote Sens. **45**(10), 3277–3288 (2007)
9. Kotwal, K., Chaudhuri, S.: Visualization of hyperspectral images using bilateral filtering. IEEE Trans. Geosci. Remote Sens. **48**(5), 2308–2316 (2010)
10. Kotwal, K., Chaudhuri, S.: An optimization-based approach to fusion of hyperspectral images. IEEE J. Sel. Top. Appl. Earth Obs. Remote Sens. **5**(2), 501–509 (2012)
11. Ruskell, G.: The Human Eye, Structure and Function Clyde W. Oyster, 766 p. Sinauer Associates, Sunderland (1999). Hardback, ISBN 0-87893-645-9, £49.95. Ophthalmic and Physiological Optics **20**(4), 349-350. http://dx.doi.org/10.1046/j.1475-1313.2000.00552.x (2000)
12. Svaetichin, G.: Spectral response curves from single cones. Acta physiol. Scand. Suppl. **39**(134), 17–46 (1956)

Real-Time Eye Detection and Tracking in the Near-Infrared Video for Drivers' Drowsiness Control

Bogusław Cyganek

Abstract This paper presents a visual system for real-time eye detection and tracking in the near-infrared (NIR) video streams for drivers' monitoring. The system starts with crude eye position estimation based on an eye model suitable for NIR processing. In the next step, eye regions are verified with the classifier operating in the higher-order decomposition of the tensor of eye prototypes. Finally, the process is augmented with the linear tracker which facilitates eye detection and allows real-time operation necessary in the automotive environment. The reported experiments show high accuracy and real-time operation of the system in the car.

Keywords Eye tracking · Real-time detection

1 Introduction

Fatigue, exhaustion, tiredness, or inattention very often leads to serious traffic accidents. Therefore, much research is conducted toward the automatic systems for driver assistance and monitoring. For this purpose computers can be used to monitor drivers' state with potential of alerting for dangerous conditions [5]. The key task of such systems is reliable and real-time detection of driver's eyes in difficult lighting conditions. In this paper, an eye recognition system operating entirely in the near-infrared conditions is described. It is a modification of our previous versions presented in [2, 3]. The main modification, which we focus mostly in this paper, is the new tracking module which greatly facilitates operation of the system and allows real-time operation on computer platforms with moderate computational power. For tracking the simplified version of the Kalman filter is used. Recently, many systems have been reported for eye recognition as well as for monitoring drivers' drowsiness. An example is the system proposed by Bergasa et al. [1]. In their approach, eye detection

B. Cyganek (✉)
AGH University of Science and Technology, Al. Mickiewicza 30,
30-059 Krakow, Poland
e-mail: cyganek@agh.edu.pl

R. Burduk et al. (eds.), *Proceedings of the 9th International Conference on Computer Recognition Systems CORES 2015*, Advances in Intelligent Systems and Computing 403, DOI 10.1007/978-3-319-26227-7_45

relies on a known effect of white spots in NIR images due to the light reflected by the retina. This requires LEDs and central placement of the camera. In the proposed system, we do not follow this idea and the cameras can be placed in a more convenient place in a car. Zhu and Ji proposed eye recognition system based on the Kalman and mean-shift trackers [19]. Their method joins appearance-based object recognition and tracking with active IR illumination, using the mentioned effect of high reflectance of the pupil. However, the system was not designed to operate in a car. Further, interesting propositions were proposed by García et al. [6], Ma et al. [13], Kawaguchi et al. [11], Wang et al. [17], as well as de Orazio et al. [5]. The paper is organized as follows. Architecture of the proposed system is presented in Sect. 2. In Sects. 3 and 4, eye detection and tracking modules are described. Sections 5 and 6 present experimental results and conclusions, respectively.

2 Architecture of the Eye Detection and Tracking System

Figure 1 depicts architecture of the presented system. Eye recognition is performed by a cascade of specialized modules, each refining output of its predecessor. Detection block consists of many processing modules. The most important is eye region candidate detection based on the novel eye model developed for processing of NIR images, described in [2, 3]. Once the positive regions are detected after the exhaustive search in the whole image, processing is facilitated with the tracker module, as shown in Fig. 1. Predicted eye positions (each tracked by a separate tracker) are fed to the Higher-Order Singular Value Decomposition (HOSVD) classifier for verification [3]. If successful, the tracking proceeds. Otherwise, the detection is started again. This switch is necessary due to appearance change of the eyes. For instance, tracking is lost in eye blinking or rapid head movements. As already mentioned, eye region candidates are classified by the tensor subspace classifier. It is a classical subspace projection method with the exception that the bases are not pattern vectors but pattern tensors [2, 16]. This change allows better representation of images which are 2D structures. The pattern space is built with help of the HOSVD decomposition of the pattern tensors.

3 Detection and Recognition by the Cascade of Classifiers

Eye detection in real car conditions cannot be achieved only with the passive sensors operating in the visible spectrum. Therefore, the central role plays acquisition of the NIR images obtained after short illumination of a driver with the LED operating in this spectrum. However, eye detection in NIR images differs from detection in the visible spectrum. This is due to different dynamics of the NIR images, as well as different reflectance properties of objects. Thus, new eye model was developed to facilitate eye detection. Details are described in [2, 3]. Eyes change their appearance

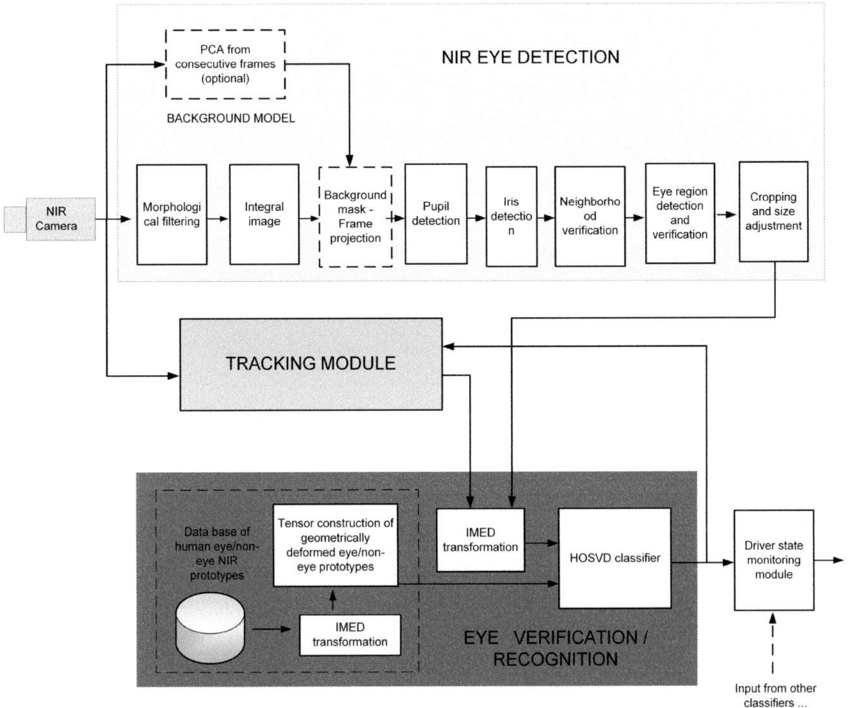

Fig. 1 Architecture of the eye recognition system in the NIR video. Detection proceeds in the whole frame. Tracking module facilitates eye positioning without exhaustive image search. Eye verification is done by the HOSVD classifier

due to driver's movements and blinking. Therefore, their proper recognition needs to be performed by a properly trained classifier [8, 12]. In the presented system, this is done with the tensor subspace method operating in the subspaces obtained from the HOSVD decomposition of the tensors containing geometrically deformed versions of the prototype eye patterns. Thanks to this, eyes can be correctly recognized in different viewing conditions. Briefly, for each normalized eye pattern $\hat{\mathbf{P}}$, and for an a priori chosen number H of tensor bases $\hat{\mathcal{T}}^{i}_{h}$, the following residual value ρ_i is computed [4]:

$$\hat{\rho}_i = \sum_{h=1}^{H} \left\langle \hat{\mathcal{T}}^{i}_{h}, \hat{\mathbf{P}} \right\rangle^2 . \tag{1}$$

In operation mode, the system returns a class i for which the corresponding $\hat{\rho}_i$ in (1) is the largest. To cope with the problem of outliers, i.e., patterns for which the system was not trained for, a minimal acceptance threshold was set on $\hat{\rho}_i$ (in our experiments it is 0.8). The number of components H in (1) was chosen from 5 to 25. Higher H allows better accuracy, at an expense of computation time.

4 Eye Tracking Module

The previously described process of eye detection starts from an exhaustive search over the whole image space. In the case of a single image, usually there is no other option. However, when processing a video stream which consists of highly correlated images, detection can be greatly facilitated with the tracking module. In other words, assuming constant motion model of an observed object, a position and a pattern of change of an object in few frames determine its possible position in the subsequent frame, and so on. One of the best known tracking methods is the Kalman filter [10]. In this case, the goal of object tracking is to use a filtering process to compute object state estimates based on the noisy observations and on the target model. The process is governed by the following equations [9, 10, 14]:

$$\mathbf{s}(k) = \Theta\, \mathbf{s}(k-1) + \Psi\, \mathbf{w}(k-1), \tag{2}$$

where $\mathbf{s}(k)$ is the tracked object state vector at time index k, Θ is the state transition matrix, $\mathbf{w}(k)$ is the unknown object maneuver, and finally Ψ denotes the maneuver/state transition matrix. The second equation of the Kalman filter concerns the object measurement (observation) related to the state and an additive noise component, as follows:

$$\mathbf{z}(k) = \mathbf{H}\,\mathbf{s}(k) + \mathbf{n}(k), \tag{3}$$

where $\mathbf{z}(k)$ is the object measurement vector, $\mathbf{H}$ denotes a state transformation matrix, and $\mathbf{n}$ is the unknown measurement noise which is assumed to be a zero-mean white stationary process. Based on the current measurement and the measurement prediction, the residual vector is computed:

$$\mathbf{r}(k) = \mathbf{z}(k) - \mathbf{H}\,\hat{\mathbf{s}}(k\ |k-1). \tag{4}$$

With the above definitions, the main goal of the Kalman filter is to compute an a posteriori state estimate $\hat{\mathbf{s}}(k\ |k)$ expressed as a linear combination of the a priori state estimate $\hat{\mathbf{s}}(k\ |k-1)$ and the weighted measurement residual $\mathbf{r}(k)$, as follows:

$$\hat{\mathbf{s}}(k\ |k) = \hat{\mathbf{s}}(k\ |k-1) + \mathbf{K}(k)\ \mathbf{r}(k). \tag{5}$$

In the last equation, which is called the smoothing, measurement update, or correction equation, $\mathbf{K}(k)$ denotes the gain matrix which, in general, changes from step to step. In the above expressions we use the notation $\hat{\mathbf{s}}(i\ |j) \equiv p\left(\mathbf{s}_i\ \middle|\mathbf{Z}_j\right)$, where $\mathbf{Z}_j$ is a sequence of all available measurements up to the time instant j, i.e., $\mathbf{Z}_j \equiv \{\mathbf{z}_i :\ i = 1 \ldots j\}$. However, a computationally simpler version of the Kalman approach is the α-β filter. It is a steady-state version of the former with constant noise and system models. In particular, the α–β filter assumes the second-order state vector containing only position and velocity of the tracked object, as follows:

$$\mathbf{s}(k) = \left[x(k) \; v(k) \right]^T, \tag{6}$$

where $x(k)$ and $v(k)$ denote the position and speed components at time index k, respectively. In this case the matrices from Eqs. (2)–(4) become as follows:

$$\Theta = \begin{bmatrix} 1 & T \\ 0 & 1 \end{bmatrix}, \tag{7}$$

where T denotes time period between state updates. In our case this is related to the frame rate, thus T=1/30 s. Now, under an assumption that in (2) maneuverability term only affects the velocity state, the object maneuverability matrix Ψ becomes

$$\Psi = \begin{bmatrix} 0 \\ T \end{bmatrix}, \tag{8}$$

and

$$\mathbf{H} = \left[1 \; 0 \right], \tag{9}$$

which means that measurement in (3) affects only spatial state x. It holds also that **w**=**h**. Finally, the filter gain **K** in (5) becomes

$$\mathbf{K} = \left[\alpha \; \tfrac{\beta}{T} \right], \tag{10}$$

where the α and β are the parameters, thus the name of this filter. These, in turn, can be determined from the sampling period T, as well as the two variances:

- σ_a, which denotes standard deviation related to the object dynamics (acceleration),
- σ_n, which denotes measurement noise standard deviation which corresponds to the object detection method.

The above three parameters can be joined into one common parameter called the tracking index, as follows:

$$\Lambda = T^2 \frac{\sigma_a}{\sigma_n}. \tag{11}$$

The tracking index, which relates object position uncertainty to the object measurement uncertainty, allows determination of the optimal tracking parameters, as shown by Kalata [9, 18]. It can be shown that the optimal values of α and β for steady-state conditions are as follows [7]:

$$\alpha = 1 - \frac{1}{16}\left(4 + \Lambda - \sqrt{\Lambda\,(8 + \Lambda)}\right)^2, \tag{12}$$

and

$$\beta = 2\,(2-\alpha) - 4\sqrt{1-\alpha}. \tag{13}$$

The α and β parameters take on small positive values less than 1. Now, considering Eqs. (6)–(10), the prediction equation (2) can be written as follows:

$$\hat{x}_k = \hat{x}_{k-1} + T\hat{\upsilon}_{k-1}, \tag{14}$$

$$\hat{\upsilon}_k = \hat{\upsilon}_{k-1}, \tag{15}$$

and the measurement

$$\hat{z}_k = x_k, \tag{16}$$

where x_k denotes (noisy) measurement at step k. Similarly, the update equation (5) can be expressed as follows:

$$\hat{x}_{k+1} = \hat{x}_k + \alpha\,\hat{r}_k, \tag{17}$$

$$\hat{\upsilon}_{k+1} = \hat{\upsilon}_k + \frac{\beta}{T}\,\hat{r}_k, \tag{18}$$

where

$$\hat{r}_k = \hat{z}_k - \hat{x}_k, \tag{19}$$

denotes the residual error between the measured and the estimated position. Equations (14)–(19) constitute a direct description of the steps of the α–β tracking algorithm for one-dimensional motion. In the case of images, two such trackers are used to track 2D object in x and y directions, respectively. The last thing is the way to determine the tracking index Λ in (11), as well as the parameters α and β that control the tracking process. However, the problem is that these parameters usually are quite different in the transient and steady-state conditions, which are given by (12) and (13). Since T is known, the tracking index Λ in (11) can be determined after estimating the variances σ_n and σ_a. The former is related to the uncertainty of object state measurement which in our case boils down to the image quantization or, in other words, to image resolution, as shown in the report by Safadi [15]. It was shown that

$$\sigma_n = \frac{1}{2\sqrt{6}r}, \tag{20}$$

where r denotes resolution of a processed image in x direction. In our system images are of resolution 640×480. Thus, $\sigma_{nx} = 32\text{e}-5$, $\sigma_{ny} = 42\text{e}-5$. To determine σ_a, which is mostly affected by changes of object acceleration, Safadi proposed to apply a separate α-β-χ filter [15]. However, we are tracking specific objects and in a specific environment. Thus, σ_a can be easily approximated by the temporal measurements of the eye movement accelerations provided by the previous detections, as follows:

$$\sigma_a = E\left[a_i^2\right] - E^2\left[a_i\right] \approx \frac{1}{N-2}\sum_{i=3}^{N} a_i^2 - \left(\frac{1}{N-2}\sum_{i=3}^{N} a_i\right)^2, \tag{21}$$

where N denotes a number of acceleration values, each computed from three consecutive frames in accordance with the following formula:

$$a_i = \frac{1}{T^2}\left[x(i) - 2x(i-1) + x(i-2)\right]. \tag{22}$$

In the paper by Kalata, it was shown that α and β change exponentially with some time constants from the initial values up to the steady-state conditions [9]. For the α–β filter, the initial values are $\alpha(0) = \beta(0) = 1$. On the other hand, for a linear motion $\sigma_a \to 0$, thus the tracking index $\Lambda = 0$ and from (12) and (13), we easily obtain $\alpha_l = 0.75$ and $\beta_l = 0.5$. In the system σ_a is computed continuously for each visible eye. Then, once the eye view is lost, the process starts again. However, when tracking is initiated, we observed that it is sufficient to change the gain parameters α and β piecewise linearly in the first frames from their initial values up to the steady-state conditions.

5 Experimental Results

The presented system operates as shown in Fig. 1. Software was implemented in C++ with the help of *DeRecLib* package [4]. The testing video streams were recorded in real car conditions, shown in Fig. 2. The experiments were conducted on a laptop with 32 GB RAM and the i7-4800QM 4-core processor. The classifier that needs to be trained is the HOSVD classifier. For this purpose a database was created as described in [2]. Stages of eye detection and tracking are presented in the block

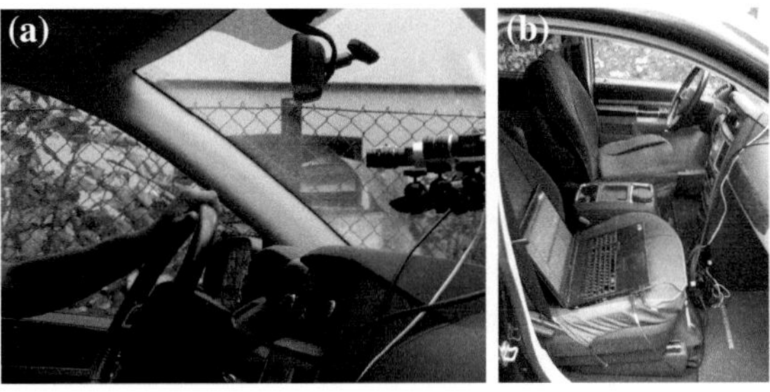

Fig. 2 Views of the driver's eye recognition system mounted in a car

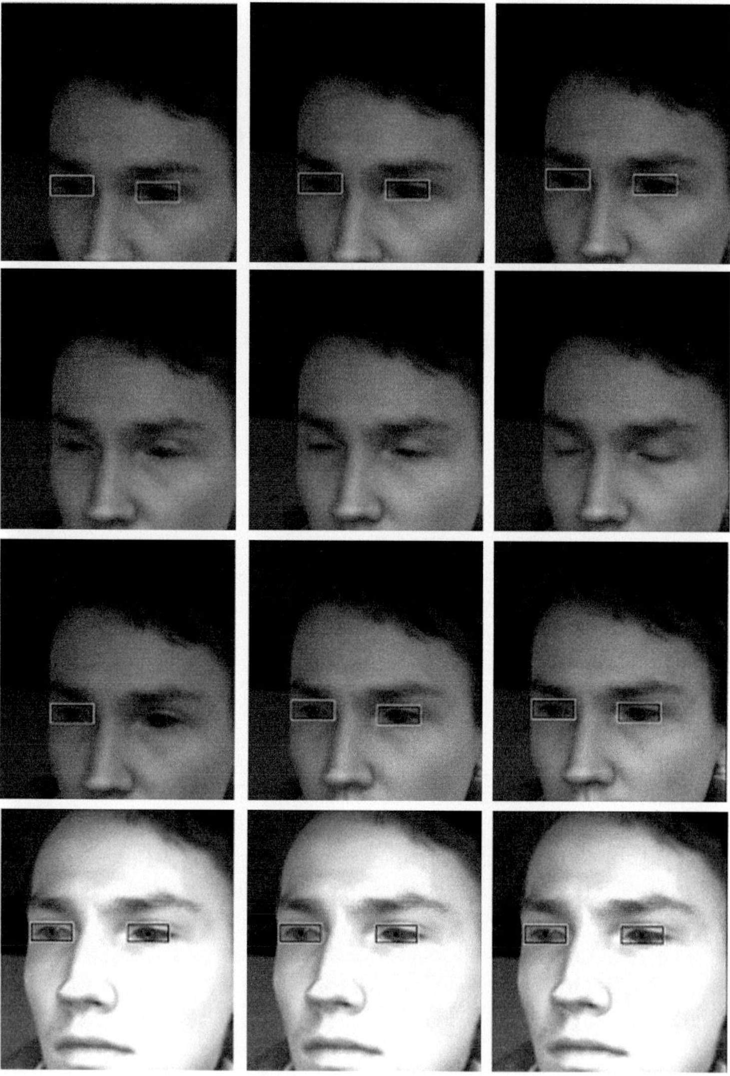

Fig. 3 Exemplary frames of driver's eye tracking in the NIR images in night conditions

diagram in Fig. 1. In the initial stage or after losing the tracking, exhaustive eye detection needs to be performed. To facilitate this process, a cascade of classifiers is used. The first one does very fast fit of the eye model, especially designed to deal with NIR conditions. Subsequent eye recognition is done with the pretrained tensor classifier. This process is augmented with the tracking module, as shown in Fig. 1. Figure 2 shows two images of the eye monitoring system mounted in a car. Results of eye tracking for one of the recorded sequences are presented in Fig. 3. Output regions are shown in black frames.

Table 1 Accuracy parameters of the eye recognition system in the NIR video

TP	FP	FN
97.5	0.1	2.4

Table 1 presents accuracy parameters of the presented system. Apart from the true-positive rate TP, the other most important parameter is the rate of false-positives FP. It needs closer examination to avoid a dangerous situation in which open eyes are reported by the system when in reality driver's eyes were closed. Driver's fatigue and drowsiness conditions can be monitored based on the output parameter of the percentage of eye closure, as described in [2]. The system allows real-time operation with the stream of 30/s images of resolution of 640×480 pixels.

6 Conclusions

In this paper, a system for eye recognition operating in the NIR spectrum is described. Its purpose is to recognize driver's eye with the purpose of detecting person's drowsiness and fatigue conditions which frequently lead to serious car accidents. The presented version of the system extends our previous propositions [2, 3] by an application of the tracking module which increased its robustness. For this purpose, the α–β filter is used which is a computationally less demanding version of the Kalman filter. Its properties, as well as details of implementation, were also presented. Thanks to this, the exhaustive search of eye candidates in each frame of the NIR video stream was eliminated. Therefore, the system easily meets requirements of real-time processing while also attains high accuracy and small ratio of false-positive responses. Properties of the system were measured in the experimental sequences acquired in real car conditions.

Acknowledgments This work was supported by the Polish National Science Center under the grant no. DEC-2013/09/B/ST6/02264 and AGH Statutory Funds no. 11.11.230.017. The author is very grateful to Mr. Marcin Bugaj, as well as to Mr. Stanisław Groński and Krzysztof Groński for their help in the experiments.

References

1. Bergasa, L.M., Nuevo, J., Sotelo, M.A., Barea, R., Lopez, E.: Visual monitoring of driver inattention. In: Prokhorov, D. (ed.) Computational Intelligence in Automotive Applications. SCI, vol. 132, pp. 25–51 (2008)
2. Cyganek, B., Gruszczyński, S.: Hybrid computer vision system for drivers' eye recognition and fatigue monitoring. Neurocomputing **126**, 78–94 (2014)

3. Cyganek, B., Gruszczyński, S.: Eye recognition in near-infrared images for driver's drowsiness monitoring. In: 2013 IEEE Intelligent Vehicles Symposium (IV), pp 397–402. Gold Coast, Australia, 23–26 June 2013
4. Cyganek, B.: Object Detection and Recognition in Digital Images. Wiley, NewYork (2013). Theory and practice
5. D'Orazio, T., Leo, M., Guaragnella, C., Distante, A.: A visual approach for driver inattention detection. Pattern Recognit. **40**, 2341–2355 (2007)
6. García, I., Bronte, S., Bergasa, L.M, Almazán, J., Yebes, J.: Vision-based drowsiness detector for real driving conditions. In: 2012 Intelligent Vehicles Symposium. Alcalá de Henares, Spain (2012)
7. Gray, E., Murray, W.: A derivation of an analytic expression for the tracking index for the alpha-beta-gamma filter. IEEE Trans. Aerosp. Electron. Syst. **29**, 1064–1065 (1993)
8. Jackowski, K., Krawczyk, B., Woźniak, M.: Improved adaptive splitting and selection: the hybrid training method of a classifier based on a feature space partitioning. Int. J. Neural Syst. **24**(3) (2014)
9. Kalata, P. R. The tracking index: A generalized parameter for α-β and α-β-γ target trackers. IEEE Transactions on Aerospace and Electronic Systems, AES -20, pp. 174–182 (1984)
10. Kalman, R.E.: A new approach to linear filtering, prediction problems. Trans. ASME J. Basic Eng. pp. 35–45 (1960)
11. Kawaguchi, T., Hidaka, D., Rizon, M.: Detection of eyes from human faces by Hough transform and separability filter. Int. Conf. Image Process. **1**, 49–52 (2000)
12. Krawczyk, B.: One-class classifier ensemble pruning and weighting with firefly algorithm. Neurocomputing **150**, 490–500 (2015)
13. Ma, Y., Ding, X., Wang, Z., Wang, N.: Robust precise eye location under probabilistic framework. IEEE Int. Conf. Autom. Face Gesture Recognit. pp. 339–344 (2004)
14. Ristic, B., Arulampalam, S., Gordon, N.: Beyond the kalman filter. Particle filters for tracking applications, Artech House (2004)
15. Safadi, R., B.: An Adaptive Tracking Algorithm for Robotics and Computer Vision Application. Technical Report MS-CIS-88-05, University of Pennsylvania (1988)
16. Savas, B., Eldén, L.: Handwritten digit classification using higher order singular value decomposition. Pattern Recognit. **40**, 993–1003 (2007)
17. Wang, P., Green M., Ji, Q., Wayman J.: Automatic eye detection and its validation. In: CVPR'05 Proceedings of the 2005 IEEE Computer Society Conference on Computer Vision and Pattern Recognition, Vol. 03, pp. 164–171 (2005)
18. Wikipedia: Alpha beta filter, http://en.wikipedia.org/wiki/Alpha_beta_filter#cite_note-Kalata-4 (2015)
19. Zhu, Z., Jib, Q.: Robust real-time eye detection and tracking under variable lighting conditions and various face orientations. Comput. Vis. Image Underst. **98**, 124–154 (2005)

Clothing Similarity Estimation Using Dominant Color Descriptor and SSIM Index

Piotr Czapiewski, Paweł Forczmański, Krzysztof Okarma, Dariusz Frejlichowski and Radosław Hofman

Abstract This paper deals with the problem of estimating the similarity of clothing for the purpose of fashion-related e-commerce systems. The images presenting fashion models are segmented and analyzed in order to detect clothing characteristics. We propose a method based on human pose estimation and body parts segmentation, followed by the analysis of dominant color and structural similarity, independently for particular body segments. The algorithm can be utilized to perform clusterization or in the simpler case—to directly search for similar outfits. The experiments performed using 1800 real-life photos proved the applicability of the proposed approach.

1 Introduction

1.1 Motivation

The fashion domain is one of the areas of e-commerce, which until recently has been largely neglected. The reason for this stems from the fact that most people are reluctant to buy clothing online, preferring to touch the garments and try them on. However,

P. Czapiewski (✉) · P. Forczmański · D. Frejlichowski
Faculty of Computer Science and Information Technology, West Pomeranian University of Technology, Szczecin, Żołnierska 49, 71-210 Szczecin, Poland
e-mail: pczapiewski@wi.zut.edu.pl

P. Forczmański
e-mail: pforczmanski@wi.zut.edu.pl

D. Frejlichowski
e-mail: dfrejlichowski@wi.zut.edu.pl

K. Okarma
Faculty of Electrical Engineering, West Pomeranian University of Technology, Szczecin, 26. Kwietnia 10, 71-126 Szczecin, Poland
e-mail: okarma@zut.edu.pl

R. Hofman
FireFrog Media sp. z o.o., Jeleniogórska 16, 60-179 Poznań, Poland
e-mail: radekh@fire-frog.pl

R. Burduk et al. (eds.), *Proceedings of the 9th International Conference on Computer Recognition Systems CORES 2015*, Advances in Intelligent Systems and Computing 403, DOI 10.1007/978-3-319-26227-7_46

during the recent years the number of fashion-related initiatives on the Internet has been rapidly growing—from fashion blogs, to fashion-focused social networking sites, to large online clothing retailers. Along with the increase of popularity within the e-commerce area, the interest in fashion has grown within the computer science research community. Various methods to analyze and process fashion-related data are sought for, including fashion recommender systems, virtual fitting rooms, clothing image retrieval systems, etc. One of the most interesting fashion-related issues is building a fashion recommender system. Such a system, among other components, must possess the capability of automatically estimating the similarity between two garments or two complete outfits, given the image of a human model. This paper deals with the issue of analyzing unconstrained images containing human silhouettes in order to assess the similarity between two presented outfits. Both colors and material texture are taken into account, separately for particular body segments. The proposed approach will allow to automatically quantify the similarity level, which can be later used in several scenarios within fashion-related systems, e.g.,

- directly searching for a similar outfit based on a user's request;
- identifying recurring fashion patterns or styles by mining the outfits database;
- clustering the outfits in order to speed up the retrieval (within the recommender system);
- classifying user's outfits to one of the styles (predefined or found through data mining).

This paper is organized as follows. The next subsection summarizes the previous works found in the literature. Section 2 presents a detailed method description. The performed experiments are described in Sect. 3. Section 4 concludes the paper.

1.2 Previous Works

The research related to the fashion domain deals with quite broad spectrum of topics, including computer vision and semantic technologies. A human–computer interaction system called "a responsive mirror" was proposed in [17]. The purpose of this interactive tool is to support shopping in a real (not online) retail store. The type and certain attributes of garments are extracted from the live image and similar clothes are searched in a database. An approach proposed in [5] allows for the recognition of clothes' attributes from images. However, the scope of analysis is extremely limited (only coats and jackets). Similar restrictions were applied in [2]—only garments for the upper body were analyzed. A different approach has been proposed in [11]. Here, the semantic description of clothing is analyzed in order to determine the similarity between items, without using any visual descriptors. The visual similarity based on image analysis was described in [3]. The authors computed weighted similarities between certain bundled features, using point features (SIFT) and local visual words. In all the above cases only single garments were analyzed, not the whole outfits. To the best of our knowledge, no research was reported aimed at retrieval or clustering of clothing imagery data dealing with complete outfits. Although the analysis of single garments is sufficient for many simple retrieval tasks (e.g., when assisting the user while looking for a particular piece of garment in online store), the assessment of the whole outfits is required in many tasks occurring within a comprehensive fashion recommender system. For example, when trying to identify user's unique style based on their own outfits (or outfits indicated by them as interesting), the composition of garments is equally or even more important than the particular pieces. The fundamental components of the approach presented in this paper have been proposed in our earlier works. Here, we integrate the approach using body segmentation and dominant color [4, 9] with the approach to compare the textile materials using structural similarity [12], building toward a comprehensive outfit similarity assessment for the purpose of the fashion recommender system.

2 Method Description

The method for assessing the similarity of two outfits comprises the following stages:

- human silhouette detection,
- body parts segmentation,
- calculation of color descriptors and color-based similarity estimation,
- structure-based similarity estimation,
- calculation of the combined similarity (or distance) measure.

Given the database of fashion model images and the algorithm in question, depending on the particular application, the obtained distance measure can be used to

- directly search for outfits similar to a given reference outfit,
- perform a clusterization of the whole dataset in order to group similar outfits together (e.g., for the purpose of style exploration or speeding up similar outfits retrieval).

2.1 *Preprocessing and Silhouette Detection*

Several assumptions have been made regarding the input images, as described in [9]. In general, no strong constraints are given (e.g., no requirement for a plain background or a specific model's pose). However, it is required that the image contains a whole human silhouette, facing toward the camera, covering at least 40 % of the total image area, and that only one figure is visible in the foreground. In the initial preprocessing phase, the above conditions are checked. Starting with the face detection (using a Viola–Jones algorithm [14]), the upper body area is estimated using several heuristic rules based on human body proportions. The upper body area gives a starting point for the next stage. The details of this step can be found in [9].

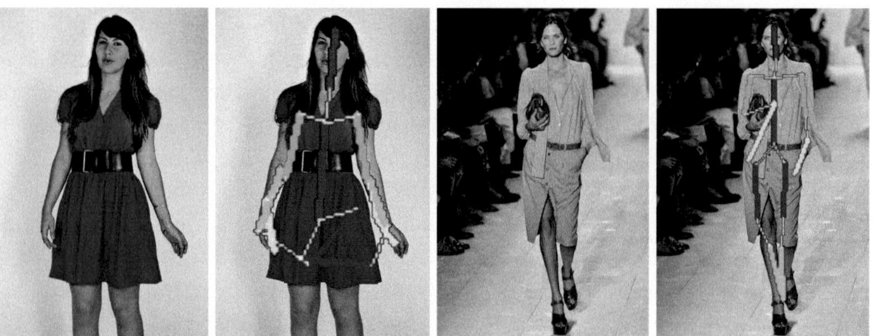

Fig. 1 Examples of body parts segmentation results

2.2 Body Parts Segmentation

Starting with the upper body area, the segmentation is performed using a watershed algorithm [13]. Next, the body part extraction is performed, following the approach proposed in [6, 8]. In our experiments we used the pose estimation software published at [7]. The output consists of 10 line segments corresponding to the following body parts: head, torso, upper and lower arms, thighs, and lower legs. The examples of segmentation results are presented in Fig. 1.[1]

2.3 Color-Based Similarity

The first aspect of similarity considered here is the distribution of colors within the extracted body segments. The applied approach was described in details in [9]. A simplified version of dominant color descriptor adopted from MPEG-7 standard [16] has been used. The color descriptor is calculated separately for each of the 10 body segments. After the conversion from RGB to HSV, color quantization is performed and a histogram with 72 bins is created. Next, all the values in the histogram except for the 8 highest ones are set to zero. Finally, the histogram is normalized. By concatenating 10 histograms, a feature vector of 720 elements is obtained, representing the distribution of dominant colors within the whole figure. The color-based com-

[1] All the photos used in the paper were obtained from Flickr.com website, from the following users: martingreffe, https://www.flickr.com/people/splinter66; icanteachyouhowtodoit, https://www.flickr.com/people/icanteachyouhowtodoit; GoToVan, https://www.flickr.com/people/gotovan; fervent-adepte-de-la-mode, https://www.flickr.com/people/51528537@N08; Christopher Macsurak, https://www.flickr.com/people/macsurak; Jessica Quirk, https://www.flickr.com/people/midwestjess; Jason Hargrove, https://www.flickr.com/people/salty_soul; Patrick Raczek, https://www.flickr.com/people/greenpatte; Frank Kovalchek, https://www.flickr.com/people/72213316@N00; Mycatkins, https://www.flickr.com/people/bigmikeyeah.

ponent of the similarity measure is calculated as follows. Given the two descriptors H_A and H_B, corresponding to the two images A and B, a histogram intersection is performed:

$$D_c(A, B) = \sum_{k=1}^{10} \sum_{i=1}^{72} \left[w_k \cdot min(H_A(i, k), H_B(i, k)) \right], \quad (1)$$

where w denotes a vector of weights. Assuming that not all body segments are equally important when comparing the outfits, the following values of weights were taken:

$$w = (0.4 \quad 0.1 \quad 0.1 \quad 0.1 \quad 0.1 \quad 0.05 \quad 0.05 \quad 0.05 \quad 0.05 \quad 0). \quad (2)$$

The subsequent weights in w correspond to torso (w_1), upper arms (w_2, w_3), thighs (w_4, w_5), lower arms (w_6, w_7), lower legs (w_8, w_9), and face (w_{10}). The resulting color-based similarity measure $D_c(A, B)$ falls within the interval of $\langle 0, 1 \rangle$, 1 denoting highly similar outfits.

2.4 *Structure-Based Similarity*

The second aspect of similarity to be considered refers to the texture of textile materials. To this end, the approach first proposed in [12] was used. As has been shown in [12], a similarity measure based on the structural similarity index (SSIM, see [15]) proved to be effective in finding similar materials, given a rectangular image segment representing the material. SSIM is calculated using a sliding window, according to the following formula:

$$SSIM(x, y) = \frac{(2\bar{x}\bar{y} + C_1) \cdot \left(2\sigma_{xy} + C_2\right)}{\left(\sigma_x^2 + \sigma_y^2 + C_1\right) \cdot \left(\bar{x}^2 + \bar{y}^2 + C_2\right)}, \quad (3)$$

where x and y denote the two compared image fragments; $\bar{x}$, σ_x, and σ_{xy} denote the local mean, local variance, and local covariance, respectively. The constants C_1 and C_2 prevent from the division by zero and were set to $C_1 = (0.01 \cdot 255)^2$ and $C_2 = (0.03 \cdot 255)^2$. The overall SSIM index for images X and Y is calculated as the average of the local similarity indexes:

$$MSSIM(X, Y) = \frac{1}{M} \sum_{j=1}^{M} SSIM(x_j, y_j), \quad (4)$$

where the iteration over j represents the sliding window calculation. The examples of finding similar textile materials in the database of image samples using SSIM are shown in Fig. 2. The results presented in [12] showed that in most cases the best performance is achieved, when calculating the SSIM in HSV space and averaging

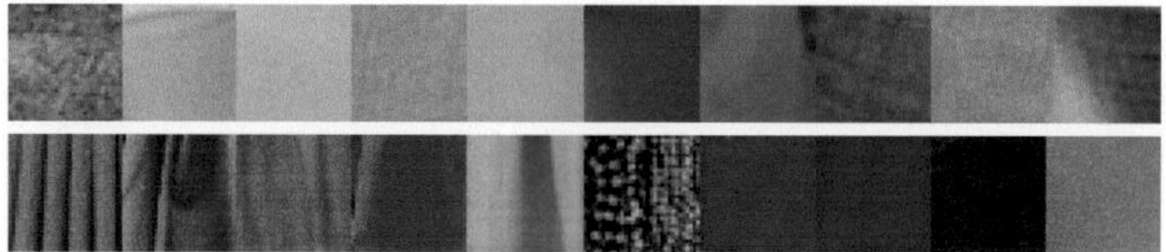

Fig. 2 Example results of retrieving similar textile materials (*first image* in each row is the reference, the rest—retrieval results in order of descending similarity)

the value obtained for each channel. However, this experiment dealt only with SSIM, without using any other means of color analysis, and in the final conclusion adding the color information separately was suggested. Here, the SSIM is used in addition to the dominant color descriptors—hence, we decided to only use the SSIM calculated over the V channel in order to prevent the prevalence of color over texture in the combined similarity measure. Given the images A and B containing a human silhouette and the segmentation results, the overall structural similarity measure is calculated in a similar way, as for the color-based similarity. The MSSIM index is calculated for each body segment and then averaged using the same weights vector w as for the color descriptor:

$$D_s(A, B) = \sum_{k=1}^{10} w_k \cdot MSSIM(a_k, b_k), \tag{5}$$

where a_k and b_k denote the square subimages of particular body segments. The images a_k and b_k are extracted from the body segments as the largest possible square inscribed in the segment.

2.5 *Combined Distance Measure*

The overall similarity measure combines both the above-described components, incorporating color and structural information. As for some of the experiments the distance measure is more suitable than the similarity measure, both indexes have been converted to distance measures and combined in a weighted average:

$$d(A, B) = w_c \cdot (1 - D_c(A, B)) + w_s \cdot (1 - D_s(A, B)), \tag{6}$$

where $d(A, B)$ denotes the distance between outfits presented in images A and B; and w_c and w_s denote the weights assigned to the color and structural components, respectively. Here we assumed $w_c = w_s = 0.5$, but finding the optimal values should be done in subsequent experiments.

3 Experiments

The experiments were performed using over 1800 images obtained from the Flickr website. All images contained one dominant human silhouette; however, the photo's composition, lighting, and model's pose varied significantly. Two main experiments were run in order to assess the suitability of created distance measure, i.e., clusterization and similar outfit extraction. The clusterization was performed according to the approach proposed in [4]. Given that no centroid-based methods like k-means could be used, the k-medoids algorithm [10] was applied. The implementation published by Brookings et al. [1] was used. Two examples of clusters found using the described approach are presented in Fig. 3. As can be seen, the outfits within the clusters are quite cohesive with respect to colors distribution and textures. The second experiment comprised in searching for similar outfits given the reference outfit. The template matching approach was used, applying the distance measure directly. Given the input outfit, the five outfits with smallest distance measure were returned. The examples of such a retrieval are presented in Fig. 4. Both experiments have shown that finding similar outfits (in terms of style) based on automatic image analysis is possible. Clusterization allowed for finding groups of outfits forming a distinguishable style. Direct retrieval based solely on a similarity measure is also possible, and could be further sped up by combining with clusterization (finding closest medoids first, then searching within one or several closest clusters). However, the quality of the results is greatly influenced by the quality of initial body part segmentations. For images, where the model's pose causes some difficulties (e.g., hands are hidden, torso is partially occluded, etc.), the resulting similarity measure is obviously wrong (e.g., mistaking the background for a part of outfit, or mixing up body segments). It should be noted, however, that those problems can be avoided in any practical, commercial

Fig. 3 Two sample clusters obtained by applying k-medoids algorithm

Fig. 4 Sample results of similar outfits retrieval (*first image* in each row is the reference outfit; the rest—outfits with the smallest distance from the reference; the value of the combined distance measure d is given underneath each retrieved outfit)

application. If the database of outfits is prepared for commercial purposes (e.g., by producers, retailers, fashion designers), all photos can be taken in the best and consistent conditions. Assuming that also the photos uploaded by users should be used (e.g., within a social networking website), several easy-to-meet constraints can be enforced (or at least advised)—standing in an upright position, face toward the camera, hands along the torso, against a plain background if possible. After incorporating several easy heuristic rules to the preprocessing stage, the system might warn users, whenever the obtained segmentation results seem suspicious.

4 Summary

The method proposed in this paper allows for estimating the similarity of two outfits presented in the images, taking into consideration both color and texture information. The whole silhouette is analyzed, not just the particular pieces of garments. Body segmentation is performed, and for each segment the dominant colors and structural similarity are analyzed. To this end, the simplified dominant color descriptor (adopted from MPEG-7 standard) and SSIM index are used. The obtained similarity measure was used in two experiments: clustering the database of outfits in search of stylistically distinguishable groups of outfits, and direct retrieval of outfits similar to the reference one. Both experiments proved the applicability of the proposed approach. The main factor restricting the applicability of the method is the quality of the analyzed photos (in terms of background homogeneity and human pose). Given that the images used in experiments are far from being perfect for the task, it can be assumed that the method will perform even better for well-suited photos. Further research is required, focusing mostly on certain perceptual experiments involving the potential users. In order to determine objectively the correspondence between automatically estimated similarity and the similarity perceived by human observers, the users will be asked to assess some outfits in terms of fashion attributes or accordance to their individual style. The results of such experiments will make it possible to evaluate our approach not only qualitatively, but also quantitatively. Furthermore, certain modifications could then be explored, including fine-tuning weights vectors (both for the body segments and the similarity components), choosing different colors or texture descriptors or including some semantic information (user-provided tags describing the outfits). Also, this foundation will allow us to work on certain further elements of the fashion recommender system, such as style classification or mining.

Acknowledgments The project "Construction of innovative recommendation based on users' styles system prototype: FireStyle" (original title: "Zbudowanie prototypu innowacyjnego systemu rekomendacji zgodnych ze stylami użytkowników: FireStyle") is the project co-founded by European Union (project number: UDA-POIG.01.04.00-30-196/12, value: 14.949.474,00 PLN, EU contribution: 7.879.581,50 PLN, realization period: 01.2013-10.2014). European funds—for the development of innovative economy (Fundusze Europejskie—dla rozwoju innowacyjnej gospodarki).

References

1. Brookings, T., Grashow, R., Marder, E.: Statistics of neuronal identification with open and closed loop measures of intrinsic excitability. Frontiers in Neural Circuits **6**(19), 2012, doi:10.3389/fncir.2012.00019
2. Chen, Huizhong, Gallagher, Andrew, Girod, Bernd: Describing Clothing by Semantic Attributes. In: Fitzgibbon, Andrew, Lazebnik, Svetlana, Perona, Pietro, Sato, Yoichi, Schmid, Cordelia (eds.) ECCV 2012, Part III. LNCS, vol. 7574, pp. 609–623. Springer, Heidelberg (2012)

3. Chen, Q., Li, J., Liu, Z., Lu, G., Bi, X., Wang, B.: Measuring clothing image similarity with bundled features. Int. J. Cloth. Sci. Technol. **25**(2), 119–130 (2013)
4. Czapiewski, P., Forczmański, P., Frejlichowski, D., Hofman, R.: Clustering-Based Retrieval of Similar Outfits Based on Clothes Visual Characteristics. In: Choraś, R.S. (ed.) Image Processing & Communications Challenges 6,Advances in Intelligent Systems and Computing, vol. 313, pp. 29–36, Springer International Publishing (2015)
5. Di, W., Wah, C., Bhardwaj, A., Piramuthu, R., Sundaresan, N.: Style finder: Fine-grained clothing style detection and retrieval. In: IEEE Conference on Computer Vision and Pattern Recognition Workshops (CVPRW), 2013, pp. 8–13. IEEE (2013)
6. Eichner, M., Ferrari, V.: Better appearance models for pictorial structures. In: Proceedings of British Machine Vision Conference (2009)
7. Eichner, M., Marín-Jiménez, M. J., Zisserman, A., Ferrari, V.: 2D articulated human pose estimation software, ETH Zurich, Visual Geometry Group, http://groups.inf.ed.ac.uk/calvin/articulated_human_pose_estimation_code/
8. Eichner, M., Marin-Jimenez, M., Zisserman, A., Ferrari, V.: Articulated human pose estimation and search in (Almost) unconstrained still images. Technical Report No. 272. ETH Zurich, D-ITET, BIWI (2010)
9. Forczmański, Paweł, Czapiewski, Piotr, Frejlichowski, Dariusz, Okarma, Krzysztof, Hofman, Radosław: Comparing clothing styles by means of computer vision methods. In: Chmielewski, Leszek J., Kozera, Ryszard, Shin, Bok-Suk, Wojciechowski, Konrad (eds.) ICCVG 2014. LNCS, vol. 8671, pp. 203–211. Springer, Heidelberg (2014)
10. Kaufman, L., Rousseeuw, P.: Clustering by means of medoids. North-Holland (1987)
11. Liu, Z., Wang, J., Chen, Q., Lu, G.: Clothing similarity computation based on tlac. Int. J. Cloth. Sci. Technol. **24**(4), 273–286 (2012)
12. Okarma, Krzysztof, Frejlichowski, Dariusz, Czapiewski, Piotr, Forczmański, Paweł, Hofman, Radosław: Similarity estimation of textile materials based on image quality assessment methods. In: Chmielewski, Leszek J., Kozera, Ryszard, Shin, Bok-Suk, Wojciechowski, Konrad (eds.) ICCVG 2014. LNCS, vol. 8671, pp. 478–485. Springer, Heidelberg (2014)
13. Roerdink, J., Meijster, A.: The watershed transform: definitions, algorithms and parallelization strategies. Fundam. Inf. **41**, 187–228 IOS Press (2001)
14. Viola, P., Jones, M.: Rapid object detection using a boosted cascade of simple features. In: IEEE Conference on Computer Vision and Pattern Recognition, CVPR 2001, pp. I-511–I-518. IEEE (2001)
15. Wang, Z., Bovik, A.C., Sheikh, H.R., Simoncelli, E.P.: Image quality assessment: from error visibility to structural similarity. IEEE Trans. Image Process. **13**(4), 600–612 (2004)
16. Yamada, A., Pickering, M., Jeannin, S., Cieplinski, L., Jens, R.O., Kim, M.: MPEG-7 Visual Part of Experimentation Model Version 9.0—Part 3 Dominant Color. ISO/IEC JTC1/SC29/WG11/N3914 (2001)
17. Zhang, W., Begole, B., Chu, M., Liu, J., and Yee, N.: Real-time clothes comparison based on multi-view vision. In: Second ACM/IEEE International Conference on Distributed Smart Cameras, 2008. ICDSC 2008, pp. 1–10. IEEE. (2008)

Determination of Road Traffic Flow Based on 3D Wavelet Transform of an Image Sequence

Marcin Jacek Kłos

Abstract This paper addresses the problem of processing data from road cameras for providing work parameters for traffic flow control systems. 3D wavelet transformation of image sequences is proposed to represent the traffic. In order to reduce the sensitivity to ambient light changes, of the road scene, a linear function of the coefficients is used to represent the traffic flow. The parameters of the linear function are determined using real traffic data by minimizing the MSE of vehicle detection functions. The developed algorithm was tested using a database of image sequences. Test results prove that it can be applied to determine traffic flow values for control systems. Instead of the usual elaborate image sequence processing, a hardware-based 3D wavelet transformation may be added to the control system.

1 Introduction

The task of detecting objects is struggling with many problems. Some of the difficulties are directly related to image quality, like noise, ambient light changes, or shadows. Another major problem is the implementation of the algorithm for calculation of the image representation in hardware. Constant technological progress brings the increase of image resolution, which causes the growth of bandwidth of the video streams. A video frame with a resolution of 4k has approximately a size of 24 MB, which gives 0.6 GB of data per second (depending on the video frame rate). The stream has to be processed with the highest accuracy, regardless of the field of application it is used for. Methods associated with digital image processing are used in video detector systems to provide data for control systems. Effective traffic control leads directly to the improvement of local traffic on analyzed intersections or global in whole road networks. Using a wavelet transform to represent video stream content, enables a simpler hardware design of vehicle video detectors and proliferates the use of video detectors in traffic control systems. The aim of the analysis is to find an

M.J. Kłos (✉)
Faculty of Transport, Silesian University of Technology, ul.Krasinskiego 8, Katowice, Poland
e-mail: marcin.j.klos@polsl.pll

R. Burduk et al. (eds.), *Proceedings of the 9th International Conference on Computer Recognition Systems CORES 2015*, Advances in Intelligent Systems and Computing 403, DOI 10.1007/978-3-319-26227-7_47

optimal way of using wavelet transformation of video streams for estimation of traffic flow. A linear function of transform coefficients is proposed to detect vehicles. The basic parameters, which describe movement on roads, are: density and traffic flow. This paper presents the results of elaborating a method that can be used to obtain traffic flow values, using wavelet coefficients representing image contents. Traffic flow, as road parameter, describes the size of the traffic stream. The traditional method, for determining traffic flow, is vehicle counting at a stop line, in 15 min intervals with 5 min steps. The obtained counts are converted to one hour traffic flow values. The major problem in obtaining accurate values of traffic flow is the appropriate thresholding of the vehicles detection function. The lack of movement in the detection field especially leads to large detection errors when bad threshold is used. This paper provides a discussion of the threshold selection. One of the first papers, which describes the idea of deriving traffic parameters directly from the transform of video streams is [1]. The author proposes to use wavelet transform coefficients of the image streams content to represent traffic flow and density, however, the detection function is only vaguely described. Another approach to traffic flow representation is proposed by Jiang et. al it is based on multiresolution analysis (MRA) with a wavelet function. A different approach to the problem is given in [2]. principal components analysis (PCA) is used to reduce the number of wavelet coefficients to represent features of objects. This paper is organized as follows: Sect. 2 provides a brief background information on wavelet transforms. This is followed by Sect. 3, which describes the methodology of detecting moving objects and the proposed algorithm. The result of using the proposed algorithm is described in Sect. 4. Section 5 of the paper contains conclusions and propositions for future work.

2 Discrete Wavelet Transformation

Wavelet transformations in image processing can be classified into two groups: discrete wavelet and continues wavelet. Video stream is a three-dimensional object, therefore spatial DWT cannot be used directly to detect objects in the stream. It is used separately on each frame of the video with some success to detect objects [3]. 3D DWT has the ability to represent the changes of the image content in time. Therefore, 3D should be useful for extracting motion parameters of moving objects. The video stream, used by videodetectors, is readily available in digital form so it can be directly diverted for DWT processing [4]. Instead of the usual complex image sequence processing, a hardware-based 3D DWT may be utilized for the image analysis. A number of wavelet functions is used in image processing. For instance the Coiflet wavelets are used to reduce noise in image frames. The Daubechies wavelet is used to filter objects with different textures. The Mexican hat is used, for example, to detect point sources [5]. The simplest computationaly, is the Haar wavelet, which is used to locate objects of different sizes. Discrete wavelet transformation based on Haar wavelets is chosen for processing the video stream data. The transform is useful to express local changes (e.g., changes in the occupancy of detection field).

However, the most important property of this transformation is its small processing requirements, which allows for very fast calculations of the transform coefficients. The FFT requires 0 (n log 2(n)), number of calculations, while DWT requires 0 (n) operations [6]. Three-dimensional Haar transformation is used, the third coordinate is the number of the frame in the video sequences. Every four frames a set of eight consecutive frames is taken for the calculation of 3D DWT. The luminance of pixels is processed. The one-dimensional DWT is given by (1):

$$\begin{aligned} d_{1,i} &= s_{0,2+1} - s_{0,2i} \\ s_{1,i} &= s_{0,2i} + [d_{0,2i}/2] \end{aligned} \tag{1}$$

where:

- $d_{1,i}$—describes the signal features at level 1 of decomposition,
- $s_{1,i}$—describes the mean values of pixel brightness at level 1 of decomposition.

Equation (2) describes the third level of wavelet decomposition and are considered sufficient to represent the image content. The decision of choosing the third level of decomposition was derived in the course of analysis of the objects size and the dynamic behavior of those. The average size of vehicles ranges from several tens to several hundred pixels and the vehicle displacement is up to several pixels between frames. Transform coefficients are given by the Eq. (3) which uses partial sums of pixel groups defined in (2) [7].

$$\begin{aligned}
S_0 &= \sum_{i=0}^{3}\sum_{j=0}^{3}\sum_{k=0}^{3} x_{ij}, & S_1 &= \sum_{i=0}^{3}\sum_{j=0}^{3}\sum_{k=0}^{3} x_{ijk+4}, \\
S_2 &= \sum_{i=0}^{3}\sum_{j=0}^{3}\sum_{k=0}^{3} x_{ij+4k}, & S_3 &= \sum_{i=0}^{3}\sum_{j=0}^{3}\sum_{k=0}^{3} x_{ij+4k+4}, \\
S_4 &= \sum_{i=0}^{3}\sum_{j=0}^{3}\sum_{k=0}^{3} x_{i+4kjk}, & S_5 &= \sum_{i=0}^{3}\sum_{j=0}^{3}\sum_{k=0}^{3} x_{i+4kjk+4}, \\
S_6 &= \sum_{i=0}^{3}\sum_{j=0}^{3}\sum_{k=0}^{3} x_{i+4kj+4k}, & S_7 &= \sum_{i=0}^{3}\sum_{j=0}^{3}\sum_{k=0}^{3} x_{i+4kj+4k+4}.
\end{aligned} \tag{2}$$

x_{ij}—values of pixels brightness located at i, j; k—frame number of the video stream; S_0, S_1, S_2, S_3, S_4, S_5, S_6, S_7—partial sums of pixel groups for the calculations of the coefficients. Coefficients map features such as image brightness, vertical, and horizontal spatial movement. 3D wavelet transformation assigns a set of eight coefficients values to each 8 frame image block.

$$
\begin{aligned}
(sss)_3 &= C0 = \tfrac{1}{512}\,(S_0 + S_1 + S_2 + S_3 + S_4 + S_5 + S_6 + S_7),\\
(ssd)_3 &= C1 = \tfrac{1}{256}\,(S_0 - S_1 + S_2 - S_3 + S_4 - S_5 + S_6 - S_7),\\
(sds)_3 &= C2 = \tfrac{1}{256}\,(S_0 - S_1 - S_2 + S_3 + S_4 - S_5 - S_6 + S_7),\\
(sdd)_3 &= C3 = \tfrac{1}{128}\,(S_0 - S_1 - S_2 + S_3 + S_4 - S_5 - S_6 + S_7),\\
(dsd)_3 &= C5 = \tfrac{1}{128}\,(S_0 - S_1 + S_2 - S_3 - S_4 + S_5 - S_6 + S_7),\\
(dds)_3 &= C6 = \tfrac{1}{128}\,(S_0 + S_1 - S_2 - S_3 - S_4 - S_5 + S_6 + S_7),\\
(ddd)_3 &= C7 = \tfrac{1}{64}\,(S_0 - S_1 - S_2 + S_3 - S_4 + S_5 + S_6 - S_7).
\end{aligned}
\tag{3}
$$

S_0, S_1, S_2, S_3, S_4, S_5, S_6, S_7,—partial sums of pixel groups for the calculations of the coefficients; $C1$, $C2$, $C3$, $C4$, $C5$, $C6$, $C7$—wavelet coefficients; $(sss)_3$, $(ssd)_3$, $(sds)_3$, $(sdd)_3$, $(dss)_3$, $(dsd)_3$, $(dds)_3$, $(ddd)_3$,—combined transformation operations of the third level of decompositions. The coefficient $C0$ describes the average level of luminance. Values $C1$, $C2$, $C3$ describe average changes in intensity in time. In contrast, coefficients $C4$, $C5$, $C6$, $C7$ describe movement values (Fig. 1).

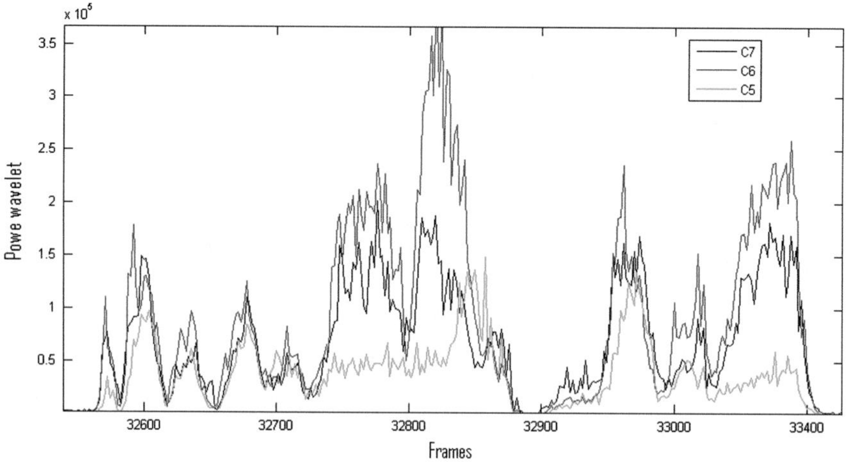

Fig. 1 Graph of 3D wavelet transformation parameters—$C5$, $C6$, $C7$

3 Implementation

Traffic flow is obtained by counting objects traveling through a defined detection field. Calculation of the occupancy of the detection field in video sequence images is shown by Eq. 4. Thresholding is used to evaluate the state of the field. Figure 2 shows a frame with a defined detection field. The detection field is placed a few meters from the stop line, to avoid the situations, when vehicles do not stop as expected. Detection function is the sum of the scaled sum of feature components ($C5$, $C6$, $C7$) and average brightness ($C1$, $C2$), which is normalized by image brightness C0. Scales w_1 and w_2 are obtained by solving a set of equations minimizing the errors of detection.

$$Y = \sum_{detectionfield} w_1 \frac{(C5 + C6 + C7)}{C0} + w_2 \frac{(C1 + C2)}{C0} > P \quad (4)$$

Y—detection function; w_1,w_2—linear function scales; $C0$, $C1$, $C2$, $C5$, $C6$, $C7$—wavelet coefficients; P—threshold value. Road traffic flow is a basic parameter, which reflects the traffic condition on the road. It represents the number of vehicles driving through the detection field. Usually, vehicle counting is conducted in 15-min intervals every 5-min.

Fig. 2 The defined detection field on the analyzed video frame

4 Results

The results of the conducted research are an algorithm for calculating traffic flow from video stream. Equation 4 with the calculated weights and appropriate threshold is used for detecting vehicles. These are counted over time and transformed to traffic flow. The graph in Fig. 3 presents the comparison of the reference traffic flow (manually counted) and the traffic flow obtained using the developed algorithm. The influence of the threshold value on the traffic flow elaboration was examined. The threshold value is referenced to the maximum values of pixel brightness. Figure 4 shows the mean square error of determining traffic flow for the range of threshold values from 0.1 to 0.5. The average error is fluctuating around 15 % for a wide range of threshold values. Based on this observation one can conclude that the detection is quite insensitivity to the threshold value.

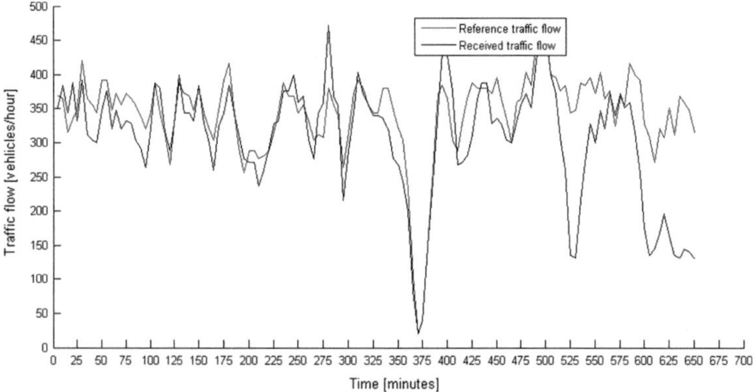

Fig. 3 Graph presenting the comparison of model traffic flow and traffic flow obtained using combination of wavelet transformation filters

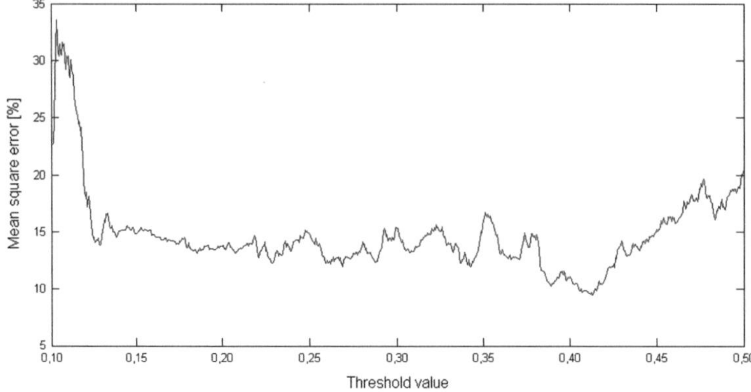

Fig. 4 The mean square error of determining traffic flow as a function of the threshold value

5 Conclusion

The elaborated method allows determining traffic flow with errors in the range of 10–15 %. Test results prove that it can be applied to determine traffic flow values for control systems. Instead of the usual elaborate image sequence processing, a hardware based on 3D wavelet transformation may be added to the control system.

References

1. Pamula, W.: Determination of road traffic parameters based on 3d wavelet representation of an image sequence. In: Bolc, L., et al. (eds.) LNCS, vol. 7594, pp. 541–548. Springer, Berlin (2012)
2. Zhu, W., Barth, M.: Vehicle trajectory-based road type and congestion recognition using wavelet analysis. In: Proceedings of the IEEE ITSC 2006 Intelligent Transportation Systems Conference Toronto, Canada, 17–20 September 2006
3. Satpute, V., Naveen, Ch., Kulat, K., Keskar, A.: Fast and memory efficient 3d-dwt based video encoding techniques. In: Proceedings of the International MultiConference of Engineers and Computer Scientists 2014, vol I, Hong Kong (2014)
4. Wu, B., Lin, Ch.: A high-performance and memory-efficient pipeline architecture for the 5/3 and 9/7 discrete wavelet transform of JPEG2000 codec. IEEE Trans. Circuits Syst. Video Technol. **15**(12), 1615–1628 (2005)
5. Gonzalez-Nuevo, J., Argueso, F., Lopez-Caniego, M., Toffolatti, L., Sanz, J., Vielva, P., Herranz, D.: The Mexican hat wavelet family: application to point-source detection in cosmic microwave background maps. Mon. Not. Roy. Astron. Soc. **369**, 1603–1610 (2006)
6. Borkowski, A., Sośnica, K.: Application of discrete wavelet transform to filtering airborne laser scanning data, Archiwum Fotogrametrii, Kartografii i Teledetekcji, vol. 20, pp. 35–45 (2009)
7. Pamula, W.: Metoda implementacji trojwymiarowej dyskretnej transformaty falkowej strumienia wideo w układach FPGA, PAK 2012 nr 07, pp. 632–634 (2012)

Part V
Medical Applications

Schmid Filter and Inpainting in Computer-Aided Erosions and Osteophytes Detection Based on Hand Radiographs

Bartosz Zieliński and Marek Skomorowski

Abstract In previous papers we presented a computer system to detect erosions and osteophytes from hand radiographs (the most common symptoms of rheumatic diseases) based on the shape analysis of the joint surfaces borders. Such borders are obtained automatically using algorithms which were also proposed in our previous articles. In this paper, we consider a new approach which analyzes patches located at the joint surfaces borders in order to determine which of them correspond to the lesions. Vectors of features which are used to classify patches are calculated by applying Schmid filter with various frequencies and scales. Additional features are obtained using inpainting. Vectors are analyzed based on Gaussian mixture model calculated with expectation maximization algorithm. The accuracy is measured with area under curve of the receiver-operating characteristic. The conducted experiments proved that, the shape approach described in our previous work can be improved by applying Schmid filter and the inpainting approach in the parsing stage, especially, in case of the lower MCP and upper PIP surfaces for which classification still remains inaccurate.

Keywords Medical imaging · Radiographs · Computer aided rheumatoid diagnosis · Erosions · Osteophytes · Inpainting · Schmid filters

B. Zieliński (✉) · M. Skomorowski
The Institute of Computer Science and Computer Mathematics,
Faculty of Mathematics and Computer Science, Jagiellonian University,
ul. Łojasiewicza 6, 30-348 Kraków, Poland
e-mail: bartosz.zielinski@uj.edu.pl
URL: http://www.ii.uj.edu.pl

M. Skomorowski
e-mail: marek.skomorowski@uj.edu.pl
URL: http://www.ii.uj.edu.pl

R. Burduk et al. (eds.), *Proceedings of the 9th International Conference on Computer Recognition Systems CORES 2015*, Advances in Intelligent Systems and Computing 403, DOI 10.1007/978-3-319-26227-7_48

1 Introduction

Within the scope of rheumatology and diagnostic radiology, it is essential to distinguish between inflammatory and noninflammatory diseases. To give a diagnosis at an early stage of a disease, radiographs are taken of the patient's hands and the symmetric joints are analyzed. The analysis is conducted in order to detect the lesions, which are taken into consideration during diagnosis, together with other tests (e.g., blood tests). However, due to the number of hand joints, such a standard analysis is exceedingly complicated and time consuming. To minimize the time spent on this analysis and to make a radiograph examination more frequent and precise, this process should be automated [17]. In previous papers, we presented a computer system to detect erosions and osteophytes from hand radiographs (the most common symptoms of rheumatic diseases) based on the shape analysis of the joint surfaces borders [3, 4, 17]. Such borders are obtained automatically using algorithms which were also proposed in our previous articles [5, 6, 16]. In this paper, we consider a different approach which analyzes patches located at the joint surfaces borders in order to determine which of them correspond to the lesions. Vectors of features which are used to classify patches are calculated by applying Schmid filter ("Gabor-like" filter resistant to rotation) with various frequencies and scales [12]. Additional features are obtained using inpainting [2]. Vectors are analyzed based on Gaussian mixture model (GMM) calculated with expectation maximization algorithm. There are a number of papers concerning the topic of computer aided rheumatoid diagnosis based on hand radiographs. Some of them focus on segmentation of the hand radiographs—see [9, 15]. Others relate to identifying the joint surface borders and detecting joint space narrowing—see [6, 14]. However, according to Peloschek et al. [11], only a few of them relate to detecting lesions in joint surfaces—see [10, 17]. In order to see more detailed state of art see [17]. This paper is organized in the following manner. First, the medical basis of the topic is outlined. Then, detecting erosions and osteophytes using a Schmid filter and inpainting approach is presented. Finally, the obtained results and the discussion are presented.

2 The Medical Basis of the Topic

Despite the fact that magnetic resonance imaging (MRI) shows the greatest sensitivity for detecting and monitoring bone erosions [8], conventional, or digital radiography of the hand is the most commonly used imaging method in diagnosis, as well as monitoring disease progression and the treatment response in case of the patients with rheumatic musculoskeletal diseases [1]. This is due to the fact that radiography is widely available, inexpensive and easy to perform, as well as valuable in differential diagnosis [8]. In general, there are two groups of rheumatic musculoskeletal diseases. The first group is described as inflammatory disorders, with rheumatoid arthritis being the most prevalent (from 0.5 to 1 % suffers from inflammatory diseases). The second group is known as noninflammatory disorders and includes the degenerative diseases of the joints, e.g., osteoarthritis, the most prevalent one (from 11 to 14 % of the population suffers from rheumatic diseases). There are several radiographic lesions [1] corresponding to these two groups. The most important ones among them are erosions and osteophytes (see Fig. 1a, b). The occurrence of those lesions in specific places may indicate, together with the other tests (e.g., blood tests), on the particular type of disease [17]. There are three joints of interest in the cases of rheumatoid arthritis and osteoarthritis (see Fig. 2): metacarpophalangeal joints (MCP, the joints between the metacarpal bones and the proximal phalanxes), proximal interphalangeal joints (PIP, the joints between the proximal and middle phalanxes), and distal interphalangeal joints (DIP, the joints between the middle and distal phalanxes). MCP joints are condyloid joints, whereas PIP and DIP joints are hinge joints. The differences in their anatomy result in differences on radiographs, therefore, the analyses in both cases has to differ [17].

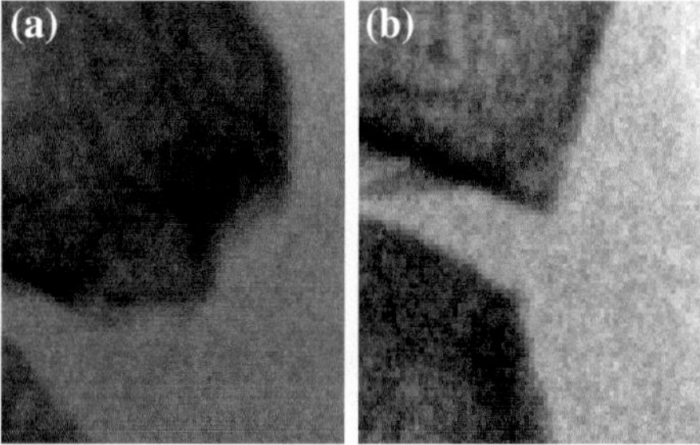

Fig. 1 An upper surface of metacarpophalangeal joint with erosion (**a**) and osteophyte (**b**)

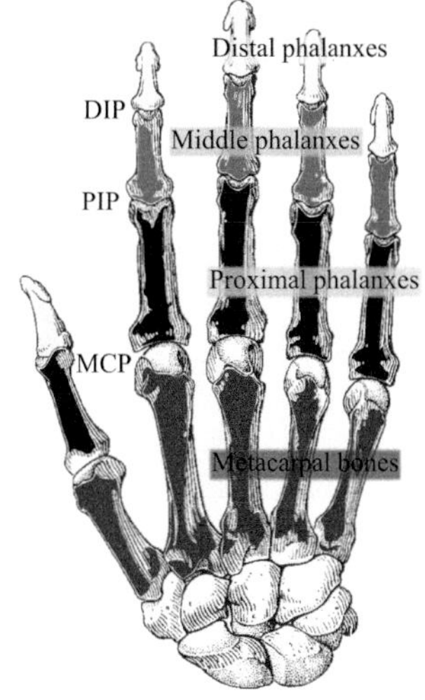

Fig. 2 Hand anatomy

3 Schmid Filter and Inpainting Approach

In our new approach, patches located at the joint surfaces borders are analyzed in order to determine which of them correspond to the lesions. As an input, it obtains points located at the border of the joint surface (see Fig. 3b). Such points can be obtained automatically using algorithms proposed in our previous articles [5, 6, 16]. For each of the points, elliptic patch is generated in such a way that the longer axis of the ellipse is perpendicular to the tangent in this point (see Fig. 3c). We use ellipses instead of circles, because only such patches can be convincingly inpainted using method described in [2].

3.1 Calculating Features Using Schmid Filter

Vectors of features which are used to classify the patches are calculated by applying Schmid filter with various frequencies and scales [12]. Schmid filter is a "Gabor-like" filter resistant to rotation of the following form:

$$F(r, \sigma, \tau) = cos\left(\frac{\pi\tau r}{\sigma}\right) exp\left(-\frac{r^2}{2\sigma^2}\right), \tag{1}$$

where τ, σ, and r correspond to frequency, scale, and radius, respectively. For our experiment we used 10 filters with pairs (σ, τ) equal $(1, 1)$, $(2, 1)$, $(2, 2)$, $(3, 1)$, $(3, 2)$, $(3, 3)$, $(4, 1)$, $(4, 2)$, $(4, 3)$, $(4, 4)$—their representations are presented in Fig. 4. We decided to use such filters, as the ellipses corresponding to patches are quite small (12×4 pixels). Each of those ten filters is convoluted with the analyzed patch and in result ten filtered patches are obtained for each point of the border. Then, mean value and standard deviation are obtained for each of the filtered patches—those 20 values are added to the patch feature vector.

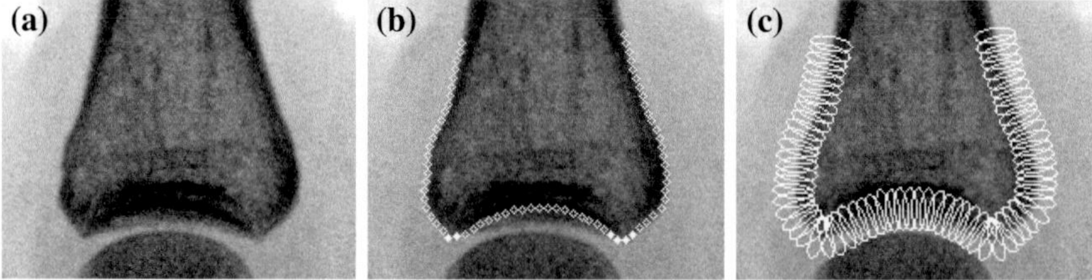

Fig. 3 Joint surfaces without and with border points (**a** and **b**, respectively) and elliptic patches which are generated in such a way that the longer axis of each ellipse is perpendicular to the tangent in the corresponding border point (**c**)

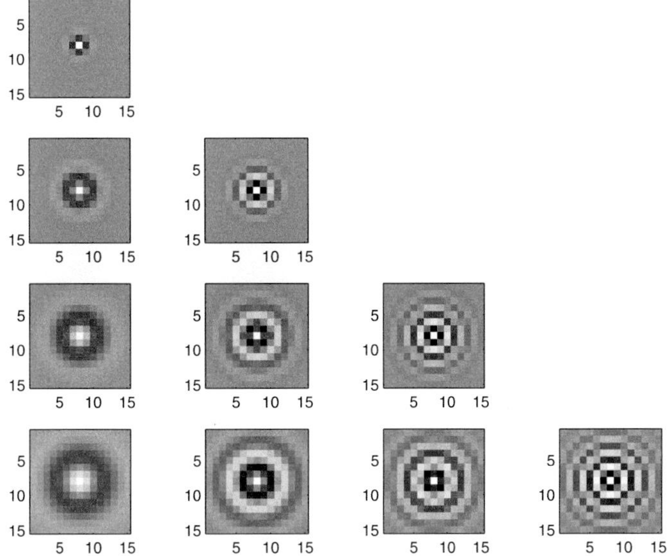

Fig. 4 Representations of the Schmid filters with (σ, τ) equal (1, 1), (2, 1), (2, 2), (3, 1), (3, 2), (3, 3), (4, 1), (4, 2), (4, 3), (4, 4). *Gray* color is represented by 0, while *black* and *white* represent negative and positive values, respectively

3.2 Calculating Features Using Inpainting

Additional features are obtained using inpainting [2]. Inpainting is the technique of modifying an image in an undetectable form. The goals and applications of inpainting are numerous, from the restoration of damaged paintings and photographs to the removal/replacement of selected objects. In [2], after the user selects the regions to be restored, the algorithm automatically fills-in these regions with information surrounding them. The fill-in is done in such a way that isophote lines arriving at the regions boundaries are completed inside. We decided to use such approach, due to the fact that in case of the joint surfaces without lesions, the isophotes are smoother than in case of the joint surfaces with lesions. Therefore, in result of applying inpainting into the patch with lesions, it should change it significantly, while patches without lesions should remain unchanged. This can be observed in Fig. 5a, b. For each border point, the patch representing difference between not inpainted and inpainted patch (see last column in Fig. 5) is calculated and then its mean value and standard deviation are obtained—they are added to the patch feature vector.

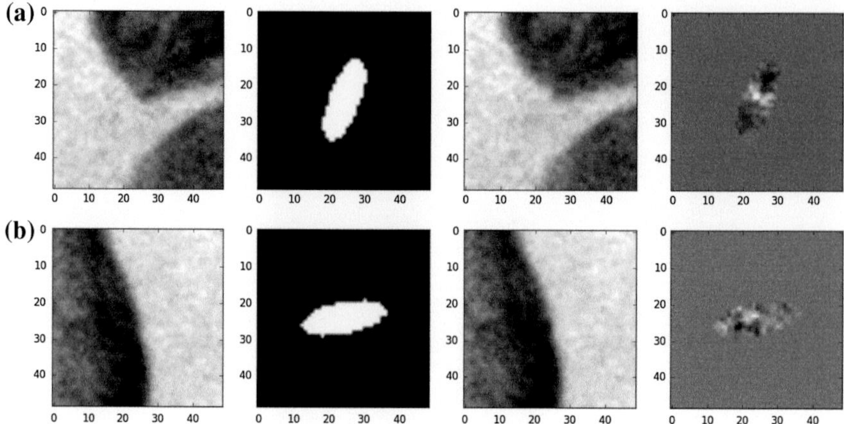

Fig. 5 The result of applying inpainting into the patch with and without lesions (**a** and **b**, respectively). First, column contains image of a patch together with its neighborhood, second column corresponds to the patch mask, third column responds to inpainted patch, while fourth column represents difference between first and third column (difference between not inpainted and inpainted patch). In the last column, *gray color* represents 0. All images are adjusted

3.3 *Classification with Gaussian Mixture Model*

Due to the fact that the number of the patches without lesions is over 40 times bigger than the number of the patches with lesions, we decided to classify vectors of features using GMM calculated with expectation maximization algorithm. Such model was build only based on the half of the patches without lesions, and then it was applied to the rest of the patches in order to detect anomalies (patches with lesions) in the data set. To obtain accurate number of components in GMM, we used Bayesian information criterion (BIC) [13].

4 Experiments and Discussion

In total, 1440 joint surfaces were analyzed in the left and the right hand radiographs of 60 patients (60 patients $\times$ 2 hands $\times$ 2 fingers of interest $\times$ 3 joints $\times$ 2 joint surfaces), acquired through the offices of the University Hospital in Kraków, Poland. Among all the patients, 20 were healthy, 20 suffered from degenerative diseases and 20 suffered from inflammatory diseases. They were selected by a rheumatologist who accepted the radiographs of the whole hand taken in the anterior posterior position and rejected the radiographs corresponding to a terminal state of disease or containing metal objects. Each radiograph was received along with the borders of the upper and the lower joint surfaces and with the corresponding locations of erosions and osteophytes.

Such data was collected in two steps. First, the joint surfaces borders and lesions were outlined by a less experienced radiologist. Then they were presented to a more experienced radiologist who could modify them. In total 92076 patches were created for 1440 joint surfaces (around 60 patches for each joint surface). Among them, 2206 patches contained lesions. The numbers of patches with and without lesions for each joint surface type are presented in Table 1. For each joint surface type, we computed GMM using vectors corresponding to half randomly chosen patches without lesions. Minimal and maximal number of components equal 1 and 21, respectively. In order to determine the significance of Schmid filter and inpainting, we decided to build three types of GMM: only for Schmid filter features, only for inpainting features and for both types of the features. Obtained models were tested using remain half of the patches without lesions and all patches with lesions. We generated receiver operating characteristic (ROC), where true positives (TP) are correctly classified patches with lesions and true negatives (TN) are correctly classified patches without lesions. Finally, the area under curve (AUC) were computed (see Table 2). In most of the cases, BIC was the best for the component number equals around 8 (see example BIC plot in Fig. 6a). In most of the cases, the results are the best for the combination of the Schmid filter and inpainting. However, in case of the lower PIP and lower DIP surfaces, the vectors containing only Schmid features and only inpainting features, respectively, turned out to be more accurate. It is probably due to the fact, that the anatomy of the lower PIP and DIP surfaces differs comparing to the other surfaces—they have sharper borders. It can be also noticed, that in most of the cases, the standardization did not improve the classification. This is not surprising, because when each column is standardized separately, we lose a bunch information about their balance. The results are the best for the upper MCP surface, similarly like in [17] (see Fig. 6b). On the other hand, the results for upper and lower DIP surfaces are surprisingly poor comparing to [17]—this will be analyzed in the feature. Nevertheless, the results for lower MCP and upper PIP surfaces are very promising (let us recall that the results for those two joint surfaces are very poor in [17]) and therefore this new approach could improve lesions detection if used together with the

Table 1 The numbers of patches with and without lesions for each joint surface type

Joint surface type	Patches with lesions	Patches without lesions
Upper MCP	462	16166
Lower MCP	245	20582
Upper PIP	368	14248
Lower PIP	175	14717
Upper DIP	503	11727
Lower DIP	331	12430
Summary	2206	89870

Table 2 The AUC for Schmid filter features, for inpainting features and for both types of the features

Joint surface type	Schmid		Inpainting		Both	
	S		S		S	
Upper MCP	0.68	0.68	0.70	0.72	**0.73**	0.72
Lower MCP	0.61	0.60	0.60	0.61	**0.66**	0.63
Upper PIP	0.62	0.62	0.61	0.61	**0.63**	0.59
Lower PIP	0.62	**0.66**	0.62	0.63	0.59	0.60
Upper DIP	0.62	0.61	0.62	0.64	**0.65**	0.60
Lower DIP	0.59	0.59	0.63	**0.65**	0.59	0.63

Letter S means that data were standardized before generating model and classifying. The best results are in bold

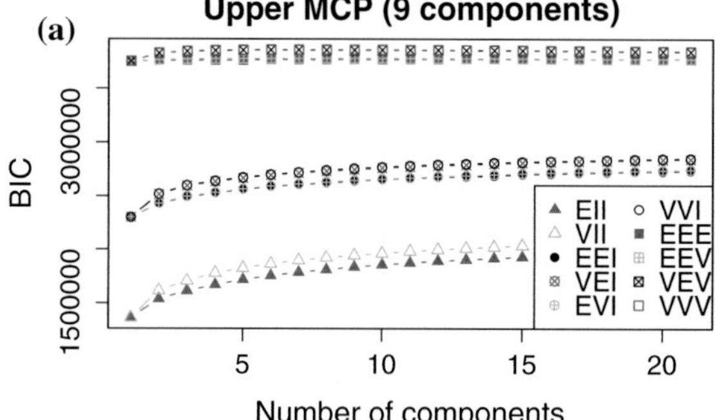

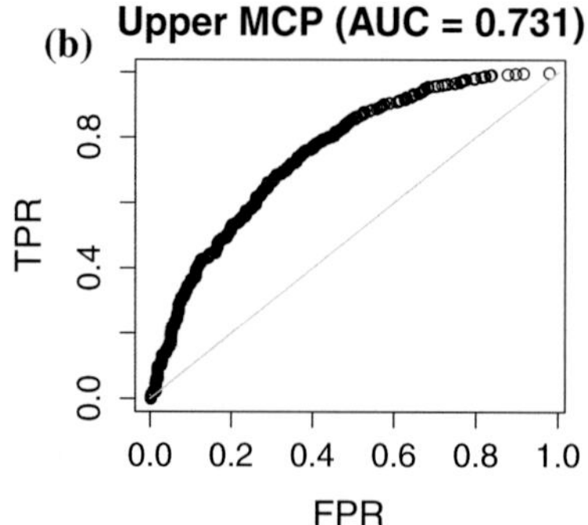

Fig. 6 BIC and AUC plots for upper MCP surface using both types of features, without standardization (BIC were computed with mclust library [7])

shape approach. In our feature work, we would like to improve the shape approach described in [17] by applying Schmid filter and inpainting approach in the parsing stage. Moreover, we would like to use more advanced methods of classification, such as random forest, and in result implement more accurate computer system to detect erosions and osteophytes from hand radiographs.

References

1. Altman, R.D., Gold, G.: Atlas of individual radiographic features in osteoarthritis, revised. Osteoarthr. Cartil. **15**, A1–A56 (2007)
2. Bertalmio, M., Bertozzi, A.L., Sapiro, G.: Navier-stokes, fluid dynamics, and image and video inpainting. In: Proceedings of the 2001 IEEE Computer Society Conference on Computer Vision and Pattern Recognition, 2001. CVPR, vol. 1, pp. I–355. IEEE (2001)
3. Bielecka, M., Bielecki, A., Korkosz, M., Skomorowski, M., Wojciechowski, W., Zieliński, B.: Application of shape description methodology to hand radiographs interpretation. In: Computer Vision and Graphics, pp. 11–18. Springer, Berlin (2010)
4. Bielecka, M., Bielecki, A., Korkosz, M., Skomorowski, M., Wojciechowski, W., Zieliński, B.: Modified jakubowski shape transducer for detecting osteophytes and erosions in finger joints. In: Adaptive and Natural Computing Algorithms, pp. 147–155. Springer, Berlin (2011)
5. Bielecki, A., Korkosz, M., Wojciechowski, W., Zieliński, B.: Identifying the borders of the upper and lower metacarpophalangeal joint surfaces on hand radiographs. In: Artificial Intelligence and Soft Computing. pp. 589–596. Springer, Berlin (2010)
6. Bielecki, A., Korkosz, M., Zieliński, B.: Hand radiographs preprocessing, image representation in the finger regions and joint space width measurements for image interpretation. Pattern Recognit. **41**(12), 3786–3798 (2008)
7. Fraley, C., Raftery, A.E.: Mclust version 3: an r package for normal mixture modeling and model-based clustering. Technical report, DTIC Document (2006)
8. Guermazi, A., Taouli, B., Lynch, J.A., Peterfy, C.G.: Imaging of bone erosion in rheumatoid arthritis. Semin. Musculoskelet. Radiol. **8**, 269–285 (2004)
9. Kauffman, J.A., Slump, C.H., Moens, H.J.B.: Segmentation of hand radiographs by using multi-level connected active appearance models. In: Medical Imaging, pp. 1571–1581. International Society for Optics and Photonics (2005)
10. Langs, G., Peloschek, P., Bischof, H., Kainberger, F.: Automatic quantification of joint space narrowing and erosions in rheumatoid arthritis. IEEE Trans. Med. Imaging **28**(1), 151–164 (2009)
11. Peloschek, P., Boesen, M., Donner, R., Kubassova, O., Birngruber, E., Patsch, J., Mayerhöfer, M., Langs, G.: Assessement of rheumatic diseases with computational radiology: current status and future potential. Eur. J. Radiol. **71**(2), 211–216 (2009)
12. Schmid, C.: Constructing models for content-based image retrieval. In: Proceedings of the 2001 IEEE Computer Society Conference on Computer Vision and Pattern Recognition, 2001. CVPR, vol. 2, pp. II–39. IEEE (2001)
13. Schwarz, G., et al.: Estimating the dimension of a model. Ann. Stat. **6**(2), 461–464 (1978)
14. Sharp, J.T., Gardner, J.C., Bennett, E.M.: Computer-based methods for measuring joint space and estimating erosion volume in the finger and wrist joints of patients with rheumatoid arthritis. Arthr. Rheum. **43**(6), 1378–1386 (2000)
15. Tadeusiewicz, R., Ogiela, M.R.: Picture languages in automatic radiological palm interpretation. Int. J. Appl. Math. Comput. Sci. **15**(2), 305 (2005)
16. Zielinski, B.: Hand radiograph analysis and joint space location improvement for image interpretation. Schedae Inform. **17**(18), 45–62 (2009)
17. Zieliński, B., Skomorowski, M., Wojciechowski, W., Korkosz, M., Sprężak, K.: Computer aided erosions and osteophytes detection based on hand radiographs. Pattern Recognit. **48**(7), 2304–2317 (2015)

Asymmetric Generalized Gaussian Mixtures for Radiographic Image Segmentation

Nafaa Nacereddine and Djemel Ziou

Abstract In this paper, a parametric histogram-based image segmentation method is used where the gray level histogram is considered as a finite mixture of asymmetric generalized Gaussian distribution (AGGD). The choice of AGGD is motivated by its flexibility to adapt the shape of the data including the asymmetry. Here, the method of moment estimation combined to the expectation–maximization algorithm (MME/EM) is originally used to estimate the mixture parameters. The proposed image segmentation approach is achieved in radiographic imaging where the image often presents an histogram with a complex shape. The experimental results provided in terms of histogram fitting error and region uniformity measure are comparable to those of the maximum likelihood method (MLE/EM) with the advantage that MME/EM method reveals to be more robust to the EM initialization than MLE/EM.

1 Introduction

The segmentation is a key step in any image analysis system where, its quality has indisputably a direct effect on the results of the next stage, i.e., the feature extraction on which are dependent the image interpretation and retrieval, the object recognition and classification, etc. For delicate applications such as radiographic imaging in both fields of industry and medicine, an accurate segmentation is more than ever required since a bad interpretation or a false diagnosis lead sometimes to irreparable harm to the human patient and the industrial plant in question. On the other hand, an accu-

N. Nacereddine (✉)
Research Center in Industrial Technologies CRTI, P.O.Box 64, 16014 Algiers, Algeria
e-mail: n.nacereddine@csc.dz; nafaa.nacereddine@enp.edu.dz

N. Nacereddine
Lab. des Math. et leurs Interactions, Centre Universitaire de Mila, Mila, Algeria

D. Ziou
Dpt. de Math. et Informatique, Université de Sherbrooke, Québec, Canada
e-mail: djemel.ziou@usherbrooke.ca

R. Burduk et al. (eds.), *Proceedings of the 9th International Conference on Computer Recognition Systems CORES 2015*, Advances in Intelligent Systems and Computing 403, DOI 10.1007/978-3-319-26227-7_49

rate modeling of unknown probability density functions *pdf* s of data, encountered in practical applications, can play an important role in the direction of simulation and design of modern signal processing systems [1]. The latter include the image gray level histogram which is, in general, multimodal and which can be approximated by a finite mixture model (FMM) [2] for the purpose of image segmentation. The most popular method to estimate FMM parameters is the expectation–maximization (EM) algorithm [3]. In most of applications, including image segmentation, the Gaussian *pdf* is used for FMM but the analyzed signals often present complex and non-Gaussian shapes. This is why, in this work, the asymmetric generalized Gaussian distribution (AGGD) has been chosen because it is shown to not only model a wide range of statistical distributions (e.g., impulsive, Laplacian, Gaussian) but also include the asymmetry [4]. This last property permits to an asymmetric generalized Gaussian mixture model (AGGMM) [5] to portray successfully a large class of signals (e.g., image gray level histogram, speech data). However, in case of maximum likelihood estimation (MLE) using EM algorithm, all the AGGD parameters in the mixture are expressed by high nonlinear equations [5] which makes the numerical solution cumbersome and sensitive to EM initial values, as we can see later in experiments. As an alternative, in this work, the moment matching estimation (MME) method [6, 7] in combination with the EM algorithm is used and newly tested on the AGG mixture model. For the aim of segmentation of images issued from the real-world applications mentioned above, MME/EM is applied and compared with MLE/EM in terms of histogram fitting, segmentation quality, and robustness to EM initialization. The remainder is organized as follows. In Sect. 2, the analytical expression of AGGD and its mixture model for image segmentation are given. Section 3 deals with the computing of AGGD moments and presents the algorithm to combine them with EM algorithm. The experiments are carried out and commented in Sect. 4. Finally, the concluding remarks are drawn in Sect. 5.

2 AGGD and Its Mixture Model

The *pdf* of a one-dimensional AGGD is defined as

$$f(x|\boldsymbol{\theta}) = \begin{cases} \dfrac{\beta}{(\alpha_1+\alpha_2)\Gamma(1/\beta)} e^{-[(-x+\mu)/\alpha_1]^\beta} & \text{if } x < \mu \\ \dfrac{\beta}{(\alpha_1+\alpha_2)\Gamma(1/\beta)} e^{-[(x-\mu)/\alpha_2]^\beta} & \text{if } x \geq \mu \end{cases} \tag{1}$$

for x ($x \in \mathbb{R}$), and where $\boldsymbol{\theta} = (\mu, \alpha_1, \alpha_2, \beta)^T$ ($\mu \in \mathbb{R}$, $\{\alpha_1, \alpha_2, \beta\} \in \mathbb{R}_+^*$) is the distribution parameter vector with the components are the left and the right scale parameters and the shape parameter, respectively. The gamma function is defined by $\Gamma(\xi) = \int_0^\infty e^{-t} t^{\xi-1} \mathrm{d}t$. The parameter β dictates the exponential rate of decay: the larger the β the flatter the *pdf*; the smaller the β the more peaked the *pdf*, as shown in

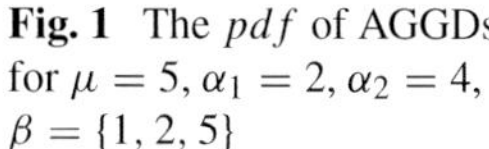

Fig. 1 The pdf of AGGDs for $\mu = 5, \alpha_1 = 2, \alpha_2 = 4$, $\beta = \{1, 2, 5\}$

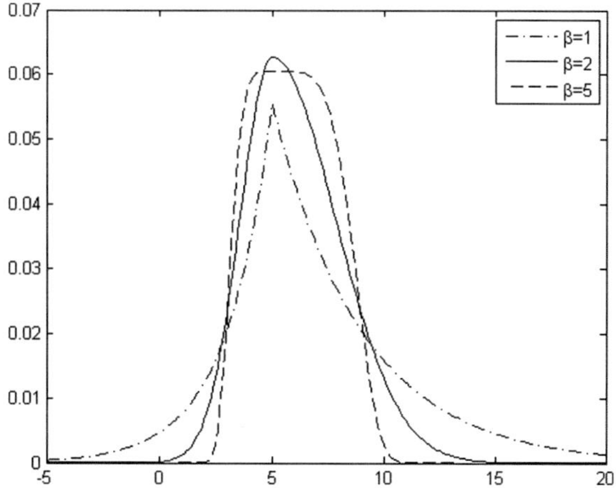

Fig. 1. Special cases of AGGD can be obtained with some values of β : asymmetric Laplacian distribution ($\beta = 1$), asymmetric Gaussian ($\beta = 2$), impulse distribution ($\beta \to 0$), and uniform distribution ($\beta \to \infty$). Also, the asymmetry is avoided if $\alpha_1 = \alpha_2$. Note that α_1 and α_2 which express the distribution width are linked to the standard deviations σ_1 and σ_2 by

$$\alpha_i = \sigma_i \sqrt{\frac{\Gamma(1/\beta)}{\Gamma(3/\beta)}}, \qquad i = 1, 2 \tag{2}$$

For the purpose of the image segmentation, let us assume that the normalized gray level histogram $h_g(g \in [0, L-1]$ with L the number of gray levels) can be approximated by an AGG mixture model as

$$f(g|\boldsymbol{\Theta}) = \sum_{m=1}^{M} \pi_m f(g|\boldsymbol{\theta}_m) \tag{3}$$

where $\boldsymbol{\Theta} = (\pi_m, \boldsymbol{\theta}_m)$ is a vector of parameters to estimate with $\boldsymbol{\theta}_m = (\pi_m, \mu_m, \alpha_{1,m}, \alpha_{2,m}, \beta_m)$, $m = 1, \ldots, M$ (M is the number of the regions to be segmented) and π_m the mth mixing parameter satisfying $\pi_m > 0$ and $\sum_{m=1}^{M} \pi_m = 1$. Thus, the complete log-likelihood function is given by

$$L_c(\boldsymbol{\Theta}) = \sum_{g=1}^{L-1} \sum_{m=1}^{M} z_{g,m} h_g \log\left[\pi_m f(g|\boldsymbol{\theta}_m)\right] \tag{4}$$

Using the MLE/EM method, the posterior probabilities $z_{g,t}$ and the mixture parameters estimates $\hat{\boldsymbol{\Theta}}$ are given in [5]. It is noted [5] that all the derived equations in M-step (AGGD parameters estimation) are highly nonlinear. In fact, there are many types of nonlinear functions, such as logarithm, power, digamma, and especially the piecewise-defined function for which, the numerical solution based on Newton–Raphson method could be instable due to its sensitivity, among other, to differentiation involved at the function breakpoint, i.e., μ_m. To map the segmented image, the optimal realization $S(u, v)$ for each pixel with (u, v) coordinates is assigned according to the Bayes decision rule

$$S(u, v) = \begin{cases} \omega_1(t_f) & \text{if} \quad p_1(t_f) = \max\limits_{m:1,\ldots,M} p_m(t_f) \\ \vdots & \\ \omega_M(t_f) & \text{if} \quad p_M(t_f) = \max\limits_{m:1,\ldots,M} p_m(t_f) \end{cases} \tag{5}$$

where, t_f is the final number of iterations for the EM algorithm convergence, $p_m(t) = \pi_m f(g|\boldsymbol{\theta}_m(t))$ and $\omega_m(t)$ is the estimated real mean parameter given by $\omega_m(t+1) = \sum\limits_g z_{g,m}(t)h_g g \Big/ \sum\limits_g z_{g,m}(t)h_g$.

3 Moment Matching Estimation (MME) on AGGMM

Proposition 1 *If X is a random variable with the AGGD density and, for any $r \in \mathbb{N}$, the r-order noncentral moment can be written as*

$$EX^r = \frac{1}{(\alpha_1 + \alpha_2)\Gamma(1/\beta)} \sum_{k=0}^{r} \binom{r}{k} \left[(-1)^k \alpha_1^{k+1} + \alpha_2^{k+1}\right] \Gamma\left(\frac{k+1}{\beta}\right) \mu^{r-k} \tag{6}$$

Proof

$$EX^r = \frac{\beta}{(\alpha_1 + \alpha_2)\Gamma(1/\beta)} \left[\int_{-\infty}^{\mu} x^r e^{-\left(\frac{-x+\mu}{\alpha_1}\right)^\beta} \mathrm{d}x + \int_{\mu}^{\infty} x^r e^{-\left(\frac{x-\mu}{\alpha_2}\right)^\beta} \mathrm{d}x \right]$$

By taking into account the following: (1) set the change of variable $y = (-x+\mu)/\alpha_1$ if $x \le \mu$ and $y = (x-\mu)/\alpha_2$ if $x > \mu$, (2) use the power series formula $(a+b)^n = \sum_{k=0}^{n} \binom{n}{k} a^{n-k} b^k$ and the definite integral $\int_0^\infty x^{\nu-1} \exp(-\eta x^p)\mathrm{d}x = \frac{1}{p}\eta^{-\frac{\nu}{p}} \Gamma\left(\frac{\nu}{p}\right)$ and (3) replace temporarily $\beta/[(\alpha_1+\alpha_2)\Gamma(1/\beta)]$ by Ω, we obtain

$$EX^r = \Omega \left[(-1)^r \alpha_1^{r+1} \int_0^\infty \left(y - \frac{\mu}{\alpha_1}\right)^r e^{-y^\beta} \mathrm{d}y + \alpha_2^{r+1} \int_0^\infty \left(y + \frac{\mu}{\alpha_2}\right)^r e^{-y^\beta} \mathrm{d}y \right]$$

$$= \Omega \sum_{k=0}^{r} \binom{r}{k} \left[(-1)^r \alpha_1^{r+1} \left(-\frac{\mu}{\alpha_1} \right)^{r-k} + \alpha_2^{r+1} \left(\frac{\mu}{\alpha_2} \right)^{r-k} \right] \int_0^{\infty} y^k e^{-y^{\beta}} \mathrm{d}y$$

$$= \frac{1}{(\alpha_1 + \alpha_2)\Gamma(1/\beta)} \sum_{k=0}^{r} \binom{r}{k} [(-1)^k \alpha_1^{k+1} + \alpha_2^{k+1}] \Gamma\left(\frac{k+1}{\beta}\right) \mu^{r-k}$$

■

Proposition 2 *Using the Proposition 1, the first four moments (mean, variance, skewness, and kurtosis) of X are deduced as*

Mean

$$me = \mu - (\alpha_1 - \alpha_2) \frac{\Gamma(2/\beta)}{\Gamma(1/\beta)} \tag{7}$$

Variance

$$\sigma^2 = \frac{\alpha_1^3 + \alpha_2^3}{\alpha_1 + \alpha_2} \times \frac{\Gamma(3/\beta)}{\Gamma(1/\beta)} - (\alpha_1 - \alpha_2)^2 \frac{\Gamma(2/\beta)^2}{\Gamma(1/\beta)^2} \tag{8}$$

Skewness

$$\gamma_1 = \frac{\alpha_1 - \alpha_2}{\sigma^3} \times \left[3 \frac{\alpha_1^3 + \alpha_2^3}{\alpha_1 + \alpha_2} \frac{\Gamma(2/\beta)\Gamma(3/\beta)}{\Gamma(1/\beta)^2} - 2(\alpha_1 - \alpha_2)^2 \frac{\Gamma(2/\beta)^3}{\Gamma(1/\beta)^3} - (\alpha_1^2 + \alpha_2^2) \frac{\Gamma(4/\beta)}{\Gamma(1/\beta)} \right] \tag{9}$$

Kurtosis

$$\gamma_2 = \frac{1}{\sigma^4} \left[\frac{\alpha_1^5 + \alpha_2^5}{\alpha_1 + \alpha_2} \times \frac{\Gamma(5/\beta)}{\Gamma(1/\beta)} - (\alpha_1 - \alpha_2)^2 \left(4(\alpha_1^2 + \alpha_2^2) \frac{\Gamma(2/\beta)\Gamma(4/\beta)}{\Gamma(1/\beta)^2} - 6 \frac{\alpha_1^3 + \alpha_2^3}{\alpha_1 + \alpha_2} \times \frac{\Gamma(2/\beta)^2\Gamma(3/\beta)}{\Gamma(1/\beta)^3} + 3(\alpha_1 - \alpha_2)^2 \frac{\Gamma(2/\beta)^4}{\Gamma(1/\beta)^4} \right) \right] - 3 \tag{10}$$

Knowing that $\{x_1, \ldots, x_n\}$ is a set of n realizations of random variable X, the empirical or sample moments denoted me_s, σ_s^2, γ_{1s}, and γ_{2s} are computed as

$$\begin{cases} me_s = \bar{x} = \sum_{i=1}^{n} x_i / n \ ; \ \sigma_s^2 = \sum_{i=1}^{n} (x_i - \bar{x})^2 / n \\ \gamma_{1s} = \sum_{i=1}^{n} (x_i - \bar{x})^3 / (n\sigma_s^3) \ ; \ \gamma_{2s} = \sum_{i=1}^{n} (x_i - \bar{x})^4 / (n\sigma_s^4) \end{cases} \tag{11}$$

In this section, the EM algorithm and the moment matching estimation method are combined for a mixture of AGGDs. In fact, here, MME has as role to estimate the AGGD parameters by equating the analytical expressions of moments given by

Algorithm 1 Pseudo-code of EM-based AGGMM moment matching method

Data: Number of modes M; initial mixture parameters $\mathbf{\Theta}^{(0)}$; convergence threshold ϵ
Result: $\hat{\mathbf{\Theta}}$, the estimate of $\mathbf{\Theta}$
$t \leftarrow 0$
Initialization: $\mathbf{\Theta}^{(t)} = \mathbf{\Theta}^{(0)}$
repeat

  E-step: Compute the a posteriori probabilities $\hat{z}_{g,m}^{(t)}$

$$z_{g,m}^{(t)} = \pi_m f(g|\boldsymbol{\theta}_m^{(t)}) \Big/ \sum_{l=1}^{M} \pi_l f(g|\boldsymbol{\theta}_l^{(t)})$$

  M-step: Maximization of $L_c(\mathbf{\Theta}^{(t)})$

$$\pi_m^{(t)} = \sum_g z_{g,m}^{(t)} h_g \text{ and } \hat{\boldsymbol{\theta}}_m^{(t)} = \boldsymbol{\theta}_{mme,m}^{(t)}$$

  $t \leftarrow t + 1$

until $\left\| \mathbb{E}\left[L_c(\mathbf{\Theta}), \hat{\mathbf{\Theta}}^{(t)}\right] - \mathbb{E}\left[L_c(\mathbf{\Theta}), \hat{\mathbf{\Theta}}^{(t-1)}\right] \right\| < \epsilon$;

Eqs. (7)–(10) with the sample moment equations given in (11) and then to solve numerically the system of nonlinear equations

$$\left\{ me = me_s \; ; \; \sigma^2 = \sigma_s^2 \; ; \; \gamma_1 = \gamma_{1s} \; ; \; \gamma_2 = \gamma_{2s} \right. \tag{12}$$

Thus, the estimated parameters $\boldsymbol{\theta}_{mme} = (\mu_{mme} \, , \alpha_{1_{mme}}, \alpha_{2_{mme}}, \beta_{mme})^T$ are obtained. So, such an estimation replaces the log-likelihood function differentiations in M-step of MLE/EM method. However, E-step remains as it is since it permits to evolve the algorithm so that, the likelihood function is maximized iteratively in order to better fit the estimated mixture to the real histogram. The combination of EM algorithm with MME for an univariate asymmetric generalized Gaussian mixture model is summarized in Algorithm 1.

4 Experiments

With the purpose to compare the performances of MME/EM with those of MLE/EM in histogram clustering for image segmentation, the experiments are carried out on real X-ray images. The first image shown in Fig. 2 represents a knee X-ray image with respectively, the lowest and the highest parts of the femur and the tibia bones. This image may be labeled a priori in three regions: the bone, the flesh, and the background. The second tested image (see Fig. 3) represents an X-ray image of a welded joint [8] and the region of interest (ROI) subjected to the segmentation. In this case, the task is to extract the weld defect indications from the regions representing other radiogram parts such as the welded joint and the base metal. We initialize the EM algorithm as follows: (1) all the mixing

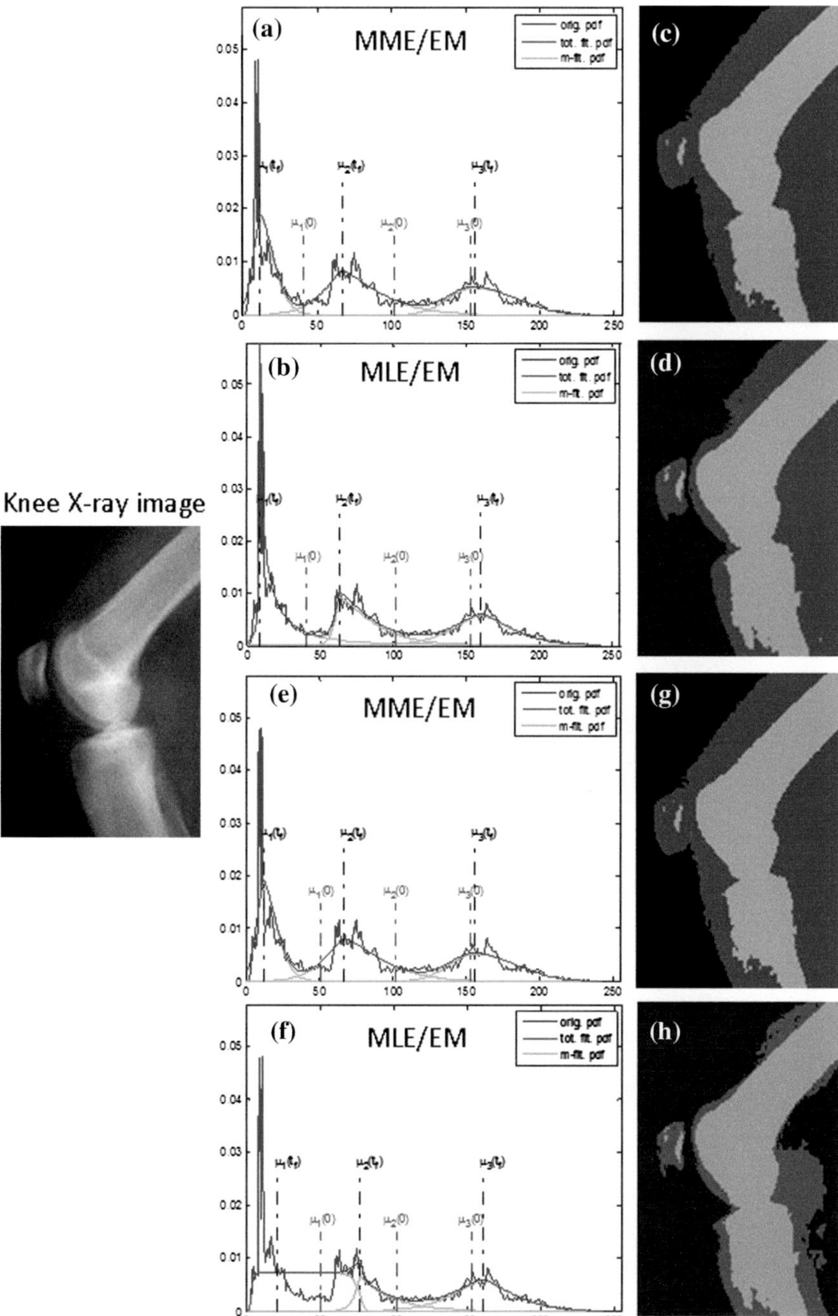

Fig. 2 Knee X-ray image. Histogram fitting and segmented regions for MME/EM in **a**, **c** and MLE/EM in **b**, **d** (with $\Delta\mu_0 = \{0\,0\,0\}$); and for MME/EM in **e**, **g** and MLE/EM in **f**, **h** (with $\Delta\mu_0 = \{10\,0\,0\}$)

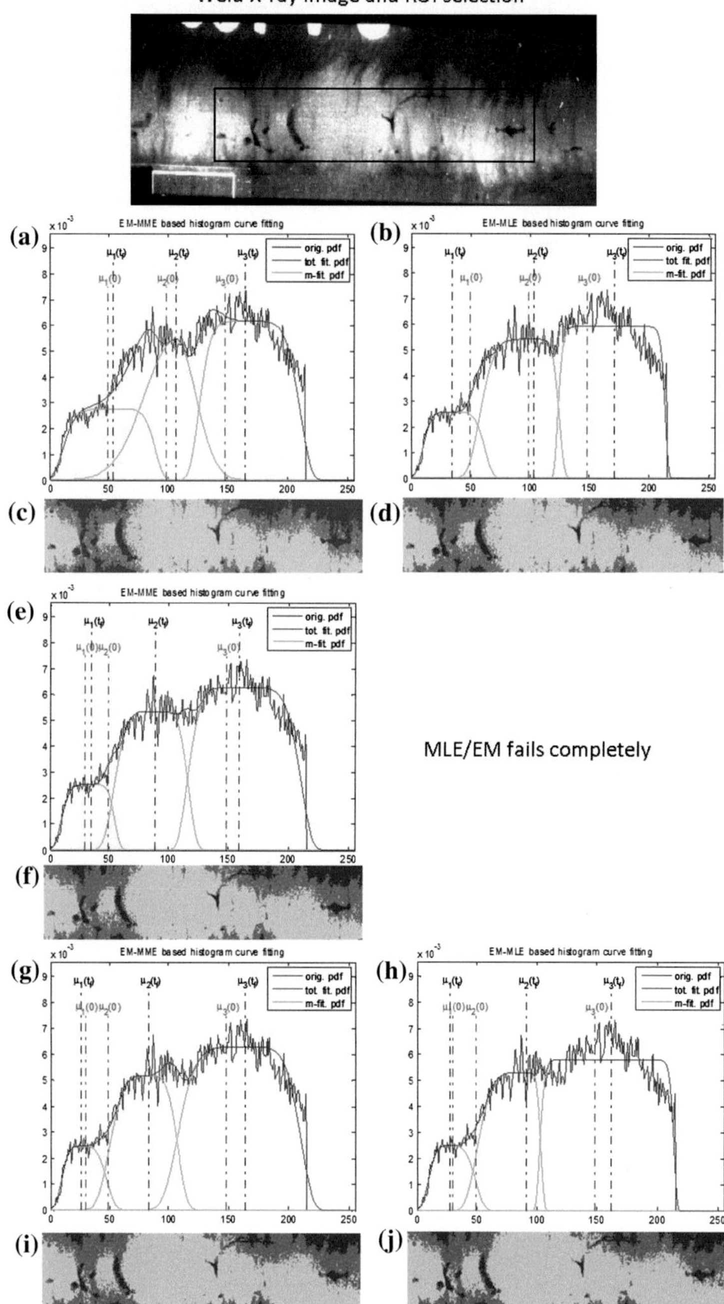

Fig. 3 Weld X-ray image. Histogram fitting and segmented regions for : (1) MME/EM in **a**, **c** and MLE/EM in **b**, **d** (with $\Delta\mu_0 = \{0\ 0\ 0\}$), (2) MME/EM in **e**, **f** (with $\Delta\mu_0 = \{-20\ \ -50\ 0\}$) and (3) MME/EM in **g**, **i** and MLE/EM in **h**, **j** (with $\Delta\mu_0 = \{-20\ \ -50\ 0\}$ and $\Pi\sigma^2_{1,0} = \Pi\sigma^2_{2,0} = \{5\ 5\ 5\}$)

parameters $\pi_m(0)$ are taken equal to $1/M$, (2) the pseudo-means $\mu_m(0)$ are chosen so that $\mu_m(0) = H^{-1}(0.05) + (2m-1)[H^{-1}(0.95) - H^{-1}(0.05)]/(2M)$ where H denotes the cumulative histogram, (3) the initial left and right variances $\sigma^2_{1,m}(0)$ and $\sigma^2_{2,m}(0)$ are equally set to $(\mu_2(0) - \mu_1(0))^2/10$ (recall that $\sigma^2_{1,2}$ is linked to $\alpha_{1,2}$ by Eq. (2)), and (4) all initial shape parameters $\beta_m(0)$ are taken equal to 1.5. Two comparison criteria, the sum of squared error (SSE) and the region uniformity measure U [9], are used to measure the histogram curve fitting and to evaluate the ensued segmentation results, respectively. They are given by

$$SSE = \sum_g \left(h(g) - f(g, \hat{\mathbf{\Theta}})\right)^2 \tag{13}$$

where h is the original histogram and $f(g, \hat{\mathbf{\Theta}})$ its estimation by AGGD mixture

$$U = 1 - \sum_{m=1}^{M} \pi_m \sigma_m^2 / \sigma_T^2 \tag{14}$$

where π_m is the area ratio of the mth segmented region, σ_m^2 its variance and σ_T^2 the total image variance. The highest is U, the better is the segmentation. In order to examine further, the influence of the EM algorithm initialization change on the estimation scores of both methods, we have introduced three parameters $\Delta\mu_0$, $\Pi\sigma^2_{1,0}$ and $\Pi\sigma^2_{2,0}$ which are, respectively, equal to $\mu_0 - \mu_0^{new}$, $\sigma^{2,new}_{1,0}/\sigma^2_{1,0}$ and $\sigma^{2,new}_{2,0}/\sigma^2_{2,0}$. For both methods MME/EM and MLE/EM and for all initializations, the fitted histograms and the images segmented accordingly, are illustrated in Figs. 2 and 3. Also, the estimated AGGMM parameters $\hat{\mathbf{\Theta}}$, the values of SSE and U are reported in Table 1. When the above-mentioned first EM initialization is used, i.e., $\Delta\mu_0 = 0$ and $\Pi\sigma^2_{10} = \Pi\sigma^2_{20} = 1$, for each of the tested Knee and Weld images, the segmentation results are given in Figs. 2a–d and 3a–d and their evaluation scores are provided by the two first rows of Table 1. In fact, the MME/EM method is as performing as the MLE/EM method in the knee X-ray image classification in bone, flesh, and background regions where U reaches 0.92 for the both. The same successfulness is observed for the weld radiographic image for both methods ($U \sim 0.83$ and 0.81) where, in spite of the bad quality of such images, characterized usually by weak contrast, blurry contours, weld overthickness and lot of artifacts and noise, the major part of the weld defects named slag inclusions, contained in the selected ROI, are well extracted. Due to the nature of the nonlinearities found in the derived AGGMM parameters by MLE/EM in [5], especially the piecewise-defined functions, the algorithm implementation works in the limits of instability and becomes, among other, very sensitive to initialization. The conclusions given above are argued by the examples given in Figs. 2e–h and 3e–j and summarized in Table 1 (Knee image: 2nd row and Weld image: 2nd and 3rd rows). In fact, when the initial mean values $\mu_{m,0}$ given at the beginning of this section are moved by amounts $\Delta\mu_{m,0}$, unlike the MME/EM method for Knee image, the MLE/EM method fails to recover correctly all the distribution modes

Table 1 AGGMM parameter estimates and fitting error via MME/EM and MLE/EM

		$\Delta\mu_0$	$\Pi\sigma^2_{1,0}$	$\Pi\sigma^2_{2,0}$	π	μ	ω	α_1	α_2	β	SSE	U
Knee X-ray image	MME	{0 0 0}	{1 1 1}	{1 1 1}	{0.32 0.39 0.29}	{12.3 66.6 156}	{15.5 77.9 164.3}	{6.9 17.7 24.4}	{12.2 33.3 39.1}	{1.6 1.4 2}	2.7×10^{-3}	**0.92**
	MLE	*id.*	*id.*	*id.*	{0.42 0.29 0.29}	{9 63 160}	{28.2 81.1 161.8}	{0.1 4.6 26.7}	{0.8 28.3 28.9}	{0.4 1.3 1.5}	**1.8** $\times 10^{-3}$	**0.92**
	MME	{10 0 0}	{1 1 1}	{1 1 1}	{0.32 0.39 0.29}	{12.3 66.4 156}	{15.4 77.7 164.2}	{7.1 17.7 24.5}	{12.3 33.4 39.2}	{1.7 1.4 2.1}	**2.7** $\times 10^{-3}$	**0.92**
	MLE	*id.*	*id.*	*id.*	{0.52 0.18 0.30}	{21.7 77 160.7}	{32.3 94.2 161.4}	{18.3 4.6 28.1}	{55.9 23.4 29}	{17.6 1.1 1.5}	4.4×10^{-3}	0.87
Weld X-ray image	MME	{0 0 0}	{1 1 1}	{1 1 1}	{0.21 0.30 0.49}	{54 107.3 164.4}	{50.2 99.4 168.5}	{44.3 37.9 39}	{36.5 23.6 47.5}	{6.3 2 6.1}	7.7×10^{-5}	**0.83**
	MLE	*id.*	*id.*	*id.*	{0.13 0.35 0.52}	{34 103.9 170.4}	{35.9 91.3 166.9}	{25.4 48.1 47.2}	{29.2 21.6 43.1}	{5.3 6 18.8}	**7.4** $\times 10^{-5}$	0.81
	MME	{−20 −50 0}	{1 1 1}	{1 1 1}	{0.11 0.33 0.56}	{34.4 89.0 159}	{32 85.5 163.2}	{26 36.7 43.8}	{21 29.5 52.5}	{5.6 5.4 6.6}	**6.8** $\times 10^{-5}$	**0.79**
	MLE	*id.*	*id.*	*id.*	Fails	Fails	Fails	Fails	Fails	Fails	–	–
	MME	{−20 −50 0}	{5 5 5}	{5 5 5}	{0.09 0.29 0.62}	{26 82.7 163.8}	{28.8 78.8 158.6}	{17.7 35 58.6}	{23.3 26.9 47.7}	{4 4.3 5.8}	**6.7** $\times 10^{-5}$	**0.76**
	MLE	*id.*	*id.*	*id.*	{0.09 0.27 0.64}	{27 91.4 161.5}	{29.4 77 157.3}	{18.5 42.8 59.7}	{23.4 12.4 52.1}	{4.1 5.4 23.6}	8.0×10^{-5}	0.74

as seen in Fig. 2f, h and which produces, as consequence, greater SSE, and lower U in comparison to MME/EM. That said, because the left part of the 2nd mode is assigned wrongly to 1st mode, we are in presence of under-segmentation where some parts of flush are confused with background. It is also noted from the fitted histogram in Fig. 2b. that MLE/EM method tracks better the 1st side of the 2nd mode despite its pronounced verticalness. However, when $\alpha_{i,m}, i = 1, 2$ get close to 0 or the new mean initial values $\mu_{1,0}^{new}$ and $\mu_{2,0}^{new}$ (see 2nd row of Table 1 for Weld and Fig. 3e) are very close to each other, the numerical solutions for EM algorithm become instable causing computation breaking as it can be seen in Fig. 3 where, MLE/EM method fails completely. Nevertheless, still for weld X-ray image, with the same new initial mean values (-20 -50 0); but now, if the initial values of the left and right variances, given in the beginning of the experiments section, are multiplied by 5 ($\Pi\sigma_{10}^2 = \Pi\sigma_{20}^2 = 5$) implying, according to Eq. (2), an increase of $\alpha_{i,m}$ by $\sqrt{5}$, the MLE/EM method becomes again stable and converges correctly with slight lower scores than MME/EM (see last row of Table 1 for Weld image).

5 Conclusion

In this paper, MME/EM method is applied on finite mixture of AGGDs to fit the histograms of images obtained from radiographic imaging. In addition to its good behavior and the reasonable agreement of its segmentation results, compared to those obtained by the MLE/EM method, the solutions given by the moment matching method, although they handled nonlinear equation system, are stable and more robust to the EM algorithm initialization. In fact, for MLE/EM, due to the nature of the nonlinearity of the derived parameter equations, especially the piecewise-defined function, the numerical solutions given by methods such as Newton–Raphson could be in the limits of stability when the scale parameters tends toward zero and in addition, very sensitive to EM initialization as shown by the experiments. That is why, in the light of the obtained results, the MME/EM method could be an interesting alternative to fit histograms approximated by this kind of distributions.

References

1. Kokkinakis, K., Nandi, A.K.: Exponent parameter estimation for generalized Gaussian probability density functions with application to speech modeling. Signal Process. **85**(9), 1852–1858 (2005)
2. McLachlan, G., Peel, D.: Finite Mixture Models. Wiley, New York (2000)
3. Dempster, A.P., Laird, N.M., Rubin, D.B.: Maximum likelihood from incomplete data via the EM algorithm. J. Royal Stat. Soc. Ser. B **39**(1), 1–38 (1977)
4. Lee, J.-Y., Nandi, A.K.: Parameter Estimation of the Asymmetric Generalised Gaussian Family of Distributions. Stat. Signal Process., IEE Colloquium, pp. 9/1–9/5 (1999)

5. Nacereddine, N., Tabbone, S., Ziou, D., Hamami, L.: Asymmetric generalized Gaussian mixture models and EM algorithm for image segmentation. In: Proceedings of 20th International Conference on Pattern Recognition (ICPR'2010), pp. 4557–4560. Istanbul (2010)
6. Lindsay, B.G., Pilla, R.S., Basak, P.: Moment-based approximations of distributions using mixtures: Theory and application. Ann. Inst. Stat. Math. **52**(2), 215–230 (2000)
7. Delicado, P., Goria, M.N.: A small sample comparison of maximum likelihood, moments and L-moments methods for the asymmetric exponential power distribution. Comput. Stat. Data Anal. **52**(3), 1661–1673 (2008)
8. Nacereddine, N., Ziou, D., Hamami, L.: Fusion-based shape descriptor for weld defect radiographic image retrieval. Int. J. Adv. Manuf. Tech. **68**(9–12), 2815–2832 (2013)
9. Ng, W.S., Lee, C.K.: Comment on using the uniformity measure for performance measure in image segmentation. IEEE Trans. PAMI **18**(9), 933–934 (1996)

Accurate Classification of ECG Patterns with Subject-Dependent Feature Vector

Piotr Augustyniak

Abstract Correct and accurate classification of ECG patterns in a long-term record requires optimal selection of feature vector. We propose a machine learning algorithm that learns from short randomly selected signal strips and, having an approval from a human operator, classifies all remaining patterns. We applied a genetic algorithm with aggressive mutation to select few most distinctive features of ECG signal. When applied to the MIT-BIH Arrhythmia Database records, the algorithm reduced the initial feature space of 57 elements to 3–5 features optimized for a particular subject. We also observe a significant reduction of misclassified beats percentage (from 2.7 % to 0.7 % in average for SVM classifier and three features) with regard to automatic correlation-based selection.

1 Introduction

In algorithmic support of electrocardiogram interpretation, the procedure of beat patterns classification plays an essential role. The resulting beat classes disclose the origin of the stimulus and represent the activation of conduction pathways, thus consistency of beats stands for repeatability of the heart action. The classification results are also used for selecting specific beats for further processing (e.g., for heart rate variability studies only normal sinus beats are taken into account [3]) and helps in rough identification of supraventricular, ventricular, and other beats from artifacts. The method applied has to cope with numerous beat types, recording condition variability and often is expected to give a repeatable result in real time. Usually, the classification uses a subset of temporal and shape-related parameters that represent irregularities in generation and conduction of the electrical heart stimulus. As a result, each beat is attributed with a morphology label given accordingly to a most corresponding physiological model. The MIT-BIH Arrhythmia Database [19], used

P. Augustyniak (✉)
AGH University of Science and Technology, 30 Mickiewicza Ave., 30-059 Krakow, Poland
e-mail: august@agh.edu.pl

R. Burduk et al. (eds.), *Proceedings of the 9th International Conference on Computer Recognition Systems CORES 2015*, Advances in Intelligent Systems and Computing 403, DOI 10.1007/978-3-319-26227-7_50

as a reference for vast number of algorithms, distinguishes 41 beat types, although in a particular record only few of them are present.

The scientific interest for classification of heartbeats arised in the first years of digital electrocardiography, and its fast and accurate implementation remains an engineering challenge till today [7]. A review of recent works starts with the paper by O'Dwyer et al. [20] who studied the different time and amplitude-based heartbeat features and their appropriacy for classification. Chang et al. [4] discussed four most common metrics to measure the similarity of two sections of the ECG signal. While Lemay et al. [11] proposed an interesting QRS classification method-based solely on ECG sample values, Mensing et al. [18] developed an alternative Poincare plot-based classification considering both temporal and voltage differences. Another sophisticated beat morphology-based classification method was presented by de Chazal et al. [5], who separately considers respective time and value features of ECG waves detected beforehand. Time-frequency domain was proposed as another feature space for ECG beats classification [12, 23]. An efficient feature space for beat classification also results from relationships of neighboring coefficients [25] in so-called cone of influence [17]. A substantial progress was made with inclusion interchannel relations present in a multi-lead ECG record (e.g., angular data) to the feature space. Several works on integrating temporal, morphological, and time–frequency features were proposed by Llamedo-Soria et al. [13–15]. Yet another alternative approach [8] was based on a syntactic model of the ECG, where selected parameters of best fitted figures were used as classification features.

Main novelty of our proposal is the use of a very small highly discriminative patient-dependent feature vector instead of a universal set of parameters. The feature selection is performed by a genetic algorithm based on interactive human-supported classification of record excerpts and the resulting feature vector can be maintained with the patient data and reused for future records taken from the same individual in similar conditions.

2 Materials and Methods

2.1 Accuracy Versus Size Optimization with a Genetic Algorithm

For solving optimization problems various methods have been implemented in machine learning procedures. One of heuristic method are genetic algorithms, inspired by the nature but recently adapted to several fields of research and used for feature selection. By the analogy to the nature, each individual encodes one subset of features, the discrimination capabilities are successively evaluated based on a learning set, and the algorithm aims to produce new individuals, represented by more distinctive features and to eliminate the others. An implemented classifier plays a pivotal role in evaluation of discriminative power of each individual, thus it determines the feature vector selected as optimal. A common approach assumes the use of a classic genetic algorithm [6], where the genes of each individual correspond to a complete feature space and represent the inclusion of a given feature to the produced feature vector. The classic evolution, however, optimizes the classification accuracy, which increases with the count of features included.

Since our aim is reduction of feature set, we applied an alternative algorithm with integer encoding of individuals in predefined small number of genes and modification of mutation scheme yielding a larger mother population. The genetic algorithm with aggressive mutation was originally proposed by Rejer [21] for building the vector of most distinctive features for EEG-based human–machine interface. Based on her future paper [22] it showed best results in competition with RelieF [10], forward selection [9], and least absolute shrinkage and selection operator (LASSO) [24]. In the aggressive mutation the mother population of M individuals is transformed into new population, but in the opposite to the classic approach, each parent of N genes yields N children with another gene mutated. Additional M individuals are created in result of a crossover operation. All individuals of this new population together with the mother population, of the total size of $M \times (N + 2)$ are then evaluated using classification accuracy as a fitness function. In result of this step, only M individuals with most discriminative features are retained as mother population for next iteration. In the worst case, the individuals from the initial mother population win the competition and the value of classification accuracy in next mother population does not decrease. Similarly to the brain pattern recognition, identification of most distinctive ECG features is needed for faster and more robust heartbeats classification. We applied the algorithm as originally proposed, but extended its applicability with individual learning step allowing for personalization of classification procedure.

2.2 The Annotated Reference Collection

The actual classification is based on subject-specific feature vector determined beforehand or calculated from the the current record. The feature selection process is performed once and and the feature vector is maintained with patient data for further use with records taken from the same individual. Although the complete feature space contains several elements, the optimized feature vector is expected to have as few elements as possible.

The learning phase is supported by a human operator. This person is asked to manually correct the initial classification of beats made for several short signal strips randomly selected in the whole record. Since this step is supervised by an expert, no particular accuracy is required at this stage, and thus, we used a conventional kNN-based algorithm using voltage difference as a feature. Keeping in mind that the strips are excerpts from different parts of the signal, we first determine the classes for each strip independently, and then repeat the classification including the beats from neighboring strips until the whole collection of beats is included. Variability of classification results represents the variability of the record content. In case of stable record, the classification result does not depend on strip selection and all beats are likely to fall in the groups created for the first strip. Otherwise, new strips contain features unexpected by the classifier and new groups are created each time when new beats are appended.

The initial classification yields annotated heartbeats originating from the patient-specific record, and provides a reference for the automatic classification of the whole record. Only few hundreds of beats from the whole record are classified in this manner, but their appropriate selection is fundamental for the overall result.

2.3 Feature Vector Optimization

The annotated reference collection of beats is randomly divided into two parts. First, containing 35 % of the total beats is used as a learning set in the procedure of feature vector optimization, while the remaining part (65 %) constitutes a testing set (Fig. 1). While the initial classification was based on the voltage difference as a feature, at this stage all considered features for all heartbeats in the learning set are calculated. Depending on the experiment setup, the feature vector consists of 3–5 items. After an initial random selection, the genetic algorithm with aggressive mutation presented

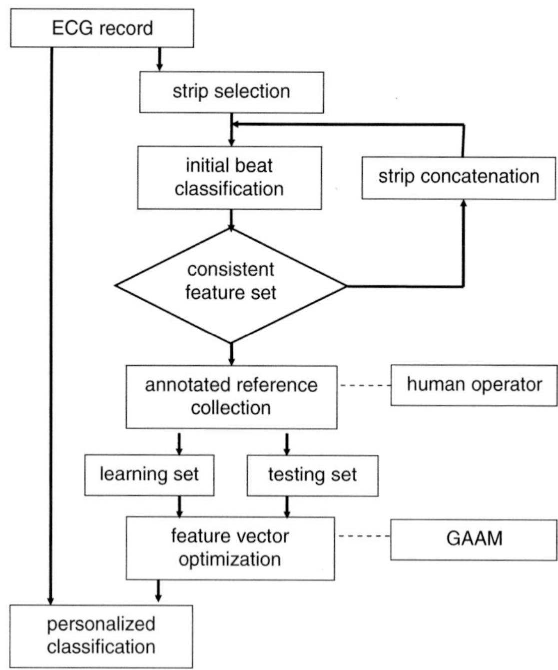

Fig. 1 Diagram flow of feature set optimization for personalized heartbeats classification

in 2.1 is applied together with the cooperating classifier to indicate the most distinctive features out of the complete feature space.

The optimized feature vector is then verified with use of testing set. The features indicated as most distinctive are calculated for all beats there and the classification is made with the classifier previously used to evaluate discriminative power of genes. Finally, the classification results are compared to the annotations for each respective beat. If the classification accuracy measured on the heartbeats from the testing set meets the quality criteria, the resulting feature vector is considered as optimal and is used for classification of the whole record. Otherwise this stage is repeated starting with new selection of learning and testing sets from the annotated reference collection of beats.

3 Test Setups

3.1 A Complete Feature Space

The proposed ECG classification method was tested with 48 double-lead half-hour records from the MIT-BIH Arrhythmia Database [19], each accompanied with annotation of time and morphology. These annotations eliminate the needs of beat detection procedure and of participation of a human expert. Implemented algorithms calculated 57 different parameters of the beat building up a complete feature space:

- 4 derived from the QRS shape,
- 7 based on the signal derivatives (acceleration),
- 3 based on the wave area,
- 12 calculated from interval lengths,
- 5 derived from angles between two leads,
- 4 derived from Poincare plots of waves,
- 16 calculated from time-frequency representation,
- 6 calculated as parameters of best fitted dual channel syntactic model.

3.2 Cooperating Classifiers

We performed two independent tests: one for a common kNN [16] and the other for linear SVM [26] as cooperating classifiers. For each record we first built the annotated reference collection out of 15 randomly selected nonoverlapping strips each of 10s length. The initial classification and human correction were not used in tests, since the reference annotations are provided by the database. The collections were then split to learning set (ca. 60 beats) and testing set (ca. 120 beats). In three attempts, the feature vector was expected to contain 3, 4, or 5 elements, so the initial feature space was reduced 11–19 times. The feature vector was initiated with random items and optimized by the aggressive mutation algorithm with 100 iterations run for 60 individuals form the learning set. Finally, the classification using optimal feature vector was validated with beats from the testing set with their annotations and used for classification of the whole record.

3.3 Methods of Feature Vector Optimization

Three different feature selection methods were applied in the test:

1. Reference classification method R1 was based on principal feature analysis, i.e., best features were selected in the whole MIT-BIH Arrhythmia database by their pairwise minimum cross correlation [1].
2. Universal classification method (UCM) was based on features globally optimized by the GAAM using a common annotated reference collection composed of beats originating from various records of the MIT-BIH Arrhythmia Database.
3. Personalized classification method (PCM) was based on features locally optimized by the GAAM using individual annotated reference collections composed of beats originating from a specific record of the MIT-BIH Arrhythmia Database.

4 Results

4.1 Classification Accuracy

Test results for classification accuracy are presented in Table 1 as the ratio of the total misclassified beats to the total count of beats. Comparing the universal classification performed with GAAM-optimized features to the reference reveals that decorrelated features (reference method) do not have the optimal discriminative power. On the other hand, comparing the classification performed with personalized feature vectors

Table 1 Percentage of misclassified heartbeats

Classifier	No. of features	R1	UCM	PCM
kNN	3	3.11	2.24	1.15
	4	2.84	2.08	1.03
	5	2.57	1.89	0.98
SVM	3	2.73	1.55	0.69
	4	2.29	1.31	0.61
	5	2.01	1.18	0.58

Table 2 Positions taken by the RR_{n+1}/RR_n ratio in the ranking of discriminative power resulting from personal optimization (the total no. of records equals 44)

Classifier	No. of features	Position				
		1st	2nd	3rd	4th	5th
kNN	3	29	7	3		
	4	29	6	2	2	
	5	28	6	3	2	1
SVM	3	32	10	2		
	4	31	9	3	1	
	5	30	8	3	2	1

to the universal method clearly demonstrates the benefit of adapting to the specific characteristics of the signal.

4.2 *Ranking the Features by Discriminative Power*

An interesting result is also the average ranking of each feature by its discrimination power. As a reference we used the globally GAAM-optimized feature vector. In case of global optimization, the RR_{n+1}/RR_n ratio was found the most discriminative feature, however for local optimization it was not always the first, and in some records not even present in the feature vector. Table 2 lists the positions taken by this parameter in the ranking of discriminative power resulting from personal optimization.

5 Discussion

Shrinking the feature vector demonstrates that a locally optimized vector with three elements (Table 1 col. 4 row 1) is more distinctive than a global minimum correlation-based vector with 5-elements (Table 1 col. 3 row 6) even using a classifier of inferior performance (kNN versus SVM). Therefore much of computational power may be saved by calculating only few most distinctive features specific for a given individual. Also in real-time classification reducing the operations is beneficial for speeding up the processing. Maintaining the same criteria of heartbeats classification in different subjects is not justified by medical knowledge.

Usually patients show a very stable ECG pattern, thus the optimization stage is performed once and remains usable until the change of recording conditions or health status. The specification of optimal feature vector is very tiny (several bytes) and may accompany the personal data. In case of using a reprogrammable personal

ECG recorder [2] or personalized interpretation software, appropriate patient-specific classification algorithms may be driven by this specification.

The originality of presented approach consists in an individual, subject-dependent selection of most distinctive ECG features, instead of using a standardized feature vector. For repeatability of the results, these features have to be accommodated by classification algorithm, e.g., programmed into a personal ECG recorder or supported by general-purpose interpretive equipment.

The proposed method requires additional data fields for storage of the specification of optimal feature vector. This may be implemented as extra patient data or by fingerprinting of the digital record. The advantage of the latter solution is that it does not require any change in current definitions of the ECG record and allows the access to this information only for authorized recipient. The proposed method also requires from the interpretation software to perform several algorithms for beats classification or to accept an external library of functions dedicated for this purpose.

Acknowledgments This scientific work is supported by the AGH University of Science and Technology in year 2015 as a research project No. 11.11.120.612.

References

1. Augustyniak, P.: The use of shape factors for heart beats classification in holterrecordings. Proc. Comput. Med. Zakop. **2–6**(05), 47–52 (1997)
2. Augustyniak, P.: Adaptive discrete ECG representation—comparing variable depth decimation and continuous non-uniform sampling. Comput. Cardiol. **29**, 165–168 (2005)
3. Augustyniak, P.: Wearable wireless heart rate monitor for continuous long-term variability studies. J. Electrocardiol. **44**(2), 195–200 (2011)
4. Chang, K.C., Lee, R.G., Wen, C., Yeh, M.F.: Comparison of similarity measures for clustering electrocardiogram complexes. Comput. Cardiol. **32**, 759–762 (2005)
5. de Chazal, P., O'Dwyer, M., Reilly, R.B.: Automatic classification of heartbeats using ecg morphology and heartbeat interval features. IEEE Trans. Biomed. Eng. **51**(7), 1196–1206 (2004)
6. Holland, J.H., Reitman, J.S.: Cognitive systems based on adaptive algorithms. ACM SIGART Bull. **63**, 43–49 (1977)
7. Jaworek, J., Augustyniak, P.: A cardiac telerehabilitation application for mobile devices. Comput. Cardiol. **38**, 241–244 (2011)
8. Jokić, S., Krčo, S., Delić, V., Sakač, D., Lukić, Z., Loncar-Turukalo, T.: An efficient approach for heartbeat classification. Comput. Cardiol. **2010**(37), 991–994 (2010)
9. Kittler, J.: Feature set search algorithms. In: Pattern Recognition and Signal Processing, pp. 41–60. Sijthoff and Noordhoff, Alphen aan den Rijn (1978)
10. Kononenko, I.: Estimating attributes: analysis and extensions of relief. In: De Raedt, L., Bergadano, F. (eds.) Machine Learning: ECML-94, pp. 171–182. Springer, Berlin (1994)
11. Lemay, M., Jacquemet, V., Forclaz, A., Vesin, J.M., Kappenberger, L.: Spatiotemporal QRST cancellation method using separate QRS and T-Waves templates. Comput. Cardiol. **32**, 611–614 (2005)
12. Llamedo-Soria, M., Martinez, J.P.: An ECG classification model based on multilead wavelet transform features. Comput. Cardiol. **34**, 105–108 (2007)
13. Llamedo-Soria, M., Martinez, J.P.: Analysis of multidoma in features for ECG classification. Comput. Cardiol. **36**, 561–564 (2009)

14. Llamedo, M., Khwaja, A., Martinez, J.P.: Analysis of 12-lead classification models for ECG classification. Comput. Cardiol. **37**, 673–676 (2010)
15. Llamedo, M., Martinez, J.: Heartbeat classification using feature selection driven by database generalization criteria. IEEE Trans. Biomed. Eng. **58**(3), 616–625 (2011)
16. MacQueen, J.: Some methods for classification and analysis of multivariate observations. In: Le Cam, L.M., Neyman, J. (eds.) Proceedings of the Fifth Berkeley Symposium on Mathematical Statistics and Probability, vol. 1, pp. 281–297. University of California Press, Berkeley (1967)
17. Mallat, S., Zhong, S.: Characterization of signals from multiscale edges. IEEE Trans. Pattern Anal. Mach. Intell. **14**(7), 710–732 (1992)
18. Mensing, S., Bystricky, W., Safer, A.: Identifying and measuring representative QT intervals in predominantly non-normal ECGs. Comput. Cardiol. **33**, 361–364 (2006)
19. Moody, G.B.: The MIT-BIH Arrhythmia Database CD-ROM, 3rd Edn. Harvard-MIT Division of Health Sciences and Technology, Cambridge (1997)
20. O'Dwyer, M., de Chazal, P., Reilly, R.I.: Beat classification for use in arrhythmia analysis. Comput. Cardiol. **27**, 395–398 (2000)
21. Rejer, I.: Genetic algorithms in EEG feature selection for the classification of movements of the left and right hand. In: Proceedings CORES2013, pp. 579–589 (2013). doi:10.1007/978-3-319-00969-8-57
22. Rejer, I.: Genetic algorithm with aggressive mutation for feature selection in BCI feature space pattern. Anal. Appl. (2014). doi:10.1007/s10044-014-0425-3
23. Rodriguez-Sotelo, J.L., Cuesta-Frau, D., Castellanos-Dominguez, G.: An improved method for unsupervised analysis of ECG beats based on WT features and J-Means clustering. Comput. Cardiol. **34**, 581–584 (2007)
24. Tibshirani, R.: Regression shrinkage and selection via thelasso. J. Stat. Soc. Ser. B **58**(1), 267–288 (1996)
25. Vansteenkiste, E., Houben, R., Pizurica, A., Philips, W.: Classifying electrocardiogram peaks using new wavelet domain features. Comput. Cardiol. **35**, 853–856 (2008)
26. Vapnik, V.N.: The Nature Of Statistical Learning Theory. Springer, New York (1995)

Environmental Microbiological Content-Based Image Retrieval System Using Internal Structure Histogram

Yan Ling Zou, Chen Li, Zeyd Boukhers, Kimiaki Shirahama, Tao Jiang and Marcin Grzegorzek

Abstract Environmental Microbiology (EM) is an important scientific field, which investigates the ecological usage of different microorganisms. Traditionally, researchers look for the information of microorganisms by checking references or consulting experts. However, these methods are time-consuming and not effective. To increase the effectiveness of EM information search, we propose a novel approach to aid the information searching work using *Content-based Image Retrieval* (CBIR). First, we use an microorganism image as input data. Second, image segmentation technique is applied to obtain the shape of the microorganism. Third, we extract shape feature from the segmented shape to represent the microorganism. Especially, we use a contour-based shape feature called *Internal Structure Histogram* (ISH) to describe the shape, which can use angles defined on the shape contour to build up a histogram and represent the structure of the microorganism. Finally, we use Euclidean distances between each ISHs to measure the similarity of different EM images in the retrieval task, and use *Average Precision* (AP) to evaluate the retrieval result. The experimental result shows the effectiveness and potential of our EM-CBIR system.

Keywords Environmental microbiology · Content-based image retrieval · Image segmentation · Internal structure histogram

---

Y.L. Zou (✉) · T. Jiang
Chengdu University of Information Technology, Chengdu, China
e-mail: zyl@cuit.edu.cn

T. Jiang
e-mail: jiang@cuit.edu.cn

Y.L. Zou · C. Li · Z. Boukhers · K. Shirahama · M. Grzegorzek
Pattern Recognition Group, University of Siegen, Siegen, Germany
e-mail: chen.li@uni-siegen.de

Z. Boukhers
e-mail: zeyd.boukhers@uni-siegen.de

K. Shirahama
e-mail: kimiaki.shirahama@uni-siegen.de

M. Grzegorzek
e-mail: marcin.grzegorzek@uni-siegen.de

R. Burduk et al. (eds.), *Proceedings of the 9th International Conference on Computer Recognition Systems CORES 2015*, Advances in Intelligent Systems and Computing 403, DOI 10.1007/978-3-319-26227-7_51

1 Introduction

Environmental microbiology is an important scientific study of microorganisms in environmental protection and treatment, which investigates the decomposing abilities of different microorganisms for pollutants and rubbishes. For example, Rotifera can eat sludge as its food and makes the water clean. And Vorticella can digest organic pollutants in wastewater and improve the quality of fresh water [14]. Traditionally, when environmental microbiological researchers obtain a new microorganism sample, they identify it by checking references or asking experts. However, these traditional methods are time-consuming and inefficient. To increase the search effectiveness of environmental microbiological information, we propose a novel *Environmental Microbiological Content-based Image Retrieval* (EM-CBIR) System using microorganism images directly. Given a query image from a user, the retrieval system starts by conducting a semi-automatic segmentation process to obtain only the region of interest, which defines the EM without any surrounding artifacts. Then, features which characterise the shape of the EM are extracted from the segmented image. To finalise the system process, we calculate the Euclidean distance between the extracted features from the query image and features extracted formerly from all images in the database. Our retrieval system outputs similar images to the query sorted from most similar to dissimilar, and gives feedback to the user. The framework of our system is shown in Fig. 1.

In this paper, we address the following problems (1) the EM image segmentation, (2) the stable shape feature extraction, and (3) the CBIR approach: (1) Image seg-

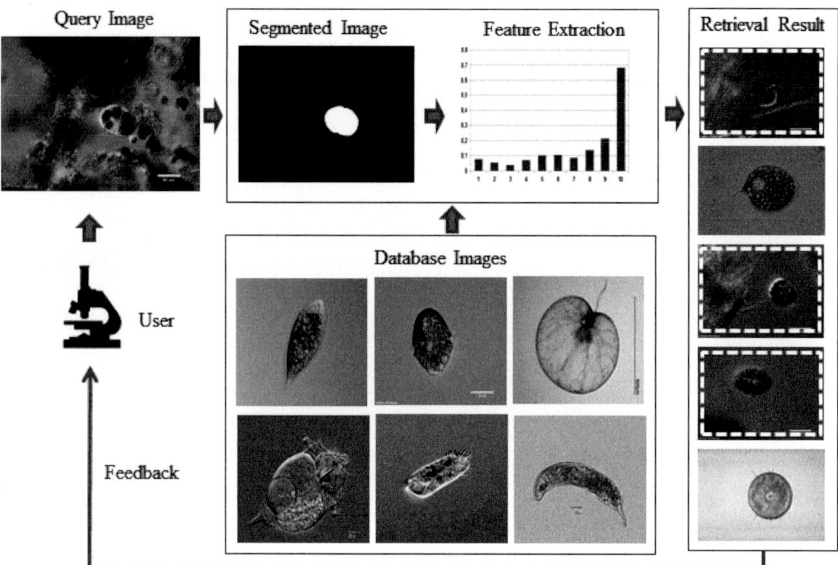

Fig. 1 Workflow of the proposed EM-CBIR system

mentation: To ensure the accuracy of removing impurities in microscopic images, we use a semi-automatic segmentation method, which purifies an EMs region boundary that is approximately specified by a user. It combines conventional manual segmentation utilities with a novel automatic approach [9]. (2) Feature extraction: Our feature extraction procedure begins with the shape description of EMs regardless of their rotation and colour. We use *Internal Structure Histogram* (ISH) to discriminate structural properties of a microorganism, which is a contour-based shape feature. ISH is extracted by equidistantly distributing sample points on the contour of an EM region. Then, we create a histogram representing the distribution of angles, each of which is defined by a combination of three sample points [9]. (3) CBIR approach: We use a CBIR method, which represents EM images by numerical feature vectors extracted automatically to represent their perceptual properties (shape, colour, texture, etc.). And, we apply *Average Precision* (AP) and mean AP (mAP) to evaluate the retrieval result.

2 Related Work

2.1 CBIR of Microorganisms

To the best of our knowledge, there is no CBIR approach which focuses on EM images. However, there are some relevant retrieval and classification methods for different types of microorganisms. We explain the novelty of our work in the next three paragraphs, which show what types of microorganisms are addressed by the existing systems.

We investigate the effectiveness of various low level image descriptors of Marine Microorganisms, which includes the colour, shape and texture features in representing the semantic categories of the marine life images. In [17], a system is developed by researchers for image retrieving and evaluating the effectiveness of various types of image descriptors for indexing marine life images, where features are represented by colour, texture and shape.

Medical microorganisms refer to the prevention, diagnosis and treatment of infectious diseases. For instance, a framework to archive and retrieve histopathology images by content is presented in [4]. This automatic image annotation framework can recognise high-level concepts after analyzing visual image contents. In [1], a multitiered content-based image retrieval system for microscopic images utilising a reference database that contains images of more than one disease is designed and developed. In this research, a multitiered method is used to classify and retrieve microscopic images involving their specific subtypes, which are mostly difficult to discriminate and classify. The system enables both multi-image query and slide-level image retrieval. In [19], an image modality classification method is proposed to classify a given set of medical images according to 18 modalities taken from five classes. Then, the most relevant images for each topic, according to a set of ad hoc information requests are retrieved.

Environmental microorganisms (EMs) are classified in [9], where a semi-automatic segmentation method, four shape features and late fusion approach are used for identifying 10 classes of EMs. But, an EM-CBIR system is still a novel topic, which constitutes a research motivation for us.

2.2 Selection of Techniques for EM-CBIR

Image segmentation: There are different segmentation methods based either on pixel intensity levels or on image context. One popular intensity-based method is Otsu thresholding [13]. Another category of methods analyse first- and second-order derivatives and the local gradients (e.g. Sobel [7], Prewitt [10], Laplacian of a Gaussian (LoG) [12], and Canny edge detectors [5]). Further, the Watershed algorithm [15] simulates the topological features of geodesy and divides the image areas considering pixel values as altitudes. As pre- or post-processing techniques, morphological operations such as erosion and dilatation are often applied [6]. Among all the methods mentioned above, we use Sobel edge detectors in our experiment since its noise suppression functionality turns it robust to noises.

Feature extraction: Since most of the microorganisms are colourless and transparent, it is practically impossible to extract their colour and texture features. Therefore, we assume shape features which capture characteristics of microorganisms based on the optic boundary between light and shade in various microscopic images. Specifically, contour-based shape features are very important descriptors for shape representation. For example, shape signature (SS), which represents a shape by a one-dimensional function that is derived from shape boundary points. Many shape signatures exist, such as centroidal profile, complex coordinates, centroid distance, tangent angle, cumulative angle, curvature, area and chord-length [16]. In [20], the author propose Fourier Descriptor (FD) to classify similarity transformed characters using Fourier transformed coefficients. The distance measurement is the weighting sum of the variance of magnitude ratios and the variance of phase difference between two sets of Fourier coefficients. A global feature, called shape context, is extracted for each corresponding point [2]. The matching between corresponding points equals the matching between the context features. To extract the shape context (ShC) at a point p, the vectors of p to all the other boundary points are found and used as the shape context of this point. ISH is a contour-based shape feature, which characterises the structures of different shapes. It is an extension of internal structure angles (ISAs) [3], each of which represents an angle defined by a combination of three points on the contour. We have extended ISAs by considering their distribution and modelling a histogram representation [9].

Image retrieval: CBIR systems [11, 18, 21] for medical images are important to deliver a stable platform to catalogue, search and retrieve images based on their content [1]. For EM-CBIR, some existing works propose related approaches to this challenging task. In [19], text- and content-based approaches are used for medical image retrieval, in which approaches of combined textual and visual features are explored. In [4], an improved content-based histopathology retrieval addresses medical imaging systems. Finally, another paper proposed a system which enables both multi-image query and slide-level image retrieval for microscopic images [1].

3 EM-CBIR Approaches

3.1 Image Segmentation

Since many EM images are very noisy and have low contrast, their full-automatic segmentation is difficult. Hence, we use a Sobel edge detector based semi-automatic segmentation approach [9]. As shown in Fig. 2(a), we input an original EM image. Then, we use a cursor operation to refine an approximate region of this microorganism in (b). Third, Sobel edge detector is applied to search the contour of this microorganism as shown in (c). Fourth, morphological operations are used to fill holes and smooth the contour in (d). Fifth, in (e) we manually select the foreground depicting the microorganism of interest. Finally, we get the segmentation result in (f).

3.2 Feature Extraction

The output of our semi-automatic segmentation method is a binary image, in which the region of an EM is separated from the background. From this binary image, shape

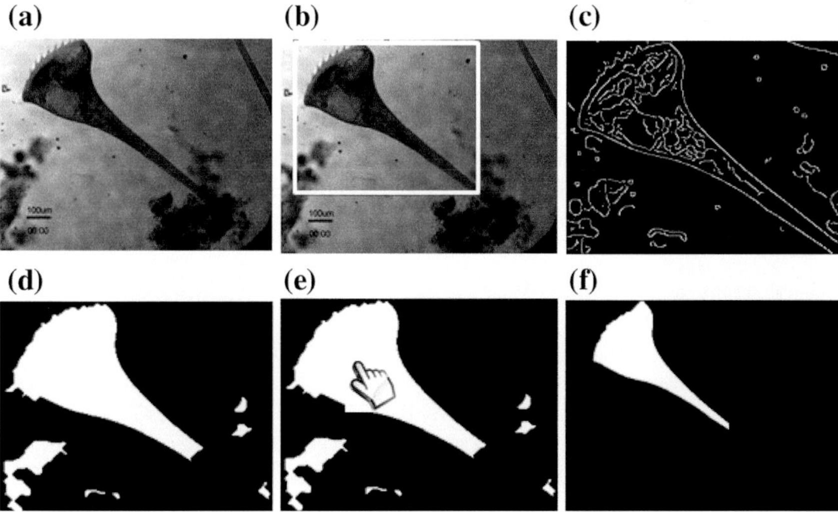

Fig. 2 An example of semi-segmentation approach

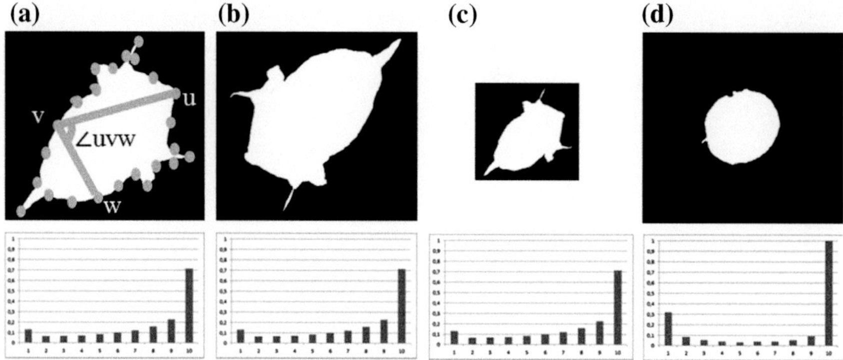

Fig. 3 An example for the distinguishing ability of ISH

feature is computed. Especially, because ISH can demonstrate shapes based on the structures effectively, we apply it in our EM-CBIR work. For determining the ISH, we first compute the internal structure angles (ISAs) of an microorganism shape. As shown in Fig. 3, V sample points ($V = 100$ in our experiments) are equidistantly distributed along the EM boundary. Then, for any possible triple of points u, v and w, an ISA is computed as their inscribed angle $\angle uvw$. Afterwards, we build a histogram by classifying the values of all the ISAs into 10 bins defined as $[0°, 36°)$, $[36°, 72°)$, ..., $[324°, 360°)$. Thus, the ISH can be represented as a 10-dimensional feature vector $\boldsymbol{c}_{\text{ISH}} = (s_1, \ldots, s_{10})^{\text{T}}$. It should be noted that the ISH deals with the frequencies (statistics) of angles defined on the EM boundary. With respect to this, one ISA $\angle uvw$ is related to the other two ISAs $\angle uwv$ and $\angle wuv$ (i.e. $\angle uvw = 180° - (\angle uwv + \angle wuv)$). However, an ISH created by two of such three ISAs loses the shape information of the EM. This is because the relation of each sample point to all the others is needed to fully represent the shape of the EM. Hence, we create the ISH that preserves frequencies of all the ISAs. In Fig. 3, there are four EM images and their ISHs are shown: (a) the original image; (b) the rotated image; (c) 50 % resized image; and (d) another microorganism. From Fig. 3, we find that because (a), (b) and (c) have the same shapes, ISH builds similar histograms for them. However, (d) has a different shape, so ISH produces a dissimilar histogram for it.

3.3 Retrieval Approach and Evaluation

We present a retrieval system that performs a similarity search in an EM image dataset for a query given by the user. Here, the similarity is defined as $1 - D$, where D is the Euclidean distance between two feature vectors. In order to evaluate performance of the proposed model and similarity measure technique, an EM-CBIR system is implemented. The challenge in EM-CBIR is that the similarity search is performed under semantic aspects. In contrast to conventional data retrieval systems, for example in relational databases, the search conditions are not explicitly known. Usually, the amount of result documents in a retrieval system is limited by the total amount of returned documents or by some threshold for similarity value. AP has been developed in the field of information retrieval, and is used as an indicator to evaluate a ranked list of retrieved samples [8]. AP is defined as follows:

$$AP = \sum\nolimits_{k=1}^{m} (P(k) \times rel(k))/(number\ of\ relevant\ EM\ images) \qquad (1)$$

where $P(k)$ is the precision computed by regarding a cut-off position as the kth position in the list, and $rel(k)$ is an indicator function which takes 1 if the EM image ranked at the kth position is relevant, otherwise 0. Thus, AP represents the average of precisions each of which is computed at the position where a relevant EM image is ranked. The value of AP increases, if the relevant EM image occurs on higher positions in the ranked list. Therefore, AP can be used to measure the retrieval performance.

4 Experimental Results

We use a real EM dataset (EMDS) as shown in Fig. 4 acquired by USTB and CUIT. EMDS contains 21 classes of EMs $\omega_1, \ldots, \omega_{21}$. Each class is represented by 20 microscopic images. In our experiments of EM retrieval, we used each EM image as a query image once and all remaining images as database images. AP is used to evaluate the result of such a retrieval process, and the mean of APs (mAP) for all 21 classes is used as an overall evaluation measure. Finally, to avoid the bias caused

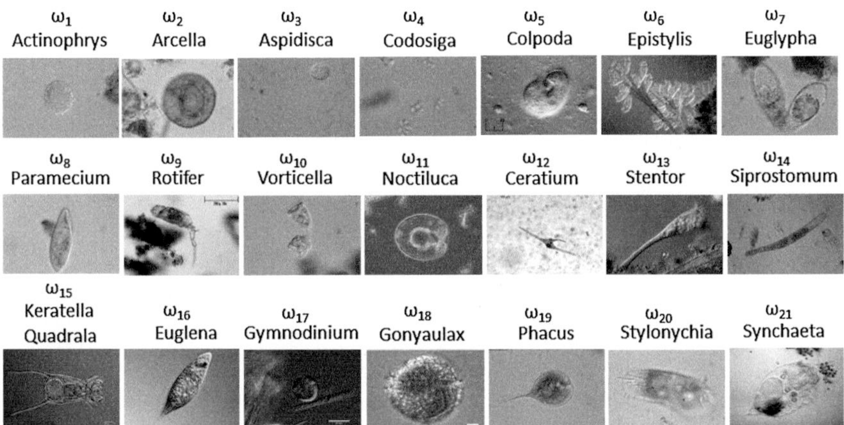

Fig. 4 Examples of original images on EMDS where $\omega_1, \ldots, \omega_{21}$ represent 21 classes of EMs

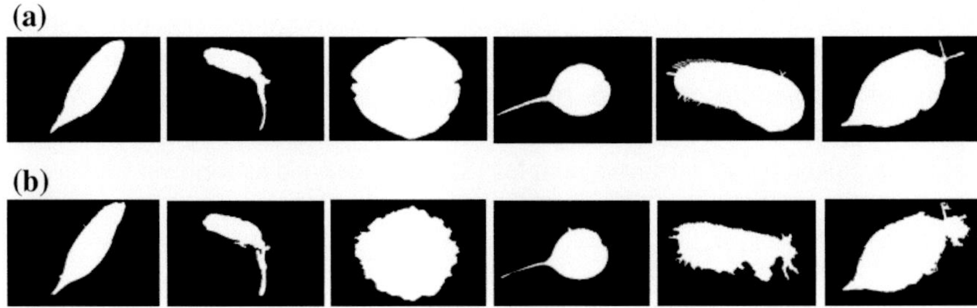

Fig. 5 Comparison of ground truth and segmented images of EMDS. **a** Examples of ground truth images of EMDS. **b** Examples of semi-automatic segmented images of EMDS

by different value ranges of different features, we normalise each feature, so that its maximum and minimum values become 1 and 0, respectively.

First, we evaluate the effectiveness of the semi-segmentation method by comparing the segmented images to ground truth images. The comparison is shown in Fig. 5. Then, to evaluate the effectiveness of our ISH feature, we compare it to three other existing contour-based shape features, namely SS, FD, and ShC in EM retrieval. For a more impartial comparison, we set five different parameters to all these features where we apply 10, 25, 50, 75 and 100 sample points along the shape contours of an EM, separately. Through the retrieval results, we can conclude that ISH has the highest mAP of 36.8 % using 50 sample points along the contours of EMs. Furthermore, we show the retrieval results visually in Fig. 6. Finally, we evaluate the usefulness of ISH on segmentation results using semi-automatic segmentation method in Fig. 7. To this end we compare the retrieval results of ISH with ground truth images and segmentation results, where we use 50 sample points on the contours of EMs.

In Fig. 7, we find that ISH obtains a mAP of 33.9 % on segmentation results, which is very close to that of ground truth images of 36.8 %. This comparison shows the robustness of ISH feature for EM-CBIR task on semi-segmentation results.

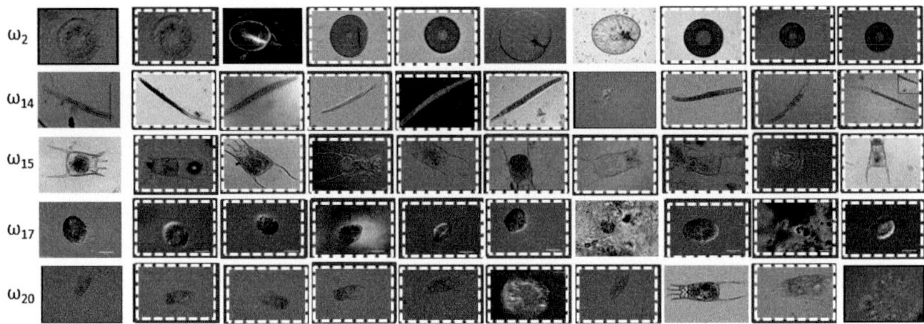

Fig. 6 Examples of EM retrieval results using ISH. The *first column* shows the query images. From the *second* to the *last columns*, the database images are sorted by their ISH similarities from high to low. The images in *white dotted line boxes* are the relevant images

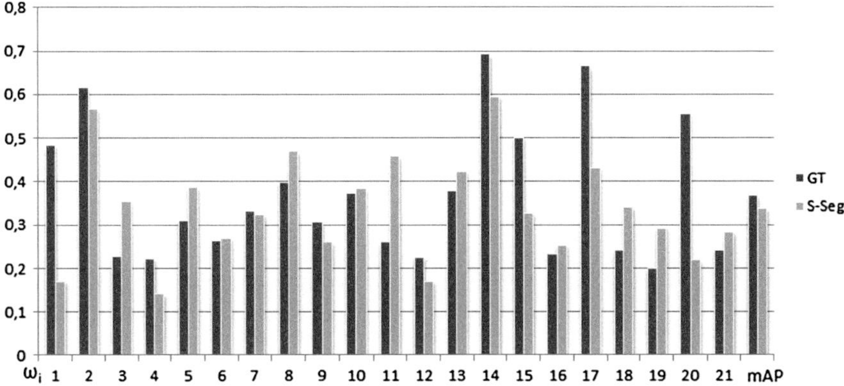

Fig. 7 Comparison of EM retrieval results using ISH on ground truth (GT) and semi-automatic segmented (S-Seg) images. $\omega_{1,\ldots,21}$ represent 21 classes of EMs. mAP is the mean AP

5 Conclusion

In this paper, we introduced an EM-CBIR system to aid the EM information searching work. Our system consists of three main phases: semi-automatic segmentation, feature extraction, as well as image retrieval. The semi-automatic segmentation based on Sobel edge detector is used to extract EM regions in noisy and low-contrast microscopic images. The characterisation of shapes in different EM regions is supported by ISH feature descriptor. In particular, we use ISH to represent the structural information based on the distribution of angles, each of which is defined by a combination of three sample points on the EM boundary. At last, we use Euclidean distance between ISHs to measure the similarity of EM images in the retrieval task. Experimental results validate the necessity and effectiveness of each of the developed methods.

Acknowledgments We thank Program of Study Abroad for Young Scholar (supported by Chengdu University of Information Technology (CUIT)), China Scholarship Council, Project (No. Y2013106) Supported by the Teaching Research Foundation of CUIT, Project (No. KYTZ201410) Supported by the Scientific Research Foundation of CUIT, and Project (No. 2015GZ0197, 2015GZ0304) Supported by Scientific Research Fund of Sichuan Provincial Science & Technology Department to support this research work. We also thank Prof. Dr. Beihai Zhou and M.Sc. Fangshu Ma from University of Science and Technology Beijing (USTB) for their great help. Also, we thank Cathrin Warnke from the University of Siegen for her proof reading.

References

1. Akakin, H., Gurcar, M.: Content-based microscopic image retrieval system for multi-image queries. Inform. Technol. Biomed. **16**(4), 758–769 (2012)
2. Belongie, S., Malik, J., Puzicha, J.: Matching shapes. In: International Conference on Computer Vision, pp. 454–461 (2001)

3. Brill, E.: Character Recognition via Fourier Descriptors. Qualitative Pattern Recognition Through Image Shaping, Los Angeles (1968)
4. Caicedo, J., Gonzalez, F., Romero, E.: Content-based histopathology image retrieval using a kernel-based semantic annotation framework. Biomed. Inform. **156**(44), 519–528 (2011)
5. Canny, J.: A computational approach to edge detection. Pattern Anal. Mach. Intell. **8**(6), 679–698 (1986)
6. Gonzalez, R.C., Woods, R.E.: Digital Image Processing, 3rd edn. Pearson International Edition, New Jersey (2008)
7. Kazakova, N., Margala, M., Durdle, N.G.: Sobel dege detection processor for a real-time volume rendering system. In: Circuits and Systems. pp. 23–26 (2004)
8. Kishida, K.: Property of Average Precision And Its Generalization: An Examination of Evaluation Indicator for Information Retrieval Experiments (2005)
9. Li, C., Shirahama, K., Grzegorzek, M.: Application of content-based image analysis to environmental microorganism classification. Biocybern. Biomed. Eng. **35**(2015), 10–21 (2015)
10. Maini, R., Sohal, J.S.: Performance evaluation of prewitt edge detector for noisy images. Graph. Vis. Image Process. **6**(3), 39–46 (2006)
11. Muller, H., Michoux, N., Bandon, D., Geissbuhler, A.: A review of content-based image retrieval systems in medical applications clinical benefits and future directions. Int. J. Med **73**(1), 1–23 (2004)
12. Neycenssac, F.: Contrast enhancement using the Laplacian-of-a-Gaussian filter. Graph. Models Image Process. **55**(6), 447–463 (1993)
13. Otsu, N.: A threshold selection method from gray-level histograms. Syst. Man Cybern. **9**(1), 62–66 (1979)
14. Pepper, I., Gerba, C., Gentry, T.: Environmental Microbiology, 3rd edn. Academic Press, San Diego (2014)
15. Roerdink, J.B.T.M., Meijster, A.: The watershed transform: definitions, algorithms and parallelization strategies. Fundam. Inform. **41**(1–2), 187–228 (2000)
16. Rui, Y., She, A., Huang, T.: A Modified Fourier Descriptor for Shape Matching in MARS. Image Databases and Multimedia Search. World Scienti1c Publishing Co, Singapore (1997)
17. Sheikh, A., H, L., Mansor, S., Fauzi, M., Anuar, F.: A content based image retrieval system for marine life images. In: International Symposium on Consumer Electronics, pp. 29–33 (2011)
18. Shyu, C., Brodley, C., Kak, A., Kosaka, A., Aisen, A., Broderick, L.: Assert: a physician-in-the-loop content-based retrieval system for HRCT image databases. Comput. Vis. Image Understand **75**(1–2), 111–132 (1999)
19. Simpson, M., Rahman, M., Phadnis, S., Apostolova, E., Demner-Fushman, D., Antani, S., Thoma, G.: Text- and content-based approaches to image modality classiffcation and retrieval for the ImageCLEF 2011 medical retrieval track. In: Information Technology in Biomedicine, pp. 758–769 (2012)
20. Zhang, D., Lu, G.: A comparative study of Fourier descriptors for shape representation and retrieval. In: Asian Conference on Computer Vision, pp. 646–651 (2002)
21. Zheng, L., Wetzel, A., Gilbertson, J., Becich, M.: Design and analysis of a content-based pathology image retrieval system. In: Information Technology in Biomedicine, pp. 249–255 (2003)

Control of a Multi-joint Hand Prosthesis—An Experimental Approach

Andrzej Wołczowski and Janusz Jakubiak

Abstract This paper presents the concept of kinematic control of prosthetic hand with 13 d.o.f., while grasping objects of different shapes and sizes. The concept refers to the process of healthy hand motion control performed by the human nervous system. Planning of grip is based on kinematic model of the hand. The parameters of subsequent phases of the gripping process were determined experimentally from the measurements carried out with a laboratory model of hand. It was assumed that the final arrangement of fingers on the workpiece is determined on the basis of information from the touch sensor system.

Keywords Hand prosthesis · Multi-joint structure control · Gripping process

1 Introduction

Contemporary prosthetic hands are controlled in bioelectric manner—by recognition of user intention, expressed by myosignals (the signals associated with the muscles activity) from the stump of amputated limb. Such control is by the nature a two-level and two-stage process. At the first stage, the decision relating the type of the desirable prosthesis movement it is elaborated, and in the second stage—the control of prosthesis mechanical structure (the motor control), performing this decision, is executed. This control structure to a small extent maps multilevel process of hand movements control performed by the human nervous system [2, 5, 6]. In the central and peripheral nervous systems of humans the flow of information is fully bidirectional: from the area of the motor cortex of the brain (where arises the motor decision), through the spinal cord and peripheral nerves, to the muscles motor end plates—by efferent

A. Wołczowski (✉) · J. Jakubiak
Faculty of Electronics Chair of Cybernetics and Robotics, Wrocław University of Technology, Wrocław, Poland
e-mail: Andrzej.Wolczowski@pwr.edu.pl

J. Jakubiak
e-mail: Janusz.Jakubiak@pwr.edu.pl

R. Burduk et al. (eds.), *Proceedings of the 9th International Conference on Computer Recognition Systems CORES 2015*, Advances in Intelligent Systems and Computing 403, DOI 10.1007/978-3-319-26227-7_52

tracts, motor commands are transferred to muscles, whereas back: from the muscle activity receptors and sensory receptors in the skin of fingers leads the afferent tracts to the area of the sensory cortex. Sensory and motor areas of the cortex are frequently linked, allowing the control flow in a closed-loop system. Also on the lower levels, information from afferent tract feedback (modifies) the efferent signals directed to the muscles, performing the reflex and stabilizing functions. This allows very precise control of the kinematic structure (musculoskeletal structure) of hand and determines its unique dexterity. In comparison to the sensory system of natural hand the prosthesis sensory system, if it exists at all, is very poor, and control decisions defining the type of desirable motion are necessarily limited to little numerous repertoire [7]. This causes that dexterity of today's active prostheses leaves much to be desired. A small step toward improving of their agility and flexibility is to provide them in touch sensors and to include this sensor information in the individual finger motor control in the process of grasping [3, 8]. The subject of the discussion of this article is the motor control capable of flexible execution of the prehensile/manipulation operations defined by movement decisions. Experimental approach has been proposed based on the determination of the successive phases of prosthesis movement operation, based on measurements carried out on a laboratory model of a prosthetic hand. It was assumed that the final posture of fingers on the held object is fixed based on the touch feedback. The organization of the paper is as follows. Section 2 presents the phases of the grasp. In Sect. 3 the hardware setup is presented followed by kinematic analysis and proposed approach to the control in Sect. 4. The preliminary test results are presented in Sect. 5 and summarized in Sect. 6.

2 Model of Grasping Process

In the description of the grasping process we denote by A_F—finger posture, F_F—finger pressure force, V_F—finger velocity, V_A—velocity of arm movement, $V(2)$—preceding stage velocity, K—human knowledge about the object being grasped (visual perception + experience), and S—the sense of grab and arm movement. In the process of grasping we can distinguish seven stages [6]:

1. rest position (starting point for the grasp preparation)—the fingers stay at rest position, and are motionless and passive ($A_F \Leftarrow K$, V_F, $F_F = 0$),
2. grasp preparation (precedes grasp closing)—the fingers are arranged depending on the shape of visually observed object and the knowledge K about the method of gripping it, with the velocity proportional to the velocity of the intended arm movement ($A_F \Leftarrow K$, $V_F \sim V_A$),
3. grasp closing (precedes holding—the fingers move with the velocity resulting from the knowledge about the object and the arm velocity during the grasp preparation stage) $(V_F \Leftarrow K) \vee (V_F \sim V_A(2))$,

4. grabbing—the fingers press on the object with a force dependent on the knowledge of the object and proportionally to the arm motion velocity in the grasp preparation phase: $(F_F \Leftarrow K) \vee (F_F \sim V_A(2))$,
5. maintaining the grasp with force adjustment—the fingers adjust the force F_F depending on the amount of squeeze and slip of the object—increase/decrease the force $(F_F \sim S)$,
6. releasing the grasp—the fingers move with a velocity dependent on the knowledge of the object behavior (e.g., small V_F force for an object with an unstable balance): $(F_F \sim K)$,
7. transition to the rest position—the fingers move with a constant velocity toward the rest position: $(V_F = \text{const})$.

Due to limitations in the exchange of information between the human nervous system and prosthesis control system (myosignals distortions and recognition process errors), the presented steps of the grasping process should be modified, replacing, as far as possible, central control by local control. Prosthesis grasping process is as follows, with additional symbols meaning A_F^0—the rest position, $y^\star$—recognized user's decision, and M—the measurement of grab parameters and arm movement.

1. rest position—fingers automatically take the rest setting A_F^0 stored in the control algorithm, remain stationary, passive $(A_F^0 \Leftarrow L,\ V_F,\ F_F = 0)$,
2. grasp preparation—the type of grasp is defined by decision $y^\star$ from the user's intent recognition level (depending on the shape of visually observed object and the knowledge K about the method of gripping it), the hand opening is max for the type of object to be gripped, the velocity proportional to the velocity of the intended arm movement $(y^\star \Leftarrow K,\ A_F \Leftarrow A_F^{\max}(L),\ V_F \Leftarrow M \vee V_F \sim V_A)$,
3. grasp closing (precedes holding—the fingers move with the velocity resulting from the knowledge about the object and the arm velocity during the grasp preparation stage) $(V_F \Leftarrow K) \vee (V_F \sim V_A(2))$,
4. grabbing—the movement of each finger member is stopped under the influence of information about touching the object to be gripped, the fingers touch/press on the object with a force established by decision $y^\star$ or proportionally to the arm motion velocity in the grasp preparation phase: $(F_F \Leftarrow K) \vee (F_F \sim V_A(2))$,
5. maintaining the grasp with force adjustment—the fingers adjust the force F_F depending on the amount of squeeze and slip of the object—increase/decrease the force $(F_F \sim M)$,
6. releasing the grasp—the fingers move with a velocity dependent on the knowledge of the object behavior (e.g., small V_F force for an object with an unstable balance): $(F_F \sim K)$,
7. transition to the rest position—the fingers move with a constant velocity toward the rest position: $(A_F,\ V_F \Leftarrow L)$.

In this paper we concentrate on rest (1) to grabbing (4) phases.

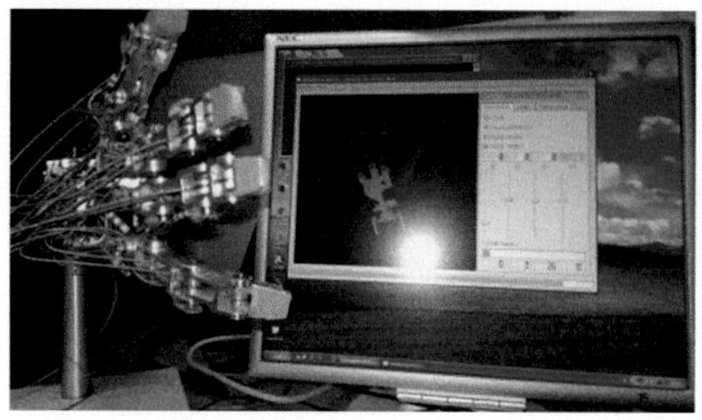

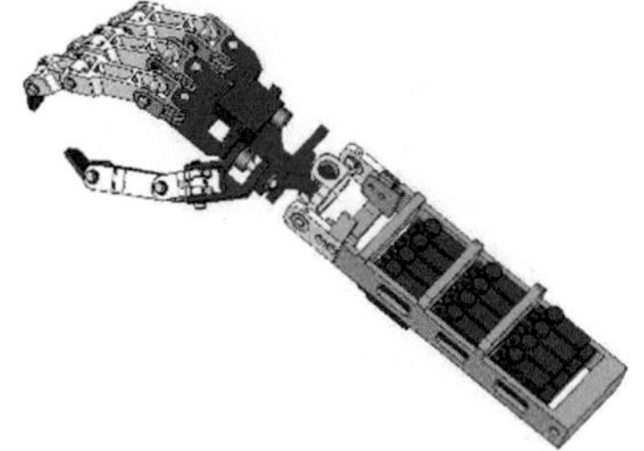

Fig. 1 Physical setup (*left*) and CAD model (*right*)

3 Setup

Laboratory model of the hand prosthesis analyzed in this paper is presented in Fig. 1. An experimental setup consists of a mechanical hand built in our department by G. Wiśniewski and a computer application to send and receive control commands and to visualize hand configuration in GUI. The mechanical hand is equipped with three fingers and a thumb. It has in total 16 rotational joints, out of which 13 are controlled independently. All three fingers are identical with 4 joints and 3 d.o.f as the last two joints are coupled. The thumb has 4 independently controlled joints. Additionally, two rotational joints are placed in a wrist. The prosthesis is actuated with 13 servo motors Futaba S3150. The one responsible for the thumb base rotation is placed in the palm and the remaining twelve in the forearm. The servo input is PWM (pulse width modulation) signal with 20 ms interval returning angle of the servo rotation as an output. The position of the shaft is transmitted to the joints with Bowden cables. The transmission ratio is 1:1, so the joint angle is proportional to the input signal. All active joints of mechanical hand are equipped with pulse encoders measuring their angular positions. Fingertips of thumb and index finger and palm are equipped with force/touch sensors detecting when a digit is in contact with a grasped object.

4 Prosthesis Control Model

4.1 Kinematics

For the purpose of describing the kinematics of the hand, we treat it as a group of four manipulators mounted in different base points located on the palm, numbered from 0 to 3 where 0 denotes the thumb. To simplify notation, we assume that the origin O is in the base of finger 1, as shown in Fig. 2. All digits are built in such a way that we can analyze them as two parts: a planar manipulator moving in plane Π_i and a rotating base which defines the orientation of the plane with respect to the

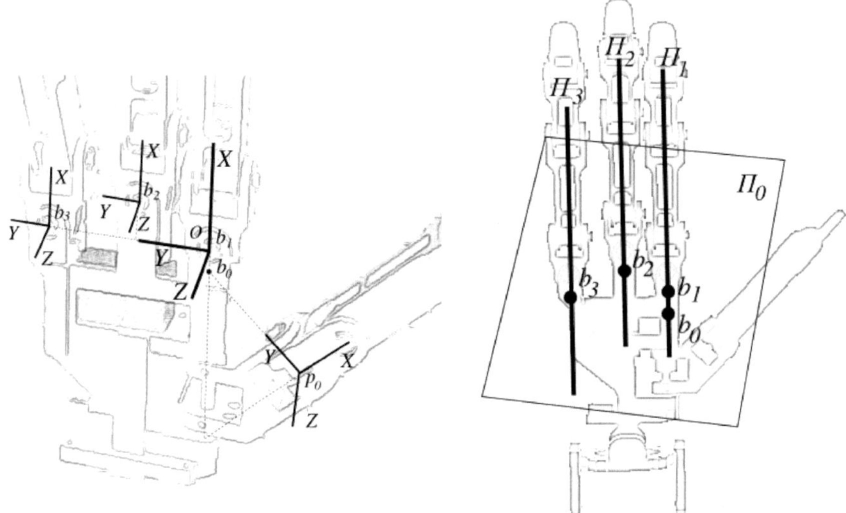

Fig. 2 Position and orientation of finger bases (*left*), finger motion planes (*right*)

palm. In the case of fingers the base rotation takes place in joint 1 and the remaining three joints move in a plane. In case of the thumb both base rotation and planar part include two joints. This observation allows us to simplify the analysis of the motion planning task by dividing them into two steps: analysis of digits' configurations in their planes and relations between the planes. A schematic plot in the palm plane is shown in Fig. 2. We denote joint angles θ_{ij}, where i is the digit number and j—joint number counted from the palm (0) to the tip (3). All joint angles form a hand state $\theta = [\theta_{ij}]$, $i, j = 0, 1, 2, 3$. It is worth noticing that all planes associated with fingers are orthogonal to the palm, while the angle between the palm and the thumb plane Π_0 depends on the configurations of joints θ_{00} and θ_{01}. Following the structure of the fingers, description of the kinematics of each digit will be divided into three parts:

- constant transformation between the origin O and the base point of the finger b_i; it is worth noting that the points b_i are invariant points of plane Π_i rotations;
- transformation in the rotational base, between finger base b_i and planar part beginning p_i;
- transformation from p_i to the tip of digit e_i.

Coordinate transformations between X and Y consist of rotation R_X^Y and translation t_X^Y and will be represented as a standard homogeneous transformation A_X^Y [4]

$$A_X^Y = \begin{bmatrix} R_X^Y & t_X^Y \\ 0 & 1 \end{bmatrix},$$

with R_X^Y being 3×3 matrix and t_X^Y—3-dimensional vector. Transformations $A_O^{b_i}$ are constant and are defined by

Table 1 Denavit–Hartenberg parameters

	fingers 1,2,3			
joint no.	θ	d	a	α
0	θ_{i0}	0	L_0	$\frac{\pi}{2}$
1	θ_{i1}	0	L_1	0
2	θ_{i2}	0	L_2	0
3	θ_{i2}	0	L_3	0

	thumb			
joint no.	θ	d	a	α
0	θ_{00}	0	0	$\frac{\pi}{4}$
1	θ_{01}	$-d_1$	L_1	$\frac{\pi}{2}$
2	θ_{02}	0	L_2	0
3	θ_{03}	0	L_3	0

$$R_O^{b_0} = \begin{bmatrix} 0 & 0 & 1 \\ 0 & -1 & 0 \\ -1 & 0 & 0 \end{bmatrix}, \qquad R_O^{b_i}|_{i=1,2,3} = I_3 \tag{1}$$

and

$$b_0 = \begin{bmatrix} -11.47 \\ 0 \\ 0 \end{bmatrix}, \quad b_1 = \begin{bmatrix} 0 \\ 0 \\ 0 \end{bmatrix}, \quad b_2 = \begin{bmatrix} 11 \\ 22 \\ 0 \end{bmatrix}, \quad b_3 = \begin{bmatrix} 0 \\ 49 \\ 0 \end{bmatrix}.$$

With the above definition of the base frames, the transformation between the base and the tip of each digit $A_{b_i}^{e_i}$ can be described with standard Denavit–Hartenberg parameters [1]. Kinematic parameters of the digits are given in Table 1. Note that for fingers the joints 2 and 3 are coupled, so the angle θ_{i2} in the left table is repeated. A transformation between the origin and p_i is given by

$$A_O^{p_i} = A_O^{b_i} A_{b_i}^{p_i}$$

and together with point b_i, it defines a plane in which the planar section of the digit i moves. For the fingers, transformations $A_{b_i}^{p_i}$ and $A_O^{p_i}$ are defined by the configuration of joint 0

$$A_{b_i}^{p_i} = \begin{bmatrix} c_{i0} & 0 & s_{i0} & L_0 c_{i0} \\ s_{i0} & 0 & c_{i0} & L_0 s_{i0} \\ 0 & 1 & 0 & 0 \\ 0 & 0 & 0 & 1 \end{bmatrix} \qquad A_O^{p_i} = \begin{bmatrix} c_{i0} & 0 & s_{i0} & L_0 c_{i0} + b_{ix} \\ s_{i0} & 0 & c_{i0} & L_0 s_{i0} + b_{iy} \\ 0 & 1 & 0 & b_{iz} \\ 0 & 0 & 0 & 1 \end{bmatrix}. \tag{2}$$

The joints 1 to 3 form a triple pendulum moving in a plane defined by point b_i and the normal vector $\hat{n}_i$ from the third column of $R_O^{p_i}$

$$p_i = b_i + \begin{bmatrix} L_0 c_{i0} \\ L_0 s_{i0} \\ 0 \end{bmatrix}, \quad \hat{n}_i = \begin{bmatrix} s_{i0} \\ c_{i0} \\ 0 \end{bmatrix}, \quad \text{for } i = 1, 2, 3. \tag{3}$$

In case of the thumb the transformation between b_0 and p_0 is obtained as a product of transformations in the first two joints and it is given by

$$R_{b_0}^{p_0} = \begin{bmatrix} c_{00}c_{01} - s_{00}s_{01}\frac{\sqrt{2}}{2} & s_{00}\frac{\sqrt{2}}{2} & c_{00}s_{01} + s_{00}c_{01}\frac{\sqrt{2}}{2} \\ s_{00}c_{01} - c_{00}s_{01}\frac{\sqrt{2}}{2} & -c_{00}\frac{\sqrt{2}}{2} & s_{00}s_{01} - c_{00}c_{01}\frac{\sqrt{2}}{2} \\ s_{01}\frac{\sqrt{2}}{2} & \frac{\sqrt{2}}{2} & c_{01}\frac{\sqrt{2}}{2} \end{bmatrix} \tag{4}$$

$$t_{b_0}^{p_0} = \left[-d_1 s_{00}\frac{\sqrt{2}}{2}, d_1 c_{00}\frac{\sqrt{2}}{2}, -d_1\frac{\sqrt{2}}{2} \right]^T$$

After transformation to the global frame we obtain a plane in which the last two segments of the thumb move

$$p_0 = t_O^{p_0} = b_0 + t_{b_0}^{p_0}, \quad \hat{n}_0 = R_O^{p_0}\begin{bmatrix} 0 \\ 0 \\ 1 \end{bmatrix} = \begin{bmatrix} c_{01}\frac{\sqrt{2}}{2} \\ -s_{00}s_{01} + c_{00}c_{01}\frac{\sqrt{2}}{2} \\ -c_{00}s_{01} - s_{00}c_{01}\frac{\sqrt{2}}{2} \end{bmatrix}. \tag{5}$$

Note that the plane of thumb planar motion is orthogonal to the palm when $c_{00}s_{01} - s_{00}c_{01}\frac{\sqrt{2}}{2} = 0$ and the planes Π_i and Π_j are parallel when $\hat{n}_i$ and $\hat{n}_j$ are.

4.2 *Control of Hand Joints*

During given grasp operation op_N joints are moved from an initial state $\theta^N(0)$ to a final configuration $\theta^N(k_N)$. Additional parameter of op_N is the grasp type which determines the planes in which the tips of digits move in the grasp closing phase. It is assumed that initial and final states and a grasp type are known and they are parameters of the grasping procedure; in the first approach, it is also assumed that orientation of the planes depends on grasp type only and not on the initial and final states.

1. Joints 0 for fingers and 0 and 1 for thumb are set to proper angles with respect to the grasp type.
2. For the joint j of the finger i we determine a pair of angles describing its initial and final configurations $(\theta_{ij}(s_0), \theta_{ij}(s_N))$.
3. For each joint we determine desired displacement $\theta_{ij}(s_N) - \theta_{ij}(s_0)$.
4. We assume number of steps k to move to the requested configuration.
5. As a result we receive single-step change required in each joint in every

$$\Delta_{ij} = \frac{\theta_{ij}(s_N) - \theta_{ij}(s_0)}{k}$$

6. The single-state goals are then sent to the joints to execute motion until one of the conditions is met: the goal angle is reached or the touch sensor detects contact with the object as presented in Fig. 3. The step is repeated until all digits are in contact with the object or the minimum angle was reached.

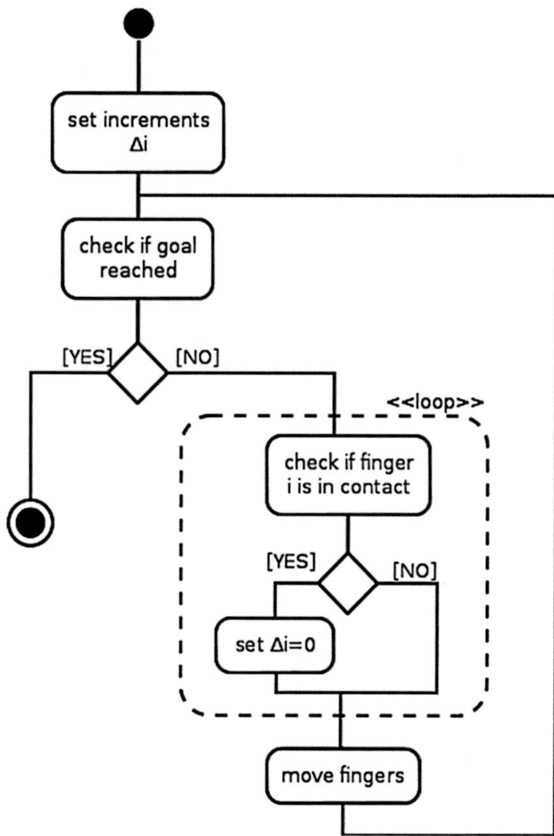

Fig. 3 Control scheme

5 Experiments

The experiments were divided into two parts. In the first one, the joint angles were measured in rest and grasp positions, for various grasp types and objects of various sizes. Tests covered cylindrical, hook, tip, and lateral grasps. During the motion operation, the initial configuration $\theta(0)$ and the final configuration $\theta(k_N)$ were defined by the sets of angles in finger joints. The first set is defined for the rest configuration, and the latter for the destination grasp while handling the object. In the experiments different types of object were handled and below we present two examples for different grasp types. The first example is a cylindrical grasp of a glass; the second is a tip grasp of a ball. The angles in the hand's rest position and grasps were collected in user-supervised procedure. For chosen objects, the user was properly placing the fingers to obtain stable grasp. Each finger was configured separately and its orientation angle was manually tuned. Next, the obtained angles for an object were recorded as the grasp patterns for those objects. The results for examplary grasps are collected in Table 2. Based on the predefined initial and final states, the single-step change was

Table 2 Joint angles in rest and grasp positions

	Fingers											
	Rest				Cylindrical				Tip			
Joint no.	0	1	2	3	0	1	2	3	0	1	2	3
1	0	13	0	0	77	11	5	0	81	16	0	0
2	0	0	0	0	1	21	39	44	2	24	0	0
3	0	0	0	26	−9	58	64	55	−4	58	0	0
4	4	0	0	26	19	58	64	55	48	58	0	0

Table 3 Planned single-step angle change for $k = 10$ and measured average velocity

	Planned increments °/step						Measured avg. velocity $^\circ/s$					
	Cylindrical				Tip		Cylindrical				Tip	
Joint no.	0	1	2	3	0	1	0	1	2	3	0	1
1	7.7	−0.2	0.5	0	8.1	0.3	37	−1	2	0	44	2
2	0.1	2.1	3.9	4.4	0.2	2.4	0.5	10	19	21	1	13
3	−0.9	5.8	6.4	2.9	−0.4	5.8	−4	28	31	14	−2	32
4	1.5	5.8	6.4	2.9	4.4	5.8	7	28	31	14	24	32

determined with an assumption that the transfer from the rest configuration to the grasp is done in 2 s. The maximum velocity of joints depends on finger and hand configurations. We determined empirically safe value for planning as $45^\circ/s$. This results in dividing the task into $k = 10$ steps. The results of planning are presented in Table 3 to the left. To the right we show the results of exemplary experiments for each grasp. The motion time was 2.07 s for the cylindrical and 1.83 s for the tip grasp. In the table we place the average velocities of the motion. It can be noted that the time of real motion differs slightly from the planned and may be either shorter or longer than planned 2 s. Configuration of the fingers during cylindrical and tip grasps for virtual model and real hand are presented in Figs. 4 and 5.

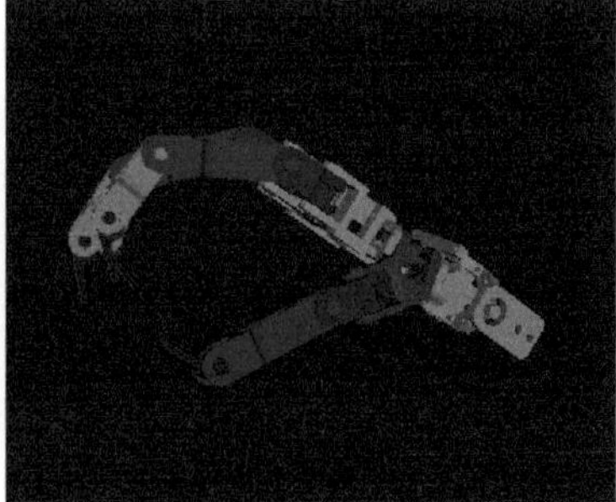

Fig. 4 A cylindrical grasp—virtual model and realization

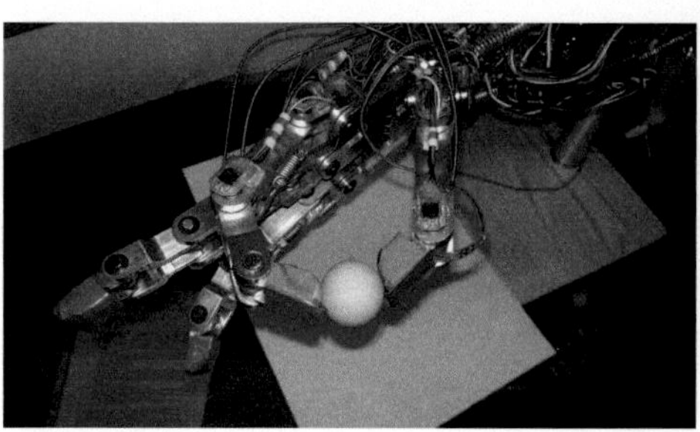
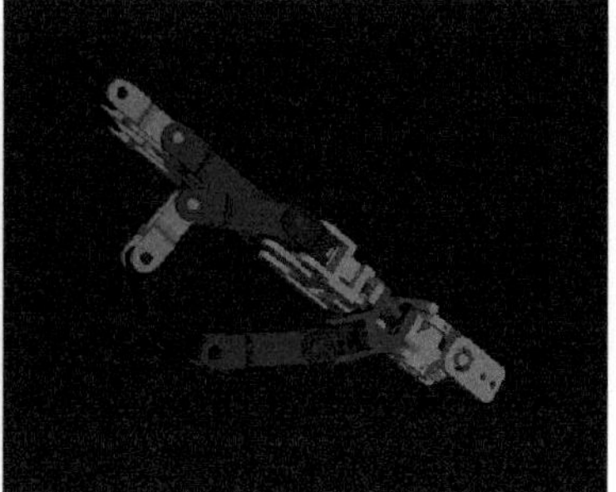

Fig. 5 A tip grasp—virtual model and realization

6 Conclusions

Experiments have shown that with the proposed method it is possible to achieve sure grasp and handling of objects with different shapes and dimensions. The objects used in the experiments were rigid and light. Apart from the grasp envelope and velocity, also the applied force influences grip safety. In the experiments touch and handling forces were set to the level which ensures nonslip grip of the handled object. The proposed algorithm allows defining maximum value of grasp force. When detected force is bigger than predefined maximum, the closing motion of hand fingers is stopped. In future, it is possible to define the higher level layer which would allow passing to the phase of maintaining the grasp. In this phase the force limit could be increased on demand to correct the grasp with a visual supervision of the user and disallow object to slip out. Non-slippery function can also be automated by additional slip sensors, which will be the subject of further research.

Acknowledgments The work of the first author was supported by National Science Center resources in 2012–2014 years as a research project No. ST6/06168. The work of the second author was supported by statutory grant No. S40142. Computational resources were provided by PL-Grid Infrastructure.

References

1. Denavit, J., Hartenberg, R.S.: A kinematic notation for lower-pair mechanisms based on matrices. Trans. ASME J. Appl. Mech. **22**, 215–221 (1955)
2. Iberall, A.: A neural model of human prehension. Technical report, University of Massachusetts (1987)
3. Kurzynski, M., Wolczowski, A.: Control system of bioprosthetic hand based on advanced analysis of biosignals and feedback from the prosthesis sensors. In: Pietka, E., Kawa, J. (eds.) Information Technologies in Biomedicine. LNCS, vol. 7339, pp. 199–208. Springer, Berlin (2012)

4. Sciavicco, L., Siciliano, B.: Modelling and Control of Robot Manipulators. The McGraw-Hill, New York (1996)
5. Taylor, C., Schwarz, R.: The anatomy and mechanics of the human hand. Artif. Limbs **2**, 22–35 (1955)
6. Wołczowski, A.: Smart hand: the concept of sensor based control. In: Proceedings of MMAR. pp. 783–790. Miedzyzdroje (2001)
7. Wolczowski, A., Kurzynski, M.: Human-machine interface in bioprosthesis control using EMG signal classification. Expert Syst. **27**(1), 53–70 (2010)
8. Wolczowski, A., Kurzynski, M., Zaplotny, P.: Concept of a system for training of bioprosthetic hand control in one side handless humans using virtual reality and visual and sensory biofeedback. J. Med. Inform. Technol. **18**, 85–91 (2011)

Hilbert–Huang Transform Applied to the Recognition of Multimodal Biosignals in the Control of Bioprosthetic Hand

Edward Puchala, Maciej Krysmann and Marek Kurzyński

Abstract This paper deals with the problem of bioprosthetic hand control via recognition of user intent on the basis of electromyography (EMG) and mechanomyography (MMG) signals acquired from the surface of a forearm. As a method of signal parameterization the Hilbert–Huang (HH) transform is applied which is an effective tool for reduction of feature space dimension. The performance of proposed recognition method based on HH transform of EMG and MMG signals was experimentally compared against an autoregressive model of dimensionality reduction using real data concerning the recognition of five types of grasping movements. The experimental results show that the HH transform approach with root mean square of amplitude feature outperforms an autoregressive method.

1 Introduction

The activity of human organisms is accompanied by physical quantities's variation which can be measured and monitored with appropriate instruments and applied to control the work of technical devices. This allows for new possibilities of using these devices by enabling a close integration of a machine and a living organism into one being. The common practice is to make use of myopotentials (called EMG signals) which are electrical signals that accompany the activity of muscles. EMG signals measured on skin are the superposition of electrical potentials generated by recruited motor units of contracting muscles. Various movements are related to the recruitment of distinct motor units, and different spatial locations of these units in

E. Puchala (✉) · M. Krysmann · M. Kurzyński
Department of Systems and Computer Networks, Wroclaw University of Technology, Wyb. Wyspianskiego 27, 50-370 Wroclaw, Poland
e-mail: edward.puchala@pwr.edu.pl

M. Krysmann
e-mail: maciej.krysmann@pwr.edu.pl

M. Kurzyński
e-mail: marek.kurzynski@pwr.edu.pl

R. Burduk et al. (eds.), *Proceedings of the 9th International Conference on Computer Recognition Systems CORES 2015*, Advances in Intelligent Systems and Computing 403, DOI 10.1007/978-3-319-26227-7_53

relation to the measuring points lead to the formation of EMG signals of differing features, e.g., with different *rms* values and different frequency spectra. The features depend on the type of executed or (in the case of an amputated limb) only imagined movement so they provide the information about the users' intention [1–3]. Bioprostheses can utilize the biosignals measured on the handicapped person's body (on the stump of a hand or a leg) to recognize a prosthesis user's intention and to control the actuators of artificial hand's fingers. Nevertheless, reliable recognition of intended movement control based on EMG signal analysis is a hard problem. The efficiency of the realized control process depends on the quality of signal representation (i.e., feature vectors) subjected to recognition, i.e., quality of biosignal acquisition and analysis process, but also on the assumed number of recognized motion classes (hand gestures) and required level of recognition reliability. The difficulty increases along with the cardinality of prosthesis movement repertoire (e.g., with prosthesis dexterity). Considering this fact, there is still a need for research in developing biosignal recognition methods that could increase the reliability of recognition of handicapped person movement decisions. According to the authors' earlier experience, increase in the efficiency of the recognition stage can be achieved through the following actions [4, 5]:

1. by introducing the concept of simultaneous analysis of different types of biosignals—authors studied the fusion of EMG signals and mechanomyography signals (MMG signals) [6];
2. by choosing the proper signal parameterization method and applying the best method to reduce the dimensionality of the feature vector;
3. through improving the recognition method.

The bioprosthesis control system developed in this study includes the abovementioned ideas. It uses both EMG and MMG signals (according to proposition (1)) which are the carrier of information about the performed hand movement. To recognize the intended hand movements (according to proposition (2)), the signal parametrization method called Hilbert–Huang (HH) [7] transform is applied. The HH transform is an effective tool for the reduction of dimension of feature space which has demonstrated its usefulness in many practical pattern recognition problems with objects described by signals, especially in different medical applications [8]. The paper presents the concept of a bioprosthesis control system which in principle consists in the recognition of a prosthesis user's intention (i.e., patient's intention) based on adequately selected parameters of EMG signal and then on the realization of the control procedure which had previously been unambiguously determined by a recognized state. The performance of proposed recognition method based on HH transform of EMG and MMG signals was experimentally compared against other method of dimensionality reduction—autoregressive model (AR) using real data concerning the recognition of five types of grasping movements. The paper arrangement is as follows. Section 2 includes the concept of prosthesis control system based on the recognition of "patient intention" and provides an insight into the nature of EMG and MMG signals, which are the source of information exploited in the recognition and control procedure. In Sect. 3 the Hilbert–Huang transformation, i.e., the key procedure for the proposed recognition method, is presented in detail. The experiments conducted and the results with discussion are presented in Sect. 4. Section 5 contains conclusions.

2 Control of Bioprosthetic Hand at the Decision Level

The bioprosthesis control based on classification of EMG and MMG signals requires the development of three stages [5]:

1. acquisition of signals;
2. reduction of dimensionality of their representation;
3. classification of signals.

The acquisition must take into account the nature of the measured signals and their measurement conditions. A quality of obtaining information depends essentially on the ratio of the measured signal power to the interfering signal power, defined as SNR (Signal-to-noise ratio). For the noninvasive methods of measurements carried out on the surface of the patient's body, to obtain a satisfactory SNR is a difficult issue [3]. Usually, the noise amplitude exceeds many times the amplitude of the measured signal. For example, for electrical signals (which include EMG signals), the amplitude of voltages induced on the patient body as a result of the influence of external electric fields may exceed more than 1000 times, the value of useful signals. This induces the need for careful design of measurement channels for different modalities, including sophisticated circuits and high-quality components. The MMG signals are mechanical vibrations propagating in the limb tissue as the muscle contracts. They have low frequency (up to 200 Hz) and small amplitude and can be registered as a "muscle sound" on the surface of the skin using microphones [9]. This sound carries essential information about individual muscle group excitation. In the case of MMG signals the basic problem is to isolate the microphone sensor from the external sound sources along with the best acquisition of the sound propagating in the patient's tissue. After the acquisition stage, the recorded signals have the form of strings of discrete samples. Their size is the product of measurement time and sampling frequency. For a typical motion, this gives a record of size between 3 and 5 thousand of samples (time of the order of 3–5 s, and the sampling of the order of 1 kHz). This "primary" representation of the signals hinders the effective classification and requires the reduction of dimensionality. This reduction leads to a representation in the form of a signal feature vector. The reduction of dimension of feature space is an essential problem in a pattern recognition because it intends to speed up the classification process by keeping the most important class-relevant features. In general, a dimension reduction can be defined as a transformation from original high-dimensional space to low-dimensional space where an accurate classifier can be constructed. There are two main methods of dimensionality reduction [10]: feature selection in which we select the best possible subset of input features and feature extraction consisting in finding a mapping to a lower dimensional space. In this study as a feature extraction algorithm the Hilbert–Huang transform is proposed [11]. The method applied is presented in the next section in detail.

3 Hilbert–Huang Transform

As it is known, one of the three processes in the bioprosthesis control based on classification of EMG and MMG signals is reduction of dimensionality of the feature vector. Reduction of dimensionality process is necessary to increase the speed of the classification algorithms. In this chapter, Hilbert–Huang transform for EMG and MMG signals will be presented. Hilbert–Huang transform consists of two fundamental processes:

1. Method of empirical mode decomposition—EMD,
2. Hilbert spectral analysis—HSA.

An empirical mode decomposition is a method with which any EMG and MMG signal sets are decomposed into finite number of intrinsic mode functions (IMF). Intrinsic mode functions should have two properties:

1. The total number of maxima and minima must be equal or differ by one from the number of zero crossing,
2. The mean value of two envelopes (defined by the local maxima and local minima) at any point of IMF should be equal to zero.

For the feature extraction process for EMG and MMG signals, the so-called fission approach (presented below) will be applied. In this case the feature vectors of EMG and MMG signals consist of the features from each intrinsic mode function. Fission approach for EMG/MMG signals is as follows:

1. Determine all maxima and minima of $EMG/MMG(t)$,
2. Using cubic spline curve connect all maxima to obtain the upper envelopes $UE(t)$,
3. Using cubic spline curve connect all minima to obtain the lower envelopes $LE(t)$,
4. Determine average signal $m_1(t)$ (Fig. 1):

$$m_1(t) = \frac{UE(t) + LE(t)}{2}, \tag{1}$$

5. Compute

$$h_1(t) = EMG/MMG(t) - m_1(t), \tag{2}$$

6. Iterate steps 1–5 for obtaining

$$h_{1,1}(t) = h_1(t) - m_{1,1}(t), \tag{3}$$

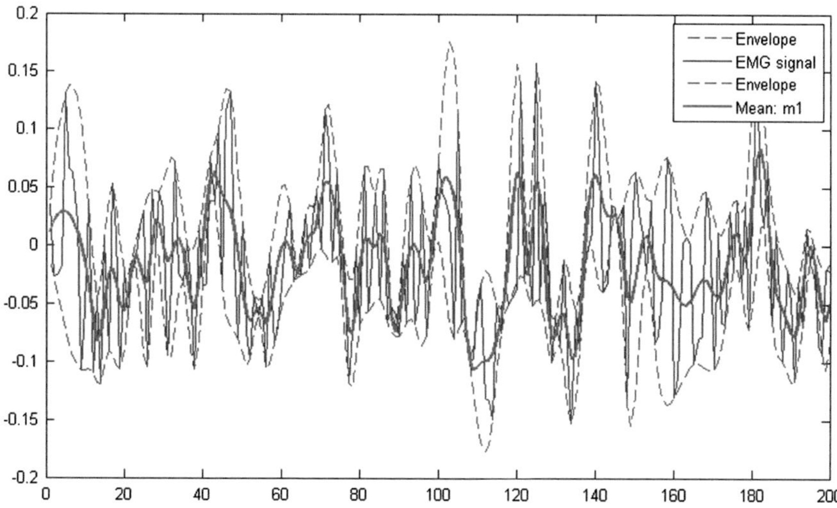

Fig. 1 EMG signal with calculated envelopes (signal duration 0.1 s)

where $m_{1,1}(t)$ denotes average signal for upper and lower envelopes of $h_1(t)$,

7. Iterate 1–5 (according to the *point 6*) until

$$h_{1,k}(t) = h_{1,k-1}(t) - m_{1,k}(t) \tag{4}$$

then, first intrinsic mode function will be defined as

$$h_{1,k}(t) = IMF_1(t) \tag{5}$$

The formula in *point 7* defines stopping criteria. Another form of stopping criteria is presented below:

$$\frac{\sum_{t=0}^{T} |(h_{1,k-1}(t) - h_{1,k}(t)) \times (h_{j,k-1}(t) - h_{1,k}(t))|}{\sum_{t=0}^{T} [h_{1,k}(t) \times h_{1,k}(t)]} < \varepsilon \tag{6}$$

where ε is a stopping parameter,

8. Compute the residual $r_1(t)$:

$$r_1(t) = EMG/MMG(t) - IMF_1(t) \tag{7}$$

and repeat whole procedure for $r_1(t)$,

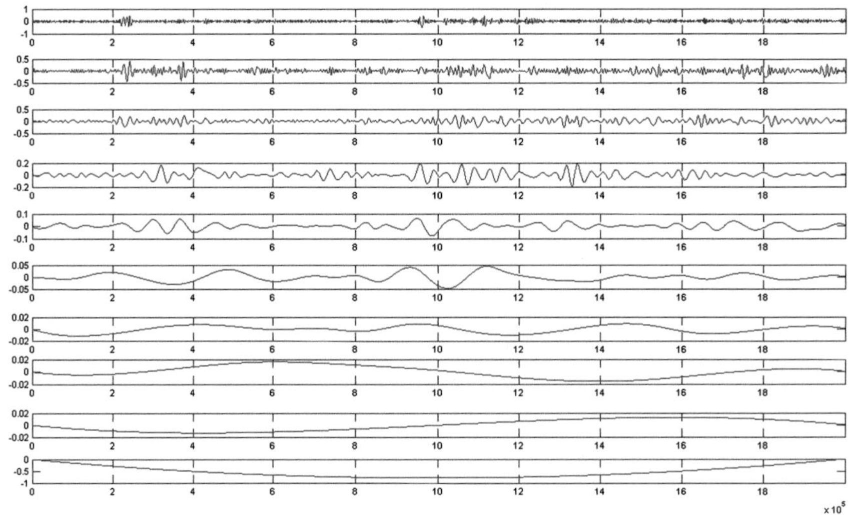

Fig. 2 EMG signal decomposition (2 s of signal)

9. Iterate points 1–8 until $r_k(t)$ will be a monotonic function and

   - the total number of maxima and minima must be equal or differ by one from the number of zero crossing,
   - the mean value of two envelopes (defined by the local maxima and local minima) at any point of IMF should be equal to zero (Fig. 2).

The last function $r_k(t)$ is the component with a smallest frequency (trend). Completely decomposed $EMG/MMG(t)$ signals($D(t)$) can be represented as the sum of all $IMF_j(t)$ and residuals $R(t)$:

$$EMG/MMG(t) \sim D(t) = R(t) + \sum_{j=1}^{n} IMF_j(t). \tag{8}$$

After completing the fission procedure, for all $IMF_j(t)$ Hilbert transform $HT[IMF_j(t)]$ should be determined. Using $HT[IMF_j(t)]$ all intrinsic mode functions are represented as an analytic signal:

$$z_j(t) = IMF_j(t) + i \times H[IMF_j(t)] = A_j(t) \times e^{-i\theta_j}(t). \tag{9}$$

To obtain information about instantaneous frequency for each $INF_j(t)$ we calculate

$$fIMF_j(t) = \frac{1}{2\pi} \times \frac{d\theta_j(t)}{dt}, \tag{10}$$

which allows determining the frequency features of INF. On the basis of information contained in [12] we propose the following components of the EMG and MMG signals' feature vectors:

- maximum amplitudes of analytic signals $z_j(t)$for each IMF—$AMAX[z_j(t)]$,
- root mean square of amplitudes for each IMF—$ARMS[z_j(t)]$,
- mean instantaneous frequency for each IMF—$MIF[z_j(t)]$.

Let us assume that the number of all IMFs for each of the EMG and MMG signals is not greater than n_{max}. Hence, it follows that the feature vectors for EMG/MMG signals have $(3 \times n_{max})$ components.

4 Experimental Investigations

The experimental investigation was divided into two parts. First was covering aspect of testing if all literature features need to be calculated from each IMF to provide best classification accuracy in case where there are 16 signals used simultaneously. The second one was performed in order to check if all IMFs carry valuable, for classification, information. In order to maintain good comparison conditions there were no signal selection methods used; all 16 channels (8 EMG and 8 MMG) were utilized to perform HH transform or AR feature extraction. For the same reason there were not used any feature selection/reduction methods.

4.1 Data Acquisition

Data was gathered using authors' acquisition system, which allows to perform simultaneous recording of EMG and MMG signals. Eight bimodal electrodes were placed

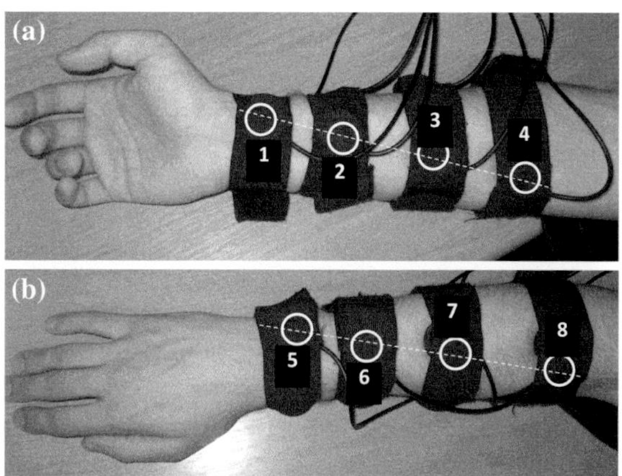

Fig. 3 Electrodes placement

as in Fig. 3. Each signal was recorded with 1000 Hz sampling frequency. Experiments were performed with 5-class problem of real-life palm grasps. There had been gathered in one session 100 repetitions of each class, which gave 500 objects. Then after one week the session was repeated in order to preserve real-life differences in muscle tiredness level. Therefore, the full database contains 1000 objects, with equal number of objects per class (200). Each object was described with 16-channel 2-second raw signals during which the gesture was made. To preserve real-life variety there were no signal trimmings were performed, so beginning of grasp could be a bit shifted in time.

4.2 Hilbert–Huang Transform

HH transform was evaluated in two ways—first to find best feature to be calculated from each IMF and the second to find best threshold value of IMFs used in feature extraction.

4.2.1 Features Evaluation

Use of all features proposed in literature could lead to very long feature vectors, which leads in many cases to a significant drop in classification accuracy. The literature proposes the use of three features per each IMF, which makes in our research vectors of minimum length of 288 features (using three features calculated from each of six IMFs per each of 16 signal channels). Therefore, each feature was evaluated separately and compared with their combinations.

4.2.2 Threshold of IMFs Number

In multisignal recognition problem, where there are several signals describing the same object, the final feature vectors tend to be very long which lead to problems in classification and requirement of more complex classifier model use. Therefore, the approach which reduces number of IMFs used for each signal was evaluated. In order to make classification possible there was a need of minimal number of IMFs per signal (across the learning set) determination. It is indispensable due to HH transform way of working. For different time objects there are different numbers of IMFs acquired and would lead to different numbers of features per signal, which would make classification problem impossible to perform. However, even the minimal number of IMFs is still around 7 (depending on a channel) in our EMG/MMG signals, which would lead to (taking into account 3 features calculated from each IMF) final feature vectors of length around 340. Therefore, investigation of implementing fixed number of IMFs used in threshold had been performed.

4.3 Autoregressive Model

Evaluated approach should be compared with the well known in literature approach. Therefore, autoregressive method was chosen, because it was also deeply evaluated in authors' earlier research [5]. The AR model is one of the linear prediction methods that predict a value y_n of a time series of data $\{y_n\}$ based on the previous values $(y_{n-1}, y_{n-2}, \ldots\ldots)$. Burg algorithm was chosen because of its many advantages [13], so it will be a good approach to compete with such method. Autoregression order, in all experiments, was set to match number of features with number of features generated by evaluated HH transform approach. It was made to have best setup for fair comparison.

4.4 Results and Discussion

Results for the comparison, using classification accuracy, were obtained from 5-nearest neighbor classifier. This classification model was used in order to provide fast classification and reduce classifier parameters' tuning bias. Simple classifier should also give better indication about feature extraction methods' differences. There had been performed 10×2-fold cross-validation approach evaluated with F test (significance level test at $\alpha = 0.1$). All differences between methods are statistically significant. Results from feature selection experiments are presented in Table 1. Classification accuracy is evaluated for the following: all 3 features; maximum amplitudes feature (AMAX); root mean square of amplitudes (ARMS); mean instantaneous frequency (MIF); maximum amplitude and least-squares amplitude features (feat. 1&2); and autoregressive features (AR). Results are divided into two groups depending on which acquisition trial they are gathered from and also they are presented combined database results (All). It can be easily seen that least-squares feature calculation method significantly outperforms all other, even autoregressive model. Moreover, this method presents invariance to differences between two acquisition sessions, which is very important in real-life use. Table 2 presents results from IMFs number threshold evaluation (Threshold = ARorder) using only root mean square of amplitudes. In other words it shows the difference in quality, of features in corresponding amount of them, between HH transform extraction and autoregressive method. It is obvious that for all cases HH transform extraction method completely

Table 1 Classification accuracy depending on used feature (in %), IMFs threshold = 7

Data	3 features	AMAX	ARMS	MIF	feat. 1&2	AR
All	36.00	89.39	**95.16**	41.58	89.85	89.58
1. 500	41.18	95.31	**98.14**	47.63	95.63	94.81
2. 500	34.29	84.31	**91.98**	37.82	84.81	83.71

Table 2 Classification accuracy depending on threshold of IMFs number

Threshold	#Features	HH transform	AR	Threshold	#Features	HH transform	AR
2	32	91.8	84.1	5	80	95.9	89.3
3	48	94.2	86.7	6	96	95.6	89.8
4	64	95.4	89.0				

outperforms AR method. Moreover, even calculation of features on base of two IMFs gives result over 91 % correct classification and the best one is for utilization of 5 IMFs which gave result almost 96 %. The HH transform approach with root mean square of amplitudes feature outperforms an autoregressive method, and moreover it is less variant to the differences between two data gathering sessions.

5 Conclusion

In the paper, we presented the concept of simultaneous analysis of two types of biosignals—EMG signals and the mechanomyography (MMG) signals. Analysis refers to the biosignal recognition processes in the control of a bioprosthetic hand. To recognize the intended hand movements the signal parametrization method called Hilbert–Huang (HH) transform was applied. The HH transform has been used as an effective tool for reduction of dimension of feature space in signal recognition tasks. The performance of proposed recognition method based on HH transform of EMG and MMG signals was experimentally compared against autoregressive method of dimensionality reduction using real data concerning the recognition of five types of grasping movements. The experimental results show that the HH transform approach with root mean square of amplitudes feature outperforms an autoregressive method.

Acknowledgments This work was financed from the National Science Center resources in 2012–2014 years as a research project No ST6/06168 and supported by the statutory funds of the Department of Systems and Computer Networks, Wroclaw University of Technology.

References

1. Englehart, K., Hudgins, B.: A robust, real-time control scheme for multifunction myoelectric control. IEEE Trans. Biomed. Eng. **50**, 848–854 (2003)
2. Wolczowski A., Kurzynski M.: Control of Artificial Hand Via Recognition of EMG Signals. Lecture Notes in Computer Science, vol. 3337, pp. 356–364. Springer, Berlin (2004)
3. Wolczowski A, Myslinski S.: Identifying the relation between finger motion and EMG signals for bioprosthesis control. In: IEEE International Conference on Methods and Models in Automation and Robotics, Miedzyzdroje, pp. 127–137 (2006)

4. Kurzynski M., Wolczowski A.: Control system of bioprosthetic hand based on advanced analysis of biosignals and feedback from the prosthesis sensors. In: Proceedings of the Third International Conference on Information Technologies in Biomedicine, pp. 199–208. Springer, Berlin (2012)
5. Kurzynski, M., Krysmann, M., Trajdos, P., Wolczowski, A.: Multiclassifier system with hybrid learning applied to the control of bioprosthetic hand. Computers in Biology and Medicine (under review)
6. Kurzynski, M., Woloszynski, A.: Multiple classifier system applied to the control of bioprosthetic hand based on recognition of multimodal biosignals. IFMBE Proc. **43**, 577–580 (2014)
7. Barnhart, B.L.: The Hilbert Huang transform: theory,applications, development University of Iowa, available at Iowa Research Online. http://ir.uiowa.edu/etd/2670
8. Huang, N.E, Shen, Z., Long, S.R., Wu, M.C., Shih, H.H., Zheng, Q., Yen, N. C., Tung, C. C., Liu, H.H.: The empirical mode decomposition and the hilbert spectrum for nonlinear and non stationary time series analysis. In: Proceedings of the Royal Society A: Mathematical, Physical and Engineering Sciences, vol. 454, no. 1971, pp. 903–995 (1998)
9. Malek, M., Coburn, J.: The utility of electromyography and mechanomyography for assesing neuromuscular functions, a noninvasive approach. Phys. Med. Rehabil. Clin. N. Am. **23**, 23–32 (2012)
10. Duda, R., Hart, P., Stork, D.: Pattern Classification. Wiley-Interscience, New York (2001)
11. Li, H., Yang, L., Huang, D.: Application of hilbert huang transform to heart rate variability analysis. In: The 2nd International Conference on Bioinformatics and Biomedical Engineering, pp. 648–651 (2008)
12. Zong, C., Chetouani, M.: Hilbert-Huang transform based physiological signals analysis for emotion recognition. In: IEEE International Symposium on Signal Processing and Information Technology (ISSPIT) (2009)
13. Schlögl, A.: A comparison of multivariate autoregressive estimators. In: Signal Processing 86, Special Section: Signal Processing in UWB Communications (2006)

Wavelet Analysis of Cell Nuclei from the Papanicolaou Smears Using Standard Deviation Ratios

Dorota Oszutowska-Mazurek, Przemysław Mazurek, Kinga Sycz and Grażyna Waker-Wójciuk

Abstract Two techniques of the image analysis of Papanicolaou stains are compared in this paper—standard deviation and standard deviation ratio for cell nuclei. The image analysis is based on diagonal details obtained from multiresolutional analysis using wavelets. Two best wavelets are presented 'coif2' and 'sym1.' Results show the importance of standard deviation ratios and smallest diagonal details for classification of cell together with cell nucleus area. Classification of cells allows rapid discrimination of cells for further analysis of them by cytoscreener.

Keywords Wavelets · Image analysis · Papanicolaou smears · Cytology

1 Introduction

Digital image analysis of cervical cytology [1] allows the detection of precancerous and cancer condition. The analysis of cervical cytology digital images could be applied in screening of precancerous and cancerous conditions. Features of atypical cells are observed in cell nuclei. Images obtained from microscope with digital camera may be processed for the purpose of computer-assisted diagnosis, although biological objects like cervical cells are considered as very complex and the large

D. Oszutowska-Mazurek
Higher School of Technology and Economics in Szczecin, Faculty of Motor Transport, Klonowica 14 St., 71244 Szczecin, Poland
e-mail: adorotta@op.pl

P. Mazurek (✉)
Department of Signal Processing and Multimedia Engineering, West-Pomeranian University of Technology, Szczecin, 26. Kwietnia 10 St., 71126 Szczecin, Poland
e-mail: przemyslaw.mazurek@zut.edu.pl

K. Sycz · G. Waker-Wójciuk
Independent Public Voivodeship United Hospital, Department of Pathomorphology, Arkońska 4 St., 71455 Szczecin, Poland

G. Waker-Wójciuk
e-mail: grazynka@blue.net.pl

R. Burduk et al. (eds.), *Proceedings of the 9th International Conference on Computer Recognition Systems CORES 2015*, Advances in Intelligent Systems and Computing 403, DOI 10.1007/978-3-319-26227-7_54

number of cells is required for the reliability of results. Computer-aided diagnosis could simplify the classification task for cytoscreener and pathomorphologist via detection of cells for further analysis. Moreover, computer-aided diagnosis could improve detection ratio and reduce screening time. Advanced scanners of microscopic slides allow the acquisition of complete slides with high resolution and multiple settings of deep of field that is important for high contrast image recording. The application of software tools for the discrimination of the objects of interest should allow fast detection of atypical cells. The main problem is the very complex content of image for conventional Papanicolaou-stained slides [3, 4, 8]. The Papanicolaou (Pap) process allows chemical segmentation of objects and adds color to some cellular structures that are important for cytoscreener. There are also liquid-based processes that use centrifuge for additional removal (mechanical) artifacts and such processes are very often used together with computer-aided diagnosis tools. Liquid-based process removes many features, which is good for simplification of algorithm, but removes some important features for general diagnosis also. The detection of atypical cells is important, but very often another kinds of atypia (like detection of viruses) could be possible using conventional Papanicolaou process. The objects of interest (cells with cells' nuclei) could be assigned to a few classes, but in this work two general classes are applied: atypical and correct due to available image database. Example images of well-separated cells are shown in Fig. 1. One

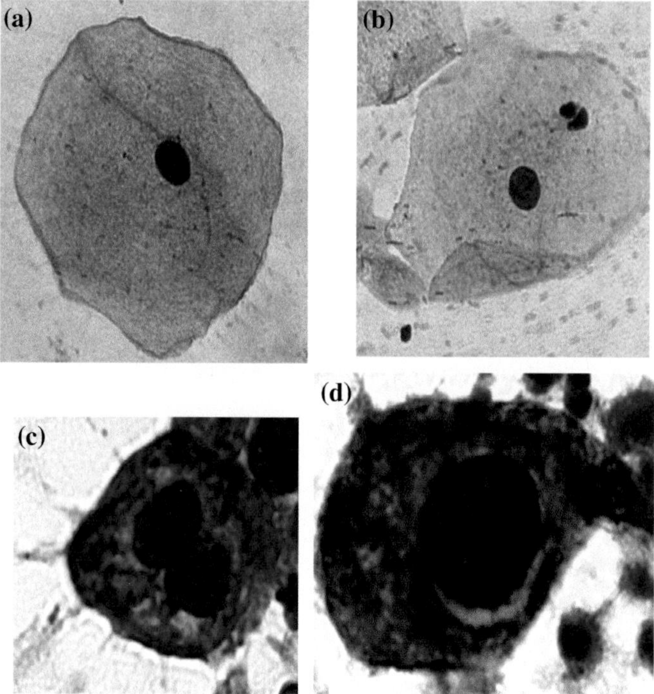

Fig. 1 Example images of cell nuclei (**a**, **b** correct; **c**, **d** atypical)

of the most important problems of databases is the resolution of images. In some researches small-resolution cameras and 100× magnification are used. Such conditions of image acquisitions are not correct for the analysis of cell nucleus details. The contour and texture analysis requires few megapixel camera and 400× magnification for example. In this work Zeiss AxioCam 5 camera (5 Mpix) was used for the database preparation.

1.1 Related Works

The proper selection of image estimators allows the classification of cells and there are two main classes: contour and texture descriptors [22]. Contour analysis of considered cells is adequate, because such contour is very often fuzzy. Texture descriptors [13] are more important and there are numerous descriptors proposed for such cells. Multiresolutional descriptors are very important because texture of cell nuclei has self-similarity properties [19]. Fractal-based descriptors are applied for numerous types of cell nuclei [1, 5, 11]. The thresholded box-counting [15], triangular prism method (TPM) [16], Tiled TPM [14], variogram [17], lacunarity [18], as well as area–perimeter method [10] are proposed for the fractal-based analysis of cervical cells. Estimators for textures of cell nuclei require support of irregular objects. The most sophisticated problem is the segmentation of cells and cell nuclei, and many works are related to this topic [6, 7, 9, 12]. The approach that is used in this paper assumes availability of the segmentation algorithm that is nonideal. The required texture descriptor should work with complete cell nucleus as well as a inner part of cell nucleus. Such descriptor should be robust to the contour segmentation. Wavelet-based estimators are also proposed for analysis of cell nuclei. In [20] is proposed AMSGLCM–Adaptive multiscale GLCM (Gray-level co-occurence matrix). In [2] 'haar' wavelet is proposed for cell nuclei and surrounding area. Multispectral images are analyzed in [21] using 'daub2,' 'daub16,' 'bior2.2,' and Gabor transform.

1.2 Content and Contribution of the Paper

Two techniques of analysis are compared in this paper. The standard deviation is computed for selected detail as reference technique. The estimator based on the ratio of two standard deviations of selected details is the proposed one (Sect. 2). The results are presented in Sect. 3 and discussion in Sect. 4. Final conclusions are presented in Sect. 5. The main problem is the selection of the proper wavelets and the provided analysis is the result of testing large number of combinations of wavelets and estimated parameters of details and approximations. The best results are provided only as well as reference.

2 Wavelets for Image Analysis

The proposed approach uses wavelets for grayscale images. The area of analysis is the largest square area with $N \times N$ size from the inner part of cell nuclei, where N is the power of two. The contrast of this square area is normalized to $\langle 0 - 1 \rangle$ range. Two wavelets are analyzed 'sym1' and 'coif2,' but similar results are obtained for other wavelets, like 'haar.' Previous works related to the analysis of cell nuclei show great importance of cell nucleus area. This parameter is the main criteria of the discrimination of cells and is known as main diagnostic tools. Large area of the cell nucleus is directly related to a kind of atypia. Such cells are in our database and considered as subclass. The second subclass is very important because it cannot be distinguished from correct cells using cell area only. This subclass of atypical cells is the most important and the proposed estimators are especially oriented to this one. Standard deviation is calculated for diagonal details (Dx) in the reference approach only. Obtained value is the second parameter that is depicted. The proposed approach uses ratios of two lower diagonal details (Dx and Dy):

$$\frac{std.dev._{Dx}}{std.dev._{Dy}} \tag{1}$$

The number of image cells is intentionally increased using morphological shrinking. The size of cell nucleus is reduced for the analysis of the influence of nonideal segmentation algorithm. Such sensitivity analysis allows better testing of the image analysis algorithms.

3 Results

The results of classification for two 'coif2' and 'sym1' wavelets are shown in Figs. 2 and 3. Atypical and correct classes are well visible, with separation of atypical subclasses. The main parameter is the cell area. Standard deviation for A0 (input image) and details: D1, D2, and D3 are calculated. The ellipsoid discriminant of 95 % confidence region is presented also, which is the main criteria of the quality of classification. This region is computed using atypical cell with small area of cell nuclei only. Such region obtained from database could be applied for the selection of cells for examination by cytoscreener. The results for standard deviation ratios as well as magnification of the confidence regions are shown in Figs. 4 and 5. Such magnifications are provided for previous two figures because ellipsoid discriminant spans over almost all cells horizontally. More sophisticated separation techniques using three estimators together but the most important results are obtained for simple variant: cell nucleus area and selected standard deviation or standard deviation ratio.

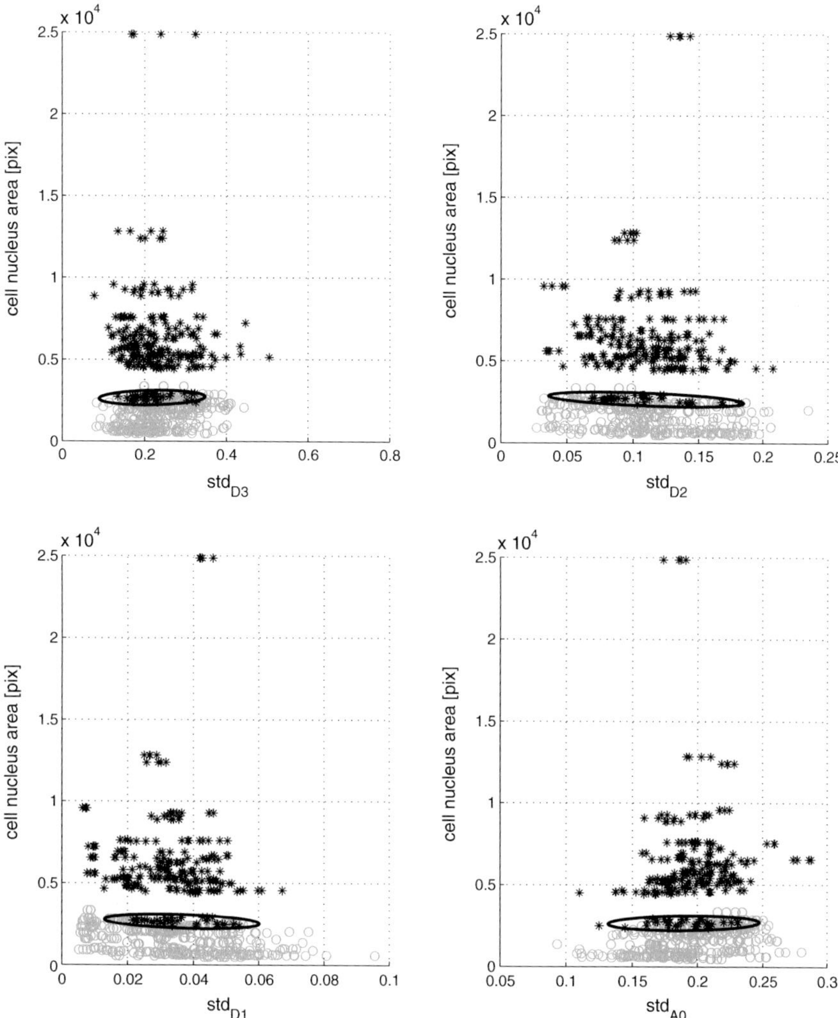

Fig. 2 'coif2' wavelet-based analysis (*circles* correct cell nuclei, *stars* atypical cell nuclei)

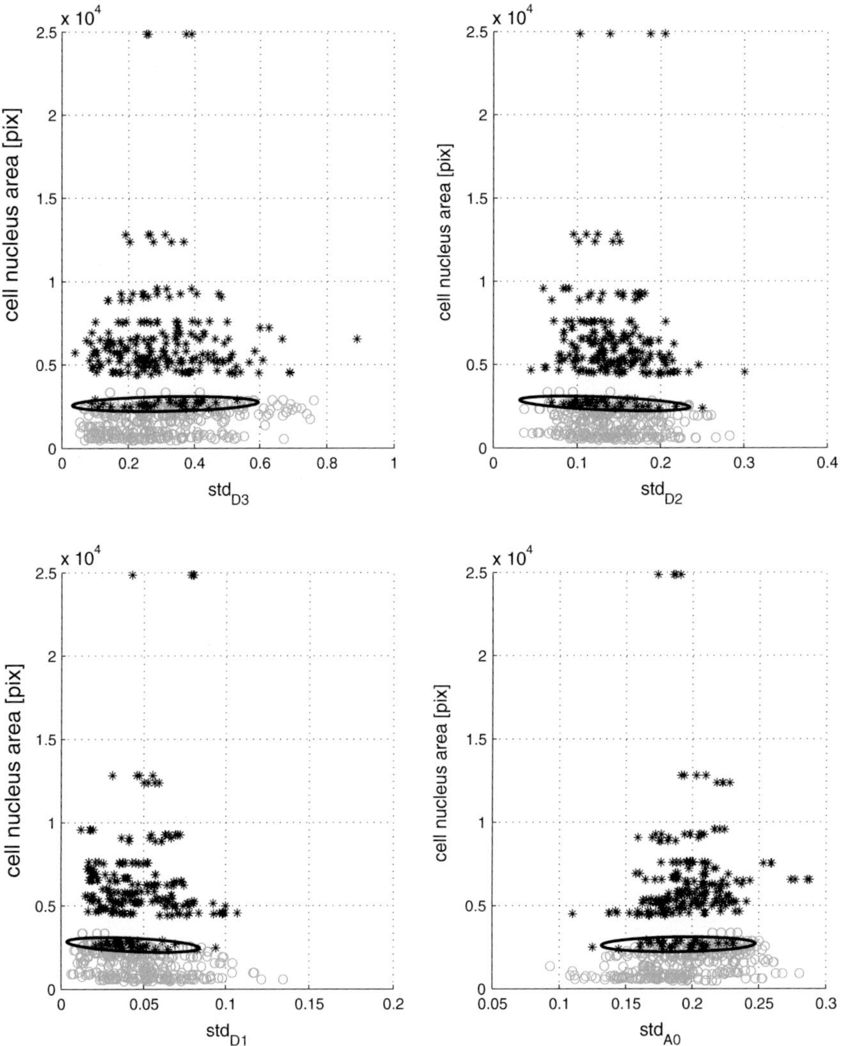

Fig. 3 'sym1' wavelet-based analysis (*circles* correct cell nuclei, *stars* atypical cell nuclei)

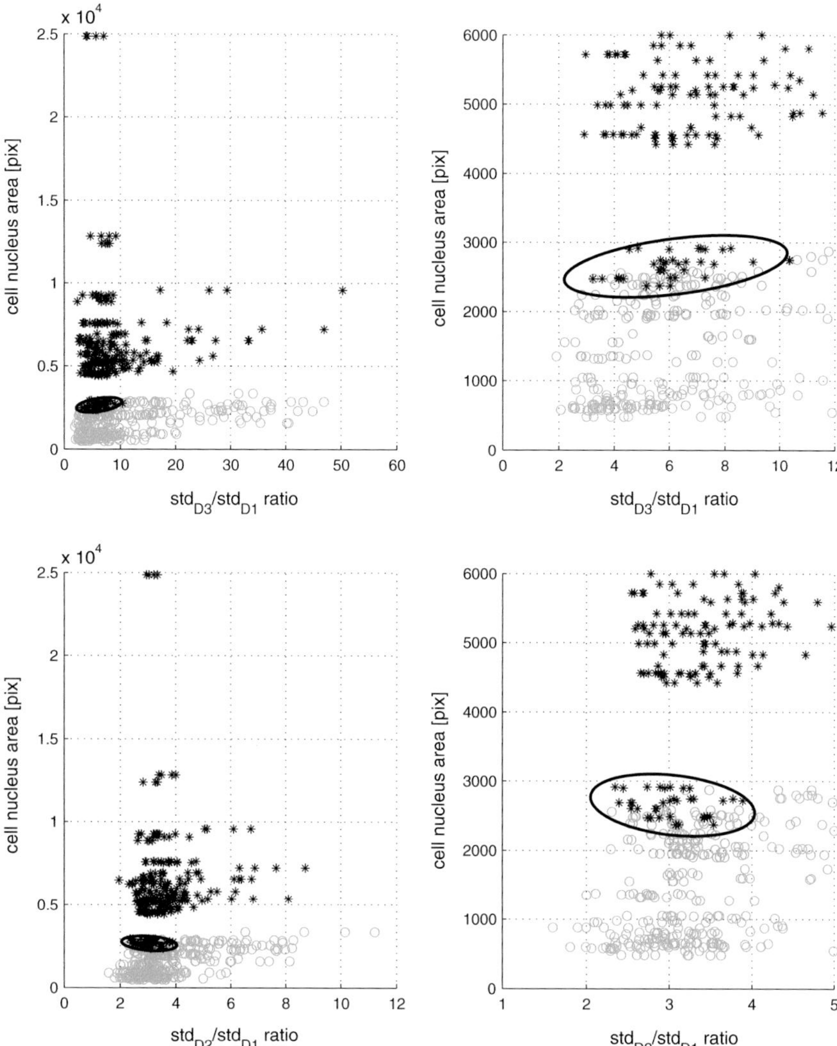

Fig. 4 'coif2' wavelet-based analysis using standard deviation ratio (*circles* correct cell nuclei, *stars* atypical cell nuclei)

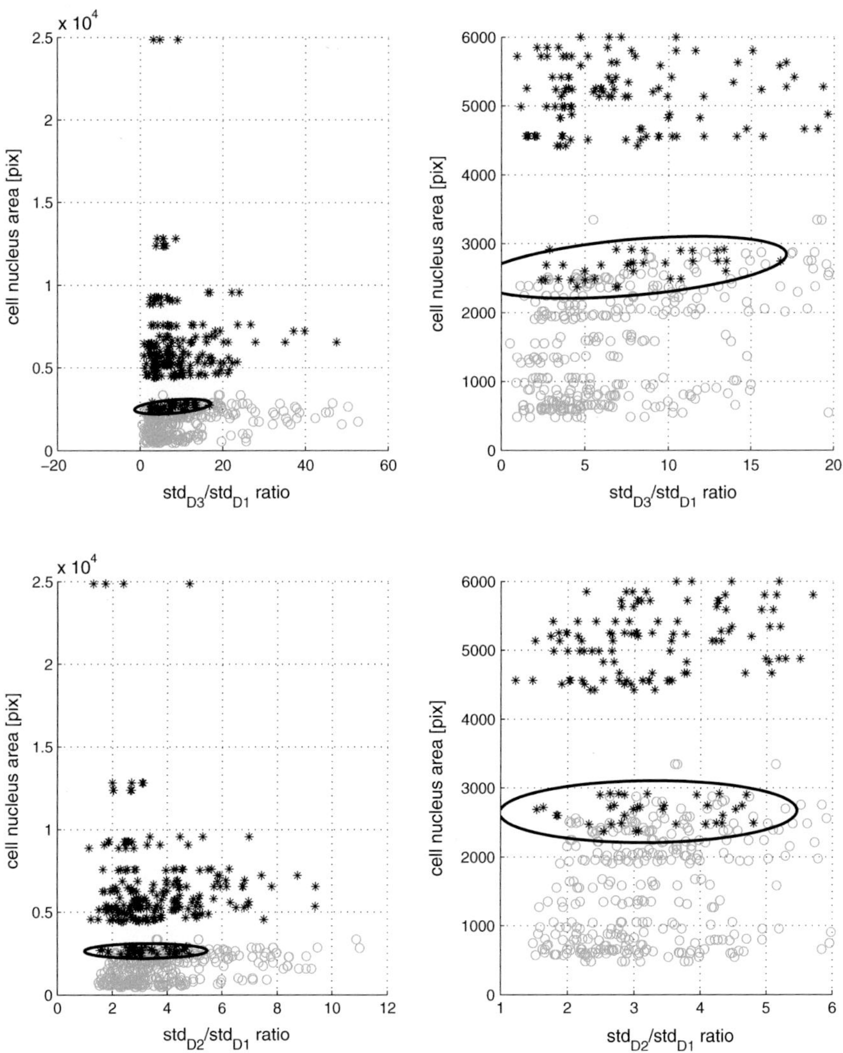

Fig. 5 'sym1' wavelet-based analysis using standard deviation ratio (*circles* correct cell nuclei, *stars* atypical cell nuclei)

4 Discussion

Ellipsoid discriminant does not allow the separation of atypical and correct cell correctly in the case of reference method (Figs. 2 and 3). Many correct cells are inside ellipsoid region, so they should be examined by cytoscreener additionally. The reduction of the number of them is possible by the application of standard deviation ratios. Two standard deviation ratios are presented (Figs. 4 and 5) and the separation is much better. The numbers of correct cells in ellipsoid region are 36 for $std.D3/std.D1$ ('coif2'), 32 for $std.D2/std.D1$ ('coif2'), 49 for $std.D3/std.D1$ ('sym1'), and 54 for $std.D2/std.D1$ ('sym1'). The best results are for 'coif2' wavelet, but the selection of other types improves separation also. The details are limited to the third level only, because many correct one cells' nuclei are small and the higher levels of analysis reduce size of the image.

5 Conclusions

Image analysis of cell nuclei is not trivial task. The provided results for both estimator types, based on the standard deviation, show importance of wavelet and selection of the proper details or approximations as well as estimated value. The standard deviation is often used as an estimator, but provided results show importance of ratios, like standard deviation ratios. The best results are obtained for the first two details. Similar results are obtained for the selection of multiresolutional levels for the fractal-based estimators. Higher scales do not provide important information for the estimation of available database. The obtained separation allows the reduction of cell that should be examined by cytoscreener, but the separation is nonideal. The expected separation based only on the wavelets is not achieved and combined classification is required. Provided results on different cell areas show low sensitivity to the selection of area of cell which is very important due to nonideal segmentation (manual or automatic).

References

1. Adam, R., Silva, R., Pereira, F., Leite, N., Lorand-Metze, I., Metze, K.: The fractal dimension of nuclear chromatin as a prognostic factor in acute precursor B lymphoblastic leukemia. Cell. Oncol. **28**, 55–59 (2006)
2. Bondarenko, A., Katsuk, A.: Extracting feature vectors of biomedical. In: KORUS'2005, pp. 52–56 (2005)
3. Chosia, M., Domagała, W.: Cytologia szyjki macicy. Fundacja Pro Pharmacia Futura (2010)
4. Cibas, E., Ducatman, B.: Cytology. Diagnostic Principles and Clinical Correlates, Saunders Elsevier, Philadelphia (2009)

5. Ferro, D., Falconi, M., Adam, R., Ortega, M., Lima, C., de Souza, C., Lorand-Metze, I., Metze, K.: Fractal characteristics of May-Grünwald-Giemsa stained chromatin are independent prognostic factors for survival in multiple myeloma. PLoS ONE **6**(6), 1–8 (2011)
6. Filipczuk, P., Wojtak, W., Obuchowicz, A.: Automatic nuclei detection on cytological images using the firefly optimization algorithm. Lecture Notes in Computer Science (including subseries Lecture Notes in Artificial Intelligence and Lecture Notes in Bioinformatics) LNBI, vol. 7339, pp. 85–92 (2012)
7. Frejlichowski, D.: Detection of erythrocyte cells in microscopy images. Electr. Rev. **88**(10b), 264–267 (2012)
8. Hoda, R., Hoda, S.: Fundamentals of Pap Test Cytology. Humana Press, Totowa (2007)
9. Hrebień, M., Korbicz, J., Obuchowicz, A.: Hough transform, (1+1) search strategy and watershed algorithm in segmentation of cytological images. Adv. Soft Comput. **45**, 550–557 (2007)
10. Mazurek, P., Oszutowska-Mazurek, D.: From slit-Island method to ising model—analysis of grayscale images. Int. J. Appl. Math. Comput. Sci. **24**(1), 49–63 (2014)
11. Metze, K.: Fractal dimension of chromatin and cancer prognosis. Epigenomics **2**(5), 601–604 (2010)
12. Obuchowicz, A., Hrebień, M., Nieczkowski, T., Marciniak, A.: Computational intelligence techniques in image segmentation for cytopathology. Stud. Comput. Intell. **151**, 169–199 (2008)
13. Okarma, K., Frejlichowski, D., Czapiewski, P., Forczmań, ski, P., Hofman, R.: Similarity Estimation of Textile Materials Based on Image Quality Assessment Methods. Lecture Notes in Computer Science, vol. 8671, pp. 478–485 (2014)
14. Oszutowska, D., Purczyński, J.: Estimation of the fractal dimension using tiled triangular prism method for biological non-rectangular objects. Electr. Rev. **88**(10b), 261–263 (2012)
15. Oszutowska-Mazurek, D., Mazurek, P., Sycz, K., Waker-Wójciuk, G.: Adaptive windowed threshold for box counting algorithm in cytoscreening applications. Advances in Intelligent Systems and Computing, vol. 233, pp. 3–12. Springer, Berlin (2014)
16. Oszutowska-Mazurek, D., Mazurek, P., Sycz, K., Wójciuk, G.W.: Estimation of fractal dimension according to optical density of cell nuclei in Papanicolaou smears. Information Technologies in Biomedicine 2012 (ITIB'2012). Lecture Notes in Computer Science, vol. 7339, pp. 456–463. Springer, Heidelberg (2012)
17. Oszutowska-Mazurek, D., Mazurek, P., Sycz, K., Wójciuk, G.W.: Variogram based estimator of fractal dimension for the analysis of cell nuclei from the Papanicolaou smears. Advances in Intelligent Systems and Computing, vol. 184, pp. 47–54. Springer, Berlin (2013)
18. Oszutowska-Mazurek, D., Mazurek, P., Sycz, K., Wójciuk, G.W.: Lacunarity Based Estimator for the Analysis of Cell Nuclei. LNCS, vol. 8671, pp. 486–493. Springer, Berlin (2014)
19. Steven, I.: Linear Richardson plots from non-fractal data sets. Dutch Math. Geol. **25**(6), 737–751 (1993)
20. Walker, R.: Adaptive multi-scale texture analysis with applications to automated cytology. Ph.D. thesis, University of Queensland (1997)
21. Zhao, T., Wachman, E., Farkas, D.: A novel scheme for abnormal cell detection in pap smear images. In: Proceedings of SPIE 5318, Advanced Biomedical and Clinical Diagnostic Systems II, vol. 151 (2004). Accessed 1 July 2004
22. Zieliński, K., Strzelecki, M.: Komputerowa analiza obrazu biomedycznego. Wstęp do morfometrii i patologii ilościowej, Wydawnictwo Naukowe PWN (2002)

EEG Patterns Analysis in the Process of Recovery from Interruptions

Izabela Rejer and Jarosław Jankowski

Abstract This paper reports the results of the experiment addressing the recovery from interruption phenomenon in terms of brain activity patterns. The aim of the experiment was to find out whether it is possible to find any significant differences in brain activity between subjects performing the task in the recovery period better or worse than the control group. The main outcome from the experiment was that the brain activity of the subjects who performed better than the control group did not change significantly during back to task period compared to interruption period. On the contrary, for subjects whose performance was worse than in the control group, the significant changes in signal power in some frequency bands were found.

Keywords EEG pattern analysis · Brain activity patterns · Interruptions · Recovery from interruption · Human–computer interaction · HCI

1 Introduction

Nowadays, any human activity is performed in a very noisy environment. Digital and traditional media continuously deliver new information and distracting stimuli. As a result, it is very difficult to focus on a primary task, which is repeatedly interrupted by incoming messages, calls, advertisements, etc. Since in our times it is impossible to avoid interruptions, the main question posed in this field is how to continue a main task effectively after the interruption has been finished. This question has been addressed in earlier research from many different points of view. For example, in the field of human–computer interaction (HCI) usually the minimal negative impact of

I. Rejer · J. Jankowski (✉)
Faculty of Computer Science and Information Technology, West Pomeranian University of Technology, Szczecin, Poland
e-mail: jjankowski@wi.zut.edu.pl

I. Rejer
e-mail: irejer@wi.zut.edu.pl

R. Burduk et al. (eds.), *Proceedings of the 9th International Conference on Computer Recognition Systems CORES 2015*, Advances in Intelligent Systems and Computing 403, DOI 10.1007/978-3-319-26227-7_55

internal messages on demanding tasks is looked for [1]. On the other hand, in multitasking environments, the analysis is focused on finding the effective mechanisms for switching a user attention between tasks [2]. Yet, another research regards the relationship between the time of interruption and the efficiency of returning to the main task [3]. Of course also the influence of the type of interruption and type of task performed before the interruption on the successful recovery is analyzed [4].

The research carried out in the field has erupted in recent years due to the development of electronic systems enabling the direct analysis of the impact of interruptions on the decision process [5], and also enabling the simultaneous analysis of many aspects of interruptions, like timing, interruptions frequency, ability to block interruptions, the relevancy to main editorial content, etc. Along with developing the interactive media, new tools and environments for conducting cognitive research aimed at attracting user attention or influencing his current cognitive processes have also evolved [6].

Interruption is defined as "an externally generated randomly occurring, discrete event that breaks continuity of cognitive focus on a primary task" [7]. Other authors define interruption as an "unanticipated issue rising up from the environment while a main action is being performed" [8]. As a result of an unpredictable and uncontrollable nature of interruptions, the stress level of a task performer increases, which can have a negative effect on performance after interruption [9]. However, the stress level induced by an interruption is not constant—it differs significantly depending on many factors. Xia and Sudharshan analyzed the influence of interruptions on natural cognitive flow in relation to online customer decision processes for both abstract and concrete goals [5]. For customers who had to deal with concrete goals, a much higher level of frustration was detected.

Not only the users' stress level but also the quality of performance after interruption depends on external factors. However, while the stress is almost always present when the interruption occurs, the performance after interruption can be not only worse, but also the same or sometimes even better. An example of research with negative outcome can be the research conducted by Edwards and Gronlund [10]. They analyzed the similarity between interruption and primary task along with memory representation of primary task and found that when interruptions were related to the primary content, memory and performance were negatively affected.

Distraction and conflict theory also discuss the relationship between performance of primary task and disruptions [11]. Results of experiments, conducted by Baron, revealed that while disruptions affect the performance of complex task, they have not any direct effect when simple tasks are performed. However, even with simple tasks psychological negative effects of interruptions are observed [12].

Usually, the performance deteriorates when interruption appears [9], but in some research the opposite phenomenon is reported. For example, Speier in 1996 conducted a study aimed at analyzing the influence of different cognitive and social characteristics such as frequency, duration, content, complexity, timing (cognitive characteristics), and form of interruption techniques used for interruption generation and social expectations (social characteristics) [13]. The author showed that differences in characteristics of interruptions, types of goals, and individual differences of customers play important role in the performance after interruption. He proved also that when interruptions are used properly, they can be an effective technique for enhancing the quality of this performance, for example by attracting customer attention. Also most computer system users can handle interruptions effectively. They have the ability to switch attention between tasks and focus on the primary task immediately after interruption [14]. "Zeigarnik Effect" shows that details of interrupted tasks can be recalled even better than those of uninterrupted tasks [15].

In view of the short review of research given above, it is apparent that interruptions can have not only negative, but sometimes also positive influence on the performance after interruption. Hence, the question is how to prepare, modulate, or format the interruption to induce this positive outcome. Currently, this question is addressed by preparing different variations of interrupting content and evaluating their influence on the performance in recovery period in many dimensions, e.g., time or accuracy. These measures provide an answer which forms of interruptions have more positive impact on the subject than others but they do not provide any evidence to address "why" question. Meanwhile, if we were able to detect why a given interruption induces a positive or negative subject's response, we would be able to prepare interruptions better suited to the subject's expectations. We believe that in order to evaluate the true influence of the interruption on the subject during the recovery period we should use a more direct approach than measuring the time or accuracy—we should look insight the subject's brain.

The paper reports the results of our preliminary experiment addressing the recovery from interruption phenomenon, conducted at West Pomeranian University of Technology in Szczecin. Two goals were posed in the experiment: first, establishing the influence of a simple 3-s visual interruption on the text reading process (in regard to text understanding and the time needed for completing the task), and second, investigating whether there is any consistency in the brain activity in the recovery after interruption period. At this stage of the survey we did not want to look for the reasons for positive or negative subject's behavior in the recovery period; we wanted only to find out if the subject's brain responded in a similar way for the similar interruptions.

The rest of the paper is organized as follows. Section 2 presents methods applied in the experiment. Two next sections, Results and Discussion, describe the output of the experiment and its analysis. And finally, the paper is summarized in the last section.

2 Experiment Setup and Methods

The experiment was performed with 14 subjects (students from the West Pomeranian University of Technology in Szczecin), 9 men and 5 women, aged between 20 and 25. All subjects were right-handed without any previous mental disorders. Before the experiment, each subject was fully informed about the experiment. The subjects were randomly assigned to two groups, called the treatment and the control groups, respectively. Before assigning subjects to groups, they were segmented according to the sex. After the assignment, the treatment group was composed of 4 male and 2 female, and the control group of 5 men and 3 women.

The experiment was composed of two stages. At both stages the subjects were presented with a text. The task was to read the text, understand it, and to answer a ten-question test, testing the level of text understanding. The text was presented in a computer screen as 10 short pages, each of the lengths of about 300 words. The decision when to display the next page was left for the subject (each page ended with a "move forward" button). No option to move back to previously read pages was available.

The difference in the experimental conditions for both groups was that while the control group was presented only with a pure text, in the treatment group, the process of reading a text was disturbed by advertisements presentation. Ten advertisements were displayed during the experiment, one per each text page. To avoid the habituation effect, the onset of each advertisement presentation was chosen randomly between 5 and 15 s after a new text page release. The period during which the advertisement was displayed on the screen was fixed (3 s).

The performance of the subjects from both groups was measured in two dimensions: the level of the text understanding and the time needed for completing the task. To test the level of text understanding subjects had to fulfill 10 yes/no questions' test. In order to draw the subject attention to the reading activity, the level of text understanding was evaluated at the end of the experiment. On the other hand, to make the subject more agitated during the advertisements presentation, the time needed to complete the whole experiment was measured. At the end of the experiment the subjects were ranked and awarded according to their results.

EEG data were recorded during the experiment, however, only from the subjects from the treatment group. Since there was any "disruption of the cognitive process" during the experiment with the control group, there was also nothing to investigate in their EEG signals. The EEG data were recorded from four monopolar channels at a sampling frequency of 250.03 Hz. Six passive electrodes were used in the experiments. Four of them were attached to subjects' scalp at Fp1, Fp2, F3, and F4 positions according to the International 10–20 system [16]. The reference and ground electrodes were placed at the right and left mastoid, respectively. The impedance of the electrodes was kept below 5 kΩ. EEG signal was acquired with EEG DigiTrack amplifier (Elmiko) and recorded with DigiTrack software.

In the signal preprocessing stage, a simple band-pass filter (1–30 Hz) was used. After filtering, the mean value was removed from each channel. Next, the epochs were extracted from the continuous signal recorded from a subject during the whole experiment. Each epoch started 3 s before the advertisement onset and ended 3 s after the advertisement offset. Hence, the epoch lasted 9 s; during the first 3 s the subject was reading the text, during the next 3 s he was looking at the advertisement, and during the final 3 s he was reading the text again. Since 10 advertisements were presented during the experiment, 10 epochs were extracted for each subject.

After extracting epochs, we inspected them visually in view of artifacts. We assumed that the data analysis would be done on the basis of at least 1 s of continuous recording. Therefore, we looked for the epochs that contained at least 1 s of artifact-free continuous data in each of the three segments (the first text reading, the advertisements presentation, and the second text reading). The visual inspection revealed that each epoch fulfilled our requirements, and hence all 10 epochs for all 6 subjects were accepted for the analysis.

In order to determine the brain activity patterns related to different stages of the experiment, we analyzed the changes in the signal power in six classic frequency bands: delta (1–4 Hz), theta (4–8 Hz), alpha (8–13 Hz), low beta (13–18 Hz), medium beta (18–24 Hz), and high beta (14–30 Hz). For each frequency band, channel, and epoch, we calculated three values—the average signal power in the period when a subject was reading the first part of a text (PT1), the average signal power in the period when an advertisement was displayed (PA), and the average signal power in the period when a subject was reading the second part of the text (PT2).

To find out whether any significant effects appeared in the cortical activity recorded from a subject after removing the advertisement from the screen, we performed the statistical analysis of 10 epochs extracted for a subject. Since we wanted to know whether the "back to task activity" brought the statistically significant difference in each frequency band and each channel separately, we performed 24 (6 frequency bands $\times$ 4 EEG channels) paired t-student tests per testing condition. We performed two types of tests. The first one tested PT2 against PA, and the second tested PT2 against PT1. Hence, in the first group of tests we tested the null hypothesis H0: Average $(PT2_{ch,f})$ = Average $(PA_{ch,f})$ against the alternative hypothesis H1: Average $(PT2_{ch,f}) \neq$ Average $(PA_{ch,f})$, and in the second group of tests we tested the null hypothesis H0: Average $(PT2_{ch,f})$ = Average $(PT1_{ch,f})$ against the alternative hypothesis H1: Average $(PT2_{ch,f}) \neq$ Average $(PT1_{ch,f})$, where ch—channel index (ch $= 1 \ldots 4$), and f—frequency band index (f $= 1 \ldots 6$). The further analysis was performed for all pairs of averages where the null hypothesis was rejected, i.e., for all pairs where both averages differed significantly. To find out the direction of the change, we calculated the difference between the average value of PT2 and the average value of PA for the tests testing Average (PT2) against Average (PA) and the difference between the average value of PT2 and the average value of PT1 for the tests testing Average (PT2) against Average (PT1).

3 Results

Table 1 presents the results of the experiment in terms of execution time and text understanding. The execution time was measured separately for each text page and then was averaged for each subject. To make the execution time comparable for both groups, the time spent for advertisement presentation (three seconds) was subtracted from the average execution time calculated for subjects from the treatment group.

Table 1 Task performance for treatment and control group

Treatment group									
Subject	S1	S2	S3	S4	S5	S6			Avg
Average time (s)	31.10	28.96	38.03	25.55	30.25	30.46			31.28
Text understanding (%)	0.80	0.90	0.80	0.90	0.90	0.70			0.83
Control group									
Subject	S7	S8	S9	S10	S11	S12	S13	S14	Avg
Average time (s)	37.77	30.74	23.10	37.13	38.25	32.54	40.88	47.96	36.05
Text understanding (%)	0.90	0.60	1.00	0.80	0.80	0.80	0.90	0.80	0.83

Table 2 Brain activity patterns for subjects S1–S6 in BTT period compared to AD period and in BTT period compared to BA period

		BTT period versus AD period				BTT period versus BA period			
		Fp1	Fp2	F3	F4	Fp1	Fp2	F3	F4
S1	Low beta: 13–18 Hz						–		
	Medium beta: 18–24 Hz							–	
	High beta: 24–30 Hz			+					
S2	Theta: 4–8 Hz								+
S3	Delta: 1–4 Hz		–						
	Theta: 4–8 Hz 18–24 Hz		–						
	Low beta: 13–18 Hz				+				
S5	Medium beta: 18–24 Hz					–			
S6	Alpha: 8–13 Hz				+				

The text understanding was measured on the basis of the outcome from 10-question yes/no questionnaires fulfilled by the subjects after completing the reading task.

Table 2 presents brain activity patterns established for individual subjects in the back to task (BTT) period compared to the ads displaying (AD) period and compared to the before ads presentation (BA) period. The signs inside the table denote the direction of the change in the average signal power calculated over all 10 epochs. Symbol '+' means that the average signal power in the given frequency band and in the given channel was greater in BTT period; symbol '–' means that the average signal power was greater in AD or BA period. Only significant results, tested with paired t-student test are presented in the table. The comparison of the signal power of significant patterns for both pairs of periods is presented in Fig. 1 (BTT vs. AD period) and in Fig. 2 (BTT vs. BA period). As it can be noticed in both figures, in general brain activity patterns found for BTT versus AD period were stronger than those found for BTT versus BA period.

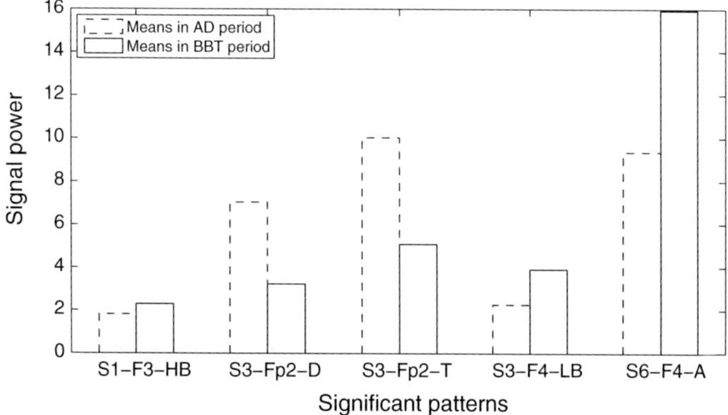

Fig. 1 The comparison of the signal power in BBT versus AD period for significant brain activity patterns; A-alpha, D-delta, T-theta, LB-low beta, MB-medium beta, HB-high beta

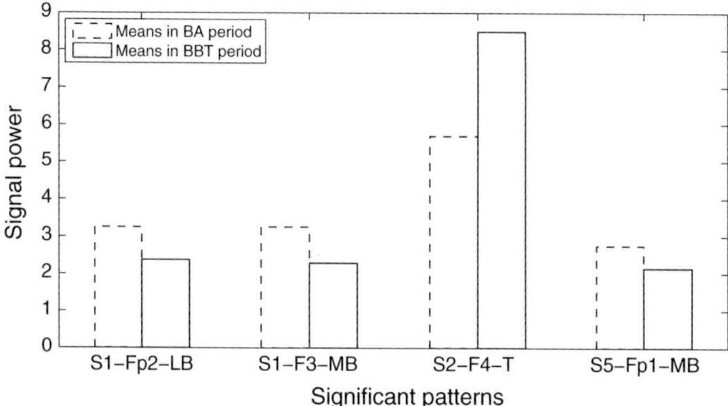

Fig. 2 The comparison of the signal power in BBT versus BA period for significant brain activity patterns; A-alpha, D-delta, T-theta, LB-low beta, MB-medium beta, HB-high beta

4 Discussion

The results presented in Table 2 can be summarized as follows:

1. The only significant difference in signal power observed for subject S1 was the increase in power in the high beta sub-band in channel F3 in BTT period compared to AD period, which suggests the approach motivation of the subject during back to task activity. At the same time, however, the drop in two beta sub-bands in both hemispheres was noted when comparing BBT period to BA period, which means that the subject was significantly less focused on the reading task in BBT period.
2. No significant changes in the brain activity over the analyzed channels were found for subjects S2, S4, and S5 when comparing BTT period and AD period. Two

significant changes observed in BTT period compared to BA period (an increase in the theta band for subject S2, and drop in the beta band for subject S5) indicate the drop in the concentration level BTT period, and the approach motivation of both subjects.

3. All three patterns found for subject S3 are consistent with each other. The drop in the signal power in two bands of the lowest frequency (delta and theta) over the right prefrontal cortex and the increase of the signal power in the beta band over the right frontal cortex clearly indicate the withdraw motivation of the subject.
4. Only one significant change in the signal power was found for subject S6—the increase in alpha band in the right frontal cortex in BTT period compared to AD period. This pattern indicates the approach motivation of the subject when he returned to the text reading task.

When we started our experiment, we believed that when the text reading process was interrupted by displaying ads, the total time needed for completing the task and the overall performance would be worse compared to the control group. However, after the analysis of related papers (discussed in Sect. 1), we abandon our first radical assumption, and started to expect that the subjects from the treatment group could achieve better performance than subjects from the control group. This assumption proved to be the correct one. As Table 1 presents, the average performance of the subjects from both groups was exactly the same. This means that interrupting of the text reading task by displaying 3-s ads did not bring any negative consequences for the task performance after interruption. Moreover, the time needed for completing the whole task was about 15 % shorter for the subjects whose concentration on the task was disrupted by displaying ads. These results are in line with those reported in [14]. Obviously, we do not state that this is the general truth because the effectiveness of backing to task depends on many factors, such as the length of the interruption, its invasiveness, the modality involved, etc.

The aim of our study was to find out whether this "back to task" activity has any reflection in the subjects' brains. In other words, we wanted to confront our time and performance metrics with the actual brain activity patterns. At first we assumed that when the subject successfully returns to his original task after interrupting it by ads presentation, the prefrontal and/or frontal parts of the brain should become more active. Hence, generally we expected the increase in the signal power in the beta sub-bands and decrease in the signal power in the alpha sub-bands. We expected also the overall drop in concentration during back to task activity, compared to BA period. After performing the experiments, it occurred, however, that none of the two assumptions were exactly valid. The first assumption, about the increase of the brain activity, was true only for two out of all six subjects (S1 and S3). For three of the remaining subjects (S2, S4, and S5) none of the significant changes were found in BTT period compared to BA period, and the only significant change found for the last subject (S6) suggested the drop, instead of the rise, of the brain activation.

What is really interesting here is that all three subjects whose brain activity remained on the same level during BTT and AD periods (S2, S4, and S5) performed significantly better than the average subject from both treatment and control groups

(Table 1). To be more precise, there was only one subject from the control group (S9) that performed better than these three subjects, and two subjects whose performance was on the same level (S7 and S13). Moreover, subjects S2, S4, and S5 also finished the whole task quicker than the average subject from both group. Also this time there was one subject from the control group who finished the task in a shorter time (S9), but the remaining 10 subjects were much slower in completing the task. In addition to lack of the significant patterns in BTT period compared to AD period, the analysis of BTT period compared to BA period performed for subjects S2, S4, and S5 revealed inactivation patterns with approach motivation for subjects S2 and S5, and no significant patterns for subject S4. One of the possible explanations why all three subjects performed better than most of the remaining subjects from both groups could be that they stayed on the same level of alertness during the whole task (S4) or even tried to improve themselves by being more agitated in BTT period than in BA period (S2 and S5).

While subjects S2, S4, and S5 performed better and quicker than most of the remaining subjects, subjects S1, S3, and S6 were less precise in their answers that the average subject from the treatment group. Their performance was also worse than the average performance of the subjects from the control group. The main difference between these three subjects and the subjects S2, S4, and S5 in terms of the brain activity patters is that all three "worse performing" subjects presented significant activity patterns in BTT versus AD period (as it can be noticed in Fig. 1 the differences in signal power in both periods were substantial). The patterns, however, were not consistent with each other. While the patterns found for two of the subjects (S1 and S6) indicated their approach motivation, the patterns found for subject S3 clearly indicated the withdraw motivation (Table 2, and Fig. 1). The approach motivation of subject S1 and S6 could mean that they were involved in the task and they wanted to proceed (and they did since their execution time was quick enough). Their alertness, however, could be too high to remember the details of the text. On the contrary, subject S3 presented negative attitude to the experiment and was rather not very interested in its continuation, which was in agreement with his achievements.

5 Conclusion

Summing up the results obtained in the experiment, it should be stated that the 3-s ads interrupting the reading task did not distract the subjects who returned to the main task with ease. Their average performance was on the same level as for the control group and they completed the whole task even quicker than subjects who were not disturbed by ads presentation. The main difference that was found in brain activity patterns between subjects whose performance was high and subjects who performed worse was the lack of significant patterns in the first group. Our assumption is that they stayed relax during the whole experiment and their brains were on the similar alertness level in all three analyzed periods. Due to this, they were able to perform the task better. In order to confirm this assumption, we plan to perform a full experiment on a much larger group of subjects. Moreover, in our future work we would like also to perform similar experiments with different forms of interruptions and other cognitive tasks to find the differences in brain activity patterns appearing during the successful and unsuccessful recoveries from interruption.

References

1. Ramchurn, S.D., Deitch, B., Thompson, M.K., De Roure, D.C., Jennings, N.R., Luck, M.: Minimising intrusiveness in pervasive computing environments using multi-agent negotiation. In: Proceedings of the First Annual International Conference on Mobile and Ubiquitous Systems: Networking and Services (MobiQuitous'04). IEEE Computer Society, Los Alamitos, pp. 364–372 (2004)
2. Appelbaum, S.H., Marchionni, A., Fernandez, A.: The multi-tasking paradox: perceptions, problems and strategies. Manag. Decis. **46**(9), 1313–1325 (2008)
3. Altmann, E.M., Trafton, J.G.: Timecourse of recovery from task interruption: data and a model. Psychon. Bull. Rev. **14**(6), 1079–1084 (2007)
4. Hodgetts, H.M., Jones, D.M.: Contextual cues aid recovery from interruption: the role of associative activation. J. Exp. Psychol.: Learn. Mem. Cogn. **32**(5), 1120–1132 (2006)
5. Xia, L., Sudharshan, D.: Effects of interruptions on consumer online decision processes. J. Consum. Psychol. **12**(3), 265–280 (2002)
6. Mandel, N., Johnson, E.: Constructing preferences online: can web pages change what you want? (Working paper). University of Pennsylvania, Wharton School, Philadelphia (1999)
7. Corragio, L.: Deleterious effects of intermittent interruptions on the task performance of knowledge workers: a laboratory investigation. Unpublished doctoral dissertation, University of Arizona, Tucson (1990)
8. Zhang, Y., Pigot, I., Mayers, A.: Attention switching during interruptions. In: Proceedings of the Third International Conference on Machine Learning and Cybernetics. Shanghai, 26–29 August 2004, pp. 276–281
9. Cohen, S.: Aftereffects of stress on human performance and social behavior: a review of research and theory. Psychol. Bull. **88**(1), 82–108 (1980)
10. Edwards, M.B., Gronlund, S.D.: Task Interruption and its effects on memory. Memory **6**(6), 665–687 (1998)
11. Baron, R.S.: Distraction-conflict theory: progress and problems. Adv. Exp. Soc. Psychol. **19**, 1–39 (1986)
12. Speier, C., Valacich, J.S., Vessey, I.: The influence of task interruption on individual decision making: an information overload perspective. Decis. Mak. **30**(2), 337–360 (1999)
13. Speier, Ch.: The effect of task interruption and information presentation on individual decision making. Unpublished doctoral dissertation, Indiana University, Bloomington (1996)
14. Zha, W., Wu, H.D.: The impact of online disruptive ads on users' comprehension, evaluation of site credibility, and sentiment of intrusiveness. Am. Commun. J. **16**(2), 15–28 (2014)
15. Van Bergen, A.: Task Interruption. North-Holland, Amsterdam (1968)
16. Jasper, H.H.: The ten-twenty electrode system of the international federation in electroencephalography and clinical neurophysiology. EEG J. **10**, 371–375 (1958)

Part VI
Application

Application of Syntactic Pattern Recognition Methods for Electrical Load Forecasting

Mariusz Flasiński, Janusz Jurek and Tomasz Peszek

Abstract Electrical load forecasting is an important problem concerning safe and cost-efficient operation of the power system. Although many techniques are used to predict an electrical load, a research into constructing more accurate methods and software tools is still being conducted over the world. In this paper an experimental application for improving an accuracy of an electrical load prediction is presented. It is based on the syntactic pattern recognition approach and FGDPLL(k) string automata. The application has been tested on the real data delivered by one of the Polish electrical distribution companies.

1 Introduction

Accurate electrical load forecasting is very important for a safe and cost-efficient operation of the power system. It is crucial for electrical distribution companies, because better forecasts mean better trade profits for a distributor. Since the problem is very difficult, a lot of methods and software tools have been developed [1, 13]. Nevertheless, there is still a need for constructing more accurate methods and systems. In this paper, a real-world example of the forecasting process in case of Tauron Polish Energy Distribution Division in Gliwice, Poland is considered. A prediction of an electrical demand of the clients one day ahead is prepared daily at 8 am. The prediction concerns the 24-h period, and the demand is defined for every hour of this period. On the basis of this prediction Tauron buys a proper amount of energy on the Polish energy market. There are many methods of forecasting applied by Tauron, for example,

M. Flasiński (✉) · J. Jurek · T. Peszek
Information Technology Systems Department, Jagiellonian University Cracow, ul. prof. St. Łojasiewicza 4, 30-348 Cracow, Poland
e-mail: mariusz.flasinski@uj.edu.pl

J. Jurek
e-mail: janusz.jurek@uj.edu.pl

R. Burduk et al. (eds.), *Proceedings of the 9th International Conference on Computer Recognition Systems CORES 2015*, Advances in Intelligent Systems and Computing 403, DOI 10.1007/978-3-319-26227-7_56

- neural networks, using such parameters as time, a type of a day (a weekday, Saturday, Sunday, or a holiday), weather (e.g., temperature or insolation), and random effects;
- the autoregressive integrated moving average (ARIMA) method, based on historic data;
- methods based on expert reports prepared by experienced specialists in the field of electrical load forecasting.

These methods are combined together to obtain the final prediction. In this paper a method of a correction of the final prediction for making it more precise is presented. The method is based on the syntactic pattern recognition approach. To make the correction a *structure* of an electrical load pattern is analyzed. Section 2 contains basic definitions of the approach. In Sect. 3 input data for the syntactic pattern recognition system as well as the output results are introduced. A description of the system architecture is included in Sect. 4. Experimental results are presented in Sect. 5, whereas conclusions are contained in the final section.

2 GDPLL(*k*) Grammars and FGDPLL(*k*) Automata

Syntactic pattern recognition [2, 5, 7] is based on the theory of formal languages, grammars, and automata. In the first phase of a recognition process a structural representation of a pattern is generated. First, a pattern is segmented in order to identify elementary components, called *primitives*. Second, a symbolic representation of the pattern in the form of a string, which consists of symbols representing primitives, is defined. This representation is treated as a word belonging to a formal language. During the next phase such a word is analyzed by a formal automaton, which is constructed on the basis of a formal grammar generating the corresponding formal language. In result a derivation of the analyzed string is obtained. The derivation is used by a syntactic pattern recognition system for describing structural features of the pattern and for recognizing (classifying) it. In our previous works [3, 8–10] we have presented a recognition method based on the so-called GDPLL(*k*) grammars. There are several advantages of the method in comparison to other well-known approaches [5]. The GDPLL(*k*) grammars are characterized by very good discriminative properties (they are able to generate a considerable subclass of context-sensitive languages). An efficient parsing algorithm for GDPLL(*k*) grammars has been constructed as well as a grammatical inference algorithm. Let us introduce two basic definitions corresponding to GDPLL(*k*) grammars [9].

Definition 1 A *generalized dynamically programmed context-free grammar* is a six-tuple $G = (V, \Sigma, O, P, S, M)$, where V is a finite, nonempty alphabet; $\Sigma \subset V$ is a finite, nonempty set of terminal symbols (let $N = V \setminus \Sigma$); O is a set of basic operations on the values stored in the memory; $S \in N$ is the starting symbol; M is a memory; P is a finite set of productions of the form: $p_i = (\mu_i, L_i, R_i, A_i)$ in which $\mu_i : M \longrightarrow \{TRUE, FALSE\}$ is the predicate of applicability of the production p_i

defined with the use of operations ($\in O$) performed over M; $L_i \in N$ and $R_i \in V^*$ are left- and right-hand sides of p_i, respectively; and A_i is the sequence of operations ($\in O$) over M, which should be performed if the production is to be applied. □

Definition 2 Let $G = (V, \Sigma, O, P, S, M)$ be a generalized dynamically programmed context-free grammar. The grammar G is called a *GDPLL(k) grammar*, if the following two conditions are fulfilled.

1. Stearns's condition of LL(k) grammars. (The top-down left-hand side derivation is deterministic if it is allowed to look at k input symbols to the right of the current position of the input head in the string.)
2. There exists a certain number ξ such that after the application of ξ productions in a left-hand side derivation we get at the "left-hand side" of a sentence at least one new terminal symbol. □

A parsing algorithm for GDPLL(k) grammars (GDPLL(k) parser) has been described in [3, 9]. We do not present it in the paper. We just notice here that the algorithm proceeds in a top-down manner. A syntactic pattern recognition approach is very convenient for analysis of patterns of a structural nature. On the other hand, its effectiveness can worsen in case of fuzzy (or distorted) patterns. Therefore, enhanced

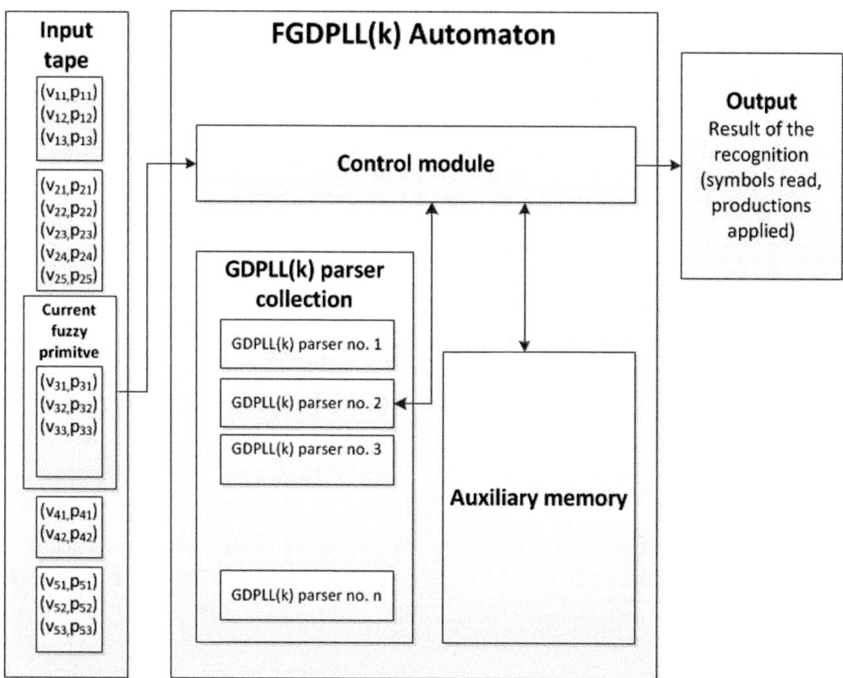

Fig. 1 Architecture of FGDPLL(k) automaton

syntactic pattern recognition models have been developed (e.g., [4, 11]). The model we apply deals with the fuzziness problem with the use of a FGDPLL(k) automaton [11]. A formal definition of the automaton is very complex, so we present here just a brief description of its work and the scheme of its architecture (see Fig. 1). The input for a FGDPLL(k) automaton is a string of *vectors*. (Instead of a string of alphabet symbols, as usual in case of syntactic pattern recognition applications.) The vectors are of the form $((a_1, p_1), \ldots, (a_n, p_n))$, where a_i is the possible symbolic value of the primitive, and p_i is the probability that the primitive is equal to a_i. The FGDPLL(k) automaton maintains a collections of GDPLL(k) parsers, which are needed to perform simultaneously several derivation processes for each possible symbolic value of an input primitive. The control module of the automaton is responsible for identifying which of the derivation processes should be continued, and which of them should be stopped. For this purpose it uses an auxiliary memory, which contains computed probability values for each derivation process. Let us notice that the work of the FGDPLL(k) automaton can be parallelized, which reduces the time needed for the recognition [6].

3 Input and Output Data

In order to model and test the syntactic pattern recognition system for electrical load forecasting, called *SPR system* later on, data concerning years 2012 and 2013 have been delivered by Tauron. (For confidentiality reasons the data have been coded.) For these two years the following information for a given day has been received:

- an area identifier (AR),
- a type of the day, i.e., a working day or a non-working day (TD),
- a day of a week: Monday–Sunday (WD),
- a season: spring, summer, autumn, or winter (SN),
- an amount of non-working days before a given day (FDB),
- an amount of non-working days after a given day (FDA),
- a "typical" or "non-typical" day (AT),
- a load forecast $(F_1, \ldots, F_{24})$ for each hour in a two-day period; prepared in a "traditional" way by Tauron, i.e., without an application of syntactic pattern recognition methods,
- an air temperature forecast $(T_1, \ldots, T_{24})$ for each hour in a two-day period,
- an insolation forecast $(I_1, \ldots, I_{24})$ for each hour in two days ahead period.

The SPR system generates a correction $(C_1, \ldots, C_{24})$ of the load forecast for a given day, i.e., the more accurate forecast is defined as $(F_1 + C_1, \ldots, F_{24} + C_{24})$. Information about a current electrical load in a given day (for each hour) is the reference data used for the testing purposes.

4 Architecture of the SPR System

A general scheme of the SPR system is presented in Fig. 2. First, the input data (mostly numeric) described in a previous section are transformed into the *symbolic form* acceptable by the FGDPLL(k) automaton. It is performed with the help of probabilistic neural networks [12]. We use two probabilistic neural networks PNN_1 and PNN_2. The first one is applied for a classification of the day, the second—for a classification of the weather. Both PNNs in our model deliver, additionally, a distribution representing the probability of belonging of the recognized primitive to proper classes. (PNN_1 and PNN_1 are trained and the GDPLL(k) grammar is inferred on the basis of historic data delivered by Tauron.) A transformation scheme of the input data into the (fuzzy) symbolic ones is shown in Table 1. The FGDPLL(k) automaton receives a fuzzy pattern describing the day and weather conditions as an input. It performs syntax analysis of the pattern (following the rules described in [11]). The result is either a set of most probable patterns (i.e., patterns being the most probable "versions" of the input fuzzy pattern) with information about probabilities of their derivation or the lack of the acceptance. Thus, the FGDPLL(k) automaton is used as a transducer in our method. The transducer translates an input fuzzy pattern into a set of output (not fuzzy) patterns, which are the basis for making the correction of the forecast in the next step. (The lack of the acceptance of a fuzzy pattern means that the "original" forecast will not be corrected.) Let us assume that the result of the analysis performed by the FGDPLL(k) automaton is the acceptance. Then, the correction module of the SPR system (cf. Fig. 2) calculates a proper correction. The

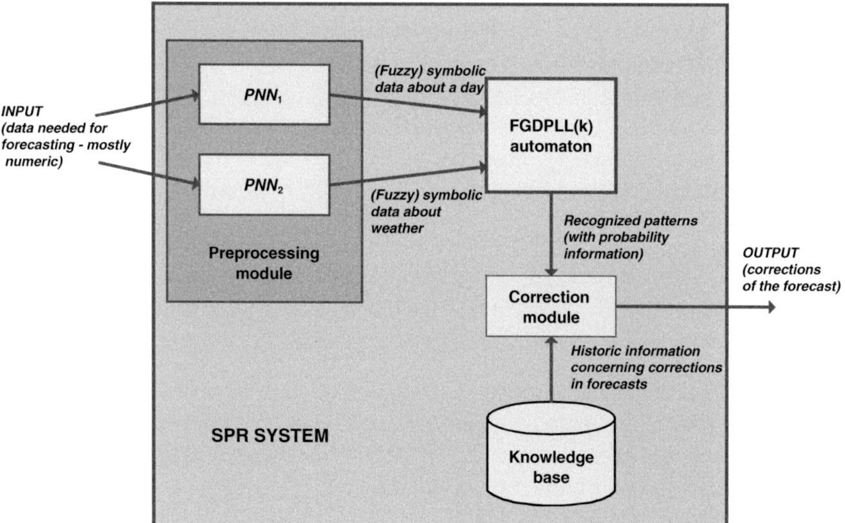

Fig. 2 General scheme of SPR system

Table 1 Data preprocessing with the help of PNN_1 and PNN_2

Input data	Preprocessed data
AR TD WD SN *FDB FDA AT*	Symbolic (fuzzy) representation of a day (PNN_1)
$F_1, \ldots, F_6$ $T_1, \ldots, T_6$ $I_1, \ldots, I_6$	Symbolic (fuzzy) representation of the weather in I period of a day (PNN_2)
$F_7, \ldots, F_{12}$ $T_7, \ldots, T_{12}$ $I_7, \ldots, I_{12}$	Symbolic (fuzzy) representation of the weather in II period of a day (PNN_2)
$F_{13}, \ldots, F_{18}$ $T_{13}, \ldots, T_{18}$ $I_{13}, \ldots, I_{18}$	Symbolic (fuzzy) representation of the weather in III period of a day (PNN_2)
$F_{19}, \ldots, F_{24}$ $T_{19}, \ldots, T_{24}$ $I_{19}, \ldots, I_{24}$	Symbolic (fuzzy) representation of the weather in IV period of a day (PNN_2)

calculation is performed with the use of the knowledge base, according to the historic data corresponding to the current (real) weather circumstances and the real electrical load. Let a result delivered by the FGDPLL(k) automaton be of the following form:

$$(w_1, p_1), \ldots, (w_n, p_n),$$

where w_i is a recognized word (one of the most probable patterns), and p_i is a probability of its recognition. For each $i = 1, \ldots, n$ the knowledge base is searched for the cases represented by w_i. The correction module uses a pattern-matching technique to find adequate cases. Let us assume that $j = 1, \ldots, k$ are indices of the found (adequate) cases stored in the knowledge base. Let d_j be a vector of differences between forecasted and current values of the electrical load (for a given period in a day) stored in the knowledge base. Let $p(j) = p_i$, if the j-case corresponds to the word w_i. A correction of the forecast *corr* is calculated according to the following formula:

$$corr = \sum_{j=1}^{k} c_j d_j, \ where \ c_j = \frac{p(j)}{\sum_{t=1}^{k} p(t)}.$$

Hence, a correction of the forecast is the weighted sum of corrections based on historic data. One of the main windows of the SPR system is presented in Fig. 3. It contains information about parameters for the forecast for a given day and results of their analysis, including a correction of the forecast.

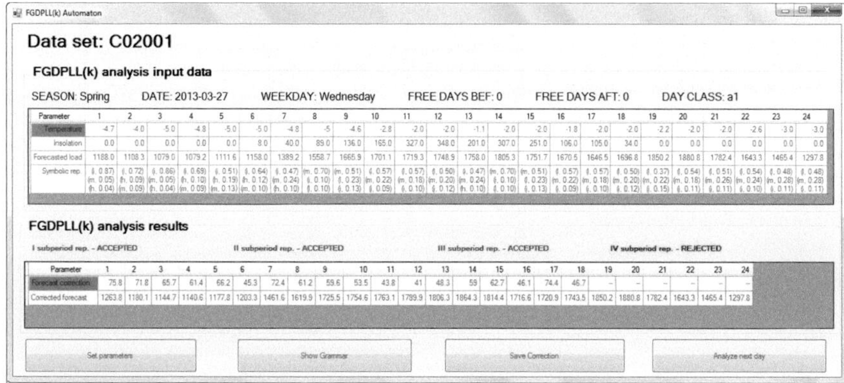

Fig. 3 SPR system: an example view

5 Experimental Results

The SPR system has been tested on data delivered by Tauron PE S.A. company. Dataset submitted to analysis consisted of 731 cases (years 2012, 2013), all of which concerned single distributional area. Since energy consumption is periodical, we use the cases of 2012 as a learning set and the cases of 2013 as a validation set. The performance of the system was measured while working on the validation set. An example plot displaying the absolute percentage error in case of "traditional" forecasts (i.e., prepared with help of neural networks, autoregression models, and expert knowledge) and forecasts with corrections made according to the SPR system results is presented in Fig. 4. As it can be easily seen, the SPR corrected forecast is better than the forecast prepared with "traditional" methods in most cases. As a tool for an evaluation of the overall forecast accuracy, *MAPE* (mean absolute percentage

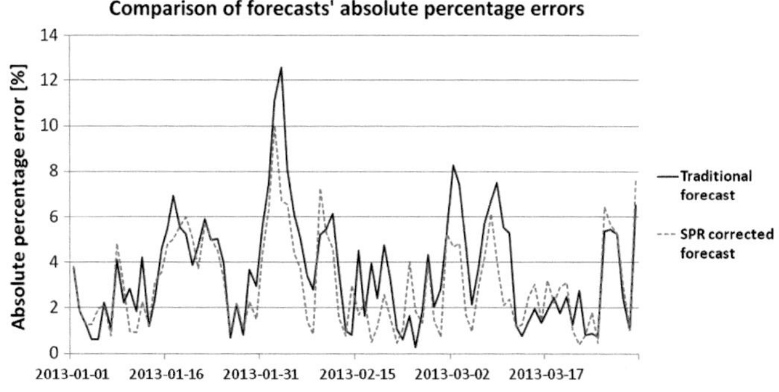

Fig. 4 Absolute percentage error of "traditional" and corrected forecasts

Table 2 Comparison between $MAPE_T$ and $MAPE_C$

Time period	Result of SPR	Number of patterns	$MAPE_T$	$MAPE_C$
I/00:00-05:59	Unrecognized	102	3.10	3.10
	Recognized	263	3.02	2.95
II/06:00-11:59	Unrecognized	94	2.64	2.64
	Recognized	271	3.20	2.93
III/12:00-17:59	Unrecognized	79	2.95	2.95
	Recognized	286	3.13	2.90
IV/18:00-23:59	Unrecognized	73	2.63	2.63
	Recognized	292	2.71	2.64

error) has been used. Let a forecast $F = (f_1, \ldots, f_n)$, and a current (real) electrical load $R = (r_1, \ldots, r_n)$. Then, *MAPE* is defined in the following way:

$$MAPE(R, F) = \frac{1}{n} \sum_{k=1}^{n} \frac{|r_i - f_i|}{r_i}$$

A comparison between *MAPE* in case of "traditional" forecasts ($MAPE_T$) and forecasts with corrections made according to the results obtained from the SPR system ($MAPE_C$) is presented in Table 2. One can easily notice that in most cases (when a pattern has been recognized by the SPR system), $MAPE_C$ is better than $MAPE_T$ by 0, 07–0, 27. Although the difference seems to be small, it could result in significant profits, if we consider energy purchasing on the Polish energy market. At the end of this section, let us mention that the method has been tested on the (limited) dataset delivered by the company. The first test results are very promising, nevertheless it is too early to state that our new method ultimately outperformed "traditional" one. Before such a conclusion, wider tests have to be done (more distribution areas, longer time periods) and the areas of effectiveness and ineffectiveness should be identified and characterized.

6 Concluding Remarks

In this paper, recent results of the research into improving electrical load forecasting with the help of syntactic pattern recognition methods have been presented. It has been shown that the model based on GDPLL(k) grammars and FGDPLL(k) automata has been successfully used in case of such an application. The main advantages of the model are as follows: GDPLL(k) grammars are of a big generating/descriptive power, which allows us to recognize even very complex patterns; FGDPLL(k) automaton is computationally efficient, which makes it useful even in online systems; FGDPLL(k) automaton is able to analyze (recognize) fuzzy/distorted patterns which are typical in real-world applications. The practical experiments with the use of FGDPLL(k) automata for electrical load forecasting are in an early stage, but the results already achieved are promising. Even a small improvement of a forecast accuracy can bring considerable profits in case of an electrical energy distribution company.

References

1. Alfares, H.K., Nazeeruddin, M.: Electric load forecasting: literature survey and classifcation of methods. Int. J. Syst. Sci. **33**, 23–34 (2002)
2. Bunke, H.O., Sanfeliu, A. (eds.): Syntactic and Structural Pattern Recognition—Theory and Applications. World Scientific, Singapore (1990)
3. Flasiński, M., Jurek, J.: Dynamically programmed automata for quasi context sensitive languages as a tool for inference support in pattern recognition-based real-time control expert systems. Pattern Recognit. **32**, 671–690 (1999)
4. Flasiński, M., Jurek, J.: On the analysis of fuzzy string patterns with the help of extended and stochastic GDPLL(k) grammars. Fundamenta Informaticae **71**, 1–14 (2006) (IOS Press, Amsterdam)
5. Flasiński, M., Jurek, J.: Fundamental methodological issues of syntactic pattern recognition. Pattern Anal. Appl. **17**, 465–480 (2014) (Springer, Berlin)
6. Flasiński, M., Jurek, J., Peszek, T.: Parallel Processing Model for Syntactic Pattern Recognition-Based Electrical Load Forecast. Lecture Notes in Computer Science, vol. 8384, pp. 338–347. Springer, Berlin (2014)
7. Fu, K.S.: Syntactic Pattern Recognition and Applications. Prentice Hall, Englewood Cliffs (1982)
8. Jurek, J.: Towards Grammatical Inferencing of GDPLL(k) Grammars for Applications in Syntactic Pattern Recognition-Based Expert Systems. Lecture Notes in Computer Science, vol. 3070, pp. 604–609. Springer, Berlin (2004)
9. Jurek, J.: Recent developments of the syntactic pattern recognition model based on quasi-context sensitive languages. Pattern Recognit. Lett. **26**, 1011–1018 (2005) (Elsevier, Amsterdam)
10. Jurek, J.: Grammatical Inference as a Tool for Constructing Self-learning Syntactic Pattern Recognition-Based Agents. Lecture Notes in Computer Science, vol. 5103, pp. 712–721. Springer, Berlin (2008)
11. Jurek, J., Peszek, T.: Model of Syntactic Recognition of Distorted String Patterns with the Help of GDPLL(k)-Based Automata. Advances in Intelligent and Soft Computing, vol. 226, pp. 101–110. Springer, Berlin (2013)
12. Specht, D.F.: Probabilistic neural networks. Neural Netw. **3**, 109–118 (1990)
13. Taylor, J., McSharry, P.: Short-term load forecasting methods: an evaluation based on European data. IEEE Trans. Power Syst. **22**, 2213–2219 (2008)

Improvements to Segmentation Method of Stained Lymphoma Tissue Section Images

Lukasz Roszkowiak, Anna Korzynska, Marylene Lejeune, Ramon Bosch and Carlos Lopez

Abstract We present the METINUS (METhod of Immunohistochemical NUclei Segmentation), which is a improved and modified version of supporting tool for pathologists from 2010. The method supports examination of immunohistochemically stained thin tissue sections from biopsy of follicular lymphoma patients. The software localizes and counts FOXP3 expression in the cells' nuclei supporting standard procedure of diagnosis and prognosis. The algorithm performs colour separation followed by object extraction and validation. Objects with statistical parameters not in specified range are disqualified from further assessment. To calculate the statistics we use the following: three channels of RGB, three channels of Lab colour space, brown channel and three layers completed with colour deconvolution. Division of the objects is done with support of watershed and colour deconvolution algorithm. Evaluation was performed on arbitrarily chosen 20 images with moderate quality of most typical tissues. We compared results of improved method with the previous version in the context of semiautomatic, pathologist controlled, computer-aided result of quantification as reference. Comparison is based on quantity of nuclei located per image using Kendall's tau-b correlation coefficient. It shows concordance of 0.91 between results of proposed method and reference, while with previous version it is only 0.71.

Keywords Immunohistochemistry · Nuclear quantification · Pathology · Follicular lymphoma · Biomedical image processing · Image segmentation

---

L. Roszkowiak (✉) · A. Korzynska
Nalecz Institute of Biocybernetics and Biomedical Engineering, Ks. Trojdena 4 Str., Warsaw, Poland
e-mail: lroszkowiak@ibib.waw.pl

M. Lejeune · C. Lopez
Molecular Biology and Research Section, IISPV, Hospital de Tortosa Verge de la Cinta, C/Esplanetes 14, Tortosa, Spain

R. Bosch
Pathology Department, Hospital de Tortosa Verge de la Cinta, C/Esplanetes 14, Tortosa, Spain

R. Burduk et al. (eds.), *Proceedings of the 9th International Conference on Computer Recognition Systems CORES 2015*, Advances in Intelligent Systems and Computing 403, DOI 10.1007/978-3-319-26227-7_57

1 Background

There are over 20 types of lymphoma cancer that develop from abnormal proliferation of lymphocytes. Follicular lymphoma is the second most common form of non-Hodgkin's lymphoma [16]. The tumour, which has at least a partially follicular pattern, arises from a germinal-centre B cell in the majority of cases. It has long median survival and rare spontaneous remission. Immune patterns present in the microenvironment of tumours have been described as predictors of patients' behaviour. In follicular lymphoma, the quantity of regulatory T cells, which can be marked using FOXP3 antigen, is related with the clinicobiologic behaviour of these patients [1]. It can be inspected through examination under microscope, once biopsy samples are specifically stained. Common practice is using indirect immunohistochemical method with primary antibody against FOXP3 and final staining with 3,3'-diaminobenzidine (DAB) and contra-stained with haematoxylin (H) [18]. The examination of antigen expression most often relies on visual scoring done by expert. As human evaluation is prone to error [17], we proposed an automatic method of T-cell pattern evaluation using digital image of tissue sample and computer-assisted image processing method. A major advantage of computer techniques is the reproducibility of achieved results. Most computer-based procedures for immunohistochemistry image analysis have limited applicability due to numerous drawbacks. Nevertheless, there are several available systems for microscopic image analysis, like EAMUS [4], system developed by Markiewicz [11] and others [6, 9, 10]. The technique based on image feature extraction is used for blood vessel extraction form tissue micro-array samples [3] and nuclei extraction [2]. Other applications suggest conversion to colour spaces, other than RGB, like HSV [8] or CMYK [14]. Moreover, specialized colour deconvolution algorithms [15] can be applied to discriminate multiple spectra. Tissue samples used for this study are stained with DAB&H, so images show brown objects among blue ones. The FOXP3 proteins are targeted by DAB, thus brown—immunopositive objects turn out to be nuclei of regulatory T cells. Stained with blue colour are mostly immunonegative B cells among other negative immune cells like stromal elements. Range in size and shape of brown objects is limited as can be expected, since they are not cancerous cells. On the contrary, immunonegative objects have various sizes and shapes, from round to elongated, depending on the overall course and severity of the disease. Most observable objects have slight but visible texture. Space between cells is filled with bright background that sometimes has slight blue tint. Unfortunately, each and every sample differs one from the other in colour intensity, range of colour and tone. In this study, we present improved and modified version of the previously described software [12]. It was originally created in 2010 to support examination of follicular lymphoma samples stained with DAB&H. From now on, this method will be called METINUS (METhod of Immunohistochemical NUclei Segmentation) and this name will be used in future publications. For test images we use digitalized samples of the tissue, prepared according to standard procedure described in detail in [7]. Images were acquired in Hospital Verge de la Cinta in Tortosa, Spain.

2 Methodology

Software presented in this study can be virtually divided into four main parts: image pre-processing, image processing, object extraction and object analysis, as presented in Fig. 1. The initial two stages were modified and adapted from the previous version [12] to fit into this software. Object extraction and analysis stages are utterly novel. Loaded image has to be prepared before applying main segmentation algorithm [13]. Initially, colour separation is applied and then the image is rescaled as it was described previously [12]. During colour separation all colours other than brown are covered with white background so the result image contains only user-defined brown colour. Because of the variability in stain intensity, white balance and other colour features, the definition of brown colour has to be manually adjusted for a given set of images. Unfortunately, dark blue colour is often included during this separation as it is similar to very dark brown in RGB colour space. The image undergoes rescaling with the Gaussian pyramid algorithm. As size and resolution of the image becomes smaller, the time needed for calculations decreases. This type of resize method keeps objects compact, which is crucial for reconstruction operation. Main image processing consists of adaptive threshold, morphological opening and morphological reconstruction. During this stage image is binarized; two categories, objects and background, are left. Adaptive threshold method adapts the threshold value for each pixel to the local image characteristics. It is not possible to use global thresholding because of fluctuations of colour and contrast observed in this type of images. With the use of morphological operations small groups of pixels and singular pixels are removed and shape of found objects is refined. Afterwards, the image is resized to fit the original image size. Entirely innovative to this procedure are two subsequent stages: object extraction and analysis. As a result of prior processing the map of objects is created, on which we can perform objects extraction. We collect not only geometrical parameters such as area, eccentricity and perimeter but also statistical data. Mean, standard deviation and median are calculated for each object not only in every layer of RGB colour space but also for Lab colour space, defined

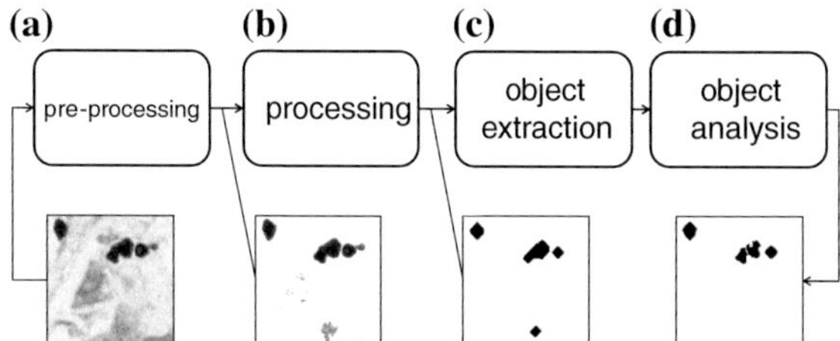

Fig. 1 Diagram of the presented algorithm with image examples. Image **a** is a section of RGB input image. Image **b** is the same fragment after colour separation. Image **c** is the result of other operations, including thresholding. Image **d** is a result of objects analysis—the lower object is removed and the central object is split in two

Table 1 Statistical parameters used to validate objects

Parameter	Validation criteria
Area	<2000
Mean of H layer (colour deconvolution)	<185
Standard deviation of H layer (colour deconvolution)	>53
Mean of DAB layer (colour deconvolution)	<108
Mean of 3rd component of colour deconvolution	>170

brown channel [19] and images obtained with the use of colour deconvolution [15]. The deconvolution algorithm lets us separate stains present in the image of immunohistochemically stained tissue sample. Particularly for this set of images we get H layer, DAB layer and third component. Based on these factors we perform validation of found objects. Those with statistical parameters not in specified (see Table 1) range are disqualified from further assessment. Taking into consideration a possible variation (of colour and tones) in the image we have to adjust those thresholds experimentally to given set of images. So far we conducted validation with the use of area parameter and statistical parameters of all three layers (separated stains) acquired after colour deconvolution. One of the main problems with segmented objects is clustering, and thus next step is dividing groups of cells with watershed segmentation. Dividing of objects is done on the single layer representing Haematoxylin dye separation obtained by colour deconvolution of the original immunohistochemistry image. Then, if the divided elements fulfil our discrimination constraints, they are assessed once more. Finally, a map to count number of objects is created. Majority of the algorithm was implemented in MATLAB. The adaptive threshold algorithm is executed in C to speed up this time-consuming operation. We compared results of improved method with the previous version and with Image-Pro Plus (version 6.2.0.424 for Windows) semiautomatic segmentation. The segmentation in Image-Pro Plus was performed by "Count/Size" procedure. Intensity range was manually adjusted by expert based on a histogram. Additionally, we filtered found objects based on their measurements. We took into account only those that had area between 50 and 2000 pixels, roundness between 0 and 3 and had no holes. To compare the results we used Kendall's tau-b correlation coefficient [5] that measures the strength of the relationship between two sets of samples. It is commonly used as rank-based correlation metric that calculates the difference between the rate of concordance and discordance of samples. The range of this coefficient is between -1 and 1, where 1 indicates the perfect agreement of samples, -1 indicates they are fully inverted and 0 for independent samples.

3 Results and Discussion

The evaluation was performed on a set of 20 arbitrarily chosen images which contain images of most typical tissues. To eliminate variances in colour introduced during staining process all chosen images are from one tissue micro-array—so they were stained simultaneously. All images are well focused. The stain expression is moderate, without weak or most intensive stains. Images with major artefacts, like ripples, wrinkles or folded tissue, were omitted. The comparison is based on quantity of nuclei located per image. Results of evaluation are shown in Table 2. Additionally, we present the number of clusters found by improved software in each image. The commonly used concordance measure Kendall's tau-b is used to show performance improvement achieved using the proposed method of quantification in comparison to the previous one in the context of Image-Pro reference method. In this investigation, the pairwise and continuous version of the test is used and the results are presented in Table 3. It can be observed that the concordance between the new version of the

Table 2 Comparison of results achieved with different softwares

#	Software		Image-Pro Plus 6.2	Clusters
	Previous	Improved		
1	78	24	36	5
2	570	325	390	15
3	17	6	11	1
4	93	44	63	3
5	387	190	228	16
6	155	31	38	2
7	235	18	18	1
8	331	162	189	10
9	575	46	54	8
10	605	261	340	19
11	858	337	316	22
12	197	58	79	1
13	573	265	312	17
14	1090	512	481	48
15	104	57	72	1
16	31	21	16	2
17	91	16	28	2
18	93	22	24	3
19	586	382	379	41
20	39	12	15	1

Table 3 Comparison of results with Kendall's tau-b coefficient

	Previous	Improved	Image-Pro
Previous	1	0.74	0.71
Improved	0.74	1	0.91
Image-Pro	0.71	0.91	1

method of segmentation and reference method from Image-Pro is higher, coefficient value of 0.91, than the previous version concordance (0.71) and is statistically significant. The results of Image-pro Plus are based on the intensity range of the image histogram modified interactively by the operator. It was manually adjusted so that brown objects are in the selected range. Expert's assessment was global and strictly visual without access to particular object evaluation. There was no manual adding or discarding of objects as well as "no border" parameter of any object was changed. As can be seen in Table 2 the highest count of nuclei for every image is achieved by the previous version of the software, described before [12]. Also, the other observable tendency is that in most cases Image-pro Plus detects more objects than our improved software, except for images number 11, 14, 16 and 19 where it is different. It occurs when brown artefacts which are not nucleus but rather deposits of stains are added to the number of nuclei. The main reason for more objects detected by our improved method than Image-Pro Plus is the presence of nuclei clusters in these four images. Image-pro Plus does not perform any division of those clusters, while our software is dividing the largest area over 400 pixels with watershed algorithm with support of information coming from results of colour deconvolution algorithm. In the previous version of the software watershed algorithm was performed on whole image, dividing every object. This gave varied results. Some objects were oversegmented into small pieces (see Fig. 2), often too small to be nucleus. Our improvements modify that problem by executing watershed algorithm only on chosen objects which size testifies to the fact that they are clusters. By this way we can distinguish clusters by their area and divide only them. For images with dispersed objects, the main problem is undetected nuclei. The improved algorithm was slightly adjusted to a new set of images with colour separation parameters. Fortunately, it seems only few objects were omitted; we hope to verify that during comparison to the human expert—pathologist. In previous experiments [12] we proved that this method is useful because of small (11 %) difference between the manual assessment and achieved results for images with dispersed objects. In the improved version of proposed algorithm found, objects undergo evaluation step, based on statistical information: mean, standard deviation and median of stains separated with the colour deconvolution. Clustered objects are divided with watershed algorithm based not only on the information of objects shape, but also on information coming from a single layer representing Haematoxylin dye separated with colour deconvolution. It seems to decrease the difference between the manual assessment and our method's results, and it will be examined in the future.

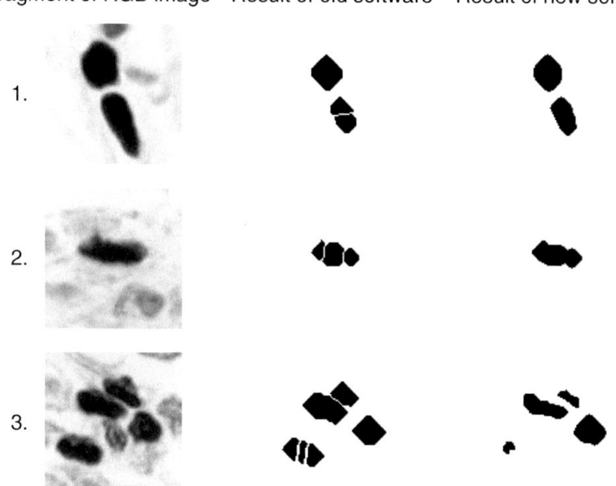

Fig. 2 The comparison of results given by the old and new, improved software. Examples *1* and *2* show unnecessary division of object but improvements cause that it is not divided. In *3* oversegmented object, in the *bottom part*, is limited to very small area but overall number of objects is kept correct. Additionally, the other objects present in example *3* have more natural shape (less defined by shape of structure element of morphology operation) in comparison to the old software's results

4 Conclusions and Future Work

Overall, the comparison of the results shows that modifications proposed in this paper:

1. extraction and evaluation of objects (according to the mean and standard deviation of colour deconvolution layers),
2. performing of watershed algorithm on each cluster separately instead of global handling

is improving results of quantification and makes it closer to the performance of semiautomatic computer-aided pathologist counting, by

1. reduction of false positive objects (as some of them are discarded during validation);
2. improved division of clustered objects;
3. reduction of oversegmentation in case of singular objects.

In the future work we plan to further improve precision and accuracy of the METINUS. We plan to use DAB layer in watershed segmentation, alongside the single layer representing Haematoxylin dye separated with colour deconvolution. It should be considered to calculate watershed algorithm for all objects as in many cases additional unnecessary fragments are taken in. Furthermore, it should improve the shape of objects' borders—keep it more natural and indifferent to shape of morphological operation structural element. Moreover, we will try to find any other significant validation threshold for better discrimination unnecessary objects. Additionally, an implementation of dispersed calculation is planned to improve performance and speed up calculations. In order to verify usefulness of improvements, we plan to perform a full spectrum evaluation in comparison to human expert assessment as next step evaluation.

References

1. Alvaro, T., Lejeune, M., Salvado, M.T., Lopez, C., Jaen, J., Bosch, R., Pons, L.E.: Immunohistochemical patterns of reactive microenvironment are associated with clinicobiologic behavior in follicular lymphoma patients. J. Clin. Oncol. **24**(34), 5350–5357 (2006)
2. Bueno, G., Gonzalez, R., Deniz, O., Garcia-Rojo, M., Gonzalez-Garcia, J., Fernandez-Carrobles, M., Vallez, N., Salido, J.: A parallel solution for high resolution histological image analysis. Comput. Methods Programs Biomed. **108**(1), 388–401 (2012). http://www.sciencedirect.com/science/article/pii/S016926071200082X
3. Fernandez-Carrobles, M., Tadeo, I., Bueno, G., Noguera, R., Deniz, O., Salido, J., Garcia-Rojo, M.: TMA vessel segmentation based on color and morphological features: application to angiogenesis research. Sci. World J. **2013**, 11 (2013)
4. Kayser, K., Radziszowski, D., Bzdyl, P., Sommer, R., Kayser, G.: Towards an automated virtual slide screening: theoretical considerations and practical experiences of automated tissue-based virtual diagnosis to be implemented in the internet. Diagn. Pathol. **1**, 1–8 (2006).http://dx.doi.org/10.1186/1746-1596-1-10
5. Kendall, M.G.: A new measure of rank correlation. Biometrika **30**(1–2), 81–93 (1938). http://biomet.oxfordjournals.org/content/30/1-2/81.short
6. Korzynska, A., Strojny, W., Hoppe, A., Wertheim, D., Hoser, P.: Segmentation of microscope images of living cells. Pattern Anal. Appl. **10**(4), 301–319 (2007). http://dx.doi.org/10.1007/s10044-007-0069-7
7. Korzynska, A., Roszkowiak, L., Lopez, C., Bosch, R., Witkowski, L., Lejeune, M.: Validation of various adaptive threshold methods of segmentation applied to follicular lymphoma digital images stained with 3,3'-diaminobenzidine& haematoxylin. Diagn. Pathol. **8**(1), 48 (2013). http://www.diagnosticpathology.org/content/8/1/48
8. Kuse, M., Sharma, T., Gupta, S.: A classification scheme for lymphocyte segmentation in h&e stained histology images. In: Unay, D., Cataltepe, Z., Aksoy, S. (eds.) Recognizing Patterns in Signals, Speech, Images and Videos. Lecture Notes in Computer Science, vol. 6388, pp. 235–243. Springer, Berlin (2010). http://dx.doi.org/10.1007/978-3-642-17711-8_24
9. Lopez, C., Lejeune, M., Salvado, M.T., Escriva, P., Bosch, R., Pons, L., Alvaro, T., Roig, J., Cugat, X., Baucells, J., Jaen, J.: Automated quantification of nuclear immunohistochemical markers with different complexity. Histochem. Cell Biol. **129**, 379–387 (2008). http://dx.doi.org/10.1007/s00418-007-0368-5
10. Lopez, C., Lejeune, M., Bosch, R., Korzynska, A., Garcia-Rojo, M., Salvado, M.T., Alvaro, T., Callau, C., Roso, A., Jaen, J.: Digital image analysis in breast cancer: an example of an automated methodology and the effects of image compression. Stud. Health Technol. Inf. **179**, 155–171 (2011)
11. Markiewicz, T.: Using matlab software with Tomcat server and Java platform for remote image analysis in pathology. Diagn. Pathol. **6**, 1–7 (2011). http://dx.doi.org/10.1186/1746-1596-6-S1-S18
12. Neuman, U., Korzynska, A., Lopez, C., Lejeune, M.: Segmentation of stained lymphoma tissue section images. In: Piętka, E., Kawa, J. (eds.) Information Technologies in Biomedicine.

Advances in Intelligent and Soft Computing, vol. 69, pp. 101–113. Springer, Heidelberg (2010). http://dx.doi.org/10.1007/978-3-642-13105-9_11
13. Neuman, U., Korzynska, A., Lopez, C., Lejeune, M., Roszkowiak, L., Bosch, R.: Equalisation of archival microscopic images from immunohistochemically stained tissue sections. Biocybern. Biomed. Eng. **33**(1), 63–76 (2013)
14. Pham, N.A., Morrison, A., Schwock, J., Aviel-Ronen, S., Iakovlev, V., Tsao, M.S., Ho, J., Hedley, D.: Quantitative image analysis of immunohistochemical stains using a CMYK color model. Diagn. Pathol. **2**(1), 8 (2007). http://www.diagnosticpathology.org/content/2/1/8
15. Ruifrok, A., Johnston, D.: Quantification of histochemical staining by color deconvolution. Anal. Quant. Cytol. Histol. **23**(4), 291–299 (2001). http://europepmc.org/abstract/MED/11531144
16. Sandeep S.D., Wright, G., Tan, B., Rosenwald, A., Gascoyne, R.D., Chan, W.C., Fisher, R.I., Braziel, R.M., Rimsza, L.M., Grogan, T.M., Miller, T.P., LeBlanc, M., Greiner, T.C., Weisenburger, D.D., Lynch, J.C., Vose, J., Armitage, J.O., Smeland, E.B., Kvaloy, S., Holte, H., Delabie, J., Connors, J.M., Lansdorp, P.M., Ouyang, Q., Lister, T.A., Davies, A.J., Norton, A.J., Muller-Hermelink, H.K., Ott, G., Campo, E., Montserrat, E., Wilson, W.H., Jaffe, E.S., Simon, R., Yang, L., Powell, J., Zhao, H., Goldschmidt, N., Chiorazzi, M., Staudt, L.M.: Prediction of survival in follicular lymphoma based on molecular features of tumor-infiltrating immune cells. New Engl. J. Med. **351**(21), 2159–2169 (2004). http://dx.doi.org/10.1056/NEJMoa041869, pMID: 15548776
17. Seidal, T., Balaton, A.J., Battifora, H.: Interpretation and quantification of immunostains. Am. J. Surg. Pathol. **25**(9), 1204–1207 (2001). http://journals.lww.com/ajsp/Fulltext/2001/09000/Interpretation_and_Quantification_of_Immunostains.13.aspx
18. Swerdlow, S.: For Research on Cancer, I.A., Organization, W.H.: WHO classification of tumours of haematopoietic and lymphoid tissues. World Health Organization classification of tumours, International Agency for Research on Cancer (2008). http://books.google.pl/books?id=WqsTAQAAMAAJ, iSBN-13: 9789283224310; ISBN-10: 9283224310
19. Tadrous, P.: Digital stain separation for histological images. J. Microsc. **240**(2), 164–172 (2010). http://dx.doi.org/10.1111/j.1365-2818.2010.03390.x

Swipe Text Input for Touchless Interfaces

Mateusz Wierzchowski and Adam Nowosielski

Abstract Swipe typing has been designed for touchscreen devices and consists of sliding a finger or stylus through consecutive letters lifting only between words. Since the tracked path contains many redundant letters the accurate recognition of intended word requires a good input path analyser and a word search engine. There are many proprietary solutions of swipe typing available on the market; however, all of them focus on touchscreen devices. On the other hand, there is a growing number of non-contact interfaces operated with gestures. These interfaces limit their operation to pointer manipulation in graphical user interface. The problem of typing in most cases is omitted. Of course touchless interfaces are not designed for text entry purposes but at least some decent possibility of text entry is necessary. In this paper the word recognition algorithm from the tracked path created by the hand movement in front of the wall screen projector is proposed. It is compared with other solutions based on individual key selection in touchless environment.

Keywords Touchless typing · Swipe typing · Text entry · Gesture interaction · Human–computer interaction

1 Introduction

Swipe typing is a method of interaction with a soft keyboard (also referred to as virtual keyboard or on-screen keyboard), originally designed for touchscreen devices. The main idea is to reduce the number of the movements done by the user in order to input a word. In traditional typing every key of a soft keyboard, representing a single character, has to be separately pressed. In the swipe typing, however, the entire word

M. Wierzchowski (✉) · A. Nowosielski
Faculty of Computer Science and Information Technology, West Pomeranian University of Technology, Szczecin, Żołnierska 52, 71-210 Szczecin, Poland
e-mail: mwierzchowski@wi.zut.edu.pl

A. Nowosielski
e-mail: anowosielski@wi.zut.edu.pl

R. Burduk et al. (eds.), *Proceedings of the 9th International Conference on Computer Recognition Systems CORES 2015*, Advances in Intelligent Systems and Computing 403, DOI 10.1007/978-3-319-26227-7_58

is entered by a single gesture. The user touches the key representing the first letter of a word, then drags the pointing object serially over every word's character to finally raise it over the last one. A path of the performed gesture consists of many redundant characters. It must be therefore analysed using a word search engine in order to determine the typed phrase. Reduction of the user movements is crucial here, since it is difficult for the user to coordinate his motion without feeling a physical feedback. Decisively, swipe typing methods provide faster and more intuitive ways of text entry with on-screen keyboards, but the final efficiency of an interface depends on the infallibility of the swipe gesture path analysis and the word search engine. The interaction in the outlined approach is performed in a haptic environment. The paper focuses on the swipe text input approach adaptation for touchless interfaces and proposes an algorithm for the task. The rest of the paper is organized as follows. In Sect. 2 the related works and context of the problem are addressed. Section 3 describes the architecture of the swipe text entry touchless interface. In Sect. 4 the interface is described and the user interactions are explained. Section 5 provides the details on swipe gesture path analysis algorithm. The paper ends with a summary where evaluation of proposed solution is reported and some conclusions drawn.

2 Related Works

The very first text entry interface based on the swipe concept was created by Swype company [1]. It was the third-party QWERTY soft keyboard application for touch-screen devices, initially available on Windows Mobile platform from 2011. Swype continued its development, bringing their product to devices running Android, Bada, MeeGo, Symbian, Windows 7 and iOS. Owing to growing popularity of the solution, similar products appeared on the market, including Android Gesture Typing, SwiftKey Flow, SwipeIt, SlideIT, TouchPal, ShapeWriter, Multiling O Keyboard and Sony Gesture Input. Due to the market competition in this field, providers of these solutions do not reveal their know-how about path analysis algorithms. In all the above proprietary solutions the interaction is performed in haptic environment. A relatively new field of application is the touchless environment. Touchless interfaces are very interesting solutions for the human–computer communication. Touchless interaction enables users to operate without additional medium and intermediate equipment. Those interfaces are particularly suitable for [2–6]: public spaces, clinical environments, consumer electronics, vision-based augmented reality, 3D CAD modelling, sign language recognition, robot control, brainstorming and visualization, etc. The non-contact interaction is usually limited to some kind of manipulation on a display in graphical user interface by the hand movements. Touchless text entry is not the most important element in the interface. Nevertheless, the possibility of writing using the same medium of communication would be valuable and the adaptation of swipe text entry approach can be a significant improvement. An interesting solution has already been presented in [8] where the user's hand and fingers are tracked. The separation of words is achieved here with a pinch gesture. The text entry in touchless environment is considered as a novel issue [7, 8]. Alternatives in the form of automatic speech to text transcription require relatively silent surroundings. Dedicated solutions for disabled people are usually slow and unintuitive. In entertainment products there is a lack of convenient method of typing with touchless interfaces. Therefore, there is a need for a new and intuitive text input interface. It could be used to type on Smart TVs, game consoles or PCs with hands free, while standing in some distance from the screen. The solution that meets these needs is presented in the paper.

3 Structure of Touchless Swipe Text Entry System

Figure 1 presents the scheme of proposed solution for swipe text entry system. The processing is performed on the classical PC equipment, which could be for instance replaced with a SmartTV or a video game console. Proprietary Kinect sensor from Microsoft is utilized for tracking the user's hand movements. The other solutions for hand tracking and gesture recognition reported in literature include [4]: gloves, body suits and marker-based techniques. There is abundant work devoted to the problem of hand gesture recognition and good literature surveys are provided in [3, 4, 9]. Most recent works in the NUI field, however, take advantage of 3D depth sensors [10] and Kinect is preferred here. It provides some sort of unified environment, ensures a similar basis for comparison purposes and finally is accepted as valuable middleware for higher level solutions. The software developer kit for Kinect is available as a free download [11] from the manufacturer's home page [12]. For the above reasons the Kinect sensor has been chosen for our interface. The hardware is complemented

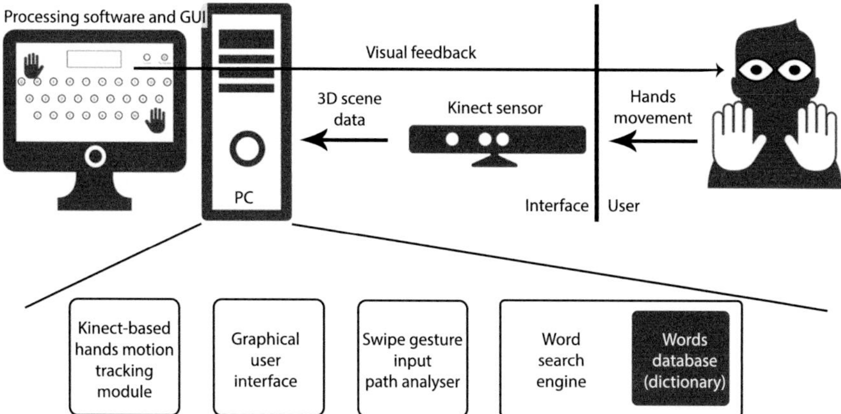

Fig. 1 The structure of the touchless swipe text entry system

with the software consisting of an input path analyser, a word search engine with corresponding database and a graphical user interface with QWERTY soft keyboard. All these components are controlled by the Kinect-based hands motion tracking module. The swipe gesture input path analyser interprets the data describing hand movements as a specific primal phrase (detected letters sequence, before the use of the word search engine). The proposed algorithm is explained in detail in Sect. 4. The word search engine is very important here, for it is able to find the closest word to the initially recognized phrase (primal phrase) using the dictionary (words database). It also provides a list of suggested phrases, close to the typed one. This allows to reduce the number of incorrectly typed words, as well as to easily replace one phrase with another.

4 Interface and Interaction

The GUI displays a well-known QWERTY layout. It contains all the elements necessary to provide comfortable and intuitive ways of interaction with the user. In order to make a visual feedback more understandable, a number of visual effects were created, such as dynamic sizes and background colour of the elements. The scheme of the interface layout is presented in Fig. 2. The window of the application displays a section of the depth map, indicating the user silhouette (displayed for some feedback purposes). The user is able to interact with the interface by moving his hands in front of the sensor. The movements are modelled using 3D skeleton model. This model has two nodes representing positions of each hand of the user in analysed 3D

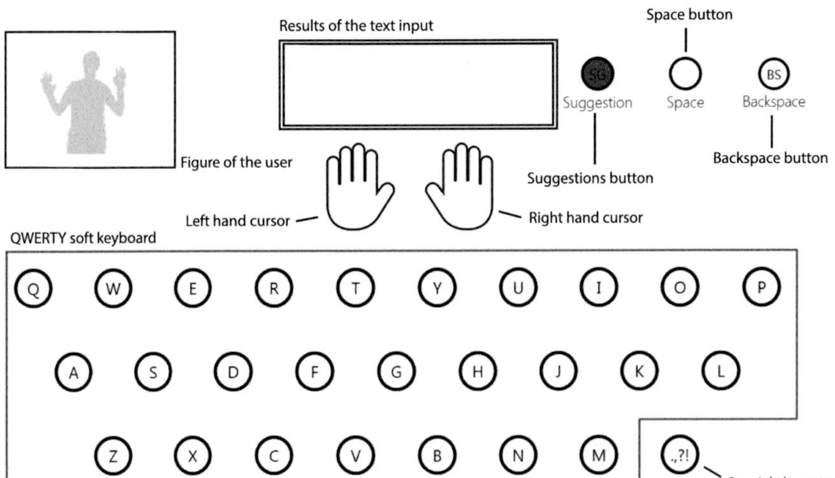

Fig. 2 The graphical user interface scheme

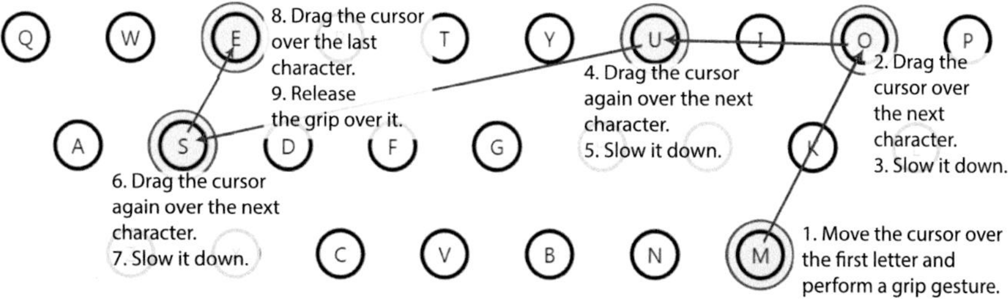

Fig. 3 The algorithm for swipe gesture on the example of the word "MOUSE"

space. These coordinates are cast on the graphical interface window and determine cursors positions. Two different ways to enter the text have been proposed. The first one is analogous to the generally accepted way of interaction with a soft keyboard. The system recognizes a push gesture performed by hand in front of the intended character. It is similar to key press in traditional solution. In addition, two modes have been proposed. In the first mode the user performs single-handed actions, while in the second the user uses both of them. The second and promoted method is the swipe text input. Its purpose is to provide a fast and easy way of touchless text entry. However, the method of the proposed solution cannot be considered as an equivalent of those in touch screen devices. It is similar with those in well-known swipe interfaces, but at the same time is modified to be better optimized for touchless systems and to simplify the swipe gesture input path analyser algorithms. It has to be performed using one hand only. Traditionally, the whole word is entered with a single swipe gesture and a space is automatically inserted after it. The detailed algorithm for swipe gesture is given as follows and schematically depicted in Fig. 3:

1. Move the cursor over the key of the first letter of the word.
2. Perform a grip gesture over that button (make a fist without releasing).
3. If the word is a single character go to the step no. 7.
4. Unhesitating, move the fist to drag the cursor over the next key.
5. Slow down the hand movement.
6. If there are any characters of the word left go back to step no. 4.
7. Open the fist, in order to release the grip over the key of the last character.

5 Proposition of Swipe Touchless Text Input Algorithm

The proposed solution consists of graphical user interface, hands motion tracking module, input path analyser and word search engine. The two last-mentioned components are described in detail in the current section.

5.1 The Swipe Gesture Input Path Analyser

The analysis of the swipe gesture input path is a complex task. First off, the input data have to be explained. During the swipe gesture input process two data structures are stored. The first one is a vector (x, y, t), where x and y are the coordinates of the cursor on the window plane (in pixels) and t corresponds to the time in milliseconds. This data set is stored per every input frame from the sensor. The second structure stores the type of the event associated with gestures: grip, grip release, pointer enter and pointer leave. The time of the beginning and the end of the interaction during the gesture action (in milliseconds), as well as the indicated character, are also recorded. Since it is possible to obtain a chronological order of the cursor–keyboard interactions, the sequence of the keys, over which the indicator was dragged, is known. However, it is very unlikely or impossible that the cursor was dragged over target letters only. In most cases it makes unintended interactions with those keys that are positioned on the path. For example, the most simple and the shortest path for the pointer to move from the letter "M" to the letter "O", on a QWERTY keyboard, is through the "K" key. That is why the stored character sequence contains many unintended letters. The sequence for the word "MOUSE", for example, could be like "**M**K**O**I**U**YFD**SE**". The whole purpose of the analyser is to automatically decide which of the stored characters were intended by the user. There is no need to analyse the first and the last interactions, for it is clear that they point to the first and the last character of the typed word. The rest of the interactions are classified as intended or not, using a value of the introduced user decision error factor (*UDEF*). It is a quotient of the average speed of the cursor, during a single interaction and the duration of this interaction (1). The higher its value is, the more unlikely is that an interaction was intended by the user. Mathematical equations explaining how to calculate *UDEF* are presented below:

$$UDEF(n) = \begin{cases} 0 & : n \in \{1, N\} \\ \frac{\overline{Vc_n}}{ti_n} & : 1 < n < N \end{cases}, \tag{1}$$

where

$$\overline{Vc_n} = \frac{\sum\limits_{1 \le p \le P} Vc_p}{P}, \tag{2}$$

$$ti_n = te_n - tb_n, \tag{3}$$

$$Vc_p = \begin{cases} 0 & : p = 1 \\ \frac{\sqrt{(Xc_p - Xc_{p-1})^2 + (Yc_p - Yc_{p-1})^2}}{t_p - t_{p-1}} & : 1 < p \le P \end{cases}, \tag{4}$$

$$t \in < tb_n, te_n >,$$

$$n \in \{1, 2, 3, \ldots, N\},$$

$$p \in \{1, 2, 3, \ldots, P\},$$

$UDEF$	—user decision error factor,
n	—nth cursor–keyboard interaction during a single swipe gesture,
N	—number of cursor–keyboard interactions during a single swipe gesture,
$\overline{Vc_n}$	—average cursor speed during nth interaction [px/s],
ti_n	—duration of nth interaction [s],
p	—pth cursor position during a single swipe gesture,
P	—number of cursor positions stored during a single swipe gesture,
Vc_p	—speed of the cursor in pth position [px/s],
tb_n	—the time of the beginning of nth interaction [s],
te_n	—the time of the end of nth interaction [s],
Xc_p, Yc_p	—coordinates of the cursor in pth position,
t_p	—the time in the pth cursor position [s].

Obtained values (according to the above calculations) of the user decision error factors for each cursor–keyboard interaction during a single swipe gesture can now be used to determine whether or not an interaction was intended by the user. In order to perform it precisely, the acceptance threshold *TH* needs to be calculated individually for every swipe gesture. This is a product of the average *UDEF* value in a swipe gesture and the tolerance level of hesitant cursor movement H. The value of H must be determined experimentally. The value of H was experimentally determined in our studies as 0.15. The logical function of acceptance of the interaction as intended by the user, denoted as *Fa*, is presented below:

$$Fa(n) = \begin{cases} 0 : UDEF_n > TH \\ 1 : UDEF_n \leq TH \end{cases}, \tag{5}$$

where

$$TH = \overline{UDEF} \cdot H, \tag{6}$$

$$\overline{UDEF} = \frac{\sum\limits_{1 \leq n \leq N} UDEF_n}{N}, \tag{7}$$

$$n \in \{1, 2, 3, \ldots, N\},$$

$$H \in < 0, 1 >,$$

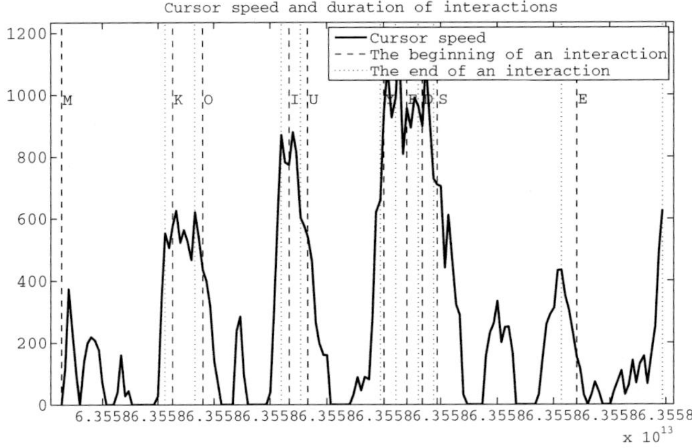

Fig. 4 Example of swipe gesture for the word "MOUSE" showing the cursor speed and the duration of the interactions

Fa	—logical function of acceptance an interaction as intended by the user,
TH	—acceptance threshold,
H	—tolerance level of the hesitant cursor movement.

Figure 3 presents an example of cursor speed and the duration of the interaction with consecutive characters laying on the performed path for the word "MOUSE", touchless entered by the user using our interface. The speed values are low for the intended letters (Fig. 4). Figure 5 presents the graph showing *UDEF* and *TH* values for swipe gesture for the word "MOUSE" and the same case as above. Hight *UDEF* values are noticed for unintended letters. Proper adjustments of the threshold value, as 0.15 coefficient of the average *UDEF* value in our case, allow the proper character eliminations. Letters in circles on the abscissa (Fig. 5) are determined to be intended by the user.

5.2 The Word Search Engine

The swipe gesture input path analyser interprets the data corresponding to hand movements. The detected letter sequence is cleared of unnecessary characters which were unintended but located on the movement path. The resultant sequence constitutes the primal phrase. The primal phrase should be matched with available database words. For this task the dictionaries provided for the OpenOffice suite are utilized. The word search engine is very important here, for it is able to find the closest word to the initially recognized phrase (primal phrase) using the dictionary. It also provides a list

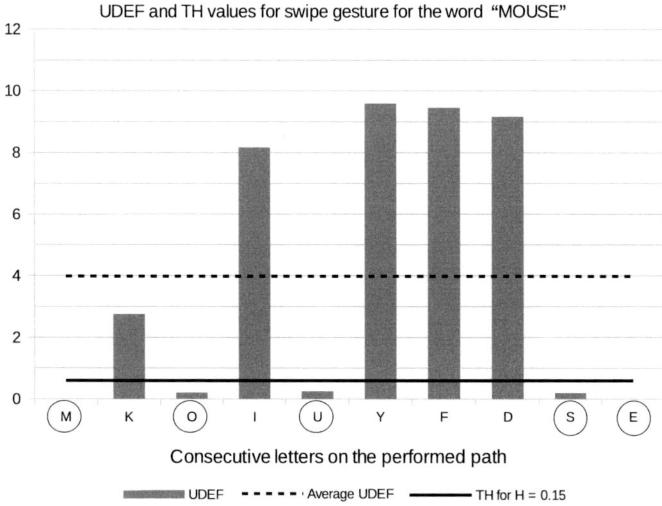

Fig. 5 The graph showing *UDEF* and *TH* values for swipe gesture for the word "MOUSE"

of suggested phrases, close to the typed one. The user can easily replace the primal phrase with one suggested at any time, no matter which typing method was used to enter the word. It allows to significantly reduce the number of mistakes made by swipe gesture input path analyser. The solution implements Hunspell Spell Checker engine [13]. Core algorithms of this engine are based on the MySpell spellchecker, which is a portable and thread-safe C++ library reimplementation of ispell created by Kevin Hendricks for the OpenOffice project. In short, MySpell and ispell spell-checking methods are focused on affixes stripping and looking up substrings in the dictionary. In addition, they use switches (in linguistic terms privative lexical features) that can eliminate some incorrect results of these heuristic methods. These approaches failed to perform efficiently in languages with rich morphology. The HunSpell method is the solution for that problem by incorporating a multistep sequential affix-stripping algorithm. As a multilingual spellchecker it is one of the most popular open source word search engines today.

6 Summary

In this paper, the swipe text input algorithm for touchless interfaces has been proposed. It has a variety of uses in spontaneous, non-contact, casual writing performed in front of some display screen. The operations are carried out with the hand gesture movements. The proposed mechanism imitates the modern way of word entry on haptic devices where words are entered using single, yet compound, movements instead of individual letter selection. Adaptation of such approach to touchless environment required some modifications to compensate for freedom of the user movements and for the distinction between intended user actions from involuntary ones. The proposed solution utilizes its own algorithm for the performed path analysis. The user decision error factor *UDEF* has been introduced to classify user actions as intended or not. We evaluated our swipe solution in comparison with individual letter selection writing performed in touchless environment using similar experimental conditions (the same middleware algorithms for gesture recognition and the same group of 44 participants). Swipe text input has been compared with push gesture typing on QWERTY layout keyboard (using our interface and basic entry techniques, see Sect. 3). One hand and double hands modes have been individually tested. The results are as follows: average of 42 s for swipe approach and surprisingly equal average typing speed of 01:14 min in both push gesture QWERTY interfaces one operated with single hand and the other with both hands. The experiments have been performed under *forced* error correction conditions [14] which guarantee the error-free resultant text and justifies the use of time performance as evaluation measure. In conclusion, the presented solution proved to be effective.

References

1. The Swype solution, http://www.swype.com/
2. Chattopadhyay, D., Bolchini, D.: Understanding Visual Feedback in Large-Display Touchless Interactions: An Exploratory Study. Research report, IUPUI Scholar Works, Indiana University (2014)
3. Sarkar, A.R., Sanyal, G., Majumder, S.: Hand gesture recognition systems: a survey. Int. J. Comput. Appl. **71**(15), 25–37 (2013)
4. Mitra, S., Acharya, T.: Gesture recognition: a survey. IEEE Trans. Syst. Man Cybern. Part C: Appl. Rev. **37**(3), 311–324 (2007)
5. Johnson, R., O'Hara, K., Sellen, A., Cousins, C., Criminisi, A.: Exploring the potential for touchless interaction in image-guided interventional radiology. In: ACM Conference on Computer-Human Interaction (CHI '11), pp. 3323–3332 (2011)
6. Dostal, J., Hinrichs, U., Kristensson, P.O., Quigley, A.: SpiderEyes: designing attention- and proximity-aware collaborative interfaces for wall-sized displays. In: Proceedings of the 19th international conference on Intelligent User Interfaces IUI '14. pp. 143–152. Haifa, Israel (2014)
7. Nowosielski, Adam: QWERTY- and 8pen-Based Touchless Text Input with Hand Movement. In: Chmielewski, Leszek J., Kozera, Ryszard, Shin, Bok-Suk, Wojciechowski, Konrad (eds.) Computer Vision and Graphics. Lecture Notes in Computer Science, vol. 8671, pp. 470–477. Springer, Heidelberg (2014)
8. Markussen, A., Jakobsen, M.R., Hornbæk, K.: Vulture: a mid-air word-gesture keyboard. In: Proceedings of the SIGCHI Conference on Human Factors in Computing Systems CHI'14, pp. 1073–1082 (2014)
9. Ibraheem, N.A., Khan, R.Z.: Survey on various gesture recognition technologies and techniques. Int. J. Comput. Appl. **50**(7), 38–44 (2012)
10. Placitelli, A.P.: Toward a framework for rapid prototyping of touchless user interfaces. In: 2012 Sixth International Conference on Complex, Intelligent, and Software Intensive Systems, pp. 539–543 (2012)
11. Giorio, C., Fascinari, M.: Kinect in Motion - Audio and Visual Tracking by Example Starting. Packt Publishing, Birmingham (2013)

12. The Kinect for Windows SDK, http://www.microsoft.com/en-us/kinectforwindows/
13. The Hunspell solution, http://hunspell.sourceforge.net/
14. Arif, A.S., Stuerzlinger, W.: Analysis of text entry performance metrics. In: 2009 IEEE Toronto International Conference Science and Technology for Humanity (TIC-STH), pp. 100–105 (2009)

Defect Detection in Furniture Elements with the Hough Transform Applied to 3D Data

Leszek J Chmielewski, Katarzyna Laszewicz-Śmietańska, Piotr Mitas, Arkadiusz Orłowski, Jarosław Górski, Grzegorz Gawdzik, Maciej Janowicz, Jacek Wilkowski and Piotr Podziewski

Abstract Defects in furniture elements were detected using data from a commercially available structured light 3D scanner. Out-of-plane deviations down to 0.15 mm were analyzed successfully. The hierarchical, iterated version of the Hough transform was used. The calculation of position of the plane could be separated from that of its direction due to the assumption of nearly horizontal location of the plane, which is natural when the tested elements lie on a horizontal surface.

Keywords Defect detection · Quality inspection · Furniture elements · Iterated · Hierarchical · Hough transform · 3D scanner · Structured light

1 Introduction

Coordinate measuring techniques, including the noncontact 3D measurement methods, are applied more and more frequently in different areas and are one of the fastest growing areas of modern metrology. Three-dimensional scanning, thanks to its high measurement precision and speed, can be an effective tool for monitoring the dimensional and shape accuracy of products manufactured in the various branches of timber industry, including the manufacturing of furniture elements. A key aspect of modern furniture production is striving to ensure high quality, in accordance with

L.J. Chmielewski (✉) · A. Orłowski · G. Gawdzik · M. Janowicz
Faculty of Applied Informatics and Mathematics (WZIM),
Warsaw University of Life Sciences (SGGW), ul. Nowoursynowska 159,
02-775 Warsaw, Poland
e-mail: leszek_chmielewski@sggw.pl
URL: http://www.wzim.sggw.pl

K. Laszewicz-Śmietańska · P. Mitas · J. Górski · J. Wilkowski · P. Podziewski
Faculty of Wood Technology (WTD),
Warsaw University of Life Sciences (SGGW), ul. Nowoursynowska 159,
02-775 Warsaw, Poland
e-mail: katarzyna_laszewicz@sggw.pl
URL: http://www.wtd.sggw.pl

R. Burduk et al. (eds.), *Proceedings of the 9th International Conference on Computer Recognition Systems CORES 2015*, Advances in Intelligent Systems and Computing 403, DOI 10.1007/978-3-319-26227-7_59

the increasing requirements of potential customers [1–4]. This implies the need for automated quality inspection of furniture parts, including the dimension and shape accuracy. One of the effective tools for this task could be machine vision understood as automatic analysis of images [5, 6]. The use of 3D scanning would be another step forward. At the moment, however, 3D scanners are used in furniture industry only occasionally, and almost exclusively in the reconstruction of missing pieces of furniture (reverse engineering), or to duplicate reference objects of complex shapes. There are no reports of any practical application in the field of technical inspection of mass-production of furniture. However, it can be expected that a fully functional and inexpensive automatic identification systems of dimensional inspection for typical furniture elements, using the universal structured light 3D scanners which are widely available, could prove to be quite interesting for furniture manufacturers who face the stringent quality requirements. Detecting the deviations from the specifications during and at the end of the technological processes can minimize the number of defects, which leads to reduction in manufacturing costs. The domain of quality control of furniture elements should be strictly distinguished from the domain of automatic analysis of anatomical defects in timber. There are many solutions to the latter problem and industrial systems for analyzing, cutting, and sorting are available on the market (see, e.g., [7–9] to name but a few). There exists extensive literature on this subject (see, e.g., [10, 11] for reviews). The reason for this situation can be probably attributed to that the production of raw timber is a mass production, while the furniture designs are varied to a much larger extent. In this paper, we shall concentrate on the detection of defects which can be defined as a deviation from a plane. For the detection of planes in 3D we shall use the Hough transform (HT). The range of the literature on HT is very wide; however, relatively little attention has been paid to the detection of planes. In [12] the classical Hough transform is used for detecting a plane in 3D parameter space, which is relatively time and memory consuming; therefore, in conclusion another method of plane detection chosen as more efficient. In [13] a number of Hough space architectures is reviewed and an original version is proposed in the form of a ball to receive even accuracy in the solution space and good rotation invariance. The three-point randomized version is finally chosen, where three parameters of the plane are calculated together from random 3-tuples of points. In [14], the classical HT formulation is used. Standard HT is used, but much attention is paid to the problem of manipulating the plane representation within the frames of the conformal geometric algebra. In [15] the observation that the scans made with the LIDAR technique have the shape of conics is used to speed up the accumulation. The curvatures found from the data are used. In [16] the data are used with the known neighborhood information, so the calculations can be started from finding the local direction in such data, so the accumulation of direction and distance can be decoupled. In [17] the points are first clustered and for the Hough transform the clusters with known centroids are used, so the direction can also be easily found. In our work we use the Hough transform for the most general case of raw data with no information on the neighborhood of a data point. Therefore, no local information on the directions or derivatives in such data is available. However, for the furniture elements which lie on a flat surface it is possible to make an assumption that the interesting plane is close to horizontal (within some margin). This will make it possible to use the hierarchical HT. To make the calculations simpler no nonlinear equations will be used. The approach used will be close to that used previously in [18] in another application. We shall also use some experience from our previous work on defect detection [19–21]. We shall investigate the potential of using the data from a 3D scanner working according to the principle of structured light. The scanners of this type available on the market attain the measurement accuracy well below 1 mm for the objects having the length over 1 m. This is a very moderate cost solution in comparison to the laser (LIDAR) scanners on one side and the professional measurement systems for timber on the other. The remainder of this paper is organized as follows. The data sets used and the problem to be solved are described in Sect. 2. The method proposed is presented in Sect. 3. Results are shown and discussed in Sect. 4 and the paper is concluded in Sect. 5.

2 Data and Problem Statement

In this study three defects with three levels of expected difficulty were taken into consideration. The difficulty was related to the size of the defect. Images of the objects with these defects are shown in Figs. 2a–d, 3a–d and 4a in the sequence according to the expected difficulty. The large defect in Fig. 2a is a *chipping* resulting form the fracture during the process of machining the cut-out. The defect in Fig. 3a is a *gap* which resulted in the process of machining and assembling a finger joint (the face view of the joint itself is not shown). The defect in Fig. 4a is an *orange peel* which emerged during the process of painting a surface of an element. The expected difficulty of detecting this defect is large due to that the depth of the irregularities is close to or below the limit of accuracy of the measuring device. The objects were scanned with the structured light scanner *Smarttech scan 3D DUAL VOLUME* [22] with a 5 Mpix sensor. The scans were taken from the distance 500 mm. At this distance the density of points is around 8 points/mm in each direction x, y and the accuracy in direction z is 0.4−0.8 mm, depending on the reflecting properties of the surface. The data for the objects are displayed in Figs. 2b–d, 3a–d and 4a, b, respectively. In this introductory study, we shall consider the defects which can be considered as the deviation of dimensions from the largest plane present in the data, which will also serve as the reference line for dimensions. The approach using the same principle can be extended to more numerous geometrical defects, with the use of the hierarchy principle. Here, we wish to concentrate upon the testing of the limits to which the dimensions can be derived from the images considered.

3 Method

The equation of a plane in the coordinate system $Oxyz$ (let Oz be vertical) is

$$(x - x_0)n_x^1 + (y - y_0)n_y^1 + (z - z_0)n_z^1 = 0, \tag{1}$$

where the six parameters are: $[n_x^1, n_y^1, n_z^1] = n^1$—normal vector (it can be a unit vector); (x_0, y_0, z_0)—a reference point belonging to the plane. Note that the components of the normal vector can be expressed in terms of the angles it forms with the coordinate system, as used in many publications, but we shall not refer to these angles to keep all the equations linear. To show explicitly that the plane is parameterized with only three parameters we shall assume that the vertical component is unit. This assumption is consistent with the assumption that we consider the nearly vertical planes. The reference point can be located anywhere on the plane, so let it lie on Oz. Therefore, Eq. (1) becomes

$$x\,n_x + y\,n_y + z - z_0 = 0. \tag{2}$$

The measurement points are $P_i = (x_i, y_i, z_i)$, $i = 1, \ldots, M$, where $M \gg 3$. The plane represented by the greatest number of these points is to be detected. Some of the points can represent other part of the scanned furniture element, including possibly the defect. This assumption, together with the assumption of near-horizontal position, makes it possible to estimate the parameter z_0 by the maximum on the projection of all the points onto the axis Oz. When z_0 is known the remaining parameters n_x, n_y can be found in a 2D Hough transform according to (2). To make the result more stable with respect to the estimate of z_0 the data are expressed in the barycentric coordinates. Now, the n_x, n_y found are substituted into (2) and new z_0 is found in a 1D transform. These steps are repeated until the parameters stabilize. The iterative HT was originally proposed in [23], in a different application. The resolution of the accumulators used was 0.01 mm for z_0 (range according to data), and 0.001 for n_x, n_y in the range $\langle -1, 1\rangle$ in the example of Fig. 2 and 0.0001 in the range $\langle -0.1, 0.1\rangle$ in the examples of Figs. 3 and 4. The stop criterion was zero change in the parameters. Now, the out-of-plane distances of points can be found. The model of the object is used to extract the part of the infinite plane which is actually present in the object (this can be seen in Fig. 2a, d where the points belonging to the cut-out in the object were removed). Then, the distances are thresholded with a set of thresholds, according to the expected size of the defect. The points for which the distance is beyond the threshold represent the defect. The way the calculations proceed, for two from the examples considered, is shown in Fig. 1. Please note that in the first iteration the accumulator for z_0 has a flat maximum. This indicates that the assumption of initial closeness to planarity is important in this approach. Please note also a characteristic circle-like shape in the accumulator for the normal vector elements. When z_0 is not accurate the normal sort of turns around the accurate position to minimize at least the part of the distances. This is the reason why it is profitable to express the measurement

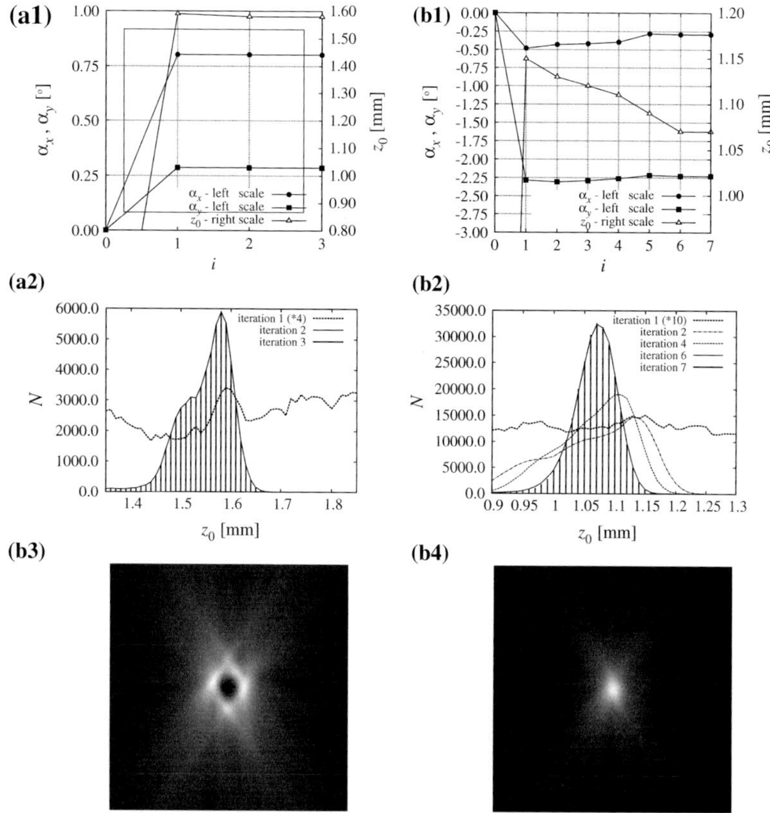

Fig. 1 Chosen results: **a** chipping, **b** gap. Subfigures **a1, b1**: results of iterations (components of the normal transformed to angles); **a2, b2**: accumulators of height z_0 for all iterations; **b3, b4**: accumulator of normal vector $[n_x, n_y]$ for first and last iteration (scaled into $\langle 0, 255\rangle$ for visualization), only central parts of tables are shown

points in the coordinate system in which the points are possibly close to the origin. The calculations for data containing several hundred thousands of data points took up to 30 s to analyze. The process could be accelerated by using the randomized approach and variable resolution in subsequent iterations [18]. The method can be extended to hierarchically find multiple shapes by eliminating the shapes already found from the data, which is a generally known technique in HT. The methods for detection of other shapes, like lines and circles in 2D or cylinders in 3D, are known and can be adapted to the problem of interest.

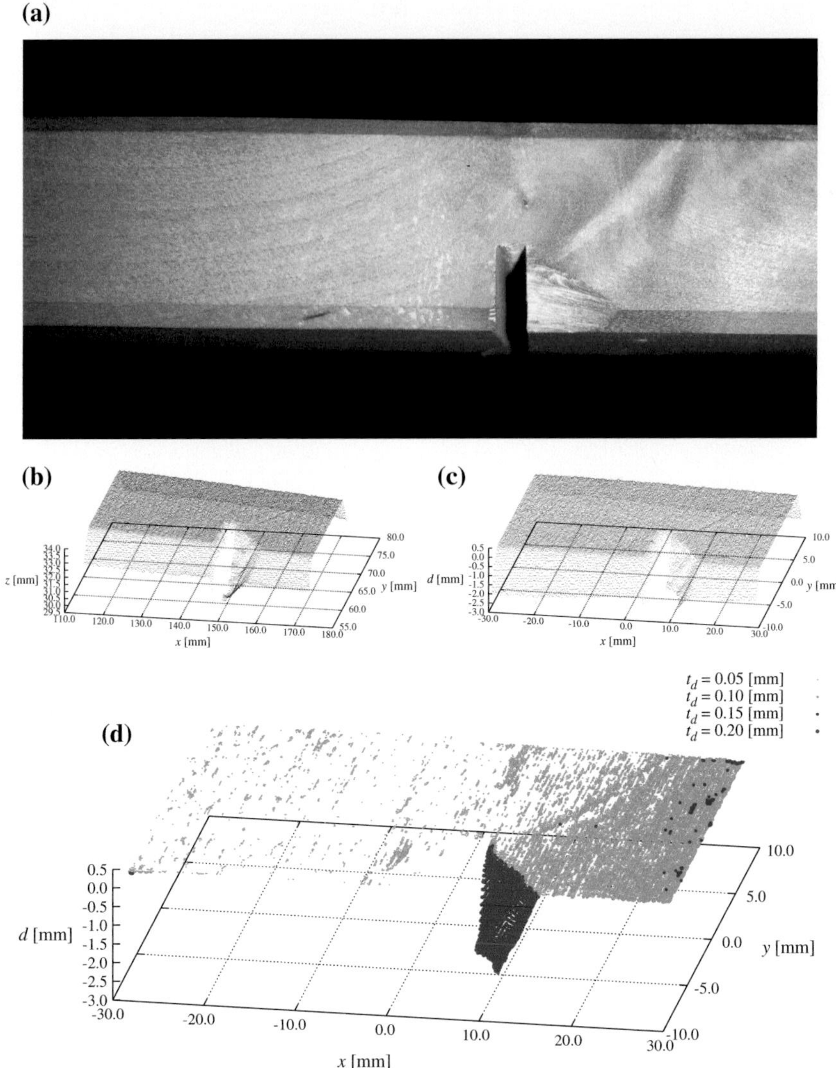

Fig. 2 Defect *chipping* as an example of an easy case. **a** View of the defect; **b** raw scanned data; **c** data with the main plane located horizontally, so the plot represents distance from the plane; **d** points having distances larger than t_d: *red color* represents the defect. In (**d**) the points belonging to the cut-out in the object were removed

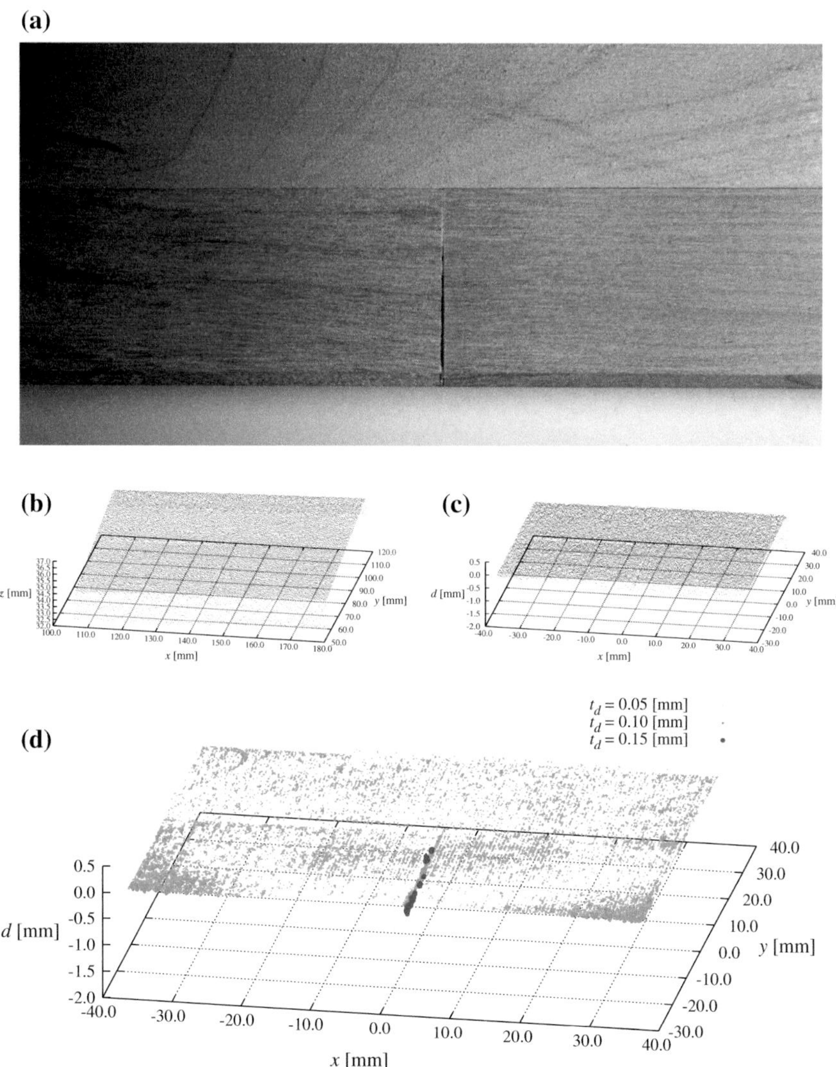

Fig. 3 Defect *gap* in a finger joint. **a** View of the defect; **b** raw scanned data; **c** data with the main plane located horizontally, so the plot represents distance from the plane; **d** points having distances larger than t_d: *red color* represents the defect

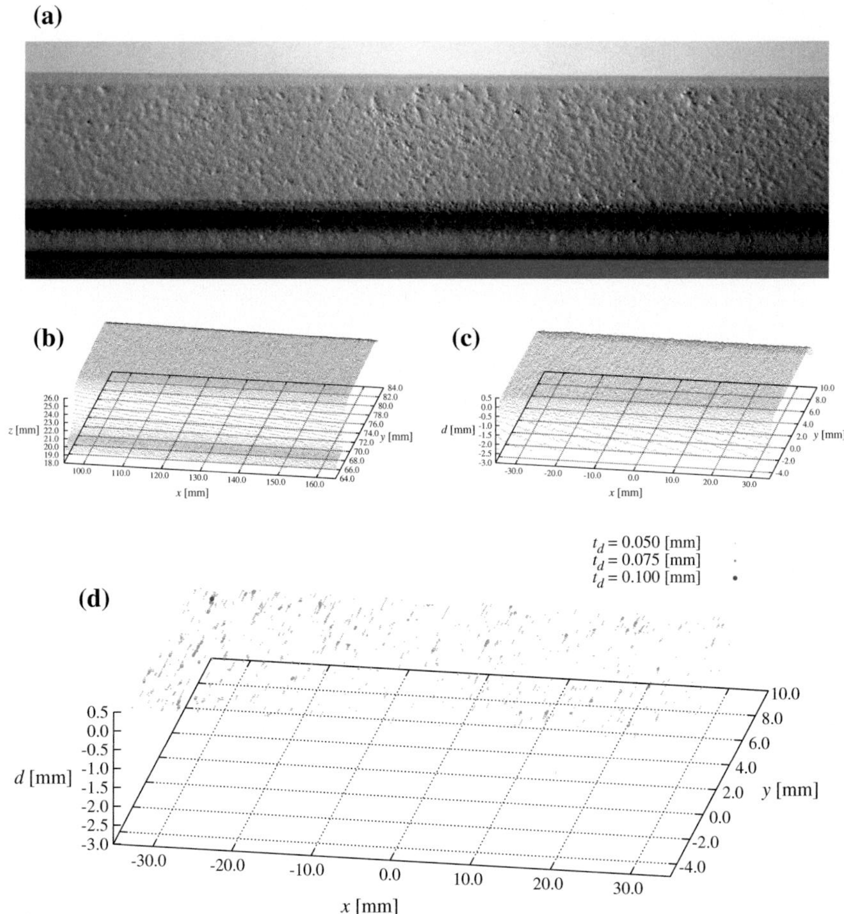

Fig. 4 Defect *orange peel* on a surface of an element as an example of a difficult case. **a** View of the defect; **b** raw scanned data; **c** data with the main plane located horizontally, so the plot represents distance from the plane; **d** points having distances larger than t_d: defect is hardly visible

4 Results and Discussion

The data shown in Figs. 2b–d, 3a–d and 4a, b transformed to show the distances from the main planar surface are shown in Figs. 2c, d, 3a–d and 4a–c, respectively. In Figs. 2d, 3a–d and 4a–d the distances thresholded at subsequent levels are shown, which represent the final results. For the *chipping* in the machined element in Fig. 2 the result is close to expected and the defect is clearly visible. However, the distribution out-of-plane distances across the object is not even—there are larger values in the right-hand side. Therefore, the resolution for the normal vector was increased in the following examples. The results for the *gap* in the finger joint in Fig. 3 show that the out-of-plane distances are evenly distributed for the object besides the defect. This indicates that the accuracy of the plane detection was high enough. The defect is also visible clearly; it is well within the capability of the measuring device. The defect *orange peel* in Fig. 4 is an example where the defect is at the border of the scanner accuracy which was 0.08 mm in this case. The defects can not be distinguished in Fig. 4d. It should be noted that better accuracy could be attained with the scanner used if the scanning mode with a smaller field of view were used. This example is shown, however, to indicate that some defects should be detected separately from the others.

5 Conclusion

To our best knowledge, automatic quality inspection of furniture elements is a subject which is virtually absent in the literature. We have performed an introductory study on the detection of defects with the use of a 3D structured light scanner. The experience gained is very promising. The measuring method allowed us to reliably find the defects of the dimension down to 0.15 mm out-of-plane deviation which is in conformity with the accuracy of the scanner used. The two-level hierarchical iterative Hough transform performed well in finding the most distinct plane in the data, provided that an assumption of closeness to horizontal location of this plane is valid, which is natural in the problem considered. The methodology used can be extended to shapes containing more than one plane and other shape patterns used in the design of furniture elements.

References

1. Young, T., Winistorfer, P., Siqun, W.: Multivariate control charts of MDF and OSB vertical density profile attributes. For. Prod. J. **49**(5), 79–86 (1999)
2. Zakrzewski, W., Staniszewska, A.: Accuracy of Woodwork Cutting (in Polish). Wydaw. Akademii Rolniczej im. Augusta Cieszkowskiego, Poznań (2002)
3. Lăzărescu, C.N., Lăzărescu, C.: Tolerances and Dimensional Control in Wood Industry. Editura Universităţii Transilvania, Braşov (2005)
4. Laszewicz, K., Górski, J.: Control charts as a tool for the management of dimensional accuracy of mechanical wood processing (in Russian). Ann. Wars. Univ. Life Sci.-SGGW, For. Wood Technol **65**, 88–92 (2008)
5. Laszewicz, K.: Machining Accuracy of MDF Boards Milling Process (in Polish). Ph.D. thesis. Warsaw University of Life Sciences, Faculty of Wood Technology, Warsaw (2011)
6. Laszewicz, K., Górski, J., Wilkowski, J.: Long-term accuracy of MDF milling process-development of adaptive control system corresponding to progression of tool wear. Eur. J. Wood Wood Prod. **71**(3), 383–385 (2013). doi:10.1007/s00107-013-0679-2
7. Innovativ vision: woodeye. http://woodeye.se/en/ (2015). Accessed 09 Mar 2015
8. WEINING: company. http://www.weinig.com (2015). Accessed 09 Mar 2015
9. MiCROTEC: company. http://www.microtec.eu (2015). Accessed 09 Mar 2015

10. Bucur, V.: Techniques for high resolution imaging of wood structure: a review. Meas. Sci. Technol. **14**(12), R91 (2003). doi:10.1088/0957-0233/14/12/R01
11. Longuetaud, F., Mothe, F., Kerautret, B., Krähenbühl, A., Hory, L., Leban, J., Debled-Rennesson, I.: Automatic knot detection and measurements from X-ray CT images of wood: a review and validation of an improved algorithm on softwood samples. Comput. Electron. Agric. **85**, 77–89 (2012). doi:10.1016/j.compag.2012.03.013
12. Tarsha-Kurdi, F., Grussenmeyer, P.: Hough-transform and extended RANSAC algorithms for automatic detection of 3D building roof planes from Lidar data. In: Proceedings of the ISPRS Workshop on Laser Scanning 2007 and SilviLaser 2007. vol. XXXVI-3/W52., Espoo, Finland, pp. 407–412 (Sep 2007)
13. Borrmann, D., Elseberg, J., Lingemann, K., Nüchter, A.: The 3D hough transform for plane detection in point clouds: a review and a new accumulator design. 3D Res. **2**(2), 1–13 (2011). doi:10.1007/3DRes.02(2011)3
14. Bernal-Marin, M., Bayro-Corrochano, E.: Integration of Hough transform of lines and planes in the framework of conformal geometric algebra for 2D and 3D robot vision. Pattern Recognit. Lett. **32**(16), 2213–2223 (2011). doi:10.1016/10.1016/j.patrec.2011.05.014
15. Grant, W., Voorhies, R., Itti, L.: Finding planes in LiDAR point clouds for real-time registration. In: Proceedings of the 2013 IEEE/RSJ International Conference Intelligent Robots and Systems IROS, pp. 4347–4354 (Nov 2013). doi:10.1109/IROS.2013.6696980
16. Hulik, R., et al.: Continuous plane detection in point-cloud data based on 3D Hough Transform. J. Vis. Commun. Image Represent. **25**(1), 86–97 (2014). doi:10.1016/j.jvcir.2013.04.001
17. Limberger, F.A., Oliveira, M.M.: Real-time detection of planar regions in unorganized point clouds. Pattern Recognit. **48**(6), 2043–2053 (2015). doi:10.1016/j.patcog.2014.12.020
18. Chmielewski, L., Orłowski, A.: Ground level recovery from terrestrial laser scanning data with the variably randomized iterated hierarchical hough transform. In: Azzopardi, G., Petnov, N., (ed.) Proceedings of the International Conference on Computer Analysis of Images and Patterns CAIP 2015, vol. 9256 of LNCS., Valletta, Malta, Springer Verlag (2–4 Sep 2015) pp. 630–641 (Part I). doi:10.1007/978-3-319-23192-1_53
19. Nieniewski, M., Chmielewski, L., Jóźwik, A., Skłodowski, M.: Morphological detection and feature-based classification of cracked regions in ferrites. Mach. Gr. Vis. **8**(4), 699–712 (1999)
20. Jóźwik, A., Chmielewski, L., Skłodowski, M., Cudny, W.: A parallel net of (1-NN, k-NN) classifiers for optical inspection of surface defects in ferrites. Mach. Gr. Vis. **7**(1–2), 99–112 (1998)
21. Mari, M., et al.: The CRASH Project: Defect detection and classification in ferrite cores. In: Bimbo, A.D., (ed.) Proceedings of the 9th International Conference on Image Analysis and Processing, vol. 1310 of LNCS., Florence, Italy, Springer Verlag (17–19 Sep 1997) pp. 781–787 (vol. II)
22. SMARTTECH Ltd.: Scanner scan3D dual volume (2015). http://smarttech3dscanner.com/3d-scanners/for-industry/scan3d-dual-volume/. Accessed 09 Mar 2015
23. Habib, A., Schenk, T.: New approach for matching surfaces from laser scanners and optical sensors. In: Csatho, B.M., (ed.): Proceedings of the Joint Workshop of ISPRS III/5 and III/2 on Mapping Surface Structure and Topography by Air-borne and Space-borne Lasers, La Jolla, San Diego, CA (9–11 Nov 1999)

Prediction of Trend Reversals in Stock Market by Classification of Japanese Candlesticks

Leszek J. Chmielewski, Maciej Janowicz and Arkadiusz Orłowski

Abstract K-means clustering algorithm has been used to classify patterns of Japanese candlesticks which accompany the approach to trend reversals in the prices of several assets registered in the Warsaw stock exchange (GPW). It has been found that the trend reversals seem to be preceded by specific combinations of candlesticks with notable frequency. Surprisingly, the same patterns appear in both "bullish" and "bearish" trend reversals. The above findings should stimulate further studies on the problem of applicability of the so-called technical analysis in the stock markets.

Keywords Clustering · K-means · Trend reversals · Japanese candlesticks

1 Introduction

The technical analysis of the stock market assets [1, 2] belongs to the most controversial approaches to investigations of data series, especially having economical and financial meaning. This is because the aim of technical analysis is, basically, no less than the approximate prediction of specific values of the data which appear as realizations of a random process. On one hand, it has been declared a kind of "pseudo-science," which, because of the incorrectness of its most important principles, cannot lead to any sustainable increase of returns above the market level [3, 4]. On the other hand, it has also been applied without any serious knowledge about the market dynamics. Considerable revision of the ultra-critical stand of many experts regarding

L.J. Chmielewski · M. Janowicz (✉) · A. Orłowski
Faculty of Applied Informatics and Mathematics (WZIM), Warsaw University of Life Sciences (SGGW), Poland, ul. Nowoursynowska 159, 02-775 Warsaw, Poland
e-mail: leszek_chmielewski@sggw.pl
URL: http://www.wzim.sggw.pl

M. Janowicz
e-mail: maciej_janowicz@sggw.pl

A. Orłowski
e-mail: arkadiusz_orlowski@sggw.pl

R. Burduk et al. (eds.), *Proceedings of the 9th International Conference on Computer Recognition Systems CORES 2015*, Advances in Intelligent Systems and Computing 403, DOI 10.1007/978-3-319-26227-7_60

technical analysis has been influenced by more recent publications, e.g. [5–8] As a part of a common knowledge about the stock market dynamics let us notice that time series generated by the prices of stocks are *not* random walks, and at least the short-time correlations *are* present. Whether they can indeed be exploited with the purpose of maximization of returns is an open question. What we investigate here is meant to be a very small contribution to answer it. It is well known that a very important part of technical analysis is the localization of possible ends of a given trend—upward or downward. Many technical-analysis indicators (like MACD—Moving Average Convergence-Divergence or RSI—Relative Strength Index) are used by technical analysts for this purpose. Another tool used to this end is the graphical patterns made by the sequences of Japanese candlesticks. A Japanese candlestick is a four-element sequence containing the opening, maximal, minimal, and closing prices of an asset during a given trade session. We do not analyze here the patterns themselves but we attempt to obtain an answer to the following rather humble question: Are the trend reversals accompanied more often by some types of candlesticks than by others? For that purpose we perform first the classification of candlesticks to have the above "types" well defined. For the purpose of classification we employ a well-established clustering algorithm called K-means [9–13]. Let us notice that clusterization of candlesticks for a given asset allows one to ascribe labels to them. This, in turn, makes it possible to investigate how their sequences with given labels have performed in the past and what is the predictive power (if any) of sequences with particular labels. The main body of this work is organized as follows. In Sect. 2 we recall and modify the definition of the Japanese candlesticks, which form our working example. In Sect. 3 we provide our results of the relation between candlestick types and trend reversals in prices of several stocks. Finally, Sect. 4 comprises some concluding remarks.

2 Japanese Candlesticks as a Representation of Value of Assets in Stock Market

The Japanese candlestick is a sequence of four numbers $(O(a,t), X(a,t), N(a,t), C(a,t))$, where O denotes the opening value of the asset a at the trading day t, X is the maximum value (*high*) reached during the trading session, N is the minimum (*low*), and C is the closing value. There exists a well-known graphical representation of the candlestick [14] often considered important in what is called the technical analysis of stock markets. In what follows below we employ a sequence of five elements (O, X, N, C, V), which we call an *augmented Japanese candlestick* where V represents the transaction volume associated with the asset and the trading day. An augmented candlestick of the asset a on the day t can be denoted as a 5-tuple

$$\mathbf{y}(a;t) = (O(a,t), X(a,t), N(a,t), C(a,t), V(a,t)). \tag{1}$$

In the following we shall call it simply a candlestick. The time series of $n + 1$ candlesticks, called otherwise a sequence, can be written as

$$S_n(a;t) = (\mathbf{y}(a;t), \mathbf{y}(a;t+1), \ldots, \mathbf{y}(a;t+n)). \tag{2}$$

Each sequence has its own starting time t and ending time $t + n$. We define the (Euclidean) distance between two candlesticks (differently from our previous work [15]) as

$$d(\mathbf{y}_1, \mathbf{y}_2) = \sum_i \left(y_{1,i} - y_{2,i}\right)^2, \tag{3}$$

where $y_{1,i}$ and $y_{2,i}$ are corresponding components of $\mathbf{y}_1$ and $\mathbf{y}_2$ respectively, i.e., they run through the elements of appropriate sets $\{O, X, N, C, V\}$. In order to consider this formula meaningful, the values of the asset and the transaction volume must be comparable. To achieve this, we normalize all time series by subtracting the closing values from the opening ones as well as from the maxima and minima, and dividing O, X, N, and C by the standard deviation of C. Similarly, the volume is also divided by its standard deviation. In this way, the standard deviations of renormalized C and V are exactly 1. All candlesticks analyzed further are normalized in the above sense. Using the K-means algorithm we have classified the Japanese candlesticks as

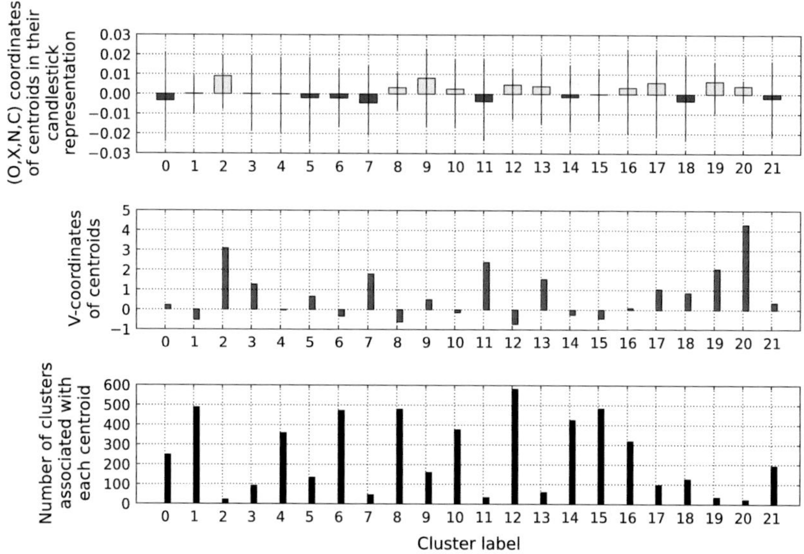

Fig. 1 Coordinates of centroids corresponding to clusters of Japanese candlesticks obtained using the K-means algorithm for the BZWBK stocks (5327 trading days with equal number of candlesticks). The first four coordinates of centroids shown in the upper part of the figure are displayed as candlesticks, the fifth coordinates (i.e., volume) are given as *red* horizontal bars in the middle part of the figure while the *black* horizontal bars in the figures illustrate the numerical amount of elements in the clusters corresponding to the centroids

they appear in the group of the 20 biggest and most powerful stocks (WIG20) in the Warsaw stock exchange. This has been done for each of the 20 stocks separately. We have assumed that there are 32 clusters. This should have corresponded to the 32 semi-quantitative features of the candlesticks: a candlestick can be "black" (close value lower than open value) or "white", its body length (i.e., absolute value of the difference between the close and open value) can be larger or smaller than average, its upper "shadow" (i.e., the difference between the maximum and the larger of open and close value) can be longer or shorter than average, the lower "shadow" can also be long or short with analogous meaning; finally, the corresponding volume can be larger or smaller than average. All this gives 2^5 features, hence 32 clusters. An example of the coordinates of the centroids associated with each cluster for BZWBK stocks is shown in Fig. 1. We have used the implementation of the K-means algorithm as given in the module `Scikit-learn` [16, 17]. To improve the presentation, for every centroid with coordinates (O, X, N, C, V) we subtracted the first coordinate from the first four to obtain $(0, X - O, N - O, C - O, V)$ and displayed its candlestick representation. This is the reason why the candles in the upper part of Fig. 1 have the same level of opening values.

3 Trend Reversals and Candlesticks

It is by no means self-evident what the trend in the data obtained from the random process really means qualitatively, even though the intuitive meaning is understandable. In particular, it is not clear when the trend actually starts and when it ends. To quantify these concepts, we have employed the following simple procedure. To every closing value of a given asset we have ascribed the slope of the straight line obtained from the preceding 10 (ten) closing values. As an indicator of the start and end of the trend we have chosen the change of sign of the above slope. A justification for using the number 10 is that the period of two trading weeks is usually considered important by the technical analysts. For every change of sign of the slope as defined above, which appears between nth and $(n + 1)$th trading sessions, we have ascribed one of the 32 cluster labels (from 0 to 31) of the candlestick which appeared in the session $n + 1$. If, however, a cluster with a given label contained less than $N/240$ sticks, where N is the number of trading days considered, it has been discarded. For instance, in the case of BZWBK stocks ($N = 5327$) we have kept only 22 clusters (and 22 labels). We have applied such a filter to exclude candlesticks that appear too rarely, less than once per year. In Fig. 2 we have plotted an example of the time series generated by a stock in GPW (BZWBK stock, trading session No. 1000–1500) together with the associated time-dependence of the slopes generated by 10 preceding closing values. In Fig. 3 we have plotted the relative frequencies of appearances of the cluster labels ($i = 0$) for the change from downward to upward trend and from the upward to downward for BZWBK stops. It has been calculated as the ratios $M_{j,du}/L_j$ and $M_{j,ud}/L_j$, where $M_{j,du}$ is the number of appearances of a candle belonging to the ith cluster near the down-to-up trend reversal, $M_{j,ud}$ is the number

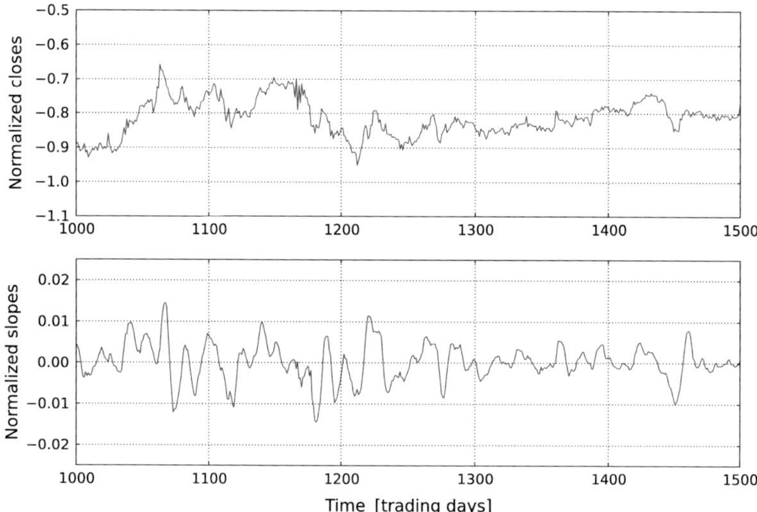

Fig. 2 Example of the time series generated by a stock in GPW (BZWBK stock, trading session no. 1000–1500)—*upper part* of the figure, together with the associated trading-session dependence of the slopes generated by 10 preceding closing values—*lower part* of the figure

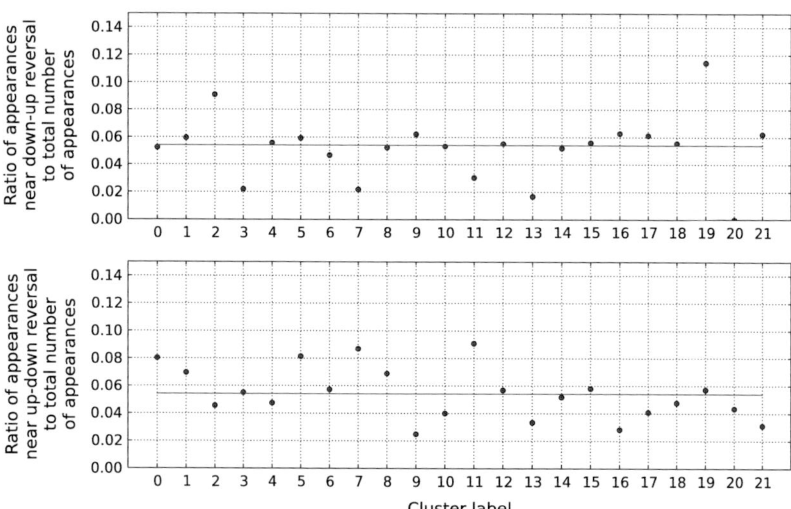

Fig. 3 Relative frequencies of appearances of the cluster labels for the change from downward to upward trend (*upper part* of the figure) and from the upward to downward (*lower part* of the figure) for BZWBK stocks. The *horizontal lines* represent the ratio of the total number of trend reversals to the number of trading days

of appearances of a candle belonging to the ith cluster near the up-to-down trend reversal, and L_j is the total amount of appearances of any candle belonging to jth cluster. This and similar figures which we have obtained for other stocks traded at GPW have been somewhat surprising to us. Indeed, it seems that there are such types of candlesticks that appear frequently close to the trend reversals (as defined above) and relatively rarely outside the regions close to the zeros of the slope series. Since the total number of these zeros has been of the order of $N/10$ (575 for BZWBK stocks), where N is the number of trading sessions, one may be tempted to say that, in fact, candlesticks of some types are concentrated in the trend-reversal regions. We have observed that this behavior is more pronounced in the stocks that are traded longer in GPW and are much less visible for stocks relatively new in the market. It is quite evident, however, that the most significant from the point of view of the trend reversals are those types of candlesticks that, in general, appear quite infrequently. This is rather intuitive from the point of view of technical analysis. What is more, as for the down-to-up reversal, the most significant candlestick is the one with relatively long and light body (i.e., closing price is larger than the opening price), relatively short "shadows" and quite large volume. This means that during the trading day there is almost a steady growth and the interest of investors to buy an asset does not diminish. The fact that, with such investors' mood, the new "bullish" trend is likely, seem to be intuitive. Similarly, the sticks that appear relatively likely near the up-to-down reversals are those with closing prices lower than the opening ones but with relatively long "shadows" and large volumes. It seems that the trade days on which the prices are characterized with such candlesticks are very turbulent, but such that the pessimistic moods finally settle down and the "bearish" reversal becomes likely. Let us notice that the candlestick that appears most frequently in the trend-reversal regions is that of the very short body (closing price very close to the opening price). It is called "doji" and, according to the technical analysis folklore, it traditionally signifies indecision in the market. The trend reversal may indeed follow, but the prediction of "doji" usually depends on further information about the context.

4 Concluding Remarks

In this work we have performed classification of the Japanese candlesticks that appear in the stocks traded in the Warsaw stock exchange using the standard K-means clustering algorithm. With each trend-reversal point we have attached a label associated with the cluster to which the candlestick appearing at that point belongs. We have found that there exist types of candlesticks that frequently tend to appear close to the trend-reversal regions and others that cannot be found in such regions. Needless to say, the above results are very preliminary and require careful reexamination. What is especially important is to find a much more convincing definition of the trend and the trend reversal than that given in this work in terms of linear regression slopes. What is more, as always stressed by technical analysts, the candlesticks are to be considered within the specific market context. This can be done using other technical-analysis indicators. We hope to report the results of such improved analysis in a forthcoming publication.

References

1. Murphy, J.: Technical Analysis of Financial Markets. New York Institute of Finance, New York (1999)
2. Kaufman, P.: Trading Systems and Methods. Wiley, New York (2013)
3. Malkiel, B.: A Random Walk Down the Wall Street. Norton, New York (1981)
4. Fama, E., Blume, M.: Filter rules and stock-market trading. J. Bus. **39**, 226–241 (1966)
5. Brock, W., Lakonishok, J., LeBaron, B.: Simple technical trading rules and the stochastic properties of stock returns. J. Financ. **47**(5), 1731–1764 (1992)
6. Lo, A., MacKinley, A.: Stock market prices do not follow random walks: evidence from a simple specification test. Rev. Financ. Stud. **1**, 41–66 (1988)
7. Lo, A., MacKinley, A.: A Non-Random Walk down Wall Street. Princeton University Press, Princeton (1999)
8. Lo, A., Mamaysky, H., Wang, J.: Foundations of technical analysis: computational algorithms, statistical inference, and empirical implementation. J. Financ. **55**(4), 1705–1765 (2000)
9. MacQueen, J.: Some methods for classification and analysis of multivariate observations. In: Cam, M.L., Neyman, J. (eds.) Proceeding of 5th Berkeley Symposium on Mathematical Statistics and Probability, vol. 1, pp. 281–297. University of California Press, Berkeley (1967)
10. Steinhaus, H.: Sur la division des corps matériels en parties. Bull. Acad. Polon. Sci. **4**(12), 801–804 (1957)
11. Lloyd, S.: Least square quantization in PCM (1957) Bell Telephone Laboratories Paper
12. Forgy, E.: Cluster analysis of multivariate data: efficiency versus interpretability of classifications. Biometrics **21**(3), 768–769 (1965)
13. Hartigan, J.: Clustering Algorithms. Wiley, New York (1975)
14. Wikipedia: Candlestick chart – Wikipedia, The Free Encyclopedia (2014) http://en.wikipedia.org/w/index.php?title=Candlestick_chart. [Online; accessed 19-December-2014]
15. Chmielewski, L., Janowicz, M., Orłowski, A.: Clustering algorithm based on molecular dynamics with Nose-Hoover thermostat. Application to Japanese candlesticks. In: Rutkowski, L., et al., (eds.) Artificial Intelligence and Soft Computing: Proceeding of International Conference ICAISC 2015. Lecture Notes in Artificial Intelligence, vol. 9120, pp. 330–340. Springer (2015)
16. Scikit-learn Community: Scikit-learn - machine learning in Python (2015) http://scikit-learn.org. [Online; accessed 10-February-2015]
17. Pedregosa, F., Varoquaux, G., Gramfort, A., et al.: Scikit-learn: machine learning in python. J. Mach. Learn. Res. **12**, 2825–2830 (2011)

Tracklet-Based Viterbi Track-Before-Detect Algorithm for Line Following Robots

Grzegorz Matczak and Przemysław Mazurek

Abstract Line following robots could be applied in numerous applications with artificial or natural line. The proposed algorithm uses tracklets and Cartesian-to-polar conversion together with Viterbi algorithm for the estimation of line. The line could be low contrast or deteriorated and Monte Carlo tests are applied for the analysis of algorithm properties. Two algorithms are presented and compared—Viterbi and proposed Tracklet-based Viterbi Track-Before-Detect algorithms. Both of them are evaluated and properties are presented. The proposed algorithm could be better for smoother lines if higher noise disturb images.

1 Introduction

Line following robots are applied in numerous industrial [3] and entertainment applications. They use vision-based systems, that could be very simple or sophisticated. Simplest line following robots have applied two optical senors for the estimation of the position over the line. More advanced approach is based on the linear senor (1D camera) located under the robot. Cameras could be applied for the forward looking with perspective view of the line. This approach improves response of overall system, because the estimation of trajectory before the robot could be possible. Larger view allows better determination of line if distortions of many types occur. Some recent line following robots use 3D depth cameras for the estimation of the line. The main problem of line following robots is the vision system that should be robust for the poor light conditions and low line-to-background contrast. Such system should fulfill requirements of the real-time processing for constant speed maintaining. High-speed

G. Matczak · P. Mazurek (✉)
Department of Signal Processing and Multimedia Engineering,
West–Pomeranian University of Technology, Szczecin,
26. Kwietnia 10 Street, 71126 Szczecin, Poland
e-mail: przemyslaw.mazurek@zut.edu.pl

G. Matczak
e-mail: grzegorz.matczak@gmail.com

R. Burduk et al. (eds.), *Proceedings of the 9th International Conference on Computer Recognition Systems CORES 2015*, Advances in Intelligent Systems and Computing 403, DOI 10.1007/978-3-319-26227-7_61

line following robots are in the area of interest of robotics enthusiasts and they are verified in numerous contests also. The importance of line following robots is not reduced by the availability of alternative navigation techniques. Satellite navigation (GPS) systems are not feasible for indoor applications. They have limited performance for outdoor scenarios and the real-time position is estimated with 2 m quality for single frequency receiver (L1) using standard (consumer grade) receivers. The position error could be reduced to 0.1 m for special civil receiver that are very costly. Moreover, limited availability of such receivers is related to political reasons. SLAM (Simultaneous Localization and Mapping) techniques are more costly due to higher computation power demands and line following robots are much simpler. This line could be artificially made or natural one.

1.1 Related Works

Line following robots are proposed in 60s of the previous century and the most notable is the Stanford Cart [14]. The area of applications related to harvesting and similar task, where the line is not specified directly, is considered in [13]. Road line estimation is necessary for automatic control of vehicle or is applied in lane departure warning system [15] using Hough Transform. Agricultural applications need natural lines estimation that could be low contrast [1] and the estimation of line is necessary even in LIDAR systems [17]. Recent works provide the approach based on the application of Track-Before-Detect algorithms [2] that improves lines estimation quality, so even lines hidden in the noise background could be detected [9, 10]. Additional trajectory filtering [4] may improve results independently on noise suppression in the spatial domain.

1.2 Content and Contribution of the Paper

The Viterbi algorithm is originally applied for time series with unknown multiple states at specific moments [16]. Replacing of such state by the 2D position is possible and the line estimation is possible for low curvatures. Directional filtering (FIR-based) that is parallel to the line direction improves the line estimation [9]. Local directional filtering is necessary with fixed direction for better estimation of curvatures. Such approach is important for the computation cost reduction also, so Viterbi algorithm is not applied for every image row. The idea of hierarchical processing of such data is introduced in [6]. Two filtering approaches are possible. First one is computed using original image and second one uses local Cartesian-to-polar transformation. The second approach is assumed in this paper and is considered in Sect. 2. Viterbi algorithm is applied for the selection of the most probable path and is considered in Sect. 3. Monte Carlo approach is applied for the numerical evaluation of algorithm performance and is considered in Sect. 4. The discussion is provided in Sect. 5 and the final conclusions are provided in Sect. 6.

2 Measurement Space Transformation

The acquired image frames are rectangular with Cartesian coordinates. Viterbi algorithm, discussed in next section, process them using small length track (tracklets). Every tracklet value is related to the average value of pixel related to the starting particular position and orientation:

$$d_{\{x,y,\phi\}} = \int_{l_\phi} I\left((x,y) \in l_\phi\right) \mathrm{d}l_\phi, \tag{1}$$

where l_ϕ is the line oriented with some angle direction ϕ and $\phi \in \langle -\alpha, \alpha \rangle$ Cartesian-to-polar transformation could be applied for limited angle range. The cone of analysis in such transformation is reduced to $\langle -\alpha, \alpha \rangle$. Multiple tracklets are processed starting from the particular pixel (x, y) with assumed cone (Fig. 1). This cone could be modified to triangle for the equalization of tracklets length in vertical direction. Example input image and the result of Cartesian-to-polar conversion with cone area are shown in Fig. 2. Such conversion changes sloped lines to straight lines that allows the application of Viterbi algorithm in more straightforward way like it is considered in [9, 10]. The cone (Fig. 2) in converted to the rectangular region corresponding with $\langle 180° - 60°, 180° + 60° \rangle$ range and this certain rectangular area is processed. Such reformulation simplifies the implementation of Viterbi algorithm, that is important in real-time processing applications. White areas present in Figs. 1 and 2 are omitted. Measurement space transformation could be applied locally with equal vertical length of tracklets what is computationally practicable.

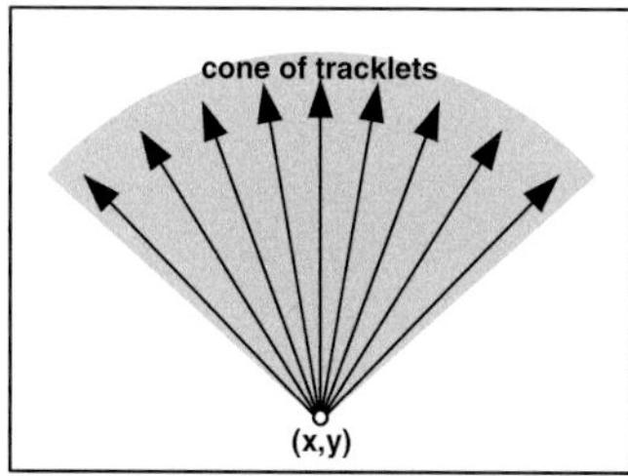

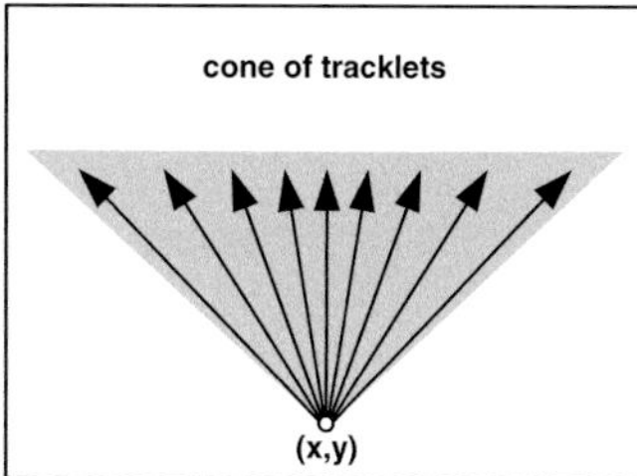

Fig. 1 Possible tracklets for particular pixel: equal length of tracklets (*left*) and equal vertical length of tracklets (*right*)

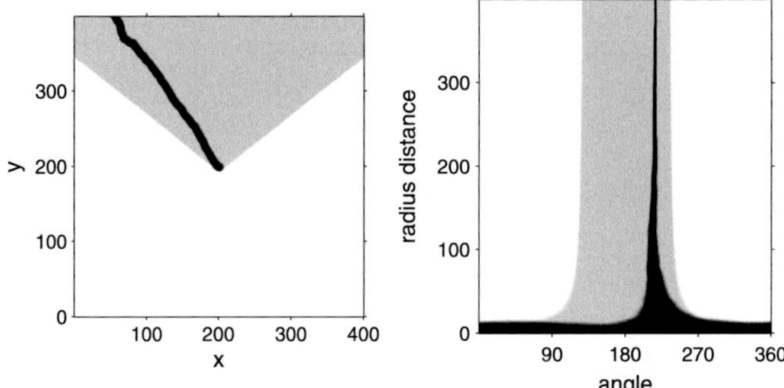

Fig. 2 Input image with line (*black*) and cone (*gray*) of analysis (*left*) and result of Cartesian-to-polar conversion (*right*)

3 Viterbi Algorithm

Viterbi algorithm could be applied for image processing using two approaches. First approach assumes direct relation between pixels and states. Second approach uses transformed image space using directional filtering for example, so tracklets could be processed. Viterbi algorithm uses two processing phases for particular starting row of pixels. The first phase uses forward processing and the second is backward for the best trajectory (best tracklets set) obtained from the first forward phase. Obtained result is the single tracklet related to the single starting pixel of the particular row. Such tracklets allows the robot motion control. Tracklets possible transitions are defined by the trellis. Example part of trellis is shown in Fig. 3 for two rows of pixels/nodes (n and $n + 1$). There are n_{max} of rows in forward direction and Cartesian-to-polar conversion is well fitted to this definition of the Viterbi algorithm. The length of tracklets could be variable, but fixed is possible especially for polar measurement space. The transition cost is assumed as a value of particular tracklet which is added to the value of node. This node is related to the 2D position of tracklet beginning.

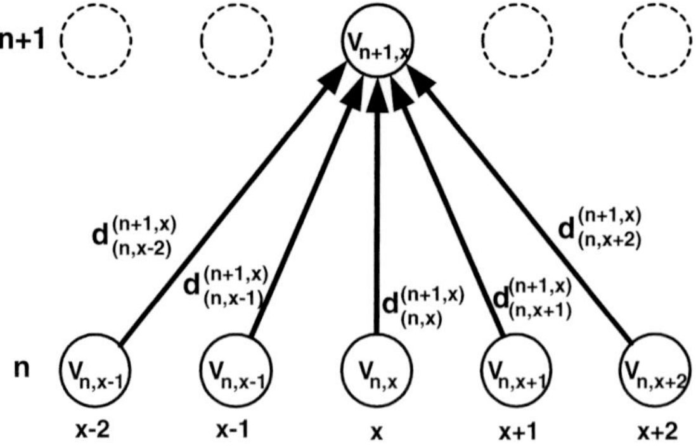

Fig. 3 Local paths in example trellis

Multiple tracklets are merged in nodes of the next row and only the maximal or minimal values are preserved. The minimal values are preserved in this paper for the following settings: grayscale image with values from the normalized $\langle 0 - 1 \rangle$ range; white color (value 1) corresponds to the background; the black color corresponds to the line (value 0), for the ideal case. The local cost for modified cone (triangle) is corrected due to different length of tracklets between nodes of two following rows:

$$d_{corrected\{.,\phi\}} = \frac{d_{\{.,\phi\}}}{|d_{\{.,\phi\}}|}. \tag{2}$$

The image is processed using trellis and the example part of the trellis is show in Fig. 3. The Viterbi algorithm process n_{max} rows of nodes starting from the bottom row ($n = 1$) and the obtained result is the transition between first and second row only. Further transitions between rows $n = 2 \cdots n_{max}$ are processed for the selection of this particular transition and are not reused in next steps (e.g., $n = 2 \cdots n_{max} + 1$). Overall process is repeated for next sets (e.g., $n = 2 \cdots n_{max} + 1$) without relation to the previous results. Initialization is the assignment of zero value to the total cost V for nodes of the first row ($n = 1$):

$$V_{n=1,.} = 0 \tag{3}$$

The local cost d (exactly $d_{corrected}$) is added to previous nodes:

$$V_{n+1,x} = \min\left(V_{n,x+g} + X_{n,x+g}\right), \tag{4}$$

but the minimal cumulative cost is preserved only. The transition is preserved in L using the following formula:

$$L_n^{n+1,x} = \arg\min_g \left(V_{n,x+g} + X_{n,x+g}\right). \tag{5}$$

The first phase of the Viterbi algorithm is computed up to n_{max}. The solution for the first phase (path) is the node with minimal value:

$$P_{opt} = \max\left(V_{n=n_{min},.}\right), \tag{6}$$

that is related to the x–position:

$$x_{n=n_{max}} = \arg\min_x \left(V_{n=n_{max},x}\right). \tag{7}$$

The Selection of the final node allows backward processing for the selection of transition between two first rows. The following recursive processing formula:

$$x_{n-1} = x_m + L_{n-1}^{n,x}, \tag{8}$$

for successively decremented row numbers:

$$n = n_{max}, \ldots, 2. \tag{9}$$

Two results are obtained—most probable node and transition to the next row.

4 Monte Carlo Test of Tracklets-Based Algorithm

The Tracklet-based algorithm reduces number of the processed row by the Viterbi algorithm (Fig. 4) and introduces directional filtering by the accumulation of values related to the path between rows. The increased number of rows n is responsible for better noise suppression. The filtering expect low probability of line direction changes and such example line is show in Fig. 4. Monte Carlo approach allows the estimation of the properties of algorithms because is unbiased. The comparison of algorithms is possible also for identical input images. The generator of images allows the testing for different controllable conditions, which is very important for the algorithm evaluation. The test scenario is based on the analysis of image that has 200×60 pixel resolution. An example results of the single case are depicted in the Fig. 5. The line has standard deviation 1.0 and image is disturbed by a few lines with standard deviation selected by additive Gaussian noise. Two algorithms are compared: Viterbi and Tracklet-based tracking for exactly the same input images. There are 1400 testing scenarios for specified Gaussian noise (std. dev. 0–0.8), with two different line direction change probabilities {0.25, 1.25}. The main test was related to different number of rows processed by the Tracklet-based algorithm. Comparative results for the Viterbi algorithm are provided also (Fig. 8). The cumulative computation time

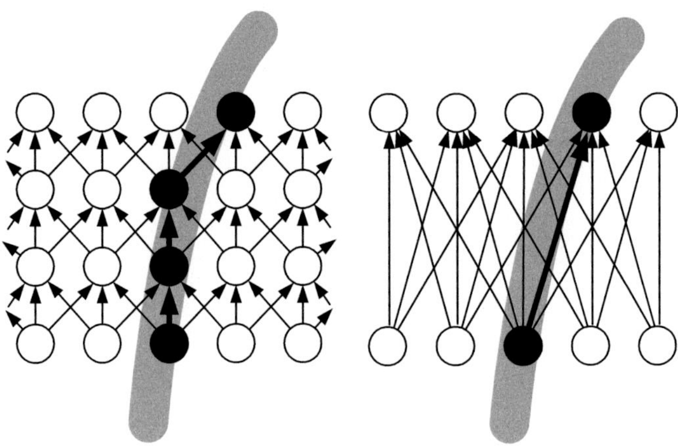

Fig. 4 Processed path in the Viterbi Algorithm (*left*) and in the tracklet-based Viterbi algorithm (*right*)

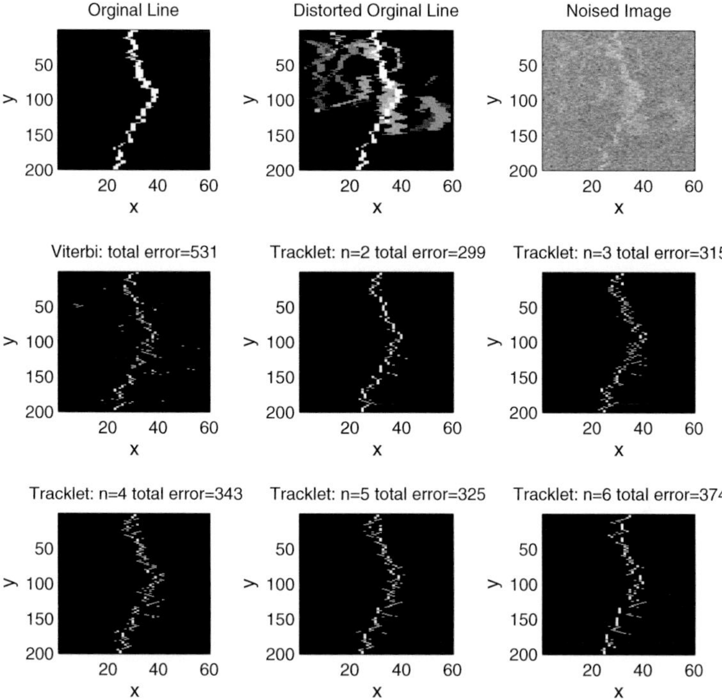

Fig. 5 Example results for specific image—std. dev. 0.45 and line direction change probability 0.25

is 500 h of 2.4 GHz core using Matlab for this test, but the real-time processing is possible for robot using optimized code. The regression line is shown in Fig. 6 for comparison both algorithms. The slope and position of this line shows the advantages of the Tracklet-based algorithm, because the mean position error for horizontal direction is smaller.

5 Discussion

Very interesting results are shown in Fig. 7. Two histograms are depicted for both algorithms and they are different. Better distribution of errors is achieved for the Tracklet-based algorithm because probability of errors is exponential (smaller errors are more probable). The errors for the Viterbi algorithm have mean value near to 5.

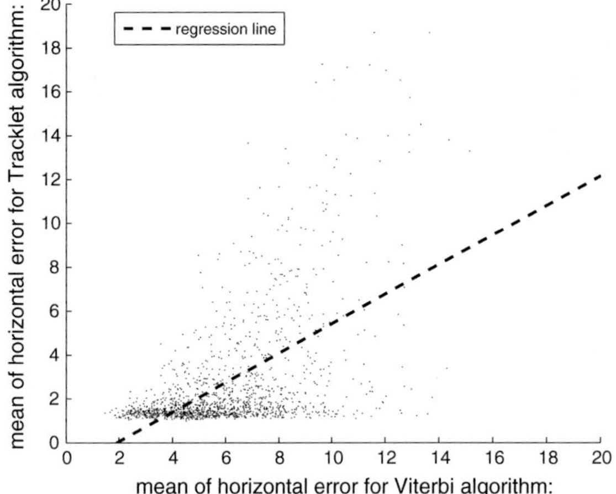

Fig. 6 Comparison of mean errors for Viterbi and tracklet-based algorithms (Monte Carlo test for std. dev. 0.45 and line direction change probability 0.25)

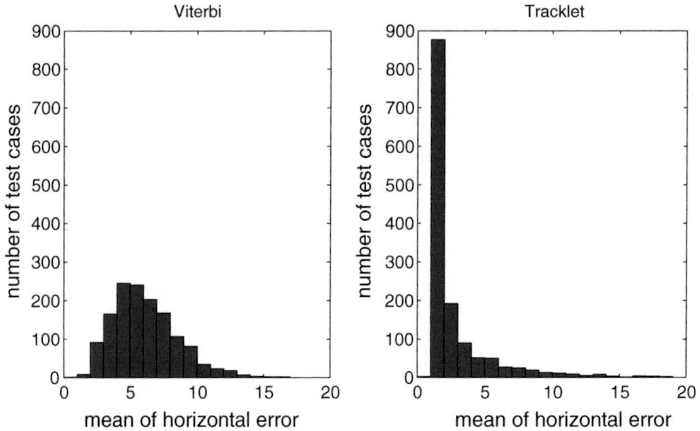

Fig. 7 Histogram of mean error for Viterbi and tracklet-based algorithm (Monte Carlo test for std. dev. 0.45 and line direction change probability 0.25)

Final test (Fig. 8) shows better performance of the Viterbi algorithm for very small noise of images. The dynamic of the line influences the results. Small n could be selected for higher probabilities of line direction changes. Longer integration (larger n) is recommended for the smooth lines.

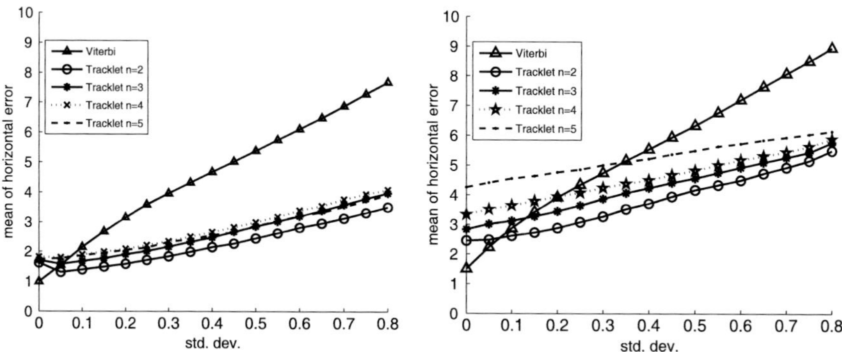

Fig. 8 Mean error for Viterbi and tracklet-based algorithms (Monte Carlo test for variable noise and line direction change probability 0.25 (*left*) and 1.25 (*right*))

6 Conclusions

Line following robots could be improved using vision systems based on 2D camera. Difficult light conditions or line deteriorations are not significant problem for advanced image processing algorithms. Track-Before-Detect (TBD) approach allows processing of weak lines, even below the noise floor. There are applications where only noise is observed, so object's signal is the noise pattern [8, 11]. Such case needs preprocessing of measurements independently on impulse noise suppression. Textures suppression algorithms [12] are also important for the reduction of spatial interferences. The cost of computation for TBD algorithms is significant, but modern available computing devices like SIMD-based CPU cores, GPGPUs and FPGA devices support real-time processing. Optimization techniques for TBD algorithms implementations are also possible [5, 7]. Viterbi algorithms could be accelerated using dedicated hardware that is very often available in modern DSPs. This algorithm is used for the digital communication applications, and there are dedicated instructions or coprocessors in DSPs. Monte Carlo approach allows testing of Viterbi and proposed Tracklet-based Viterbi algorithms and fitting parameters for the best performance and computational cost reduction. This approach allows the comparison of algorithms and the selection of better if variable conditions related to estimated background noise occurs. Such technique of image analysis allows the design better line estimation. Further works will be related to the implementation of mentioned algorithms for the line following robot.

Acknowledgments This work is supported by the UE EFRR ZPORR project Z/2.32/I/1.3.1/267/05 "Szczecin University of Technology—Research and Education Center of Modern Multimedia Technologies" (Poland).

References

1. Astrand, B., Baerveldt, A.: A vision-based row-following system for agricultural field machinery. Mechatronics **15**(2), 251–269 (2005)
2. Blackman, S., Popoli, R.: Design and Analysis of Modern Tracking Systems. Artech House, Norwood (1999)
3. Horan, B., Najdovski, Z., Black, T., Nahavandi, S., Crothers, P.: Oztug mobile robot for manufacturing transportation. In: Proceedings of the IEEE International Conference on Systems, Man and Cybernetics (SMC 2011). pp. 3554–3560 (2011)
4. Marchewka, A.: Crack detection on asphalt surface image using local minimum analysis. Adv. Intell. Soft Comput. **84**, 353–359 (2010)
5. Mazurek, P.: Optimization of bayesian track-before-detect algorithms for GPGPUs implementations. Electr. Rev. R. **86**(7), 187–189 (2010)
6. Mazurek, P.: Hierarchical track-before-detect algorithm for tracking of amplitude modulated signals. Adv. Intell. Soft Comput. **102**, 511–518 (2011)
7. Mazurek, P.: Code reordering using local random extraction and insertion (LREI) operator for GPGPU-based track-before-detect systems. Soft Comput. **18**(6), 1095–1106 (2013)
8. Mazurek, P.: Track-before-detect filter banks for noise object tracking. Int. J. Electron. Telecommun. **59**(4), 325–330 (2013)
9. Mazurek, P.: Directional filter and the viterbi algorithm for line following robots. In: Chmielewski, L., Kozera, R., Shin, B.S., Wojciechowski, K. (eds.) Computer Vision and Graphics. Lecture Notes in Computer Science, vol. 8671, pp. 428–435. Springer, Berlin (2014)
10. Mazurek, P.: Line estimation using the viterbi algorithm and track–before–detect approach for line following mobile robots. In: Proceedings of the 19th International Conference on Methods and Models in Automation and Robotics. pp. 788–793 (2014)
11. Mazurek, P.: Preprocessing using maximal autocovariance for spatio-temporal track-before-detect algorithm. Adv. Intell. Syst. Comput. **233**, 45–54 (2014)
12. Okarma, K., Frejlichowski, D., Czapiewski, P., Forczmański, P., Hofman, R.: Similarity estimation of textile materials based on image quality assessment methods. Lecture Notes in Computer Science **8671**, 478–485 (2014)
13. Ollis, M.: Perception Algorithms for a Harvesting Robot. CMU-RI-TR-97-43. Carnegie Mellon University, Pittsburgh (1997)
14. Schmidt, R.J.: A Study of the Real-time Control of a Computer-Driven Vehicle, Ph.D. thesis. Stanford University, Stanford (1971)
15. Taubel, G., Yang, J.S.: A lane departure warning system based on the integration of the optical flow and Hough transform methods. In: Proceedings of the 2013 10th IEEE International Conference on Control and Automation (ICCA) Hangzhou, China, June 12–14, 2013. pp. 1352–1357 (2013)
16. Viterbi, A.: Error bounds for convolutional codes and an asymptotically optimum decoding algorithm. IEEE Trans. Inf. Theory **13**(2), 260–269 (1967)
17. Zhang, J., Chambers, A., Maeta, S., Bergerman, M., Singh, S.: 3D perception for accurate row following: Methodology and results. In: Proceedings of the 2013 IEEE/RSJ International Conference on Intelligent Robots and Systems (IROS) November 3-7, 2013. Tokyo, Japan. pp. 5306–5313 (2013)

Evolutionary Algorithms for Fast Parallel Classification

Tomáš Ježowicz, Petr Buček, Jan Platoš and Václav Snášel

Abstract The classification tries to assign the best category to given unknown records based on previous observations. It is clear that with the growing amount of data, any classification algorithm can be very slow. The learning speed of many developed state-of-the-art algorithms like deep neural networks or support vector machines is very low. Evolutionary-based approaches in classification have the same problem. This paper describes five different evolutionary-based approaches that solve the classification problem and run in real time. This was achieved by using GPU parallelization. These classifiers are evaluated on two collections that contains millions of records. The proposed parallel approach is much faster and preserve the same precision as a serial version.

Keywords Classification · Parallelization · Speedup · Cuda · GPU

1 Introduction

Classification algorithms are used in many fields, e.g., text categorization and spam filtering [1, 17], optical character recognition [14], machine vision [10], bioinfor-

T. Ježowicz (✉) · P. Buček · J. Platoš · V. Snášel
Department of Computer Science, FEECS, VŠB-Technical University of Ostrava, 17. listopadu, 708 33 Ostrava-Poruba, Czech Republic
e-mail: tomas.jezowicz@vsb.cz

P. Buček
e-mail: petr.bucek.2@vsb.cz

J. Platoš
e-mail: jan.platos@vsb.cz

V. Snášel
e-mail: vaclav.snasel@vsb.cz

T. Ježowicz · P. Buček · J. Platoš · V. Snášel
IT4Innovations, VŠB-Technical University of Ostrava, 17. listopadu 15/2172, 708 33 Ostrava-Poruba, Czech Republic

R. Burduk et al. (eds.), *Proceedings of the 9th International Conference on Computer Recognition Systems CORES 2015*, Advances in Intelligent Systems and Computing 403, DOI 10.1007/978-3-319-26227-7_62

matics (classify proteins according to their functions) [3]. The classification can be solved with numerous methods. Current state-of-the-art methods include artificial neural networks (ANN) [9, 21], Support vector machines (SVM) [19] or deep learning (DL) [2, 7].

In general, one of the main disadvantages of the classification is that it needs a lot of labeled data. The more data are available the better accuracy is usually achieved. But with growing amount of data the algorithms require more computation time and therefore they are very slow.

The common approach chosen by many users is to accelerate algorithms by using GPUs. The use of graphic cards (GPUs) for general purpose (GPGPU) computations has become a new trend. CUDA[1] has provided a new standard that allows to use hundreds of cores that can concurrently run thousands of computing threads. The classification algorithms are no exception and they uses GPUs to accelerate their computations on many different levels. Although, the acceleration with GPU may not be sufficient, e.g., see paper [2], where authors were not able to use the whole dataset due to the computational costs of DL.

In this paper, we are dealing with evolutionary-based classification approach that does not often reach quality of the state-of-the-art algorithms (depending on the dataset), however we show that we are able to train the classifiers in a matter of seconds even for millions of records. The classifiers have also a great advantage in the testing phase. Once the classifier have been trained, the classification of a new sample is a simple scalar product. We also compare the quality of results of five different bio-inspired methods.

2 Related Work

There are many approaches where authors have tried to improve speed of the classification algorithms on GPU. A high-performance traffic classifier on GPU based on a decision-trees was presented in [28]. Paper [4] proposed a model for classification evaluation on GPU. The authors used genetic programming (introduced in [12]) and showed that the proposal is much more efficient than current Java implementation of the algorithm. The same authors published papers [5, 6] where they improved their model based on genetic programming.

Sarkar [16] in 2011 successfully used genetic algorithms for a classification task. Authors of paper [15] in 2012 used the PSO-based classification algorithm for document classification on GPU and they reached 7–10× speedup.

However, all above-mentioned papers dealt only with hundreds or thousands of records in datasets (except for [6] where they used Poker dataset with 1 million records). Much work has been done in paper [29], where authors used nonparametric Naive Bayes classifier designed with no training phase. This classifier is able to

[1]http://www.nvidia.com/object/cuda_home_new.html.

classify images from 80 million tiny images dataset [20]. They are able to classify images in a few seconds without the training phase. However, the training phase may be important as shown in this paper.

3 Classification and Classifiers

The main goal of classification algorithms is to predict the category of previously unseen data accurately. These algorithms build a model which is based on past observations, then they make predictions or decisions. The past observations are labeled vectors $v_i \in V_r$ (training set). Every vector then belongs to one or more predefined category $c_i \in C$. Then, the trained classifier tries to assign a class label to an unknown vector $v_i \in V_t$ (testing set).

The well-known metric in classification and information retrieval is precision (Pr) and recall (Re). Both precision and recall are based on the understanding and measure of relevance. Mathematical definitions are as follows: $Precision = \frac{TP}{TP+FP}$, $Recall = \frac{TP}{TP+FN}$ .Where TP is a count of correctly classified vectors (true positives), FP is a number of vectors that were incorrectly assigned to a given category (false positives), and FN (false negatives) is a number of vectors that were not assigned into given category, although they should be. Usually both precision and recall are combined into a weighted harmonic mean of precision and recall (F-score) $F_1 = \frac{2 \times Pr \times Re}{Pr+Re}$.

4 Evolutionary-Based Classification

The classification with the evolutionary-based algorithm (EBA) was introduced in [22] where they used particle swarm optimization algorithm (PSO). The authors defined the classifier as follows: EBA tries to find a vector and a threshold for each category. The goal is to find a vector that best describes a category. A document is considered to be in the category if it is similar to a particular particle in evolutionary algorithm. To measure similarity between vectors they used the standard cosine similarity $sim(u, v) = \frac{u \cdot v}{\|u\| \|v\|}$. So a cost function for one particle requires to compute the similarities between a particle vector and all documents in dataset. The document is in the category if the cosine similarity is smaller than some predefined threshold.

In this paper, we also compare the quality of other evolutionary-based algorithms. The cost function is same for all evolutionary algorithms (F-score).

4.1 Particle Swarm Optimization

Particle swarm optimization (PSO) is algorithm inspired by the behavior of the swarm of birds and other animals that uses a collective behavior. PSO was created by Kennedy and Eberhart in 1995 [8, 11]. Like other optimizing algorithms, PSO search the optimal solution of the cost function.

PSO creates the set of particles that are involved in finding solution. Each particle remembers its position and velocity. Algorithm updates particle positions step by step, influenced by the best known position.

4.2 *Bat Algorithm*

Bat algorithm (BA) is another meta-heuristic algorithm by Xin-She Yang from 2010. It is based on the movement of bats with the help of echolocation. Each bat (particle) remembers its position, frequency and velocity. For local search, bats move randomly [25].

4.3 *Cuckoo Search*

Cuckoo search (CS) was developed by Xin-She Yang and Suash Deb and published in 2009 [23]. It is inspired by the behavior of cuckoo species, which use brood parasitism for reproduction. The important key part of CS is the application of the Lévy flights. CS uses idealizing rules, especially:

- Cuckoo lays one egg per step in a random nest.
- The best rated nests are used for next generation.
- There is fixed number of host nests and each nest with cuckoo egg can be discovered by the host with probability $p_a \in [0, 1]$ [13].

If a host bird found the nest with a cuckoo egg, it can Build a new nest or Abandon nest or Remove the cuckoo egg.

4.4 *Firefly Algorithm*

Firefly algorithm (FA) was published in 2008 [24]. The algorithm is inspired by flashing characteristics of fireflies. Light intensity (also affected by the distance) represents attractiveness of firefly [27].

In standard FA, each firefly (particle) is an idealized and simplified version of real firefly. There are three rules [18]:

- All fireflies are unisex (attracted to all the others).
- Firefly will move to a brighter one. If there is no one, it will move randomly.
- Brightness of a firefly is affected or determined by the cost function.

There are two key parts of FA: the variation of light intensity and formulation of the attractiveness.

4.5 *Flower Pollination Algorithm*

The main inspiration of the FPA (2012) was the transfer of pollen, which is provided by pollinators, such as birds, bees, and other animals. The process of FPA is driven by following rules [26]:

- Biotic and cross-pollination is a global process (via Lévy flights).
- Abiotic and self-pollination is a local process.
- Probability of reproduction is proportional to the similarity of the flowers.
- Both types of pollination are defined with probability $p \in [0, 1]$.

5 Classification on GPU

Paper [15] showed the speedup of PSO-based classifier algorithm by computing F-score using GPU. We propose a new solution with more optimized F-score computation which uses not only PSO-based classifier but also four other nature-inspired meta-heuristics. The description of a new optimized F-score computation is proposed in the following sections.

5.1 *Parallel Cost Function on Multiple GPUs*

The classifier uses the cosine of the angle between two vectors as a similarity measure. We simplified this calculation by normalizing all vectors to have unit length, so the denominator is equal to one. The similarity measure is then computed by simple scalar product of two vectors. This is one of the most important observations due to performance reasons on GPU. Note that this does not affect the algorithm at all (except for some negligible rounding errors) because the angle between vectors a and b is equal to the angle between vectors $|a|$ and $|b|$. Normalized vectors were used in all experiments described in this paper.

The cost function is computed for all particles at the same time. The computation can be separated into two parts.

The first part of the computation requires to measure similarity between each particle and each record in the dataset. Since all vectors were normalized we used matrix–matrix multiplication. Providing that the matrix A represents the dataset and matrix B represents the particles. The matrix product C then represents the similarity between each particle and each record in the dataset. In other words the scalar product between row vector of the matrix A and column vector of the matrix B computes the similarity between these two vectors (the record and the particle). The highly optimized cuBlas library from the CUDA Toolkit was used for matrix–matrix multiplication.

Each particle requires to compute the F-score based on its similarity to every other documents in the dataset. The cuBlas library stores the matrices in column-major format. So each row of the resulting matrix C corresponds to the particle's result. Each column of the matrix corresponds the documents, e.g., the C[i][j] represents the similarity between the ith particle and jth document in the dataset.

The second part of the computation was done by parallel reduction of each row. Once the similarities were computed, the quantity of correctly and not correctly classified documents (true positive, true negative, and false positive) needs to be summed together. So the final F-score can be computed. Only one block for each particle was allocated on the GPU. The block then performed reduction in shared memory with the maximal number of threads. Each thread computed two values. The single block of threads reduces (sums) all true positives, false positives, and false negatives and at the end only one thread computes the F-score for a single particle and writes it to the global memory.

The proposed algorithm uses the combination of MPI[2] and CUDA and it is primarily designed for clusters. However, it can be used even on a single computer with single GPU (see Table 2). The more nodes with GPU are used the more particles can be computed at the same time. There is one master node which handles selected evolutionary algorithm. Other nodes are waiting until it is required by the master node to compute the cost functions for the particles. Then, the particles are distributed (MPI Scatter) to the nodes where they are normalized and the F-score is computed as described above. he algorithm is described in Algorithm 1.

Algorithm 1 Train EBA on GPUs

1. Allocate the required amount of memory (every node)
2. Copy whole dataset to GPU memory (every node)
3. Evolutionary algorithm initialize (master node)

for 1..*N* **do**

 4.1 Compute evolution algorithm part 1 (master node)
 4.2 Distribute particles to the nodes (master node)
 4.3 Compute F-score (every node)
 4.4 Gather results from the nodes (master node)
 4.5 Compute evolution algorithm part 2 (master node)

end for

5. Finalize evolutionary algortithm (master node)

6 Experiment Results

This section contains the description of the performed experiments on two data sets. First part of the experiments tested efficiency of the classification. Second part of the experimental results contains the description of the speedup factor on the different GPUs.

[2]http://www.open-mpi.org/.

In our experiments we used Anselm cluster[3] with up to 23 nodes equipped with 2x Intel Sandy Bridge E5-2470, 2.3 GHz, 96GB RAM, NVIDIA Kepler K20 (2496 processor cores, 5 GB memory).

We used two different datasets. Both datasets have been produced using Monte Carlo simulations and both try to distinguish between a signal process (produces supersymmetric particles) and a background process. So the datasets have two classes and the classification is binary. We used Higgs[4] and Sussy dataset[5] [2]. Higgs dataset contains 11,000,000 records, 28 attributes and has 53 % positive examples. Susy dataset contains 5,000,000 records, 18 attributes and has 46 % positive examples. The complete set of features has been used in our experiments. We used 500,000 last records as a test set.

6.1 Efficiency of the Classification

Classifiers (PSO, BA, CS, FA, FPA) were tested on Higgs and Susy datasets. First, they were trained on 10,500,000 examples (Higgs) and 4,500,000 examples (Susy). All algorithms used complete attribute set (low-level and high-level). Then, they were tested on the last 500,000 examples in each dataset. For our primary metric we chose the same metric as in [15, 22]. We selected best threshold, precision, recall, and F-score values that algorithms produced. Each classifier was tested 100 times for different thresholds. The best threshold for each classifier was selected. Since the mathematics behind the parallel and serial versions of the EBAs is the same the results were the same. Table 1a shows the results for Higgs dataset. All classifiers F-scores for this dataset were almost identical (around 0.7). The best result was produced by PSO algorithm (0.7075). The Susy results (Table 1b) were more interesting. The best F-scores were reached by PSO and FA algorithms (0.8194 and 0.8168).

6.2 Performance Test

The last experiments were focused on the performance (speedup) of the proposed algorithms. Figure 1 shows average speedup on 1, 4, 8, and 16 GPUs compared to serial CPU version of the algorithms. The average was computed from all five algorithms. The detailed computation times and speedup is shown in Table 2 for Susy dataset and Table 3 for Higgs dataset. Each GPU is able to effectively compute only certain amount of particles at once. The maximal amount of particles can be computed as follows: $n = \frac{GPU_{mem} - bytes \times m \times k}{bytes \times (m+k)}$, where n is the number of particles, GPU_{mem} is available GPU memory, m is the number of elements, k is the number of attributes

[3] https://docs.it4i.cz/anselm-cluster-documentation.

[4] http://archive.ics.uci.edu/ml/datasets/HIGGS.

[5] http://archive.ics.uci.edu/ml/datasets/SUSY.

Table 1 Efficiency (Higgs, Susy)

Alg.	Thr.	Precision	Recall	F-score
(a) Efficiency (Higgs)				
PSO	−0.36	0.5628	0.9524	0.7075
BA	−0.28	0.5292	0.9031	0.6672
CS	−0.06	0.5382	0.9583	0.6810
FA	−0.22	0.5311	0.9577	0.6833
FPA	−0.10	0.5308	0.9508	0.6813
(b) Efficiency (Susy)				
PSO	−0.12	0.7559	0.8946	0.8194
BA	−0.58	0.5426	1.0000	0.7035
CS	−0.10	0.6308	0.9326	0.7525
FA	−0.24	0.7381	0.9142	0.8168
FPA	−0.46	0.6427	0.8628	0.7367

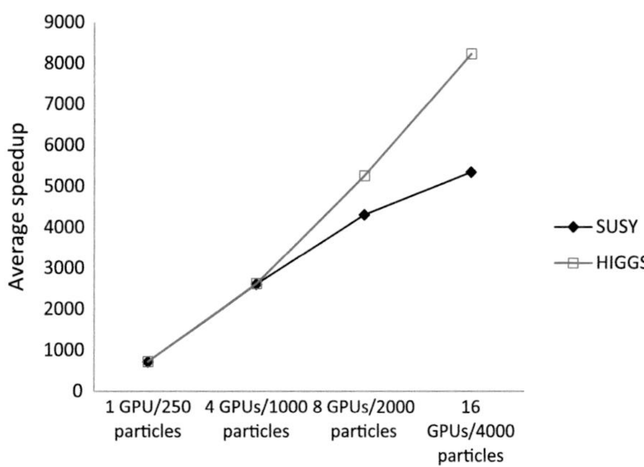

Fig. 1 Average speedup on Higgs and Susy dataset

in the collection and *bytes* is size of each element. Similarly the maximum number of records and attributes can be computed. Required memory is used for matrix multiplication. We selected 80 particles per GPU for Higgs collection and 250 particles per GPU for Susy collection to provide enough memory space for the particle's results and the category array. The results requires $n \times bytes$ memory and the category $m \times bytes$ memory. The most interesting is the speedup for 16 GPUs which ranges from 10 to 11 thousands if compared to our own optimized CPU version. The slowest speedup was reached using FA algorithm, because it compares every individual with all other individuals.

Table 2 Iteration time for the Susy dataset (milliseconds)

Algorithm	1 GPU/250 particles			8 GPU/2,000 particles			16 GPU/4,000 particles		
	CPU	GPU	Speedup	CPU	GPU	Speedup	CPU	GPU	Speedup
PSO	101,090	142.55	709.20	829,540	144.15	5,754.76	1,614,640	145.36	11,108.23
BA	103,990	142.73	728.63	829,090	145.36	5,703.57	1,651,640	147.81	11,174.32
CS	102,680	143.17	717.19	828,240	150.02	5,520.79	1,654,290	156.63	10,561.73
FA	101,190	146.11	692.60	810,750	373.60	2,170.09	1,666,490	931.12	1,783.32
FPA	103,290	142.77	723.46	829,890	146.02	5,683.56	1,660,270	149.37	11,115.14

Table 3 Iteration time for the Higgs dataset (milliseconds)

Algorithm	1 GPU/80 particles			8 GPU/640 particles			16 GPU/1,280 particles		
	CPU	GPU	Speedup	CPU	GPU	Speedup	CPU	GPU	Speedup
PSO	105,470	146.55	719.66	865,255	147.43	5,869.04	1,695,520	164.20	10,326.23
BA	106,346	146.54	725.72	828,224	152.79	5,420.81	1,690,450	149.16	11,332.98
CS	105,629	146.87	719.21	848,136	150.29	5,643.46	1,694,150	152.93	11,078.20
FA	106,501	146.83	725.34	848,189	150.29	5,643.81	1,696,580	397.49	4,268.27
FPA	105,760	146.64	721.24	849,215	148.14	5,732.57	1,702,270	149.69	11,372.20

7 Discussion and Conclusion

The authors in paper [15] and [22] showed that the PSO algorithm may reach a high F-score and may be comparable to state-of-the-art algorithms for some datasets. Tables 2 and 3 seem to derive the similar performance. This may mean that any other bio-inspired algorithms have a potential for providing the similar performance.

It is well known fact that the state-of-the-art algorithms are not usually able to train classifiers in a matter of seconds for large datasets, e.g., back propagation computation etc. As shown in this paper, EBAs may be able to train classifier in a matter of seconds for relatively large datasets. Such speedup is reached by using GPUs to compute the F-score. EBAs have another large advantage. Once, the vector that best describes a category is found, the classification of an unknown document is very unexpensive operation. It is simple a scalar product of a category vector and an unknown document vector (almost negligible time).

The main disadvantage of EBAs is probably a high bias (underfitting). A category is described only by one vector that have a same dimension as an input data. This does not have to be a problem for some datasets, but it may be interesting to introduce some bias-variance tradeoff in the future.

This paper describes evolutionary-based classification problem implemented using cluster of GPUs. We present the architecture of the algorithm as well as the optimization of the computation tasks. We tested five different bio-inspired algorithms and evaluated their precision and computation efficiency.

The average achieved speedup of the algorithm using cluster of nodes with GPUs was 5,000 $\times$ and 9,000 $\times$ for the datasets. Therefore, we are able to deal with millions of documents in seconds.

Acknowledgments This work was supported by the IT4Innovations Centre of Excellence project (CZ.1.05/1.1.00/02.0070), funded by the European Regional Development Fund and the national budget of the Czech Republic via the Research and Development for Innovations Operational Programme and by Project SP2015/105 "DPDM—Database of Performance and Dependability Models" of the Student Grand System, VŠB—Technical University of Ostrava and by Project SP2015/146 "Parallel processing of Big data 2" of the Student Grand System, VŠB—Technical University of Ostrava.

References

1. Androutsopoulos, I., Koutsias, J., Chandrinos, K.V., Paliouras, G., Spyropoulos, C.D.: An evaluation of naive bayesian anti-spam filtering. arXiv preprint arXiv:cs/0006013 (2000)
2. Baldi, P., Sadowski, P., Whiteson, D.: Searching for exotic particles in high-energy physics with deep learning. Nat. Commun. **5** (2014)
3. Brun, C., Chevenet, F., Martin, D., Wojcik, J., Guénoche, A., Jacq, B., et al.: Functional classification of proteins for the prediction of cellular function from a protein-protein interaction network. Genome Biol. **5**(1), R6 (2004)
4. Cano, A., Zafra, A., Ventura, S.: Solving classification problems using genetic programming algorithms on gpus. In: Hybrid Artificial Intelligence Systems, pp. 17–26. Springer (2010)

5. Cano, A., Zafra, A., Ventura, S.: A parallel genetic programming algorithm for classification. In: Hybrid Artificial Intelligent Systems, pp. 172–181. Springer (2011)
6. Cano, A., Zafra, A., Ventura, S.: Speeding up the evaluation phase of gp classification algorithms on gpus. Soft Comput. **16**(2), 187–202 (2012)
7. Deng, L., Yu, D.: Deep learning: Methods and applications. Found. Trends Signal Process. 7(34), 197–387 (2013). http://dx.doi.org/10.1561/2000000039
8. Eberhart, R., Kennedy, J.: A new optimizer using particle swarm theory. In: Proceedings of the Sixth International Symposium on Micro Machine and Human Science, MHS '95, pp. 39–43, 4–6 Oct 1995
9. Hagan, M.T., Demuth, H.B., Beale, M.H., et al.: Neural Network Design. Pws Publishers, Boston (1996)
10. Jain, R., Kasturi, R., Schunck, B.G.: Machine Vision, vol. 5. McGraw-Hill, New York (1995)
11. Kennedy, J., Eberhart, R.: Particle swarm optimization. In: Proceedings of IEEE International Conference on Neural Networks, vol. 4, pp. 1942–1948, Nov–Dec 1995
12. Koza, J.R.: Genetic Programming: On the Programming of Computers By Means of Natural Selection, vol. 1. MIT press, Cambridge (1992)
13. Manikandan, P., Selvarajan, S.: Data Clustering Using Cuckoo Search Algorithm (CSA), pp. 1275–1283 (2012)
14. Mori, S., Nishida, H., Yamada, H.: Optical Character Recognition. Wiley, New York (1999)
15. Platos, J., Snasel, V., Jezowicz, T., Kromer, P., Abraham, A.: A pso-based document classification algorithm accelerated by the cuda platform. In: IEEE International Conference on Systems, Man, and Cybernetics (SMC), pp. 1936–1941 (2012)
16. Sarkar, B.K., Chakraborty, S.K.: Classification system using parallel genetic algorithm. Int. J. Innov. Comput. Appl. **3**(4), 223–241 (2011)
17. Sebastiani, F.: Machine learning in automated text categorization. ACM Comput. Surv. **34**(1), 1–47 (2002). http://doi.acm.org/10.1145/505282.505283
18. Srivatsava, P.R., Mallikarjun, B., Yang, X.S.: Optimal test sequence generation using firefly algorithm. Swarm Evol. Comput. **8**(2013), 4453 (2012)
19. Suykens, J.A., Vandewalle, J.: Least squares support vector machine classifiers. Neural Process. Lett. **9**(3), 293–300 (1999)
20. Torralba, A., Fergus, R., Freeman, W.T.: 80 million tiny images: A large data set for nonparametric object and scene recognition. IEEE Trans. Pattern Anal. Mach. Intell. **30**(11), 1958–1970 (2008)
21. Wang, S.C.: Artificial neural network. In: Interdisciplinary Computing in Java Programming, pp. 81–100. Springer, New York (2003)
22. Wang, Z., Zhang, Q., Zhang, D.: A pso-based web document classification algorithm. In: Proceedings of the Eighth ACIS International Conference on Software Engineering, Artificial Intelligence, Networking, and Parallel/Distributed Computing, SNPD '07, vol. 03, pp. 659–664. IEEE Computer Society, Washington, D.C., (2007). http://dx.doi.org/10.1109/SNPD.2007.84
23. Yang, X.S., Deb, S.: Cuckoo search via lvy flights. In: Proceedings of World Congress on Nature and Biologically Inspired Computing, pp. 210–214 (2009)
24. Yang, X.S.: Nature-inspired Metaheuristic Algorithms, 2nd edn, pp. 81–89. Luniver Press, Frome (2010)
25. Yang, X.S.: A new metaheuristic bat-inspired algorithm. In: Nature Inspired Cooperative Strategies for Optimization, pp. 65–74 (2010)
26. Yang, X.S.: Flower pollination algorithm for global optimization. In: Unconventional Computation and Natural Computation, pp. 240–250. Springer, Berlin (2012)
27. Yang, X.S.: Nature-Inspired Optimization Algorithms. School of Science and Technology, Middlesex University, London (2014)
28. Zhou, S., Nittoor, P.R., Prasanna, V.K.: High-performance traffic classification on gpu. In: IEEE 26th International Symposium on Computer Architecture and High Performance Computing (SBAC-PAD) IEEE, pp. 97–104 (2014)
29. Zhu, L., Jin, H., Zheng, R., Feng, X.: Effective naive bayes nearest neighbor based image classification on GPU. J. Supercomput. **68**(2), 820–848 (2014)

A Hardware Architecture for Calculating LBP-Based Image Region Descriptors

Marek Kraft and Michał Fularz

Abstract In this paper, an efficient hardware architecture, enabling the computation of LBP-based image region descriptors is presented. The complete region descriptor is formed by combining individual local descriptors and arranging them into a grid, as typically used in object detection and recognition. The proposed solution performs massively parallel, pipelined computations, facilitating the processing of over two hundred VGA frames per second, and can easily be adopted to different window and grid sizes for the use of other descriptors.

Keywords Programmable logic · Hardware accelerator · LBP · Region descriptor · Object recognition · Object detection

1 Introduction

Object detection and recognition based on images is a field of active study. Most of the contemporary solutions use a following processing scheme: a detection window is moved through all the locations in the image and a decision is made on whether or not the currently analyzed image patch contains an instance of the object. The process is usually performed in multiple scales to account for possible scaling of detected objects. The key elements of such object detection and recognition systems are therefore region descriptor extractor and classifier. The problem is usually formed as binary classification. The most common choices of classifier are boosted cascade classifier [19], support vector machines (SVM) [5], or decision trees and forests [4]. While the choice of classifier has a significant impact on the quality of results, the choice of input data is no less important. Throughout the years,

M. Kraft (✉) · M. Fularz
Institute of Control and Information Engineering, Poznan University of Technology, Piotrowo 3A, 60-965 Poznan, Poland
e-mail: marek.kraft@put.poznan.pl

M. Fularz
e-mail: michal.fularz@put.poznan.pl

R. Burduk et al. (eds.), *Proceedings of the 9th International Conference on Computer Recognition Systems CORES 2015*, Advances in Intelligent Systems and Computing 403, DOI 10.1007/978-3-319-26227-7_63

numerous approaches have been proposed, including histogram of orientation gradients (HOG) first proposed in [6, 18], Haar-like features [13, 17] and point feature descriptors [2, 10]. Another popular method for neighborhood description in object detection and recognition applications are the local binary patterns (LBP) [16]. While originally developed for texture description, their beneficial properties, such as conceptual and computational simplicity, robustness to illumination and descriptive capability made them a useful tool for a range of other applications. Examples of such applications include face detection and recognition [1], car and text detection [21] or human silhouette detection [22] or even action recognition [11].

In this paper, we present a specialized hardware architecture that calculates LBP-based descriptors of image regions of arbitrary size. An extended LBP variant, called the nonuniform, redundant local binary pattern, is used as a base for the implementation [14, 15]. The architecture employs a sliding window approach. The window is divided into regular grids of square cells, and the final region descriptor is constructed by concatenating LBP histograms from individual contributing cells.

2 Goal and Motivation

Even though algorithmically and conceptually simple, computation of a features as well as their distributions in the form of the local occurrence histograms for the whole image is a compute-intensive task and the requirement of real-time input image stream processing may be difficult to satisfy. The use of modern high performance multicore microprocessors and graphics processing units may be a way to deal with computational performance issues, but only when the energy consumption is not critical [3]. However, applications such as video monitoring and surveillance or robotics typically require low power consumption. Implementation of image processing algorithms on programmable hardware may therefore be regarded as another viable alternative. Using programmable hardware as a computational platform allows to take advantage of the parallel processing capability and flexibility to freely form the internal architecture on one hand and to keep the power consumption at a reasonable level on the other hand, especially when local image processing is performed [7].

With the above remarks in mind, the decision was made to implement the feature vector extractor using a hybrid system-on-chip comprised of a dualcore CPU and a pool of programmable logic as a hardware platform.

3 The Implemented Algorithm

LBPs are image features computed for every image pixel. They have the form of a binary vector, computed from grayscale images based on the intensity values of the image function in the processed pixel's neighborhood. The most common form of LBP uses a 3×3 neighborhood and is denoted $LBP_{8,1}$. The first index term denotes the number of neighboring pixels contributing to the descriptor construction and corresponds directly to the descriptor's bit length, and the second index term denotes the radius (in pixels) of these neighboring pixels. If we denote the value of the image function of the central pixel by i_c, and the value of the image function of the eight neighboring pixels by i_0 to i_7, the $LBP_{8,1}$ is formulated as given in Eqs. (1) and (2).

$$s(x) = \begin{cases} 1 \text{ if } & x \geq 0 \\ 0 \text{ if } & x < 0 \end{cases} \tag{1}$$

$$LBP_{8,1} = \sum_{p=0}^{7} = s(i_p - i_c)2^p \tag{2}$$

Although the results of the comparison operations are essentially binary, the descriptor is interpreted as an 8-bit unsigned integer value and has 256 possible values. As stated in [16], detailed analysis reveals that up to 36 of these values make up to over 90 called uniform local binary patterns (ULBP) and are identified by the fact that their binary representation has at most two 0/1 transitions. Accounting for these patterns and classifying all the remaining patterns from the complete, 256-element set to a separate, single class makes the description with ULBP more compact than description with LBP. In such case, the number of unique descriptors is reduced from 256 to 59 with no significant impact on distinguishability. Further modification of ULBP was proposed in [14] and further developed as a region descriptor in [15]. The rationale behind the modification was the fact that two objects with similar structure but with different foreground–background contrast are represented using different ULBPs. For example, detection windows containing an image of a black car and an image of a white car, both with the same gray background would be described by a set of different ULBPs, even though the class and shape of the objects is the same. To overcome this issue, the ULBPs and their binary negation (e.g., 0b00000110 and 0b11111001) are considered to be the same, making the description more robust and reducing the number of possible binary patterns to 30. Such patterns are called nonredundant uniform library patterns (NRULBP). For object recognition and detection, the most common approach is to compute the histograms of features over some arbi-

Fig. 1 The illustration of complete window descriptor formation based on local feature histograms

trary pixel regions, and then arrange a grid of such regions. The complete detection window comprised of such regions slides over the whole image to perform detection. The input data of the detector or the classifier are the concatenated feature histograms of the regions, forming a the region descriptor as shown in Fig. 1.

4 Description of the Implemented Architecture

The block diagram of the datapath for the coprocessor is given in Fig. 2. The coprocessor is fully pipelined and systolic—for each incoming image pixel a complete descriptor vector is returned. The data is first organized into a 3×3 pixel window for simultaneous access using the internal dual-port memories and registers. Based on the values of the image function of these pixels, the raw, basic LBP is computed as given in Eqs. (1) and (2). Raw LBP is then converted into NRULBP with a lookup table operation. The block diagram of the block performing these operations is shown in Fig. 3. The stream of NRULBPs computed for each pixel is then organized into a 50×1 window, enabling simultaneous access for rapid computation of individual local histograms, and combining them into the complete sliding window descriptor. The computation of the local histogram is based on the observation that the histogram for a rectangular region is modified by the NRULBP values entering and exiting the region. Therefore, as shown in Fig. 4, the entry and exit histograms of 5 pixels entering and exiting the currently processed region are computed and combined into a modification histogram. The modification histogram is combined with the previous

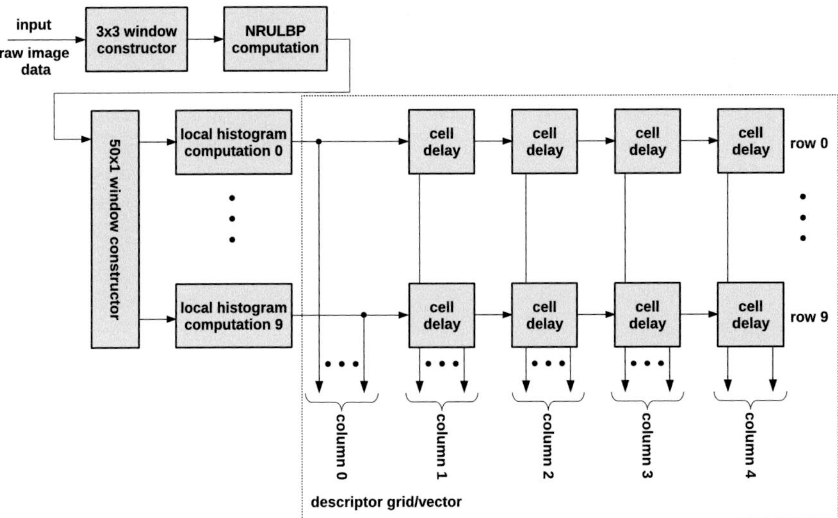

Fig. 2 The block diagram of the coprocessor

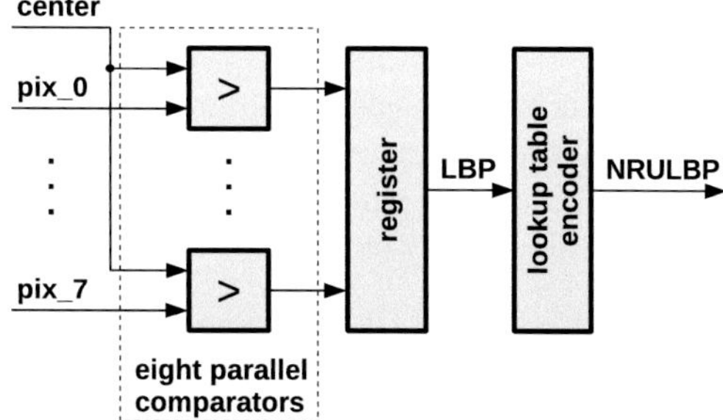

Fig. 3 The block diagram of the module performing NRULBP computation

histogram value to form the new histogram. As shown in Fig. 2, 10 such circuits working in parallel are connected to the 50×1 window, computing a single column of the descriptor in a single step. Additional cell delay blocks are connected to the outputs of the local histogram computation blocks to form the columns of the of the grid in the processed window. Each individual cell histogram is delayed by the number of cycles equal to the horizontal spacing of the columns of the cell grid (in this case the number is five) using cell delay blocks as show in Fig. 2. Such arrangement allows to compute and dispatch a complete descriptor for each one of the image pixels in a single clock cycle with a delay resulting from the length of the processing pipeline. By feeding successive pixels to the stream processor one assures that the search composed of the grid of cells slides over the whole image. The results can be directly used as an input to a massively parallel classifier implementation. Please

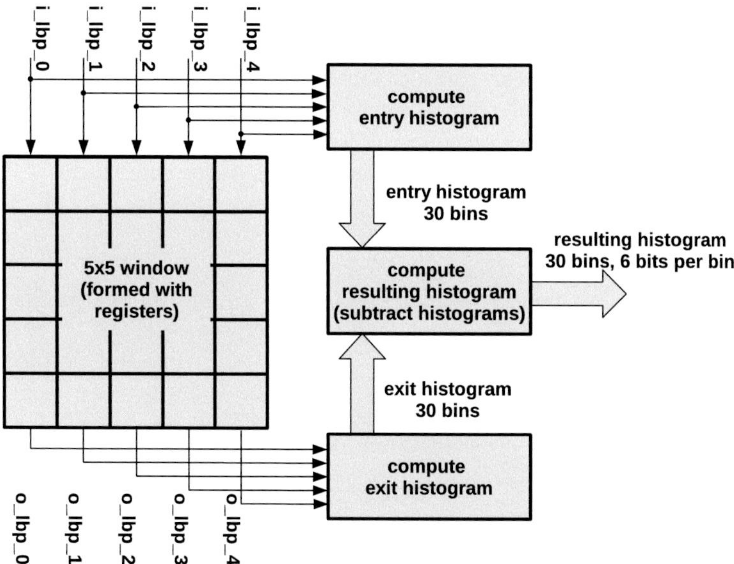

Fig. 4 The block diagram of the module computing the moving histogram

note that changing the size of the cell grid is relatively easy. The 50×1 window size needs to be adjusted to match the new described window size, and the circuit consists of two basic blocks (local histogram computation and cell delay) that are replicated as needed. Moreover, exchanging the functional block for NRULBP computation for another block using a similar processing scheme for descriptor computation such as HOG is also relatively easy.

5 Results and Discussion

The presented solution was implemented using the low-cost Zed-board evaluation board. It hosts Xilinx Zynq-7000 SoC (XC7Z020-CLG484-1), 512 MB of DDR3 RAM, Gigabit Ethernet port, USB host, HDMI output, and FMC connector for image sensor connection and has the ability to boot Linux from the SD card. Resource usage for different variants of the proposed coprocessor is given in Table 1. It should be noted that this is just an accelerator without any additional communication modules like DMA or other interface controllers. The designs differ in the size of the window used to form the descriptor and the cell grid. Increasing the vertical size requires buffering more image lines which results in higher BRAM usage, while increasing the horizontal size uses LUT and FF elements. Each of the cell delay requires 150 LUTs and 150 FFs. To help facilitate the development of different variants of the module, a Python script was developed to automate the process of different size window generation. The presented solution is capable of achieving over 80 MHz pixel clock in Xilinx Zynq 7020 device, which uses slower logic elements than some larger devices (eg., Zynq 7045, Kintex-7 or Virtex-7 devices). Despite this, thanks to its fully pipelined architecture, it is capable of processing over 250 VGA frames per second. Detailed results are given in Table 2. For comparison, the pure software implementation of presented method used for verification was capable of processing approximately 1, 5 frame per second for VGA resolution. It should be noted that it was straightforward, single-threaded implementation coded with OpenCV library. Even with heavy usage of multi-threading and optimization the software version

Table 1 Resource usage of the implemented design for 640×480 image size (designations: FFs—flip-flops, LUTs—lookup tables, BRAMs—block RAM memory blocks)

	LUT	% LUT	FF	%FF	BRAM	% BRAM
10×5	23,223	43.65	13,816	12.98	26.0	18.57
5×5	12,789	24.04	6966	6.55	13.0	9.29
6×3	12,163	22.86	6536	6.14	15.5	11.07
XC7Z020	53,200		106,400		140	
XC7Z045	218,600		437,200		545	

The values in percent are given with respect to all corresponding resources available in XC7Z020 device

Table 2 Number of frames per second that the proposed hardware accelerator can process for different image sizes

Resolution	Processing time (ms)	Frames per second
640 × 480	3.84	260.4
1280 × 720	11.52	86.8
1920 × 1080	25.92	38.6

will be an order of magnitude slower than the presented hardware solution. The main advantage of the design is its ability to easily blend with additional modules—for example, a color to grayscale converter could be attached to camera output to transform the image data to a compatible format. Since the modules implemented in programmable hardware can operate independently in parallel, additional modules will not impact the processing speed as long as they can be clocked at at least 80 MHz. Making the processing pipeline longer would only influence the associated delay. The system is designed as a source of input data for a classifier module. In [12] such solution was presented and achieved 40 MHz, which allowed to process 60 VGA frames per second. In [3] different implementations, both software and hardware, of LBP calculation modules were presented. While in some cases they achieved higher processing speed than presented design, they were much more specialized—only raw LBP values were calculated. If the whole block of histogram computation and descriptor formulation were to be dropped, the presented coprocessor can achieve the pixel clock frequency of over 250 MHz, which translates to the processing speed of 271.3 frames per second for 720HD resolution.

6 Conclusions and Future Work

Thanks to the use of programmable logic, the proposed circuit can be integrated with other dedicated hardware coprocessors. A good example would be the functional block for moving object detection [8], enabling the reduction of the number of false positives. Range of possible applications includes, e.g., smart cameras and surveillance [9] and automotive driver assistance systems [20]. Future work will be focused on the integration of the described hardware coprocessor with functional blocks implementing complete classifiers and other image and video processing realizations in programmable hardware.

Acknowledgments This research was financed by the Polish National Science Centre grant funded according to the decision DEC-2011/03/N/ST6/03022, which is gratefully acknowledged.

References

1. Ahonen, T., Hadid, A., Pietikainen, M.: Face description with local binary patterns: Application to face recognition. IEEE Trans. Pattern Anal. Mach. Intell. **28**(12), 2037–2041 (2006)
2. Bay, H., Ess, A., Tuytelaars, T., Gool, L.V.: Speeded-up robust features (SURF). Comput. Vis. Image Underst. **110**(3), 346–359 (2008). Similarity Matching in Computer Vision and Multimedia
3. Bordallo López, M., Nieto, A., Boutellier, J., Hannuksela, J., Silvén, O.: Evaluation of real-time lbp computing in multiple architectures. J. Real-Time Image Process. pp. 1–22 (2014). http://dx.doi.org/10.1007/s11554-014-0410-5
4. Breiman, L.: Random forests. Mach. Learn. **45**(1), 5–32 (2001)
5. Burges, C.J.: A tutorial on support vector machines for pattern recognition. Data Min. Knowl. Discov. **2**(2), 121–167 (1998)
6. Dalal, N., Triggs, B.: Histograms of oriented gradients for human detection. In: IEEE Computer Society Conference on Computer Vision and Pattern Recognition, CVPR 2005, vol. 1, pp. 886–893 (2005)
7. Fowers, J., Brown, G., Cooke, P., Stitt, G.: A performance and energy comparison of fpgas, gpus, and multicores for sliding-window applications. In: Proceedings of the ACM/SIGDA International Symposium on Field Programmable Gate Arrays, FPGA '12, pp. 47–56. ACM, New York (2012). http://doi.acm.org/10.1145/2145694.2145704
8. Fularz, M., Kraft, M., Kasinski, A., Acasandrei, L.: A hybrid system on chip solution for the detection and labeling of moving objects in video streams. In: Signal Processing: Algorithms, Architectures, Arrangements, and Applications (SPA), pp. 94–99 (2013)
9. Fularz, M., Kraft, M., Schmidt, A., Kasiński, A.: The architecture of an embedded smart camera for intelligent inspection and surveillance. In: Szewczyk, R., Zieliński, C., Kaliczyńska, M. (eds.) Progress in Automation, Robotics and Measuring Techniques, Advances in Intelligent Systems and Computing, vol. 350, pp. 43–52. Springer International Publishing (2015)
10. Hamdoun, O., Moutarde, F., Stanciulescu, B., Steux, B.: Person re-identification in multi-camera system by signature based on interest point descriptors collected on short video sequences. In: Second ACM/IEEE International Conference on Distributed Smart Cameras, ICDSC 2008, pp. 1–6 (2008)
11. Kellokumpu, V., Zhao, G., Pietikäinen, M.: Recognition of human actions using texture descriptors. Mach. Vis. Appl. **22**(5), 767–780 (2011)
12. Kryjak, T., Komorkiewicz, M., Gorgon, M.: Fpga implementation of real-time head-shoulder detection using local binary patterns, svm and foreground object detection. In: 2012 Conference on Design and Architectures for Signal and Image Processing (DASIP), pp. 1–8 (2012)
13. Lienhart, R., Maydt, J.: An extended set of haar-like features for rapid object detection. In: Proceedings of International Conference on Image Processing, vol. 1, pp. I–900–I–903 (2002)
14. Nguyen, D.T., Zong, Z., Ogunbona, P., Li, W.: Object detection using non-redundant local binary patterns. In: 17th IEEE International Conference on Image Processing (ICIP), pp. 4609–4612 (2010)
15. Nguyen, D.T., Ogunbona, P.O., Li, W.: A novel shape-based non-redundant local binary pattern descriptor for object detection. Pattern Recognit. **46**(5), 1485–1500 (2013)
16. Ojala, T., Pietikainen, M., Maenpaa, T.: Multiresolution gray-scale and rotation invariant texture classification with local binary patterns. IEEE Trans. Pattern Anal. Mach. Intell. **24**(7), 971–987 (2002)
17. Pavani, S.K., Delgado, D., Frangi, A.F.: Haar-like features with optimally weighted rectangles for rapid object detection. Pattern Recognit. **43**(1), 160–172 (2010)
18. Pedersoli, M., Vedaldi, A., Gonzàlez, J., Roca, X.: A coarse-to-fine approach for fast deformable object detection. Pattern Recognit. **48**(5), 1844–1853 (2015)
19. Schapire, R.: The boosting approach to machine learning: An overview. In: Denison, D., Hansen, M., Holmes, C., Mallick, B., Yu, B. (eds.) Nonlinear Estimation and Classification, Lecture Notes in Statistics, vol. 171, pp. 149–171. Springer New York (2003)

20. Sun, H., Wang, C., Wang, B., El-Sheimy, N.: Pyramid binary pattern features for real-time pedestrian detection from infrared videos. Neurocomputing **74**(5), 797–804 (2011)
21. Zhang, H., Gao, W., Chen, X., Zhao, D.: Object detection using spatial histogram features. Image Vis. Comput. **24**(4), 327–341 (2006)
22. Zheng, Y., Shen, C., Hartley, R., Huang, X.: Pyramid center-symmetric local binary/trinary patterns for effective pedestrian detection. In: Kimmel, R., Klette, R., Sugimoto, A. (eds.) Computer Vision - ACCV 2010, Lecture Notes in Computer Science, vol. 6495, pp. 281–292. Springer, Berlin (2011)

Unified Process Management for Service and Manufacture System—Material Resources

Marek Krótkiewicz, Marcin Jodłowiec, Krystian Wojtkiewicz and Katarzyna Szwedziak

Abstract The paper proposes the Unified Process Management for Service and Manufacture (UPM$^{\mathrm{Srv}}_{\mathrm{Mnf}}$), the intent of which is to constitute integrated platform for modelling and exchange of information in the field of production processes. The UPM$^{\mathrm{Srv}}_{\mathrm{Mnf}}$ consists of a theoretical basis, formal notation, modeling language, algorithms, and methods of processing information as well as modeling methodology. This article focuses on the passage concerning the most theoretical foundations in the identification of selected categories and relations such as *generalization-specialization*, *part–whole*, *class-feature*, *class-instance*, and *property-instance*. The platform used to build UPM$^{\mathrm{Srv}}_{\mathrm{Mnf}}$ is the Semantic Knowledge Base (SKB), which is the result of a research project on methods of representation and processing of knowledge.

1 Introduction

Existing approaches in the field of production engineering and business processes often focus on specific fragments or aspects of modeling type of issues [1, 2]. They are either notational issues (languages), notations, formalisms, management approach focused on processes and their management, engineering approach combined on resources and their management and financial approach characterized by the aspects of budgeting and financial flows. The aim of the research is to develop a theoretical basis to implement it and introduce as a complete set of tools both theoretically and practically covering all phases of the production process management cycle. Approach proposed by the authors is intended to be a complementary system covering all relevant aspects of the production processes management. System components

M. Krótkiewicz (✉) · M. Jodłowiec
Institute of Control and Computer Engineering, Opole University of Technology, Opole, Poland
e-mail: mkrotki@mkrotki.com

K. Wojtkiewicz · K. Szwedziak
Department of Biosystem Engineering, Opole University of Technology, Opole, Poland
e-mail: kwojt@math.uni.opole.pl

R. Burduk et al. (eds.), *Proceedings of the 9th International Conference on Computer Recognition Systems CORES 2015*, Advances in Intelligent Systems and Computing 403, DOI 10.1007/978-3-319-26227-7_64

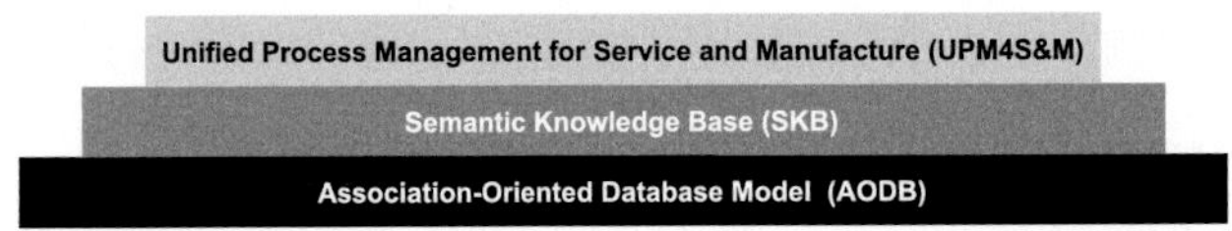

Fig. 1 Layered architecture of the system

are consistent, they support and complement each other and benefit from synergies. The proposed solution Unified Process Management for Service and Manufacture (UPM^{Srv}_{Mnf}) includes:

- theoretical basis (full identification of categories and relations between them),
- formal notation,
- modeling language and methodology,
- algorithms and processing methods.

The above-mentioned components are arranged in space consisting of material resources, time, finances, and location dimensions.

In addition, two basic interpenetrating key aspects of modeling should be emphasized: structural and behavioral. In this article, the authors focus on selected elements of the proposed system, namely a fragment of structural aspect to define the basic concepts and their relationships, what can be described as a kind of ontology. The concept is here understood in the context of SKB [3] representing reference plane for the modeled system [4]. The article focuses on the concepts of resources, namely the material resources and methods to define them. The lowest layer of modeling in which Semantic Knowledge Base (SKB) has been defined is an *Association-Oriented Database Model*—AODB (Fig. 1).

2 Resources

At the stage of the production process definition, one of the steps is to define the resources necessary for its implementation [5], i.e., their types and specific instances. For example, a resource may be an employee or more specifically an employee at a defined position or with the appropriate privileges or powers. While what is considered to be a particular resource is a person physically available for allocation in the implementation of a specific process [6]. In publications and in practice, the solutions to define resources are very different, in particular in the frame of the detail description of the types of resources and their specific instances. The proposed solution aims at organizing and standardizing the approach to define the types of resources and their instances. In particular, Sect. 3 of the paper, will discuss relationships of the following types: *class-instance* (type-instance), *class-feature* (type-attribute), *property-instance* (instance-property), *generalization-specialization* (generalization-specialization), and *part–whole* (part–whole). These relationships are the result of the solutions adopted in the Semantic Knowledge

Base (SKB), and more specifically in the Ontological Core Module (OCM^{SKB}) [3], implementation of which is presented in Sect. 3. The classic division into human resources, time, materials, etc., with subcategories is nothing but a taxonomy of resource types based on the *generalization-specialization* relationship. Unfortunately, in practice, there are solutions that confuse *generalization-specialization* relationship, with *part–whole* relationship. An example might be a method of describing the complexity concerning a specific work stations. Let this description apply to stations, the types of which are determined by their equipment; specific stations for welding, milling, drilling, painting, assembling.[1] These stations are characterized by a high capacity to adapt to specific needs (configuration of the production line to particular series of orders). Therefore, taxonomic characteristics based on the *generalization-specialization* relationship, dividing the stations into primary and backup types, for example, metallurgical processing station and its subtypes: station for milling, punching, machining, etc., has a number of disadvantages. From the modeling point of view of manufacturing processes much more adequate for the reality is the description of such resources with the use of *part–whole* relationship. These stations may be equipped with various tools and devices, and this is what determines them. In this example, it does not seem appropriate to create taxonomies based on *generalization-specialization* relationship, as it generates a large number of taxonomic tree nodes that has a large dynamic range. It results from the already mentioned mobility of equipment and tools, which would cause the need for continuous updating of specific types of stations. It should be emphasized that the types, and also the whole tree of types (taxonomy), should be possibly invariant in time. On the other hand, part–whole relation from the functional point of view is intuitively perceived as much more dynamic and can naturally be continually updated. This amounts to a process called production centers configuration, or customization of tools, equipment, stations, workshops, or entire factories, which can be configured for the purposes of a specific order.[2] To sum up this topic, methodology of modeling the production processes should take into account the differences and the methods to use the *generalization-specialization* relationship with a particular focus on the dynamics and refer to the actual specifics of the case.

3 Resources in Terms of Semantic Knowledge Base

Semantic Knowledge Base (SKB) [3] is a project which aims to develop theoretical assumptions and the implementation of a system capable of storing and processing of knowledge. Knowledge within the meaning of SKB should be treated as information along with the ability to use it. It is also assumed that the information contained in SKB has defined semantics. This semantics is defined on several levels.

[1]Based on SUMER Ltd. table of stations before the implementation of the Production Process Management System.

[2]The company Zakłady ODRA—Brzeg.

The lowest of these is the structure of the system, together with the interpretation of the meaning of the individual elements of the system. Semantic Knowledge Base has been developed in *Association-Oriented Database Model*—AODB, so all the structural elements will be presented in the form of a formal record of AODB or AML^{AO}_{DB}—Association-Oriented Modeling Language. The next level is constituted by algorithms and methods of processing this information, and the highest by the interfaces providing the ability to communicate with humans, including language interfaces, as well as specialized graphical user interfaces (GUI). Semantic Knowledge Base has a modular construction. Individual modules are characterized by a specific structure for their purpose, semantics, and links with the other modules. These relationships are particularly important because of the fact that SKB is assumed to be a hybrid system, i.e., a system based on several methods of knowledge representation, such as *ontology*, *object-oriented modeling*, *semantic networks*, *the representation of time and space* and *linguistics*. Semantic Knowledge Base is a large project, and therefore items directly used in Unified Process Management for Service and Manufacture (UPM^{Srv}_{Mnf}) will be clarified at the beginning. The basic and most essential module for the SKB is Ontological Core Module (OCM^{SKB}). It is also very important from the point of view of the subject discussed in this article, because it allows to define key terms, including material resources for production processes and the most fundamental relationship between them: *class-feature*, *property-instance*, and *class-instance*. Furthermore, what will be mentioned are *generalization-specialization* and *part–whole* relationships, which have been implemented in Relationships Module (RM^{SKB}), while in Ontological Core Module (OCM^{SKB}) the areas able to define these relations, as well as any relationships between concepts will be indicated. Ontological Core Module (OCM^{SKB}) does not provide the ability to store information about the relationships, in particular *generalization-specialization* and *part–whole* relationships type. This issue will be the subject of a separate study.

3.1 class-instance Relationship

The basic way to describe the complexity is the *class-instance* relationship. It is widely known modeling method based on the categories known in various fields of knowledge. The philosophy discusses, for instance *man* and *man at all*. In computer science what is defined are objects and classes. Management science uses the terms of types or kinds and specific items. In SKB what corresponds to those categories are accordingly INSTANCE as a specific item, object, etc., and CLASS, as a type, kind, etc. An instance is an item representing the class. There are two approaches. The first is the object-oriented approach strictly taken from the *object-oriented programming*, which treats objects as artifacts created on the basis of the class, that is, according to some kind of template that this class represents. In particular, the objects have the ability to store the values of attributes and a list of these attributes is in the class definition. In addition, the objects do not belong to the class, that class is not

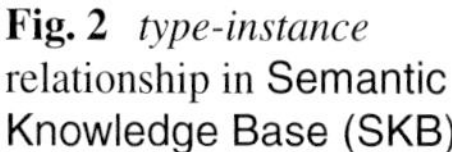

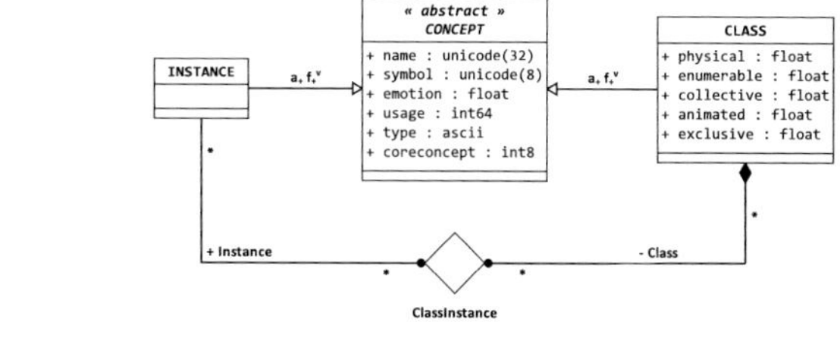

Fig. 2 *type-instance* relationship in Semantic Knowledge Base (SKB)

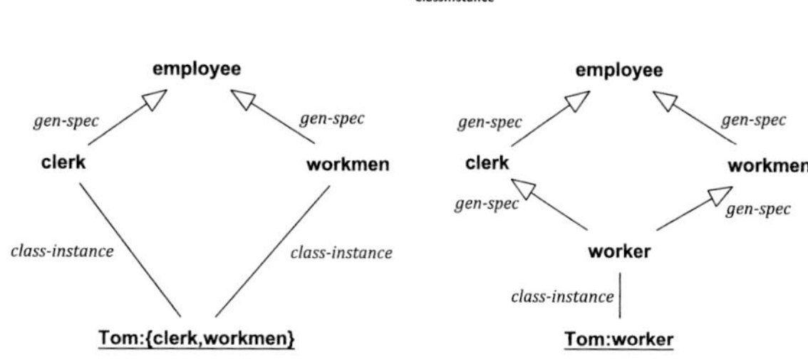

Fig. 3 Multiple inheritance of classes and attribution of the instance to multiple classes

a set, or any other collection of objects [7]. The second approach is based on the idea that class determines not only the structure and behavior of objects, but also it is a container for them. Here, in UPM_{Mnf}^{Srv} the approach based on the allocation of functions was chosen to define objects and their grouping in terms of storage.[3] In addition, the adopted solution provides the ability to assign a particular instance to several classes, as shown in Fig. 2.

From the point of resource definition view CLASS is a resource type, and INSTANCE is a particular resource. Even though this relationship itself is rather intuitive and does not require an example, it is a matter of defining instance through a few classes, what may seem semantically more complex issue. This mechanism is used only to improve the modeling, as it could easily be replaced by artificial creation of an auxiliary class (Fig. 3).

3.2 *class-feature* Relationship

This relationship allows to specify which properties a certain class has. CLASS as a description of the construction of the object in Semantic Knowledge Base consist of FEATURE. Each type represented in SKB by the CLASS has a specific set of features that enable objects to have the ability to store the values of these attributes. A simple example would be the resource being the type of a device described by a set of parameters, which will be in UPM_{Mnf}^{Srv} as the CLASS, for which there is a FEATURE set will be specified. However, information about the values of characteristics

[3] Semantic Knowledge Base provides the ability to create a collection of containers through the use of SET and SetInstance association.

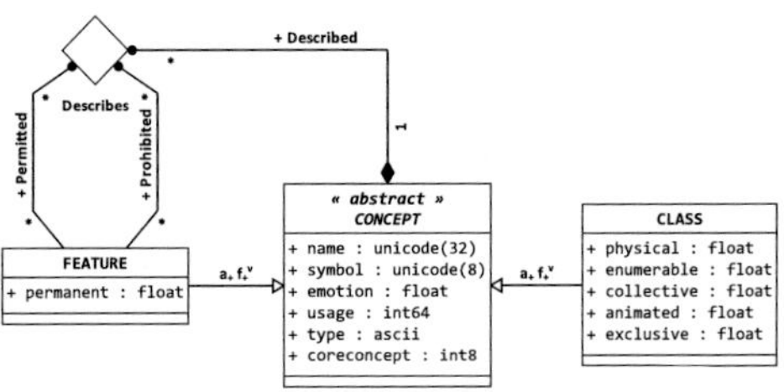

Fig. 4 *class-feature* relationship in Semantic Knowledge Base (SKB)

of the particular INSTANCE constitutes another relationship—*property-instance*.[4] Where a specific INSTANCE is assigned to two or more number of CLASS it has a FEATURE set representing the sum of FEATURE sets assigned to each of those classes respectively. In some systems of knowledge representation, it is possible to directly assign FEATURE to INSTANCE, but in Unified Process Management for Service and Manufacture this solution was abandoned. *class-feature* relationship is closely associated with the *property-instance* relationship, since, as mentioned above, INSTANCE can store values of only those FEATURE, that have been assigned to CLASS, to which the INSTANCE belongs. It is worth noting that when using the SKB mechanisms is possible to define not only the features describing INSTANCE of a specific CLASS, but also can to indicate what FEATURE should not be used for this (role: *Permitted* and *Prohibited*) (Fig. 4).

3.3 property-instance Relationship

Property is a container that holds the value of the characteristics of a specific INSTANCE, where INSTANCE is assigned to a specific CLASS. So as an INSTANCE constitutes and artifact for CLASS, a Property is an artifact for the FEATURE. Figure 5 shows a section of the structure of the Semantic Knowledge Base implemented in AODB responsible for the storage of information on the value of instance features of a particular class. This is implemented by the *n*-ary association Property, which combines the values of FEATURE defined in the VALUE or LOCALVALUE$^{\varnothing}$ to a particular INSTANCE. E.g., in the manufacture of rehabilitation tables, one of the necessary resources is a steel angle bracket, used for the construction of the table. CLASS definition of steel angle bracket includes, among others, its name, thickness and type. In terms of a specific angle bracket, which will be delivered to the factory, we can talk about the values of the features defined above. In terms of the Semantic Knowledge Base implementation of the above example requires the following:

[4]The relationship is described in detail in Sect. 3.3.

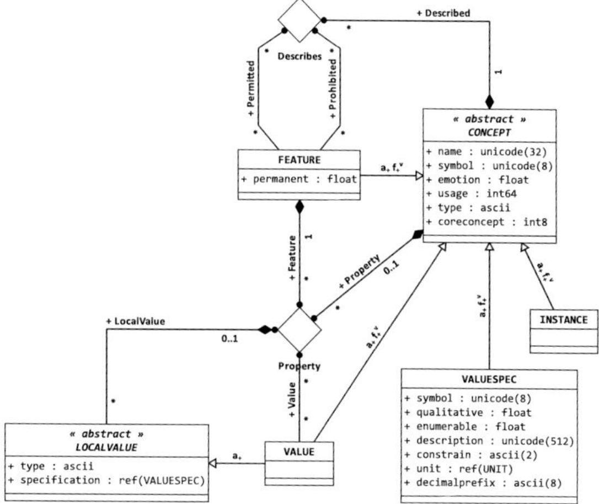

Fig. 5 *property-instance* relationship in Semantic Knowledge Base (SKB)

1. defining a CLASS—*angle_bracket*
2. defining the FEATURE: name, thickeness, type
3. determining the Describes relationships between the angle bracket CLASS, and the name, thickness and type FEATURE
4. defining INSTANCE: angle_bracket1, as *angle_bracket* CLASS instance
5. creation of Property, which will store the value of the corresponding FEATURE for angle_bracket1 INSTANCE

An important aspect, from the production processes modeling point of view is the fact that it assumes that the production center is a part of an active element featuring, among others, the possibility of initiating events involving the request of starting the flow of resources in sufficient quantity and of the relevant parameters. Expected parameters are nothing but the restrictions imposed on the values of the characteristics of resources. All INSTANCEs, in this case angle brackets, taking part in the flow of material resources will, therefore, have to comply with certain conditions. For the present instance, this restriction may apply to, for example, the type of bracket, or its minimum or maximum thickness. The issue of defining the limits focuses on behavioral aspects of manufacturing processes definition and it will not be described further in this article.

3.4 *part-whole and generalization-specialization Relationships*

Relationships named *generalization-specialization* and *part–whole* are the basis for the description of the modeled fragment of reality. However, the common descriptions of management processes, production in particular, often ignore the dependencies of this type. They are sometimes represented as unnamed relationships or episodically appear in the form of verbal descriptions, informal, intended rather to be interpreted by a person modeling or implementing. *Part–whole* relationship describes the dependency, which from the semantic point of view is to be regarded as connection of element constituting the *whole* with one or with a greater number of elements serving as *parts*. At the same time, it is important to determine the properties of the relationship. What is distinguished is the *part–whole* relationship called *week* and *strong*. In object-oriented modeling they are often referred to as *aggregation* and *composition*. *Aggregation* is a very general conceptual relationship saying that some elements are the *part* of some *whole*. But this not entails any consequences in terms of restrictions. In contrast, the composition imposes restrictions on the properties of this relationship, namely:

- from the extensional point of view:
  - part may not belong to several wholes (at the same time). This means that a specific (extensional) element constituting a part may not simultaneously be assigned to several wholes. It can be reassign to other wholes (dynamics),
  - a part cannot be the whole to itself and generally speaking the *part–whole* dependency graph must be acyclic,
- from the intensional point of view:
  - a particular type (class) can be a *part* of several *wholes*,
  - there is no restriction for a particular type (class) to stand on both sides of the *part–whole* relationship, and generally speaking cycles in the dependency graphs are allowed,
- there is a limit on the lifetime of instances that are *part* of a *whole*, namely, when a *whole* is removed from the system entirely, all the associated parts should be removed as well. It should be stressed that this relationship is unidirectional, i.e., at the time of the removal of this parts it has no effect on the *whole*, the part item may not exist independently, i.e., it must be assigned to a *whole* (Fig. 6).

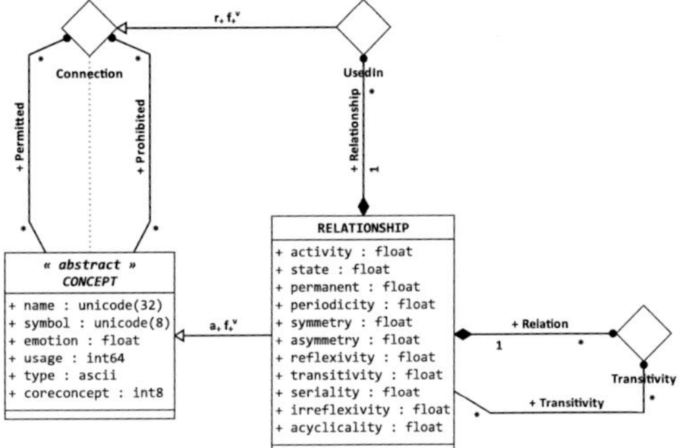

Fig. 6 *part–whole* and *generalization-specialization* relationships in Semantic Knowledge Base (SKB)

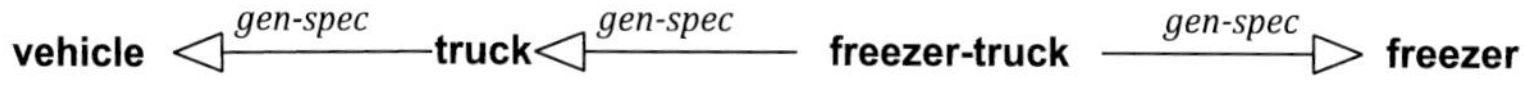

Fig. 7 Multiple inheritance of classes in Semantic Knowledge Base (SKB)

UPM$^{Srv}_{Mnf}$ has adopted above semantics. An example of a *weak* relationship, that is aggregation, can be a team member relationship. A team member can belong to several teams. Dissolution of a team does not simultaneously remove the members of the team as independent instances. Member of the team as an instance has the right to exist even when not belonging to any team. An example of a strong *part–whole* relationship, that is a *composition*, is the relationship of the furnace and the heating elements.[5] Removal of the furnace as a whole means automatic removal of all components, in this case, the heating element of the furnace. Furnace heating element cannot belong, in the sense of being a part of, to several furnaces. Furnace heating element cannot exist by itself, without the furnace, because it will not be treated as the heating element, but as a spare part, which it ceases to be at the time of installation. It then becomes an integral part (a component). Semantics of the *generalization-specialization* describes it as the relationship between types (classes) saying that *specialization* is a special case of *generalization*. What is understood as a special case is taking over all the characteristics of a base class (type) (*generalization*) and the possibility of having their own distinctive characteristics of the type of *specialization* and *generalization* and other system components. *Generalization* has all the features that combine the entire collection of *specialization*. Thanks to this, a *generalization* type contains features which do not need to be repeated in all *specializations* since the relationship is closely related to the mechanism of inheritance. This mechanism is simply based on taking over by *specialization* characteristics of *generalization*. Which greatly simplifies the definition of types (classes). The graph *generalization-specialization* (taxonomy) is a tree in the case when multiple inheritance, or inheritance from more than one base class, is not permitted. Generalizing the issue and taking into account the possibility of multiple inheritance taxonomy takes the form of an acyclic directed graph. This is due to the fact that any type (class) cannot directly or indirectly be a *specialization* or *generalization* of itself. UPM$^{Srv}_{Mnf}$ adopts the approach to the *generalization-specialization* relationship consistent with the semantics described above. For example, by defining the resources of the machinery it is possible to create a taxonomy of vehicles similar to a pattern shown in Fig. 7.

4 Summary

This article provides an introduction to the description of the Unified Process Management for Service and Manufacture being built, the elements of which are listed in Sect. 1. It is a complementary system allowing modeling, design, and

[5]Only one component of a *whole* was presented for simplification.

management of production processes in an integrated environment, dedicated to the needs of medium and large manufacturing and service companies. The main focus were aspects in the field of elementary concepts and relationships between them. A special emphasis was put on the fundamental building blocks, which are CLASS, INSTANCE, FEATURE and Property. What was explained are the relationships between them and the semantics based on Semantic Knowledge Base. The proposed system is to provide a platform that enables sharing and transfer of structures, methods, algorithms, processes, and resources not only within a single entity, but also between independent economic operators. Unification will provide model-level compatibility, as well as data structures and protocols on the low level point of view. It will also ensure consistency in terms of set of categories. Lack of unification is the main problem in system integration, merging into a common mechanism for the exchange of information, goods and services, both at the design and operation of industrial production management systems, equally in terms of production and services. Currently, the works involve the construction of information structures and definition of the semantics of the other SKB components and modules in terms of Unified Process Management for Service and Manufacture.The next step will be to develop a range of information processing methods and algorithms in the $\mathrm{UPM}_{\mathrm{Mnf}}^{\mathrm{Srv}}$ —Unified Process Management for Service and Manufacture. Accordingly, Unified Process Management for Service and Manufacture components are defined in SKB categories, which are defined in categories of the Association-Oriented Database (AODB) Model. The multilayer system architecture provides a robust process of building it, using the mechanism of encapsulation and reusability of specialized algorithms and patterns.

References

1. Liu, X., Bo, H., Ma, Y., Meng, Q.: Integrated production planning management model in iron & steel group. Comput. Integr. Manuf. Syst.-Beijing **14**(1), 24 (2008)
2. Muniz, J., Batista Jr, E.D., Loureiro, G.: Knowledge-based integrated production management model. J. Knowl. Manag. **14**(6), 858871 (2010)
3. Krótkiewicz, M., Wojtkiewicz, K.: Conceptual ontological object knowledge base and language. In: Kurzyński, M., Puchała, E., Woźniak, M., Żołnierek, A. (eds.) Computer Recognition Systems. Advances in Soft Computing, vol. II, pp. 227–234. Springer, Berlin (2005)
4. Hovey, M.: Model Categories. American Mathematical Society, Providence (1999)
5. Schwartz, J.D., Rivera, D.E.: A process control approach to tactical inventory management in production-inventory systems. Int. J. Prod. Econ. **125**(1), 111–124 (2010)
6. Kuhlmann, T., Lamping, R., Massow, C.: Agent-based production management. J. Mater. Process. Technol. **76**(1), 252–256 (1998)
7. Yang, J., Balakrishnan, G., Maeda, N., Ivani, F., Gupta, A., Sinha, N., Sankaranarayanan, S., Sharma, N.: Object model construction for inheritance in C++ and its applications to program analysis. In: Lecture Notes in Computer Science (including subseries Lecture Notes in Artificial Intelligence and Lecture Notes in Bioinformatics), 2012, vol. 7210, pp. 144–164. LNCS (2012)

Point Image Representation for Efficient Detection of Vehicles

Zbigniew Czapla

Abstract This paper presents method of image conversion into point image representation. Conversion is carried out with the use of small image gradients. Layout of binary values of point image representation is in accordance with the edges of objects comprised in the source image. Vehicles are detected through analysis of the detection field state. The state of the detection field is determined on the basis of the sum of the edge points within the detection field. The proposed method of vehicle detection is fast and simple computationally. Vehicle detection with the use of image conversion into point image representation is efficient and can by used in real-time processing. Experimental results are provided.

1 Introduction

In traffic systems image data are utilized for traffic monitoring and determination of traffic parameters [1–3]. Image data are obtained as a result of image analysis. In image analysis segmentation into regions or objects, description and classification of objects are applied. Before image analysis often image processing is carried out. An important image processing technique is edge detection. The most popular methods of edge detection are gradient methods based on discrete convolution and used convolution masks, e.g. Roberts, Sobel, Prewitt masks [4, 5]. There are also applied other methods of edge detection [6, 7]. Determination of traffic parameters requires application of vehicle detection. Sensors such as loop detectors, radars, ultrasonic sensors, infrared sensors, and cameras are utilized for vehicle detection. Traffic systems using cameras are often attractive for lower cost compared to other systems; however, traffic systems based on image data obtained from cameras are usually complex. In traffic systems based on image data various, often multistage, processing methods are applied. These methods usually contains several stages such

Z. Czapla (✉)
Faculty of Transport, Silesian University of Technology,
ul. Krasinskiego 8, 40-019 Katowice, Poland
e-mail: zbigniew.czapla@polsl.pl

R. Burduk et al. (eds.), *Proceedings of the 9th International Conference on Computer Recognition Systems CORES 2015*, Advances in Intelligent Systems and Computing 403, DOI 10.1007/978-3-319-26227-7_65

as segmentation, recovery of vehicle parameters, vehicle identification, vehicle tracking [8], background updating, shadow elimination, Kalman filtering, feature extracting [9], foreground estimation, foreground segmentation, object verification, vehicle tracking [10], description of virtual detection lines, edge detection, morphological operations, and feature extraction [11]. Proposed point image representation is based on small gradients. A set of binary values are obtained as a result of conversion into point image representation. Layout of binary values in point image representation is in accordance with objects contained in the image. Conversion into point image representation is fast and computationally simple. Proposed point image representation is intended for traffic systems and is suitable for efficient vehicle detection.

2 Image Gradients

In a continues domain an image is described by the function $f(x, y)$. The gradient along the line normal to the edge of $f(x, y)$ on a slope of α to the horizontal axis in terms of derivatives along the orthogonal axes is given by the equation

$$g(x, y) = \frac{\partial f(x, y)}{\partial x} \cos(\alpha) + \frac{\partial f(x, y)}{\partial y} \sin(\alpha). \tag{1}$$

The magnitude of the gradient $g(x, y)$ is expressed by the equation

$$|g(x, y)| = \left[\left(\frac{\partial f(x, y)}{\partial x} \right)^2 + \left(\frac{\partial f(x, y)}{\partial y} \right)^2 \right]^{\frac{1}{2}}. \tag{2}$$

In a discrete domain an image is described by the function $f(m, n)$, with m increasing horizontally and n increasing vertically, and then the edge gradient can be expressed by the row gradient $g_r(m, n)$ and the column gradient $g_c(m, n)$ as follows:

$$g(m, n) = [g_r(m, n), g_c(m, n)]. \tag{3}$$

The magnitude of the gradient $g(m, n)$ is given by

$$|g(m, n)| = \left[g_r(m, n)^2 + g_c(m, n)^2 \right]^{\frac{1}{2}}. \tag{4}$$

The small neighbourhood of gradients can be defined in orthogonal and diagonal directions. In the orthogonal directions gradients are defined in the horizontal and vertical neighbourhoods. In the horizontal neighbourhood the small row gradient is given by the equation

$$g_r(m, n) = f(m, n) - f(m - 1, n) \tag{5}$$

and in the vertical neighbourhood the small column gradient by the equation

$$g_c(m, n) = f(m, n) - f(m, n - 1). \tag{6}$$

In the diagonal neighbourhood the small gradients $g_d(x, y)$ ("diagonal down") and $g_u(x, y)$ ("diagonal up") are expressed, respectively, as follows:

$$\begin{aligned} g_d(m, n) &= f(m, n) - f(m - 1, n - 1), \\ g_u(m, n) &= f(m, n) - f(m - 1, n + 1). \end{aligned} \tag{7}$$

Small image gradients can be determined in the image by appropriate calculation of running differences of pixels along orthogonal and diagonal directions.

3 Conversion into Point Image Representation

Image conversion into point image representation is carried out on the basis of small image gradients. The source image is the greyscale digital image of intensity resolution k bits per pixel and of the size of $M \times N$ (columns $\times$ rows) of pixels. The image function $f(m, n)$ assigns a non-negative integer value of pixel intensity to each pair of non-negative integer coordinates (m, n). The coordinates satisfy conditions $0 \leq m \leq M - 1$ and $0 \leq n \leq N - 1$. In the process of image conversion into point image representation two image matrices are used: image matrix $\mathbf{A}$

$$\mathbf{A} = \left[a_{n,m}\right], \quad 0 \leq n \leq N - 1, \quad 0 \leq m \leq M - 1, \tag{8}$$

for pixel values of the source image, and binary image $\mathbf{B}$

$$\mathbf{B} = \left[b_{n,m}\right], \quad 0 \leq n \leq N - 1, \quad 0 \leq m \leq M - 1 \tag{9}$$

for target point values obtained as the result of conversion. Elements of matrix $\mathbf{A}$ are read by rows successively and for each read element, except border elements ($1 \leq n \leq N - 2$ and $1 \leq m \leq M - 2$), the magnitude of small gradients is determined. The small gradient magnitudes in the horizontal and vertical directions are determined, respectively, by the equations

$$\begin{aligned} |g_r(m, n)| &= \left|a_{n,m} - a_{n,m-1}\right|, \\ |g_c(m, n)| &= \left|a_{n,m} - a_{n-1,m}\right| \end{aligned} \tag{10}$$

and in the diagonal directions, respectively, by the equations

$$\begin{aligned} |g_d(m, n)| &= \left|a_{n,m} - a_{n-1,m-1}\right|, \\ |g_u(m, n)| &= \left|a_{n,m} - a_{n+1,m-1}\right|. \end{aligned} \tag{11}$$

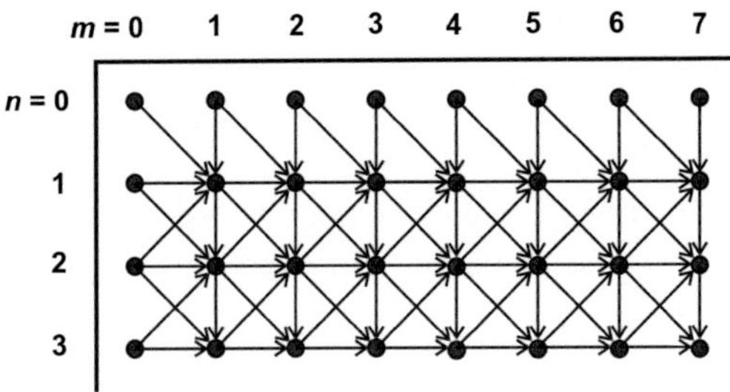

Fig. 1 Position of the image small gradients

The position of the considered small gradients in the source image is shown in Fig. 1. For each processed element of matrix **A** the maximum value of the gradient magnitude is determined

$$|g(m,n)|_{\max} = \max\left[\,|g_r(m,n)|\,,\,|g_c(m,n)|\,,\,|g_d(m,n)|\,,\,|g_u(m,n)|\,\right]. \tag{12}$$

Logical values written into binary matrix **B** are determined according to values of maximum small gradient magnitude and the set value of the threshold T as follows:

$$b_{n,m} = \begin{cases} 0 & \text{for} \;\; |g(m,n)|_{\max} \le T, \\ 1 & \text{for} \;\; |g(m,n)|_{\max} > T. \end{cases} \tag{13}$$

Elements of matrix **B**, which satisfy equation $b_{n,m} = 1$, are called the edge points. Layout of the edge points in matrix **B** corresponds to the edges contained in the image. Elements of matrix **B** form target point image representation.

4 Experimental Results

The input image sequence is obtained from the video stream of 30 frames per second. The input image sequence consists of greyscale images of intensity resolution 8 bits per pixel and of the size of 256×256 of pixels. Images of the test input image sequence are converted into point image representation. Each image of the input image sequence is labelled by the ordinal number in the sequence denoted by i. The selected images from the input image sequence are shown in Fig. 2. All images of the input image sequence are converted into point image representation. The threshold value is set $T = 8$. Layout of the edge points in the selected images from the input image sequence is shown in Fig. 3. In the figure, the edge points of point image representation are denoted by black points. The position of the edge points in point image representation properly corresponds to edges of objects in the source image and is suitable for object detection.

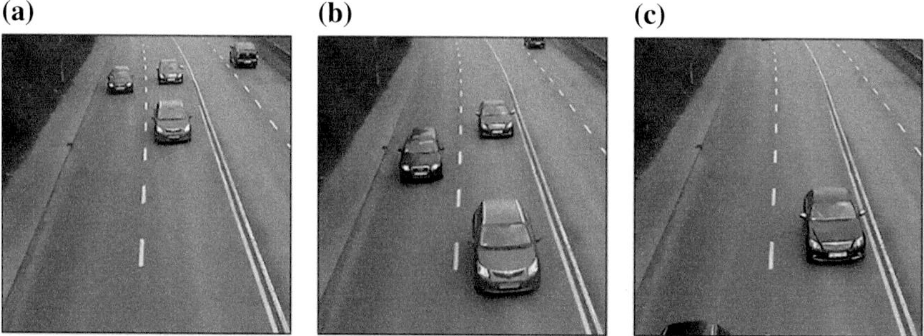

Fig. 2 Selected images from the input image sequence, **a** $i = 0$, **b** $i = 22$, **c** $i = 44$

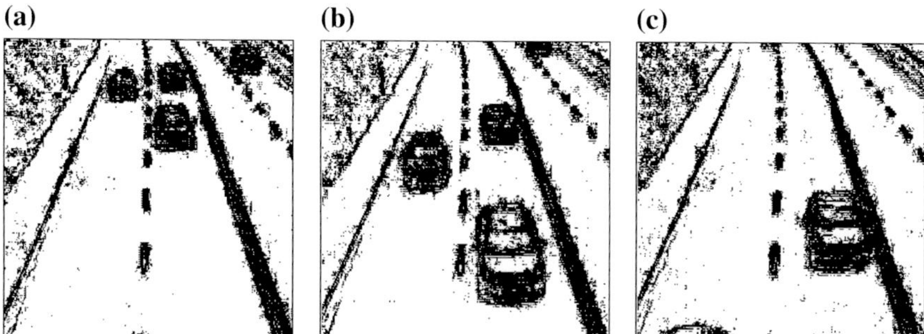

Fig. 3 Layout of the edge points in point image representation of selected images from the input image sequence, **a** $i = 0$, **b** $i = 22$, **c** $i = 44$

5 Vehicle Detection

Each image of the input image sequence is converted into point image representation. For all images of the input image sequence the rectangular detection field is defined. After image conversion into point image representation, the sum of the edge points in the detection field is calculated and considered for consecutive images of the input image sequence [12]. The detection field denoted by D is defined by the set of four points which coordinates determine the location of detection field vertices:

$$D = \{p_{LU}(m_L, n_U), p_{RU}(m_R, n_U), p_{LB}(m_L, n_B), p_{RB}(m_R, n_B)\}. \tag{14}$$

The individual vertices are defined as follows: the left, upper vertex by the point $p_{LU}(m_L, n_U)$, the right, upper vertex by the point $p_{RU}(m_R, n_U)$, the left, bottom vertex by the point $p_{LB}(m_L, n_B)$ and the right, bottom vertex by the point $p_{RB}(m_R, n_B)$. Selected images from the input image sequence with the marked detection field are shown in Fig. 4. After conversion into point image representation, the same detection field D is defined for all converted images from the input image sequence. Selected images after conversion into point image representation, with the marked detection

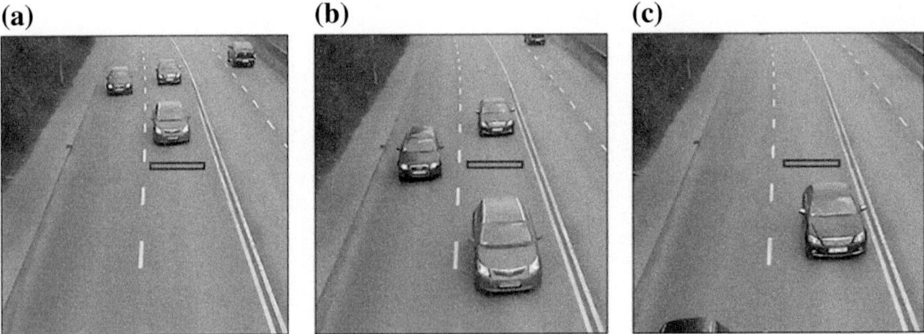

Fig. 4 Selected images from the input image sequence with the marked detection field: **a** $i = 0$, **b** $i = 22$, **c** $i = 44$

Fig. 5 Layout of the edge points in point image representation of selected images from the input image sequence with the marked detection field: **a** $i = 0$, **b** $i = 22$, **c** $i = 44$

field, are shown in Fig. 5. In the figure, the edge points of point image representation are denoted by black points. For each converted image i the arithmetic sum of the edge points (equal to 1) is calculated within the detection field as follows:

$$S_i = \sum_{n=n_U}^{n_B} \sum_{m=m_L}^{m_R} b_{n,m} : b_{n,m} = 1 \tag{15}$$

A vehicle passing through the detection field changes the sum of the edge points within the detection field. Changes of the sum of the edge points in the detection field for the input image sequence are shown in Fig. 6. The state of the detection field is determined on the basis of the average sum of the current image i and P previous images in the input image sequence. The average sum for image i is expressed by the equation

$$R_i = \frac{1}{P+1} \sum_{j=i-P}^{i} S_j \tag{16}$$

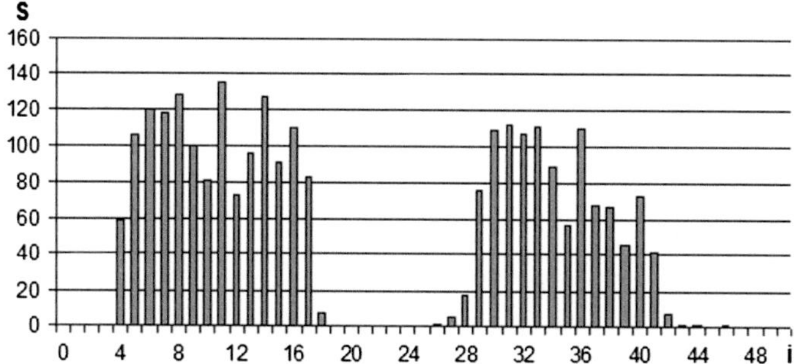

Fig. 6 Changes of the sum of the edge points in the detection field for the input image sequence

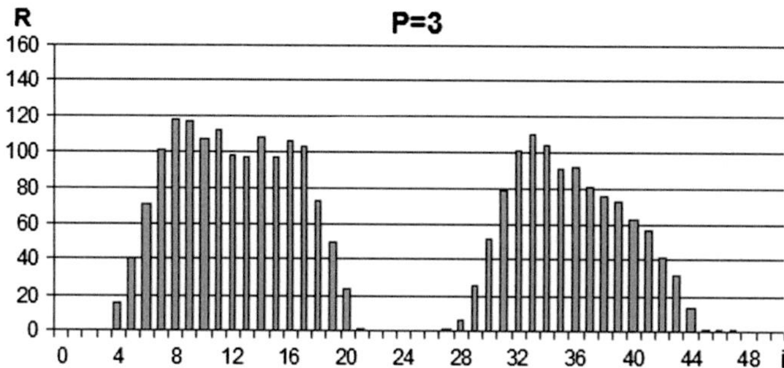

Fig. 7 Changes of the average sum of the edge points in the detection field for the input image sequence

Changes of the average sum of the edge points in the detection field for the input image sequence are shown in Fig. 7. The detection field can be in the state "detection field free" (the detection field state pointer $K = 1$) or "detection field occupied" (the detection field state pointer $K = 0$). The state of the detection field changes from "detection field free" ($K = 0$) to "detection field occupied" ($K = 1$) if the following conditions are satisfied:

$$R_i > R_O \ \wedge \ K = 0 \tag{17}$$

where R_O denotes the minimum sum of the edge points for the occupied detection field. The state of the detection field changes from "detection field occupied" ($K = 1$) to "detection field free" ($K = 0$) if the conditions

$$R_i < R_F \ \wedge \ K = 1 \tag{18}$$

are satisfied. The quantity R_F denotes the maximum sum of the edge points for the free detection field. The vehicle driving into the detection field changes its state from "detection field free" to "detection field occupied". The vehicle leaving the detection field changes its state from "detection field occupied" into "detection field

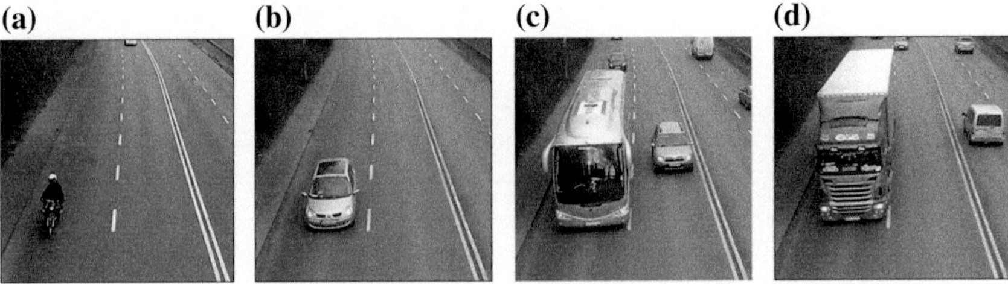

Fig. 8 Source images of various vehicles: **a** motorcycle, **b** passenger car, **c** bus, **d** articulated lorry

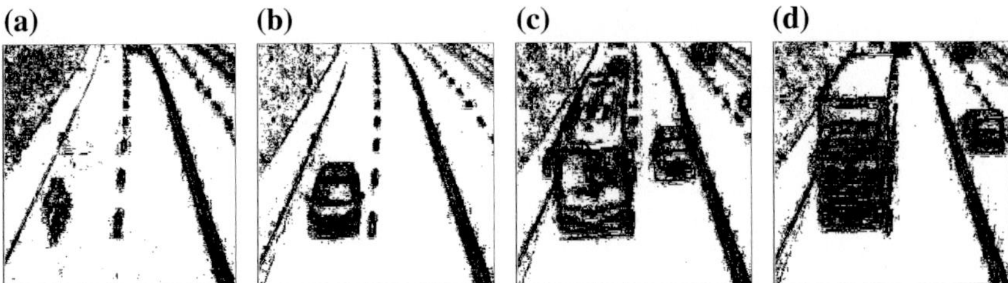

Fig. 9 Point image representation of the source images: **a** motorcycle, **b** passenger car, **c** bus, **d** articulated lorry

free". Conversion of images into point image representation allows to detect vehicles of various types such as bicycles, motorcycles, passenger cars, trucks, buses, and articulated lorry. The source images presenting motorcycle, passenger car, bus, and articulated lorry are shown in Fig. 8. The result of conversion of the source images into point image representation is shown in Fig. 9. Conversion of images into point image representation allows efficient detection of vehicles of various types.

6 Comparison to Methods of Edge Detection

The proposed method of image conversion into point image representation is comparable to methods of edge detection using discrete convolution (e.g. Sobel masks, Prewitt masks). Methods of edge detection based on discrete convolution apply repeated operations with the use of convolution masks, and therefore these methods are computationally exacting. For comparison, edge detection with the use of Sobel masks has been performed for the same selected images of the input sequence. Sobel masks for edge detection in horizontal and vertical directions are shown in Fig. 10. The result of processing of selected images from the input image sequence, after edge detection and inversion, with the marked detection field is shown in Fig. 11. The set threshold value ($T = 160$) allows to obtain the sum of the edge points within the detection field similar to the sum of the edge points obtained using image conversion into point image representation. For each pixel of the source image execution of at

1	2	1
0	0	0
-1	-2	-1

-1	-2	-1
0	0	0
1	2	1

1	0	-1
2	0	-2
1	0	-1

-1	0	1
-2	0	2
-1	0	1

Fig. 10 Sobel masks for edge detection in *horizontal* and *vertical* directions

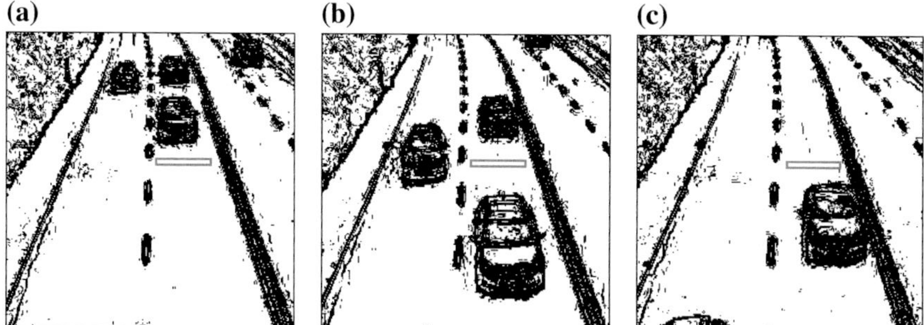

Fig. 11 Selected images from the input image sequence, after edge detection and inversion, with the marked detection field: **a** $i = 0$, **b** $i = 22$, **c** $i = 44$

least ten operations is required (eight partial sum operations, one total sum operation and one threshold operation) for one mask. Typically, four masks are applied. Quality enhancement can involve operations performed on floating-point numbers or the application of the LUT tables which additionally increases computational complexity.

7 Conclusions

Conversion of an image into point image representation is fast and simple computationally. Layout of the edge points well corresponds to the edges of objects contained in the source image and allows efficient vehicle detection. Detection of vehicles passing through the detection field is carried out by the analysis of changes of the detection field state. The state of the detection field is determined through simple calculation of the sum of the edge points within the detection field. The proposed method of image conversion into point image representation is significantly simpler than the well-known edge detection methods using discrete convolution (e.g. Sobel masks, Prewitt masks). Implementations of image conversion into point image representation can be applied instead of mask edge detection (with the use of discrete convolution) in a wide range. The proposed method is faster and uses only integer and logic operations. Vehicle detection with the use of image conversion into point image representation utilizes the small number of operations which makes it efficient and attractive computationally. The small number of operations causes that conversion into point image representation can be useful for real-time processing.

References

1. Coifman, B., Beymer, D., McLauchlan, P., Malik, J.: A real-time vision system for vehicle tracking and traffic surveillance. Transp. Res. Part C **6C**, 271–288 (1998)
2. Fernandez-Caballero, A., Gomez, F.J., Lopez-Lopez, J.: Road traffic monitoring by knowledge-driven static and dynamic image analysis. Expert Syst. Appl. **35**, 701–719 (2008)
3. Kamijo, S., Matsushita, Y., Ikeuchi, K., Sakauchi, M.: Traffic monitoring and accident detection at intersections. IEEE Trans. Intell. Transp. Syst. **1**(2), 108–118 (2000)
4. Di Zeno, S.: A note on the gradient of a multi-image. Comput. Vis., Graph. Image Process. **33**, 116–125 (1986)
5. Muthukrishnan, R., Radha, M.: Edge detection techniques for image segmentation. Intern. J. Comput. Sci. Inf. Technol. (IJCSIT) **3**(6), 259–267 (2011)
6. Kang, C.-C., Wang, W.-J.: A novel edge detection method based on the maximizing objective function. Pattern Recognit. **40**, 609–618 (2007)
7. Qian, R.J., Huang, T.: Optimal edge detection in two-dimensional images. IEEE Trans. Image Process. **5**(7), 1215–1220 (1996)
8. Gupte, S., Masoud, O., Martin, R.F.K., Papanikolopoulos, N.P.: Detection and classification of vehicles. IEEE Trans. Intell. Transp. Syst. **3**(1), 37–47 (2002)
9. Hsieh, J., Yu, S.H., Chen, Y.S., Hu, W.F.: An automatic traffic surveillance system for vehicle tracking and classification. IEEE Trans. Intell. Transp. Syst. **7**(2), 175–187 (2006)
10. Lien, C.C., Tsai, M.H.: Real-time flow analysis without background modeling. J. Inf. Technol. Appl. **5**(1), 1–14 (2011)
11. Mithun, N.C., Rashid, N.U., Rahman, S.M.M.: Detection and classification of vehicles from video using multiple time-spatial images. IEEE Trans. Intell. Transp. Syst. **13**(3), 1215–1225 (2012)
12. Czapla, Z.: Video based vehicle counting for multilane roads. Logistyka, no. **4**, 2709–2717 (2014)

Modelling Dental Milling Process with Machine Learning-Based Regression Algorithms

Konrad Jackowski, Dariusz Jankowski, Héctor Quintián, Emilio Corchado and Michał Woźniak

Abstract Control of dental milling processes is a task which can significantly reduce production costs due to possible savings in time. Appropriate setup of production parameters can be done in a course of optimisation aiming at minimising selected objective function, e.g. time. Nonetheless, the main obstacle here is lack of explicitly defined objective functions, while model of relationship between the parameters and outputs (such as costs or time) is not known. Therefore, the model must be discovered in advance to use it for optimisation. Machine learning algorithms serve this purpose perfectly. There are plethoras of competing methods and the question is which shall be selected. In this paper, we present results of extensive investigation on this question. We evaluated several well-known classical regression algorithms, ensemble approaches and feature selection techniques in order to find the best model for dental milling model.

Keywords Dental milling process · Machine learning · Regression · Ensemble of predictors · Feature selection

K. Jackowski (✉) · D. Jankowski · M. Woźniak
Department of Systems and Computer Networks, Wroclaw University of Technology, Wyb. Wyspianskiego 27, 50-370 Wroclaw, Poland
e-mail: konrad.jackowski@pwr.edu.pl

D. Jankowski
e-mail: dariusz.jankowski@pwr.edu.pl

H. Quintián · E. Corchado
Department. de Informática y Automática, Universidad de Salamanca, Salamanca, Spain
e-mail: hector.quintian@usal.es

E. Corchado
e-mail: escorchado@usal.es

R. Burduk et al. (eds.), *Proceedings of the 9th International Conference on Computer Recognition Systems CORES 2015*, Advances in Intelligent Systems and Computing 403, DOI 10.1007/978-3-319-26227-7_66

1 Introduction

Over recent years, there have been a high increase in the use of artificial intelligence and soft computing methods to solve real-world problems [5, 7, 16].

It is known that the complexity inherited in most of the new real-world problems increases with the computing capabilities [18]. Higher performance requirements with a lower amount of data examples are needed due to the costs of gathering new instances, especially in those processes where new technology arises.

The optimisation process of machine parameters could significantly help to increase companies' efficiencies and substantially contributes to cost reductions in preparation and setting machines processes and it also helps in the production process using new materials. Nevertheless, the variables and parameters setting processes are a well-known problem that has not been fully resolved yet. Several different techniques are proposed in the literature. In [8] the influence of operating parameters of ultrasonic machining is studied using Taguchi and F-test method.

Also, ANN has been used to find relationships between the mechanical properties of different real-world problems [2, 7] and they have been also applied for identification of the parameters for operating conditions [10]. Conventional methods can be greatly improved through the application of soft computing techniques [6], by generating soft computing models with high accuracy that are used to optimise machine parameters. This research is focused on creating the best model of a dental milling machine applying several soft computing techniques, comparing all these techniques in terms of RMSE (Root Mean Square Error).

This study is a consequence of a preliminary study [17], where several models were generated using ANN; however, these models did not get the expected accuracy, so in these research other softcomputing techniques have been applied using a bigger dataset than in previous study, in order to get models with lower errors.

The rest of this paper is organised as follows. Section 2 introduces all the algorithms used in this study for modelling. The experiments and commented results are presented in Sect. 3. Finally, conclusions are drawn and future work outlined.

2 Machine Learning-Based Regression

One of the method for creating model of given process is the application of regression algorithms. Their objective is to create the best possible mapping between inputs and outputs. In regression tasks, it is assumed that the output $\hat{y}$ is the realisation of real variable which depends on input X [1]. Task of finding optimal input–output mapping (1) means that one have to select model of function F, which reflect real relationships between $\hat{y}$ and X, and choose the function parameters β.

$$\hat{y} = F(X, \beta). \tag{1}$$

The task is not easy as there are several issues which have to addresses. Let us discuss the most important ones. First, in many real situations we do not have any information on model of relation (1) or its parameters. In this case one can select the model arbitrarily in hope that it allows for acceptable approximation of the relation. Alternatively, several models can be chosen and assessed experimentally by comparison of their accuracies achieved on available testing set. Lists of available models are wide. Among the other one can list linear or polynomial regression, non-parametric regression–smoothing algorithms [9]. Alternatively, one can try to employ machine learning algorithms [11], which prove their usefulness for regression and classification tasks. Advantage of this sort of systems is that most of them are very adaptable and can create mapping model which is able to adjust to current tasks in more flexible manner that regression algorithm based on mapping described in previous paragraph. Some of the machine learning-based options are described later on in Sect. 1. Second, input variables can have different types and formats, which causes problems with their unification. Nonetheless, in many cases, their transformation into numerical values can be easily done. Some problems can appear while casting nominal variables. Usually, this process requires designing tailored distance measures which are very often used by machine learning algorithms. Nonetheless, without loosing ability to generalisation, in further consideration we assume that input to the system consists of the set of numerical variables only. Next, regardless which model of F is selected, its parameters have to be adjusted to form appropriate mapping. For this purpose a learning set is used (2). It consists of set of sample, i.e. pairs of input and output.

$$LS = \{(x_1, y_1), (x_2, y_2), \ldots, (x_N, y_N)\}. \tag{2}$$

There is one essential issue related with forming the learning set. Its content has to be representative, i.e. distribution of collected samples should reflect real distribution of population in feature space. It is hard to meet this requirement as it requires collecting relatively large number of samples, what would be costly, time consuming or impractical. Therefore, usually we have in hand only limited learning set and which does not quarantine that resulting regression model approximates the real relationships acceptably. Finally, one has to select a measure for model evaluation. The same measure can be also used for regression algorithm training procedure which aims at setting optimal model parameter values. In case of regression tasks, mean root squared error measure (3) can be used for that purpose:

$$RMSE(F, LS) = \sqrt{\frac{1}{N}\sum_{n=1}^{N}(F(x_n) - y_n)^2}. \tag{3}$$

Formula (3) can be also used for comparative analysis of different regression algorithms. Due to lack of knowledge on model of dental milling process we decided to test possible wide range of regression algorithms. For our purposes we decided to select several classical regression algorithms and some ensemble methods.

2.1 Classical Algorithms

2.1.1 Linear Regression

This is supposedly the simplest predictive model [1]; nonetheless, it features surprisingly high accuracy in many practical problems. It uses simple weighted combination of input variables to calculate the output. The weights are set in a course of training procedure aiming at minimising RMSE. More advanced version of linear regression named Pace regression was proposed by Wang [19].

2.1.2 Multilayer Neural Network

This is the implementation of classical multilayer neural network [1] with nodes arranged in sequentially organised layers. Layers are fully interconnected and input signal is transferred forward from the first to the last layer. Each node performs weighted fusion of its input signals and maps the response with sigmoid transfer function. There is only one output neuron which returns the system response. Entire system is trained using back-propagation algorithm.

2.1.3 SMOreg

SMOreg stands for sequential minimal optimisation for support vector machine (SVM) for regression. SVM is one of the most popular machine learning algorithms proposed by Smola [14]. SVM computes linear regression in a high-dimensional feature space. Input data are mapped using a non-linear kernel function optimisation of the SVM that is performed by algorithm which is an extension of the SMO algorithm proposed by Platt [12].

2.2 Feature Selection

In many real situations, results obtained by classical algorithms are not satisfactory or even acceptable. Usually, it is hard to find the reason, as there are many factors which affect the results. One can list wrong selection of regression model, and poor representation of samples in learning set. There are few options how to improve the performance. One of them is feature selection. It is based on the assumptions that there can exist input variables which are not relevant to outputs. Such inputs presented to the system can negatively affect overall performance.

Removing such an irrelevant or noised feature can improve the accuracy. There are many possible ways how to do that. The optimal one is exhaustive selection, i.e. comparative evaluation of all possible combinations of features. This method is feasible only for limited number of inputs due to computational complexity. More effective (from this point of view) are heuristic algorithms such as backward or forward feature selection, or selection based on genetic approach.

2.3 Ensemble Techniques

The last presented approach to improve regression accuracy covers a wide range of ensemble techniques. All of them share the same principle, i.e. it is assumed that there is a set $\Pi = \{F_1, F_1, \ldots, F_K\}$ which consists of several elementary (classical) regression algorithms F_k (predictors).

The predictors in Π are trained independently. Ensemble system fuses responses of all elementary predictors while the fusion function varies depending on the selected model. In the most simplest form result can be obtained by calculating the average of responses (4)

$$\hat{F}(x) = \frac{1}{K}\sum_{k=1}^{K} F_k(x). \tag{4}$$

The model (4) is very intuitive and can be calculated straightforward as there are no additional parameters which have to be computed in a course of ensemble learning procedure. On the other hand, this model does not take into consideration the quality of the elementary predictors. It means that all the predictors contribute to final response in the same degree which can lead to spoiling results by week or irrelevant predictors. Much better results can be obtained by incorporating weighting which reflects the quality of the predictors (5):

$$\hat{F}(x) = \frac{1}{K}\sum_{k=1}^{K} w_k F_k(x). \tag{5}$$

While selecting model (5) one must remember that $\sum_{k=1}^{K} w_k = 1$. Important issue that must be addressed is the method of weight selection. In the simplest implementation, the weights shall be counter proportional to RMSE of the given elementary predictor. In more sophisticated case, they can be adjusted in the course of learning process which aims at minimising RMSE of the ensemble according to (3).

2.3.1 Diversity of the Ensemble

Collecting elementary predictors which shall form ensemble system is not a trivial task. Common sense tells that set of the same or similar predictors cannot help to improve quality of regression as all the responses are almost the same. Therefore, there is a necessity to inject some diversity into ensemble. There are several methods possible to do this. According to Brown et al. [4] diversity can be ensured using implicitly or explicitly diversity maintaining algorithms. In the following paragraphs, we provide some insight into selected diversity maintaining methodologies.

2.3.2 Bagging

Bagging was originally proposed by Breiman [3]. In this method ensemble consists of predictors of the same type, i.e. having the same mapping model. Diversity is generated by partitioning empirical dataset used for training individual predictors in the ensemble. Each of them is trained with its own subset created using bootstrapping, i.e. drawing sample randomly from original dataset with replacement.

2.3.3 Boosting

More sophisticated approach is used in boosting introduced by Schapire [13]. Forming homogeneous ensemble is an iterative process. At each repetition recently created predictor is weighted according to its accuracy. The better one gets higher weights in the ensemble. Samples in learning set are also weighted proportionally to accuracy of their estimation. Smaller prediction error means that weights of sample are decreased in next iteration.

2.3.4 Heterogeneous Ensemble

The last ensemble method used in our researches is creating heterogeneous ensemble, i.e. such that consists of predictors which have different mapping models. There are many techniques which can be used for setting fusion weights (5). In our tests we decided to use evolutionary algorithms for that purpose. In this method, weights are encoded in a form of chromosome (i.e. vector containing weights stacked together). Setting their values is optimisation process which aims at minimising regression error (3). Population of possible solution (weights values) are processed using standard genetic operators (i.e. mutation and crossover), which introduce some diversity into population of solutions and exchange the parts of chromosomes between selected individuals.

3 Experimental Evaluation

We set the following objectives for our research and experiments:

1. Examining wide range of classical machine learning-based regression algorithms,
2. Investigating possibility of increasing of the regression by application of ensemble methods such as bagging, boosting and heterogeneous ensembles.

3.1 Experimental Framework

The following regression algorithms were implemented and tested:

1. Linear Regression (*Linear Reg.*),
2. Pace Regression (*Pace Reg.*),
3. Multilayer Neural Network with one hidden layer consisting of 5 neurons (*MLP Reg.*),
4. SVM for regression optimised with SMO algorithm (*SMO Reg.*).

Additionally, several listed below techniques were also tested.

1. Feature selection algorithm which uses genetic algorithms. Methods were applied to all elementary predictors.
2. Homogeneous ensembles. Four ensembles were created based on four listed classical predictors, respectively. Two ensemble creating techniques were implemented:
   a. Bagging,
   b. Boosting.
3. Heterogeneous ensemble which consisted of four different classical predictors. Three methods were tested for calculating output of the system:
   a. Simple mean of predictors outputs,
   b. Weighted mean with weights set proportional to predictor accuracy,
   c. Weighted mean with weights set by evolutionary algorithm which minimised RMSE of the ensemble.

Classical predictors, bagging and boosting algorithms were modelled in KNIME (an open source data mining framework [15] available at[1]) using WEKA nodes.[2] MATLAB optimisation toolbox was for our own implementation of evolutionary ensemble.

3.2 Dataset

In this paper we use empirical material collected during manufacturing of dental pieces (see Fig. 1). A dynamic high-precision machining centre with five axes was applied in this research. This real industrial use case is described by a dataset of 218 samples obtained by a dental scanner in the manufacturing of dental pieces characterised by 12 input variables. The input variables (see Table 1) are the following: the type of work, the thickness, the number of pieces, the radius of the tool, the revolutions of the drill, the feed rate in each of the dimensions (X, Y and Z), the initial tool

[1]https://www.knime.org/.

[2]http://www.cs.waikato.ac.nz/ml/weka/.

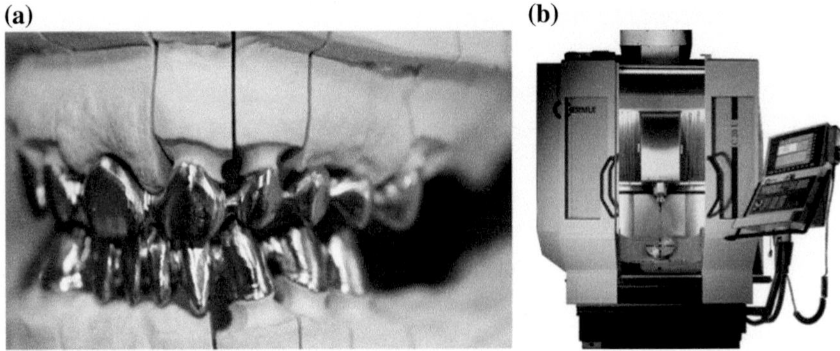

Fig. 1 Metal pieces drilled by the drilling machine. **a** Metal pieces manufactured by a dynamic high-precision machining centre with five axes. **b** Milling centre of HERMLE type-C 20 U (iTNC 530)

Table 1 Different features from the process, their units and ranges

Variable (Units)	Number of feature	Range of values
	INPUTS	
Type of work	1	1–7
Thickness (mm)	2	8–18
Number of pieces	3	1–4
Radius (mm)	4	0.15–2
Tool	5	1–4: toric, spherical, plain, drill
Revolutions per minute (RPM)	6	7,500–39,000
X-axis feed rate (mm per minute)	7	0–3,000
Y-axis feed rate (mm per minute)	8	0–3,000
Z-axis feed rate (mm per minute)	9	50–2,000
Initial diameter tool (mm)	10	91.0608–125.56
Initial temperature (°C)	11	21.6–31
Estimate work time (s)	12	6–6,318
	OUTPUTS	
Time error for manufacturing (s)	1	−682–1017

diameter, the initial temperature and the estimated duration of the work. The main parameter to estimate is the time error for manufacturing, which is the difference between the estimated time by the machine itself and real work time—negative values indicate that real exceeds estimated time. All the variables and their ranges are presented in Table 1.

3.3 Results and Discussion

Table 2 presents results obtained in test procedure. For elevating test confidence, 10-fold cross-validation methods were used. Therefore, the table shows average of RMSE (3) obtained for 10 repetitions of the experiments for 10-folds, respectively. Target signal was normalised before tests to get a value from range $< 0, 1 >$. Analysis of the results from Table 2 allows to make the following observations:

1. All elementary predictors obtained very similar results. The only one that stands out is SMO algorithm. Three other got almost the same RMSE. Therefore, it is hard to firmly state that any one can be nominated as the winner.
2. Feature selection did not bring any advantage as the results for respective algorithms remain also almost the same.
3. Bagging and Boosting did not help neither. Only small improvement can be noticed for SMO predictor as its RMSE was reduced from 0.000977 to 0.000447 by application of boosting techniques.
4. No improvement was get by ensemble techniques which use heterogeneous set of predictors. However, it is worthy to mention that simple average, which is probably one of the most popular techniques, spoiled the quality of the system as it increased the error to 0.007888. More advanced weighted fusion methods preserve the same quality. Nonetheless, as it can be seen, they are useless, as there is no error reduction.

Finally, no one ensemble technique allows to bring any significant error deduction. All tested systems got very similar results. Therefore, it can be clearly seen that there are cases for which classical regression algorithms are the best options and their improvement cannot be done by more advanced and sophisticated ensemble techniques. Nonetheless, it has to be underlined that this conclusion cannot be generalised.

Table 2 RMSE for tested regression algorithms

		Linear Reg.	MLP Reg.	Pace Reg.	SMO Reg.
Single one predictor	All inputs	0.000425	0.000424	0.000430	0.000977
	Selected features	0.000428	0.000428	0.000426	0.000949
Homogeneous ensembles	Bagging	0.000419	0.000420	0.000421	0.000693
	Boosting	0.000425	0.000411	0.000430	0.000447
Heterogeneous ensembles	Mean fusion	0.007888			
	Weighted fusion	0.000434			
	Evolutionary based fusion	0.002079			

Authors are conscious that for any other cases results could be quite different. Even more, we would like to underline that this conclusion cannot be done without making all the reported tests.

4 Conclusion

Generation of dental milling models with high accuracy allows to optimise machine parameters and a reduction in costs and time. The extensive investigation done in this paper about the modelling of dental drilling machine let us to conclude that all classical regression techniques used in this paper are the best options for modelling the dataset used in this research.

Results obtained with the soft computing techniques applied in this paper are better than results obtained in previous studies; however, the reason is not the usage of more complex modelling techniques. Based on the current results and previous one obtained in other studies, it can be concluded that the dataset used in previous studies was smaller and not so much informative than this one; so for this reason results are better using simpler softcomputing techniques.

Future lines of research include modelling the temperature difference and the erosion difference (between diameters of the tool before and after the manufacturing), which helps to measure the accuracy of the dental milling process. Additionally, it will investigate about optimisation algorithms in order to optimise the dental milling machine parameters.

Acknowledgments This work was supported by the Polish National Science Centre under the grant no. DEC-2013/09/B/ST6/02264.

References

1. Alpaydin, E.: Introduction to Machine Learning, 2nd edn. The MIT Press, Boston (2010)
2. Bosch, J., López, G., Batlles, F.: Daily solar irradiation estimation over a mountainous area using artificial neural networks. Renew. Energy **33**(7), 1622–1628 (2008). http://www.sciencedirect.com/science/article/pii/S0960148107002881
3. Breiman, L.: Bagging predictors. Mach. Learn. Boston **24**, 123–140 (1996)
4. Brown, G., Wyatt, J., Harris, R., Yao, X.: Diversity creation methods: a survey and categorisation. J. Inf. Fusion **6**, 5–20 (2005)
5. Calvo-Rolle, J.L., Casteleiro-Roca, J.L., Quintián-Pardo, H., del Carmen Meizoso-Lopez, M.: A hybrid intelligent system for PID controller using in a steel rolling process. Expert Syst. Appl. **40**(13), 5188–5196 (2013). http://dx.doi.org/10.1016/j.eswa.2013.03.013
6. Chang, H.H., Chen, Y.K.: Neuro-genetic approach to optimize parameter design of dynamic multiresponse experiments. Appl. Soft Comput. **11**(1), 436–442 (2011). http://www.sciencedirect.com/science/article/pii/S1568494609002567
7. Diz, M.L.B., Baruque, B., Corchado, E., Bajo, J., Corchado, J.M.: Hybrid neural intelligent system to predict business failure in small-to-medium-size enterprises. Int. J. Neural Syst. **21**(4), 277–296 (2011). http://dx.doi.org/10.1142/S0129065711002833

8. Kumar, V., Khamba, J.S.: Statistical analysis of experimental parameters in ultrasonic machining of tungsten carbide using the taguchi approach. J. Am. Ceram. Soc. **91**(1), 92–96 (2008). http://dx.doi.org/10.1111/j.1551-2916.2007.02107.x
9. Liero, H., Härdle, W.: Applied nonparametric regression (Biometric society monographs no. 19) Cambridge University Press 1990, p. 333. Biom. J. **33**(6), 704–704 (1991). http://dx.doi.org/10.1002/bimj.4710330610
10. Liu, Y.H., Liu, C.L., Huang, J.W., Chen, J.H.: Neural-network-based maximum power point tracking methods for photovoltaic systems operating under fast changing environments. Sol. Energy **89**(0), 42–53 (2013). http://www.sciencedirect.com/science/article/pii/S0038092X12004082
11. Mitchell, T.M.: Machine Learning, 1st edn. McGraw-Hill Inc, New York (1997)
12. Platt, J.: Probabilistic outputs for support vector machines and comparison to regularize likelihood methods. In: Smola, A., Bartlett, P., Schoelkopf, B., Schuurmans, D. (eds.) Advances in Large Margin Classifiers, pp. 61–74. MIT Press, Cambridge (2000). http://citeseer.ist.psu.edu/platt99probabilistic.html
13. Schapire, R.E.: The boosting approach to machine learning: an overview. In: Proceedings of the MSRI Workshop on Nonlinear Estimation and Classification (2001)
14. Scholkopf, B., Smola, A.J.: Learning with Kernels: Support Vector Machines, Regularization, Optimization, and Beyond. MIT Press, Cambridge (2001)
15. Silipo, R., Mazanetz, M.P.: The KNIME Cookbook: Recipes for the Advanced User. KNIME Press, Switzerland (2012)
16. Torreglosa, J., Jurado, F., García, P., Fernández, L.: PEM fuel cell modeling using system identification methods for urban transportation applications. Int. J. Hydrog. Energy **36**(13), 7628–7640 (2011). http://www.sciencedirect.com/science/article/pii/S0360319911007233 hysydays
17. Vera, V., Corchado, E., Redondo, R., Sedano, J., Garcia, A.E.: Applying soft computing techniques to optimise a dental milling process. Neurocomputing **109**, 94–104 (2013). http://dx.doi.org/10.1016/j.neucom.2012.04.033
18. Villar, J.R., González, S., Sedano, J., Corchado, E., Puigpinós, L., Ciurana, J.: Meta-heuristic improvements applied for steel sheet incremental cold shaping. Memetic Computing **4**(4), 249–261 (2012). http://dx.doi.org/10.1007/s12293-012-0100-4
19. Wang, Y., Witten, I.H.: Modeling for optimal probability prediction. In: Proceedings of the Nineteenth International Conference in Machine Learning. pp. 650–657. Sydney, Australia (2002)

Rough Sets and Fuzzy Logic Approach for Handwritten Digits and Letters Recognition

Marcin Majak and Andrzej Żołnierek

Abstract This paper presents the hybrid approach using fuzzy logic and rough sets used as a pattern recognition framework. Both fuzzy and rough sets have been introduced to deal with vagueness and uncertain data in artificial intelligence applications. In general, fuzzy logic can be related to vagueness, while rough sets deal with indiscernibility. In our work we propose two-stage algorithm. At the first stage, an optimization procedure is applied to reduce the number of features for fuzzy membership functions and to find the optimal granulation for rough sets, respectively. In the second stage, two-step classifier is used. We tested our attempt using Handprinted Forms and Characters Database containing the full page binary images of 3699 handwriting sample forms. For any segmented image classification, a crucial part lies in the proper feature extraction method. In our work, cross corner feature algorithm was used as a main tool.

1 Introduction

In the literature, one can find different examples of handwritten pattern recognition applications [3, 9]. There are two, main challenging tasks in this problem. The first one relates to proper feature extraction methods which affect the final classifier accuracy [10, 11]. The second important issue comes to classifier selection. Very often neural networks are applied as an effective tool, but recently a new branch is emerging in which hybrid approaches or classifier ensembles are implemented [6, 8]. The results of such attempts in different recognition tasks were very promising, so we decided to apply our hybrid method [13] to the problem of handwritten digits and letters recognition. The novelty of our approach consists in applying

M. Majak (✉) · A. Żołnierek
Faculty of Electronics, Department of Systems and Computer Networks, Wroclaw University of Technology, Wybrzeze Wyspianskiego 27, 50-370 Wroclaw, Poland
e-mail: marcin.majak@pwr.edu.pl

A. Żołnierek
e-mail: andrzej.zolnierek@pwr.edu.pl

R. Burduk et al. (eds.), *Proceedings of the 9th International Conference on Computer Recognition Systems CORES 2015*, Advances in Intelligent Systems and Computing 403, DOI 10.1007/978-3-319-26227-7_67

simulated annealing as a preprocessing step. It is used for feature extraction and an initial parameter setting for rough sets and fuzzy logic algorithms. Our experiments show that it significantly reduces the number of features for classification and simplifies construction of IF-THEN rules. Proposed algorithm was tested with two methods of feature extraction: simple pixel intensity ratio from a grid and more sophisticated approach called cross corner feature extraction. The organization of this paper is as follows. Section 1.2.1 describes major steps applied to character recognition systems, while in Sect. 1.2.2 one can find information about feature extraction methods. Implemented algorithm and classification results are presented in Sects. 2 and 3, respectively. Whole paper is finished with conclusions and future improvements proposed in Sect. 4.

1.1 Problem Statement

This paper deals with pattern recognition task in which we assume that the pattern (in out case segmented image) is in the state $j \in M$, where M is an m-element set of possible states numbered with the successive natural numbers ($j \in M = \{1, 2, \ldots, m\}$) [5, 12]. The state j is unknown and does not undergo our direct observation. What we can only observe are the features or tokens by which a state manifests itself. We will denote a d-dimensional measured feature vector by $x \in X$ (thus X is the feature vector space). In order to classify unknown patterns, as usual in practice, we assume that we have at our disposal a so-called training set, which in the investigated decision task consists of N training patterns:

$$S = (x_1, j_1), \ldots, (x_N, j_N)$$

x_α, j_α denote d-dimensional pattern and its true classification, respectively. In general, the decision algorithm with learning should use every time as well observed data, i.e. the feature vector x as the knowledge included in the training set S. In consequence, the general classification algorithm with learning is of the following form:

$$i = \Psi(S, x), i \in M$$

1.2 Handwritten Digits and Letters Pattern Recognition

1.2.1 Character Recognition System

Whole handwritten character recognition system can be broken down into few steps.

- Training
  - Preprocessing phase—in this step different filters are applied to the image, e.g. crop, rescale and noise removal. Additionally, slant correction for letter is required to obtain unified set.
  - Feature extraction—this allows to reduce the amount of data and reveals relevant information to classifier. At this stage normalization is needed.
  - Model/classifier construction using available training data set.
- Testing—during this stage similar steps are applied as in training
  - Preprocessing—noise removal, smoothing borders and image segmentation.
  - Feature extraction.
  - Classification—compare feature vectors to various models and find the most suitable one.

1.2.2 Feature Extraction

The main goal of feature extraction is to speed up algorithm processing and increase accuracy of the classifier for pattern recognition. Features of the character are extracted in such a way so that the whole portion of binary image is covered and each portion is distinctive. This process plays a key role in a proper pattern recognition system and even a sophisticated classifier system cannot deal with overlapping and poor quality features. Depending on the language and character types different approaches can be found in the literature. In general, we can distinguish local (in case of zones) or global (related to the whole image) characteristics. The most commonly used are region- and frequency-based methods in which a priori knowledge about image characteristic is required to properly divide image into dense and scattered foreground pixel regions [1, 2]. In some cases, diagonal feature extraction algorithms are favourable over standard grid approaches, but their implementation is rather complex with an additional preprocessing step, because we need filter processing. Image centroid and zone centroid methods work well for these characters having maximum curves, while geometry-based methods are specific to a language, i.e. it is good for such character formed by straight line and simple curves [3, 9, 11].

2 Algorithm Construction

2.1 Feature Extraction Method

In this paper two approaches for feature extraction are tested, but before features are generated, an input image is processed by sequence of digital filters. At the beginning, each image is segmented with a proper threshold mask. This output is later cropped according to calculated bounding box. After getting an uniform size image, in the third step of preprocessing, slant removal method is applied to convert all character images to a standard form. Last step is a morphological filter which deletes the redundant pixel of the image ensuring that the contours of the character image do not change. Additionally, this step removes noise from the segmented image. In this work, the following feature extraction methods were used:

- Pixel intensity ratio from selected image region
  - Whole binary image is divided into five equal intervals for both width and height of the image. This creates 25 rectangles in which we calculate ratio between white and black pixels.
- Cross corner feature extraction
  - In the first step a binary image is processed using Sobel filter with kernels given by Eq. 1:

$$g_x = \begin{bmatrix} 1 & 0 & -1 \\ 2 & 0 & -2 \\ 1 & 0 & 1 \end{bmatrix} \quad g_y = \begin{bmatrix} 1 & 2 & 1 \\ 0 & 0 & 0 \\ -1 & -2 & -1 \end{bmatrix} \tag{1}$$

  - Two calculated gradient components representing horizontal and vertical edges are later decomposed into eight chain code directions, which gives as eight new images. In each one 25 measures are taken, which results in direction feature vector comprising 200 samples ($8 \times 25 = 200$).

2.2 *Classifier Construction*

In this section we will present two-stage hybrid classifier in greater details. This paper extends algorithms implemented in [4, 13] and here its performance is checked with a real-life problem of handwritten pattern recognition task. Input to the algorithm is a set of features extracted from segmented images. Our hybrid algorithm is built as a two-stage classifier. First, a rough sets algorithm is used and if it fails then fuzzy logic classifier is applied. We have divided our implementation into two steps. In the first one, an optimization procedure is applied using simulated annealing algorithm. Its purpose is to extract valuable features for pattern recognition. In the second one, a proper classification is performed using simulated annealing solutions for granulation step in case of rough sets algorithm and the best positions of membership functions for fuzzy logic classifier. Below, algorithm construction is described in greater details:

1. **Stage 1—Rough sets algorithm implementation**. This algorithm is constructed according to the following steps:
   a. In the first step, we look for the optimal cuts of each attribute in rough sets algorithm with the help of the simulated annealing algorithm. This step

is crucial, because granulation strongly affects the classification accuracy. At the beginning, an initial solution is generated and it is a vector containing number which defines the number of intervals (potential cuts) for each attribute. The length of this vector is the same as the number of attributes characterizing single segmented image. To enable feature extraction forgetting factor named as *don't use* is introduced. It tells that a given attribute is totally excluded from the formulation of IF-THEN rules. For simplicity, let us consider an object x which is described by three features. In this case, one potential solution could be |*don't use*|4|*don't use*. It means that the first and third attributes are excluded and the second attribute is granulated with four intervals. The fitness function for assessing current solution is given by (2)

$$F = w_1 \cdot NR^2 - (1 - w_1) \cdot NNR + NR \cdot \left(\frac{1}{NOA}\right) \quad (2)$$

where w_1 is a selected weight, NR is the number of correctly classified handwritten patterns using partition from encoded individual during validation phase, NNR is the number of misclassified objects and NOA represents the number of available attributes after removing noisy features. An individual with the maximum fitness indicator is used as the final solution.

b. Next step is based on granulation preprocessing for those attributes found by optimization procedure in step 1. After the granulation procedure, value of each pattern's attribute is discretized and represented by the appropriate number of interval in which this attribute is included. Let us denote the lth attribute and its p_lth value or interval ($p_l = 1, \ldots, K_l$) by $v^l_{p_l}$.
c. In this step, using available training data we generate the set *For(C)* of all decision formulas according to the granulation from the previous stage:

$$\text{IF } (x^1 = v^1_{p_1}) \text{ AND } \ldots \text{ AND } (x^d = v^d_{p_d}) \text{ THEN } \Psi(S, x) = j \quad (3)$$

During the learning procedure, each rule is given the strength factor which is the fraction of correctly classified patterns over all objects which activated this rule.
d. In the next step for the set of formulas *For(C)*, for every $j = 1, \ldots, m$ we calculate C-lower approximation $C_*(P_j)$, C-upper approximation $C^*(P_j)$ and boundary region $CN_B(P_j)$.
e. In the last step, we can classify an incoming handwritten pattern by looking for matching rules in the set *For(C)*. At this step, three possible actions can be undertaken. If there is only one matching rule, then we classify this pattern to the class which is indicated by its decision attribute j, because for sure such a rule belongs to the lower approximation of all rules indicating j. If there is more than one matching rule in the set *For(C)*, it means that the recognized pattern should be classified by the rules from the boundary region and in this case the final decision is determined by index of boundary region for which the strength of corresponding rules is maximal. When none rule is activated

or there are two or more classes with the same strength factor, then pattern is rejected and will be classified in the next stage by fuzzy logic classifier.

2. **Stage 2—Fuzzy logic implementation**. In the fuzzy logic algorithm construction the key point lies in defining correct membership functions set for each attribute. In the literature one can find many examples of fuzzy logic algorithm construction, but here the simulated annealing algorithm is applied to generate proper membership functions in terms of their shapes and location in the feature space. In case of rough sets, simulated annealing was used to find the best granulation step for each feature, while here an optimization algorithm is used to find a proper membership function's shape and its location for each attribute. Below, the steps for constructing fuzzy logic algorithm are presented:

    a. In the first step, each attribute is divided into the same number of intervals and for each interval new membership function is assigned from the set of L using typical triangular functions. For each membership function from L, a unique linguistic variable is attached plus one additional variable *don't use* to indicate that the given attribute is not used in rule generation. Of course, at this stage we have too many membership functions per attribute and construction of IF-THEN rules would be a tedious and very complex task. It is natural that some functions can be totally removed while shapes of the others can be adjusted.
    b. In the next step to produce fuzzy rules, simulated annealing is started. This time single solution (vector) stores information about set of rules which are in the form represented by Eq. (4):

$$R_q : IF(x^1 = A^1_{q_1}) \text{AND } (x^2 = A^2_{q_2}) \dots \text{ AND } (x^d = A^d_{q_d}) \text{THEN class} = \text{j} \tag{4}$$

    It was decided that solution contains 20 fuzzy rules $R_q, q \in (1, \dots, 20)$ and their classification performance is treated as a fitness function F from Eq. (2). At this point we accomplish two goals. In the first one, we significantly decrease the number of IF-THEN rules and additionally, some features are fully removed from classification procedure. When new solution is obtained through simulated annealing procedure, in the first place each rule from encoded vector must be assigned with a proper class label and its corresponding strength. This procedure is as follows:

    - In the consequent step, we assign class label to a single rule and compute its strength. For each training pattern x_p let us calculate the compatibility grade of a single rule connected with its antecedent part $A_r = (A_{r1}, A_{r2}, \dots, A_{rd})$ using the product operator of each membership function $\mu_{A_{ri}}$ determined for A_{ri}:

$$\mu_{A_r}(x_p) = \mu_{A_r1}(x_p) \cdot \mu_{A_r2}(x_p) \cdot \dots \cdot \mu_{A_rd}(x_p) \tag{5}$$

      If we know how to calculate the compatibility grade of each training pattern, then we can determine C_r and CF_r for each rule. The fuzzy probability $P(class\ j|A_r)$ of class j, $j = (1, \ldots, m)$ indicating how pattern x can be associated with class j is shown below [7]:

$$Pr(class\ j|A_r) = \frac{\sum\limits_{x_p \in class\ j} \mu_{A_r}(x_p)}{\sum\limits_{p=1}^{m} \mu_{A_r}(x_p)} \tag{6}$$

For the rth rule R_r the label of class is assigned according to the winning rule, which means that the label with maximal probability is chosen:

$$R_r : C_r = max_{j \in \{1,\ldots,m\}} \{Pr(class\ j|A_r)\} \tag{7}$$

In the learning phase it can happen that rule R_r can be activated by patterns coming from different classes. To ensure the proper classification, each rule has a strength factor which tells how precisely rule R_r predicts the consequent class j.

$$R_r : CF_r = Pr(class\ j|A_r) - \sum_{j=1, j \neq C_r}^{m} Pr(class\ j|A_r) \tag{8}$$

If CF_r in (8) is negative, then rule R_r is denoted as *dummy* and is not taken for further reasoning; otherwise, it is used in defuzzification process to determine the final class label. Let us assume that $N_{rule} = 20$ fuzzy rules are generated with indicators C_r, CF_r determined by (7) and (8).

- Then in the last step the process of classification can be done as follows:

$$\Psi(S, x_p) = C_q \leftarrow max_{j \in \{1,\ldots,m\}} \{\mu_{A_q}(x_p) \cdot CF_r\} \tag{9}$$

The label of the class for an unknown pattern is determined by a winner rule R_w that has the maximum compatibility grade and the rule strength CF_r. If multiple fuzzy rules have the same maximum product μ_{A_r} but different consequent classes then the classification is rejected. The same action is taken if no fuzzy rule is compatible with the incoming pattern x_p.

3 Experiment Results

3.1 Input Database

To test the classifier performance Handprinted Forms and Characters Database was used. It contains the full page binary images of 3699 handwriting sample forms (HSFs) and 814,255 segmented handprinted digits and alphabetic characters from

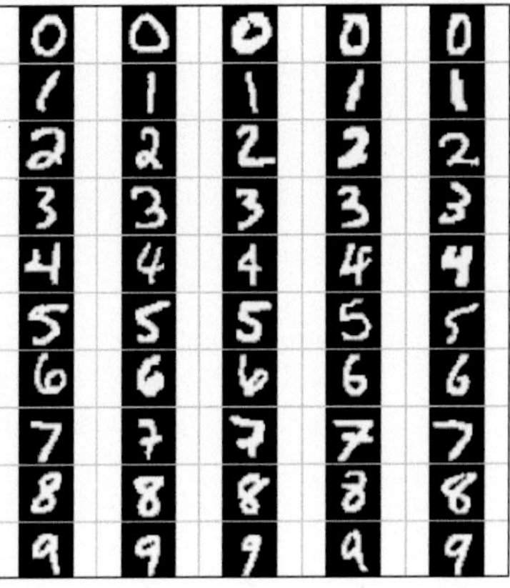

Fig. 1 Segmented characters from NIST database

those forms. Each segmented character occupies a 128 × 128 pixel raster image labelled by one of the 62 classes corresponding to "0–9", "A–Z" and "a–z". The classes of each segmentation have been manually checked such that the residual character classification error is about 0.1 %. Examples of segmented images for different writers are presented in Fig. 1.

3.2 Investigations

For the comparison purposes three different classifiers were trained and tested on the same datasets: 3-KNN (weighted distance measure), neural network (2-layer NN, 300 hidden units, mean square error) and SVM classifier (rbf kernel). Notation used in results table is as follows: AP—classifier error rate in percentage for testing set using pixel intensity ratio as a feature extraction method, BP—classifier error rate in percentage for testing set using cross corner feature extraction algorithm, S—standard deviation of algorithm accuracy in percentage, NA—averaged number of attributes used in the classification procedure. Due to the random nature of simulated annealing procedure, classification with our hybrid algorithm was repeated 50 times and the final results are averaged. To obtain training and testing sets whole database was divided using tenfold cross-validation method. Final classifier performances for two methods of feature selection (described in the Sect. 2.1) are presented in Table 1. Settings, for our hybrid classifier, were as follows:

Table 1 Classification performance for whole NIST handwritten database for four classifiers

Classifier	AP	S	NA	BP	S	NA
SVM	5.58	1.32	25/25	3.39	1.24	200/200
3-KNN	8.92	1.04	25/25	6.69	0.98	200/200
NN	5.99	1.34	25/25	3.52	1.30	200/200
Hybrid	5.52	2.03	6/25	3.38	2.01	59/200

- Simulated annealing parameters:
  - Boltzmann condition for accepting solutions,
  - Initial temperature: $T = 1000.0$,
  - Cooling rate: $C_r = 0.999$,
  - Absolute temperature: $T_t = 0.0001$,
  - Number of simulated annealing iterations: $I_{sa} = 100{,}000$.
- Fitness function weight $w_1 = 0.78$,
- Maximum number of membership functions for single attribute in fuzzy logic $f_{fl} = 10$,
- Maximum number of intervals for single attribute in rough sets $f_{rs} = 8$.

After analyzing results in Table 1, it can be seen that our hybrid approach obtains satisfactory results compared to other classifiers. In general, for the first segmentation method we obtained worse classification performance. Simple pixel intensity ratio is prone to noise and the number of intervals in which image is divided also affects the final results. For English alphabet better approach is to use diagonal methods for feature extraction. The second experiment shows that our cross corner feature extraction algorithm improves classification accuracy.

4 Conclusions

This paper presents the hybrid approach for handwritten pattern recognition using fuzzy logic and rough sets algorithms. Due to high dimensionality of the feature sets describing segmented image we have proposed simulated annealing as an optimization procedure. It allows to reduce the number of features and additionally gives information about rough sets granulation and fuzzy logic membership functions shape. Additionally, simpler and easy to read IF-THEN rules can be created. Another important issue from this paper relates to feature extraction of binary images. Two methods were compared: the first with simple pixel intensity ratio, and the second one which uses edge detection with diagonal chain coding. It was shown that more sophisticated method increases classification accuracy even if the feature space dimension is quite high. Our hybrid classifier accuracy gives satisfactory results and can compete with other algorithms. In the future we would like to implement rejection method to initially divide patterns into groups. We have noticed that in some cases an initial threshold rejection would speed up classification and give better results. This reject option can be based on the posterior density estimation for each group, for example, two groups: letters and digits.

References

1. Cheng-Lin, L., Kazuki, N., Hiroshi, S., Hiromichi, F.: Handwritten digit recognition: investigation of normalization and feature extraction techniques. Pattern Recognit. **37**, 265–279 (2004)
2. Hossain, M., Amim, M., Hong, Y.: Rapid feature extraction for bangla handwritten digit recognition. In: International Conference on Machine Learning and Cybernetics (ICMLC), pp. 1832–1837 (2011)
3. Impedovo, S., Pirlo, G., Mangini, F.: Handwritten digit recognition by multi-objective optimization of zoning methods. In: 2012 International Conference on Frontiers in Handwriting Recognition (ICFHR), pp. 675–679 (2012)
4. Majak, M.: Universal segmentation framework for medical imaging using rough sets theory and fuzzy logic clustering. Information Technologies in Biomedicine 3 (2014)
5. Majak, M., Zolnierek, A.: Rough sets approach to the problems of classification. In: Proceedings of International Conference MOSIS X, pp. 109–114 (2010)
6. Nabiha, A., Nadir, F.: New dynamic ensemble of classifiers selection approach based on confusion matrix for arabic handwritten recognition. In: International Conference on Multimedia Computing and Systems (ICMCS), pp. 308–313 (2012)
7. Qinghua, H., Zongxia, X.: Hybrid attribute reduction based on a novel fuzzy-rough model and information granulation. Pattern Recognit. **40**, 3509–3521 (2007)
8. Singh, P., Verma, A.: An experimental evaluation of feature selection based classifier ensemble for handwritten numeral recognition. In: International Conference on Electronics and Communication Systems (ICECS), pp. 1–8 (2014)
9. Wang, J., Fang-Chen, C.: An accelerometer-based digital pen with a trajectory recognition algorithm for handwritten digit and gesture recognition. In: IEEE Transactions on Industrial Electronics, pp. 2998–3007 (2012)
10. Wang, Q., Yang, A., Dai, W.: An improved feature extraction method for individual offline handwritten digit recognition. In: 8-th World Congress on Intelligent Control and Automation (WCICA), pp. 6327–6330 (2010)
11. Yuan, H., Wang, P.: Handwritten digits recognition using multiple instance learning. In: IEEE International Conference on Granular Computing (GrC), pp. 408–411 (2013)
12. Zolnierek, A., Majak, M.: Rough sets approach to the classification task with modification of decision rules. In: Proceedings of the 11th WSEAS International Conference on Systems Theory and Scientific Computation, pp. 53–56 (2011)
13. Zolnierek, A., Majak, M.: Hybrid approach using rough sets and fuzzy logic to pattern recognition task. Lecture Notes in Computer Science 8073 (2013)

Environmental Sounds Recognition Based on Image Processing Methods

Tomasz Maka and Paweł Forczmański

Abstract The article presents an approach to environmental sound recognition that uses selected methods from the field of digital image processing and recognition. The proposed technique adopts the assumption that an audio signal can be converted into a visual representation, and processed further, as an image. At the first stage the audio data are converted into rectangular matrices called feature maps. Then a two-step approach is applied: the construction of a representative database of reference samples and the identification of test samples. The process of building the database employs two-dimensional linear discriminant analysis. Then the recognition operation is carried out in a reduced feature space that has been obtained by two-dimensional Karhunen–Loeve projection. At the classification stage, a minimum distance classifier is applied to different features. As it is shown, the results are very encouraging and can be a base for many practical audio applications.

Keywords Audio classification · Image recognition · Image processing · Audio features · Linear discriminant analysis · Distance metrics

1 Introduction

The effectiveness of speech-based services is dependent on the properties of acoustic environment. Such environment may contain many static or moving sound sources. These sources produce quite complex audio structure which deteriorates the quality of input signal and has direct influence on the signal processing path. Therefore, it is important to improve the robustness of such services in the presence of environmental noise. There are many approaches to determine the acoustic scene and its

T. Maka (✉) · P. Forczmański
Faculty of Computer Science and Information Technology, West Pomeranian University of Technology, Szczecin, Żołnierska Str. 52, 71–210 Szczecin, Poland
e-mail: tmaka@wi.zut.edu.pl

P. Forczmański
e-mail: pforczmanski@wi.zut.edu.pl

R. Burduk et al. (eds.), *Proceedings of the 9th International Conference on Computer Recognition Systems CORES 2015*, Advances in Intelligent Systems and Computing 403, DOI 10.1007/978-3-319-26227-7_68

components [3, 11, 14, 19]. The environmental sound recognition (ESR) systems require a robust feature extraction stage in order to identify the background noise and the occurrences of audio events [3, 19]. Typical feature sets include features calculated in various domains [11]. However, in some cases the feature set contains properties of audio signal calculated at different timescales. The characteristics of the selected features should capture time–frequency structure of the signal. At the parameterization stage, the features are calculated in a frame-based manner in the most cases. The size of frame and time spans used in the analysis process of source signal influence on the final representation of the acoustic scene [1]. Most of feature extraction techniques use physical characteristics of sound, thus low-level feature vectors consisting of a set of coefficients calculated in the time domain, frequency, or cepstrum. Their recognition and classification are based primarily on a one-dimensional approach to data. However, there exist approaches where two-dimensional representation of audio data is employed in the speech and audio signals' analysis process [5, 15, 17, 20]. In work [13] a method based on statistical analysis of acoustic images generated from time–frequency distributions was presented. The authors used audio images based on spectrogram and the Choi–Williams energy distribution to classify acoustic sport events and different types of gunshots. An algorithm for audio classification which uses sound spectrograms as texture images was presented in [20]. It was reported that proposed method based on matching time–frequency blocks in the spectrogram achieves good performance in musical instrument classification task. The spectrogram image feature (SIF) was proposed in [5]. The approach uses normalized spectrogram, whose dynamic range is quantized into different regions and then mapped to a form of image. The distribution statistics of partitioned image was used as the feature. Since the noise influences only the limited part of image dynamic range, the approach yields to robust classification in mismatched conditions. This article focuses on a two-dimensional approach to audio signal presented in the form of a matrix. So it is natural to use algorithms from the field of digital image processing and recognition. The proposed approach is based on the observation that the visual representation of audio signal carries much more information that in case of standard and limited vector representation [6]. In contrast to the most of existing methods, it uses the feature maps instead of the spectrogram as the form of time–frequency representation, which is assumed to capture more sound characteristics in terms of classification.

2 Audio Data Representation

The feature space dedicated to audio classification tasks should include the time–frequency structure of the signal. Therefore, in typical audio and speech analysis tasks, many features calculated in several domains are employed. In our approach we have decided to use feature space created from a set of feature vectors arranged into a feature map. The whole process of feature map construction is shown in

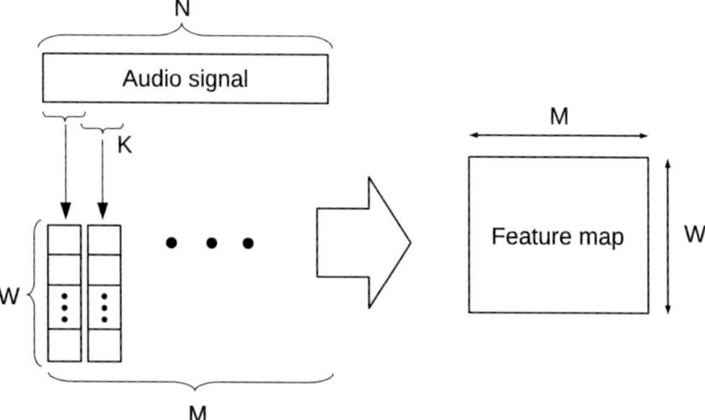

Fig. 1 Feature map calculation scheme

Fig. 1. The source signal of length N is split into a set of M overlapping frames (the overlapping between consecutive frames is equal to half of the frame size). For each frame, a feature vector of length W is calculated and a feature map is formed. The number of columns of feature map is equal to the number of frames of single feature contour and is determined by the following equation:

$$M = \left\lfloor \frac{F_s \cdot (N - K)}{\lfloor F_s \cdot R \cdot K \rfloor} \right\rfloor + 1, \tag{1}$$

where N is the total size of the signal [s], K denotes frame size [s], R is the frame overlapping $R \in (0, 1)$, and F_s is the sampling rate [Hz]. At the audio parametrization stage we have employed feature maps constructed on the basis of the following feature sets:

- *Bark Frequency Filter Bank* (BFB):
  Feature vector consists 24 energies calculated at the outputs of filter bank. Each filter is defined in the Bark frequency scale. The scale from 1 to 24 Barks reflects 24 critical bands of hearing [18].
- *Mel-Frequency Cepstral Coefficients* (MFCC):
  The MFCC vector is a form of cepstrum representation. It is calculated by mapping the power spectrum onto Mel-frequency scale using a set of filters with critical band spacing. From each band a logarithm of energy is calculated and then DCT transform is executed to obtain final MFCC representation [4].
- *Linear Prediction Coefficients* (LPC):
  The coefficients are obtained in the process of predicting a signal sample based on n past samples (n is the order of LPC). The calculation is performed by solving the linear normal equations from the autocorrelation function of input signal [16].
- *Linear-Frequency Cepstral Coefficients* (LFCC):
  The calculation process is the same as in case of MFCC. Instead of Mel-frequency scale mapping, the filters are equally spaced with constant bandwidth.

- *Linear Prediction Cepstral Coefficients* (LPCC):
  The cepstral coefficients are calculated from LPC coefficients using the recursion procedure [16]. The representation describes the envelope of log magnitude spectrum.
- *Line Spectral Frequencies* (LSF):
  The LSF is another feature derived from LPC representation. It has lower sensitivity to quantization process and better stability in comparison to LPC coefficients [16].

The size of feature vector extracted for each audio frame was constant and equals to $W = 24$. An illustration of feature maps calculated for example audio clip is shown in Fig. 2.

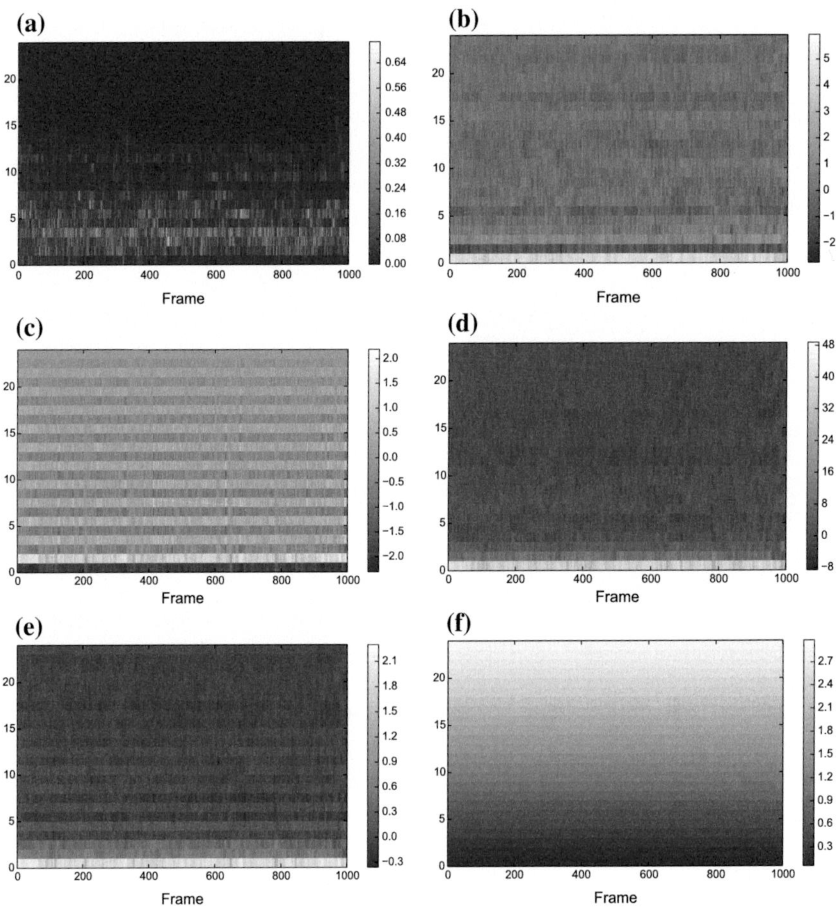

Fig. 2 Feature maps calculated using the following features, for example, audio clip ($W = 24$): BFB (**a**), MFCC (**b**), LPC (**c**), LFCC (**d**), LPCC (**e**), LSF (**f**)

3 Algorithm Description

An image of feature map captures many important low-level characteristics of audio signal. However, in order to be useful for in-depth analysis, it should be represented using rather large matrix. Processing of such large matrices, in terms of classification or recognition, causes several problems, namely large processing power overhead and storage requirements. Hence, it is necessary to reduce their size. Algorithm for the classification process of audio samples is shown in Fig. 3. It consists of three main components: data preparation (pre-processing), reduction of dimensionality (projection to the subspace features), and classification. At the stage of developing a database containing reduced form of image patterns (feature maps) we use a method employed, among others, in recognizing facial images, namely, two-dimensional linear discriminant analysis (2DLDA) [10], while at the recognition stage of test samples—two-dimensional Karhunen–Loeve transform (2DKLT) [10]. The analysis of the research showed that linear discriminant analysis may be successfully applied in dimensionality reduction stage. Since one-dimensional variant of LDA requires much more space to store covariance matrices and also does not cope with small-sample-size problem, we propose to use a 2DLDA. It was successfully applied in our previous works, especially for face, stamps, and textures recognition [7, 10, 12]. At the stage of classification, we propose to use a set of simple distance-based classifiers (k-Nearest neighbor with different distance metrics). According to the assumed methodology, all of the samples from the learning part of the database are subject to analysis (2D analysis, 2DLDA), which leads to forming the transformation matrices. They are later used during the construction phase of the reference database to project input images (2D projection, 2DKLT). Feature maps calculated for audio samples are grouped according to G classes corresponding to the labeled content. It is assumed that each input feature map is represented in shades of gray as a matrix of dimensions $W \times M$ elements, where g is the number of class ($l = 1, \ldots, L$). In the first step an average matrix of all the matrices X is calculated:

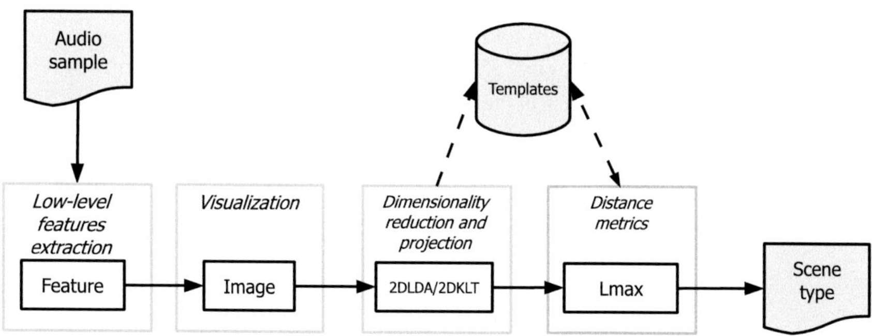

Fig. 3 General scheme of audio sample processing and classification

$$\bar{X}_{W \times M} = \frac{1}{GL} \sum_{g=1}^{G} \sum_{l=1}^{L} X_{W \times M}^{(g,l)} \tag{2}$$

and a mean matrix in each class $g = 1, \ldots, G$:

$$\bar{X}_{W \times M}^{(g)} = \frac{1}{L} \sum_{l=1}^{L} X_{W \times M}^{(g,l)}. \tag{3}$$

Due to the large size of input images (in experimental studies—even 24×1001 pixels) it is not possible to apply directly the one-dimensional LDA method, therefore, it was decided to use a two-dimensional variant—2DLDA. In this method, two covariance matrices are calculated, each one for row and column representation of feature map [10]. The covariance matrices are subject to the generalized eigenvalue problem, which creates matrices Λ and V for both representations. In the next step, from the diagonals of $\Lambda^{(Row)}$ and $\Lambda^{(Col)}$ s and p maximal elements are selected, respectively, and their positions are recorded. From $\left(V^{(Row)}\right)^T$ s rows corresponding to selected elements are extracted, and from $V^{(Col)}$ p columns in the same way. Then, the two transformation matrices are constructed: $F^{(Row)}$ containing $s \times W$ elements and $F^{(Col)}$ having $M \times p$ elements. Projection of lth feature map from the gth class $X^{(g,l)}$ into a subspace is performed by matrix multiplication [10]:

$$Y_{s \times p}^{(g,l)} = F_{s \times W}^{(Row)} \left(X_{W \times M}^{(g,l)} - \bar{X}_{W \times M} \right) F_{M \times p}^{(Col)}. \tag{4}$$

For optimal degree of reduction it was assumed that the information contained in feature maps, arranged vertically (in columns of each image), is more important than horizontal information (stored in rows of image), so we used such relation that $s \leq p$. As it was mentioned, audio pattern classification is based on the operations in the reduced feature space. Reference database contains reduced visual representations of specific features calculated for all learning samples. We employed Chebyshev distance metrics at the stage of classification using k-nearest neighbor classifier (for $k = 1$) as it is simple to compute and gives acceptable accuracy. In comparison to more complex methods, employing SVMs [8, 9] its simple implementation and low dimensionality of feature space is an obvious advantage. The Chebyshev distance (L_{max} metric) between two feature matrices $Y^{(p)}$ and $Y^{(q)}$, with standard coordinates $y_i^{(p)}$ and $y_i^{(q)}$, respectively, is calculated as follows [2]:

$$D_{\text{Cheb}}(Y^{(p)}, Y^{(q)}) := \max_{i,j}(|y_{i,j}^{(p)} - y_{i,j}^{(q)}|). \tag{5}$$

4 Experimental Evaluation

The audio data used in the evaluation stage consists of $G = 6$ classes of the environmental sounds. Every audio clip is monaural, $N = 10\,\mathrm{s}$ long with $F_s = 22050\,\mathrm{Hz}$ sampling rate—total length of audio data is 42 min long. The brief characteristics of the classes are depicted in Table 1. The frame size equal to $K = 20\,\mathrm{ms}$ with the 50 % overlapping ($R = 0.5$) between adjacent frames was used in the feature map generation stage. In order to demonstrate the complexity of the problem, Fig. 2 shows the sample feature maps for each of the classes. It should be noted that their distinguishability in original attributes space is quite limited. Each class includes 42 audio clips where 30 items are used in the training phase and 12 items in testing phase. We performed a fivefold cross-validation to avoid overfitting and make the results objective. In order to evaluate the performance of the proposed algorithm of classification, we investigated six above-presented audio features represented in a form of feature maps. The reduction parameters at the stage of 2DLDA/2DKLT are $s = 5$, $p = 10$; hence, we classified objects using 50 features, which is lower than in case of other known methods. The averaged results are presented in Table 2. In the result, the best performance is obtained for LFCC features. In case of more classes, the similar performance is given by LFCC and LPCC. The lowest accuracy was obtained for LPC and LSF features. Exemplary confusion matrices for different numbers of classes and LFCC feature map (where $G = 6$ is the most complex case and $G = 3$ is the easiest one) are provided in Table 3. Besides typical number of actual versus predicted samples, the accuracy for each class is given. A closer look at these tables unveils that the third class (ambient sounds in a restaurant) is the one with the highest classification rate. This class is probably well clustered by LDA, while other classes have more outliers caused by the presence of audio events.

Table 1 Characteristics of audio classes in our database

Class	Description
c_1	Passing trucks and cars by on the road or highway. Variable traffic intensities—various vehicles at different speeds
c_2	Sounds of rainy days. Various intensities of the rain
c_3	Ambience sounds inside the restaurant or bar. Babble noises, sounds of cutlery, coffeemaker, etc.
c_4	Inside the shopping center. Sounds of chatting people, moving trolleys, and cash register checkouts. Sometimes distant music or commercials
c_5	The sounds of waves and wind at the beach. Human voices blended with the beach sounds from time to time
c_6	Ambience sounds in the factories. Various sound sources, different types of machinery, and many periodic sounds

Table 2 Classification accuracy for different feature maps and the number of classes, where for $G = 3$ classes $c_1 - c_3$ are used, for $G = 4$, $c_1 - c_4$; for $G = 5$, $c_1 - c_5$ and for $G = 6$, $c_1 - c_6$, respectively

G	Feature type					
	BFB	MFCC	LPC	LFCC	LPCC	LSF
3	0.70	0.77	0.62	0.88	0.83	0.65
4	0.60	0.59	0.47	0.75	0.70	0.55
5	0.50	0.51	0.37	0.64	0.63	0.44
6	0.41	0.44	0.36	0.53	0.54	0.42

Table 3 Confusion matrices for $G = 6$ (a), $G = 5$ (b), $G = 4$ (c) and $G = 3$ (d)

(a)

Actual	Predicted						Accuracy
	c_1	c_2	c_3	c_4	c_5	c_6	
c_1	7	3	0	1	0	1	0.58
c_2	1	7	1	1	0	2	0.58
c_3	0	0	8	1	2	1	0.67
c_4	0	2	3	1	3	3	0.08
c_5	2	2	1	2	3	2	0.25
c_6	2	0	4	2	0	4	0.33

(b)

Actual	Predicted					Accuracy
	c_1	c_2	c_3	c_4	c_5	
c_1	8	3	0	1	0	0.67
c_2	1	8	1	1	1	0.67
c_3	0	0	9	1	2	0.75
c_4	0	0	3	1	8	0.08
c_5	1	2	1	2	6	0.5

(c)

Actual	Predicted				Accuracy
	c_1	c_2	c_3	c_4	
c_1	8	3	0	1	0.67
c_2	1	8	1	2	0.67
c_3	0	0	11	1	0.92
c_4	0	0	6	6	0.5

(d)

Actual	Predicted			Accuracy
	c_1	c_2	c_3	
c_1	8	3	1	0.67
c_2	1	9	2	0.75
c_3	0	0	12	1.0

5 Summary

The approach presented in the article shows that using selected methods from the field of digital image processing and recognition to the ESR problem may lead to promising results. In proposed algorithm a visual representation of audio data called feature map is built using six typical low-level audio features. At the next stage the feature map is projected to the reduced feature space using 2DLDA. Classification in resulting subspace is done using 1NN method. The conducted experiments on a benchmark database show the good efficiency of developed method in many cases. The main drawback of the proposed technique is quite low classification accuracy in case of occurrence of many single audio events in the audio data. Presented approach can be applied to various practical problems like acoustic surveillance, background noise suppression, and context awareness. In future work, we plan to apply more feature maps using various properties of human hearing system. Also, a dedicated audio event detection and grouping module is planned.

References

1. Abe, M., Matsumoto, J., Nishiguchi, M.: Content-based classification of audio signals using source and structure modelling. In: Proceedings of the IEEE Pacific Conference on Multimedia, pp. 280–283 (2000)
2. Cantrell, C.D.: Modern Mathematical Methods for Physicists and Engineers. Cambridge University Press, Cambridge (2000)
3. Clavel, C., Ehrette, T., Richard, G.: Events detection for an audio-based surveillance system. IEEE Int. Conf. Multimed. Expo, ICME **2005**, 1306–1309 (2005)
4. Davis, S., Mermelstein, P.: Comparison of parametric representation for monosyllabic word recognition in continuously spoken sentences. IEEE Trans. ASSP **28**(4), 357–366 (1980)
5. Dennis, J., Tran, H.D., Li, H.L.: Spectrogram image feature for sound event classification in mismatched conditions. IEEE Signal Process. Lett. **18**(2), 130–133 (2011)
6. Forczmański, P.: Evaluation of singer's voice quality by means of visual pattern recognition. J. Voice. doi:10.1016/j.jvoice.2015.03.001 (2015, in press)
7. Forczmański, P., Frejlichowski, D.: Classification of elementary stamp shapes by means of reduced point distance histogram representation. Mach. Learn. Data Min. Pattern Recognit., LNCS **7376**, 603–616 (2012)
8. Geiger, J.T., Schuller, B., Rigoll, G.: Large-scale audio feature extraction and SVM for acoustic scene classification. In: IEEE Workshop on Applications of Signal Processing to Audio and Acoustics (WASPAA), pp. 1–4 (2013)
9. Jiang, H., Bai, J., Zhang, S., Xu, B.: SVM-based audio scene classification, natural language processing and knowledge engineering. In: Proceedings of 2005 IEEE International Conference on IEEE NLP-KE'05, pp. 131–136 (2005)
10. Kukharev, G., Forczmański, P.: Face recognition by means of two-dimensional direct linear discriminant analysis. In: Proceedings of the 8th International Conference PRIP 2005 Pattern Recognition and Information Processing. Republic of Belarus, Minsk, pp. 280–283 (2005)
11. Maka, T.: Environmental background sounds classification based on properties of feature contours. In: 26th International Conference on Industrial, Engineering and Other Applications of Applied Intelligent Systems, IEA/AIE, Amsterdam, LNCS, vol. 7906, pp. 602–609 (2013)
12. Okarma, K., Forczmański, P.: 2DLDA-based texture recognition in the aspect of objective image quality assessment. Ann. Univ. Mariae Curie-Sklodowska. Sectio AI Informatica **8**(1), 99–110 (2008)
13. Paraskevas, I., Chilton, E.: Audio classification using acoustic images for retrieval from multimedia databases. In: 4th EURASIP Conference on Video/Image Processing and Multimedia Communications. IEEE, vol. 1, pp. 187–192 (2003)
14. Paraskevas, I., Potirakis, S.M., Rangoussi, M.: Natural soundscapes and identification of environmental sounds: a pattern recognition approach. In: 16th International Conference on Digital Signal Processing, pp. 5–7, 1–6 July 2009
15. Pinkowski, B.: Principal component analysis of speech spectrogram images. Pattern Recognit. **30**(5), 777–787 (1997)
16. Rabiner, L., Schafer, W.: Theory and Applications of Digital Speech Processing. Prentice-Hall, Englewood Cliffs (2010)
17. Rafii, Z., Coover, B., Han, J.: An audio fingerprinting system for live version identification using image processing techniques. In: IEEE International Conference on Acoustic, Speech and Signal Processing (ICASSP), pp. 644–648 (2014)

18. Smith III, J.O.: Spectral Audio Processing. W3K Publishing, Stanford (2011)
19. Wichern, G., Xue, J., Thornburg, H., Mechtley, B., Spanias, A.: Segmentation, indexing, and retrieval for environmental and natural sounds. IEEE Trans. Audio Speech Lang. Process. **18**(3), 688–707 (2010)
20. Yu, G., Slotine, J.: Audio classification from time-frequency texture. In: IEEE International Conference on Acoustics, Speech and Signal Processing, ICASSP. Taipei, Taiwan, pp. 1677–1680 (2009)

Investigating Combinations of Visual Audio Features and Distance Metrics in the Problem of Audio Classification

Paweł Forczmański and Tomasz Maka

Abstract The article addresses a problem of audio signal classification employing image processing and recognition methods. In such an approach, vectorized audio signal features are converted into a matrix representation (feature map), and then processed, as a regular image. In the paper, we present a process of creating a low-dimensional feature space by means of two-dimensional Linear Discriminant Analysis and projecting input feature maps into this subspace using two-dimensional Karhunen–Loeve Transform. The classification is performed in the reduced feature space by means of voting on selected distance metrics applied for various features. The experiments were aimed at finding an optimal (in terms of classification accuracy) combination of six feature types and five distance metrics.The found combination makes it possible to perform audio classification with high accuracy, yet the dimensionality of resulting feature space is significantly lower than input data.

Keywords Audio classification · Feature extraction · Image recognition · 2DLDA · 2DKLT · Distance metrics

1 Introduction

Analysis of audio signals is one of the most interesting and challenging tasks of multimedia systems. While there are many issues that are fairly good solved, e.g., the identification of persons on the basis of the registered voice and speech recognition in terms of content; there are also many problems that are still left for analysis. This includes the automatic classification of audio scene in terms of recognizing background sounds in certain environments, e.g., restaurant, factory, street, etc. The

P. Forczmański (✉) · T. Maka
Faculty of Computer Science and Information Technology, West Pomeranian University of Technology, Szczecin, Żołnierska Str. 52, 71–210 Szczecin, Poland
e-mail: pforczmanski@wi.zut.edu.pl

T. Maka
e-mail: tmaka@wi.zut.edu.pl

R. Burduk et al. (eds.), *Proceedings of the 9th International Conference on Computer Recognition Systems CORES 2015*, Advances in Intelligent Systems and Computing 403, DOI 10.1007/978-3-319-26227-7_69

sensitivity, which characterizes human sense of hearing is already achievable for machines—each of the physical quantities characterizing audio signal can now be specified much more precisely using computerized analyzers than using human sense of hearing. On the other hand, human beings are still able to use the acquired information from the audio signal in a more effective manner. It should be noted also that background sounds are extremely complex, when it comes to the formal description. The automation of audio scene classification process can have multiple purposes, e.g., improving speech-based services that depend on properties of environment, content-based audio retrieval in large multimedia databases, intelligent surveillance, etc. The audio signal can be described by a number of parameters [19, 21]. So, it seems that there is no need to seek for new methods of representation of audio signal, but we should focus on the selection and use of existing ones. As it was shown in [16, 21], most of the developed methods use physical characteristics of sound in a form of low-level feature vectors. The classification of such data is based mainly on a one-dimensional approach [1, 15]. In contrast, developed algorithm adopts a two-dimensional approach to audio signal and is based on the observation that the visual representation of audio signal may carry more useful information that in case of one-dimensional vector representation [5]. In such case, certain methods aimed at image processing may obtain higher accuracy of classification in comparison to established methods [18]. Recently, many applications of Support Vector Machines to audio scene classification have been proposed (e.g., [10, 11]). The results are promising, however, such methods are complex in terms of training. Therefore, this work is devoted to the task of evaluation of the possibility of application of pattern recognition methods in the context of automated classification of audio scene. The main focus is put on finding a combination of different audio features and adequate, simple distance metrics (used as classifiers), that gives the highest possible classification accuracy.

2 Feature Maps

In typical audio analysis task, various features calculated in several domains can be employed. The feature space should be able to capture time-frequency structure of the signal. Therefore, in our approach we use feature space created from a set of feature vectors arranged into a feature map. In such an approach, the source audio signal of length N is decomposed into a set of M overlapping frames (the overlap between consecutive frames is equal to half of the frame length). For each frame a feature vector of length W is calculated and added as a new column to the feature matrix (map). The number of columns in the feature map is equal to $M = \lfloor (F_s \cdot (N - K)) / \lfloor F_s \cdot 0.5 \cdot K \rfloor \rfloor + 1$, where: N is the total length of input signal [s], K is a frame size [s] and F_s the sampling rate [Hz]. In our investigations, the size of feature vector extracted for each audio frame was constant and equal to $W = 24$. Resulting feature maps calculated for exemplary audio clip are shown in Fig. 1. In order to capture various characteristics of audio we have employed feature

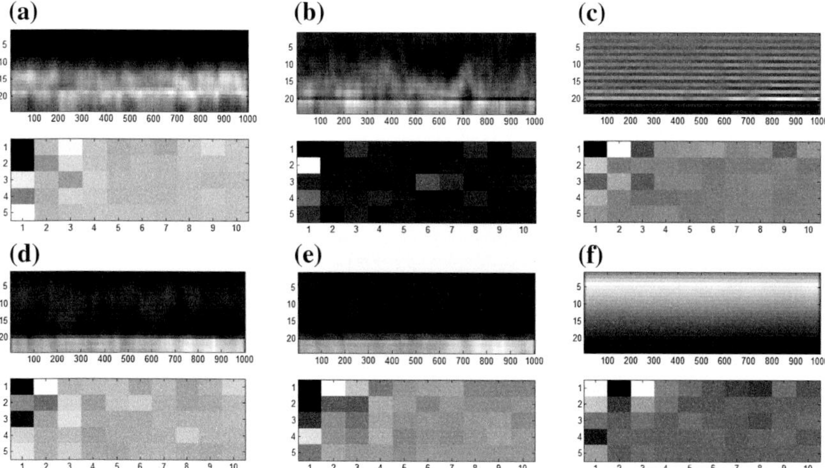

Fig. 1 Feature maps of exemplary audio signal for the following features ($W = 24$) and their 2DLDA-based representation ($s = 5, p = 10$): BFB (**a**), MFCC (**b**), LPC (**c**), LFCC (**d**), LPCC (**e**), LSF (**f**)

maps constructed on the basis of state-of-the-art feature sets, namely *Bark-Frequency Filter Bank* (BFB) [22], *Mel-Frequency Cepstral Coefficients* (MFCC) [3], *Linear Prediction Coefficients* (LPC) [19], *Linear-Frequency Cepstral Coefficients* (LFCC) [19], *Linear Prediction Cepstral Coefficients* (LPCC) [19], *Line Spectral Frequencies* (LSF) [12].

3 Algorithm Description

Created feature map is intended to preserve important low-level characteristics of audio signal. Its informative potential depends on the resolution of the image. Unfortunately, processing large image matrices requires large time and memory overhead and is not desirable. Therefore, in many similar tasks, certain dimensionality reduction is applied. Although, simple spatial resampling could be useful, it would lead to the elimination important details. Hence, we apply a more sophisticated approach, namely dimensionality reduction preceded by a data analysis stage. Our algorithm includes three components: data preparation (feature map creation), reduction of dimensionality (projection to the subspace), and classification (see Fig. 2). At the offline stage of building a reference database containing reduced forms of feature maps, we use a method employed, among others, in recognizing facial images, stamps and textures [6, 13, 17], namely Two-dimensional Linear Discriminant Analysis (2DLDA) [13], while at the recognition stage of test samples—Two-dimensional Karhunen–Loeve Transform (2DKLT) [7, 8, 13]. In order to obtain such reduc-

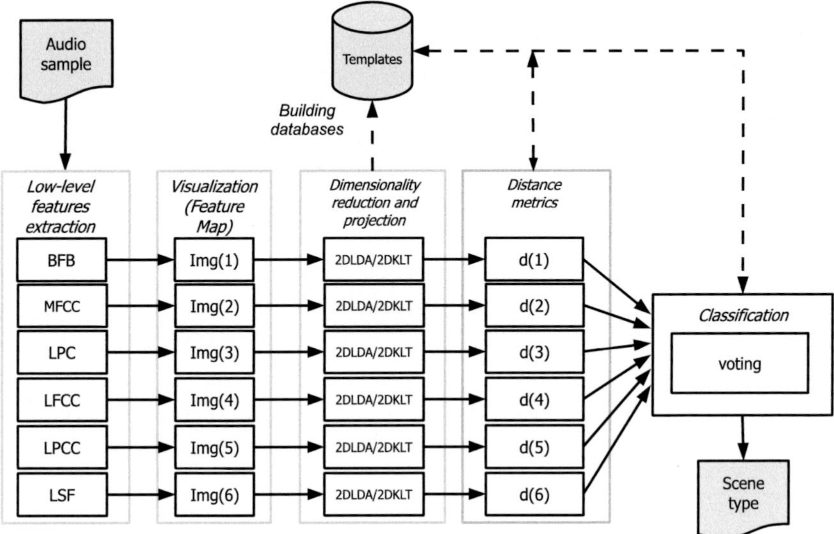

Fig. 2 Scheme of audio sample processing and classification

tion effect, together with clustering improvement, we propose to use a dimensionality reduction stage. The analysis of the research showed that Linear Discriminant Analysis may be successfully applied in this case. Since one-dimensional variant of LDA requires much more space to store covariance matrices and also does not cope with small-sample-size problem, we propose to use a two-dimensional Linear Discriminant Analysis. At the stage of classification, we use a combination of simple distance-based classifiers (1-Nearest Neighbor with different distance metrics).

3.1 Dimensionality Reduction

The aim of dimensionality reduction is to limit the volume of data describing input samples by selecting dimensions that represent them with a acceptably low error. It is typical for many methods involving subspace projection that all samples from the learning part of the database are subject to analysis (2D analysis), which leads to forming the transformation matrices. They are later used to project input images (2D projection) into a subspace forming a reference database. Feature maps calculated for audio samples are grouped according to G classes corresponding to the content. In our approach, each input feature map $X^{(g,l)}$ is represented as a grayscale image represented using matrix of dimensions $W \times M$ elements, where g is the class number and l is a respective feature map in each class ($l = 1, \ldots, L$). This approach is repeated for each feature type, which number, in our case is equal to six. In the first step, for each feature type, an average matrix $\bar{X}_{W \times M}$ of all the matrices X is calculated and a mean matrix in each class $g = 1, \ldots, G$: $\bar{X}^{(g)}_{W\times M}$. Due to the large size of input feature maps (in experimental studies—even 24 × 1001 pixels) it is not possible to apply directly the one-dimensional LDA method, therefore, it was decided to use a two-dimensional variant—2DLDA. In this method, the covariance matrices for within-class scatter (Sw) and between-class scatter (Sb) are calculated, each two for row and column representation of feature map, respectively, [13]: $Sw^{(Row)}_{W\times W}$ and $Sb^{(Row)}_{W\times W}$, $Sw^{(Col)}_{M\times M}$ and $Sb^{(Col)}_{M\times M}$. Then, we calculate the corresponding matrices H, determining the total distribution of the classes in the feature space: $H^{(Row)}_{W\times W} = \left(Sw^{(Row)}_{W\times W}\right)^{-1} Sb^{(Row)}_{W\times W}$ and $H^{(Col)}_{M\times M} = \left(Sw^{(Col)}_{M\times M}\right)^{-1} Sb^{(Col)}_{M\times M}$. They are used to maximize the so-called Fisher criterion, the aim of which is to increase the dispersion between-class scatter in relation to the intra-class [9]. This gives the improvement of clustering and significantly increases the effectiveness of later classification. For the matrices $H^{(Row)}$ and $H^{(Col)}$, we solve the task of searching for the eigenvalues $\{\Lambda^{(Row)}, \Lambda^{(Col)}\}$ and eigenvectors $\{V^{(Row)}, V^{(Col)}\}$. In the final step of analysis, from the diagonals of $\Lambda^{(Row)}$ and $\Lambda^{(Col)}$ s and p maximal elements are selected, respectively, and their positions are recorded. From $\left(V^{(Row)}\right)^T$ s rows corresponding to selected elements are extracted, and from $V^{(Col)}$ p columns in the same way. Then, the two transformation matrices are constructed: $F^{(Row)}$ containing $s \times W$ elements and $F^{(Col)}$ having $M \times p$ elements. Projection of lth feature map from the gth class $X^{(g,l)}$ into a subspace is performed by matrix multiplication [13] $Y^{(k,l)}_{s\times p} = F^{(Row)}_{s\times W}\left(X^{(g,l)}_{W\times M} - \bar{X}_{W\times M}\right) F^{(Col)}_{M\times p}$. In order to balance between reduction degree and classification performance, it was assumed that the information contained in feature maps, arranged vertically (in columns of each image) is more important than horizontal information (stored in rows of image), so we used such relation as $s \leq p$.

3.2 Distance Metrics

As it was mentioned, audio pattern classification is based on the operations in the reduced feature space. Reference database contains reduced feature maps calculated for all learning samples. The class assignment of a query feature map (its reduced form $Y^{(q)}$) is done using a distance-based Nearest-Neighbor Classifier on all reference objects $Y^{(r)}$. We applied five popular distances, described below. The Euclidean distance (L_2 norm) [4] is given as

$$d_{\text{Euclid}}(Y^{(q)}, Y^{(r)}) = \sqrt{\sum_{s,p}\left(y^{(q)}_{s,p} - y^{(r)}_{s,p}\right)^2}. \tag{1}$$

The Chebyshev distance (L_{max} norm) is defined as [2]:

$$d_{\text{Cheb}}(Y^{(q)}, Y^{(r)}) = \max_{s,p}\left(\left|y^{(q)}_{s,p} - y^{(r)}_{s,p}\right|\right). \tag{2}$$

The taxicab distance (cityblock/Manhattan metric or L_1 norm) is calculated as [20]:

$$d_{\text{City}}(Y^{(q)}, Y^{(r)}) = \sum_{s,p} \left| y_{s,p}^{(q)} - y_{s,p}^{(r)} \right| . \tag{3}$$

The Minkowski distance of order m (in our case $m = 5$) is defined as:

$$d_{\text{Mink}}(Y^{(q)}, Y^{(r)}) = \left(\sum_{s,p} \left| y_{s,p}^{(q)} - y_{s,p}^{(r)} \right|^m \right)^{\frac{1}{m}} . \tag{4}$$

The Canberra distance (a weighted version of Manhattan distance) is given as follows [14]:

$$d_{\text{Canb}}(Y^{(q)}, Y^{(r)}) = \sum_{s,p} \frac{\left| y_{s,p}^{(q)} - y_{s,p}^{(r)} \right|}{\left| y_{s,p}^{(q)} \right| + \left| y_{s,p}^{(r)} \right|} . \tag{5}$$

Above distances have been chosen on a basis of computational simplicity and proved efficiency in many pattern recognition problems.

3.3 Voting Schemes

We employed two schemes of joining elementary classifiers' results. Both of them are based on voting. The first one comes from the observation that different distant metrics applied for a single feature can have significantly different accuracy. Thus, we classify an unknown sample in parallel using five distances presented above and remember the results. When we join these five results and use majority voting scheme, we may obtain higher classification accuracy. It is also known as a mode, when we chose value that appears most often in a set of data. In case when all values appear with the same frequency or there is no possibility to chose one single *winner*, we choose arbitrarily the first result. The second approach is based on the similar assumption, that each feature may have a preferred distance metrics. It should be noted that this relation is unknown to us. In such case, during a classification of an unknown sample, we collect 30 individual results (combination of five distances and six features). After a majority voting, we have the final classification result. Both approaches are slightly different, since in the first case we set a feature and calculate distances for this particular feature, while in the second one we calculate all the combinations of features and distances. Although, the first approach is less computationally intensive, its accuracy depends on the selected features.

4 Experiments

4.1 Database Characteristics

The benchmark dataset used for the evaluation of algorithm's performance contains six classes ($g = 1, \ldots, 6$) of the environmental sounds. The sounds represent the following acoustic scenes called: 'passing cars' ($g = 1$), 'rain' ($g = 2$), 'restaurant' ($g = 3$), 'shopping centre' ($g = 4$), 'beach' ($g = 5$), and 'factory' ($g = 6$). Each class consists of 42 audio recordings divided into 30 items used for learning phase and 12 items—for testing. Every audio recording is monaural, 10 s long with 22050 Hz sampling rate—total length of audio data is 42 min long. In order to demonstrate the complexity of the problem, Fig. 1 shows exemplary feature maps for each of the classes. It should be noted that their distinguishability in original attributes space is very limited. In the same figure, reduced representations of each feature map are also given. As it can be seen, most of the energy is condensed in the upper left corner of each matrix.

4.2 Experimental Results

In order to evaluate the performance of the proposed algorithm of classification, we investigated combinations of features and classifiers on the same dataset (classification schemes were presented above). The reduction parameters at the stage of 2DLDA/2DKLT is $s = 5$ and $p = 10$, hence we classified objects using 50 features, which is lower than in case of other known methods. The results are presented in Tables 1 and 2. We provide results for different number of classes ($G = \{3, 4, 5, 6\}$), since the first three classes are rather easy to distinguish, i.e. for $G = 3$ we take into consideration classes $g = \{1, 2, 3\}$ and for $G = 6$, classes $g = \{1, 2, 3, 4, 5, 6\}$, respectively. As it can be seen, the recognition accuracy decreases with the increase of number of classes. The other observation is that it is impossible to select one pair of feature-distance for which the accuracy is the highest for all the cases. In general, LPCC and LFCC perform better, no matter which distance metric we choose. Moreover, the voting on five distances calculated on the same feature gives slight increase in the classification accuracy over the individual case. The second voting scheme was also investigated. The results of experiments are provided in Table 3. This table presents recognition accuracy for different number of classes in two variants: maximum accuracy is the highest accuracy from Tables 1 and 2, while the voted accuracy means an accuracy for voting involving provided set of six pairs feature/distance (all features have been used), e.g., BFB/d_{City}, MFCC/d_{Cheb}, etc. As it can be seen, there is a significant progress in comparison to the first scheme. After applying the second scheme we are able to classify the simplest case of $G = 3$ classes with 100 % accuracy. On the other hand, the case of $G = 6$ classes still causes problems in terms of efficient classification. This is due to the characteristics of sound clips gathered in

Table 1 Classification accuracy as a function of distance metrics and feature type, for $G = 3$ and $G = 4$

Distance	$g = \{1, 2, 3\}$						$g = \{1, 2, 3, 4\}$					
	BFB	MFCC	LPC	LFCC	LPCC	LSF	BFB	MFCC	LPC	LFCC	LPCC	LSF
d_{Cheb}	0.81	0.72	0.64	0.81	0.75	0.5	0.63	0.44	0.31	0.69	0.6	0.38
d_{Euclid}	0.81	0.69	0.67	0.78	0.81	0.61	0.52	0.5	0.33	0.5	0.58	0.42
d_{City}	0.83	0.69	0.58	0.81	0.81	0.67	0.56	0.46	0.4	0.58	0.58	0.48
d_{Mink}	0.78	0.69	0.64	0.83	0.78	0.5	0.58	0.5	0.35	0.69	0.65	0.4
d_{Canb}	0.75	0.58	0.58	0.72	0.67	0.83	0.6	0.4	0.44	0.44	0.54	0.48
Voted	0.81	0.75	0.67	0.83	0.81	0.61	0.56	0.56	0.33	0.56	0.56	0.42

Table 2 Classification accuracy as a function of distance metrics and feature type, for $G = 5$ and $G = 6$

Distance	$g = \{1, 2, 3, 4, 5\}$						$g = \{1, 2, 3, 4, 5, 6\}$					
	BFB	MFCC	LPC	LFCC	LPCC	LSF	BFB	MFCC	LPC	LFCC	LPCC	LSF
d_{Cheb}	0.47	0.42	0.22	0.53	0.52	0.25	0.32	0.38	0.15	0.42	0.4	0.25
d_{Euclid}	0.42	0.48	0.27	0.48	0.5	0.35	0.36	0.36	0.26	0.33	0.38	0.26
d_{City}	0.45	0.43	0.33	0.55	0.62	0.37	0.38	0.32	0.26	0.42	0.44	0.31
d_{Mink}	0.47	0.42	0.28	0.53	0.53	0.28	0.36	0.38	0.21	0.43	0.36	0.25
d_{Canb}	0.43	0.42	0.33	0.38	0.55	0.32	0.4	0.32	0.26	0.32	0.33	0.29
Voted	0.45	0.48	0.32	0.5	0.58	0.32	0.38	0.33	0.25	0.38	0.43	0.22

Table 3 Classification accuracy for voting scheme with different features/distances

Number of classes (G)	Maximum accuracy	Voted accuracy	Features					
			BFB	MFCC	LPC	LFCC	LPCC	LSF
3	0.83	1.0	d_{City}	d_{Cheb}	d_{Cheb}	d_{Canb}	d_{Mink}	d_{Canb}
4	0.69	0.77	d_{Mink}	d_{Euclid}	d_{Canb}	d_{Mink}	d_{Mink}	d_{Canb}
5	0.62	0.67	d_{Canb}	d_{Euclid}	d_{Canb}	d_{Cheb}	d_{City}	d_{Canb}
6	0.44	0.53	d_{Canb}	d_{Euclid}	d_{Euclid}	d_{Mink}	d_{Cheb}	d_{Canb}

Table 4 Combination of feature/distance metric giving the highest possible classification accuracy for $G = 6$ and majority voting

No.	1	2	3	4	5	6
Feature	LFCC	MFCC	BFB	LPC	LPCC	LPCC
Distance	d_{Mink}	d_{Euclid}	d_{Canb}	d_{City}	d_{City}	d_{Canb}

the benchmark database and the presence of various, nonstationary sound events in the clips. An additional experiment was devoted to finding a combination of six pairs of feature/distance metric which give the highest possible accuracy. In this experiment, we assumed that not all features are used and there may be also situations, when the same feature is used more than one time (with different distance metrics). In order to find it, 30^6 combinations were examined using brute-force strategy. The classification accuracy found for $G = 6$ is close to 0.59, and the combination is as follows (see Table 4).

5 Summary

An algorithm for automatic audio scene classification was presented. The experiments were aimed at finding an optimal combination of features and distance metrics in terms of classification accuracy. Conducted experiments on a benchmark database showed the high efficiency of developed method. In comparison to other more complex methods (e.g., [10, 11]), the presented approach has good recognition accuracy together with simple implementation and low dimensionality of feature space, especially in comparison to other techniques. In future, certain more sophisticated classifiers and distance matrices may be used, which could lead to the further increase in classification accuracy, especially in case of higher number of audio classes.

References

1. Abe, M., Matsumoto, J., Nishiguchi, M.: Content-based classification of audio signals using source and structure modelling. In: Proceedings of the IEEE Pacific Conference on Multimedia, pp. 280–283 (2000)
2. Cantrell, C.D.: Modern Mathematical Methods for Physicists and Engineers. Cambridge University Press, Cambridge (2000)
3. Davis, S., Mermelstein, P.: Comparison of parametric representation for monosyllabic word recognition in continuously spoken sentences. IEEE Trans. ASSP **28**(4), 357–366 (1980)
4. Deza, E., Deza, M.M.: Encyclopedia of Distances. Springer, Berlin (2009)
5. Forczmański, P.: Evaluation of singer's voice quality by means of visual pattern recognition. J. Voice. doi:10.1016/j.jvoice.2015.03.001 (in press, 2015)
6. Forczmański, P., Frejlichowski, D.: Classification of elementary stamp shapes by means of reduced point distance histogram representation. Mach. Learn. Data Min. Pattern Recognit., LNCS **7376**, 603–616 (2012)
7. Forczmański, P., Labedz, P.: Recognition of occluded faces based on multi-subspace classification. In: 12th IFIP TC8 International Conference on Computer Information Systems and Industrial Management Applications (CISIM), LNCS, vol. 8104, pp. 148–157 (2013)
8. Forczmański, P., Kukharev, G., Shchegoleva, N.: Simple and robust facial portraits recognition under variable lighting conditions based on two-dimensional orthogonal transformations. In: 17th International Conference on Image Analysis and Processing (ICIAP), LNCS, vol. 8156, pp. 602–611 (2013)
9. Fukunaga, K.: Introduction to Statistical Pattern Recognition, 2nd edn. Academic Press, New York (1990)
10. Geiger, J.T., Schuller, B., Rigoll, G.: Large-scale audio feature extraction and SVM for acoustic scene classification. In: IEEE Workshop on Applications of Signal Processing to Audio and Acoustics (WASPAA), pp. 1–4 (2013)
11. Jiang, H., Bai, J., Zhang, S., Xu, B.: SVM-based audio scene classification, natural language processing and knowledge engineering. In: Proceedings of 2005 IEEE International Conference on IEEE NLP-KE'05, pp. 131–136 (2005)
12. Kleijn, W., Backstrom, T., Alku, P.: On line spectral frequencies. IEEE Signal Process. Lett. **10**(3), 75–77 (2003)
13. Kukharev, G., Forczmański, P.: Face recognition by means of two-dimensional direct linear discriminant analysis. In: Proceedings of the 8th International Conference PRIP 2005 Pattern Recognition and Information Processing. Republic of Belarus, Minsk, pp. 280–283 (2005)
14. Lance, G.N., Williams, W.T.: Computer programs for hierarchical polythetic classification ("similarity analysis"). Comput. J. **9**(1), 60–64 (1966)
15. Maka, T.: Attributes of audio feature contours for automatic singing evaluation. In: 36th International Conference on Telecommunications and Signal Processing (TSP), pp. 517–520. Rome, Italy, 2–4 July 2013
16. McKinney, M., Breebaart, J.: Features for audio and music classification. In: Proceedings of the International Symposium on Music Information Retrieval, pp. 151–158. Baltimore, Maryland (USA), 26–30 Oct 2003
17. Okarma, K., Forczmański, P.: 2DLDA-based texture recognition in the aspect of objective image quality assessment. Ann. Univ. Mariae Curie-Sklodowska. Sectio AI Informatica **8**(1), 99–110 (2008)
18. Paraskevas, I., Chilton, E.: Audio classification using acoustic images for retrieval from multimedia databases. In: 4th EURASIP Conference focused on Video/Image Processing and Multimedia Communications. EC-VIP-MC, pp. 187–192. Zagreb, Croatia, 2–5 July 2003
19. Rabiner, L., Schafer, W.: Theory and Applications of Digital Speech Processing. Prentice-Hall, Englewood Cliffs (2010)
20. Sammut, C., Webb, G.: Encyclopedia of Machine Learning. Springer, Berlin (2010)

21. Schuller, B., Wimmer, M., Moesenlechner, L., Kern, C., Arsic, D., Rigoll, G.: Brute-forcing hierarchical functionals for paralinguistics: a waste of feature space? In: IEEE International Conference on Acoustics, Speech and Signal Processing, ICASSP, pp. 4501–4504 (2008)
22. Smith III, J.O.: Spectral Audio Processing. W3K Publishing, Stanford (2011)

Enhancing Tracking Capabilities of KDE Background Subtraction-Based Algorithm Using Edge Histograms

Piotr Kowaleczko and Przemyslaw Rokita

Abstract The paper presents a method which allows to improve tracking abilities of conventional background subtraction-based algorithm. The presented algorithm which is a result of the studies is a hybrid method consisting of the Kernel Density Estimation (KDE) background subtraction tracking method and the Edge Histograms Displacement Calculation (EHDC) algorithm. Tracking ratios before and after merging with EHDC have been measured and presented. The paper also describes an algorithm eliminating cyclic changes in image's intensities values, which have significant influence on the input data for the hybrid algorithm. The influence of moving-camera video specificity on the output data has been pointed out.

Keywords Edge histogram · Background subtraction · Tracking · KDE · EHDC

1 Introduction

All the research presented in this paper is the continuation of the research described in [1, 2], where the Edge Histograms Displacement Calculation (EHDC) algorithm has been thoroughly described. In essence, values of horizontal and vertical edge histograms are calculated according to the formulas (1) and (2), respectively.

$$H_v(m) = \sum_{n=0}^{N-1} I(m, n). \tag{1}$$

P. Kowaleczko (✉)
C4ISR Systems Integration Division, Air Force Institute of Technology,
Ksiecia Boleslawa 6, 01-494 Warsaw, Poland
e-mail: piotr.kowaleczko@itwl.pl

P. Rokita
Faculty of Cybernetics, Military University of Technology,
Gen. S. Kaliskiego 2, 00-908 Warsaw, Poland
e-mail: pro@ii.pw.edu.pl

R. Burduk et al. (eds.), *Proceedings of the 9th International Conference on Computer Recognition Systems CORES 2015*, Advances in Intelligent Systems and Computing 403, DOI 10.1007/978-3-319-26227-7_70

$$H_h(n) = \sum_{m=0}^{M-1} I(m, n). \tag{2}$$

where:
$0 \le m < M$, and M—image width
$0 \le n < N$, and N—image height
I—canny edge-filtered binary image.

Common (the most similar) fragments of histograms of two consecutive frames are determined, and EHDC values are calculated basing on the displacement between them. Implemented in ITWL WH-1 electro-optical gimbal, the algorithm has two variants. The difference between them lies in different measurement of likeliness of the histograms' fragments: the normalized cross-correlation method and the differential method. All the results presented in this paper have been obtained using the normalized cross-correlation EHDC method [2, 3]. The algorithm of determining vertical histogram displacement value d_{vert} (and so the image displacement in horizontal direction) has been presented below.

1. Calculate vertical histogram f_{vert} of $(t-1)$th frame.
2. Calculate vertical histogram g_{vert} of tth frame.
3. Determine the width of a search window—m, where $m < M$ and M are the number of vertical histogram's values.
4. Calculate values of K_{vert} array, which are the values of correlations between the search window (values of f_{vert}) and the analyzed windows (values of g_{vert}) panned in left and in right direction from g_{vert} median.
5. Determining index of the maximum K_{vert} array value—d_{vert}.

The formulas for calculating the values of K_{vert} array and d_{vert} have been presented below (3 and 4). Similarly, the image displacement in vertical direction (d_{hor}) can be computed using horizontal edge histograms f_{hor} and g_{hor}.

$$d_{vert} = argmax_i(K_{vert}[i]). \tag{3}$$

$$K_{vert}[l] = \frac{\sum_{i=a}^{b}\left((f_{vert}[i] - \bar{f})(g_{vert}[i+l] - \bar{g}_l)\right)}{\sqrt{\sum_{i=a}^{b}(f_{vert}[i] - \bar{f})^2 \sum_{i=a}^{b}(g_{vert}[i+l] - \bar{g}_l)^2}}. \tag{4}$$

where:
$a = \frac{M}{2} - \frac{m}{2}$, and $b = \frac{M}{2} + \frac{m}{2}$,
$\bar{f}$—the mean value of the f_{vert} search window,
$\bar{g}_l$—the mean value of the l position-panned g_{vert} analysis window.

It has been proved in [2] that the optimal value of search window's width is 20 % of frame's width (height for horizontal displacements). However, EHDC in itself cannot be used to localize moving object on the video material. This was the reason to investigate the possibility of using Edge Histograms-based methods to support the already existing detection/tracking methods [4].

2 Assisting KDE Background Subtraction-Based Tracking Algorithm with Edge Histograms Methods

If there is a necessity to filter moving objects out of a static background, the simplest way to obtain very good results is to use differential background subtraction algorithm. The algorithm works properly only if there is a significant difference between the color of tracked object and background's color. Unfortunately, the simplest calculation of pixel difference between two successive frames, and applying threshold operation does not give satisfying results because in most cases, the input data (two consecutive frames) does not fulfill basic background subtraction (BS) method requirements which are: static camera location and lack of noise in the background image.

In most real video materials, slight changes of intensity values are present, as well as the background can vary because of the movements of little objects composing it (e.g., leaves on tree). Also small movements of the camera can efficiently make the basic BS method useless. To make the algorithm robust to these kind of interferences, different types of BS-based models have been worked out. If input images are noisy or the background is slightly changing, the best algorithms to use are KDE, 1-G and GMM. It is also important to choose the distance measure properly for the BS method selected. The comparison of all algorithms mentioned above and the results obtained can be found in [5–7].

In the proposed hybrid algorithm, a KDE BS method has been implemented. In this case, the algorithm is **not** used to detect **differences between matrixes of image pixels, but** to analyze **differences between edge histograms** of subsequent frames. In order to investigate how merging BS methods with EHDC affects tracking results, it was necessary to determine optimal values of KDE BS algorithm's parameters first. Optimal values of parameters have been set basing on the first 460 frames of analyzed video material, when there was no significant movement of camera.

2.1 Determining Optimal Values of Parameters for KDE Background Subtraction-Based Algorithm

Kernel Density Estimation algorithm is one of many variations of a conventional background subtraction method. It is often used when input video material is noisy, or the background is composed of many little objects. Working of the KDE model is based on continuous monitoring of image parameters and having them up-to-date in order to detect sudden changes in the background. In most cases slight variations of background exist, so that the estimation of probability density function for each pixel's value is necessary. These function are most frequently used together with Parzen-Window method, which has been exhaustively described in [8].

In case of using KDE BS method to detect changes in edge histograms, only histograms of last T frames are analyzed. The probability of belonging the ith value of histogram to the "not-changed subset" is calculated according to (5) formula.

$$P(h[i]_t) = \frac{1}{T} \sum_{a=t-T}^{t} \frac{1}{\sqrt{2\pi\sigma^2}} e^{-\frac{(h[i]_t - h[i]_a)^2}{2\sigma^2}}. \tag{5}$$

where:
$h[i]_a$—ith value of edge histogram for ath frame,
t—current time (pointing the most recent video frame),
σ^2—variance of histogram's values.

The procedure of determining optimal values of parameters for the KDE algorithm is composed of three steps:

1. determination of the mean value of variance $\bar{\sigma}^2$
2. determination of optimal threshold P^* for probability P (according to (5))
3. determination of optimal number of previously analyzed values: T^*.

It should not be forgotten that all the analysis described in this paper refer to values of edge histograms, and not to pixel intensities' values. The calculated P defines the probability, with which a single histogram's value in t moment can be considered as matching to T previous (in time) values. Exceeding calculated optimal threshold P^* results in classifying the analyzed histogram's value as not matching to the pattern, and so it could be assumed that a significant change has been noted in this position (an object has appeared/disappeared in this image fragment).

2.1.1 Determining Mean Value of Variance $\bar{\sigma}^2$

During video material analysis, slight changes of edge (vertical) histogram's columns have been noted. They were caused by changes in illumination level [9] and the video device itself. A series of measurements have been made in order to provide the mean variance σ^2 for the KDE algorithm as the value is used in calculations of P value. All computation have been made for a series of 70 first video frames (when the position of camera was fixed), and for all the columns apart from these, representing image columns with a silhouette of men on them. Although the men does only slight movements of his legs on the movie, he causes quite big changes in histogram's values. The variance for each column in the domain of 70 first frames was calculated basing on the following formula:

$$\sigma^2 = \frac{1}{70} \sum_{i=0}^{69} (h_i - \bar{h})^2 \tag{6}$$

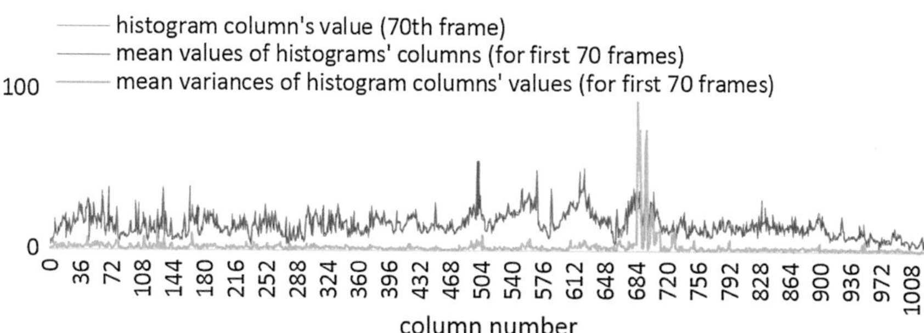

Fig. 1 Values of each histogram's column for 70th video frame; mean values and mean variances calculated for the first 70 frames

where:
h_i—value of histogram's column in ith video frame,
$\bar{h}$—mean value of histogram's column calculated for all 70 video frames.

Figure 1 shows values of each histogram's column in 70th video frame as well as mean values and mean variances calculated for all first 70 frames. The analysis of the plot proves that distances between histogram's values for 70th frame and calculated mean values are very small. High values of variances between 60th and 75th image column were caused by slight movements of men's legs. The σ^2 value used in (5) formula was calculated as the mean value of histogram column's variances for the first 70 frames, excluding variances for columns representing men's silhouette (from 650 to 750).

$$\bar{\sigma^2} = \frac{1}{923}\left(\sum_{i=0}^{649}\sigma_i^2 + \sum_{i=751}^{1023}\sigma_i^2\right) = 1.632 \tag{7}$$

where:
σ_i^2—value of the variance of ith video frame column

2.1.2 Determining Optimal Threshold for the Probability and Optimal Number of Previously Analyzed Histogram's Values

The determination of P^* and T^* values is subjective and related to each another. The measurements were based on the analysis of differences between P values for columns in which significant changes have been noted. The analysis have been made for different values of T. The maximum noted value of P differences indicates, that also the recognition capabilities of algorithm are maximal. In this case, the threshold value for P as well as T value, for which analyzed P values was calculated, were denoted as optimal values P^* and T^*. All the computations were made for 650th column of histogram. Figure 2 shows probability values P calculated for 650th histogram's column for a series of consecutive frames, and for $T = 7$.

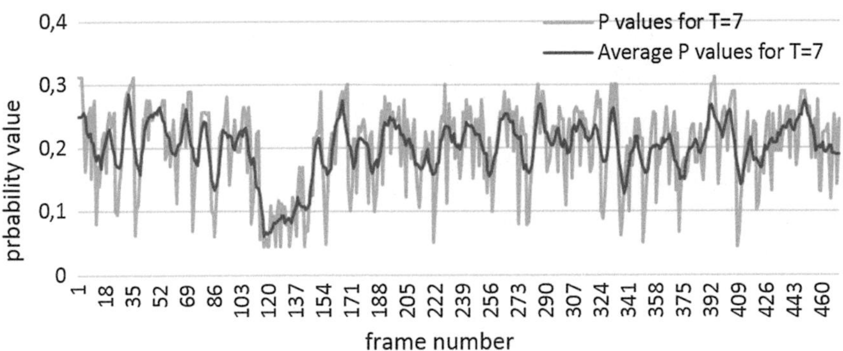

Fig. 2 Probability values P calculated for 650th histogram's column for a series of consecutive frames; $T = 7$

2.1.3 Eliminating Cyclic Changes of Probability Values

Cyclic changes of P values were noted when analyzing the plot. The analysis of the video material proved, that the reason of interferences lays in the video coding method used (DivX). Based on MPEG, the codec keeps full image information only in so-called "intra frames", which in this case appear every 12 frames. The accumulation of illumination changes in the scene during the period between two I-frames is the reason of interferences. In order to facilitate the analysis, a technique of averaging was used. Originally used in signal processing, an averaging filter with 7 values-width analysis window has been implemented according to the formula (8).

$$\bar{P} = \frac{1}{7}\sum_{i-3}^{i+3} P_i \tag{8}$$

where: P_i—value of the probability calculated for ith video frame.

Window's width of 7 values have been chosen because of good quality of the results and not losing too many details of the graph (Fig. 2). As may be seen around 100th and 150th frame, a significant drop of probability values takes place. It is caused by an object (a men) passing through the analyzed histogram/image column. The efficiency and robustness of the algorithm depends on the proper choice of P threshold value, and of the number of previously analyzed values T. Choosing optimal P threshold value: P^* and optimal T: T^* should be based only on the biggest difference noted between P values for 100–150th frames and P calculated for the rest of video material. All the measurements have been made for $T \in \langle 2, 20 \rangle$. Juxtaposition of graphs for every T value showed that they are very similar each to another, and so the choice of T^* was completely subjective. It seems that for $T = 7$ the difference between P values with, and without men's silhouette included is most noticeable. The graph of P values for $T = 7$ have been presented on Fig. 2. The best object detection results can be obtained in this case by setting the optimal probability

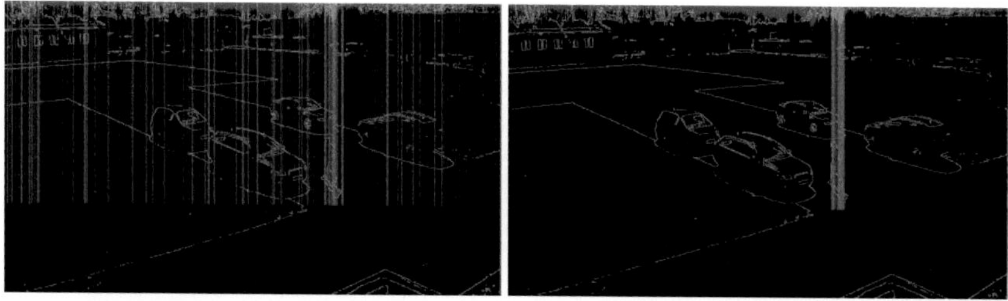

Fig. 3 108th frame of analyzed video with (*right*) and without (*left*) probability-averaging filter implemented in the KDE algorithm used in the analysis of edge histograms

threshold value to 0.11, so finally:

$$T^* = 7 \tag{9}$$

$$P^* = 0.11 \tag{10}$$

A verification of usefulness of the probability-values-averaging technique has been made. An OpenCV [10] application has been created in order to visualize the results. Figure 3 shows output data for both versions of the algorithm: with, and without probability-averaging filter implemented. Columns of edge histogram in which a change has been noted are indicated with green lines. As it can be seen, implementing the averaging filter for probability values gives a significant improvement of results' quality. The number of green lines has been drastically reduced. The only changes noted were these, connected with a moving object (a man).

2.2 *Using Displacements Calculated Using EHDC Method in KDE Background Subtraction-Based Algorithm*

The hybrid algorithm which is a subject matter of this paper has been created in order to be able to detect moving objects both for static and for moving-camera videos. Working of the algorithm is based on two steps

1. Using camera displacements calculated using EHDC method to determine common fragments of the seven most recent edge histograms
2. Using KDE BS algorithm (with averaging filter implemented) to detect changes between common fragments of frames' histograms (Fig. 4).

The most difficult part in determining common fragments for histograms of seven consecutive frames was to take into account the possibility of changing the direction of camera's movement/rotation. The algorithm which calculates common fragments is based on determination of the maximal image displacements in each direction, and

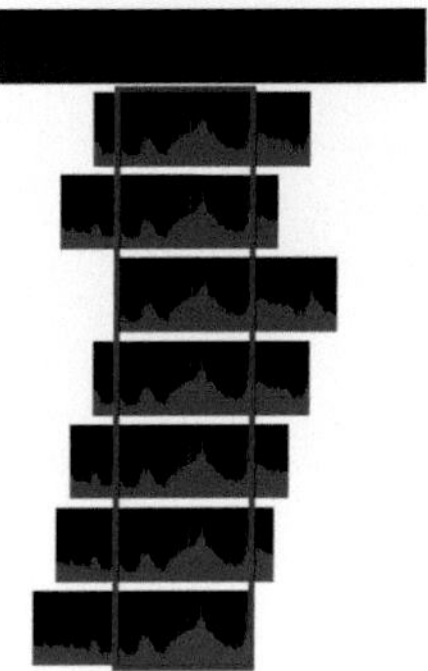

Fig. 4 A common fragment for 7 consecutive vertical edge histograms

then deleting groups of extreme left and extreme right histogram columns. The number of columns deleted depends on camera displacements in left, and in right direction during seven analyzed frames. Extreme columns are deleted from the histogram of the oldest frame. The set of columns derived in this manner is a common set for all seven histograms. In theory, the KDE analysis of seven corresponding vertical histogram fragments should lead to detect only a few changed columns (these representing a moving person) as the majority of image would be treated as background. In Fig. 5 (presenting the results for algorithms with and without EHDC correction implemented), the columns in which a change has been noted are indicated with green lines. The difference in the number of changed columns between both versions of the algorithm is significant. The results obtained prove that merging KDE BS method with EHDC displacement corrections, gives a significant amelioration of results' quality. For video fragments in which camera moves/rotates, the number of detected changes in histograms' columns decreases more than two times after implementing the proposed method. Figure 6 shows the number of histogram's columns that have changed when using algorithm with (*blue*) and without (*green*) EHDC modification.

Each significant increase of the number of changed columns (Fig. 6) coincides with the moments, when camera moves. The error limit's value (*red color*) was dynamically adjusted, depending on the number of moving persons and their distance from

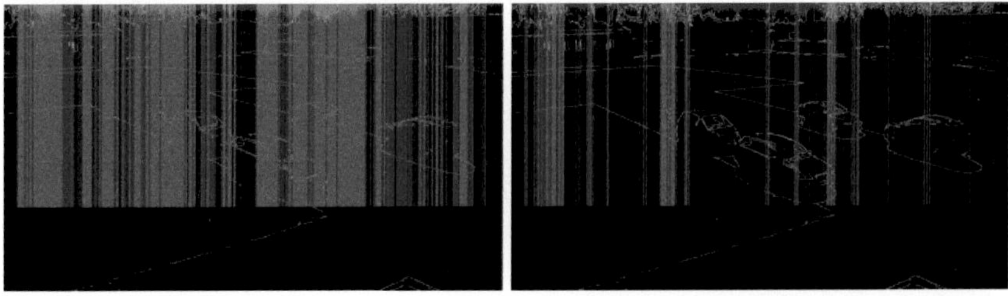

Fig. 5 Changes of histogram columns' values for the same frame of moving-camera video fragment for the conventional KDE BS algorithm (*left*) and for KDE BS algorithm with EHDC correction implemented (*right*)

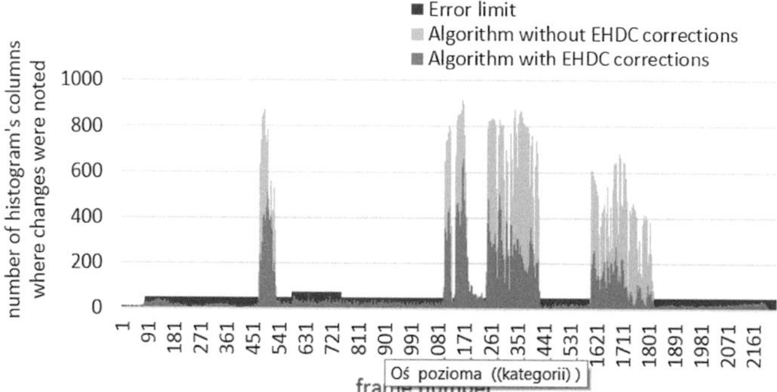

Fig. 6 Number of histogram's columns that have changed when using KDE BS algorithm with (*blue*) and without (*green*) EHDC modification implemented

the camera. It does not allow to determine if an object has been found or not. It only helps to answer the question if it is possible to find it. In theory, after implementing EHDC displacement correction, the error limit should never be exceeded, both when camera is moving/rotating, and when it's position is fixed. As it can be seen, this assumption is false when considering fragments of video when camera's position is not fixed. Even after determining and analyzing only corresponding fragments of seven consecutive histograms, many columns have changed enough to exceed the threshold. This situation is caused by inaccuracy of the video camera device. It has been proved (Fig. 7), that when camera moves, objects' edges become blurry and stretched. Therefore, when using BS methods, instead of detecting change on one-pixel-wide edge, changes are noted in several histogram columns/lines. In the case of analyzed video sequence, only vertical edges were stretched, because the camera was rotating mainly in horizontal direction. The example of significant change of object's shape has been presented on (Fig. 7). The edges are sharp for static camera image, while for the moving one, a blur can be seen (especially for the rear bumper and the shadow).

Fig. 7 Fragments of a frame for static camera (*left*), and for moving-camera (*right*) video

3 Conclusions

The purpose of the research presented in this paper was to use EHDC-calculated displacement values [2] together with the KDE background subtraction-based algorithm in order to detect image changes both on static and moving-camera video sequences. The KDE BS method did not analyze image pixels' intensities, but values of its edge histogram columns. Merging EHDC displacement corrections with the KDE algorithm, resulted in a great (more than two times) reduction of falsely detected changes for video material with a moving/rotating camera. Nevertheless, the results were not sufficient to allow the user, or any other algorithm to determine moving object's position. Measured values for video fragments in which camera's position was not fixed prove, that too many changes have been noted. The reason of the failure is that the algorithm is not robust enough when stretching/blurring occurs in moving-camera videos. In this case, the hybrid algorithm cannot be used for tracking purposes. However a great diminution of false detections allows to believe, that creating a solution which eliminates the influence of blur/stretch could make the algorithm robust for all kinds of video input data. This, as well as merging EHDC with other tracking algorithms, would be the subject matter of future research.

References

1. Rokita, P.: Fast tracking using edge histograms. SPIE. Real-Time Imaging II, vol. 3028, p. 91, 3 Apr 1997
2. Kowaleczko P., Rokita P.: Wykorzystanie algorytmow opartych na metodzie Edge Histograms do okreslenia parametrow ruchu kamer glowicy optoelektronicznej WH-1 (in Polish), KNTWE 2014 conference materials (2014)
3. Lewis, J.P.: Fast normalized cross-correlation. Vis. Interface **10**(1), 120–123 (1995)
4. Yilmaz, A., Javed, O., Shah, M.: Object tracking—a survey. Acm Comput. Surv. (CSUR) **38**(4), 13 (2006)
5. Benezeth Y., Jodoin P., Emile B., Laurent H., Rosenberger C.: Review and Evaluation of Commonly-Implemented Background Subtraction Algorithms. In: Pattern Recognition, ICPR 2008, 19th International Conference on Pattern Recognition (2008)
6. Piccardi, M.: Background subtraction techniques: a review. 2004 IEEE Int. Conf. Syst. Man Cybern. **4**, 3099–3104 (2004)
7. Elgammal A., Harwood D., Davis L.: Non-parametric model for background subtraction, ECCV (2000)
8. Babich, G.A., Camps, O.I.: Weighted Parzen windows for pattern classification. IEEE Trans. Pattern Anal. Mach. Intell. **18**(5), 567–570 (1996)
9. Kovac J., Peer P., Solina F.: Eliminating the Influence of Non-Standard Illumination from Images, Technical Report, (2003)
10. OpenCV documentation, http://docs.opencv.org

Implicit Links-Based Techniques to Enrich K-Nearest Neighbors and Naive Bayes Algorithms for Web Page Classification

Abdelbadie Belmouhcine and Mohammed Benkhalifa

Abstract The web has developed into one of the most relevant data sources and becomes now a broad knowledge base for almost all fields. Its content grows faster, and its size becomes larger every day. Due to this big amount of data, web page classification becomes crucial since users encounter difficulties in finding what they are seeking, even though they use search engines. Web page classification is the process of assigning a web page to one or more classes based on previously seen labeled examples. Web pages contain a lot of contextual features that can be used to enhance the classification's accuracy. In this paper, we present a similarity computation technique that is based on implicit links extracted from the query-log, and used with K-Nearest Neighbors (KNN) in web page classification. We also introduce an implicit links-based probability computation method used with Naive Bayes (NB) for web page classification. The new computed similarity and probability help enrich KNN and NB respectively for web page classification. Experiments are conducted on two subsets of Open Directory Project (ODP). Results show that: (1) when applied as a similarity for KNN, the implicit links-based similarity helps improve results. (2) the implicit links-based probability helps ameliorate results provided by NB using only text-based probability.

1 Introduction

The World Wide Web's size is growing faster, making the task of finding relevant information on the web much more difficult. The Internet contains billions of web pages, making the need for performant automatic web page categorization techniques more noticeable. Web page categorization creates new problematic research issues

---

A. Belmouhcine (✉) · M. Benkhalifa
Computer Science Laboratory (LRI), Sciences Faculty, Computer Science Department, Mohammed V University, Rabat, Morocco
e-mail: belmouhcine@gmail.com

M. Benkhalifa
e-mail: khalifa@fsr.ac.ma

R. Burduk et al. (eds.), *Proceedings of the 9th International Conference on Computer Recognition Systems CORES 2015*, Advances in Intelligent Systems and Computing 403, DOI 10.1007/978-3-319-26227-7_71

due to abundant information enclosed within hypertext. Text classifiers, usually, work on textual features but are not concerned with distinctive characteristics possessed by web pages.

Automatic web page classification is a supervised learning process in which a classifier is trained using some manually labeled examples of predefined classes and then applied to future unseen examples to predict categories for them. Web page classification is essential for many tasks such as focused crawling, construction and expansion of web directories, topic specific web link analysis, improvement of the web search's quality, assisted web browsing, web content filtering, and question/answer systems.

To reduce the effect of the curse of dimensionality problem [1], that prevents satisfactory results when K-Nearest Neighbors (KNN) is applied and dimensions are high, we introduce a similarity based on clicks frequencies called implicit links-based similarity. For two web pages p_1 and p_2, this similarity corresponds to the weight of the edge linking p_1 and p_2, if p_2 is among the direct neighbors of p_1. Also, we modify the probability computation mechanism of Naive Bayes (NB) in order to use weights of edges relating web pages and involve users' intuitive judgments in the process of probabilities computation. Instead of using solely the probability of a category c given a web page p, we employ the probability of c given a web page p and its neighbors.

We test the approach on two subsets of the Open Directory Project (ODP) [2]. We also used AOL query-log [3, 4] to extract implicit links and build a web pages' implicit graph.

Experimental study shows that the implicit links-based similarity helps improve results obtained by KNN. Also, implicit links-based probability helps improve results given by NB. The main contributions of this work are:

1. The introduction of a new similarity based on clicks frequencies. This similarity uses neighborhood information and helps reduce the effect of the curse of dimensionality faced when using KNN based on the text only.
2. The use of clicks frequencies in the process of probabilities calculation. Those frequencies involve users' intuitive judgments and neighboring information in probabilities computation.

Section 2 includes a review of previous work on web page classification. Section 3 gives details of the proposed techniques. Section 4 shows the experimental setting adopted and contains a discussion of obtained results. The final section concludes our work and cites some of our future perspectives.

2 Related Work

Many researchers tried to modify NB and KNN using web pages' contextual information. Kwon and Lee [5] supplemented KNN approach with a feature selection method and a term-weighting scheme using markup tags. They also reformed the inter-document similarity measure. They showed that their method improved the performance of KNN classification. He and Liu [6] proposed a new method of web categorization using the output from an improved feature selection based on Independent Component Analysis as input for NB classifier. Their experiments demonstrated that their method provided acceptable classification accuracy. Youquan et al. [7] contributed to NB algorithm improvement by considering the semi-structure nature of web pages. They divided HTML tags into "supervise tags" groups to supervise the classification. They introduced a new weighting approach based on those "supervise tags" groups. They concluded that NB classifier performance benefited from their weighting policy. Fernandez et al. [8] used NB classifier for new enriched web page representations. They tried six term-weighting functions. Some of them used only text of web pages and were based on frequencies. Others combined several criteria including information from HTML tags. They also compared two different models of Bayesian prior probabilities: the event and the Gaussian model, applied to representations based on frequencies and relevance, respectively. Their experiments showed that, in general, the Gaussian model obtained better results than event model when enriched representation was considered.

Besides, many works built artificial links between web pages instead of hyperlinks [9–12]. However, those approaches use features of neighboring web pages and their labels to classify target web pages. They do not use frequencies of appearance of two web pages together in the query-log to compute similarities between them or to calculate the probability that a web page belongs to a category.

All those cited works focus on contextual features of web pages and their neighbors and involve them in representations of web pages to improve performances of classifiers. In this paper, we inspect the use of implicit links to calculate inter-document similarity and membership probabilities.

3 Proposed Techniques

In this paper, we construct a graph using implicit links extracted from the query-log and use them to enrich KNN and NB for web page classification. KNN often suffers from the curse of dimensionality in web page classification due to the significant size of the vocabulary. Clicks frequencies can indicate the degree of relationship between two web pages based on users' intuitive judgments and are independent of the feature space dimension. Thus, they can be used to compute similarities between web pages. On the other hand, using the text alone to compute probabilities used by NB is not enough, since sometimes words are shared between many categories. Thus, clicks frequencies can be used to compute probabilities based on neighboring web pages' labels.

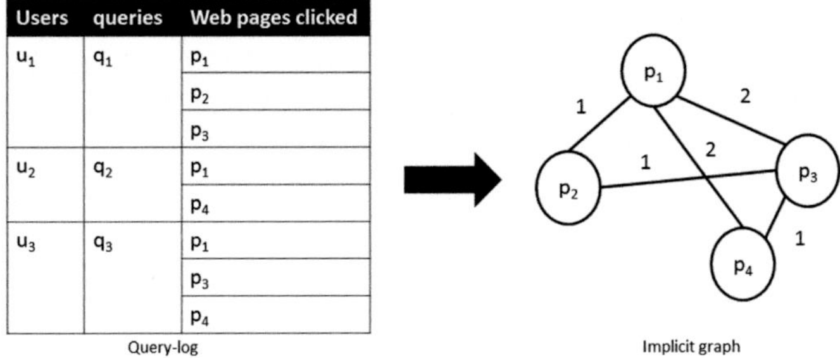

Users	queries	Web pages clicked
u_1	q_1	p_1
		p_2
		p_3
u_2	q_2	p_1
		p_4
u_3	q_3	p_1
		p_3
		p_4

Fig. 1 Example of a web pages' implicit graph constructed from a query-log

3.1 *Web Pages' Implicit Graph*

As it was done previously [9, 12], we create a graph of web pages called web pages' implicit graph, using implicit links built by leveraging information from the query-log. We construct a graph whose vertices are web pages, and edges relate web pages clicked together by a user through a query. Those edges are weighted using frequencies of clicks on their both extremities together as provided by the query-log. Figure 1 illustrates a graph constructed from a query-log.

3.2 *Implicit Links-Based Similarity Computation Technique*

Instead of using a text-based similarity alone, we introduce a novel similarity that leverages clicks frequencies. The new similarity is computed according to the following formula:

$$Sim(p_i, p_j) = \begin{cases} w_{ij} & \text{if } p_j \in N(p_i) \\ 0 & \text{otherwise} \end{cases} . \tag{1}$$

Where p_i and p_j are two web pages, w_{ij} is the weight of the edge relating p_i to the web page p_j in the implicit graph (number of clicks), and $N(p_i)$ denotes the set of neighbors of p_i.

The proposed similarity is based on implicit links. Thus, it is a semantic similarity since it is based on users' intuitive judgments. This similarity will be used with KNN to help it reduce the effect of the problem of the curse of dimensionality because weights are independent of the feature space dimension.

To classify a web page p_i, KNN calculates similarities between p_i and each web page in the training set. Then, it ranks web pages in the training set based on those similarities. In our case, KNN does a two level ranking. First, it ranks web pages using the implicit links-based similarity. Then, web pages having this similarity equal to zero are ranked again using the cosine similarity.

3.3 Implicit Links-Based Probability Computation Technique

To compute the probability of a particular class c for a web page p_i, we consider both its text and its neighbors. Thus, instead of calculating the probability of a category c given a web page p_i ($P(c|p_i)$), we calculate the probability of p_i by looking at its text along with its neighbors belonging to the class c. If none of the neighbors of p_i belongs to c, the probability is computed by considering only p_i's text. The probability of a class c given a web page p_i and its neighbors $N(p_i)$ is computed as follows:

$$P(c|p_i, N(p_i)) = \frac{P(p_i, N(p_i)|c) \times P(c)}{P(p_i, N(p_i))}. \tag{2}$$

We make a naive assumption that the text of web page p_i is independent of labels of its neighbors. Hence, the probability of a class c given a web page p_i and its neighbors $N(p_i)$ will be computed according to the following formulas:

$$P(c|p_i, N(p_i)) = \frac{P(p_i|c) \times P(N(p_i)|c) \times P(c)}{P(p_i, N(p_i))}. \tag{3}$$

$$P(N(p_i)|c) = \frac{score(p_i, c)}{Z(c)}. \tag{4}$$

$$score(p_i, c) = 1 + \sum_{p_j \in N(p_i) \& class(p_j) = c} w_{ij}. \tag{5}$$

$$Z(c) = \sum_{p_i \in V} score(p_i, c). \tag{6}$$

Where w_{ij} is the weight of the link relating p_i and p_j in the implicit graph, $Z(c)$ is normalization constant ensuring that probabilities range between 0 and 1, $class(p_j)$ is the class of the web page p_j, V is the set of all web pages and $N(p_i)$ is the set of neighbors of p_i.

Let p_i be a web page and $N(p_i)$ its neighbors, the likelihood of p_i's neighbors given a class c corresponds to the sum of weights of edges linking p_i to web pages of category c. We add 1 to this sum to avoid probabilities that are equal to 0, in the case where no web page among p_i's neighbors belongs to the class c.

To do NB classification we only need the numerator, since $P(c|p_i, N(p_i))$ are proportional to $P(p_i|c) \times P(N(p_i)|c) \times P(c)$.

4 Experiments

In this section, we present experiments used to validate the advantage of our proposed similarity and probability. We describe conducted experiments, i.e. pre-processing techniques, classifiers, datasets, query-log, performance evaluation metrics, and obtained results with the discussion.

4.1 Experimental Setup

4.1.1 Pre-processing

We applied some pre-processing techniques to each web page in datasets. The aim of those techniques is cleaning and normalizing the raw text contained in these web pages. In tokenization step, we turn all terms to lower case. Then, we removed some special characters, punctuation marks and numbers. Also, we removed all scripts, styles, mimes headings and HTML tags. For the stemming process, we used the well-known Porter method [13]. After the pre-processing stage, we build the dictionary that consists of words resulting from pre-processing. Thus, we consider web pages as bags of words. We represent our web pages using the conventional Vector Space Model [14]. We map each web page p onto its vector $v_p = (n_{1p}, n_{2p}, \ldots, n_{mp})$; where n_{ip} denotes the weight of the ith term in the web page p. We adopted the Term Frequency-Inverse Document Frequency (TF-IDF) [15, 16]-based weighting model to obtain the weights.

4.1.2 Classifiers Used

Naive Bayes

We used the implicit links-based probability with NB which is an easy and very popular probabilistic classification algorithm [17, 18]. It uses joint probabilities of features and categories to estimate the probabilities of classes given a document and makes the assumption that features are conditionally independent of each other to make the computation of joint probabilities simple. Our experiments were conducted using multinomial NB.

K-Nearest Neighbors

We used the implicit links-based similarity with KNN which is the most simple classification algorithm [19]. It is a lazy learner that predicts the category of an instance based on its K nearest training examples in the feature space using an inter-instance similarity. This algorithm does not generate a model from training instances but rather stores all those training examples directly and uses them to determine the class of a new instance. Our experiments were conducted using k equals to 10.

4.1.3 Datasets

Datasets used in this article are extracted from the ODP [2]. ODP is a tremendous repository containing around 4.6 million web pages and is organized into 765,282 categories and subcategories [20]. For the first dataset, we extracted data from 6 categories: "Arts" (2823 web pages), "Computers" (1221 web pages), "Health" (2129 web pages), "Science" (909 web pages), and "Sports" (1262 web pages). For the second dataset, we extracted web pages from two categories "Adult" (830 web pages), and "KidsAndTeens" (817 web pages). We also created a category named "Other" (782 web pages) which contains web pages from categories other than the two ones cited previously.

We used in this work AOL query-log [4], which contains a collection of around 20 million web queries collected from 650,000 users over three months.

For both datasets, all experiments were conducted using ten folds cross-validation to reduce the uncertainty of data split between training and test data.

4.1.4 Evaluation Measures

To evaluate the performance obtained, we used the standard metrics: recall, precision, and F1, which are commonly used to evaluate the classification task. Additionally, to measure global averages across multiple categories, we used two averaging techniques: Micro-averaging and Macro-averaging.

4.2 Results and Discussion

In this section, we present experiments' results obtained using implicit links-based similarity with KNN and using implicit links-based probability with NB on two subsets of ODP. We discuss our approach according to five points. The two first points are related to the performance achieved by KNN. The third point discusses the influence of the sparseness of the implicit graph on KNN's performance. The two last points concern results obtained by NB.

- As shown in Table 1 and for both datasets, F1 scores obtained using KNN along with the implicit links-based similarity (KNN(Text+Implicit Links)) are better than those obtained using KNN along with cosine similarity (KNN(Text)). Those results prove that clicks frequencies can be used as a web page similarity, which is

Table 1 Performance obtained by KNN when using implicit links-based similarity

(a) On the first subset of ODP

	Arts			Computers			Health			Science			Sports		
	P	R	F1	P	R	F1	P	R	F1	P	R	F1	P	R	F1
KNN1	0.714	0.887	0.791	**0.727**	0.703	0.715	0.836	0.828	0.832	0.866	**0.505**	0.638	0.89	0.717	0.794
KNN2	**0.722**	**0.894**	**0.799**	0.724	**0.711**	**0.717**	**0.85**	**0.842**	**0.846**	**0.881**	0.503	**0.64**	**0.896**	**0.723**	**0.8**

(b) On the second subset of ODP

	Adult			KidsAndTeens			Other		
	P	R	F1	P	R	F1	P	R	F1
KNN1	0.759	0.964	0.849	0.892	0.739	0.808	0.875	**0.767**	0.817
KNN2	**0.773**	**0.982**	**0.865**	**0.904**	**0.788**	**0.842**	**0.911**	0.761	**0.829**

KNN1: KNN(Text)
KNN2: KNN(Text+Implicit Links)

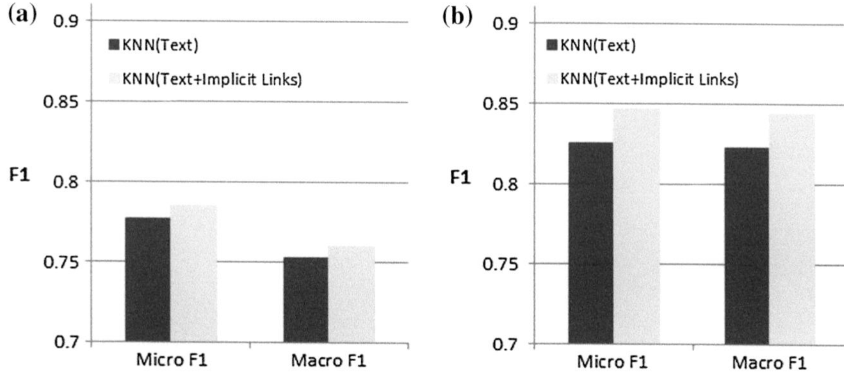

Fig. 2 A comparison of micro F1 and macro F1 averages given by KNN using text-based similarity and implicit links-based similarity. **a** On the first subset of ODP. **b** On the second subset of ODP

Table 2 Improvement obtained by KNN that uses implicit links-based similarity over KNN based on the text alone

	Improvement on the first subset of ODP		Improvement on the second subset of ODP	
	Using only linked web pages (%)	Using all web pages (%)	Using only linked web pages (%)	Using all web pages (%)
Micro F1	**2.41**	1.03	**3.6**	2.55
Macro F1	**2.26**	0.93	**3.62**	2.55

the basis for KNN classification, and hence, they can represent reliable indicators on the similitude of web pages.

- As shown in Fig. 2 and for both datasets, macro F1 and micro F1 averages obtained when using KNN with the implicit links-based similarity (KNN(Text+Implicit Links)) are better than those achieved when using KNN with cosine similarity (KNN(Text)). Those results prove that the proposed similarity brings improvement in both rare and common categories.
- The first dataset contains 8344 web pages, 3661 of them are related to at least one other web page. The second dataset contains 2429 web pages, 1799 of them are related to at least one other web page. Thus, the implicit graph is very sparse. If we consider only, web pages linked to at least one web page, the improvement of macro F1 and micro F1 over the use of only the text will increase as shown in Table 2. Hence, the sparseness of the implicit graph has a negative influence on the performance of KNN that uses implicit links-based similarity.
- As shown in Table 3 and for both datasets, the performance obtained using NB with implicit links-based probability (NB(Text+Implicit Links)) is higher than the one obtained when using NB with text-based probability (NB(Text)). This improvement is achieved because neighbors and users' intuitive judgments carry information about the relevance of a class to a web page. Sometimes, the text of a web page does not reflect its category since some words are shared by many classes

Table 3 Performance obtained by NB when using implicit links-based probability

(a) On the first subset of ODP

	Arts			Computers			Health			Science			Sports		
	P	R	F1	P	R	F1	P	R	F1	P	R	F1	P	R	F1
NB1	0.82	0.792	0.806	0.582	0.769	0.663	0.891	0.817	0.852	0.668	0.7	0.684	0.899	0.784	0.838
NB2	**0.836**	**0.795**	**0.815**	**0.586**	**0.778**	**0.668**	**0.897**	**0.834**	**0.864**	**0.673**	**0.701**	**0.687**	**0.902**	**0.793**	**0.844**

(b) On the second subset of ODP

	Adult			KidsAndTeens			Other		
	P	R	F1	P	R	F1	P	R	F1
NB1	0.938	0.911	0.924	0.881	0.886	0.883	0.861	0.88	0.87
NB2	**0.963**	**0.917**	**0.939**	**0.889**	**0.904**	**0.896**	**0.868**	**0.896**	**0.882**

NB1: NB(Text)
NB2: NB(Text+Implicit Links)

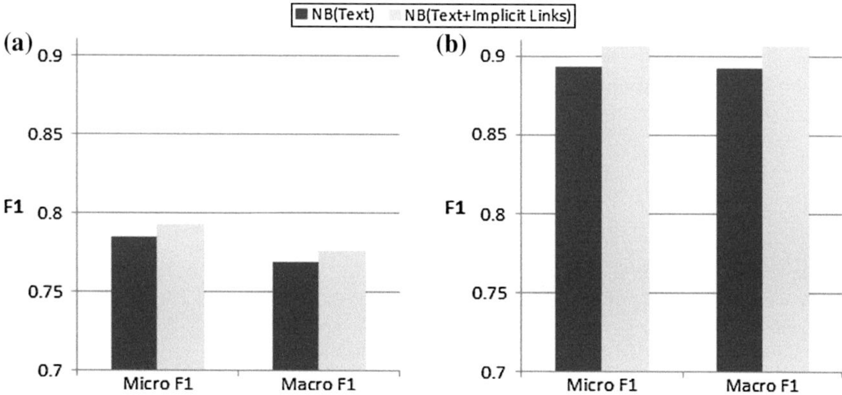

Fig. 3 A comparison of micro F1 and macro F1 averages given by NB using text-based probability and implicit links-based probability. **a** On the first subset of ODP. **b** On the second subset of ODP

(which is a general problem not related only to NB). Thus, the multiplication of probabilities by a score that is based on frequencies of occurrences of categories among the web page's neighbors helps highlight its subject and changes initial text-based ranks of categories.

- As shown in Fig. 3 and for both datasets, micro F1 and macro F1 averages obtained when using implicit links-based probability with NB (NB(Text+Implicit Links)) are higher than those obtained when using NB with text-based probability (NB(Text)). Those results prove that involving neighbors and users' intuitive judgments in the process of probabilities computation has a positive impact on the performance. This impact is the same for both common and rare categories.

5 Conclusion

In this paper, we have presented two computation techniques based on implicit links. The first one is a similarity computation technique for KNN web classifier, and the second one is a probability computation method for NB web classifier. The experimental results show that, within an appropriate experimental setting, the implicit links-based similarity helps ameliorate results obtained by KNN that uses only the cosine similarity. Also based on our experiments, the proposed probability helps ameliorate results obtained using NB with text-based probability. Our findings include:

1. The use of KNN along with the similarity based on cosine similarity and clicks frequencies gives better results than the use of KNN with only cosine similarity.
2. The use of NB along with the probabilities based on clicks frequencies gives better results than the use of NB along with text-based probabilities.

In the future, we will try to use clicks frequencies with similarities other than cosine. Additionally, we will use hyperlinks rather than implicit links, in order to evaluate the impact of the probability and the similarity with hyperlinks.

References

1. Beyer, K.S., Goldstein, J., Ramakrishnan, R., Shaft, U.: When Is Nearest Neighbor Meaningful?. In: Proceedings of the 7th International Conference on Database Theory, pp. 217–235, London, UK (1999)
2. ODP—Open Directory Project (Online). http://www.dmoz.org/
3. AOL Search Query Logs—RP (Online). http://www.researchpipeline.com/mediawiki/index.php?title=AOL_Search_Query_Logs
4. AOL search data mirrors (Online). http://gregsadetsky.com/aol-data/
5. Kwon, O.-W., Lee, J.-H.: Web page classification based on k-nearest neighbor approach. In: Proceedings of the fifth international workshop on Information retrieval with Asian languages, pp. 9–15, New York, NY, USA (2000)
6. He, Z., Liu, Z.: A novel approach to naive bayes web page automatic classification. In: Fifth International Conference on Fuzzy Systems and Knowledge Discovery (FSKD), vo. 2, pp. 361–365 (2008)
7. Youquan, H., Jianfang, X., Cheng, X.: An improved naive bayesian algorithm for web page text classification. In: Eighth International Conference on Fuzzy Systems and Knowledge Discovery (FSKD), vo. 3, pp. 1765–1768 (2011)
8. Fernandez, V.F., Herranz, S.M., Unanue, R.M., Rubio, A.C.: Naive Bayes web page classification with HTML Mark-Up enrichment. In: International Multi-Conference on Computing in the Global Information Technology (ICCGI), pp. 48–48 (2006)
9. Shen, D., Sun, J.-T., Yang, Q., Chen, Z.: A comparison of implicit and explicit links for web page classification. In: Proceedings of the 15th international conference on World Wide Web, pp. 643–650, New York (2006)
10. Xue, G.-R., Yu, Y., Shen, D., Yang, Q., Zeng, H.-J., Chen, Z.: Reinforcing web-object categorization through interrelationships. Data Min. Knowl. Discov. **12**(2–3), 229–248 (2006)
11. Kim, S.-M., Pantel, P., Duan, L., Gaffney, S.: Improving web page classification by label-propagation over click graphs. In: Proceedings of the 18th ACM Conference on Information and Knowledge Management, pp. 1077–1086, New York, NY, USA (2009)
12. Belmouhcine, A., Benkhalifa, M.: Formal concept analysis based corrective approach using query-log for web page classification. J. Emerg. Technol. Web Intell. **6**(2) (2014)
13. Porter, M.F.: In: Sparck Jones, K., Willett, P. (eds.) Readings in Information Retrieval, pp. 313–316. Morgan Kaufmann Publishers Inc., San Francisco (1997)
14. Salton, G., McGill, M.J.: Introduction to Modern Information Retrieval. McGraw-Hill Inc, New York (1986)
15. Salton, G., Buckley, C.: Term-weighting approaches in automatic text retrieval. Inf. Process. Manage. **24**(5), 513–523 (1988)
16. Jones, K.S.: A statistical interpretation of term specificity and its application in retrieval. J. Doc. **28**, 11–21 (1972)
17. Mitchell, T.M.: Machine Learning, 1st edn. McGraw-Hill Science/Engineering/Math (1997)
18. McCallum, A., Nigam, K.: A comparison of event models for Naive Bayes text classification (1998)
19. Aha, D., Kibler, D.: Instance-based learning algorithms. Mach. Learn. **6**, 37–66 (1991)
20. Henderson, L.: Automated Text Classification in the DMOZ Hierarchy (2009)

Semi-unsupervised Machine Learning for Anomaly Detection in HTTP Traffic

Rafał Kozik, Michał Choraś, Rafał Renk and Witold Hołubowicz

Abstract Currently, the growing popularity of publicly available web services is one of the driving forces for so-called "web hacking" activities. The main contribution of this paper is the semi-unsupervised anomaly detection method for HTTP traffic anomaly detection. We made the assumption that during the learning phase (for the captured volume of HTTP traffic), only small friction of samples is labelled. Our experiments show that the proposed method allows us to achieve the ratios of true positive and false positive errors below 1 %.

1 Introduction

Hypertext Transfer Protocol (HTTP) is one of the most frequently used protocols of application layer in TCP/IP model. This is because the fact that nowadays the significant part of ICT solutions rely on web servers or web services. HTTP protocol has been proven to be reliable mean of communication between computers in distributed network. The growing popularity of publicly available web services is one of the driving forces for so-called "web hacking" activities. According to Symantec report [1], the number of web attacks blocked per day has increased by 23 %,

R. Kozik (✉) · M. Choraś · W. Hołubowicz
Institute of Telecommunications and Computer Science, UTP University of Science and Technology, Bydgoszcz, Poland
e-mail: rkozik@utp.edu.pl

M. Choraś
e-mail: mchoras@itti.com.pl

R. Kozik · M. Choraś · R. Renk
ITTI Ltd., Poznan, Poland

R. Renk · W. Hołubowicz
Adam Mickiewicz University, UAM, Poznan, Poland
e-mail: renk@amu.edu.pl

R. Burduk et al. (eds.), *Proceedings of the 9th International Conference on Computer Recognition Systems CORES 2015*, Advances in Intelligent Systems and Computing 403, DOI 10.1007/978-3-319-26227-7_72

compared to previous years. Moreover, the number and importance of web sites, web services and web applications are constantly growing. It can be noticed that more and more sophisticated applications provided over the HTTP(S) protocol are currently developed to address wide range of users needs including entertainment, business, social activities, etc. Another factor drawing the attention of hackers is the fact that HTTP is the most popular protocol allowed by Internet firewalls and since operating systems are constantly better and better protected; it is easier to focus on web-based applications that are more likely to have security exploits. Protection against attack targeting application layer can be difficult. According to [2], the cross-site scripting remains the number one in 2014. Authors of this report also stated that that kind of security exploits may allow the attacker to access sensitive data or overtake the site and users accounts. According to White Hat Security Report [2], it may take 180 days (in average) to have security vulnerability patched. Therefore, is is important to have effective and efficient tools to counter that kind of attack. In the market, there are plenty of solutions called WAF (Web Application Firewall). Some of them apply signatures (regular expressions, patterns, etc.) in order to detect an attack. The patterns (or rules) are typically matched against the content of the packet (e.g. packet header or payload). Commonly, the signatures (in form of reactive rules) of an attack for a software like Snort [3] are provided by experts form the cyber community. Typically, for deterministic attacks it is fairly easy to develop patterns that will clearly identify particular attack. However, the problem of developing new signatures becomes more complicated when it comes to to attacks patterns that can be easily obfuscated. This paper is structured as follows. First, we provide and overview of anomaly based methods for cybersecurity purposes. Afterwards, an overview of the proposed method is given. In following section, the experimental setup is described. Results and conclusions are given after.

2 Overview of Anomaly Based Solutions

The anomaly based methods for cyber attacks detection build a model that describes normal and abnormal behaviour of network traffic. Commonly, such methods use two types of algorithms from machine-learning theory, namely unsupervised and supervised approach. For unsupervised learning commonly, [4–10] clustering approaches are used that usually adapt algorithms like k-means, fuzzy c-means, QT and SVM. The clustered network traffic established using the mentioned approaches commonly requires decision whenever given cluster should be indicated as a malicious or not. Pure unsupervised algorithms use a majority rule telling that only the biggest clusters are considered normal. That means that network events that happen frequently have no symptoms of an attack. In practice, it is a human role to tell which cluster should be considered as abnormal one. The supervised machine-learning techniques require at least one phase of learning in order to establish the model traffic. The learning is typically done off-line and is conducted on specially prepared (cleaned) traffic traces. One of the exemplar approaches to supervised machine learning for cyber attack detection is the autoregression stochastic process (AR) [11–13]. In the literature, there are also methods using Kalman filters [14]. Recently, the solutions adapting SVM [15], neural networks [16] and ID3-established decision trees [17] are gaining popularity.

3 Method Overview

The proposed algorithm operates on a server side of the web application. As it is shown in Fig. 1, the HTTP traffic generated by client web browser is going trough proxy server and relayed to HTTP server and our anomaly detection (AD) algorithm. Each time the AD algorithm detects anomaly, the system administrator is notified. The HTTP is a text-based protocol. Therefore, the information is first encoded with feature vectors. The procedure is described in Sect. 3.2. Afterwards, we apply clustering method to associated similar HTTP requests with the same clusters.

3.1 Data Extraction

The proposed algorithm works on HTTP requests, which have structured textual form. The procedure of HTTP stream segmentation is shown in Fig. 2. First, the request is split into request method (e.g. GET, POST, PUT, etc.), URL, request arguments and payload. Second, each request is checked against the whitelist. After that, the content of the request is analysed (see "Validate structure and content" block in Fig. 2). During the analysis, we encode parameters using procedure presented in Sect. 3.2. Using the methodology presented in Sect. 3.3, we indicate whenever given sample is normal or anomalous.

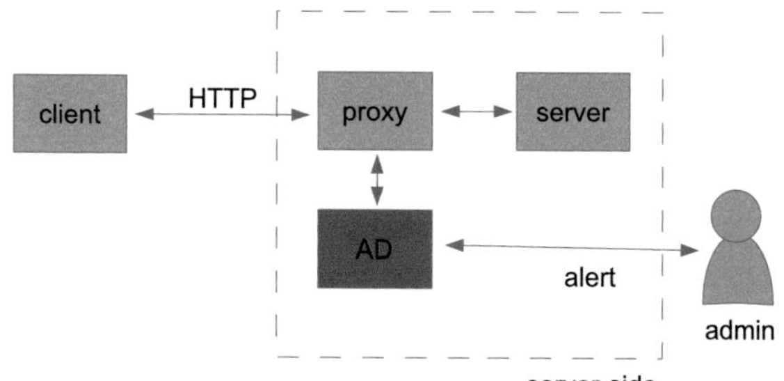

Fig. 1 Information flow of the proposed anomaly detection (AD) method

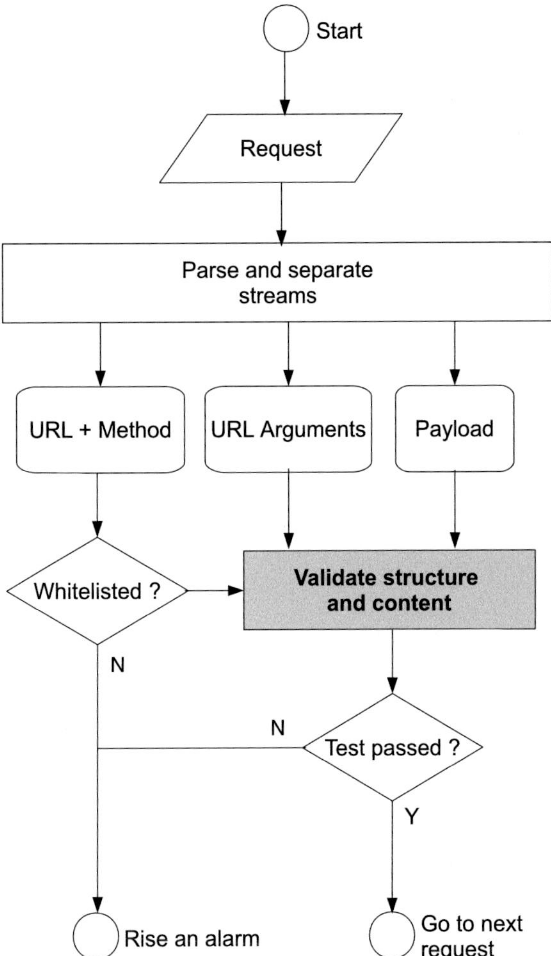

Fig. 2 HTTP stream segmentation procedure

3.2 *Parameters Encoding*

As parameters we understand HTTP request variables (present both in URL address and in request body) that have the following form: "param1=value1 & param2= value2". In this approach, we have used statistical features to encode above-mentioned HTTP parameters in order to obtain feature vectors. We count number of characters that the decimal value in an ASCII table belongs to following ranges: $< 0, 31 >, < 32, 47 >, < 48, 57 >, < 58, 64 >, < 65, 90 >, < 91, 96 >, < 97, 122 >, < 123, 127 >< 128, 255 >$. It may look heuristically, but different

Table 1 ASCII code ranges and corresponding characters

Group	Range	Characters
1	$< 0, 31 >$	Control characters
2	$< 32, 47 >$	SPACE!"#$%&'()*+,-./
3	$< 48, 57 >$	0123456789
4	$< 58, 64 >$	:;<=>?@
5	$< 65, 90 >$	ABCDEFGHIJKLMNOPQRSTUVWXYZ
6	$< 91, 96 >$	[\]^_'
7	$< 97, 122 >$	abcdefghijklmnopqrstuvwxyz
8	$< 123, 127 >$	{~}|
9	$< 128, 255 >$	Special characters

ranges represent different types of symbols like numbers, quotes, letters or special characters (see Table 1). In result, our histogram will have 9 bins. This encoding schema is used separately for each HTTP request parameter.

3.3 HTTP Traffic Model

As mentioned in Sect. 3.2, each of HTTP parameters value is encoded yielding set of feature vectors. This set is clustered using mixture of Gaussian function described by Eq. (1), where x indicates the feature vector, N number of clusters c, π probability that given feature vector belongs to cluster c, ϕ is the multivariate Gaussian (μ_g, Σ_g) and Ψ indicates the name of HTTP parameter.

$$f(x, \Psi) = \sum_{c=1}^{N} \pi_c \phi(x|\mu_g, \Sigma_g, \Psi) \tag{1}$$

The mixture of models is fitted using state-of-the-art expectation maximisation (EM) algorithm [18]. Each of the Gaussian function indicates a cluster in the analyzed dataset. In order to identify the clusters, we assume that the data used for learning is weakly labelled. For example, having some feature vectors indicating normal HTTP requests, we identify which cluster those belong to. In this manner, we assign the labels to clusters based on data samples we have. The procedure is shown in Fig. 3. The labels are obtained from additional tools (e.g. firewalls) or using expert knowledge. In this approach, we used (on average) 10 labelled samples (which is less than 0.5 % of all data samples). Whenever the new HTTP request is being tested against the attacks, first the request parameters are encoded in order to extract feature vectors. Then, the feature vectors are assigned to the closest cluster. The request is flagged as anomaly whenever it belongs to unknown or anomalous cluster.

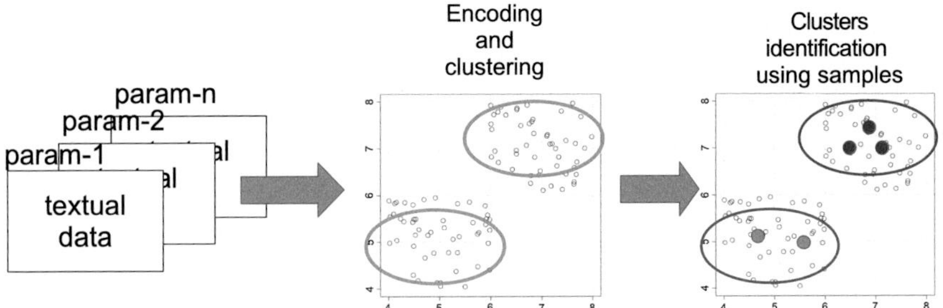

Fig. 3 Procedure for building HTTP traffic model. The cluster is identified using known HTTP traffic samples (either normal or anomalous)

4 Experimental Setup

In this section, our evaluation methodology is described. In the experiments, we have considered situation where the network administrator has huge volume of unlabelled network traces (in our case HTTP server logs) and only limited number of them are labelled. Several approaches to gather such labelled data samples are possible. For instance, data gathered during system tests (unit, deployment, integration, etc.) can be used. Moreover (as described in previous section), it is possible to use tools like sqlmap [19] or ZAP [20] tools to generate additional traces. Other approach to obtain some strong examples of attacks is to use log analysis tools like Apache Scalp [21]. In our experiment, attacks are conducted on php-based web service with the state-of-the-art tools for services penetration and SQL injection. The traffic generated by attacking tools are combined together with normal traffic (genuine queries) in order to estimate the effectiveness of the proposed methods. The web service used for penetration test is so-called LAMP (Linux + Apache + MySQL + PHP) server with MySQL back-end. It is one of the most commonly used servers, and therefore it was used for validation purposes. The server was deployed on Linux Ubuntu operation system. For penetration tests examples, services developed in PHP scripts and shipped by default with the server are validated. Attack injection methodology is based on the known SQL injection methods, namely: Boolean-based blind, time-based blind, error-based, UNION query and stacked queries. For that purpose sqlmap tool is used. It is an open source penetration and testing tool that allows the user to automate the process of validating the tested services against the SQL injection flaws. In order to avoid double-counting the same attack patterns during the evaluation process, we decided to gather first the malicious SQL queries generated by sqlmap and ZAP (several hundreds of different injection trials). After that genuine traffic (generated by crawlers and during the normal web service usage) is gathered. Such prepared data is used during the evaluation test the results are presented in Sect. 5.

5 Results

In order to compare different clustering methods, we have used above-described approach to obtain the labelled data. Our purpose was not to use that data to learn an algorithm. In our first experiments, we have compared following algorithms: k-means, fuzzy c-means, Bayesian Hierarchical Clustering and Gaussian Mixture Models. Typically, an administrator is not able to tell in advance that given cluster obtained with particular algorithm is normal or anomalous. Therefore, in this experiment we assumed the majority rule. An cluster is considered normal if certain amount of data samples is assigned to it. In Table 2, the percentage distribution of data samples among clusters extracted with different algorithms is presented. It can be noticed that for GMM algorithm there is a small number of clusters that contain both normal and anomalous samples. For instance, cluster 4 contain 73 % of all normal samples while having 4.9 % of anomalous. The second experiment was aimed at evaluating the effectiveness of attack detection in our testbed. We used tenfold cross-validation approach to obtain results presented in Table 3. As we have expected, the distribution of normal an anomalous data samples between different clusters presented in Table 2 is correlated with the effectiveness shown in Table 3. It can be assumed that due to the nature of HTTP traffic (the distribution of samples) transformed to proposed feature space, makes the GMM method most suitable for that problem.

Table 2 Percentage distribution of data samples among clusters extracted with different algorithms

	k-means		c-means		b-clustering		GMM	
cluster	N	A	N	A	N	A	N	A
1	32,4%	70,7%	11,2%	14,3%	11,2%	14,3%	0,0%	21,8%
2	0,0%	8,8%	2,4%	7,2%	2,4%	7,2%	20,8%	3,3%
3	11,3%	0,0%	30,0%	73,9%	30,0%	73,9%	6,2%	0,0%
4	0,0%	11,1%	45,1%	0,0%	45,1%	0,0%	73,0%	4,9%
5	0,0%	9,4%	0,1%	3,6%	0,1%	3,6%	0,0%	37,5%
6	56,3%	0,0%	11,3%	1,0%	11,3%	1,0%	0,0%	32,6%

Table 3 Effectiveness of the proposed method evaluated for different clustering techniques (FP-false positives, FN-false negatives, CE-classification error)

	FP [%]	FN [%]	CE [%]
GMM	0.71 ± 0.28	0.23 ± 0.1	0.63 ± 0.23
c-means	30.17 ± 3.20	3.82 ± 1.31	26.03 ± 7.30
k-means	9.21 ± 0.22	0.1 ± 0.33	7.79 ± 0.22
b-clustering	0.73 ± 0.30	0.39 ± 0.1	0.67 ± 0.31

6 Conclusions

In this paper, an semi-unsupervised anomaly detection method for HTTP traffic anomaly detection is proposed. We have evaluated our method against effectiveness of web application cyberattacks detection. In our experiments, we have assumed the problem of weakly labelled data. Obtained results prove that the proposed method can be considered as valuable information source for network administrator.

Acknowledgments This work was partially supported by Applied Research Programme (PBS) of the National Centre for Research and Development (NCBR) funds allocated for the Research Project number PBS1/A3/14/2012 (SECOR).

References

1. Symantec, Internet Security Threat Report, vol. 19. http://www.symantec.com/security_response/publications/threatreport.jsp (2014)
2. WhiteHat Website Security Statistics Report. https://www.whitehatsec.com/resource/stats.html
3. SNORT project homepage. http://www.snort.org/
4. Sharma, M., Toshniwal, D.: Pre-clustering algorithm for anomaly detection and clustering that uses variable size buckets. In: 1st International Conference on Recent Advances in Information Technology (RAIT), pp. 515–519, 15–17 March 2012
5. Adaniya, M.H.A.C., Lima, M.F., Rodrigues, J.J.P.C., Abrao, T., Proenca, M.L.: Anomaly detection using DSNS and firefly harmonic clustering algorithm. In: IEEE International Conference on Communications (ICC), pp. 1183–1187, 10–15 June 2012
6. Mazel, J., Casas, P., Labit, Y., Owezarski, P.: Sub-space clustering, inter-clustering results association and anomaly correlation for unsupervised network anomaly detection. In: 7th International Conference on Network and Service Management (CNSM), pp. 1–8, 24–28 Oct 2011
7. Yang, C., Deng, F., Yang, H.: An unsupervised anomaly detection approach using subtractive clustering and hidden markov model. In: Second International Conference on Communications and Networking in China. CHINACOM'07, pp. 313–316, 22–24 Aug 2007
8. Liang, H., Wei-wu, R., Fei, R.: An adaptive anomaly detection based on hierarchical clustering. In: 1st International Conference on Information Science and Engineering (ICISE), pp. 1626–1629, 26–28 Dec 2009
9. Pons, P., Latapy, M.: Computing communities in large networks using random walks. J. Graph Algorithms Appl. **10**(2), 191–218 (2006)
10. Liao, Q., Blaich, A., Van Bruggen, D., Striegel, A.: Managing networks through context: graph visualization and exploration. Comput. Netw. **54**, 2809–2824 (2010)
11. Ricciato, F., Fleischer, W.: Bottleneck detection via aggregate rate analysis: a real casein a 3G network. In: Proceedings of the IEEE/IFIP NOMS (2004)
12. Thottan, M., Ji, C.: Anomaly detection in IP networks. IEEE Trans. Signal Process. [Special Issue of Signal Processing in Networking], **51**(8): 2191–2204 (2003)
13. Rish, I., Brodie, M., Sheng, M., Odintsova, N., Beygelzimer, A., Grabarnik, G., Hernandez, K.: Adaptive diagnosis in distributed systems. IEEE Trans. Neural Netw. **16**(5), 1088–1109 (2005)
14. Soule, A., Salamatian, K., Taft, N.: Combining filtering and statistical methods for anomaly detection. In: Proceedings of IMC Workshop (2005)
15. Ma, J., Dai, G., Xu, Z.: Network anomaly detection using dissimilarity-based one-class SVM classifier. In: International Conference on Parallel Processing Workshops. ICPPW'09, pp. 409–414, 22–25 Sept 2009

16. Ma, R., Liu, Y., Lin, X., Wang, Z.: Network anomaly detection using RBF neural network with hybrid QPSO. In: IEEE International Conference on Networking, Sensing and Control. ICNSC, pp. 1284–1287, 6–8 April 2008
17. Gaddam, S.R., Phoha, V.V., Balagani, K.S.: K-Means+ID3: a novel method for supervised anomaly detection by cascading K-Means clustering and ID3 decision tree learning methods. IEEE Trans. Knowl. Data Eng. **19**(3), 345–354 (2007)
18. Fraley, C., Raftery, A.E.: Model-based clustering, discriminant analysis, and density estimation. J. Am. Stat. Assoc. **97**, 611–631 (2002)
19. Automatic SQL injection and database takeover tool. http://sqlmap.org/
20. OWASP Zed Attack Proxy Project. https://www.owasp.org/index.php/OWASP_Zed_Attack_Proxy_Project
21. Apache log analyzer for security. https://code.google.com/p/apache-scalp/

Sentiment Classification of the Slovenian News Texts

Jože Bučar, Janez Povh and Martin Žnidaršič

Abstract This paper deals with automatic two class document-level sentiment classification. We retrieved textual documents with political, business, economic and financial content from five Slovenian web media. By annotating a sample of 10,427 documents, we obtained a labelled corpus in the Slovenian language. Five classifiers were evaluated on this corpus: multinomial naïve Bayes, support vector machines, random forest, k-nearest neighbour and naïve Bayes, out of which the first three were used also in the assessment of the pre-processing options. Among the selected classifiers, multinomial naïve Bayes outperforms the naïve Bayes, k-nearest neighbour, random forest and support vector machines classifier in terms of classification accuracy. The best selection of pre-processing options achieves more than 95 % classification accuracy with Naïve Bayes Multinomial and more than 85 % with support vector machines and random forest classifier.

Keywords Sentiment analysis · Document classification · Machine learning · Slovenian language · Corpus

J. Bučar (✉) · J. Povh
Faculty of Information Studies, Laboratory of Data Technologies, Ulica talcev 3, 8000 Novo mesto, Slovenia
e-mail: joze.bucar@fis.unm.si
URL: http://datalab.fis.unm.si

J. Povh
e-mail: janez.povh@fis.unm.si

M. Žnidaršič
Department of Knowledge Technologies, Jožef Stefan Institute, Jamova cesta 39, 1000 Ljubljana, Slovenia
e-mail: martin.znidarsic@ijs.si
URL: http://kt.ijs.si

R. Burduk et al. (eds.), *Proceedings of the 9th International Conference on Computer Recognition Systems CORES 2015*, Advances in Intelligent Systems and Computing 403, DOI 10.1007/978-3-319-26227-7_73

1 Introduction

The unprecedented growth of the Web in the past two decades has created an entirely new way to retrieve and share information. Between 2000 and 2014, the number of web users increased more than seven times and already surpassed three million in June 2014, which represents more than 42 % of the world's population [10]. Web is the largest publicly available data source in the world and there is not enough human resources to inspect this data. People strive to detect and obtain relevant information from this chaotic cluster of data, with a hope to better understand our world and the quality of our lives. An increasing number of blogs, newsgroups, forums, chat rooms and social networks attracts web users to share their feelings and ideas about products, services, events, etc. The popularity of social media has escalated the interest in sentiment analysis [23]. Sentiment analysis, also known as opinion mining, is an emerging area of research for efficient analysis of informal, subjective, opinionated web content in source materials by applying natural language processing (NLP), computational linguistics and text analytics [17]. Sentiment classification is a supervised learning problem with usually three classes: negative, neutral and positive opinion or sentiment. Its commercial use can be noticed when

- analysing consumer habits, behaviour, trends, competitors and market buzz,
- carrying out quality control to prevent negative viral effects,
- estimating and evaluating responses to company-related events and incidents in multiple languages,
- collecting and extracting user opinions of products and services.

When classifying documents, we deal with: document representation, feature selection and document modelling [9]. In the first two phases, we choose a feature set to represent a document. The whole content of a document is then represented with a document feature vector. In the last phase, we build a document model where we use a selection of features. In our research, we explore which of the existing and frequently used classifiers based on our data set achieves the best results in terms of time consumption and performance. Also, we study the impact of various pre-processing options on classification performance. The rest of the paper is organised as follows: Sect. 2 briefly introduces the background and related work. In Sect. 3, we describe a proposed approach and implemented solutions. Section 4 presents the evaluation of our experimental results. Finally, we discuss the obtained results and present our prospects in section conclusions and future work.

2 Background and Related Work

The first studies in sentiment document classification were made around the year 2000; even a bit earlier. They were mainly focused on financial news [5, 8], movie [18] and product reviews [21], especially on the popular internet movie database

(IMDb) and product reviews downloaded from Amazon [6]. McCallum and Nigam performed a comparison of event models for Naïve Bayes (NB) text classification and different vocabulary sizes on the newsgroups data set [15]. The Naïve Bayes Multinomial (NBM) achieved 86 % accuracy. Pang et al. [18] compared the performance of various classifiers when determining the sentiment of a document and found out that support vector machines (SVM) yielded the best results in most cases. When selecting features they tried unigrams, bigrams, part-of-speech (POS) tags and term positions; however, they produced the best results using unigrams alone (82.9 % accuracy). Godbole et al. [7] worked on sources like newspapers and blog posts at the level of words and achieved accuracy 82.7–95.7 %, while others, who worked on documents, such as blog or twitter posts [19], and full web pages, have in general accuracy of around 65–85%. Despite the fact that we may be able to build comprehensive lexicons of sentiment-annotated words, there is still an issue how to detect it in a given text correctly. Data scientists use several tools and methods to determine and classify the sentiment of digital text automatically. Sentiment classification has been studied by numerous researchers subsequently and might be the most widely studied problem in the field of sentiment analysis (see a survey in [16]). Most techniques apply supervised learning where a bag of individual words (unigrams) is the most commonly used documents representation. Large set of features have been tried by researchers such as terms frequency (TF), term frequency–inverse document frequency (TF-IDF) weighting schemes, POS tags, opinion words and phrases, negations, syntactic dependency [14]. In the 1980s an article was published which found out that the average amount of negative news on ABC, CBS and NBC was 46.8 % [20]. Since then the proportion of negative news has increased in most media. Negative news is often cheap and easy to produce; moreover, it makes profit to the media. Some media are obligated to regulate the proportion of positive and negative news. More than 55 % of all the web pages, whose content language is known, are in English. The Slovenian language is with 0.1 % on the 35th place among the most common languages on the Web [22]. This is the main reason for the lack of research on sentiment classification methods suited to the Slovenian language.

3 Methodology

We retrieved 198, 186 textual documents such as news articles from the digital archive of five different Slovenian web media (Žurnal24, Rtvslo, 24ur, Dnevnik and Finance). All retrieved documents were enriched with political, business, economic and finanical content and were published between 1st September 2007 and 31st December 2013. Initially, we removed spelling mistakes in textual content. This was followed by the annotation process where six annotators manually annotated a random sample of 10,427 documents independently (approximately 2,000 documents per web media). All annotators are native speakers and were told to specify sentiment from the perspective of an average Slovenian web user. We used a five-level Likert scale [13], in which a sentiment was given a number from 1 to 5 (1—very negative,

2—negative, 3—neutral, 4—positive and 5—very positive). Annotation of documents was carried out on three levels independently, i.e. document-level, paragraph-level and sentence-level. To evaluate the process of annotation, Spearman correlation and Cronbach's alpha [4] were calculated. This labelled corpus was used as a training set to train, test and evaluate the classification techniques.

3.1 Sentiment Analysis Algorithm Selection

The classification of texts is one of the key tasks in text mining. The automation of procedures for the purpose of classification of texts has thus become an important activity, which has contributed to more efficient work. Data miners use a variety of tools and a wide range of learning algorithms [21] to tackle this problem. We carried out a classification and performance assessment of machine learning algorithms by applying the stratified tenfold cross-validation method. In order to efficiently predict the category within two class (negative and positive) document-level sentiment classification of 5002 documents, we chose to test the performance of the following classification algorithms: NB [12], NBM [15], SVM [3], k-nearest neighbour (KNN) [1] and random forest (RF) [2]. The performance of selected classifiers was measured using the accuracy:

$$Accuracy = \frac{Number\ of\ Correctly\ Classified\ Examples}{Number\ of\ All\ Examples} \tag{1}$$

3.2 Pre-processing Options

Data pre-processing is an important step in sentiment analysis. We apply standard text processing techniques that include text tokenization (splitting text into individual words/terms), upper-case letters replacement with lower-case, stop word removal (removing words that do not hold relevant information) and n-gram construction (concatenating 1 to n words appearing consecutively in a document). Various pre-processing options that were used in our experiment:

- Terms with minimum frequency 2 (to eliminate terms that appear only once),
- TF or TF-IDF weighting scheme,
- Optional: replace upper-case letters with lower-case,
- Optional: stop words removal (stop words list contains nearly 1800 words),
- N-gram tokenizer (unigrams, bigrams, trigrams and combinations).

4 Experiments and Evaluation

In the experiment six annotators labelled 10,427 documents independently, which took almost a whole year. We calculated Cronbach's alpha and Spearman coefficients (see Table 1) to describe internal consistency of annotation process. Cronbach's alpha (0.87) indicates an excellent internal consistency. Spearman coeficients also show good correlations of annotation. The maximum value of the Spearman coefficients (0.74) was found between annotator 4 and 6, while the minimum was found between annotator 3 and 5.

4.1 *Corpora*

On the basis of averaging annotations, we labelled 5,425 (52 %) documents as neutral, 3,337 (32 %) as negative and 1,667 (16 %) documents as positive. A negative sentiment was assigned to an article if its average score was less than or equal to 2.4, neutral if its average score was greater than 2.4 and less than 3.6, and positive sentiment if its average score was greater than or equal to 3.6. We can notice that 24ur publishes the biggest proportion of negative news per medium, while Finance publishes the most positive news articles. Interestingly, the proportion of negative news is twice as large as positive in all web media, with the exception of Finance, which seems to have a more balanced content. We generated two corpora, the first containing labelled documents as either negative or positive, while the other contains neutral documents as well. Table 2 shows statistic information about corpora (Fig. 1).

Table 1 Spearman coefficients between annotators

	Ann1	Ann2	Ann3	Ann4	Ann5	Ann6
Ann1	1	0.72	0.58	0.65	0.58	0.70
Ann2	0.72	1	0.57	0.62	0.60	0.69
Ann3	0.58	0.57	1	0.55	0.54	0.62
Ann4	0.65	0.62	0.55	1	0.61	0.74
Ann5	0.58	0.60	0.54	0.61	1	0.66
Ann6	0.70	0.69	0.62	0.74	0.66	1

Table 2 Corpora statistic information

Corpus	Neg & pos	Neg & neu & pos
# instances	5,002	10,427
# words	1,486,430	3,142,877
# unique words	94,770	132,658

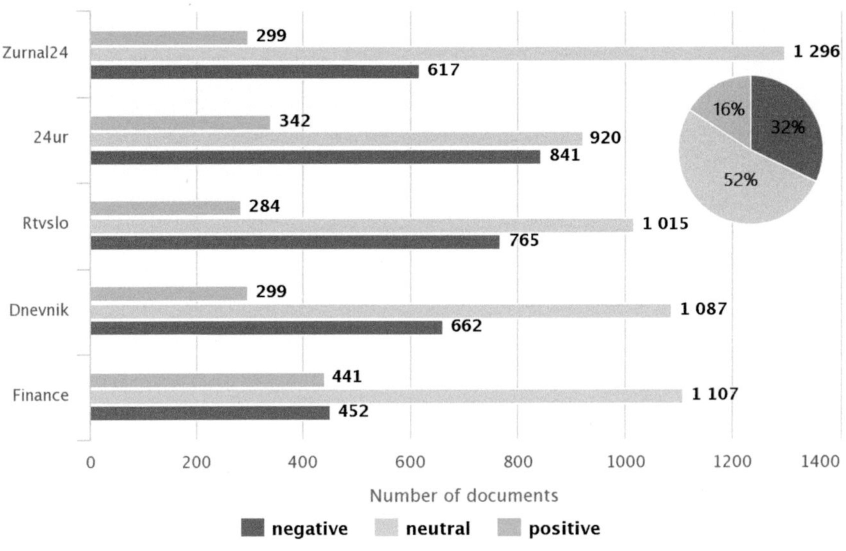

Fig. 1 Sentiment proportion per web media

4.2 *Document Classification Using Various Pre-processing Options*

In Sect. 3.2, we discussed several pre-processing options. In the first experiment, we applied standard text pre-processing (text tokenization, stop word removal, upper-case to lower-case letters transformation, unigram, bigram and trigram construction) on corpus with negative and positive documents (5002 instances). The resulting terms were used as features in the construction of feature vectors representing the documents, where feature vector construction was based on the TF-IDF feature weighting scheme. We also added the condition that a given term has to appear at least twice in the entire corpus. We evaluated performance using five classifiers (NB, NBM, SVM, KNN and RF) by applying the stratified tenfold cross-validation method (Table 3). In these experiments, we achieved the accuracies of 91.84 % for NBM, 86.71 % for SVM, 85.19 % for random forest with 100 trees (RF-100), 77.57 % for NB and 63.35 % for k-nearest neighbour with k = 10 neighbours (KNN-10) on the test set. We also tuned two parameters:

- number of nearest neighbours at k-NN, where we applied k = 1 (KNN-1), k = 5 (KNN-5) and k = 10 (KNN-10) nearest neighbours,
- number of trees at Random Forest classifier, where we applied 10 (RF-10), 50 (RF-50) and 100 trees (RF-100) (Table 3).

Table 3 Evaluation performance and time taken to train and test models for various classifiers by applying the stratified tenfold cross-validation method

Classifier	NB	NBM	SVM	KNN-1	KNN-5	KNN-10	RF-10	RF-50	RF-100
Accuracy (%)	77.57	91.84	86.71	60.24	61.38	63.35	79.69	84.59	85.19
Time (s)	6.55	0.01	28.16	4.04	4.21	4.34	8.62	43.88	86.41

Our results show that the NBM classifier outperforms other classifiers significantly from the perspective of the classification accuracy and time taken to train and test model. SVM and RF-100 produce satisfactory results, while NB and KNN perform much worse. We also showed that tuning the number of nearest neighbours at k-nearest neighbour classifier and number of trees at random forest classifier affects the evaluation performance. The next experimental study, combination of the discussed pre-processing options, is the best one in terms of the classification accuracy. Forty combinations of pre-processing options were tested using three classifiers (NBM, SVM and RF-100) on corpus with negative and positive documents (5002 instances). We applied the stratified tenfold cross-validation method. In each iteration of cross-validation, we trained classifiers on the same training partition of the data and then evaluated them on the same test partition of the data. Special care was taken not to observe the testing partition even in the pre-processing processes; thus the pre-processing is always conducted only on the learning partition of the data in each iteration of the stratified tenfold cross-validation method. Comparisons between the applied algorithms were performed using t-paired tests with significant level set at 95 %.

We present accuracies for all the tested pre-processing options and three classifiers (NBM, SVM and RF-100) in Table 4. The best accuracy is obtained by NBM when using bigrams and TF-IDF scheme. Results of NBM classifier confirm that standard TF-IDF approach to feature vector construction performs better than TF weighting scheme [11]. Moreover, using bigrams and the combinations of unigrams, bigrams and trigrams return best results. SVM performs best when applying the combinations of unigrams, bigrams and trigrams together. It can be noticed that stop words removal does not contribute to a better performance as well as transforming upper-case letters to lower-case. RF-100 performs best when applying TF-IDF scheme, removing stop words, transforming upper-case to lower-case letters and constructing unigrams, bigrams and trigrams. Relative to each pre-processing option and classifier, + (−) sign in the first column means that the accuracy of this classifier is significantly better (worse) than NBM. However, these results show, once more, that NBM is better overall.

Table 4 Evaluation performance (accuracy and standard deviation) for NBM, SVM and RF-100 using various pre-precessing options by applying the stratified tenfold cross-validation method

ID	Pre-processing options				Avg. acc. ± std. dev. (%)				
	TF and TF-IDF	Lower-case	Stop words	Ngrams (1, 2, 3)	NBM		SVM		RF-100
1	TF			1	89.08 ± 1.26	–	84.87 ± 1.07	–	82.17 ± 1.84
2	TF-IDF			1	91.40 ± 1.62	–	84.87 ± 1.07	–	81.93 ± 1.45
3	TF			2	92.06 ± 1.04	–	87.09 ± 1.50	–	80.23 ± 1.20
4	TF-IDF			2	**95.10 ± 0.92**	–	87.09 ± 1.50	–	80.05 ± 1.19
5	TF			3	89.44 ± 1.00	–	86.63 ± 1.68	–	80.51 ± 1.62
6	TF-IDF			3	93.82 ± 0.91	–	86.63 ± 1.67	–	80.33 ± 1.55
7	TF			1 + 2	89.18 ± 0.95	–	85.84 ± 1.72	–	82.91 ± 1.42
8	TF-IDF			1 + 2	91.78 ± 1.06	–	85.84 ± 1.72	–	82.39 ± 1.15
9	TF			1 + 2 + 3	89.40 ± 0.59	–	86.19 ± 1.94	–	83.19 ± 1.50
10	TF-IDF			1 + 2 + 3	91.98 ± 0.67	–	86.19 ± 1.94	–	83.17 ± 1.47
11	TF	x		1	89.34 ± 1.64	–	85.17 ± 1.69	–	82.51 ± 2.10
12	TF-IDF	x		1	91.12 ± 1.23	–	85.17 ± 1.69	–	82.45 ± 1.52
13	TF	x		2	92.08 ± 0.84	–	86.33 ± 1.14	–	80.97 ± 1.57
14	TF-IDF	x		2	94.60 ± 0.98	–	86.33 ± 1.14	–	80.57 ± 1.02
15	TF	x		3	89.34 ± 0.82	–	**87.54 ± 0.87**	–	81.03 ± 1.47
16	TF-IDF	x		3	94.10 ± 0.85	–	**87.54 ± 0.87**	–	81.07 ± 1.37
17	TF	x		1 + 2	89.24 ± 1.11	–	86.53 ± 1.41	–	83.49 ± 1.28
18	TF-IDF	x		1 + 2	91.84 ± 0.93	–	86.53 ± 1.41	–	82.87 ± 1.53
19	TF	x		1 + 2 + 3	88.96 ± 0.87	–	86.54 ± 1.65	–	83.55 ± 1.69
20	TF-IDF	x		1 + 2 + 3	91.54 ± 0.85	–	86.54 ± 1.65	–	83.59 ± 2.13
21	TF		x	1	89.06 ± 1.47	–	85.01 ± 1.52	–	84.73 ± 1.67
22	TF-IDF		x	1	91.36 ± 1.65	–	85.01 ± 1.52	–	84.99 ± 1.82

(continued)

Table 4 (continued)

ID	Pre-processing options				Avg. acc. ± std. dev. (%)				
	TF and TF-IDF	Lower-case	Stop words	Ngrams (1, 2, 3)	NBM		SVM		RF-100
23	TF		x	2	92.06 ± 1.04	–	87.09 ± 1.50	–	80.23 ± 1.20
24	TF-IDF		x	2	**95.10 ± 0.92**	–	87.09 ± 1.50	–	80.05 ± 1.19
25	TF		x	3	89.44 ± 1.00	–	86.63 ± 1.67	–	80.51 ± 1.62
26	TF-IDF		x	3	93.82 ± 0.91	–	86.63 ± 1.67	–	80.33 ± 1.55
27	TF		x	1 + 2	89.78 ± 0.95	–	85.77 ± 2.11	–	84.55 ± 1.50
28	TF-IDF		x	1 + 2	92.22 ± 1.02	–	85.77 ± 2.11	–	84.41 ± 1.81
29	TF		x	1 + 2 + 3	89.88 ± 0.56	–	86.68 ± 1.91	–	84.59 ± 1.86
30	TF-IDF		x	1 + 2 + 3	92.26 ± 0.55	–	86.68 ± 1.91	–	84.87 ± 1.64
31	TF	x	x	1	89.22 ± 1.41	–	85.25 ± 1.47	–	84.55 ± 1.43
32	TF-IDF	x	x	1	91.50 ± 1.40	–	85.25 ± 1.48	–	84.93 ± 1.65
33	TF	x	x	2	92.08 ± 0.84	–	86.33 ± 1.14	–	80.97 ± 1.57
34	TF-IDF	x	x	2	94.60 ± 0.98	–	86.33 ± 1.14	–	80.57 ± 1.02
35	TF	x	x	3	89.34 ± 0.82	–	**87.54 ± 0.87**	–	81.03 ± 1.47
36	TF-IDF	x	x	3	94.10 ± 0.85	–	**87.54 ± 0.87**	–	81.07 ± 1.37
37	TF	x	x	1 + 2	90.02 ± 0.92	–	86.73 ± 1.77	–	84.73 ± 1.52
38	TF-IDF	x	x	1 + 2	92.04 ± 0.89	–	86.73 ± 1.77	–	84.69 ± 1.67
39	TF	x	x	1 + 2 + 3	89.50 ± 1.20	–	86.71 ± 1.79	–	84.39 ± 1.76
40	TF-IDF	x	x	1 + 2 + 3	91.84 ± 0.87	–	86.71 ± 1.79	–	**85.19 ± 1.13**

5 Conclusions and Future Work

In this paper, we introduced an automatic two class (negative and positive) document-level sentiment classification on the obtained corpus of 5,002 textual documents with political, business, economic and financial content published between 1st September 2007 and 31st December 2013 in five Slovenian web media. This corpus is useful for both academia and industry; however, it will be publicly available under CC BY license for further research. We evaluated five classifiers, especially NBM, SVM and RF-100 which were evaluated using various pre-processing options. We also showed that the NBM classifier outperforms other classifiers significantly in terms of the classification accuracy as well as time that is needed to train and test a model. Considering two class document-level sentiment classification, we are also interested in three class (negative, neutral and positive) document-level sentiment classification. The annotation of documents was carried out on three levels (document, paragraph and sentence) independently; therefore we are eager to try whether fragmentation on paragraph and sentence-level can contribute to achieve a better performance. The next perspective is to study negation and how we can detect and incorporate it into the set of features. In the future, we will explore the use of feature selection, and we will apply other feature selection methods to create a sentiment lexicon for the Slovenian language.

Acknowledgments Work supported by Creative Core FISNM-3330-13-500033 'Simulations' project funded by the European Union, The European Regional Development Fund and Young Researcher Programme by Slovenian Research Agency. The operation is carried out within the framework of the Operational Programme for Strengthening Regional Development Potentials for the period 2007–2013, Development Priority 1: Competitiveness and research excellence, Priority Guideline 1.1: Improving the competitive skills and research excellence.

References

1. Aha, D.W., Kibler, D., Albert, M.A.: Instance-based learning algorithms. Mach. Learn. **6**, 37–66 (1991)
2. Breiman, L.: Random forests. Mach. Learn. **45**, 5–32 (2001)
3. Cortes, C., Vapnik, V.: Support vector networks. Mach. Learn. **20**, 273–297 (1995)
4. Cronbach, L.J.: Coefficient alpha and the internal structure of tests. Psychometrika **16**, 297–334 (1951)
5. Das, S.R., Chen, M.Y.: Yahoo! for amazon: Extracting market sentiment from stock message boards. In: Proceedings of the Asia Pacific Finance Association Annual Conference (APFA) (2001)
6. Dave, K., Lawrence, S., Pennock, D.M.: Mining the peanut gallery: Opinion extraction and semantic classification of product reviews. In: Proceedings of the World Wide Web Conference (2003)
7. Godbole, N., Srinivasaiah, M., Skiena, S.: Large-scale sentiment analysis for news and blogs. Proc. Int. Conf. Weblogs Soc. Media **2**, 1–4 (2007)
8. Hatzivassiloglou, V., McKeown, K.R.: Predicting the semantic orientation of adjectives. In: Proceedings of the 8th Conference on European Chapter of the Association for Computational Linguistics, pp. 174–181 (1997)

9. Hrala, M., Král, P.: Evaluation of the document classification approaches. In: Proceedings of the 8th International Conference on Computer Recognition Systems CORES, pp. 877–885 (2013)
10. Internet World Stats, World Internet Users and 2014 Population Stats (2014), http://www.internetworldstats.com/stats.htm. Accessed 10 Mar 2015
11. Joachims, T.: Text categorization with support vector machines: Learning with many relevant features. In: Proceedings of the European Conference on Machine Learning, pp. 137–142 (1998)
12. Lewis, D.D.: Naïe (Bayes) at forty: the independent assumption in information retrieval. Mach. Learn.: ECML **98**, 4–15 (1998)
13. Likert, R.: A technique for the measurement of attitudes. Arch. Psychol. **22**, 1–55 (1932)
14. Liu, B.: Web Data Mining: Exploring Hyperlinks, Contents, and Usage Data, pp. 469–492. Springer, New York (2011)
15. McCallum, A., Nigam, K.: A comparison of event models for Naïve Bayes text classification. In: AAAI-98 Workshop on Learning for Text Categorization, vol. 752, pp. 41–48 (1998)
16. Paliouras, G., Papatheodorou, C., Karkaletsis, V., Spyropoulos, C.: Discovering user communities on the internet using unsupervised machine learning techniques. Interact. Comput. **14**, 761–791 (2002)
17. Pang, B., Lee, L.: Opinion mining and sentiment analysis. Found. Trends Inf. Retr. **2**, 1–135 (2008)
18. Pang, B., Lee, L., Vaithyanathan, S.: Thumbs up?: Sentiment classification using machine learning techniques. In: Proceedings of the ACL-02 Conference on Empirical Methods in Natural Language Processing, pp. 79–86 (2002)
19. Smailović, J., Grčar, M., Lavrač, N., Žnidaršič, M.: Predictive sentiment analysis of tweets: A stock market application. In: Human-Computer Interaction and Knowledge Discovery in Complex, Unstructured, Big Data, pp. 77–88 (2013)
20. Stone, G.C., Grusin, E.: Network TV as the Bad News Bearer. Journal. Q. **61**, 517 (1984)
21. Turney, P.D.: Thumbs up or thumbs down? Semantic orientation applied to unsupervised classification of reviews. In: Proceedings of the 40th Annual Meeting on Association for Computational Linguistics, Association for Computational Linguistics, pp. 417–424 (2002)
22. Web Technology Surveys Usage of content languages for websites (2011), http://w3techs.com/technologies/overview/content_language/all. Accessed 08 Mar 2015
23. Wright, A.: Mining the Web for Feelings, Not Facts. New York Times 24 (2009)

A Snoring Sound Analysis Application Using K-Mean Clustering Method on Mobile Devices

Thakerng Wongsirichot, Nantanat Iad-ua and Jutatip Wibulkit

Abstract Patients with chronic diseases are increasing around the globe. Healthcare professionals attempt to find possible causes of the chronic diseases. One of the most possible causes is the sleep disorder. Sleep apnea, OSA and CSA, may be an evidence of chronic diseases. In order to detect the sleep apnea, the polysomnography (PSG) or the sleep test is required for patients. A number of parameters will be collected on patients whilst they are asleep. However, due to the limitation of the PSG test in some countries, researchers attempt to find other available alternative approaches. In this research work, a mobile application has been constructed to perform a screening test of OSA. With our initial experiment test, 74.70 % instances have been correctly classified. An application of SMOTE into a minority class is performed and achieves up to 80.10 % correctly classified instances. Limitations of the mobile application and our technique have also discussed.

Keywords Snoring sound · PSG · Sleep apnea

1 Introduction

A number of patients with chronic diseases such as hypertension and cardiovascular diseases are increasing around the world. The chronic diseases are caused by various factors such as lack of exercises, genetic inheritances and mutations, prolonged use of some medications, risk behaviours, etc. [1]. Conditions of patients with chronic

T. Wongsirichot (✉) · N. Iad-ua · J. Wibulkit
The Innovative Information Technology for Health Science and Society Research Unit (INTACT), Information and Communication Technology Programme, Faculty of Science, Prince of Songkla University, Songkhla 90112, Thailand
e-mail: thakerng.w@psu.ac.th

N. Iad-ua
e-mail: mate_narak@hotmail.com

J. Wibulkit
e-mail: tip-love-04@hotmail.com

R. Burduk et al. (eds.), *Proceedings of the 9th International Conference on Computer Recognition Systems CORES 2015*, Advances in Intelligent Systems and Computing 403, DOI 10.1007/978-3-319-26227-7_74

diseases are steadily degrading proper medications or treatments are not provided. In order to discover a proper medication or treatment, deepen diagnostic techniques are possibly required. With advancements of medical diagnostic tools and technologies, a number of chronic diseases and symptoms are deeply studied in order to accomplish their causes. Snoring is one of the initial symptoms that possibly provokes chronic diseases. Since 2002, there are a number of researchers interested in studies of snoring that related to health issues. An initial report shows approximately 20–40 % of adults snore whilst they are asleep. However, not all of the snorers have or will have chronic diseases. Snoring with apnea, a suspension of external breathing, is a key determination of current and future chronic diseases in patients [2].

2 Sleep Apnea

Sleep apnea (SA) is one of the sleep disorder symptoms that a series of stopped breathing or infrequent breathing episodes. There are two categories of SA that are obstructive sleep apnea (OSA) and central sleep apnea (CSA). Both of them share some common properties including stopped breathing, reduction of airflows (Hypopnea), etc. However, the OSA is caused by abnormal organs in the respiratory systems. It contains a series of external breathing pauses Hypopnea within a sleeping interval. OSA patients attempt to recover their stopped breathings to normal parameters. On the other hand, the CSA is caused by brain-related abnormalities such as a tumour in the brain. CSA patients are not able to recover their stopped breathing at all. In addition to the OSA and the CSA, a mixed sleep apnea between OSA and CSA can be happened. Statistically, OSA is the most discovered cases for all of the SA, which is 84 %, and the rest are CSA and mixed SA [3]. Chronic diseases such as hypertension, angina, myocardial infarction, arrhythmias and even stroke may be diagnosed in patients with SA especially the long-term SA [4, 5]. Potential OSA patients may be advised by their healthcare professionals to appoint for a sleep test so-called the polysomnography (PSG). The PSG test usually performs in a controlled environment in a hospital. The patients are require to stay in a hospital for at least a night. In unsuccessful laboratory interpretation cases, the patients are required to re-attend the test. The PSG test gathers a set of parameters in order to detect sleep disorders. After a full test has been performed, sleep technicians analyse the gathered data and interpret them into summarised readable results. The PSG test is not available in some hospitals due to the cost of the PSG equipments and their set up costs. Therefore, some researchers attempt to find alternative methods in order to identify SA. For example, a group of researchers uses electrocardiogram (ECG) and electroencephalogram (ECG) together with specific data mining techniques to identify SA [6, 7]. OSA patients have a series of stopped breathing episodes. In those episodes, their snoring sound is stopped because their airways have been partially or fully blocked by their respiratory organs. Additionally, after a short period of time, it can be seconds or even minutes, the patients will shortly awake (some may not feel that they are awake but the EEG shows their awaking sleep stage). In addition to the consciousness, a presence of snore with a louder noise is usually recognised in order to immediately breathe in air. With this phenomenon, a preliminary assumption of a snoring episode is addressed as follows.

- Stage 1: Normal Snoring—A patient snores.
- Stage 2: Stopped Breathing—The patient temporarily stops snore.
- Stage 3: Return to Normal Snoring—The patient retakes air. Louder noise or a strong cough presents at the very first second after an interval of temporary stopped breathing.

3 Experimental Design

Five sets of snoring sounds (average 8 hours recording for each set) have been collected from the Songklanagarind hospital with a formal permission under the affirmation of patients' confidentiality concerns. Due to patient confidentiality, the experiments are not able to perform on real patients. Experiment tests have been set up in a computer laboratory. All snoring sounds have been continuously played in a computer equipped with a speaker. We construct a mobile application to analyse

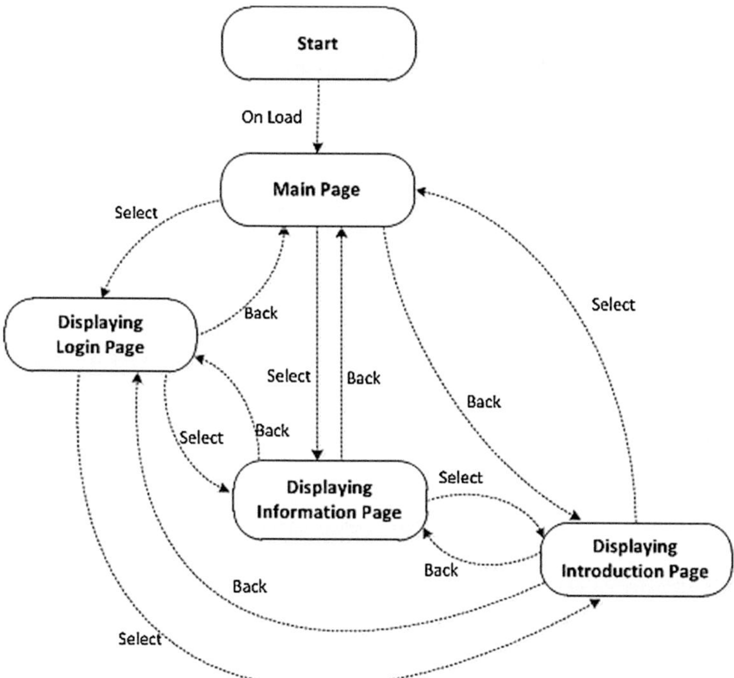

Fig. 1 State diagram of the application

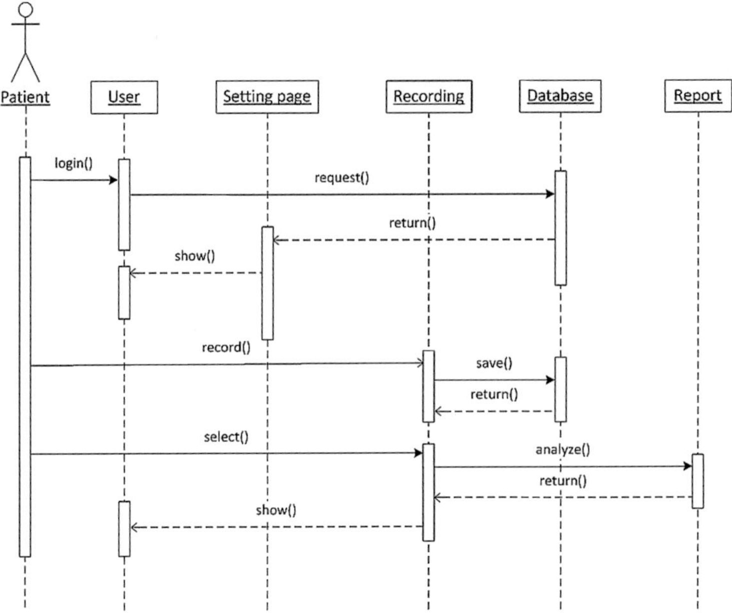

Fig. 2 Sequence diagram of the application

the snoring sounds in order to discover patterns of SA. The mobile application is designed with the following menu actions.

Figures 1 and 2 represent a state diagram and sequence diagram of the application, respectively. Figure 3 represent a set of GUIs of the application. Users are able to press the start button in order to conduct the sleep analysis. The application provides

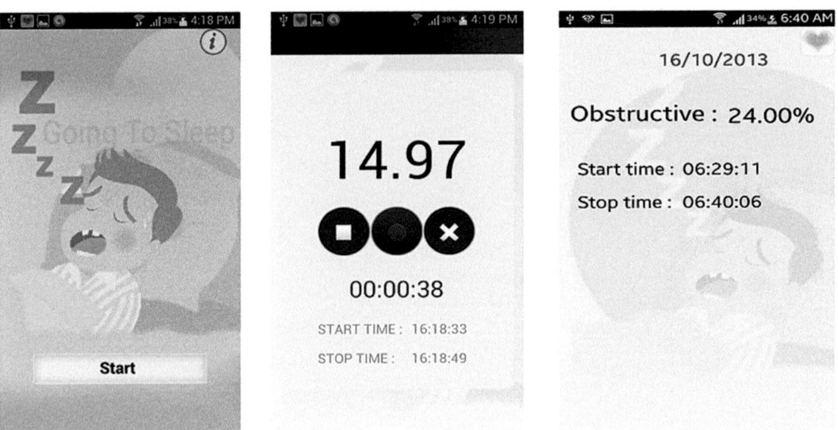

Fig. 3 Sample GUIs of the application

Algorithm 1 Noise reduction

Require: *AMP* is a set of amplitude of snoring sound signals.

```
for all A_i ∈ AMP do
  SumSecof(A_i) = SumSecof(A_i) + A_i
  if i = 60 then
    SumMinofx = SumMinofx + (SumSecof(A_i) div 60
    return SumMinofx
    x++
  end if
  SumSecof(A_i) ← 0
  SumMinofx ← 0
end for
```

predefined settings such as an idle time interval after pressing the start button, alerting when the local storage is reduced to a predefined value. The recording process shows a current amplitude detected. Users can press the cross sign to stop the recording. Finally, it also shows the output of the analysis result. It identifies the OSA episodes detected from the recording.

4 Algorithms and Techniques

One of the most vital techniques that is required in all sound analysis works is the noise reduction process. There is no recommendation for the noise reduction process in snoring sounds especially on mobile devices. Therefore, we have designed a simple noise reduction process as presented in Algorithm 1. After a full process of the noise reduction has been employed, the K-mean clustering method is performed in order to divide two classes, OSA and Non-OSA. Algorithm 2 is an adjusted K-mean clustering method that is used in our experiment using Weka [8].

5 Experimental Results and Discussions

We have collected and designed four snoring sound features, which have originally been extracted from the PSG test, for the classification using the K-Mean clustering method.

1. Beginning_Amp: Beginning amplitude of an OSA or Non-OSA interval.
2. Ending_Amp: Ending amplitude of an OSA or Non-OSA interval.
3. Max_Amp: Maximum amplitude between the beginning and ending amplitude.
4. OSA/NOSA: OSA or Non-OSA, diagnosed by sleep technicians.

Algorithm 2 Adjusted K-Mean Clustering Method [9]

```
Require: E is a set of noise-reduced amplitude of snoring sound series.
K ← 0 // two clusters: OSA and non-OSA
MaxIter ← 20 // a number of iterations
for all C_i ∈ C do
  C_i ← E_j ∈ E
end for
for all E_i ∈ E do
  l(E_i) ← argminDistance(E_j, C_i) j ∈ {1..k}
end for
chg ← false
iter ← 0
repeat
  for all C_i ∈ C do
    UpdateCluster(C_i)
  end for
  for all E_i ∈ E do
    minDist ← argminDistance(E_j, C_i) j ∈ {1..k}
    if minDist ≠ l(E_i) then
      l(E_i) ← minDist
      chg ← true
    end if
  end for
  iter++
until (chg = true)and(iter <= MaxIter)
```

Table 1 represents a confusion matrix of the classification model. Table 2, the average accuracy is 0.747 with F-measure of 0.430. However, there is one of the obvious abnormalities that require an attention. Table 1 shows the total number of cases is 22292, which are 252 OSA and 22040 Non-OSA. It is a sign of imbalanced dataset. The synthetic minority over-sampling technique (SMOTE) technique is selected for

Table 1 Confusion matrix of the classification model

	Classified OSA	Classified non-OSA	Total
OSA	43	209	252
Non-OSA	5410	16,507	22,040
Total	5453	16,716	22,292

Table 2 Performance evaluation results of the classification model

	Classification of OSA	Classification of non-OSA	Overall
Accuracy	0.747	0.746	0.747
Precision	0.007	0.987	0.498
Sensitivity	0.171	0.753	0.462
Specificity	0.753	0.169	0.462
F-Measure	0.020	0.850	0.430

Table 3 Confusion matrix of the classification model with SMOTE

	Classified OSA	Classified non-OSA	Total
OSA	21,478*	3722*	25,200*
Non-OSA	5672	16,245	21,917
Total	5453	16,716	47,117

*100 times SMOTE

Table 4 Performance Evaluation Results of the Classification Model with SMOTE

	Classification of OSA	Classification of non-OSA	Overall
Accuracy	0.801	0.801	0.801
Precision	0.791	0.813	0.802
Sensitivity	0.852	0.741	0.797
Specificity	0.741	0.852	0.797
F-Measure	0.820	0.780	0.800

eliminating the issue. It increases the amount of instances in the minority class (OSA) [10]. A result of the classification model with an application of 100 times SMOTE into the OSA class is shown in Table 3. Table 4 represents the performance evaluation results of the classification model after applied the SMOTE into the minority class. A classification improvement in both classes is shown. Specifically, the OSA classification result has been dramatically improved from the initial experiment as shown in Table 2. A possible explanation for the scenario may be due to the minority class effect. However, in terms of medical diagnostics, the mobile application is not constructed for the PSG substitution. It is more likely a rough screening device that can be used at home. A repetition of screening tests using the mobile application may be a valued guideline for a patient. The results may assist a doctor to decide whether the patient is required to attend the PSG test. Technically, there are a number of possible factors that affect the result.

1. The noise reduction algorithm: an adjusted simple moving average technique has been used in this experiment. However, other possible noise reduction techniques will be selected for further experiments in order to compare the classification performances.
2. Mobile devices: each mobile device has different capabilities in detecting noises and collecting sounds. Additionally, limitations of local storages are also addressed.
3. Positions of patients: a distance between a mobile device and a patient may affect the performance of the screening test.

6 Conclusions

The sleep apnea is a hidden symptom that can provoke future chronic diseases. The PSG test is the best diagnostic tool in order to detect the SA. However, due to its high costs and availabilities, some researchers are looking for alternative diagnostic tools. We have constructed an alpha version of mobile application to detect the OSA whilst a patient is asleep. The K-Mean clustering method has been selected as a core classification of OSA and Non-OSA. With our initial experiment test, 74.70 % instances have been correctly classified. A further investigation has been performed and discovered a possibility of the minority class effect on the OSA class. The SMOTE technique has been employed to increase the number of instances in the OSA class. The application of the SMOTE achieves upto 80.10 % correctly classified instances. Possible factors that affect the result include the noise reduction algorithm, quality of recording capabilities mobile devices and positions of patients. Further research is required to improve the classification performance in order to fulfil acceptable diagnostic capacities.

References

1. Centers for Disease Control and Prevention, http://www.cdc.gov/
2. Hoffstein, V.: Apnea and snoring: State of the art and future directions. Otorhinolaryngol. Belg. **56**, 205–236 (2002)
3. Morgenthaler, T.I., Kagramanov, V., Hanak, V., Decker, P.A.: Complex sleep apnea syndrome: is it a unique clinical syndrome? Sleep **29**, 1203–1209 (2006)
4. Flemons, W.W., Remmers, J.E., Gillis, A.M.: Sleep apnea and cardiac arrhythmias, Is there a relationship? Am. Rev. Respir. Dis. **148**, 618–621 (1993)
5. Guilleminault, C., Connolly, S.J., Winkle, R.A.: Cardiac arrhythmia and conduction disturbances during sleep in 400 patients with sleep apnea syndrome. Am. J. Cardiol. **52**, 490–494 (1983)
6. Kocak O., Bayrak T., Erdamar A., Ozparlak L., Telatar Z., Erogul O.: Automated detection and classification of sleep apnea types using electrocardiogram (ECG) and electroencephalogram (EEG) features. Adv. Electrocardiogr.-Clin. Appl. 212–230 (2012)
7. Khandoker, A.H., Palaniswami, M., Karmakar, C.K.: Support vector machines for automated recognition of obstructive sleep apnea syndrome from ECG recordings. IEEE Trans. Inf. Technol. Biomed. **13**, 37–48 (2009)
8. Weka, The University of Waikato, http://www.cs.waikato.ac.nz/ml/weka/
9. Han J., Kamber M.: Data mining: concepts and techniques. Elsevier, USA (2011)
10. Chawala, N.V., Bowyer, K.W., Hall, L.O., Kegelmeyer, W.P.: SMOTE: Synthetic minority over-sampleing technique. J. Artif. Intell. Res. **16**, 321–357 (2002)

DDoS Attacks Detection by Means of Statistical Models

Tomasz Andrysiak and Łukasz Saganowski

Abstract In this article we present a network traffic DDoS attacks detection method based on modeling the variability with the use of conditional average and variance in examined time series. Variability predictions of the analyzed network traffic are realized by estimated statistical models ARFIMA and FIGARCH. We propose simple parameter estimation models with the use of maximum likelihood function. The choice of sparingly parameterized form of the models is realized by means of information criteria representing a compromise between brevity of representation and the size of the prediction error. In the described method we propose using statistical relations between predicted and analyzed network traffic in order to detect abnormal behavior possibly being a result of a network attack. Performed experiments confirmed effectiveness of the analyzed method and cogency of the statistical models. *abstract* environment.

Keywords DDoS attacks · Anomaly detection · Statistical models

1 Introduction

At present, the biggest challenge for information systems is providing proper protection against threats. Growing number of attacks, their spreading range, and complexity enforce a dynamic development of network protection systems. This is realized by mechanisms of supervising and monitoring security of computer networks. They are implemented as IDS/IPS Intrusion Detection/Prevention Systems.

T. Andrysiak (✉) · Ł. Saganowski
Institute of Telecommunications, UTP University of Science and Technology, ul. Kaliskiego 7, 85-789 Bydgoszcz, Poland
e-mail: tomasz.andrysiak@utp.edu.pl

Ł. Saganowski
e-mail: lukasz.saganowski@utp.edu.pl

R. Burduk et al. (eds.), *Proceedings of the 9th International Conference on Computer Recognition Systems CORES 2015*, Advances in Intelligent Systems and Computing 403, DOI 10.1007/978-3-319-26227-7_75

They detect attacks directed onto widely understood network resources of information systems [1]. The techniques used in IDS systems based on statistical methods can be divided into two groups. The first one consists of methods using threshold analysis examining the frequency of events and exceeding their limits in the described time unit. The information about an attack is achieved when the examined units exceed certain threshold values. A crucial drawback of those methods is their susceptibility to errors connected with temporary violent rise in legal network traffic and problems connected with setting reference levels causing an alarm [2]. The second group consists of methods detecting statistical anomalies on the basis of estimated specific parameter profiles of a network traffic. The profiles are characterized by average quantity values, i.e., the number of IP packages, the number of newly dialed connections per time unit, ratio of packages of individual network protocols, etc. It can also be observed that there are some statistical dependences resulting from the part of the day (for instance a greater network traffic strictly after starting work). It is also possible to keep statistics for individual network protocols (for example, quantity ratio of SYN and FIN packages of TCP protocol). IDS systems based on those methods are able to learn a typical network profile—this process lasts from few to several weeks and then compare the current network activity with the memorized profile. The comparison of these two profiles will provide a basis for determining whether there is something disturbance occurring in the network (for instance an attack) [3]. The primary advantage of methods based on anomaly detection is their ability to identify unknown attack types, because they do not depend on information how a particular attack looks like, but on what does not correspond to regular norms of the network traffic. Therefore, IDS/IPS systems detecting anomalies are more effective than systems using signatures in case of identifying new, unknown attack types. Anomaly detection methods have been a topic of numerous surveys and review articles [4]. In works describing the methods there were used techniques consisting in machine learning, neural networks, and expert systems. At present, anomaly detection methods that are particularly intensively developed are those based on statistical models describing the analyzed network traffic. The most often used models are autoregressive ARMA or ARIMA, and Conditional Heteroscedastic Models ARCH and GARCH, which allow to estimate profiles of a normal network traffic [4, 5]. In the present article, we propose using estimation of statistical models ARFIMA and FIGARCH for defined behavior profiles of a given network traffic. The process of anomaly detection (a network attack) is realized by comparison of parameters of a normal behavior (predicted on the basis of tested statistical models) and parameters of real network traffic. This paper is organized as follows. After the introduction, in Sect. 2, the overview of DDoS attacks is presented. In Sect. 3 the ARFIMA and FIGARCH model for data traffic prediction is described in details. Then, in Sect. 4, the anomaly detection system based on ARFIMA—FIGARCH model estimation is shown. Experimental results and conclusion are given thereafter.

2 Overview of DDoS Attacks

Currently, DoS and DDoS attacks have become an important issue of broadly defined IT infrastructure security. Victims of the attacks are often single personal computers as well as supercomputers and vast networks. The outcomes of such activities are experienced by regular Internet users, biggest companies dealing in new technologies that often provide mass services, and powerful governmental organizations of many countries. Despite substantial effort and funds directed to enhancing IT security procedures, at present, we are not able to protect effectively against such attacks.

Attacks such as distributed denial of service (DDoS) use already known techniques of denial of service (DoS) realized with new technology. DoS attack has two crucial restrictions. First, it is performed from a single computer whose Internet connection bandwidth is too low compared to the bandwidth of the victim. Second, while performing the attack from one computer, the attacker may be subjected to a faster detection. Therefore, DoS attack is often conducted on smaller servers containing WWW sites. Attacks on bigger objects, for instance a portal or DNS server, require using a more sophisticated method DDoS, i.e., Distributed Denial of Service, which was created as a response to DoS limitations. The main difference between both methods concerns quantity factor. In DDoS, an attack is performed not from a single computer, but simultaneously from numerous overtaken machines. The sole idea of DDoS attack is therefore simple. However, what constitutes a challenge is its preparation which sometimes lasts many months. The reason is obvious, it is necessary to take over so many computers that will make the attack successful. The period of preparations is the longer, the more powerful are the victims system resources.

Why are the DDoS attacks so dangerous? Most of all, they are difficult to deter due to the fact that their source is greatly distributed. What is worse, the hosts administrators most often do not realize that they are actively participating in the attacks. The statistics are appalling a survey carried out by University of California, San Diego, point that monthly there are performed approximately 15,000 DDoS attacks.

There are a number of methods for conducting a DDoS attack. First, every operational system requires free memory space. If the attacker succeeds in allocating the whole available memory, theoretically, the system will stop functioning, or at least its performance will fall drastically. Such a brutal attack is able to block normal work of even the most efficient IT systems. The second method is based on the use of restrictions of file systems. The third means consists in using malfunctioning network applications or the kernel or errors in the operating system configuration. It is much easier to protect against the above-mentioned kind of attack by proper configuration of such a system. Most of all, it is characteristic for DoS method, which in contrast with DDoS, usually is not based on sending a great number of requests. Errors in TCP/IP stacks of different operational systems constitute an example here. In extreme cases, sending a few packages will be enough to remotely hang the server. The last method is generating a sufficiently big network traffic so that routers or servers cannot handle it [6].

Attacks of this kind are becoming a more and more serious problem. According to quarterly reports published by Prolexic company, within the last 12 months the number of DDoS attacks has risen by 22 %. Campaigns last longer-not 28.5 h as previously, but 34.5 h (a rise by 21 %). The average traffic generated during the attack is approximately 2 GB/s and is more or less 25 % greater than in 2013. The record so far was an attack on the Spamhaus, an organization dedicated to the fight against spam. In March 2013, a hostile network traffic was directed toward servers of that organization with the speed of 300 GB/s. However, according to Arbor company, most of attacks (over 60 %) still do not exceed 1 GB/s. Nevertheless, they still constitute a serious threat [7].

The reason for DDoS attacks being so problematic is that nowadays there are no effective means and methods allowing to protect the IT systems from them. It is only possible to limit the outcomes of those attacks by early identification. One of such solutions is detection of network traffic anomalies that are aftermath of a DDoS attack.

3 Statistical Models for Network Traffic Prediction

Most research on statistical analysis of time series concerns processes characterized by lack of or poor connection between variables which are separated by some time period. Nevertheless, in numerous uses there is a need for modeling processes whose autocorrelation function is slowly decreasing, and the relation between distant observations—even though it is not big—is essential. An interesting approach toward properties of long-memory time series was applying the autoregression with moving average in the process of fractional diversification. As a result, ARFIMA model (Fractional Differenced Noise and Auto Regressive Moving Average) was obtained and implemented by Grange, Joyeux, and Hosking [8, 9]. ARFIMA is a generalization of ARMA and ARIMA models. Another approach describing time series was taking into account the dependence of the conditional variance of the process on its previous values with the use of ARCH model (Autoregressive Conditional Heteroskedastic Model) introduced by Engel [10]. Generalization of this approach was FIGARCH model (Fractionally Integrated GARCH) introduced by Baillie et al. [11].

3.1 ARFIMA Model

The autoregressive fractional integrated moving average model called ARFIMA (p, d, q) is a combination of fractional differenced noise and auto regressive moving average which is proposed by Grange, Joyeux, and Hosking, in order to analyze the long-memory property [8, 9]. The ARFIMA (p, d, q) model for time series y_t is written as

$$\Phi(L)(1-L)^d y_t = \Theta(L)\varepsilon_t, \quad t = 1, 2, \ldots, \Omega, \tag{1}$$

where y_t is the time series, $\varepsilon_t \sim (0, \sigma^2)$ is the white noise process with zero-mean and variances σ^2, $\Phi(L) = 1 - \phi_1 L - \phi_2 L^2 - \cdots - \phi_p L^p$ is the autoregressive polynomial and $\Theta(L) = 1 + \theta_1 L + \theta_2 L^2 + \cdots + \theta_q L^q$ is the moving average polynomial, L is the backward shift operator, and $(1 - L)^d$ is the fractional differencing operator given by the following binomial expansion:

$$(1-L)^d = \sum_{k=0}^{\infty} \binom{d}{k} (-1)^k L^k \tag{2}$$

and

$$\binom{d}{k}(-1)^k = \frac{\Gamma(d+1)(-1)^k}{\Gamma(d-k+1)\Gamma(k+1)} = \frac{\Gamma(-d+k)}{\Gamma(-d)\Gamma(k+1)}, \tag{3}$$

where $\Gamma(*)$ denotes the gamma function and d is the number of differences required to give a stationary series and $(1 - L)^d$ is the dth power of the differencing operator. When $d \in (-0, 5\ , 0, 5)$, the $ARFIMA(p, d, q)$ process is stationary, and if $d \in (0\ , 0, 5)$ the process presents long-memory behavior. Forecasting ARFIMA processes are usually carried out using an infinite autoregressive representation of (1), written as $\Pi(L)\, y_t = \varepsilon_t$,

$$y_t = \sum_{i=1}^{\infty} \pi_i y_{t-i} + \varepsilon_t, \tag{4}$$

where $\Pi(L) = 1 - \pi_1 L - \pi_2 L^2 - \cdots = \Phi\,(L)\,(1-L)^d\,\Theta\,(L)^{-1}$. In terms of practical implementation, this form needs truncation after k lags, but there is no obvious way of doing it. This truncation problem will also be related to the forecast horizon considered in predictions (see [12]). From (4), it is clear that the forecasting rule will pick up the influence of distant lags, thus capturing their persistent influence. However, if a shift in the process occurs, this means that pre-shift lags will also have some weight on the prediction, which may cause some biases for post-shift horizons [13].

3.2 FIGARCH Model

The model enabling description of long-memory in variance series is FIGARCH (p, d, q) (fractionally integrated GARCH) introduced by Baillie, Bollerslev, and Mikkelsen et al. [11]. The FIGARCH (p, d, q) model for time series y_t can be written as

$$y_t = \mu + \varepsilon_t, \quad t = 1, 2, \ldots, \Omega, \tag{5}$$

$$\varepsilon_t = z_t\sqrt{h_t}, \quad \varepsilon_t|\Theta_{t-1} \sim N(0, h_t), \tag{6}$$

$$h_t = \alpha_0 + \beta(L)h_t + \left[1 - \beta(L) - [1 - \phi(L)](1 - L)^d\right]\varepsilon_t^2, \tag{7}$$

where z_t is a zero-mean and unit variance process, h_t is a positive time-dependent conditional variance defined as $h_t = E\left(\varepsilon_t^2|\Theta_{t-1}\right)$ and Θ_{t-1} is the information set up to time $t-1$. The FIGARCH (p, d, q) model of the conditional variance can be motivated as ARFIMA model applied to the squared innovations

$$\varphi(L)(1 - L)^d \varepsilon_t^2 = \alpha_0 + (1 - \beta(L))\vartheta_t, \quad \vartheta_t = \varepsilon_t^2 - h_t, \tag{8}$$

where $\varphi(L) = \varphi_1 L - \varphi_2 L^2 - \cdots - \varphi_p L^p$ and $\beta(L) = \beta_1 L + \beta_2 L^2 + \cdots + \beta_q L^q$ and $(1 - \beta(L))$ have all their roots outside the unit circle, L is the lag operator and $0 < d < 1$ is the fractional integration parameter. If $d = 0$, then FIGARCH model is reduced to GARCH; for $d = 1$ though, it becomes IGARCH model. However, FIGARCH model does not always reduce to GARCH model. If GARCH process is stationary in broader sense, then the influence of current variance on its forecasting values decreases to zero in exponential pace. In IGARCH case, the current variance has indefinite influence on the forecast of conditional variance. For FIGARCH process, the mentioned influence decreases to zero far more slowly than in GARCH process, i.e., according to the hyperbolic function [11, 14]. Rearranging the terms in (8), an alternative representation for the FIGARCH (p, d, q) model may be obtained as

$$[1 - \beta(L)]h_t = \alpha_0 + [1 - \beta(L) - \varphi(L)](1 - L)^d \varepsilon_t^2. \tag{9}$$

From (10), the conditional variance h_t of y_t is given by

$$h_t = \alpha_0[1 - \beta(1)]^{-1} + \lambda(L)\varepsilon_t^2, \tag{10}$$

where $\lambda(L) = \lambda_1 L + \lambda_2 L^2 + \cdots$ Of course, for the FIGARCH (p, d, q), for (8) to be well-defined, the conditional variance in the $ARH(\infty)$ representation in (10) must be non-negative, *i.e.*, $\lambda_k = 0$ for $k = 1, 2, \ldots$ Solving the problem of forecasting using Eq. (10) may be obtained as

$$h_{t+1} = \alpha_0[1 - \beta(1)]^{-1} + \lambda_1\varepsilon_t^2 + \lambda_2\varepsilon_{t-1}^2 + \cdots \tag{11}$$

The one-step ahead forecast of h_t is given by

$$h_t(1) = \alpha_0[1 - \beta(1)]^{-1} + \lambda_1\varepsilon_t^2 + \lambda_2\varepsilon_{t-1}^2 + \cdots \tag{12}$$

By analogy, the two-step ahead forecast is given by

$$h_t(2) = \alpha_0[1 - \beta(1)]^{-1} + \lambda_1 h_t(1) + \lambda_2\varepsilon_t^2 + \cdots \tag{13}$$

In general, the n-step ahead forecast is can be written as

$$h_t(n) = \alpha_0 [1 - \beta(1)]^{-1} + \lambda_1 h_t(n-1) + \cdots + \lambda_{n-1} h_t(1) + \lambda_n \varepsilon_t^2 + \lambda_{n+1} \varepsilon_{t-1}^2 + \cdots \tag{14}$$

In practical application, we stop at a large N and this leads to the forecasting equation

$$h_t(n) \approx \alpha_0 [1 - \beta(1)]^{-1} + \sum_{i=1}^{n-1} \lambda_i h_t(n-i) + \sum_{j=0}^{N} \lambda_{n+j} \varepsilon_{t-j}^2. \tag{15}$$

The parameters will have to be replaced by their corresponding estimates [14].

4 Parameters Estimation and the Choice of Model

The most often used methods of estimation of autoregressive models parameters are: maximum likelihood method (MLE) and quasi-maximum likelihood method (QMLE). This is due to the fact that estimation of the parameters by means of both methods is relatively simple and effective. The basic problem of computing with MLE method is finding a solution to the equation

$$\frac{\partial \ln(L_\Omega(\rho))}{\partial \theta} = 0, \tag{16}$$

where θ is the estimated set of parameters, $L_\Omega(\rho)$ is the likelihood function, and Ω is a number of observations. Mostly, in general case the analytic solution to the Eq. (16) is impossible and then numerical estimation is employed. The basic problem occurring while using the maximum likelihood method is necessity to define the whole model, and consequently the sensitivity of the resulting estimator for any errors in the specification of the AR and MA polynomials responsible for the dynamics of the process [15, 16]. There is no universal criterion for the choice of the model. Usually, the case is as follows: the more complex model, the greater is the value of the likelihood function. As a result, adjusting the model to the data is more effective. However, estimation of a higher number of parameters is connected with bigger errors. Therefore, it is crucial to find a compromise between the quantity of parameters occurring in the model and the value of likelihood function. The choice of the economic form of the model is often based on information criteria such as Akaike (AIC) or Schwarz (SIC). Values of the mentioned criteria can be estimated on the basis of the following formulas:

$$AIC(\rho) = -2\ln(L_\Omega(\rho)) + 2\rho, \tag{17}$$

$$SIC(\rho) = -2\ln(L_\Omega(\rho)) + \rho\ln(\Omega), \tag{18}$$

where ρ is the number of the model's parameters. From different forms of the model, the one that is chosen has the smallest information criterion value [12, 17]. In our article, we proposed the maximum likelihood method for parameters estimation and the choice of the form of the model. The method was chosen due to its relative simplicity and computational efficiency. For ARFIMA model, we used HR estimator (described in Haslett and Raftery [18]) and automatic model selection algorithm based on the information criteria (see Hyndman and Khandakar [19]). For FIGARCH model estimation, we used methodology described in the present article [14].

5 Experimental Results

In this section, we presented some results in case of ARFIMA and FIGARH statistical model usage for DDoS attack detection. We simulated real-world DDoS and application specific DDoS attacks for single LAN test network. As a network sensor we used SNORT IDS [20]. SNORT in our case is responsible for traffic capture and extracting network traffic features (see Table 1). Additionally we also used traffic testbed that contains DDoS attacks [21]. Twelve traffic features were used for evaluation of presented ARFIMA and FIGARH statistical models. Obviously, not all traffic features were sufficient for detecting all simulated attacks because they

Table 1 Network traffic features used for experiments

Traffic feature	Traffic feature description
f_1	Number of TCP packets
f_2	In TCP packets
f_3	Out TCP packets
f_4	Number of TCP packets in LAN
f_5	Number of UDP datagrams
f_6	Number of UDP datagrams in LAN
f_7	Number of ICMP packets
f_8	Out ICMP packets
f_9	Number of ICMP packets in LAN
f_{10}	Number of TCP packets with SYN and ACK flags
f_{11}	Out TCP packets (port 80)
f_{12}	In TCP packets (port 80)

Table 2 Detection Rate DR (%) and False Positive FP (%) for a given network traffic features

Traffic feature	FIGARH	ARFIMA	Traffic Feature	FIGARH	ARFIMA
f_1	5.26	8.26	f_1	5.46	4.23
f_2	5.26	12.52	f_2	5.17	4.84
f_3	0.00	12.52	f_3	5.45	4.22
f_4	15.78	10.52	f_4	5.44	4.02
f_5	10.52	14.52	f_5	5.64	4.23
f_6	25.22	35.24	f_6	5.24	4.24
f_7	90.73	98.43	f_7	7.68	6.12
f_8	83.68	96.43	f_8	1.22	0.32
f_9	80.42	85.95	f_9	6.34	4.20
f_{10}	10.52	14.22	f_{10}	5.23	4.56
f_{11}	0.00	8.26	f_{11}	4.58	3.26
f_{12}	0.00	14.22	f_{12}	4.86	3.52

Table 3 Evaluation of proposed method with the use of real world network traffic testbed [21] for 4 days of traffic

Trace date	2008-05-21	2008-08-20	2008-11-15	2009-01-15
ARFIMA DR (%)	85	80	95	82
FIGARH DR (%)	80	70	85	75

Table 4 Evaluation of proposed method with the use of real world network traffic testbed [21] for 4 days of traffic

Trace date	2008-02-20	2008-02-21	2009-05-21	2009-02-15
ARFIMA DR (%)	82	81	92	81
FIGARH DR (%)	79	75	84	77

have not got impact on entire set of traffic features presented in Table 1. In Table 2 we presented detection rate DR and false positive FP values for 12 traffic features. Additionally in Tables 3 and 4, there are results for external testbed for 4 days of network traffic, respectively. We can conclude that ARFIMA model gives us better results in case of DR and FP for the used testbed in our experiments.

6 Conclusion

Cybersecurity of information systems is contemporarily a key research factor. The growing number of DDoS attacks, their expending reach, and the level of complexity stimulate the dynamic development of network defensive systems. The techniques of statistical anomaly detections are recently the most commonly used for monitoring as well as detecting the attacks. In the present article, the construction of statistical autoregressive models, ARFIMA and FIFARCH, has been described. The above-mentioned models present the statistic variability of modeled parameters by means of the average or conditional variance. For estimation of parameters and identification of models the maximum likelihood method together with information criteria were used. As a result of their work the satisfying statistic measurements for researched signals of network traffic were obtained. The process of anomaly (attacks) detecting consist in comparison of estimated behavior parameters with real network traffic factors. The obtained results outstandingly signify that the anomalies included in the network traffic signal can be detected by suggested methods.

References

1. Jackson, K.: Intrusion Detection Systems (IDS). Product Survey. Los Alamos National Library, LA-UR-99-3883 (1999)
2. Esposito, M., Mazzariello, C., Oliviero, F., Romano, S.P., Sansone C.: Evaluating pattern recognition techniques in intrusion detection systems. In: PRIS, pp. 144–153 (2005)
3. Lakhina, A., Crovella, M., Diot, C.H.: Characterization of network-wide anomalies in traffic flows. In: Proceedings of the 4th ACM SIGCOMM Conference on Internet Measurement, pp. 201–206 (2004)
4. Chondola, V., Banerjee, A., Kumar, V.: Anomaly detection: a survey. ACM Comput. Surv. **41**(3), 1–72 (2009)
5. Rodriguez, A., Mozos, M.: Improving network security through traffic log anomaly detection using time series analysis. In: Computational Intelligence in Security for Information Systems, pp. 125–133 (2010)
6. Liang H., Xiaoming B.: Research of DDoS attack mechanism and its defense frame, computer research and development (ICCRD). In: 3rd International Conference, pp. 440–442 (2011)
7. Atak i Obrona 2013 Raport, Ataki i metody obrony w internecie w Polsce (2013)
8. Granger, C.W.J., Joyeux, R.: An introduction to long-memory time series models andfractional differencing. J. Time Ser. Anal. **1**, 15–29 (1980)
9. Hosking, J.: Fractional differencing. Biometrika **68**, 165–176 (1981)
10. Engle, R.: Autoregressive conditional heteroskedasticity with estimates of the variance of UK inflation. Econometrica **50**, 987–1008 (1982)
11. Baillie, R., Bollerslev, T., Mikkelsen, H.: Fractionally integrated generalized autoregressive conditional heteroskedasticity. J. Econom. **74**, 3–30 (1996)
12. Crato, N., Ray, B.K.: Model selection and forecasting for long-range dependent processes. J. Forecast. **15**, 107–125 (1996)
13. Gabriel, V.J., Martins, L.F.: On the forecasting ability of ARFIMA models when infrequent breaks occur. Econom. J. **7**, 455–475 (2004)
14. Tayefi, M., Ramanathan, T.V.: An overview of FIGARCH and related time series models. AUSTRIAN J. Stat. **41**(3), 175–196 (2012)
15. Box, G., Jenkins, G., Reinsel, G.: Time Series Analysis. Holden-day, San Francisco (1970)
16. Brockwell, P., Davis, R.: Introduction to Time Series and Forecasting. Springer, Berlin (2002)
17. Beran, J.A.: Statistics for Long-Memory Processes. Chapman and Hall, New York (1994)
18. Haslett, J.: Raftery AE space-time modelling with long-memory dependence: assessing Ireland's wind power resource (with discussion). Appl. Stat. **38**(1), 1–50 (1989)
19. Hyndman, R.J., Khandakar, Y.: Automatic time series forecasting: the forecast Package for R. J. Stat. Softw. **27**(3), 1–22 (2008)
20. SNORT—Intrusion Detection System, https://www.snort.org/
21. The CAIDA Dataset, http://www.caida.org/data (2006–2009)

Part VII
RGB-D Perception: Recent Developments and Applications

Infrared Image-Based 3D Surface Reconstruction of Free-Form Texture-Less Objects

Karolina Przerwa, Włodzimierz Kasprzak and Maciej Stefańczyk

Abstract The analysis of infrared (IR) images obtained from a robot-mounted camera is presented, with the purpose to reconstruct the 3D surface of texture-less objects located in close range to the camera. The prospective application of this approach is object's pose recognition with the aim of object grasping by a robot hand. Algorithms are developed that rely on the analysis of the luminance distribution in the IR image (the so-called *shape-from-shading* approach), followed by a depth-map approximation. Laboratory tests were carried out in order to evaluate the quality of obtained depth maps on some reference objects and to compare the estimated depth maps with corresponding point clouds acquired by a MS-Kinect device.

1 Introduction

In industrial applications often a linear laser scanner is used to measure the surface of an nearby object. However, in some robot grasping problems, it is important to recognize from a close range (i.e., from a camera mounted on the manipulator arm) the 3D pose of a free-form object, preferably without any texture, like a door handle, fruits, etc. We expect, that the application of integrated RGB-D image sensors, such as the MS-Kinect, for surface analysis of close-range objects will be possible, when the technology is ready for miniaturization and when it allows close-range measurements [1]. However, dealing with small object-to-sensor distances, the single camera-based

K. Przerwa · W. Kasprzak (✉) · M. Stefańczyk
Institute of Control and Computation Engineering, Warsaw University of Technology, ul. Nowowiejska 15/19, 00-665 Warsaw, Poland
e-mail: W.Kasprzak@elka.pw.edu.pl
URL: http://ia.pw.edu.pl/

K. Przerwa
e-mail: K.Przerwa@stud.elka.pw.edu.pl

M. Stefańczyk
e-mail: M.Stefanczyk@elka.pw.edu.pl

R. Burduk et al. (eds.), *Proceedings of the 9th International Conference on Computer Recognition Systems CORES 2015*, Advances in Intelligent Systems and Computing 403, DOI 10.1007/978-3-319-26227-7_76

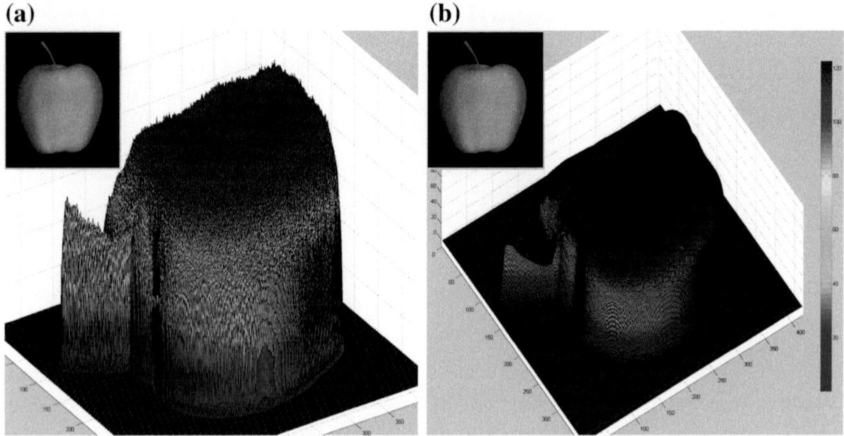

Fig. 1 Steps of depth map reconstruction. **a** Raw depth map. **b** Map approximation

solution may be still efficient [2]. The recognition of 3D objects can be conducted on the basis of color or monochrome images obtained by camera vision [3], when a 3D model of the object's type is known in advance. Even in the absence of a 3D model, the 3D surface can be reconstructed from an image, while applying specific algorithms and constraints. For example, in [4] the depth map is obtained based on prior information about the lighting and due to analysis of luminance distribution in the image. This approach solves the so-called *shape-from-shading* problem [5–7]. In this paper, two algorithms for single object surface reconstruction are presented, based on single infrared image analysis (IR). The most important steps, i.e., depth map estimation and approximation with surface feature extraction, are described in Sects. 2 and 3 (Fig. 1). Test results are provided in Sect. 4. There is also a comparison given of estimated depth maps and corresponding MS-Kinect point clouds, obtained for the same objects. The paper is completed by conclusions (Sect. 5).

2 Depth Map Estimation

2.1 The Lambertian Model

The Lambertian model of brightness distribution is applied, that is approximately true for light sources far away from the camera (Fig. 2), leading to unidirectional lighting. In this model, the intensity distribution in the acquired image corresponds to local surface characteristics, assuming the direction and nature of the illumination source are known or can be estimated. Let us denote: S—the direction of light propagation, N—normal vector to surface, x—depth axis, (x, y)—image coordinates of a pixel, τ—the *incidence* angle between the source direction S and the image axis

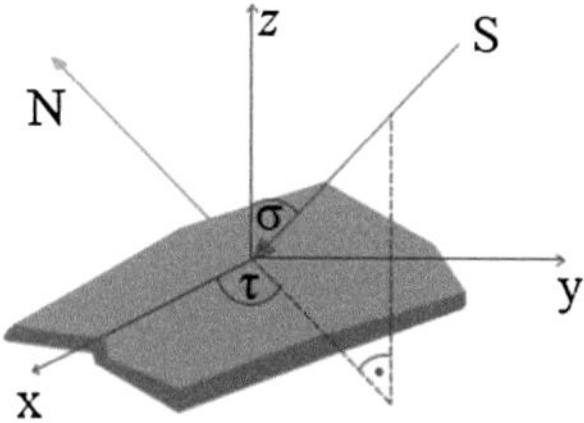

Fig. 2 The Lambertian model of light propagation

x, σ—the *inclination* angle between source orientation S and the depth axis z. A correspondence is assumed between the incidence and inclination angles of source, the normal vector to surface element, the location of observer and the reflected intensity value. Let $I(x, y)$ represents the intensity image, where (x, y) are pixel coordinates. The Lambertian model is expressed by the so-called *reflectance function* [4, 7]:

$$I(x, y) \simeq R(p, q) = \frac{1 + pp_s + qq_s}{\sqrt{1 + p^2 + q^2}\sqrt{1 + p_s^2 + q_s^2}} = \frac{\cos\sigma + p\cos\tau\sin\sigma + q\sin\tau\sin\sigma}{\sqrt{1 + p^2 + q^2}}, \tag{1}$$

where: $p = \frac{\partial z}{\partial x}$ and $q = \frac{\partial z}{\partial y}$—the unknown object surface gradient along the x axis and y axis, respectively; $p_s = \frac{\cos\tau\sin\sigma}{\cos\sigma}$ and $q_s = \frac{\sin\tau\sin\sigma}{\cos\sigma}$—the source direction coefficients.

2.2 Iterative Gradient-Based SFS

In this method, the unknown gradient values p and q are approximated using discrete differences:

$$p = \frac{\partial z}{\partial x} \simeq z(x, y) - z(x - 1, y) \tag{2}$$

$$q = \frac{\partial z}{\partial y} \simeq z(x, y) - z(x, y - 1) \tag{3}$$

Now, the reflectance equation takes the form:

$$0 = f\left(I(x, y), z(x, y), z(x - 1, y), z(x, y - 1)\right) = I(x, y) - R\left(z(x, y) - z(x - 1, y), z(x, y) - z(x, y - 1)\right) \tag{4}$$

After expanding the function $f(.)$ into Taylor series, limiting the equation to second-order components, and applying the iterative Jacobi method (substitution of previous estimates for $z(.)$ in the neighborhood of (x, y): $z(x-1, y) \to z^{n-1}(x-1, y)$, $z(x, y-1) \to z^{n-1}(x, y-1)$), an iterative formula appears, that is used for depth estimation:

$$z^n(x, y) = z^{(n-1)}(x, y) + \frac{-f\left(z^{(n-1)}(x, y)\right)}{\frac{d}{dz(x,y)} f\left(z^{(n-1)}(x, y)\right)} \tag{5}$$

where:

$$\frac{df\left(z^{(n-1)}(x, y)\right)}{dz(x, y)} = -\left(\frac{(p_s + q_s)}{\sqrt{p^2 + q^2 + 1} * \sqrt{p_s^2 + q_s^2 + 1}} - \frac{(p + q)(pp_s + qq_s + 1)}{\sqrt{(p^2 + q^2 + 1)^3} * \sqrt{p_s^2 + q_s^2 + 1}}\right)$$

$$z^{(0)}(x, y) = 0.$$

2.3 Fourier-Domain SFS

Pentland has proposed another SFS algorithm [7] that solves the problem in the Fourier domain. Let us expand the reflectance function into a Taylor series around some point $p = p_0,\ q = q_0$, and limit it to first-order components:

$$I(x, y) = R(p_0, q_0) + (p - p_0)\frac{\partial R}{\partial p}(p_0, q_0) + (q - q_0)\frac{\partial R}{\partial q}(p_0, q_0)$$

Because of the definition of $R(p_0, q_0)$ (1) for the point, $p_0 = q_0 = 0$, the above equation reduces to

$$I(x, y) = \cos\sigma + p \cdot \cos\tau \cdot \sin\sigma + q \cdot \sin\tau \cdot \sin\sigma \tag{6}$$

Next, both sides are transformed by a 2D Fourier transform into the 2D frequency domain. The Fourier transforms of partial derivatives p, q, i.e., $F_p(f, \theta)$ and $F_q(f, \theta)$ are related to the Fourier transform of the depth map $z(x, y)$, i.e., $F_z(f, \theta)$ as follows:

$$F_p(f, \theta) = F_z(f, \theta)\left(2\pi\cos(\theta) f \cdot e^{i\pi/2}\right),$$
$$F_q(f, \theta) = F_z(f, \theta)\left(2\pi\sin(\theta) f \cdot e^{i\pi/2}\right)$$

where f is radial frequency and θ its orientation. By observing, that the Fourier transform of $\cos\sigma$ is a single value, it will be ignored, what finally leads to the following correspondence:

$$F_I(f, \theta) = F_z(f, \theta)\, T(f, \theta) \tag{7}$$

where: F_z—Fourier transform of the depth map $z(x, y)$, F_I—Fourier transform of the IR image, $T(f, \theta) = 2\pi f \cdot e^{i\pi/2} [\cos\tau \sin\sigma \cos(\theta) + \sin\tau \sin\sigma \sin(\theta)]$. From above Eq. (7), one gets the Fourier transform of the depth map $z(x, y)$ as:

$$F_z(f, \theta) = \frac{F_I}{2\pi f \cdot e^{i\pi/2} [\cos\tau \cdot \sin\sigma \cdot \cos\theta + \sin\tau \cdot \sin\sigma \cdot \sin\theta]} \tag{8}$$

Assuming, that both components of the lighting direction, τ, σ are known, the depth map's Fourier components can be estimated. Even when only τ is known (the tilt component), while σ is not, the Fourier components may be recovered up to a multiplicative constant. Finally, the depth map is estimated by the inverse discrete Fourier transform (IDFT) of above equation:

$$z(x, y) = IDFT\{F_z(f, \theta)\} \tag{9}$$

3 Depth-Map Approximation and Feature Extraction

3.1 *Second-Order Polynomials*

The depth map obtained by a SFS analysis, usually contains large amount of noise. The map will be smoothed by approximating it by a second-order function of two variables, with respect to a local neighborhood of every pixel. Polynomial orthogonal base functions are used for this purpose:

$$z(x, y) \approx g(x, y) = k_1 + k_2x + k_3y + k_4x^2 + k_5xy + k_6y^2 \tag{10}$$

$$g(x, y) = \sum_{i=1}^{6} K_i(x, y)h_i(x, y) \tag{11}$$

where $h(x, y) = [1,\ x,\ y,\ x^2 - 2,\ xy,\ y^2 - 2]$ is the orthogonal polynomial base.

3.1.1 Closed-Form Estimation of Coefficients

The coefficients K_i are estimated from the depth information in the neighborhood of given pixel, (x, y), as follows:

$$K_i(x, y) = \frac{\sum_{x',y'} h_i(x', y')z(x + x', y + y')}{\sum_{x',y'} h_i^2(x', y')} \tag{12}$$

where $x' = \{-W, \ldots, -1, 0, 1, \ldots, W\}$, $y' = \{W, \ldots, -1, 0, 1, \ldots, W\}$, $i \in \{1, 2, \ldots, 6\}$, and (x', y')—pixel in local neighborhood of size parameterized by W.

3.1.2 Coefficients by Convolution Filters

An alternative way to obtain the coefficients K_i is to pass the depth map through different 2D filters, which are defined by appropriate convolution kernels. In our experiments kernels of size 5×5 are used:

$$K_2 = \begin{bmatrix} -2 & -2 & -2 & -2 & -2 \\ -1 & -1 & -1 & -1 & -1 \\ 0 & 0 & 0 & 0 & 0 \\ 1 & 1 & 1 & 1 & 1 \\ 2 & 2 & 2 & 2 & 2 \end{bmatrix} = (K_3)^T \qquad K_4 = \begin{bmatrix} 2 & 2 & 2 & 2 & 2 \\ -1 & -1 & -1 & -1 & -1 \\ -2 & -2 & -2 & -2 & -2 \\ -1 & -1 & -1 & -1 & -1 \\ 2 & 2 & 2 & 2 & 2 \end{bmatrix} = (K_6)^T$$

3.2 Surface Features

The coefficients K_i are responsible for the object's surface type related to the image point (x, y). In particular, the second-order coefficients, K_4 and K_6, represent the convexity and concavity of the function.

4 Results

In several test series, the proposed approach has been evaluated by observing:

1. the quality of depth map estimation by different methods and parameter setting;
2. the time complexity of particular processing steps;
3. the comparison of depth map with corresponding point clouds generated by a MS-Kinect.

In preliminary tests, synthetic images have been used, generated under a strict directional lighting of the object without reflected and distributed light. Then the main tests have been performed on real images, acquired by an IR camera, while the scene has been covered by a "shadeless tent" in order to limit the disturbances of lighting.

4.1 Test Images

The synthetic image set consisted of two objects (torus and tea pot), lighted by directional light coming alternatively from 10 different places (Fig. 3). The coordinate system (xyz) is consistent with Fig. 2, where the z axis is perpendicular to the image plane.

Fig. 3 Three synthetic teapot scenes for light sources at direction: **a** $\tau = 30°, \sigma = 60°$, **b** $\tau = 60°, \sigma = 30°$, **c** $\tau = 315°, \sigma = 225°$

4.2 Dependence from Lighting Direction

The dependance of the two proposed SFS algorithms from the lighting direction is observed by analyzing synthetic images. The depth estimation error (mean and standard deviation) is calculated in dependance on the lighting direction. Before evaluation, the depth maps were always normalized to the interval of [0, 1]. The error statistics is shown in Table 1 for the "teapot." The mean value over 10 directions of the MSE is around 0.0317 (for iterative SFS) and 0.0148 (for Fourier-based SFS). The iterative SFS performed worse for large incidence angle values, whereas the Fourier-based SFS lost against its competitor only for one setting: $\tau = 45°, \sigma = 45°$. The above results can be summarized as follows:

- The MSE of depth estimation for synthetic data (perfect lighting) is in the range of 0.001–0.030 (for Fourier SFS) or 0.002–0.067 (for the gradient SFS), depending on the lighting direction. This can be assumed to be the lowest possible error for SFS-based depth maps estimated from real images;
- The iterative gradient-based SFS algorithm performs much worse than the Fourier-based SFS, specially for large light incidence angles (cases 6–10 in Table 1).

Three of the obtained depth maps for the teapot object are illustrated in Fig. 4.

4.3 SFS in Real Conditions

Some real test images are shown in Fig. 5. The lighting orientation was restricted to a pair of angular values: $\tau = 45°, \sigma = 45°$. This is sufficient for the identification of abilities of particular algorithms. The estimated depth maps are illustrated in Fig. 6.

Table 1 Statistics of depth maps estimated for the synthetic 'teapot' scene

Original map		Lighting direction	Iter. SFS map			Fourier SFS map		
Mean	Var		Mean	Var	MSE	Mean	Var	MSE
0.5317	**0.0938**	1: $\tau = 45° \ \sigma = 45°$	0.5894	0.0937	0.0033	0.6090	0.0883	0.0067
		2 : $\tau = 30° \ \sigma = 30°$	0.6120	0.0620	0.0127	0.6971	0.0657	0.0369
		3: $\tau = 60° \ \sigma = 60°$	0.4863	0.0920	0.0019	0.4990	0.0873	0.0008
		4: $\tau = 30° \ \sigma = 60°$	0.4602	0.0996	0.0058	0.4781	0.0972	0.0032
		5: $\tau = 60° \ \sigma = 30°$	0.6859	0.0749	0.0293	0.7072	0.0654	0.0409
		6: $\tau = 315° \ \sigma = 225°$	0.2369	0.0157	0.0499	0.5416	0.0688	0.0015
		7: $\tau = 330° \ \sigma = 330°$	0.2347	0.0148	0.0504	0.5838	0.0716	0.0055
		8: $\tau = 300° \ \sigma = 300°$	0.2067	0.0231	0.0674	0.3968	0.0790	0.0151
		9: $\tau = 330° \ \sigma = 300°$	0.2166	0.0240	0.0630	0.4276	0.0789	0.0085
		10: $\tau = 300° \ \sigma = 330°$	0.2792	0.0105	0.0331	0.6547	0.0487	0.0293
		1–10 : Aver. MSE	–	–	0.0317	–	–	0.0148

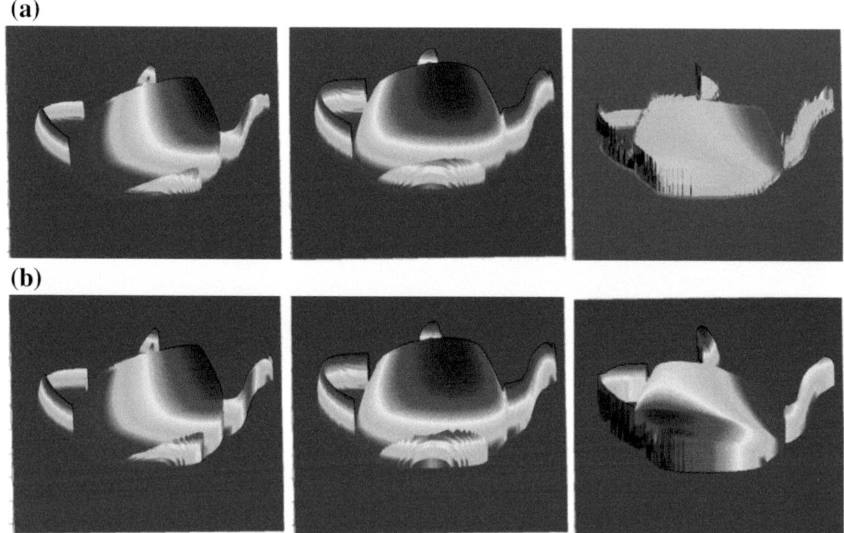

Fig. 4 Examples of depth maps obtained for the teapot object using the iterative and Fourier-based SFS algorithms **a** Iterative SFS. **b** Fourier-based SFS

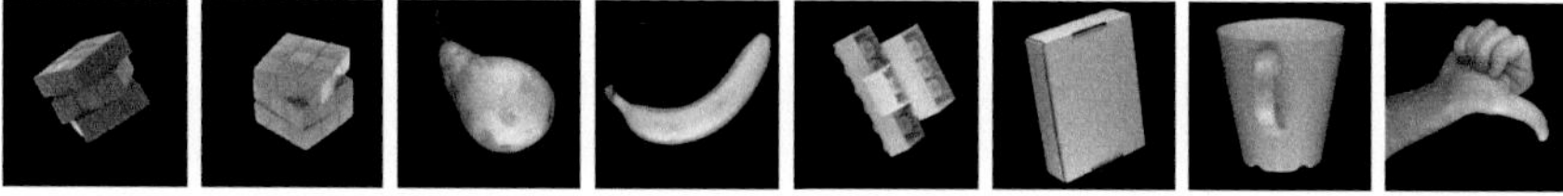

Fig. 5 Examples of real images

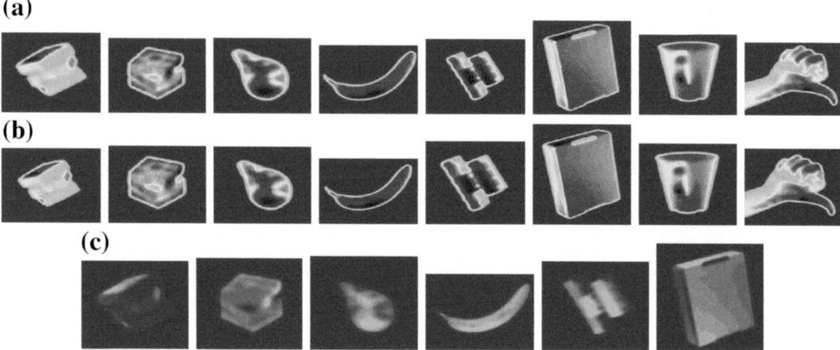

Fig. 6 Example of depth maps obtained by the iterative and Fourier-based SFS algorithms **a** Iterative SFS. **b** Fourier SFS. **c** Depth maps after approximation

Table 2 Average processing times of main steps

Step	Algorithm	Average time (ms)
SFS	Iterative SFS with discrete gradient	163
	Fourier-based SFS	40
Approximation	By local neighborhood analysis	60
	By convolution kernels	27

4.4 Time Complexity

Processing times have been measured on a Notebook with procesor Intel Core i7-3160QM 2.3GHz, 4 processor cores, 16GB RAM, OS Windows 7. Average times for the main steps when processing images with resolution 400×400 are shown in Table 2. There were only up to 30 MB requirements for RAM storage.

4.5 Comparison with Point Clouds

The third set of images is point clouds of real objects acquired by the MS-Kinect device. As for real objects no reference data is available, a comparison of general statistical features of the depth map (obtained by a SFS algorithm) and the point cloud (acquired by the Kinect device) is made (Table 3). One can also make a visual comparison of the maps (Fig. 7). The comparison can be summarized as follows:

- Kinect depth map suffers from the existence of many outliers, which suppress the relative depths of useful points to zero. Only after canceling these outliers depth measurements corresponding to real existing surfaces can be unveiled.

Table 3 Statistics of depth maps (SFS) and point clouds (Kinect)

Object	Fourier SFS		Kinect (outliers)		Kinect (NO outliers)	
	Mean	Std	Mean	Std	Mean	Std
1: Spun	0.19	0.18	0.395	0.140	0.467	0.162
2: Rubik	0.22	0.18	0.057	0.143	0.359	0.331
3: Tomato	0.20	0.17	0.049	0.122	0.464	0.371
4: Lego	0.22	0.18	0.039	0.105	0.391	0.373
5: Cup	0.16	0.16	0.065	0.140	0.386	0.355
6: Palm	0.20	0.18	0.107	0.229	0.265	0.255
7: Banana	0.15	0.14	0.034	0.035	0.440	0.232
8: Gamepad	0.24	0.20	0.040	0.069	0.439	0.340

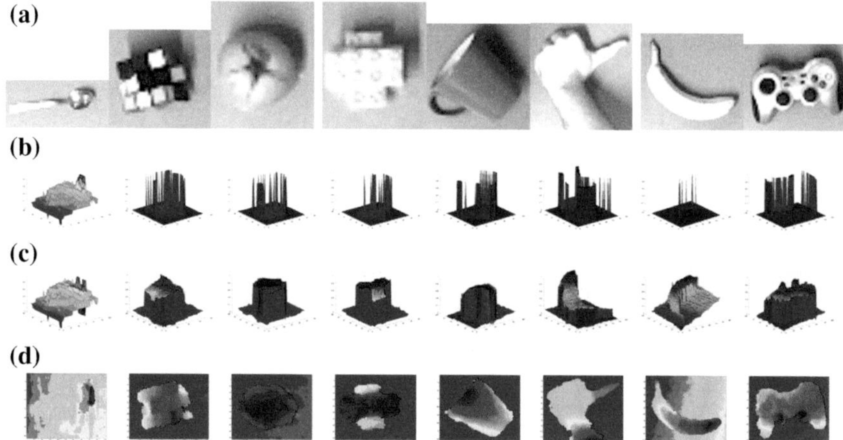

Fig. 7 Visual inspection of Kinect results for 8 objects: **a** Objects. **b** Point clouds with outliers. **c** Point clouds after outlier elimination. **d** Final depth maps

- Still, for close-range objects, the Kinect is working under critical conditions. The depth value is nearly constant (the normalized mean is usually around 0.5 and standard deviation values are quite high as a result of small depth intervals being mapped to the range of [0, 1]).
- The SFS-based depth maps show a much differentiated distribution of mean values and much narrower standard deviation values. This corresponds to a well-mapped surface structure of the tested objects.
- Visual inspection confirms, that the surface structure of the object is much detailed represented in the SFS-depth map than in the Kinect's point cloud.

5 Conclusions

It was shown, that close-range object pose recognition can well be based on depth-maps obtained from single IR images. A map of relative depth values is sufficient, as it represents the surface structure, enables surface feature extraction and classification into object parts. In perfect lighting conditions, the expected MSE of (relative) depth estimation by the two SFS algorithms is around 0.001–0.070, depending on the lighting direction. In a real environment, slightly worse results are obtained, but still being of higher quality than the point clouds generated by a Kinect device for close-range objects.

Acknowledgments The authors gratefully acknowledge the financial support of the National Centre for Research and Development (Poland), grant no. PBS1/A3/8/2012.

References

1. Stefańczyk, M., Kasprzak, W.: Multimodal segmentation of dense depth maps and associated color information. Lect. Notes Comput. Sci. **7594**, 626–632 (2013)
2. Jiang, L., et al.: 3D surface reconstruction and analysis in automated apple stem-end/calyx identification. Trans. ASAE **52**(5), 1775–1784 (2009)
3. Kasprzak, W.: Image and speech recognition. Rozpoznawanie obrazow i sygnałow mowy. Warsaw University of Technology, WUT publishing house, Warsaw (2009)
4. Ping-Sing, T., Shah, M.: Shape from shading using linear approximation. Image Vis. Comput. **12**(8), 487–498 (1994)
5. Horn, B.K.P.: Obtaining shape from shading information. In: Winston, P.H. (ed.) The Psychology of Computer Vision. McGraw-Hill, New York (1975)
6. Ikeuchi, K., Horn, B.K.P.: Numerical shape from shading and occluding boundaries. Artif. Intell. **17**, 141–185 (1981)
7. Pentland, A.P.: Linear shape from shading. Int. J. Comput. Vis. **4**(2), 153–162 (1990)

Utilization of Colour in ICP-based Point Cloud Registration

Marta Łępicka, Tomasz Kornuta and Maciej Stefańczyk

Abstract Advent of RGB-D sensors fostered the progress of computer vision algorithms spanning from object recognition, object and scene modelling to human activity recognition. This paper presents a new flavour of ICP algorithm developed for the purpose of pair-wise registration of colour point clouds generated from RGB-D images. After a brief introduction to the registration problem, we analyze the ICP algorithm and survey its different flavours in order to indicate potential methods of injecting of colour into it. Our consideration led to a solution, which we validate experimentally on colour point clouds from the publicly available dataset.

1 Introduction

In order to operate in unstructured and dynamically changing environments, service robots must possess the ability to recognize diverse objects. RGB-D sensors provide colour images supplemented with depth maps, which significantly facilitate the object recognition [11]. One of typical solutions to object recognition depends on three-dimensional models of objects, stored in the form of point clouds, supplemented with sparse clouds of features. In such a case, the recognition relies on matching and spatial grouping of correspondences between model and features extracted from the image, as presented in, e.g. [14]. Object models can be extracted straight from CAD model, however, in the case of objects of everyday use, those models are typically not existing. Hence, this raises a problem of building models straight from the physical objects, having diverse shapes, colours and textures. Of course a collection of images

M. Łepicka (✉) · T. Kornuta · M. Stefańczyk
Institute of Control and Computation Engineering, Warsaw University of Technology, Nowowiejska 15/19, 00-665 Warsaw, Poland
e-mail: marta.lepicka@gmail.com

T. Kornuta
e-mail: T.Kornuta@ia.pw.edu.pl

M. Stefańczyk
e-mail: M.Stefanczyk@ia.pw.edu.pl

R. Burduk et al. (eds.), *Proceedings of the 9th International Conference on Computer Recognition Systems CORES 2015*, Advances in Intelligent Systems and Computing 403, DOI 10.1007/978-3-319-26227-7_77

of a given object taken from different views can play a role of a multi-view model of the object itself (e.g. organized in an object-pose tree [7]), what bypasses the problem of integration of views into a 3D model. We, however, believe that autonomous modelling of unknown objects is one of the crucial steps towards enabling robots to adapt to new, unconstrained environments. In computer vision the problem of combination of multiple views into a single, consistent 3-D model is known as registration. In this paper, we focus on one of the RGB-D registration subproblems—on the so-called pair-wise registration of point clouds. This is typically performed in two steps: first, on the basis of the correspondences found between features extracted from the two merged point clouds an initial transformation between them is estimated , which is subsequently refined in several iterations, by minimization of a distance between selected pairs of points. The former step is known under the name visual odometry (VO) [9] (due to the fact that it estimates the motion of the sensor), whereas the latter is known as the iterative closest point (ICP) [2]. In this paper, we investigate the possible methods of modification of the classical ICP, relying on spatial coordinates of points, using the colour of those points. In the following section, we analyze the ICP algorithm and indicate two possible places where the colour can be utilized. Next we present a discussion leading to an ICP flavour, subsequently validated on a publicly available dataset.

2 Iterative Closest Point

Despite the name iterative closest point appeared first in the work of Besl and McKay [1], the same subject, being a part of a problem of registration of multiple range images, was addressed earlier, e.g. by Chen and Medioni [2]. Since then it was investigated by many researchers and as reported in one of the most current surveys on ICP [10] in 2011 there were more then 400 papers related to this subject, proposing diverse modifications and improvements of the algorithm. Searching for the proper method of utilization of colour, we had to distinguish and analyze the major steps of the algorithm. For this purpose, we followed the taxonomy of ICP variants proposed by Rusinkiewicz and Levoy [12] and updated it by steps described in [10]. This resulted in the ICP algorithm presented in Fig. 1. The input of the algorithm consist of two point clouds (named **reference** and **reading** clouds after [10]), that first typically undergo preprocessing in the **Data filtering** blocks. They might consist of several operations, extending the input cloud (e.g. by adding normal vectors) or decreasing its size (selection of points which will be matched, e.g. random removal of some of them). Next, the filtered reading cloud is rotated and translated using the initial transformation (e.g. estimated by VO) or transformation found in the previous iteration. The resulting reading cloud and filtered reference cloud are next associated together in **Data association**, which find correspondences between points from both clouds, for example with the use of k-Nearest Neighbours (kNN) algorithm.

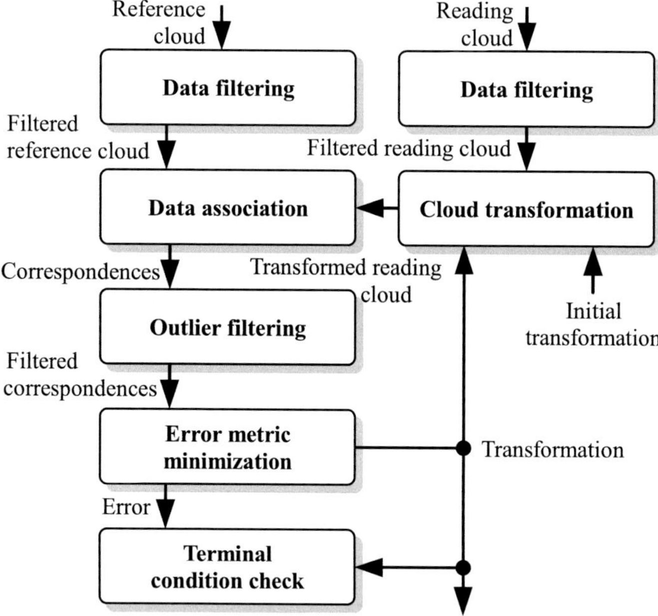

Fig. 1 Data flow diagram of ICP framework

The subsequent **Outlier filtering** is responsible for rejecting certain pairs and/or by adding different weights to them. The exemplary criteria include maximum authorized distance between corresponding point or distances from the centres of cloud masses. The remaining correspondences form the input to the optimization procedure **Error metric minimization**, which minimizes the error being sum of distances between corresponding points by changing the transformation applied to the reading cloud. This is typically done with the use of singular value decomposition (SVD) or Levenberg–Marquardt algorithm (LMA). Finally, the **Terminal condition check**s whether the procedure should end or repeat once again, by monitoring the number of performed iterations, convergence of resulting error or adjustment of transformation.

3 Injection of Colour into ICP

Analysis of the presented algorithm indicates two possible steps, where colour can be utilized: during finding of correspondences in **Data association** or used in computation of distances between points in **Error metric minimization**. The most elementary error metric used in ICP (i.e. objective function of the ICP minimization procedure) utilizes Euclidean distance (L2) between Cartesian coordinates of corresponding points. Assuming that $s_i = (s_i^x, s_i^y, s_i^z, 1)^T$ and $t_i = (t_i^x, t_i^y, t_i^z, 1)^T$ are coordinates of a point of reference cloud and corresponding point from reading cloud respectively, the refined transformation in the form homogeneous matrix M_{opt} is found by minimizing the function:

$$M_{opt} = \arg\min_{M} \sum_{i} d(s_i', t_i), \tag{1}$$

where $s_i' = M \cdot s_i$ is the projection of the point using current estimation of transformation M, whereas distance is defined as

$$d(s_i, t_i) = \sqrt{(s_i^x - t_i^x)^2 + (s_i^y - t_i^y)^2 + (s_i^z - t_i^z)^2}. \tag{2}$$

This metric was proposed in [1] and currently is known as *point-to-point* metric. An alternative yet quite popular metric was proposed in [2]. It introduces utilization of normal vectors. Assuming that $n_i = (n_i^x, n_i^y, n_i^z, 0)^T$ is the normal vector of the reading cloud, the objective function is defined as:

$$M_{opt} = \arg\min_{M} \sum_{i} d(s_i', t_i, n_i), \tag{3}$$

where $s_i' = M \cdot s_i$ and distance is given by the equation:

$$d(s_i, t_i, n_i) = |(s_i - t_i) \bullet n_i|. \tag{4}$$

As it gives good results for aligning planes, hence was named *point-to-plane*, in contrast to *point-to-point* metric from 1. The idea of injection of colour into the ICP error metric is not new, e.g. more than two decades ago Godin et al. [3] proposed an ICP falvour named ICCP (iterative closest compatible point), enhancing the range data with additional intensity information. However, the major question appeared: how to normalize the impact of two parts of the objective function, being partially measured in metres (Cartesian part) and in pixel intensity (colour part). For example, Johnson and Kang [4] proposed a simple colour/shape L2 distance definition:

$$d(s_i, t_i) = \sqrt{(s_i^x - t_i^x)^2 + (s_i^y - t_i^y)^2 + (s_i^z - t_i^z)^2 + \alpha_1(s_i^Y - t_i^Y)^2 + \alpha_2(s_i^I - t_i^I)^2 + \alpha_3(s_i^Q - t_i^Q)^2}, \tag{5}$$

with s_i^Y, s_i^I and s_i^Q representing Y, I, Q components of colour of a given point, and α_1, α_2 and α_3 being hand-crafted normalization coefficients. A slightly different approach was described in [8], where an ICP flavour called Hue-Assisted ICP was proposed. The authors used *point-to-point* distance given 2 during the optimization, but decided to utilize colour during the data association stage by definition of distance:

$$d(s_i, t_i) = \sqrt{(s_i^x - t_i^x)^2 + (s_i^y - t_i^y)^2 + (s_i^z - t_i^z)^2 + (s_i^H - t_i^H)^2}, \tag{6}$$

where s_i^H represents hue from the HSV colour space. As the hue values were normalized to the range 0–1, the authors proposed a method of normalizing the point coordinates to the same range by using the maximum range of the utilized LIDAR. In our research, first we have injected the colour to error metric in the **Error metric minimisation** step, analogically to [4], hence used error metric similar to 5, but with RGB channels instead of YIQ. However, we disapproved both of the aforementioned normalization methods—the hand-crafted parameters are typically optimized for a given dataset (hence must be fine-tuned for a different one), whereas utilization of maximum range of sensors (~3.5 m for Kinect) in registration of articulated objects (of size ranging from a few to a few tens of cm) would result in predominant influence of the colour similarity over Cartesian distance. Thus we decided to use the the maximal radius of the neighbouring sphere (the parameter of the k-NN used in **Data association**) instead, which in fact was the maximum Cartesian distance between found correspondences. Sadly, despite it was quite logical, the solution had very low convergence and the majority of found transformations was improper. After analysis, we have concluded that such an optimization could properly work only in the case of surfaces with smoothly changing colours, whereas in the case of complex textures colour can change abruptly. For this reason, we decided not to inject colour into the goal function and focus on **Data association**. But instead of merging together spatial and colour information, we decided to split this process into two: first we find k neighbours relying only on Cartesian coordinates of points according to the 2, and in the second step select the nearest one on the basis of colour similarity, i.e.:

$$d(s_i, t_i) = \sqrt{(s_i^R - t_i^R)^2 + (s_i^G - t_i^G)^2 + (s_i^B - t_i^B)^2}. \quad (7)$$

This results from another conclusion that it makes no sense to search for points with similar colours when they belong to the totally different fragments of the cloud. But when we already know that a given subset lie within a neighbouring sphere, the colour information might facilitate the selection of the most proper candidate for a correspondence. In the following section, we compare four flavours of ICP, two of which utilize the presented two-step data association. The combinations of metrics used in **Data association** and **Error metric minimisation** are presented in Table 1.

Table 1 The combinations of metrics used in comparison of ICP

Step/ICP flavour	Standard	Normals	Colour	Colour + normals
Data association	2	2	2 + 7	2 + 7
Error metric minimisation	2	4	2	4

4 Experimental Verification

Comparison of different ICP flavours requires a suitable methodology, including a proper dataset and meaningful performance metrics. In the early works on ICP (e.g. in [12]), the authors typically used artificially generated scenes, with ground truth consisting of known overlapping and transformation, and focused on the rapidity of convergence and final accuracy, the former being the number of iterations and the latter being measured as root mean squared error (RMSE) between the computed and known transformation. In our work, instead of relying on artificial data, we decided to follow the work [10] and use real-world dataset. Thus we decided to present results of the experiments performed on the sequences of images of objects taken from publicly accessible Washington RGB-D Object Dataset [6], enabling other researches interested in this subject to compare their results with ours. Each sequence contains different views of a given object, examples of each one of them are presented in Fig. 2. We selected those objects because they represent different shapes, i.e. bag of chips is an irregular object, cereal box represents a cuboidal objects (with flat surfaces), whereas can represents cylindrical objects (with convex surfaces), thus are to some extent similar to the test scenes from [12]. During the experiments, we processed each point cloud constituting the segmented object from each sequence separately, by transforming it with known homogeneous matrix and adding a gaussian noise to each point coordinates. The resulting cloud was next aligned with the original one. In the following, we present the results for transformation consisting of rotation of 0.1° around Z and translation along X, Y and Z equal to 5 mm, so the point clouds before registration were only partially overlapping each other. At the start, we collected the following metrics: the number of iterations, RMSE for the translational part of transformation (computed on the basis of difference between the given and computed translation) and RMSE for the orientation (computed as geodesic distance on a unit sphere as suggested in [10]). This, however, appeared to be not sufficient to effectively compare the results. Hence, we started to collect another RMSE, i.e. the mean Euclidean distances between all of the corresponding points from both clouds, and present them grouped into bins, separately for each of the sequences.

As some of them appear to be bigger than 0.01 m, they clearly indicated that they were incorrectly aligned. Hence we added a threshold enabling us to determine

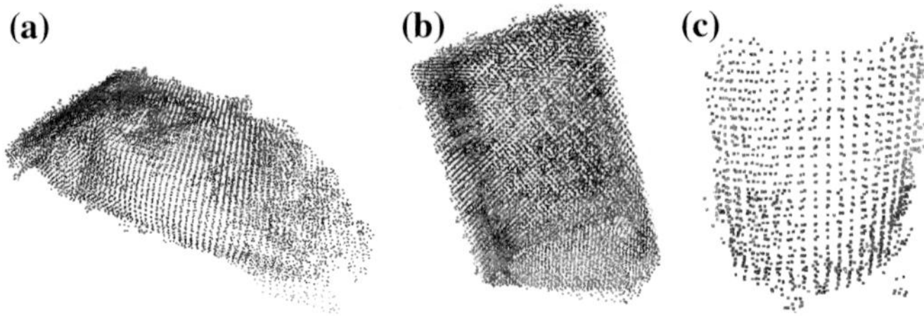

Fig. 2 Exemplary point clouds from sequences containing selected objects. **a** Bag of chips. **b** Cereal box. **c** Soda can

Table 2 Results of convergence for the sequence containing bag of chips

ICP flavour	Not converged (%)	Bad convergence (%)	Success rate (%)	Mean number of iterations (–)
Standard	0	2.78	97.22	35.72
Normals	1.64	18.61	79.75	22.49
Colour	0	2.03	97.97	19.7
Colour + normals	0.51	14.81	84.68	22.5

whether the found alignment is incorrect, as well as we started to record the number of convergence failures. Because we used only part of the point clouds from each sequence (but more than 100 for each) and each cloud consisted of slightly different number of points constituting the object, we decided to present everything as percentages. Table 2 presents the accumulated results of convergence for the sequence of images of bag of chips (object named *food_bag_6* in [6], sequence of 790 images). As it possess an irregular shape and a heterogeneous texture, the standard and colour ICP gave better results than the ICP normal flavour. Although the standard ICP had really good convergence, the RMSE was in this case rather big (Fig. 3) and the injection of colour greatly improved both the convergence speed (less number of iterations) and the final alignment quality (RMSE lower than 10^{-5} m). The flavour using both colour and normals appeared to have slightly better RMSE and convergence success rate than ICP Normals, but under all criteria it is worse than ICP colour. The results for point clouds containing boxes are presented in Table 3 (object *food_box_11*, 782 images). As those clouds contain mainly flat, textured surfaces, the ICP Standard had the worst results, being overwhelmed by the other three flavours, taking into account the convergence success rate, speed and mean RMSE (Fig. 4). However, despite the obvious supremacy of ICP Colour + normals in convergence success rate and lowest RMSE, it is worth noting that the ICP colour converged much faster than the others and had results similar to those from aligning bag of chips, being only slightly worse than the two flavours using normals. In yet another opposition,

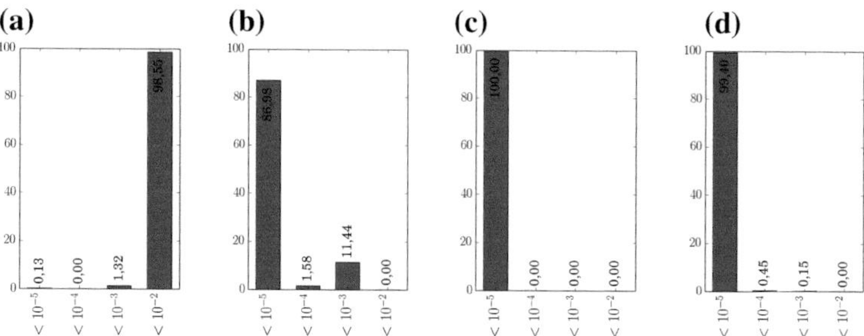

Fig. 3 Histograms containing mean RMSE for the bag of chips sequence. **a** ICP standard. **b** ICP normals. **c** ICP colour. **d** ICP colour + normals

Table 3 Results of convergence for the sequence containing cereal box

ICP flavour	Not converged (%)	Bad convergence (%)	Success rate (%)	Mean number of iterations (−)
Standard	0	51.15	48.85	27.12
Normals	6.39	4.99	88.62	10.84
Colour	0	11.25	88.75	21.92
Colour + normals	1.92	6.01	92.07	11.16

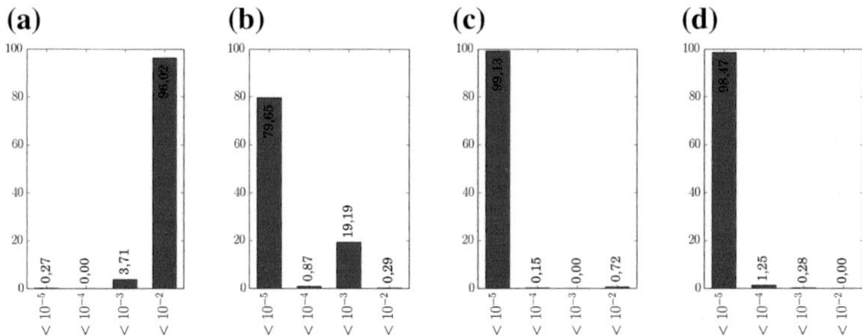

Fig. 4 Histograms containing mean RMSE for the cereal box sequence. **a** ICP standard. **b** ICP normals. **c** ICP colour. **d** ICP colour + normals

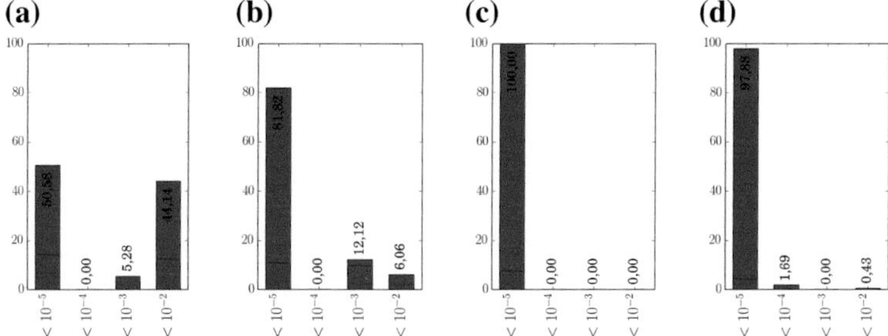

Fig. 5 Histograms containing mean RMSE for the can sequence. **a** ICP standard. **b** ICP normals. **c** ICP colour. **d** ICP colour + normals

the results for aligning the sequence of cloud containing can (object *food_can_8*, 789 images), i.e. a shape convex, present a failure of the normal-based ICP flavours, where the convergence success rate was reduced to almost 4 % for pure normals and to 56 % for normals combined with colour, recognizing the ICP Colour as the best with 100 % success rate similarly to ICP Standard, but with significantly (i.e. three orders) smaller RMSE than the latter. Summing up the presented results, we conclude that the ICP colour flavour is the best choice, with high convergence success rate, small number of required iterations and mean RMSE in majority of cases smaller than 10^{-4} m. But the most important is that it was the most stable, whereas the ICP standard and the two flavours using normals failed to converge very often (Fig. 5 and Table 4).

Table 4 Results of convergence for the sequence containing soda can

ICP flavour	Not converged (%)	Bad convergence (%)	Success rate (%)	Mean number of iterations (−)
Standard	0	0	100	46.02
Normals	95.69	0.13	4.18	16.79
Colour	0	0.38	99.62	15.77
Colour + normals	65.65	4.44	29.91	16.91

5 Conclusions

The classical ICP relies on spatial coordinates of points, whereas RGB-D sensors offer also colour associated with each of those points. For this reason, we investigated and discussed possibilities of injecting colour in ICP and implemented a new ICP flavour. We also analyzed methods of experimental verification of our approach, as well as metrics that can be used. We implemented different flavours of ICP with the use of the DisCODe framework [13], with components encapsulating selected functions and classes from the PCL and OpenCV libraries. The experimental results confirm that our solution behaves correctly and gives better results in comparison to the other, selected ICP flavours; however, we point out that this method is developed for rich-textured objects and scenes. The presented research on ICP is in fact a spin off project from our ongoing work on recognition of articulated objects for the purpose of their grasping and manipulation [5], where we encountered the problem of acquisition of three-dimensional models of the objects of interest. Aside of pair-wise point cloud alignment, we are also trying to use the SLAM-based loop closure methods (i.e. LUM and ELCH) to refine the generated object models. Recently, we also created a MongoDB-based repository for object models and acquired views of objects and scenes, which constitutes a knowledge base for our robotic system at the one hand, and a benchmark for registration and recognition algorithms at the other. In particular, we use it in our further research on ICP, including blurring of the colour of neighbouring points (which might foster the convergence) and on methods of automatic rejection of improper alignments (i.e. without a priori known ground truth).

Acknowledgments The authors acknowledge the financial support of the National Centre for Research and Development grant no. PBS1/A3/8/2012.

References

1. Besl, P., McKay, N.: A method for registration of 3-d shapes. IEEE Trans. Pattern Anal. Mach. Intell. **14**(2), 239–256 (1992)
2. Chen, Y., Medioni, G.: Object modeling by registration of multiple range images. In: Proceedings of the 1991 IEEE International Conference on Robotics and Automation, pp. 2724–2729. IEEE (1991)

3. Godin, G., Rioux, M., Baribeau, R.: Three-dimensional registration using range and intensity information. In: Photonics for Industrial Applications. International Society for Optics and Photonics, pp. 279–290 (1994)
4. Johnson, A.E., Kang, S.B.: Registration and integration of textured 3d data. Image vis. comput. **17**(2), 135–147 (1999)
5. Kasprzak, W., Kornuta, T., Zieliński, C.: A virtual receptor in a robot control framework. In: Recent Advances in Automation, Robotics and Measuring Techniques. Advances in Intelligent Systems and Computing (AISC), Springer (2014)
6. Lai, K., Bo, L., Ren, X., Fox, D.: A large-scale hierarchical multi-view RGB-D object dataset. In: 2011 IEEE International Conference on Robotics and Automation (ICRA), pp. 1817–1824. IEEE (2011)
7. Lai, K., Bo, L., Ren, X., Fox, D.: A scalable tree-based approach for joint object and pose recognition. In: Proceedings of the AAAI Conference on Artificial Intelligence, pp. 1474–1480 (2011)
8. Men, H., Gebre, B., Pochiraju, K.: Color point cloud registration with 4d icp algorithm. In: 2011 IEEE International Conference on Robotics and Automation (ICRA), pp. 1511–1516. IEEE (2011)
9. Nistér, D., Naroditsky, O., Bergen, J.: Visual odometry. In: Proceedings of the 2004 IEEE Computer Society Conference on Computer Vision and Pattern Recognition, CVPR 2004, vol. 1, I-652 p. IEEE (2004)
10. Pomerleau, F., Colas, F., Siegwart, R., Magnenat, S.: Comparing icp variants on real-world data sets. Auton. Robot. **34**(3), 133–148 (2013)
11. Ren, X., Fox, D., Konolige, K.: Change their perception: RGB-D for 3-D modeling and recognition. IEEE Robot. Autom. Mag **20**(4), 49–59 (2013)
12. Rusinkiewicz, S., Levoy, M.: Efficient variants of the icp algorithm. In: Proceedings of the Third International Conference on 3-D Digital Imaging and Modeling, 2001, pp. 145–152. IEEE (2001)
13. Stefańczyk, M., Kornuta, T.: Handling of asynchronous data flow in robot perception subsystems. In: Simulation, Modeling, and Programming forAutonomous Robots. Lecture Notes in Computer Science, vol. 8810, pp. 509–520. Springer (2014)
14. Xie, Z., Singh, A., Uang, J., Narayan, K.S., Abbeel, P.: Multimodal blending for high-accuracy instance recognition. In: 2013 IEEE/RSJ International Conference on Intelligent Robots and Systems (IROS), pp. 2214–2221. IEEE (2013)

Range Sensors Simulation Using GPU Ray Tracing

Karol Majek and Janusz Bedkowski

Abstract In this paper the GPU-accelerated range sensors simulation is discussed. Range sensors generate large amount of data per second and to simulate these high-performance simulation is needed. We propose to use parallel ray tracing on graphics processing units to improve the performance of range sensors simulation. The multiple range sensors are described and simulated using NVIDIA OptiX ray tracing engine. This work is focused on the performance of the GPU acceleration of range images simulation in complex environments. Proposed method is tested using several state-of-the-art ray tracing datasets. The software is publicly available as an open-source project SensorSimRT.

Keywords Ray tracing · RGB-D sensors · Simulation

1 Introduction and Related Work

Mobile robotics uses multiple robotic platforms with many types of sensors. There are multiple sensors attached to one platform. The sensors are gathering more and more data with higher frame rates resulting increase of the number of measurements per second. Such increase results in higher computational load and rises hardware requirements for simulators to provide the proper data throughput. To provide real-time simulation of such sensors in very complex large environments, the high-performance GPU-accelerated ray tracing is needed. The Optix engine is the natural choice for this type of application. The implementation of the simulation framework is available on GitHub (https://github.com/LIDER-MSAS/SensorSimRT). In subsequent sections, we present state of the art in robotic simulation and ray tracing.

K. Majek (✉)
Institute of Mathematical Machines, Krzywickiego 34,
02-078 Warsaw, Poland
e-mail: karolmajek@gmail.com

J. Bedkowski
e-mail: januszbedkowski@gmail.com

R. Burduk et al. (eds.), *Proceedings of the 9th International Conference on Computer Recognition Systems CORES 2015*, Advances in Intelligent Systems and Computing 403, DOI 10.1007/978-3-319-26227-7_78

We present multiple depth sensors which are described and simulated. The paper concludes with the results section where the experimental results are presented and discussed in conclusions.

Ray Tracing

The ray tracing algorithms are very popular in computer graphics because of their capabilities. They can render high-quality shadows, lighting, reflections, and effects impossible or difficult with other techniques [1]. The algorithms must compute intersections for all rays with the scene. The complexity of the scenes is increasing to achieve more realistic environments and results in increase of the computational load. To accelerate ray tracing for the scene, the data partitioning structures are used such as

- Grid [2],
- Octree [3, 4],
- Kd-tree [5, 6],
- Bounding Volume Hierarchy (BVH) [7].

The range sensors simulation especially the laser range finders (LRFs) can be simulated using the basic ray tracing without external lighting tracing only the rays from LRF to objects.

Depth Sensors

Kornuta and Stefańczyk compared the wide range of RGB-D sensors in [8]. They described stereo sensors (Bumblebee 2, nDepth, Bumblebee XB3), structural light (Microsoft Kinect, Primesense, Structure), time of flight (SICK LMS-1xx, Hokuyo UTM, Velodyne HDL-64E, SwissRanger). The Kinect v1 sensor is widely used in robotics. Its depth sensor has 300×200 resolution. The full resolution depth image (640×480) is acquired using hardware interpolation. The depth range is from 1.8 to 3.5 m. The field of view is 57° horizontal and 43° vertical. The Kinect v2 is the new sensor which can be used in robotics. It is based on time of flight (TOF) principle instead of structured-light technique in Kinect v1. In 2014 Amon and Fuhrmann evaluated the spatial resolution of v1 and v2 Kinect sensors [9]. The purpose of the evaluation was to compare the spatial resolution accuracy of the face tracking system. The depth range starts from 1.3 to 3.5 m. The resolution of the depth image is 512×424 and the horizontal and vertical field of view is $70 \times 60°$, respectively. The Velodyne HDL-64E lidar has 4 groups of 16 lasers in a rotating unit. There are also 2 groups of 32 laser receivers. The field of view is 360° horizontal and 26.8° vertical. The Velodyne HDL-32E has 32 laser/detector pairs. The vertical field of view is -30.67–$+10.67°$. The frame rate is user selectable from 5 to 20 with default set to 10 Hz. Maximum number of points per second is 700,000. The Velodyne VLP-16 lidar is newly introduced and does not have full specification widely available. The product leaflet claims for 16 channels, $\pm$ 15° vertical field of view, range over 100 meters and 300,000 point per second. This scanner will be simulated with assumption of 10 Hz frequency of gathering full scan. The MultiSense-SL is a sensor from Carnegie Robotics [10] that combines multiple data sources. It consists of the Hokuyo UTM-30LX-EW LRF and two RGB cameras for stereo vision. The Hokuyo is axially rotated on a spindle. The vendor claims that the speed is user-specified, but do not say what is the minimum and the maximum speed. The laser has scan angle of 270° and 1080 measurements with angular resolution of 0.25°. The maximum distance is 30 meters. The device performs scan in every 25 ms. Sensor documentation lacks of information about the rotation speed or number of points per scan, so this sensor will not be evaluated in experiments section. The Mandala 3D Unit [11] is a new rotating unit which supports SICK LMS1xx and SICK TIM5xx laser scanners. The rotating unit has a customizable speed. In experiments for this article, the SICK LMS100 version is considered. This laser generates 541 rays in 270° field of view. To acquire full 270 × 360° scan, only half revolution of the unit is needed. The default frequency of the SICK scanner is 10 Hz and it will be used for the simulation. The Z + F Imager 5010 is a geodetic laser scanner which can be used in mobile robotics tasks. It has the range of 187.3 m and 320 × 360° field of view. The scanner generates point clouds from 1 to 10,000 million points. The middle 5000 × 5000 resolution, which gives 25 million points, will be used in experiments to simulate this scanner.

Simulators

Microsoft Robotics Developer Studio (MSRDS) is a freely available programming environment for building robotics applications. The simulator supports SICK LMS200 and Kinect simulation [12]. According to the datasheet, the MSRDS 4 supports up to 4 Kinect sensors simulated simultaneously. MORSE [13] is an open-source robotics simulator. It is designed as a "Software In the Loop." The simulator is built on top of blender using python scripts. The simulations are executed in Game Engine mode to ensure high-quality rendering and to use Bullet physics engine. The simulator is primarily focused on real-time simulation of whole robots and less on accurate simulation of sensors. The Defense Advanced Research Projects Agency (DARPA) Robotics Challenge (DRC) of 2012–2015 used the CloudSim tool in the Virtual Robotics Challenge (VRC) [14]. The simulator was providing simulation of the scenes and ATLAS robot physics and its sensors. Robot was equipped with stereo cameras and the MultiSense-Sl sensor. All the sensors where simulated in real time. The CloudSim is a well-known Gazebo simulator running in the cloud on a OpenStack instance. Unified System for Automation and Robot Simulation (USARSim) can simulate laser rangefinders (LRF) in a smoky environment as described in [15]. USARSim is built on the commercially available Unreal Engine 3.0 by Epic Games. Range-Image can be simulated using the Z-buffer from the rendering pipeline. This approach is used in [16]. Authors use OpenGL library to find the distances to the closest object in each pixel. Color information is also simulated. The simulated depth image is compared with measured real environment and the result is used to produce a likelihood for the particle pose to use in Monte Carlo Localization. In [17], authors have shown a GPU-accelerated framework for simulation of TOF-sensors based on the PMD (photo mixing detector) technology. It uses the vertex and fragment programs in shading language to compute the depth image.

2 Simulation of Range Sensors

In this paper, we introduce GPU-accelerated simulation of several range sensors. We use state-of-the-art parallel GPU ray tracing implemented in the NVIDIA Optix engine. The idea of simulation is to shoot rays from the sensor to the scene and find the closest hits. The simulation pipeline is shown in Fig. 1. We use OptiX rtuTraversal API which provides information only about ray hits without shading or recursion capabilities. This API works both on CPU or on GPU. To perform the ray tracing, the scene and sensor rays must be prepared in GPU memory. The scene model is loaded from .obj file as a set of triangle and is copied to the GPU memory. The ray data are generated for the type of sensor as an array of sources and ray directions. Currently, the ray data are prepared in the host memory and then copied to the GPU. The number of rays corresponds to the number of points in a profile multiplied by the number of profiles in a full point cloud. The resulting depth image is copied back to the host memory. To measure the performance, several types of environments and scanners are used. The measurement is performed using 1000 iterations of the ray tracing for the specified pair model-scanner. In the listings below, we present the code for the traversal. To initialize the RTUtraversal structure (t), the rtuTraversalCreate must be used. It can be used for different queries, here it computes closest hit. The CPU ray tracing is performed if the number of CPU threads *cpu_threads* is more than 0. The initialization code for the ray traversal:

```
RTUtraversal t;
rtuTraversalCreate(&t, RTU_QUERY_TYPE_CLOSEST_HIT,
RTU_RAYFORMAT_ORIGIN_DIRECTION_INTERLEAVED,
RTU_TRIFORMAT_MESH,
RTU_OUTPUT_NONE, cpu_threads ? RTU_INITOPTION_CPU_ONLY :
RTU_INITOPTION_NONE, context);
```

For the CPU ray tracing the number of CPU threads must be set:

```
rtuTraversalSetOption(t, RTU_OPTION_INT_NUM_THREADS,
(void*)(&cpu_threads));
```

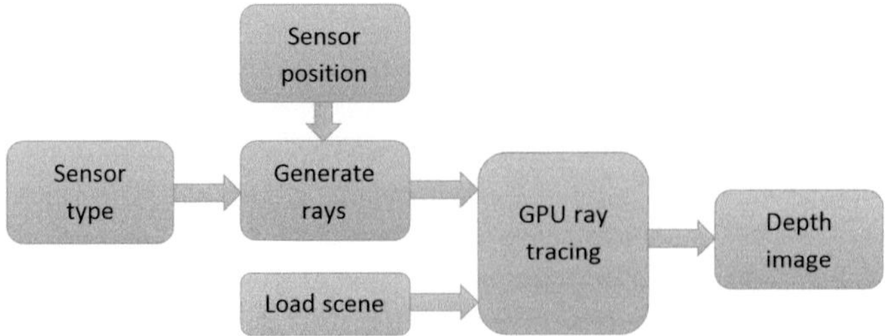

Fig. 1 The simulation pipeline

The second step is to map the ray data. The data containing information about rays is an array *rays* of floating point values. Size of the array is six times *num_rays*—the rays count. The ray is represented by the 3D position of the source and the direction. The array contains interleaved origin and direction coordinates.

```
float* rays_mapped_ptr;
unsigned int num_rays;
float* rays;    \\x0,y0,z0,dirx0,diry0,dirz0,\ldots
//Generate rays
rtuTraversalMapRays(t, num_rays, &rays_mapped_ptr);
memcpy(rays_mapped_ptr, rays, num_rays*sizeof(float3)*2);
rtuTraversalUnmapRays(t);
```

Next step is to load the mesh containing the geometrical information about the environment. We use the triangle mesh format—an array of vertices and array of indices for all triangles. The rays can be set also as triangles soup, an array of coordinates of the triangles, using *rtuTraversalSetTriangles* function.

```
unsigned int num_tris, num_verts;
float* verts;     //v0x,v0y,v0z,\ldots
unsigned int* indices;  //tr0v0,tr0v1,tr0v2,\ldots
//Load mesh
rtuTraversalSetMesh(t, num_verts, verts, num_tris, indices);
```

Before the traversal, the preprocessing function *rtuTraversalPreprocess(t)* can be called. It performs preprocessing such as the acceleration structure building or OptiX context compilation. This is not necessary to call this function as *rtuTraversal Traverse*, will call this internally as necessary. It takes about 1.4 for the *Conference*, 52 for the *San Miguel*, and about 80 s for the *Power Plant* scene. It is performed only once to optimize the ray tracing process in a complex scene. To correctly measure the performance, the 50 *warm up* iterations is done. After that 1000 iterations of the ray tracing are performed to measure the performance. In each iteration, the following code is executed:

```
rtuTraversalMapRays(t, num_rays, &rays_mapped_ptr);
memcpy(rays_mapped_ptr, rays,  num_rays*sizeof(float3)*2);
rtuTraversalUnmapRays(t);
rtuTraversalTraverse(t);
rtuTraversalMapResults(t, &results_mapped_ptr);
memcpy(results, results_mapped_ptr, num_rays *
sizeof(RTUtraversalresult);
rtuTraversalUnmapResults(t);
```

The sensors simulated using GPU ray tracing are listed in the Table 1. There can be found an information about the rays count used in the simulation. The GPU simulation

Table 1 Information about the simulated sensors

Sensor	Rays count	Scan time in milliseconds
LMS100	541	100
Velodyne VLP-16	$16 \cdot 1{,}875 = 30{,}000$	100
Kinect 1	$300 \cdot 200 = 60{,}000$	33.3
Velodyne HDL-32E	$32 \cdot 2{,}187 = 69{,}984$	100
Mandala 3D Unit LMS100	$541 \cdot 250 = 135{,}250$	25,000
Kinect 2	$512 \cdot 424 = 217{,}088$	33.3
Velodyne HDL-64E	$64 \cdot 4{,}000 = 256{,}000$	67–200
ZF Imager 5010	$5{,}000 \cdot 5{,}000 = 25{,}000{,}000$	202,000

of each sensor is several times faster as shown in experiments section, so multiple sensors can be simulated at the same time. The simulation of the Microsoft Kinect sensor is simplified. Instead of simulating the IR emitter and detector, the time of flight method is used. This approximation eliminates emitter/receiver shadows so it is only for the performance check. Simulation of the structured-light is considered as a future work.

3 Results

To compare the performance of range sensors simulation, we have done experiments using different environments. The environments are listed in Table 2. We report the results of our experiments on the Conference, San Miguel, and Power Plant [18] models. The simulation was running on a modern laptop with the NVIDIA GeForce GTX970M using OptiX 3.7.0 and Intel Core i7-4710HQ 2.5GHz. The GTX970M has 10 Streaming Multiprocessors with 128 CUDA cores each. Each multiprocessor can run 2048 threads simultaneously. The time was measured including ray data copying to GPU and copying the results back in each iteration. The used sensors with information about number of rays and the scanning time by the real hardware are listed in Table 1. The example outputs for the both Kinect sensors, Mandala 3D unit, and ZF 5010 in three scenes are shown in Table 3. The depth images from the rest of sensors are not presented because of the aspect ratio and height of depth images. The

Table 2 Environments used in the experiments

Environment	Triangles count	Vertices count
Conference	282,759	166,940
San miguel	7,880,512	4,488,339
Power plant	12,759,246	5,984,083

Table 3 Simulated depth images for Kinect 1 and 2, Mandala sensor, and ZF 5010 in conference, San miguel, and power plant scenes

Conference scene	San miguel scene	Power plant scene
Kinect 1		
Kinect 2		
Mandala 3D Unit		
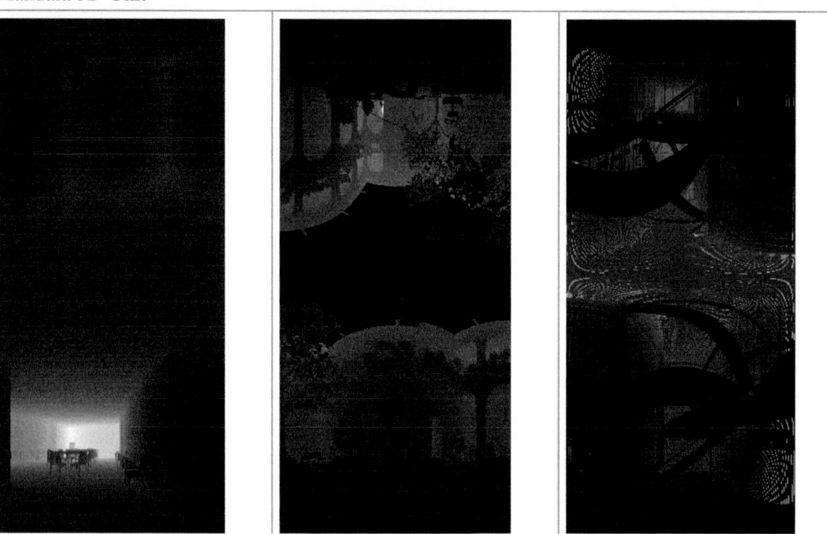		

(continued)

Table 3 (continued)

Conference scene	San miguel scene	Power plant scene
ZF 5010		

Table 4 Time of the ray tracing iteration measured during 1000 iterations of ray traversal in scenes. The values are in milliseconds

Sensor	Conference					San miguel	Power plant
	GPU	1 CPU	2 CPUs	4 CPUs	8 CPUs	GPU	GPU
LMS100	0.73	0.28	0.22	0.16	0.30	1.03	2.16
VLP-16	1.41	15.44	8.15	4.23	3.63	1.62	3.19
Kinect 1	1.84	53.80	28.08	14.58	11.80	2.73	3.96
HDL-32E	2.01	34.35	18.78	10.03	8.42	2.21	4.47
Mandala	2.13	47.42	25.70	14.23	11.74	4.11	6.23
Kinect 2	2.99	53.80	28.08	14.58	11.80	5.08	8.27
HDL-64E	3.87	129	70.06	37.73	32.19	5.55	12.23
ZF 5010	185	7149	4047	2790	2078	406	832

San Miguel scene model used in simulation has some artifacts. Some objects appear as small spheres which can be seen in presented depth images. The experimental results are shown in the Table 4. Simulation of the Velodyne sensors uses 1875, 2187, and 4000 for VLP-16, HDL-32E, and HDL-64E, respectively (Fig. 2). CPU implementation is tested only in the conference scene because of the performance of the single/multicore implementation.

4 Conclusions and Future Work

The experimental results in this paper show that the GPU ray tracing can be efficiently used to simulate depth sensors. Ray tracing algorithms are the natural choice for such sensors. The performance of the simulation enables simulation of the multiple sensors in complex environments. The complexity of the environment corresponds to the

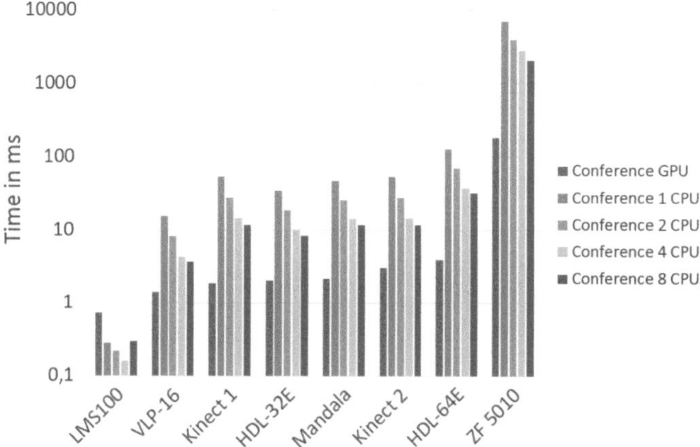

Fig. 2 Comparison between computation times using GPU and multiple CPUs on the conference scene

number of triangles in the scene. The power plant scene has 45 times more triangles than the conference. Such increase of triangles count results in 2–5 times increase of the ray traversing time depends on sensor. GPU-enabled ray tracing outperforms the CPU implementation if the number of rays is large enough. Simulation of a single SICK LMS100 scanner is done faster using serial implementation because of the time needed for data copy to the GPU and the results copy from it. Usage of the multiple CPUs decreases the simulation time. It can be used on machines without the CUDA support. The high performance of the simulation is an advantage because there is time for additional processing such as adding noise or discretization of the range values. The implementation of the GPU-accelerated simulation framework is available on GitHub (https://github.com/LIDER-MSAS/SensorSimRT). The future work will be integration of the simulator with robotics frameworks such as the ROS (robot operating system). The simulation can be improved adding the noise simulation, structured-light support, or multiple echo returns. Noise will be simulated using the NVIDIA CUDA Random Number Generation library (cuRAND) which is a part of the CUDA Toolkit. It can support multiple distributions such as uniform, normal, or Poisson. Adding the structured-light support needs to add the light rays tracing into the simulation pipeline. This is not supported by the fast ray tracing provided by the OptiX rtuTraversal API. The regular Optix API supports the simulation of the light sources, so it can be use for that purpose. The performance of this solution must be tested. The multiple echo returns can be implemented as a custom intersection program which will store multiple values of depth into ray's buffer. The experiments with dynamic environments should be done to find how it will affect the simulation performance. The ray generation functions can be done in parallel in the GPU memory directly using the CUDA technology and CUDA—Optix integration.

Acknowledgments This work is done with the partial support of NCBiR (Polish National Center for Research and Development) project: "Research of Mobile Spatial Assistance System" Nr: LIDER/036/659/L-4/12/NCBR/2013 and from the European Community's Seventh Framework Programme (FP7/2007-2013) under grant agreement nr:285417—project ICARUS Integrated Components for Assisted Rescue and Unmanned Search operations.

References

1. Whitted, T.: An Improved Illumination Model For Shaded Display. ACM Siggraph 2005 Courses. ACM, New York (2005)
2. Amanatides, J., Woo, A.: A fast voxel traversal algorithm for ray tracing. Eurographics **87**(3), 3–10 (1987)
3. Hanan, Samet: Implementing ray tracing with octrees and neighbor finding. Comput. Graph. **13**(4), 445–460 (1989)
4. Marc, Levoy: Efficient ray tracing of volume data. ACM Trans. Graph. (TOG) **9**(3), 245–261 (1990)
5. Donald, Fussell, Subramanian Kalpathi, R.: Fast Ray Tracing Using kd Trees. Department of Computer Sciences, University of Texas at Austin, Texas (1988)
6. Zhou, K., et al.: Real-time kd-tree construction on graphics hardware. ACM Trans. Graph. (TOG) **27**(5), 126 (2008)
7. Gunther, J.: Realtime ray tracing on GPU with BVH-based packet traversal. In: IEEE Symposium on Interactive Ray Tracing RT'07 (2007)
8. Maciej, Stefańczyk, Tomasz, Kornuta: Akwizycja obrazów RGB-D: metody. Pomiary, Automatyka, Robotyka **18**, 82–90 (2014)
9. Amon, C., Fuhrmann, F.: Evaluation of the spatial resolution accuracy of the face tracking system for kinect for windows v1 and v2. In: Proceedings of the 6th Congress of the Alps Adria Acoustics Association (2014)
10. Carnegie Robotics. http://carnegierobotics.com/multisense-sl/
11. Mandala Robotics. http://mandalarobotics.com/
12. Microsoft Robotics Developer Studio. http://www.microsoft.com/robotics/
13. Echeverria, G., et al.: Modular open robots simulation engine: Morse. In: Proceedings of the 2011 IEEE International Conference on Robotics and Automation (ICRA) (2011)
14. Aguero, C.E., et al. Inside the Virtual Robotics Challenge: Simulating Real-timeRobotic Disaster Response
15. Formsma, O., Dijkshoorn, N., van Noort, S., Visser, A.: Realistic simulation of laser range finder behavior in a smoky environment. In: Ruiz-del-Solar, J. (ed.) Lecture Notes in Computer Science, vol. 6556, pp. 336–349. Springer, Heidelberg (2010)
16. Fallon, M.F., Johannsson, H., Leonard, J.J.: Efficient scene simulation for robust Monte Carlo localization using an RGB-D camera. In: Proceedings of the 2012 IEEE International Conference on Robotics and Automation (ICRA) (2012)
17. Keller, M., et al.: A simulation framework for time-of-flight sensors. In: Proceedings of the IEEE 2007 International Symposium on Signals, Circuits and Systems (ISSCS 2007) vol. 1 (2007)
18. Computer Graphics Archive. http://graphics.cs.williams.edu/data/

View Synthesis with Kinect-Based Tracking for Motion Parallax Depth Cue on a 2D Display

Michał Joachimiak, Mikołaj Wasielica, Piotr Skrzypczyński, Janusz Sobecki and Moncef Gabbouj

Abstract Recent advancements in 3D video generation, processing, compression, and rendering increase accessibility to 3D video content. However, the majority of 3D displays available on the market belong to the stereoscopic display class and require users to wear special glasses in order to perceive depth. As an alternative, autostereoscopic displays can render multiple views without any additional equipment. The depth perception on stereoscopic and autostereoscopic displays is realized via a binocular depth cue called stereopsis. Another important depth cue, that is not exploited by autostereoscopic displays, is motion parallax which is a monocular depth cue. To enable the motion parallax effect on a 2D display, we propose to use the Kinect sensor to estimate the pose of the viewer. Based on pose of the viewer the real-time view synthesis software adjusts the view and creates the motion parallax effect on a 2D display. We believe that the proposed solution can enhance the content displayed on digital signature displays, kiosks, and other advertisement media where many users observe the content during move and use of the glasses-based 3D displays is not possible or too expensive.

Keywords 3D video · View synthesis · Motion parallax · Tracking · Kinect

1 Introduction

Human beings perceive the neighboring environment in 3D thanks to the interpretation of binocular and monocular depth cues [12]. Binocular depth cues can be seen only with use of both eyes, and are based on stereopsis—a displacement between the

M. Joachimiak (✉) · M. Gabbouj
Tampere University of Technology, Tampere, Finland
e-mail: michal.joachimiak@tut.fi

M. Wasielica · P. Skrzypczyński
Poznan University of Technology, Poznan, Poland
e-mail: mikolajwasielica@gmail.com

J. Sobecki
Wroclaw University of Technology, Wroclaw, Poland

R. Burduk et al. (eds.), *Proceedings of the 9th International Conference on Computer Recognition Systems CORES 2015*, Advances in Intelligent Systems and Computing 403, DOI 10.1007/978-3-319-26227-7_79

corresponding points in images of the same scene, delivered to the left and right eye. The differences in horizontal position between corresponding points in the left and right images are called binocular disparities and used by the visual cortex to sense depth. Monocular depth cues can be perceived with single eye, creating the depth sensation by perspective, texture, shadows, and motion parallax [5]. The stereopsis and motion parallax are the two most important sensory cues for depth perception [5]. An important advantage of the motion parallax depth cue is the absence of visual fatigue and discomfort [11] induced by vergence-accommodation conflict. The conflict occurs on 3D displays, that exploit binocular depth cue, due to inaccuracies between vergence and accommodation that make the fusion of binocular stimulus difficult.

The majority of 3D displays available on the market render stereoscopic images that comprise two views delivered separately to each eye. The views can be temporally or spatially separated by shutter or polarized glasses, correspondingly. In case of the shutter glasses, only one view at a time is displayed and corresponding eye is uncovered by deactivating the rolling shutter built into the glasses. The shutter that corresponds to the other eye is activated to block the view. This solution requires the display that can render video stream with at least double-frame rate. In polarized glasses-based technology the polarizing filters separate the views. The polarizing filters with corresponding orientation are used on top of the light source units on the display. In case of polarized displays, the spatial separation of the views imposes decreased spatial resolution of the 3D display. Even though, in the case of time sequential displays the resolution of the rendered view is not decreased, the frame rate of the video is limited. Imperfect view separation with use of glasses makes a small proportion of one eye's image visible in the other eye. This phenomenon is known as a 3D crosstalk [16] and, in both cases, leads to decreased subjective quality of the 3D percept caused by *ghosting artifacts*.

Autostereoscopic 3D displays can render more than two views, simultaneously, and do not require the user to wear any additional viewing gear. The state-of-the-art autostereoscopic displays (ASD) use light-directing mechanisms consisting of either lenticular lenses or parallax barriers, aligned on the surface of the display to provide different view for each eye. Due to the fact that ASDs render multiple views at a time they share the problem of decreased resolution caused by lower amount of cells corresponding to a single view. The autostereoscopic displays also exhibit ghosting artifacts caused by imperfect separation of the images intended for the left and right eye [4]. The presence of the aforementioned artifacts is one of the factors that limits application of such displays in public-space areas, such as advertising or information kiosks. Another two factors that slow down the widespread adoption of ASDs are high production cost and limited spatial resolution compared to the classic 2D displays.

The former research showed that the presence of head motion parallax, obtained by tracking of the observer's head and rendering the view accordingly, improves the ability to discriminate depth on a 3D display. In the experiments described in [10], the subjects were asked to judge the depth of the random dot sinusoidal gratings rendered on a 3D display equipped with the head tracking system. The results, corresponding to the cases in which the head motion parallax was used, showed consistently lower error rate in subjective depth assessment.

While stereoscopic and autostereoscopic 3D displays rely on the stereopsis to enable depth perception, it is impossible to recreate this effect on a conventional 2D display without special eyewear. Considering this limit and the fact that the motion parallax is one of the most important depth cues [13], we propose a cost-effective solution that enables depth sensation on a typical 2D display without enforcing the user to wear any glasses or other devices.

2 Related Work

Systems that track eyes, face, or head and render the content according to the viewpoint of the observer to simulate the motion parallax effect have been reported in literature mainly in the context of virtual reality, teleconferencing, and human–computer interaction with use of 3D display. However, they can be applied in numerous digital signage applications [2] which encompass installations based on information or advertising screens, as well as other digital communication solutions in public places. In one of the seminal works [14] a visual operating system, that uses autostereoscopic display, is proposed. In this solution, the head motion and gaze direction are tracked with use of the color cameras. However, the view is not adjusted according to the head position. Instead, the head motion and gaze tracking are used to enable interaction of the user with the operating system. Another work in which face detection works on the video signal, captured from the camera assembled in the computer, is presented in [6]. The authors propose to render stereoscopic images with respect to the detected face position. They use face detection and tracking to detect the change of the head position and alter the viewpoint of the virtual camera. Based on the face position, the virtual scene can be rendered with use of two different image rendering methods, diffuse-based or raytracing. Due to changes in light conditions, the quality of the captured video stream can cause noise in the head position measurement. Another disadvantage of the proposed solution is the usage of anaglyph glasses that cause distortions in color perception. Solutions like the one described in [6], even if affordable, are very inconvenient in applications that involve occasional viewers, such as information kiosks in public places. The nature of these places and type of the content to be delivered, which is mostly for advertising purposes, limits the possibility to distribute glasses to the viewers.

3D video teleconferencing systems deal with a problem more similar to our application, which is video delivery. The use of motion parallax effect in a teleconferencing application that utilizes a standard camera embedded in the computer is presented in [9]. The proposed solution employs head tracking to adjust a pseudo-3D view of the remote location according to the viewer's position. The system segments out the person with use of background subtraction, similarly to [14]. The measurement of the viewer's head position is used to tilt the image in which the foreground layer changes position with respect to the background layer. In this way, the pseudo-3D effect of looking through the window is achieved. Similarly to [14], changing lighting conditions impact the accuracy of the head tracker and segmentation. Thus, operation in dim or dark conditions is not possible.

Another system that utilizes motion parallax for teleconferencing was presented in [19]. The authors propose to use two methods to enhance depth perception, namely box framing and layered video. In case of box framing, the image is rendered in an artificial box which produces the window-like effect. The layered video requires the segmentation of the person similarly to [9] and the system adjusts the virtual view according the head position of the viewer. Even though the time of flight (TOF) camera is used, the depth information helps only with the segmentation and is not used for the estimation of the head position. This solution requires computationally complex face tracking that works on the RGB camera input, vulnerable to environmental conditions such as light changes or moving background objects.

Some systems utilize head tracking to enable continuous parallax for passive multiview displays. An exemplary solution [3], that removes the negative effect of the visibility zone change, uses face detection and tracking to adjust the views according to the observation angle of the viewer. Nevertheless, the calibration between the face position and the observation angle is required. The system proposed in [3] has similar weak points to previous solutions since the input from RGB camera is used for head tracking.

When a 3D perception relies on the motion parallax the accuracy of head pose estimation, robustness of the tracker, and response time to the changing head position become very important [19]. While there are many possible means to accomplish robust tracking of a single user in this type of application, most of them face practical drawbacks related to the use of regular color camera in constantly changing lighting conditions and dynamic background in uncontrolled environments, like public spaces. We propose a system designed in similar principles, but employing only off-the-shelf hardware components, such as the Kinect sensor, regular 2D display, and a commodity PC. The RGB-D sensor ensures fast and robust tracking of the user—not only the head motion, but also the whole body position, which is important in the context of our target applications.

3 Virtual View Synthesis with Motion Parallax

In difference to the previous solutions, the proposed one enables rendering a multitude of virtual views from a 3D video sequence. In result, the 3D effect is not simulated like in [19] or [9] and not only originally captured views are rendered on display like in [3]. The system takes on input a 3D video sequence stored in the multiview video and depth (MVD) format. We selected the MVD format since it is supported by the recent multiview-and-depth coding extension (MVC+D) of the advanced video coding (H.264/AVC) standard [1]. The selection of test sequences and views for input is chosen according to the common test conditions (CTC) [15]. Since the desired view change has to be as smooth as possible, the amount of interpolated views is

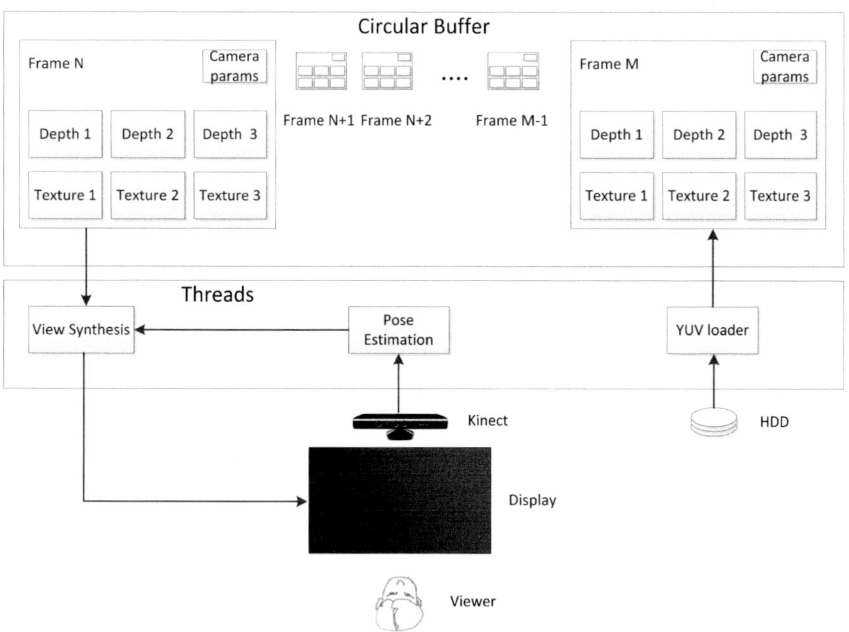

Fig. 1 Outline of the proposed system

increased comparing to CTC. Since the depth signal is embedded in this format, the depth image-based rendering (DIBR) [7] is possible at dense virtual camera positions. The outline of the system is presented in Fig. 1. To enable real-time operation, the processing is split into three threads that run independently. The view synthesis (VS) and YUV loader (YUVL) modules operate on the circular buffer in order to minimize thread stall time. Each data block of the circular buffer consists of camera parameters, three texture, and three corresponding depth views of a single frame. The frames of the sequence in the MVD format and corresponding camera parameters are loaded sequentially, from hard disk into the circular buffer memory, by the YUV loader thread. At any operating point, the viewer can freely move in front of the display. To enable operation in a longer distance range from the display, the tracking of the viewer is realized in two modes. In the first mode, the human pose tracking is utilized. The depth map captured by the Kinect sensor is processed by the pose estimation (PE) module that utilizes the Microsoft Kinect SDK v1.8 library. In this step the human silhouette is extracted from a single input depth image. Then, a per-pixel body part distribution is inferred. Finally, local modes of each part distribution are estimated to give a confidence-weighted proposal for the 3D location of body joints [17]. The algorithm is able to track up to two users. Since the motion parallax effect can be simulated for only one viewer the PE module selects the one which is closer to the display. The skeleton measurement data (Fig. 2a) is obtained in meters. The 3D coordinate system whose origin is located in the center of the Kinect sensor is used

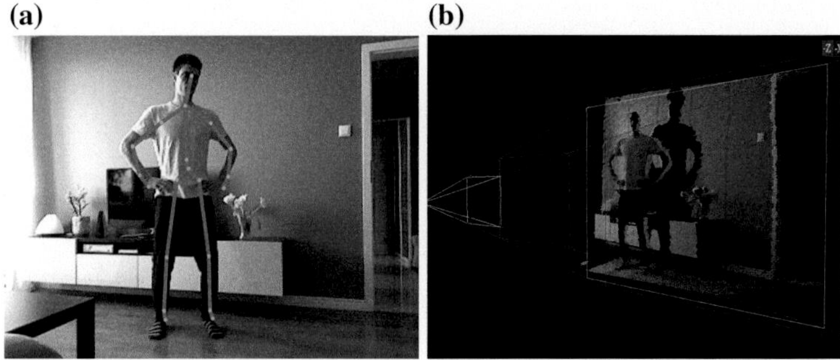

Fig. 2 Kinect-based user tracking: estimated skeleton joints (**a**), Kinect's field of view overlaid on a pseudo-color-coded depth image (**b**)

(Fig. 2b). In our setup the optical axis of the sensor and display's normal are parallel and the sensor is situated in the symmetry plane of the screen. The vertical position and the elevation angle of the Kinect are unrestricted. The only requirement is to capture the whole user's silhouette. Since our solution is intended for information kiosks and similar publicly accessible places, we assume that the user stands in front of the display, rather than sitting at a desk or table. In the second mode the PE module works in close-range distance and tracks only the face. The scene in front of the display is captured by the depth sensor and RGB camera of the Kinect sensor in order to perform face tracking. The modification of active appearance model that uses depth is utilized [18]. The PE module uses the position of the user's head in the Cartesian coordinates (Fig. 4a) and calculates its angular position in the transverse plane relatively to the Kinect optical axis (Fig. 4b). To calculate the angular position, we use the equation: $\alpha = \text{atan2}(x, y)$, where x and y are the head coordinates. Thanks to the conversion to angular coordinates we achieve invariance of the view while the user moves closer to the screen. The angular spread is limited to $\pm 10°$ measured from the optical axis at the center of the display to achieve adequacy of the head movements and displayed view. This value was adjusted experimentally to provide an optimal user experience. Then, the head angular position is normalized from $< -10; 10 >$ to $< 0; 18 >$ and discretized to integers in order to obtain the actual view index used in the view synthesis process. Based on the pose of the viewer the VS module decides whether to interpolate the view or display the originally captured one. The VS module utilizes the DIBR [7] process to interpolate the required view. The layout of interpolated views is presented in Fig. 3. The virtual views, represented by cameras in black, are interpolated from the captured views, marked as cameras in white and their corresponding depth data. The system assumes that the input data is generated by the shift-sensor camera setup [7] and is rectified [8]. Since in the shift-sensor camera setup the optical axes of the camera array are parallel the keystone effect can be avoided and vertical disparities are not present. Thanks to the rectification the corresponding points lay on the corresponding horizontal lines of neighboring views. This approach simplifies the 3D image warping which is essentially a reprojection from the source view to the world coordinates followed by a projection from world coordinates to the target virtual view. The shift-sensor algorithm [7] used in our approach, simplifies the 3D warping to the horizontal shift which decreases the computational complexity. Thanks to this approach, the interpolation for subpixel image generation is also simplified to one-dimensional. As depicted in Fig. 3 the viewer can move on the baseline that is perpendicular to the camera axes. When the viewer moves outside of the limited range the utmost, original view is displayed. The amount of interpolated views is limited to 18, therefore the algorithm discretizes the angular head position to this number, and as the result, we obtain the actual view number to render (Fig. 4c). Two exemplary views obtained from our system are presented in Fig. 5. These views represent the two most distant images, presented to the user at the left and the right side of the field of view.

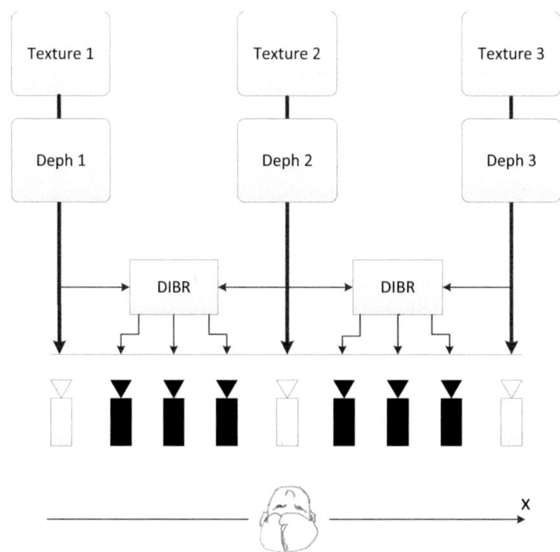

Fig. 3 The layout of interpolated views with respect to the captured ones

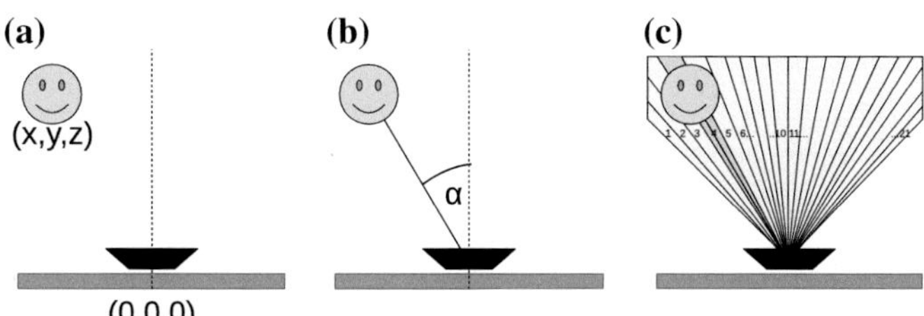

Fig. 4 Concept of the head tracking and view switching method

Fig. 5 Two example images presented to the user at the limits of the field of view

4 Experiments in the Application Context

The 3D perception that relies on the motion parallax by estimating the observer head or body pose may be used in numerous digital signage applications [2]. The feasibility of our solution in the mentioned application scenarios has been investigated in simple experiments accomplished by an expert user. The involvement of few other persons in these experiments enabled to simulate operation in a public place. In the first experiment, the human body detection range in the proposed system was investigated. The Kinect sensor was situated 1 m above the floor and leveled. The user tracking system was able to reliably detect the user's body posture up to 3.6 m from the sensor regardless of whether the person was standing or sitting (Fig. 6a). Persons situated farther were ignored by the system. Then, the minimal tracking distance was determined. This distance depends on the user's height. For the 1.8 m tall expert user the minimal detection distance was 1.9 m when standing, and 0.9 m when sitting (Fig. 6b). In the sitting mode only the head and arm joints were tracked, while in the default mode all the body joints were tracked. In the proposed system both modes are possible, since the view synthesis module requires only the head's position. The detection area was tested in both the daylight and in a darkened room. Unlike the camera-based head/face tracking systems, the Kinect-based solution does not suffer from degraded performance under low light conditions. In the second experiment, the influence, of other persons that may enter the scene, on the user's 3D perception

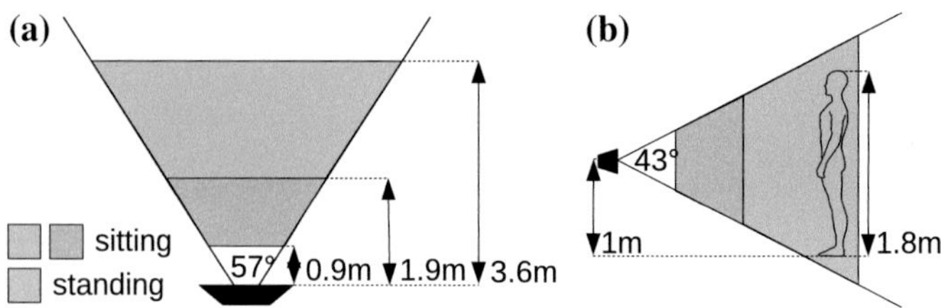

Fig. 6 The field of view for user's body detection by the Kinect sensor

comfort was investigated. The proposed system implements two tracking modes: the first one tracks the person closest to the sensor, while the second one always tracks the same person, until the active user is explicitly changed by a gesture command. In the first mode, the user was able to see comfortably in 3D till he was the closest person to the display/sensor. People walking behind him had no influence to the detection algorithm. Only when someone came closer to the sensor than the user, he took control of the displayed view, thus becoming the user. In such a situation, the first user could not continue watching in 3D. This mode provides a reasonable compromise between the viewing comfort and the flexibility of application in public places, where the persons most interested in the content usually come closer to the display. In the second mode, the tracking is fixed on the person who performed a proper gesture command recognized by the system. The distance from the sensor has no influence to the users's viewing comfort, because all other persons in the field of view are ignored. When the user was obscured to the sensor by a person passing by, the system was able to resume tracking. However, this feature worked only if the number of persons in the field of view was lower than six, because the Kinect SDK is able to track simultaneously positions of up to 6 persons. While the user was completely obscured by a person passing by, the system had no information of the head position and displayed the last view, till the occluding person changed position. Obviously, this could cause a viewing discomfort, but it should be noted that a person who was able to completely obscure the user to the sensor, blocked also the user's view of the display. Thus, he could not observe the improper view. As the system always computes the viewing angle from the current position of the user's head, the proper motion parallax view was recovered immediately after the occluding person has moved. However, in some situations the person passing in front of the display was tall enough to obscure the Kinect view of the user, but shorter than the user, who could notice the viewing discomfort. Therefore, we placed the Kinect sensor at the elevation of 2 m to provide better scene view. This solved the problem with partial occlusions. When the person passing by was much shorter than the user (e.g., a child), the system was able to work normally.

5 Conclusions

We propose to employ the active RGB-D sensor for tracking the head and body position to detect the viewpoint of the 3D video viewer. The Microsoft Kinect comes with a sophisticated software API for gaming that enables relatively precise and robust tracking of body parts in real-time. We employ this API and demonstrate that the availability of the observer's head and body pose estimate enables view rendering on a standard PC for motion parallax effect. The proposed solution utilizes a computationally optimized shift-sensor view synthesis [7] that enables real-time virtual view rendering with motion parallax depth cue. Since the observer's head position estimation is based on the overall posture detection, our system achieves the unique property of being able to track the user with partly or even completely covered faces, i.e., with sunglasses or beard. Since the human body area is much larger than the area of the face, we achieved a better detection reliability from farther distances, comparing to the standard face detection algorithms, which need close position of the face to the camera. Therefore, tracking the whole body by the Kinect is able to detect observer's head positions in realistic scenarios related to signage and advertisement applications in public places.

Acknowledgments This work was partially funded by the Poznan University of Technology 04/45/DSPB/0104 grant. The authors express gratitude to the ENGINE project under the EU 7th Framework Programme for research, grant agreement no 316097 for support during the preparation of the publication.

References

1. Advanced video coding for generic audiovisual services. ITU-T Recommendation H.264 and ISO/IEC 14496–10 (MPEG-4 AVC) (2013)
2. Anisiewicz, J., Jakubicki, B., Sobecki, J., Wantuła, Z.: Configuration of Complex Interactive Environments. In: Proceedings of the New Research in Multimedia and Internet Systems, pp. 239–249. Springer International Publishing, Berlin (2015)
3. Boev, A., Raunio, K., Georgiev, M., Gotchev, A., Egiazarian, K.: OpenGL-based control of semi-active 3D display. In: Proceedings of the 3DTV-Conference: The True Vision-Capture, Transmission and Display of 3D Video, Istanbul, pp. 125–128 (2008)
4. Boev, A., Gotchev, A., Egiazarian, K.: Stereoscopic artifacts on portable auto-stereoscopic displays: what matters?. In: Proceedings of the Workshop on Video Processing and Quality Metrics for Consumer Electronics, Scottsdale, pp. 24–29 (2009)
5. Cutting, J.E., Vishton, P.M.: Perceiving layout and knowing distances: the integration, relative potency and contextual use of different information about depth. Perception of Space and Motion, pp. 69–117. Academic Press, San Diego (1995)
6. Dąbała, Ł., Rokita, P.: Simulated holography based on stereoscopy and face tracking. In: Chmielewski, Leszek J., Kozera, Ryszard, Shin, Bok-Suk, Wojciechowski, Konrad (eds.) Computer Vision and Graphics. Lecture Notes in Computer Science, vol. 8671, pp. 163–170. Springer, Heidelberg (2014)
7. Fehn, C.: Depth-image-based rendering (DIBR), compression and transmission for a new approach on 3D-TV. In: Proceedings of the SPIE Conference Stereoscopic Displays and Virtual Reality Systems XI, USA, pp. 93–104 (2004)
8. Fusiello, A., Trucco, E., Verri, A.: A compact algorithm for rectification of stereo pairs. J. Mach. Vis. Appl. **12**(1), 16–22 (2000)
9. Harrison, C., Hudson, S.E.: Pseudo-3D video conferencing with a generic webcam. In: Proceedings of the 10th IEEE International Symposium on Multimedia, Berkeley, USA, pp. 236–241 (2008)
10. Lackner, K., Boev, A., Gotchev, A.: Binocular depth perception: does head parallax help people see better in depth?. In: Proceeding of the 3DTV-Conference: The True Vision-Capture, Transmission and Display of 3D Video, Budapest, Hungary (2014)
11. Lambooija, M.T., Ijsselsteijn, W., Heynderickx, I.: Visual discomfort in stereoscopic displays: a Review. In: Proceedings of the SPIE 6490, Stereoscopic Displays and Virtual Reality Systems XIV (2007)
12. Mikkola, M., Boev, A., Gotchev, A.: Relative importance of depth cues on portable autostereoscopic display. In: Proceedings of the 3rd Workshop on Mobile Video Delivery, pp. 63–68 (2010)

13. Nawrot, M.: Depth from motion parallax scales with eye movement gain. J. Vis. **3**(11), 17 (2003)
14. Pastoor, S., Liu, J., Renault, S.: An experimental multimedia system allowing 3-d visualization and eye-controlled interaction without user-worn devices. IEEE Trans. Multimed. **1**(1), 41–52 (1999)
15. Rusanovskyy, D., Muller, K., Vetro, A.: Common test conditions of 3DV core experiments (2013). ISO/IEC JTC1/SC29/WG11 JCT3V-E1100
16. Seuntiens, P.J., Meesters, L.M., Ijsselsteijn, W.A.: Perceptual attributes of crosstalk in 3D images. J. Disp. **26**(4), 177–183 (2005)
17. Shotton, J., Fitzgibbon, A., Cook, M., Sharp, T., Finocchio, M., Moore, R., Kipman, A., Blake, A.: Real-time human pose recognition in parts from single depth images. In: Proceedings of the IEEE Conference on Computer Vision and Pattern Recognition, Colorado Springs, pp. 1297–1304 (2011)
18. Smolyanskiy, N., Huitema, C., Liang, L., Anderson, S.E.: Real-time 3D face tracking based on active appearance model constrained by depth data. Image and Vision Computing **32**(11), 860–869 (2014)
19. Zhang, C., Yin, Z., Florencio, D.: Improving depth perception with motion parallax and its application in teleconferencing. In: Proceedings of the IEEE International Workshop on Multimedia Signal Processing, Rio de Janeiro, pp. 1–6 (2009)

Author Index

R. Burduk et al. (eds.), *Proceedings of the 9th International Conference on Computer Recognition Systems CORES 2015*, Advances in Intelligent Systems and Computing 403, DOI 10.1007/978-3-319-26227-7